Electrical Wiring

RESIDENTIAL

EINHIPLE @9CSU. BUFFALO, EDU

Based on the 2002
NATIONAL ELECTRICAL CODE®

Electrical Wiring

RESIDENTIAL

Fourteenth Edition

RAY C. MULLIN

DELMAR

THOMSON LEARNING

Australia   Canada   Mexico   Singapore   Spain   United Kingdom   United States

DELMAR
THOMSON LEARNING

Electrical Wiring Residential, Fourteenth Edition
by Ray C. Mullin

Business Unit Director:
Alar Elken

Executive Editor:
Sandy Clark

Acquisitions Editor:
Mark Huth

Editorial Assistant:
Dawn Daugherty

Development Editor:
Jennifer A. Thompson

Executive Marketing Manager:
Maura Theriault

Channel Manager:
Fair Huntoon

Marketing Coordinator:
Brian McGrath

Executive Production Manager:
Mary Ellen Black

Production Coordinator:
Toni Hansen

Project Editor:
Barbara L. Diaz

Art/Design Coordinator:
Cheri Plasse

Cover Design:
Charles Cummings, Advertising

For permission to use material from this text or product,
contact us by
Tel   (800) 730-2214
Fax   (800) 730-2215
www.thomsonrights.com

Library of Congress Cataloging-in-Publication Data

Mullin, Ray C.
     Electrical wiring residential : residential / Ray C. Mullin.—
14th ed.
        p. cm.
     "Based on the 2002 National Electrical Code."
     ISBN 0-7668-3284-8 (alk. paper)
     1. Electric wiring, Interior.   I. Title.
TK3285 .M84 2001
621.319'24—dc21   2001028888

NOTICE TO THE READER

CONTENTS

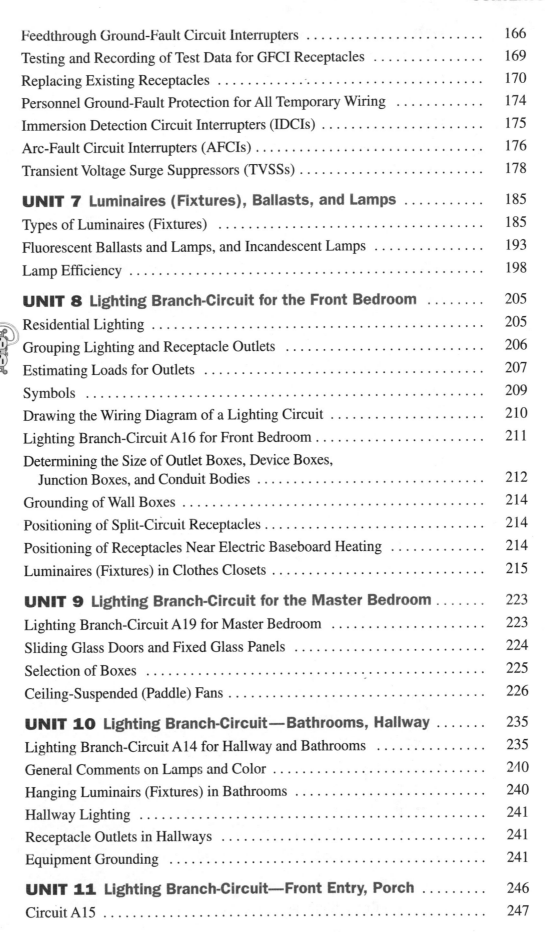

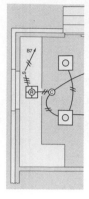

FOREWORD

THE ROLE OF THE ELECTRICAL INSPECTOR
AND THE IMPORTANCE OF PROPER TRAINING

When one considers how different groups within the electrical industry affect the end product used by consumers, it is easy to see the importance of those groups working together in a coordinated manner. From the time electrical products are designed to the time when electrical installations are completed or when electrical equipment is purchased by the consumer, many steps must be taken to make it all possible. During this process, close attention to safety standards must be given to the design, installation, and use of electrical products. Product safety standards include testing guidelines that can be used to help ensure that products perform in a safe manner. Products so evaluated by appropriate third-party testing laboratories may be labeled with authorized markings as a means of verifying that the product complies with the minimum requirements of the safety standard.

The *National Electrical Code*® is an electrical safety standard containing rules affecting products, installation procedures, and use of electrical equipment. It is the most widely recognized safety standard in this country. The *NEC*® is usually adopted through a legislative process of a governmental agency and becomes enforceable as a law within that jurisdiction. The *NEC*® is amended every three years in response to such things as the development of new products, the recognition of different installation and use procedures, and the discovery of potential hazards to persons and property. Electrical inspectors have the responsibility of enforcing the safety requirements of the *National Electrical Code*.®

Electrical inspectors are an important part of the safety program. They are involved in the development of safety rules as well as in the application of those rules to electrical installations. The responsibility borne by electrical inspectors is significant. To be responsible for knowing safety rules, making proper interpretation of those rules, and properly applying them to a job site is a major endeavor. In order for the electrical inspector to properly do his or her job, knowledge and skill must be developed. This is achieved through adequate education and experience. It is vital that electrical inspectors be knowledgeable of the rules in the *National Electrical Code*® and know how to apply those rules.

The need for adequate training and experience is not limited to the electrical inspector. All who are a part of the electrical construction industry can benefit by it. In order for one to maintain an acceptable level of proficiency in a profession or trade, one must stay current with the ever-changing electrical field. That proficiency is achieved through continuing education and training. There is no substitute for such training and experience. To emphasize this point, I refer to a philosophy taught by Ray Mullin: "The cost of education is small when compared to the price paid for ignorance."

Philip H. Cox
Executive Director
International Association
of Electrical Inspectors

INTRODUCTION

The fourteenth edition of *Electrical Wiring—Residential* is based on the 2002 *National Electrical Code* (*NEC*).® The *NEC*® is used as the basic standard for the layout and construction of residential electrical systems. In this text, thorough explanations are provided of Code requirements as they relate to residential wiring. To gain the greatest benefit from this edition, the student must use the *National Electrical Code*® on a continuing basis. Please note that throughout this text, the *National Electrical Code*® is represented by the abbreviation *NEC*.®

Why Is This Different from Other House Wiring Books?

It is extremely difficult to learn the *NEC*® by merely reading it. This text brings together the rules of the *NEC*® and the wiring of an actual house. You will study the rules from the *NEC*,® and apply those rules to a true-to-life house wiring installation.

Take a moment to look at the Table of Contents. It is immediately apparent that you will not learn such things as how to drill a hole, tape a splice, fish a cable through a wall, use tools, or repair broken plaster around a box. These things you already know, or are learning on the job. The emphasis of this text is to teach you how to wire a house that "meets Code." Doing it right the first time is far better than having to do it over because the electrical inspector turned down your job.

The first seven chapters in this book concentrate on basic electrical code requirements that apply to house wiring. This includes safety when working with electricity; construction symbols, plans, and specifications; wiring methods; conductor sizing; circuit layout; wiring diagrams; numerous ways to connect switches and receptacles; how to wire recessed luminaires (lighting fixtures); ground-fault circuit-interrupters (GFCIs); arc-fault circuit-interrupters (AFCIs); and surge suppressors.

The remaining chapters are devoted to the wiring of an actual house—room by room and circuit by circuit. All of these circuits come together for calculating the size of the main service. Since proper grounding is a key safety issue, the subject is covered in detail.

You will also learn about security systems, fire and smoke detectors, low-voltage remote-control wiring, and swimming pools, and you will be introduced to structured wiring for home automation.

You will find this text unique in that you will use the text, an actual set of plans and specifications, and the *NEC*®—all at the same time. It is perfect to learn house wiring and makes an excellent reference source for looking up specific topics relating to house wiring.

Also included are many recommendations that are above and beyond the basic *NEC*® requirements.

The *NEC*® (NFPA 70) becomes mandatory only after it has been adopted by a city, county, state, or other governing body. Until officially adopted, the *NEC*® is merely advisory in nature. State and local electrical codes may contain modifications of the *NEC*® to meet local requirements. In most cases, local codes will adopt certain more stringent regulations than those found in the *NEC*.® For example, the *NEC*® recognizes nonmetallic-sheathed cable as an acceptable wiring method for house wiring. Yet, the city of Chicago and surrounding counties do not

---

National Electrical Code® and *NEC*® are registered trademarks of the National Fire Protection Association, Inc., Quincy, MA 02269.

permit nonmetallic-sheathed cable for house wiring. In these areas, all house wiring is done with electrical metallic tubing (thinwall).

There are also instances where a governing body has legislated action that waives specific *NEC®* requirements, feeling that the *NEC®* was too restrictive on that particular issue. Such instances are very rare. The instructor is encouraged to furnish students with any local variations from the *NEC®* that would affect this residential installation in a specific locality.

Electrical wiring is a skilled trade. Wiring should not be done by anyone not familiar with the hazards involved. It is a highly technical skill that requires much training. This material provides all of the electrical codes and standards information needed to approach house wiring in a safe manner. ▶The *NEC®* defines a **qualified person** as "*one who has skills and knowledge related to the construction and operation of the electrical equipment and installation, and has received safety training on the hazards involved.*"◀

Do not work on live circuits! Always de-energize the system before working on it! There is no compromise when it comes to safety! Many injuries and deaths have occurred when individuals worked on live equipment. The question is always: "Would the injury or death have occurred had the power been shut off?" The answer is "No!"

All mandatory safety-related work practices are found in the Federal Regulation Occupational Safety and Health Act (OSHA), Title 29, Subpart S—Electrical, Sections 1910.331 through 1910.360.

Electrical Wiring—Residential is the most comprehensive text available anywhere. It covers residential wiring topics from the most advanced *NEC®* standpoint to the fundamental requirements necessary for someone wanting to learn the basics. Every Code rule is covered by text, illustrations, examples, and wiring diagrams.

Most electrical inspectors across the country are members of the International Association of Electrical Inspectors. This organization publishes one of the finest technical bimonthly magazines devoted entirely to the *NEC®* and related topics. This organization is open to individuals that are not electrical inspectors. Electrical instructors, vo-tech students, apprentices, electricians, consulting engineers, contractors, and distributors are encouraged to join the IAEI so they can stay up to date on all *NEC®* issues, changes, and interpretations. An application form that explains the benefits of membership in the IAEI can be found in the Appendix of this text.

THE ELECTRICAL TRADE

Electricity is safe when it is handled properly. If electrical installations are made in an unsafe manner, great potential exists for damage to property as a result of fire. But more importantly, electrical installations that do not meet the minimum requirements of applicable codes and standards can and do result in personal injury and/or death.

Most communities, cities, counties, and states adopt regulations by reference to the *NEC®*. These regulations also specify any exceptions to the Code to cover local conditions.

Most building codes and standards contain definitions for the various levels of competency of workers in the electrical industry. Here are some examples of typical definitions.

Apprentice shall mean a person who is required to be registered, who is in compliance with the provisions of this article, and who is working at the trade in the employment of a registered electrical contractor and is under the direct supervision of a licensed master electrician, journeyman electrician, or residential wireman.

Residential Wireman shall mean a person having the necessary qualifications, training, experience, and technical knowledge to wire for and install electrical apparatus and equipment for wiring one-, two-, three-, and four-family dwellings. A residential wireman is sometimes referred to as a *Class B Electrician*.

Journeyman Electrician shall mean a person having the necessary qualifications, training, experience, and technical knowledge to wire for, install, and repair electrical apparatus and equipment for light, heat, power, and other purposes, in accordance with standard rules and regulations governing such work.

Master Electrician means a person having the necessary qualifications, training, experience, and technical knowledge to properly plan, lay out, and supervise the installation and repair of wiring apparatus and equipment for electric light, heat, power, and other purposes, in accordance with standard codes and regulations governing such work, such as the *NEC*.®

Electrical Contractor means any person, firm, copartnership, corporation, association, or combination thereof who undertakes or offers to undertake for another the planning, laying out, supervising and installing, or the making of additions, alterations, and repairs in the installation of wiring apparatus and equipment for electrical light, heat, and power.

THE FOURTEENTH EDITION

Continuing in the tradition of previous editions, this edition thoroughly explains how Code changes affect wiring installations. New and revised full-color illustrations supplement the explanations to ensure that electricians understand the new Code requirements. New photos reflect the latest wiring materials and components available on the market. Revised review questions test student understanding of the new content. New tables that summarize Code requirements offer a quick reference tool for students. Another reference aid is the tables reprinted directly from the 2002 edition of the *NEC*.® The extensive revisions for the fourteenth edition make *Electrical Wiring—Residential* the most up-to-date and well-organized guide to house wiring. Coverage of the *NEC*® has been expanded to well over 950 Code references.

A modern residence blueprint serves as the basis for the wiring schematics, cable layouts, and discussions provided in the text. Additional plans may be obtained by contacting Delmar. Each unit dealing with a specific type of wiring is referenced to the appropriate plan sheet. All wiring systems are described in detail—lighting, appliance, heating, service entrance, and so on.

The house selected for this edition is scaled for current construction practices and costs. Note, however, that the wiring, luminaires (lighting fixtures), appliances, number of outlets, number of circuits, and track lighting are not all commonly found in a home of this size. The wiring may incorporate more features than are absolutely necessary. This was done to present as many features and Code issues as possible to give the student more experience in wiring a residence.

This text does *not* focus on such basic skills as drilling a hole, splicing two wires together, stripping insulations from wire, or fishing a wire through walls. These skills are certainly necessary for good workmanship, but it is assumed that the student (electrical apprentice or journeyman electrician) has mastered the mechanical skills on the job.

This text *does* focus on the technical skills required to perform electrical installations. It covers such topics as calculating conductor sizes, calculating voltage drop, determining appliance circuit requirements, sizing service, connecting electric appliances, grounding service and equipment, installing recessed luminaires (fixtures), and much more. These are critical skills that can make the difference between an installation that "meets Code" and one that does not. The electrician must understand the reasons for following Code regulations to achieve an installation that is essentially free from hazard to life and property.

NEW FOR THE 2002 EDITION

Note: Symbols have been added to indicate changes in the 2002 National Electrical Code vs. the 1999 National Electrical Code. ▶ ◀

1. All of the comprehensive renumbering and relocation of sections and articles in the *NEC*® has been updated in this edition. This edition of the *NEC*® has seen the most changes ever.

2. Because the *NEC*® has become an international standard, metric values (SI Units) appear first in the *NEC*,® followed by inch-pound values in parenthesis. In addition, the *NEC*® has rounded off many of the metric measurements. For example, in the *NEC*,® 1.83 m (6 ft) becomes 1.8 m (6 ft).

   For ease in learning and to continue to be user friendly, this text shows inch-pound first, followed by metrics in parentheses, for example, 6 ft (1.8 m).

3. The term *luminaire* has been used internationally for years. It means the same as *lighting fixture*. The term *luminaire* includes the fixture, lamps, and ballast (if used) as one unit. The 2002 edition of the *NEC*® uses the term *luminaire* followed by the word *fixture* in parenthesis, for example, *luminaire (fixture)*. This is to ease the transition of the terms. It is expected that in the 2005 *NEC*,® only the term *luminaire* will be used.

4. The *NEC*® has adopted the international standard for showing conduit sizes. In the *NEC*,® both metric designator number and trade size are shown. In this text, again to keep things simple and easy to understand, conduit, tubing, and raceway sizes are shown as *trade sizes*. For example, electrical metallic tubing is referred to as trade size ½ EMT.

5. The word "No." to indicate wire size has been deleted from the *NEC*.® In the past, a conductor might have been referred to in the *NEC*® as a "No. 12," or "No. 12 AWG." Beginning with the 2002 edition of the *NEC*,® this conductor will simply be referred to as "12 AWG."

6. The *NEC*® now uses a decimal system for numbering. For example, *210-52* appears as *210.52*. This change has been reflected throughout *Electrical Wiring—Residential*.

7. The term "Section" has been removed throughout the *NEC*.® For example, *Section 210-52* now appears as *210.52*. These are still "Sections," but it is not necessary to repeat this word over and over. In this text, all Code references are shown in italics.

8. The "Appendix" in the *NEC*® is now referred to as the "Annex." The Appendix of *Electrical Wiring—Residential* will continue to be called the Appendix.

9. A gray insulated conductor is no longer permitted to be used as an ungrounded "hot" conductor. Its use is now limited to being a grounded conductor. A grounded neutral conductor may now be white or gray.

10. Arc-fault circuit interrupters are now required for protecting all 125-volt, single-phase, 15- and 20-ampere outlets installed in dwelling unit bedrooms. This requirement became effective January 1, 2002. Meeting this requirement is as simple as installing an AFCI circuit breaker in the panel for the branch-circuit that supplies the bedroom.

11. Type MC Metal Clad cable has been added.

12. A comparison chart has been added to show differences betweeen Type AC and MC cable.

13. A comparison chart has been added to show features and uses for all types of service-entrance cables.

14. An expanded list of web sites for electrical and electronics manufacturers, as well as many resource web sites for home automation (structured wiring systems) information, are found in the Appendix of this text.

15. There has been a complete rewrite of the swimming pool unit and a revision of the large fold-out plan.

16. There has been a complete rewrite of the service entrance and grounding unit.

17. The lengthy unit covering service-entrance equipment, panelboards, grounding, bonding, fuses, and circuit breakers has been divided into two units, making them more manageable and easier to understand.

18. A section on how to establish a proper ground at a detached garage or other outbuilding has been added, complete with a detailed wiring diagram.

19. Information on providing a properly sized branch-circuit for microwave ovens has been added.

20. Information for connecting foreign-made equipment has been added.

21. The home automation unit has been totally revised to reflect the state of the art relative to this type of structured wiring.

22. By popular demand, a new unit has been added on standby generators.

23. More has been added about the symptoms of potential problems in homes wired with aluminum conductors.

24. Additional data has been added relating to timers.

SUPPLEMENTS

An Instructor's Guide is available and consists of the following information: selected text references, answers to unit-end reviews, a final examination covering the content of the entire text, a blank service-entrance calculation form, answers to the final examination, and over 120 transparency masters.

ABOUT THE AUTHOR

This text was written by Ray C. Mullin, former electrical instructor for the Wisconsin Schools of Vocational, Technical, and Adult Education. He is a former member of the International Brotherhood of Electrical Workers. He is a member of the International Association of Electrical Inspectors, the Institute of Electrical and Electronic Engineers, and the National Fire Protection Association, Electrical Section, and has served on Code Making Panel 4 of the *National Electrical Code*.

Mr. Mullin completed his apprenticeship training and worked as a journeyman and supervisor for residential, commercial, and industrial installations. He has taught both day and night electrical apprentice and journeyman courses, has conducted engineering seminars, and has conducted many technical Code workshops and seminars at International Association of Electrical Inspectors Chapter and Section meetings, and has served on their Code panels.

He has written many technical articles that have appeared in electrical trade publications. He has served as a consultant to electrical equipment manufacturers regarding conformance of their products to industry standards, and on legal issues relative to personal injury lawsuits resulting from the misuse of electricity and electrical equipment. He has served as an expert witness.

Mr. Mullin presents his knowledge and experience in this text in a clear-cut manner that is easy to understand. This presentation will help students to fully understand the essentials required to pass the residential licensing examinations and to perform residential wiring that "meets Code."

Mr. Mullin is the author of *House Wiring with the NEC®*—a text that focuses entirely on the *National Electrical Code®* requirements for house wiring. He is co-author of *Electrical Wiring —Commercial, Illustrated Electrical Calculations,* and *The Smart House.* He has contributed technical material for Delmar's *Electrical Grounding* and to the International Association of Electrical Inspectors' texts *Soares Book On Grounding* and *Ferm's Fast Finder.*

He served on the Executive Board of the Western Section of the International Association of Electrical Inspectors, on their Code Clearing Committee, and, in the past, served as Secretary/ Treasurer of the Indiana Chapter of the IAEI.

Mr. Mullin is past Director, Technical Liaison for a major electrical manufacturer. In this position, he was deeply involved in electrical codes and standards as well as contributing and developing technical training material for use by this company's field engineering personnel.

Mr. Mullin attended the University of Wisconsin, Colorado State University, and the Milwaukee School of Engineering.

IMPORTANT NOTE

This edition of *Electrical Wiring—Residential* was completed after all normal steps of revising the *National Electrical Code® NFPA 70* were taken, and before the actual issuance and publication of the 2002 edition of the *NEC.®*

These steps include: NFPA solicits proposals for *2002 NEC,®* interested parties submit proposals to NFPA, proposals sent to Code Making Panels (CMPs), proposals reviewed by CMPs and Technical Correlating Committee, Report on Proposals document published, interested parties submit comments on the proposals to NFPA, review of Comments by CMPs and Technical Correlating Committee, Report on Comments document published, review of all Proposals and Comments at NFPA Annual Meeting, new motions permitted to be made at the NFPA Annual Meeting. Finally, the Standard Council meets to review actions made at the NFPA Annual Meeting, and to authorize publication of the *NEC.®*

Every effort has been made to be technically correct, but there is always the possibility of typographical errors, or appeals made to the NFPA Board of Directors after the normal review process that could result in reversal of previous decisions by the CMPs.

If changes in the *NEC®* do occur after the printing of this text, these changes will be incorporated in the next printing.

The National Fire Protection Association has a standard procedure to introduce changes between *NEC®* Code cycles after the actual *NEC®* is printed. These are called "Tentative Interim Amendments," or TIAs. TIAs and typographical errors can be downloaded from the NFPA website, www.nfpa.org, to make your copy of the Code current.

ACKNOWLEDGMENTS

I must again give thanks to my wife Helen for her patience while I devoted unlimited time to attending meetings and working many hundreds of hours revising this full-color edition of *Electrical Wiring—Residential*. This happens for each Code cycle—every three years. And for this Code cycle, the renumbering and relocation of Sections and Articles has been a real challenge.

The team at Delmar, as always, has done an outstanding job in bringing this edition to press on a timely basis. Their drive, dedication, and attention to detail keeps *Electrical Wiring—Residential* the country's leading text on house wiring. This book has reached the highest state of excellence in the textbook industry!

Special thanks to my longtime friend Phil Cox, Executive Director of the International Association of Electrical Inspectors, for his foreword to this text. Phil emphasizes the need for proper education in the electrical industry, particularly about the *National Electrical Code.*®

I would also like to thank my many other friends in the electrical industry across the country; my personal friends and outstanding Code experts Ron O'Riley, Phil Simmons, and Charlie Trout; the electrical staff at NFPA headquarters, and the many Code Making Panel members that clarified specific code proposals for the 2002 *NEC.*® All offered excellent comments and suggestions that helped make this edition of *Electrical Wiring—Residential* the most comprehensive and code-compliant ever. Special thanks to Robert Boiko for his continual technical input on water heaters and their related controls.

I wish I could name all of these friends, but there are so many, it would take pages to include all of their names. Thanks to all of you for your support.

The author and publisher would like to thank the following reviewers for their contributions:

DeWain Belote
Pinellas Tech Educational Center
St. Petersburg, FL

Les Brinkley
Ashtabula County JVC
Ashtabula, OH

Larry Catron
Scott County Vo-Tech School
Gate City, VA

Robert Coy
KY Tech Elizabethtown
Elizabethtown, KY

Warren DeJardin
Northeast Wisconsin Technical College
Oneida, WI

Edward Dwyer
Triangle Tech
Pittsburgh, PA

Charles Eldridge
Indianapolis Power & Light
Indianapolis, IN

Greg Fletcher
Kennebec Valley Tech College
Fairfield, ME

David Gehlauf
Tri-County Vocational School
Glouster, OH

Fred Johnson
Champlain Valley Tech
Plattsburg, NY

Thomas B. Lockett
Vatterott College
Quincy, IL

Cliff Redigar
Independent Electrical Contractors
Denver, CO

Gary Reiman
Dunwoody Institute
Minneapolis, MN

Max Schroeder
Barton County Community College
Hoisington, KS

Special thanks to Mike Forister, one of the most knowledgeable Code experts in the country, for his thorough technical review of the Code content.

The author and Publisher wish to thank the following companies for their contributions of data, illustrations, art work, photos, and technical information:

AFC Cable Systems

American Home Lighting Institute

BRK Electronics (First Alert), A Division of Sunbeam Corporation

Carlon, A Lamson & Sessions Company

Certified Ballast Manufacturers

Chromalox

Cooper Industries, Bussmann Division

Cooper Industries, Cooper Lighting

Electri-Flex Co.

Heyco Molded Products, Inc.

Honeywell

Hubbell Incorporated, Wiring Devices Division

International Association of Electrical Inspectors

Juno Lighting, Inc.

The Kohler Co.

Leviton Manufacturing Co., Inc.

Lutron Electronics Co., Inc.

Midwest Electric Products, Inc.

Milbank Manufacturing Co.

Motorola

NuTone, Inc.

Paragon Electric Company, Inc.

Progress Lighting

RACO, A Division of Hubbell Incorporated

Reiker Enterprises, Inc.

Rheem Manufacturing Co.

Seatek Co., Inc.

Square D, Groupe Schneider

Superior Electric Co.

THERM-O-DISC

Thomas & Betts Corporation

Wiremold Co.

Applicable tables and section references are reprinted with permission from NFPA 70-2002, *National Electrical Code,*® copyright © 2002, National Fire Protection Association, Quincy, MA 02269. This reprinted material is not the complete and official position of the NFPA on the referenced subject, which is represented only by the standard in its entirety.

General Information for Electrical Installations

OBJECTIVES

After studying this unit, the student should be able to

- understand the basic safety rules for working on electrical systems.
- explain how electrical wiring information is conveyed to the electrician at the construction or installation site.
- demonstrate how the specifications are used in estimating cost and in making electrical installations.
- explain why symbols and notations are used on electrical drawings.
- list the Nationally Recognized Testing Laboratories (NRTL) that are responsible for establishing electrical standards and ensuring that materials meet the standards.
- develop a better understanding of the metric system of measurement.
- begin to refer to the latest edition of the *NEC*.®
- have a better understanding of the new *Residential Electrical Maintenance Code for One- and Two-Family Dwellings*.

SAFETY IN THE WORKPLACE

Before venturing into the study of residential wiring, let's discuss safety. The electrician might be involved in new construction where the only electrical power is temporary power, or in remodel work where the home already has electrical power.

Serious injury can result if the power has not been properly turned off before beginning to work on an electrical problem. Electricity is dangerous. The voltage levels in a home are 120 volts between one "hot" conductor and the "neutral" conductor or a grounded surface. Between the two "hot" conductors, the voltage is 240 volts. These are the *line-to-ground* or *line-to-line* voltages. From basic electrical theory, *line voltage appears across an open in a series circuit*. Getting caught "in series" with a

120-volt circuit will give you a 120-volt shock. For example, open-circuit voltage across the two wires of a single-pole switch on a lighting circuit is 120 volts, when the switch is in the "OFF" position and the lamp(s) are in place.

Likewise, getting caught "in series" with a 240-volt circuit will give you a 240-volt shock. Working on electrical equipment with the power turned on can result in death or serious injury either as a direct result of electricity (electrocution or burns), or from an indirect secondary reaction such as falling off a ladder or jerking away from the "hot" conductor into moving parts of equipment, such as into the turning blades of a furnace fan. Dropping a metal tool onto live parts, allowing metal shavings from a drilling operation to fall onto live parts c electrical equipment, cutting into a "live" conductor and a "neutral" conductor at the same time, or touching the "live" wire and the "neutral" wire or a grounded surface at the same time can cause injury. A short-circuit or ground fault can result in an *arc blast* that can cause serious injury or death. The heat of an electrical arc has been determined to be hotter than the sun. Tiny hot "balls" of copper can fly into your eye or onto your skin. Working on switches, receptacles, luminaires (fixtures), or appliances with the power turned on is dangerous.

Figure 1-1 shows a disconnect switch that has been locked and tagged.

Dirt, debris, and moisture can also set the stage for equipment failures and personal injury. Neatness and cleanliness in the workplace are a must.

It's the Law!

Not only is it a good idea to use proper safety measures as you work on and around electrical systems, it is *required* by law. As an electrician or electrical contractor, you need to be aware of these regulations.

According to *NFPA 70E Electrical Safety Requirements for Employee Workplaces*, circuits and conductors are not considered to be in an electrically safe condition until all sources of energy are removed, the disconnecting means is under lockout/tagout, and the absence of voltage is verified by an approved voltage testing device.

The federal regulations in the **Occupational Safety and Health Act (OSHA)** Number 29, Subpart S, in Part 1910.332, discuss the training needed for those who face a risk of electrical injury. Proper training means being trained in and familiar with the safety-related work practices required by paragraphs 1910.331 through 1910.335. Numerous texts are available that cover the OSHA requirements in greater detail.

▶The *NEC*® defines a **qualified person** as "one who has skills and knowledge related to the construction and operation of the equipment and has specific safety training on the hazards involved."◀ Merely telling someone or being told to "be careful" does not meet the definition of proper training, and does not make the person qualified.

Only qualified persons are permitted to work on or near exposed energized equipment. To become

Figure 1-1   A typical disconnect switch with a lock and a tag attached to it. In the OSHA, ANSI, and NFPA standards, this is referred to as the *lockout/tagout* procedure.

qualified, a person must have the skill and techniques necessary to distinguish exposed live parts from the other parts of electrical equipment, must be able to determine the voltage of exposed live parts, and must be trained in the use of special precautionary techniques such as personal protective equipment, insulations, shielding material, and insulated tools.

Subpart S, 1910.333 requires that safety-related work practices shall be employed to prevent electric shock or other injuries resulting from either direct or indirect electrical contact. Live parts to which an employee may be exposed shall be de-energized before the employee works on or near them, unless the employer can demonstrate that de-energizing introduces additional or increased hazards.

Working on equipment "live" is acceptable only if there would be a greater hazard if the system were de-energized. Examples of this would be hospital life support systems, some alarm systems, certain ventilation systems in hazardous locations, and the power for critical illumination circuits. Working on energized equipment requires proper insulated tools, proper nonflammable clothing, rubber gloves, protective shields and goggles, and in some cases, rubber blankets. None of these reasons to work on electrical equipment "hot" are found in residential wiring.

OSHA regulations allow only qualified personnel to work on or near electrical circuits or equipment that have not been de-energized. The OSHA regulations provide rules regarding "lockout and tagging" to make sure that the electrical equipment being worked on will not inadvertently be turned on while someone is working on the supposedly "dead" equipment. As the OSHA regulations state, "a lock and a tag shall be placed on each disconnecting means used to de-energize circuits and equipment."

Safety training is an important part of all recognized electrical apprenticeship and journeyman training programs across the country.

The **National Fire Protection Association (NFPA)** publications *NFPA 70E Standard for Electrical Safety Requirements for Employee Workplaces* and *NFPA 70B Recommended Practices for Electrical Equipment Maintenance* present much of the same text regarding electrical safety as does the OSHA regulation.

Safety cannot be compromised! The rule is *turn off and lock off the power, then properly tag the disconnect with a description of exactly what that particular disconnect controls.* With safety utmost in our minds, let us begin our course on the wiring of a typical residence.

LICENSING AND PERMITS

Most communities, counties, and/or states require electricians and electrical contractors to be licensed. This usually means they have taken and have passed a test. To maintain a valid license, many states require electricians and electrical contractors to attend and satisfactorily complete approved continuing education courses consisting of a specified number of classroom hours over a given period of time. Quite often, a community will have a "Residential

Only" license for electricians and contractors that limits their activity to house wiring.

Permits are a means for a community to permanently record electrical work to be done, who is doing the work, and to schedule inspections during and after the "rough-in" stage and in the "final" stages of construction. Usually, permits must be issued prior to starting an electrical project. In most cases, homeowners are allowed to do electrical work in their own home where they live but not in other properties they might own.

Figure 1-2 is a simple application for an electrical permit form. Some permit application forms are much more detailed.

If you are not familiar with licensing and permit requirements in your area, it makes sense to check this out with your local electrical inspector or building department before starting an electrical project. Not to do so could prove to be very costly. Many questions can be answered: You will find out what tests, if any, must be taken; which permits are needed; which electrical code is enforced in your community; minimum size electrical service; and so on. Generally, the electrical permit is taken out by an electrical contractor who is licensed and registered as an electrical contractor in the jurisdictional area.

For new construction or for a main electrical service change, you will also need to contact the electric utility.

Temporary Wiring

There is an ever-present electrical shock hazard on construction sites. The *NEC®* addresses this in *Article 527*. This is covered in Unit 6.

PLANS

An architect or electrical engineer prepares a set of drawings that shows the necessary instructions and details needed by the skilled workers that are to build the structure. These are referred to as **plans**, **prints**, **blueprints**, **drawings**, **construction drawings**, or **working drawings**. The sizes, quantities, and locations of the materials required and the construction features of the structural members are shown at a glance. These details of construction must be studied and interpreted by each skilled construction craftsperson—masons, carpenters, electricians, and others—before the actual work is started.

APPLICATION FOR ELECTRICAL PERMIT
VILLAGE OF ANYWHERE, USA   1-234-567-8900, EXT. 1234

Date _____   Permit No. _____

Owner _____   Job Address _____

Telephone No. _____   Job Start Date _____

CONTRACTOR INFORMATION AND SIGNATURE

Electrical Contractor _____   Tel. No. _____

Address _____ City _____ State _____ Zip _____

Registration No. _____ City of Registration _____

Supervising Electrician (Please Print Name) _____

Supervising Electrician's Signature _____

Insurance Bond _____ Village Business License _____

SERVICE INSPECTIONS OR REVISIONS

Existing Service Size: Amps_____ Volts_____ No. of Circuits_____ No. Added _____

New Service Size: Amps_____ Volts_____ No. of Circuits_____

Type: Overhead_____ Underground_____

Service Installation Fees: 100-200 amps $50_____ over 200 amps $75_____

MOTORS AND AIR CONDITIONING EQUIPMENT

No. of Motors: up to 1 HP @ $50_____   Over 1-10 HP @ $50_____

                11-25 HP @ $25_____   Over 25 HP @ $25_____

Air Conditioner/Heat Pump:

No. of Tons_____ ($20 for first ton, $5 for each additional ton)   _____

Furnace (electric):   kW_____ Amps_____ ($25)   _____

Dryer (electric):   kW_____ Amps_____ ($10)   _____

Range, oven, cooktop (electric): Total kW_____ Amps_____ ($10 each)   _____

Water Heater (electric): kW_____ Amps_____ ($10)   _____

TYPE OF MISCELLANEOUS ELECTRIC WORK

_____

_____

Minimum Inspection Fee: $40.00   _____

Escrow Deposit, if applicable   _____

TOTAL DUE   _____

Figure 1-2   A typical Application for Electrical Permit.

The electrician must be able to: (1) convert the two-dimensional plans into an actual electrical installation, and (2) visualize the many different views of the plans and coordinate them into a three-dimensional picture, as shown in Figure 1-3.

The ability to visualize an accurate three-dimensional picture requires a thorough knowledge of blueprint reading. Because all of the skilled trades use a common set of plans, the electrician must be able to interpret the lines and symbols that refer to the electrical installation and also those used by the other construction trades. The electrician must know the structural makeup of the building and the construction materials to be used

Plans might be **black line prints**, which are simply photocopies, or they might be computer-aided drawings (CAD). Although pretty much a thing of the past, we still have blueprints. Blueprints are

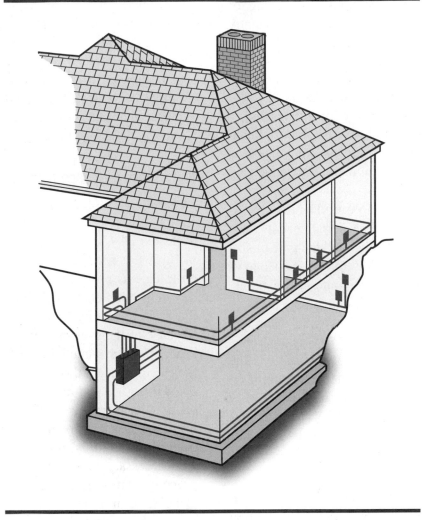

Figure 1-3   Three-dimensional view of house wiring.

created by running yellowish, light-sensitive paper and transparent Mylar® with black images on it through a blueprint machine, where the two sheets are exposed to a bright light for a short period of time. The light-sensitive paper turns white in all places except where the black images are. Where the black images are, the light-sensitive paper turns blue. That is why it is called a blueprint.

A tear-out/fold-out set of full-size plans for the residence referred to throughout this text is included in the back of the book.

SPECIFICATIONS

The **specifications** (specs, for short) for the electrical work indicated on the plans for the residence discussed throughout this text are found in the back of the book, before the Appendix.

Working drawings are usually complex because of the amount of information that must be included. To prevent confusing detail, it is standard practice to include with each set of plans a set of detailed written specifications prepared by the architect.

These specifications provide general information to be used by all trades involved in the construction. In addition, specialized information is given for the individual trades. The specifications include information on the sizes, the type, and the desired quality of the standard parts to be used in the structure.

Typical specifications include a section on "General Clauses and Conditions," which is applicable to all trades involved in the construction. This section is followed by detailed requirements for the various trades—excavating, masonry, carpentry, plumbing, heating, electrical work, painting, and others.

In the electrical specifications, the listing of standard electrical parts and supplies frequently includes the manufacturers' names and the catalog numbers of the specified items. Such information ensures that these items will be of the correct size, type, and electrical rating, and that the quality will meet a certain standard. To allow for the possibility that the contractor will not always be able to obtain the specified item, the phrase "or equivalent" is usually added after the manufacturer's name and catalog number.

The specifications are also useful to the electrical contractor in that all of the items needed for a specific job are grouped together and the type or size of each item is indicated. This information allows the contractor to prepare an accurate cost estimate without having to find all of the data on the plans.

If there is a difference between the plans and specifications, the specifications will take preference. The electrical contractor should discuss the matter with the homeowner, architect, and engineer. The cost of the installation might vary considerably because of the difference(s), so obtain any changes to the plans and/or specifications in writing.

SYMBOLS AND NOTATIONS

The architect uses **symbols** and **notations** to simplify the drawing and presentation of information concerning electrical devices, appliances, and equipment. For example, an electric range outlet symbol is shown in Figure 1-4.

Most symbols have a standard interpretation throughout the country as adopted by the American National Standards Institute (ANSI). Symbols are described in detail in Unit 2.

A notation will generally be found on the plans (blueprints) next to a specific symbol calling attention to a variation, type, size, quantity, or other necessary information. In reality, a symbol might be considered to be a notation because symbols represent words, phrases, numbers, quantities, and so on.

Figure 1-4　An electric range outlet symbol.

Another method of using notations to avoid cluttering up a blueprint is to provide a system of symbols that refer to a specific table. For example, the written sentences on plans could be included in a table referred to by a notation. Figure 2-10 is an example of how this could be done. The special symbols that refer to the table would have been shown on the actual plan.

NATIONAL ELECTRICAL CODE® (NEC®)

The *National Electrical Code®* is published by the National Fire Protection Association, and is referred to as NFPA 70. The *NEC®* was first published in 1897. It is revised every three years so as to be as up-to-date as possible. The *NEC®* does not become law until adopted by official action of the legislative body of a city, municipality, county, or state. Because of the ever-present danger of fire or shock hazard through some failure of the electrical system, the electrician and the electrical contractor must use listed materials and must perform all work in accordance with recognized standards. The *NEC®* is the basic standard that governs electrical work. The purpose of the Code is to provide information considered necessary for the safeguarding of people and property against electrical hazards. It states that the installation must be essentially hazard-free, but that such an installation is not necessarily efficient, convenient, or adequate, *90.1(B).* (Note: All *National Electrical Code®* section references used throughout this text are printed in italics.) It is the electrician's responsibility to ensure that the installation meets these criteria. In addition to the *NEC®,* the electrician must also consider local and state codes.

Code Arrangement

It is important to understand the arrangement of the *NEC®.*

- Administration and Enforcement,
  Article 80　　A comprehensive *suggested* typical electrical ordinance that could be adopted on a local level. It covers all the basics found in most electrical ordinances.
  Article 80 is not enforceable, it is merely a suggested electrical ordinance.

- Introduction, *Article 90*   An introduction to the *National Electrical Code.* Explains what is and what is not covered in the *NEC,* the arrangement of the *NEC,* who has the authority to enforce the Code, what mandatory rules are, what permissive rules are, and the basics of metric vs. inch-pound measurements found throughout the *NEC.*

- Chapter 1   General
Article 100: Definitions
Article 110: Requirements for Electrical Installations
Applies to all electrical installations.

- Chapter 2   Wiring and Protection
Articles 200–299
Applies to all electrical installations.

- Chapter 3   Wiring Methods
Articles 300–399
Applies to all electrical installations.

- Chapter 4   Equipment for General Use
Articles 400–499
Applies to all electrical installations.

- Chapter 5   Special Occupancies
Articles 500–599

- Chapter 6   Special Equipment
Articles 600–699

- Chapter 7   Special Conditions
Articles 700–799

- Chapter 8   Communications Systems
Articles 800–899
The articles in Chapter 8 are not subject to the requirements of Articles 1 through 7 unless specifically stated.

- Chapter 9   Tables showing dimensional data for raceways and conductors, resistance and reactance values of conductors, Class 2 and Class 3 circuit power limitations.

- Annex A   Provides a comprehensive list of product safety standards from Underwriters Laboratories (UL).

- Annex B   Provides data for determining conductor ampacity where there is engineering supervision.

- Annex C   Has tables showing the maximum number of conductors permitted in various types of raceways.

- Annex D   Examples of load calculations.

- Index   The alphabetical index for the *NEC.*

Note: Annexes and Fine Print Notes (FPN) are informational only. They are not mandatory.

Important: ▶In past editions of the *NEC,* requirements appeared as *Sections*. The word *Section* has been removed in most places in the *NEC.* This edition of *Electrical Wiring—Residential* reflects this change. When you see a stand-alone number in italics, such as *310.15,* you will know that this is referencing a section from the *NEC.*

In past editions, a Code section was referred to like this:

See *Section 310-15* for ampacity ratings of conductors.

In this latest edition, a Code section is referred to like this:

See *310.15* for ampacity ratings of conductors. ◀

Who Writes the Code?

For each Code cycle, the NFPA solicits proposals from anyone interested in electrical safety. Anyone may submit a proposal to change the *NEC* using the Proposal Form found in the back of the *NEC.* Proposals received are then given to a specific Code Making Panel (CMP) to accept, reject, accept in part, accept in principle, or accept in principle in part. These actions are published in the Report on Proposals (ROP). Individuals may send in their comments on these actions using the Comment Form found in the ROP. The CMPs meet again to review and take action on the comments received.

These actions are published in the Report on Comments (ROC). Final action (voting) on proposals and comments is taken at the NFPA annual meeting.

Individuals that serve on Code Making Panels are electrical inspectors, electrical contractors, electrical engineers, individuals from utilities, manufacturers, testing laboratories, the Consumer Product Safety Commission, insurance companies, and similar organizations. CMP members are appointed by the NFPA. The CMPs have 10 to 20 principal members, plus a similar number of alternate members. All of the CMPs have a good balance of representation.

Which Edition of the *NEC*® to Use?

This text is based on the 2002 edition of the *NEC.*® Some municipalities, cities, counties, and states have not yet adopted this edition and may still be using older editions. Check with your local electrical inspector to find out which edition of the Code is in force.

Copies of the latest edition of the *NEC-NFPA 70* may be ordered from the following:

Delmar, a division of Thomson Learning
P.O. Box 15015
Albany, New York 12212-5015
1-800-347-7707

NFPA
11 Tracy Drive
Avon, MA 02322-9908
1-800-344-3555

International Association of Electrical Inspectors
901 Waterfall Way, Suite 602
Richardson, TX 75080-7702
1-800-786-IAEI

Residential Electrical Maintenance Code for One- and Two-Family Dwellings

This code is published by the National Fire Protection Association, and is referred to as *NFPA 73*. It is a rather brief seven-page code that provides requirements for evaluating installed electrical systems within and associated with *existing* one- and two-family dwellings to identify

safety, fire, and shock hazards—such as improper installations, overheating, physical deterioration, and abuse. This code lists most of the electrical things to inspect in an existing dwelling that could result in a fire or shock hazard if not corrected. It only points out things to look for that are visible. It does not get into examining concealed wiring that would require removal of permanent parts of the structure. It also does not get into calculations, location requirements, and complex topics as does the *National Electrical Code® NFPA 70.*

This code uses simple statements such as: *The service shall be adequate to serve the connected load; overcurrent protective devices shall be properly rated for conductor ampacities; conductors shall be properly terminated and supported at panelboards, boxes, and devices; raceways shall be securely fastened in place; cables and cable assemblies shall be properly secured and supported; fixture canopies shall be in place and properly secured; covers shall be in place and properly secured;* and so forth.

This code can be an extremely useful guide for electricians doing remodel work, and for electrical inspectors wanting to bring an existing dwelling to a reasonably safe condition. Many localities require that when a home changes ownership, the wiring must be brought up to some minimum standard, but not necessarily as extensively as would be the case for new construction. For example, a local code might require that a minimum 100-ampere, 120/240-volt service be installed, that the service ground be brought up to the requirements of the latest edition of the *NEC,*® that 20-ampere appliance circuits be installed in the kitchen and dining room, that **ground fault circuit interrupter (GFCI)** receptacles be installed in bathrooms and other specified areas, and that a separate 20-ampere branch-circuit be installed for the laundry equipment (clothes washer).

Homes for Physically Challenged People

The *NEC*® does not address the wiring of homes for the physically or mentally challenged, or disabilities associated with the elderly.

Each installation for the physically challenged must be based upon the specific need(s) of the individual(s) who will occupy the home. Some physically challenged people are bedridden, require the mobility of a wheelchair, have trouble reaching, have trouble bending, etc. There are no hard and fast rules that *must* be followed—only many suggestions to consider.

Some of these are:

- Install more ceiling luminaires (lighting fixtures) instead of switching receptacles. Cords and lamps are obstacles.

- Install luminaires (lighting fixtures) having more than one bulb.

- Go "overboard" in the amount of lighting for all rooms, entrances, stairways, stairwells, closets, pantries, bathrooms, and so on.

- Use higher wattage bulbs—not to exceed the wattage permitted in the specific fixture.

- Consider installing luminaires (lighting fixtures) in certain areas (such as bathrooms and hallways) to be controlled by motion detectors.

- Consider installing exhaust fans in certain areas (such as laundries, showers, and bathrooms) that turn on automatically when the humidity reaches a predetermined value.

- Consider the height of switches and thermostats, usually 42 in. (1.0 m) or lower, instead of the standard 46–52 in. (1.15 to 1.3 m).

- Consider rocker-type switches instead of toggle type.

- Install pilot light switches.

- Consider "jumbo" switches.

- Be sure stairways and stairwells are well lit.

- Consider stair tread lighting.

- Position lighting switches so as not to be over stairways or ramps.

- Locate switches and receptacles to be readily and easily accessible—not behind doors or other hard-to-reach places.

- Consider installing wall receptacle outlets higher 24–27 in. (600 mm to 675 mm) than normal 12 in. (300 mm) height.

- Install lighted doorbell buttons.

- Chimes: consider adding a strategically located "dedicated" lamp(s) that will turn on when door bell buttons are pushed. The wiring diagram for this is found in Figure 25-25.

- Telephones: consider adding visible light(s) strategically located that will flash at the same time the telephone is ringing.

- Consider installing receptacle outlets and switches on the face of kitchen cabinets. Wall outlets and switches can be impossible for the physically challenged person to reach.

- Consider fire, smoke, and security systems directly connected to a central office for fast response to emergencies that do not depend upon the disabled to initiate the call.

- Consider installing the fuse box or breaker panel on the first floor instead of in the basement.

- Consider the advanced home systems concept of remote control of lighting, receptacles, appliances, television, telephones, and so on. The control features can make life much easier for a disabled person.

The American National Standards Institute (ANSI) publication *ANSI A117.1–1998, Guidelines for Accessible and Usable Buildings and Facilities,* contains many suggestions and considerations for buildings and facilities for the physically challenged. This and other standards can be ordered from:

American National Standards Institute
11 West 42nd Street
New York, NY 10036
212-642-4900
www.ansi.com

The same standard is available from the International Code Council (ICC) under the publication number *ICC/ANSI A117.1-1998.*

Another virtually unlimited source of standards is:

Global Engineering Documents
15 Inverness Way East
Englewood, CO 80112

Building Codes

The majority of the building departments across the country have for the most part adopted the *NEC®* rather than attempting to develop their own electrical codes. As you study this text, you will note numerous references to the *NEC*,® the electrical inspector, or the authority having jurisdiction.

An authority's level of knowledge varies. The electrical inspector may be full-time or part-time and may also have responsibility for other trades, such as plumbing or heating. The heads of the building departments in many communities are typically called the Building Commissioners or Directors of Development. Regardless of title, they are responsible for ensuring that the building codes in their communities are followed.

Communities, rather than writing their own codes, belong to one of the major building officials' organizations, and have adopted the building codes of that particular organization. These organizations are:

- **BOCA** Building Official & Code
  Administrators International, Inc.
  4051 W. Flossmoor Road
  Country Club Hills, Illinois 60478-5795
  Phone: 708-799-2300

- **ICBO** International Conference of
  Building Officials
  5360 Workman Mill Road
  Whittier, California 90601-2298
  Phone: 562-699-0541

- **SBCCI** Southern Building Code Congress
  International, Inc.
  900 Montclair Road
  Birmingham, Alabama 35212-1206
  Phone: 205-591-1853

- **IAPMO** International Association of
  Plumbing and Mechanical Officials
  5032 Alhambra Avenue
  Los Angeles, CA 90032-3490
  Phone: 213-223-1471

- **ICC** International Code Council
  5203 Leesburg Pike, Suite 600
  Falls Church, VA 22041-3401
  Phone: 703-931-4533

ICC is an organization that provides technical, educational, and informational products and services in support of international codes. BOCA, ICBO, and SBCCI are members of this organization. The ICC publishes a building code, mechanical code, plumbing code, fire code, energy code, residential code, and an electrical code. The ICC electrical code is similar in many respects to the *NEC*,® but there are differences. This text, *Electrical Wiring—Residential*, is based on the *NEC*.®

Because local electrical codes may differ from the *NEC*,® you should check with the local inspection authority to determine which edition of the *NEC*® is enforced, and what, if any, local requirements or amendments take precedence over the *NEC*.®

AMERICAN NATIONAL STANDARDS INSTITUTE

The **American National Standards Institute (ANSI)** is an organization that coordinates the efforts and results of the various standards-developing organizations, such as those mentioned in previous paragraphs. Through this process, ANSI approves standards that then become recognized as American national standards. One will find much similarity between the technical information found in ANSI standards, the Underwriters Laboratories standards, the International Electronic and Electrical Engineers standards, and the *NEC*.®

International Association of Electrical Inspectors

The International Association of Electrical Inspectors (IAEI) is a nonprofit organization. The IAEI membership consists of electrical inspectors, building officials, electricians, engineers, contractors, and manufacturers throughout the United States and Canada. The major goal of the IAEI is to improve the understanding of the *NEC*.® Representatives of this organization serve as members of the various Code Making Panels of the *NEC*,® and share equally with other members in the task of reviewing and revising the *NEC*.®

The IAEI publishes a bimonthly magazine— *The IAEI News*. It is devoted entirely to electrical code topics. Anyone in the electrical industry is welcome to join the IAEI. An application form is found after the Appendix of this text. Their address is:

International Association of Electrical Inspectors
901 Waterfall Way, Suite 602
Richardson, TX 75080-7702
Phone: 1-800-786-IAEI
www.iaei.com

Code Definitions

The electrical industry uses many words (terms) that are unique to the electrical trade. These terms need clear definitions to enable the electrician to understand completely the meaning intended by the Code.

Article 100 of the *NEC®* is a "dictionary" of these terms. *Article 90* also provides further clarification of terms used in the *NEC.®* A few of the terms defined in the Code follow.

Ampacity: The current in amperes a conductor can carry continuously under the conditions of use without exceeding its temperature rating.

Approved: Acceptable to the authority having jurisdiction.

►**Authority Having Jurisdiction:** The organization, office, or individual responsible for approving equipment, materials, an installation, or a procedure. See *90.4.*◄

Dwelling Unit: One or more rooms for the use of one or more persons as a housekeeping unit with space for eating, living, and sleeping, and permanent provisions for cooking and sanitation. (Note: The terms *dwelling* and *residence* are used interchangeably throughout this text.)

Fine Print Notes (FPN): Fine Print Notes (FPNs) are found throughout the Code. FPNs are "explanatory" in nature, in that they refer to other sections of the Code. FPNs also define things where further description is necessary. See *90.5.*

Identified (as applied to equipment): Recognizable as suitable for a specific purpose, function, use, environment, or application, where described in a particular Code requirement. Suitability of use, marked on or provided with the equipment, may include labeling or listing.

Identified (as applied to conductors and terminals): The grounded circuit, feeder or service-entrance conductor is "identified" by its white or gray color. Terminals on wiring devices, panels, and disconnect switches have their terminals for the grounded circuit conductor "identified" with silver colored screws. In the electrical industry,you will hear the terms *grounded circuit conductor, neutral conductor,* or *the white wire.* For all practical purposes, these terms all mean the same thing. They refer to the grounded circuit conductor. An exception to this is that a neutral conductor is always a grounded conductor, but a grounded circuit conductor is not always a neutral, as in the case of industrial wiring where a three-phase, grounded "B" phase system might be used. See *250.20.*

Labeled: Equipment or materials to which has been attached a label, symbol, or other identifying mark of an organization acceptable to the authority having jurisdiction and concerned with product evaluation, that maintains periodic inspection of production of labeled equipment or materials and by whose labeling the manufacturer indicates compliance with appropriate standards or performance in a specified manner.

►**Listed:** Equipment, materials, or services included in a list published by an organization that is acceptable to the authority having jurisdiction and concerned with evaluation of products or services, that maintains periodic inspection of production of listed equipment or materials or periodic evaluation of services, and whose listing states that either the equipment, material, or services meets appropriate designated standards or has been tested and found suitable for a specified purpose.◄

Shall: Indicates a mandatory rule, *90.5.* As you study the *NEC,®* think of the word "shall" as meaning "must." Some examples found in the *NEC®* where the word "shall" is used in combination with other words are: *shall be, shall have, shall not, shall be permitted, shall not be permitted,* and *shall not be required.*

One of the most far-reaching *NEC®* rules is *110.3(B).* This section states that the installation and use of listed or labeled equipment must conform to any instructions included in the listing or labeling. This means that an entire electrical system and all of the system's electrical equipment must be *installed* and *used* in accordance with the *NEC®* and the numerous standards to which the electrical equipment has been tested.

Metrics (SI) and the *NEC®*

►The United States is the last major country in the world not using the metric system as the primary system. We have been very comfortable using

English (United States Customary) values. But this is changing. Manufacturers are now showing both inch-pound and metric dimensions in their catalogs. Plans and specifications for governmental new construction and renovation projects started after January 1, 1994 have been using the metric system. You may not feel comfortable with metrics, but metrics are here to stay. You might just as well get familiar with the metric system.

The *NEC*® and other NFPA standards are becoming international standards. All measurements in the 2002 *NEC*® are shown with metrics first, followed by the inch-pound value in parentheses. For example, 600 mm (24 in.).

In *Electrical Wiring—Residential*, ease in understanding is of utmost importance. Therefore, inch-pound values are shown first, followed by metric values in parentheses. For example, 24 in. (600 mm).

A *soft metric conversion* describes a product already designed and manufactured to the inch-pound system, with its dimensions converted to metric dimensions. The product does not change in size.

A *hard metric measurement* describes a product designed to SI metric dimensions. No conversion from inch-pound measurement units is involved. A *hard conversion* describes an existing product redesigned into a new size.

In the 2002 edition of the *NEC*,® existing inch-pound dimensions did not change. Metric conversions were made, then rounded off. Where rounding off would create a safety hazard, the metric conversions are mathematically identical.

For example, if a dimension is required to be six ft, it is shown in the *NEC*® as 1.8 m (6 ft). Note that the 6 ft remains the same, and the metric value of 1.83 m has been rounded off to 1.8 m. This edition of *Electrical Wiring—Residential* reflects these rounded-off changes, except that the inch-pound measurement is shown first, i.e., 6 ft (1.8 m).

Trade Sizes

A unique situation exists. Strange as it may seem, what electricians have been referring to for years has not been correct!

Raceway sizes have always been an approximation. For example, there has never been a ½-in. race-

TRADE SIZE	INSIDE DIAMETER (I.D.)
½ Electrical Metallic Tubing	0.622 in.
½ Electrical Nonmetallic Tubing	0.560 in.
½ Flexible Metal Conduit	0.635 in.
½ Rigid Metal Conduit	0.632 in.
½ Intermediate Metal Conduit	0.660 in.

Table 1-1 Comparison of trade size vs. actual inside diameters.

way! Measurements taken from the *NEC*® for a few types of raceways are shown in Table 1-1.

You can readily see that the cross-sectional areas, critical when determining conductor fill, are different. It makes sense to refer to conduit, raceway, and tubing sizes as *trade sizes*. The *NEC*® in *90.9(C)(1)* states that "where the actual measured size of a product is not the same as the nominal size, trade size designators shall be used rather than dimensions. Trade practices shall be followed in all cases." This edition of *Electrical Wiring—Residential* uses the term *trade size* when referring to conduits, raceways, and tubing. For example, past editions may have referred to a ½-in. EMT. In this edition, it is referred to as trade size ½ EMT.

The *NEC*® also uses the term *metric designator*. A ½-in. EMT is shown as *metric designator 16 (½)*. A 1-in. EMT is shown as *metric designator 27 (1)*. The numbers 16 and 27 are the *metric designator* values. The (½) and (1) are the *trade sizes*. The metric designator is the raceway's inside diameter—in rounded off millimeters (mm). Table 1-2 shows some of the more common sizes of conduit, raceways, and tubing. A complete table is found in the *NEC*,® *Table 300.1(C)*. Because of possible confusion, this text uses only the term *trade size* when referring to conduit and raceway sizes.

Conduit knockouts in boxes do not measure up to what we call them. Table 1-3 shows some examples.

Outlet boxes and device boxes use their nominal measurement as their *trade size*. For example, a 4 in. × 4 in. × 1½ in. does not have an internal cubic-inch volume of 4 in. × 4 in. × 1½ in. = 24 in.³ *Table 314.16(A)* shows this size box as having 21 cubic-in. volume. This table shows *trade sizes* in two columns—millimeters and inches.

METRIC DESIGNATOR AND TRADE SIZE	
Metric Designator	**Trade Size**
12	⅜
16	½
21	¾
27	1
35	1¼
41	1½
53	2
63	2½
78	3

Table 1-2   Trade sizes of raceways and their metric designator identification.

TRADE SIZE KNOCKOUT	ACTUAL MEASUREMENT
½	⅞ in.
¾	1³⁄₃₂ in.
1	1⅜ in.

Table 1-3   Comparison of knockout trade size vs. actual measurement.

In this text, a square outlet box is referred to as trade size 4 × 4 × 1½. Similarly, a single-gang device box would be referred to as a trade size 3 × 2 × 3 box.

Trade sizes for construction material will not change. A 2 × 4 is really a *name*, not an actual dimension. A 2 × 4 will still be referred to as a 2 × 4. This is its *trade size*.

In this text, measurements directly related to the *NEC®* are given in both inch-pound and metric units. In many instances, only the inch-pound units are shown. This is particularly true for the examples of raceway and box fill calculations, load calculations for square foot areas, and on the plans (drawings).

Because the *NEC®* rounded off most metric conversion values, a computation using metrics results in a different answer when compared to the same computation done using inch-pounds. For example, load calculations for a residence are based on 3 volt-amperes per square foot or 33 volt-amperes per square meter.

For a 40 ft × 50 ft dwelling: 3 VA × 40 ft × 50 ft = 6000 volt-amperes.

In metrics, using the rounded off values in the *NEC®*: 33 VA × 12 m × 15 m = 5940 volt-amperes.

The difference is small, but nevertheless, there is a difference.

To show calculations in both units throughout this text would be very difficult to understand and would take up too much space. Calculations in either metrics or inch-pounds are in compliance with the *NEC®, 90.9(D)*. In *90.9(C)(3)* we find that metric units are not required if the industry practice is to use inch-pound units. ◀

It is interesting to note that the examples in *Chapter 9* of the *NEC®* use inch-pound units, not metrics.

Guide to Metric Usage

The metric system is a "base-10" or "decimal" system in that values can be easily multiplied or divided by ten or "powers of ten." The metric system as we know it today is known as the International System of Units (SI), derived from the French term "le Système International d'Unités."

In the metric system, the units increase or decrease in multiples of 10, 100, 1000, and so on. For instance, one megawatt (1,000,000 watts) is 1000 times greater than one kilowatt (1000 watts).

By assigning a name to a measurement, such as a *watt*, the name becomes the unit. Adding a prefix to the unit, such as *kilo*, forms the new name *kilowatt*, meaning 1000 watts. Refer to Table 1-4 for prefixes used in the metric system.

The prefixes used most commonly are *centi*, *kilo*, and *milli*. Consider that the basic unit is a meter (one). Therefore, a centimeter is 0.01 meter, a kilometer is 1000 meters, and a millimeter is 0.001 meter.

Some common measurements of length and equivalents are shown in Table 1-5.

Electricians will find it useful to refer to the conversion factors and their abbreviations shown in Table 1-6.

Refer to the Appendix of this text for a comprehensive metric conversion table. The table includes information on how to "round off" numbers for practical use on the job.

PREFIX	SYMBOL	MULTIPLIER	SCIENTIFIC NOTATION (POWERS OF TEN)	VALUE
tera	T	1 000 000 000 000	10^{12}	one trillion (1 000 000 000 000/1)
giga	G	1 000 000 000	10^{9}	one billion (1 000 000 000/1)
mega	M	1 000 000	10^{6}	one million (1 000 000/1)
kilo	k	1 000	10^{3}	one thousand (1 000/1)
hecto	h	100	10^{2}	one hundred (100/1)
deka	da	10	10^{1}	ten (10/1)
unit		1	—	one (1)
deci	d	0.1	10^{-1}	one tenth (1/10)
centi	c	0.01	10^{-2}	one hundredth (1/100)
milli	m	0.001	10^{-3}	one thousandth (1/1 000)
micro	μ	0.000 001	10^{-6}	one millionth (1/1 000 000)
nano	n	0.000 000 001	10^{-9}	one billionth (1/1 000 000 000)
pico	p	0.000 000 000 001	10^{-12}	one trillionth (1/1 000 000 000 000)

Table 1-4   Metric prefixes, symbols, multipliers, powers, and values.

one inch	=	2.54	centimeters
	=	25.4	millimeters
	=	0.025 4	meter
one foot	=	12	inches
	=	0.304 8	meter
	=	30.48	centimeters
	=	304.8	millimeters
one yard	=	3	feet
	=	36	inches
	=	0.914 4	meter
	=	914.4	millimeters
one meter	=	100	centimeters
	=	1 000	millimeters
	=	1.093	yards
	=	3.281	feet
	=	39.370	inches

Table 1-5   Some common measurements of length and their equivalents.

inches (in) × 0.025 4	=	meters (m)
inches (in) × 0.254	=	decimeters (dm)
inches (in) × 2.54	=	centimeters (cm)
centimeters (cm) × 0.393 7	=	inches (in)
inches (in) × 25.4	=	millimeters (mm)
millimeters (mm) × 0.039 37	=	inches (in)
feet (ft) × 0.304 8	=	meters (m)
meters (m) × 3.280 8	=	feet (ft)
square inches (in²) × 6.452	=	square centimeters (cm²)
square centimeters (cm²) × 0.155	=	square inches (in²)
square feet (ft²) × 0.093	=	square meters (m²)
square meters (m²) × 10.764	=	square feet (ft²)
square yards (yd²) × 0.836 1	=	square meters (m²)
square meters (m²) × 1.196	=	square yards (yd²)
kilometers (km) × 1 000	=	meters (m)
kilometers (km) × 0.621	=	miles (mi)
miles (mi) × 1.609	=	kilometers (km)

Table 1-6   Useful conversions (English/SI-SI/English) and their abbreviations.

NATIONALLY RECOGNIZED TESTING LABORATORIES

How does one know if a product is safe to use? Manufacturers, consumers, regulatory authorities, and others recognize the importance of independent, "third-party" testing of products in an effort to reduce safety risks. Unless you have all of the necessary test equipment and knowledge of how to properly test a product for safety, the surest way is to accept the findings of a third-party testing agency. In other words, "look for the label." **Nationally recognized testing laboratories (NRTL)** have the knowledge and wherewithal to test and evaluate products for safety. Make sure the product has a listing marking on it from a nationally recognized testing laboratory. If the product is too small to have a listing mark on the product itself, then look for the marking on the carton the product came in. A listed product must also be used and installed properly to assure safety. The basic rule is found in *110.3(B)*, which states that *"listed or labeled equipment shall be installed and used in accordance with any instructions included in the listing or labeling."*

There are a number of nationally recognized laboratories. The following laboratories do a considerable amount of testing and listing of electrical equipment.

Underwriters Laboratories, Inc. (UL®)

Underwriters Laboratories, Inc. (UL), founded in 1894, is a highly qualified, nationally recognized testing laboratory. UL develops standards, and performs tests to these standards. Most reputable

Figure 1-5   The Underwriters Laboratories mark.

manufacturers of electrical equipment submit their products to UL where the equipment is subjected to numerous tests. These tests determine if the product can perform safely under normal and abnormal conditions to meet published standards. After UL tests and evaluates a product, and determines that the product complies with the specific standard, the manufacturer is then permitted to *label* its product with the UL Mark (see Figure 1-5). The products are then *listed* in a UL Directory.

It should be noted that UL *does not approve* a product. Rather, UL *lists* those products that conform to a specific safety standard. A UL Listing Mark on a product means that representative samples of the product have been tested and evaluated to nationally recognized safety standards with regard to fire, electric shock, and related safety hazards.

Do not confuse a UL marking in a circle with the markings found on *recognized components*. Recognized components that have passed certain tests are marked with the letters *RU* printed backwards (see Figure 1-6). By themselves, recognized components are not to be field-installed. They are intended for use in end-use products or systems that would ultimately be tested and "listed," with the

final assembly becoming a UL listed product. Some examples of recognized components are relays, ballasts, insulating materials, special switches, and so on.

Useful UL publications are:

Electrical Construction Equipment Directory (Green Book)

Electrical Appliance and Utilization Equipment Directory (Orange Book)

General Information for Electrical Equipment Directory (White Book)

It is extremely useful for an electrician, electrical contractor, and/or electrical inspector to refer to these directories when looking for specific requirements, permitted uses, limitations, and so on, for a certain product. These directories can be purchased from Underwriters Laboratories, Inc. You may not be able to purchase all of the previous directories because of cost restraints. In this case, choose what is probably the best companion to the *NEC,*® which is the White Book.

Many times the answer to a product-related question cannot be found in the *NEC.*® It is quite possible that the answer might be found in the UL Directories.

The Green and Orange Directories provide technical information regarding a particular product, plus they list the names and addresses of manufacturers and the manufacturers' identification numbers. The White Book provides the technical information by which a product is tested, but does not show manufacturers' names and addresses.

The previous directories can be obtained by writing or calling:

Underwriters Laboratories, Inc.
333 Pfingsten Road
Northbrook, Illinois 60062-2096
847-272-8800

Underwriters Laboratories, Inc. and the Canadian Standards Association (CSA) have worked out an agreement whereby either agency can test, evaluate, and list equipment for the other agency. For example, UL might test and list an air-conditioner unit to the requirements of UL Standard

Figure 1-6   The *recognized components* mark.

Figure 1-7 The mark indicating that UL has tested and listed a product for compliance *only* with Canadian requirements.

1995 (Heating and Cooling Equipment) because the Canadian Standard C22.2 No. 236-M90 is a mirror image of UL Standard 1995. One by one, the UL and CSA Standards are becoming similar.

When UL tests and lists products that comply to the requirements of a particular CSA standard, the UL logo shown in Figure 1-7 will appear with a "C" outside of and to the left of the circle.

This means the product has been tested and evaluated for compliance *only* with Canadian requirements.

This harmonization of Codes and Standards is going on in the world today as a result of the North American Free Trade Act (NAFTA). Discussions are also going on with Mexico. When all of this is finalized, electrical equipment standards may be the same in the United States, Canada, and Mexico.

CSA International

CSA International, formerly known as the Canadian Standards Association, is the Canadian counterpart of Underwriters Laboratories, Inc. in the United States. CSA International is the source of the *Canadian Electrical Code (CEC)* and of the Canadian Standards for the testing, evaluation, and listing of electrical equipment in Canada. The *Canadian Electrical Code* is quite different from the *NEC.* A Canadian version of *Electrical Wiring— Residential* is available in Canada.

The *Canadian Electrical Code* and CSA standards can be obtained by contacting:

CSA International
178 Rexdale Blvd.
Etobicoke, Ontario, Canada
M9R 1W3
Tel: 1-800-463-6727
416-747-4000
e-mail: certinfo@csa.ca

Intertek Testing Services ITS

Formerly known as Electrical Testing Laboratories, ITS is a nationally recognized testing laboratory. They provide testing, evaluation, labeling, listing, and follow-up service for the safety testing of electrical products. This is done in conformance to nationally recognized safety standards, or to specifically designated requirements of jurisdictional authorities.

Information can be obtained by contacting:

Intertek Testing Services, NA, Inc.
Certification Center
24 Groton Avenue
Cortland, NY 13045-2014
Phone: Headquarters: 978-263-2662
Verify Listings: 1-800-DIR-LIST
Certification: 607-758-6290
Engineering: 607-758-6286

National Electrical Manufacturers Association NEMA

The National Electrical Manufacturers Association (NEMA) represents nearly 600 manufacturers of electrical products. NEMA develops electrical equipment standards, which, in many instances, are very similar to UL and other consensus standards. NEMA has representatives on the Code Making Panels for the *NEC.* Additional information can be obtained by contacting:

National Electrical Manufacturers Association
1300 North 17th Street, Suite 1847
Rosslyn, VA 22209
Phone: 703-841-3200

World Wide Web Sites

A comprehensive list of World Wide Web sites for manufacturers, organizations, and inspection and testing agencies is provided in the back of this text.

REVIEW

Note: Refer to the *NEC*® or the plans in the back of this text where necessary.

1. What is the purpose of specifications? _____

   _____

2. In what additional way are the specifications particularly useful to the electrical
   contractor? _____

   _____

3. What is done to prevent a plan from becoming confusing because of too much detail?

   _____

   _____

4. Name three requirements contained in the specifications regarding material.

   a. _____   c. _____

   b. _____

5. The specifications state that all work shall be done _____

   _____

6. What phrase is used when a substitution is permitted for a specific item? _____

   _____

7. What is the purpose of an electrical symbol? _____

   _____

8. What is a notation? _____

   _____

9. Where are notations found? _____

10. List at least twelve electrical notations found on the plans for this residence. Refer to
    the plans at the back of the text. _____

    _____

    _____

    _____

    _____

    _____

11. What three parties must be satisfied with the completed electrical installation?

    a. _____   b. _____   c. _____

12. What Code sets standards for electrical installation work? _____

    _____

13. What authority enforces the standards set by the Code? _____

    _____

14. Does the Code provide minimum or maximum standards? _____

15. What do the letters *UL* signify? _____

_____

16. What section of the Code states that all listed or labeled equipment shall be installed or used in accordance with any instructions included in the listing or labeling?

_____

17. When the words "shall be" appear in a Code reference, they mean that it (must)(may) be done. (Underline the correct word.)

18. What is the purpose of the *NEC®*?

_____

_____

_____

_____

_____

19. Does compliance with the Code always result in an electrical installation that is adequate, safe, and efficient? Why? _____

_____

_____

_____

_____

20. Name two nationally recognized testing laboratories. _____

_____

_____

21. a. Do Underwriters Laboratories and the other recognized testing laboratories "approve" products? _____

   b. What do these testing laboratories do? _____

_____

_____

_____

22. a. Has the *NEC®* been officially adopted by the community in which you live? _____

   b. By the state in which you live? _____

   c. If your answer is YES to a. or b., are there amendments to the *NEC®*? _____

   d. If your answer is YES to c., list some of the more important amendments.

_____

_____

_____

_____

23. Does the *NEC®* make suggestions about how to wire a house that will be occupied by handicapped persons? _____

24. A junction box on a piece of European equipment is marked 200 cm^3. Convert this to cubic inches. _____

_____

25. Convert 4,500 watts to Btu/hour. _____

_____

26. You will learn in Unit 3 that residential lighting loads are based on 3 volt-amperes per ft^2 (33 volt-amperes per m^2). Determine the minimum lighting load required for an area of 186 m^2. _____

_____

Electrical Symbols and Outlets

OBJECTIVES

After studying this unit, the student should be able to

- identify and explain the electrical outlet symbols used in the plans of the single-family dwelling.

- discuss the types of outlets, boxes, luminaires (fixtures), and switches used in the residence.

- explain the methods of mounting the various electrical devices used in the residence.

- understand the meaning of the terms *receptacle outlet* and *lighting outlet*.

- understand the preferred way to position receptacles in wall boxes.

- understand issues involved in remodel work.

- know how to position wall boxes in relation to finished wall surfaces.

- know how to make surface extensions from concealed wiring methods.

- understand how to determine the number of wires permitted in a given size box.

- discuss interchangeable wiring devices.

- understand the concept of fire resistance rating of walls and ceilings.

ELECTRICAL SYMBOLS

Electrical symbols used on an architectural plan show the location and type of electrical device required. A typical electrical installation as taken from a plan is shown in Figure 2-1.

The *NEC®* describes an **outlet** as "a point on the wiring system at which current is taken to supply utilization equipment."

A **receptacle outlet** is "an outlet where one or more receptacles are installed," Figure 2-2.

A **lighting outlet** is "an outlet intended for the direct connection of a lampholder, a luminaire (lighting fixture), or a pendant cord terminating in a lampholder," Figure 2-3.

The *NEC®* defines a *device* as "a unit of an electrical system which is intended to carry but not utilize electric energy."

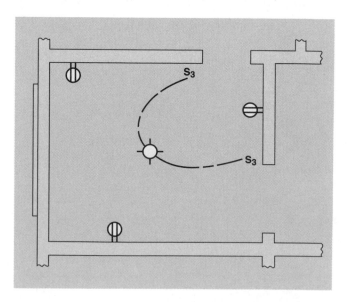

Figure 2-1 Use of electrical symbols and notations on a floor plan.

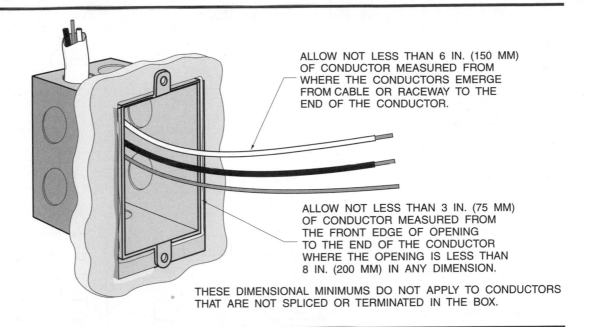

ALLOW NOT LESS THAN 6 IN. (150 MM) OF CONDUCTOR MEASURED FROM WHERE THE CONDUCTORS EMERGE FROM CABLE OR RACEWAY TO THE END OF THE CONDUCTOR.

ALLOW NOT LESS THAN 3 IN. (75 MM) OF CONDUCTOR MEASURED FROM THE FRONT EDGE OF OPENING TO THE END OF THE CONDUCTOR WHERE THE OPENING IS LESS THAN 8 IN. (200 MM) IN ANY DIMENSION.

THESE DIMENSIONAL MINIMUMS DO NOT APPLY TO CONDUCTORS THAT ARE NOT SPLICED OR TERMINATED IN THE BOX.

Figure 2-2 Where a receptacle is connected to the branch-circuit wires, the outlet is called a *receptacle outlet*. *NEC® 300.14* states the minimum length of conductors at an outlet, junction, or switch point for splices or the connection of luminaires (fixtures) or devices. Leaving the conductors too long can cause crowding of the wires in the box, leading to possible short circuits and/or ground faults.

By definition, toggle switches and receptacles are devices because they carry current, but do not consume power.

The term *opening* is widely used by electricians and electrical contractors when estimating the cost of an installation. The term *opening* covers all lighting outlets, receptacle outlets, junction boxes, switches, etc. The electrician and/or electrical contractor will estimate a job at "X dollars per lighting outlet," "X dollars per switch," "X dollars per receptacle outlet," and so on. These estimates will include the time and material needed to complete the job. Each type of electrical *opening* is represented on the electrical plans as a symbol. The electrical openings in Figure 2-1 are shown by the symbols in Figure 2-4.

ANSI recently published a totally revised standard entitled *Symbols for Electrical Construction Drawings*. This is the first revision in over 20 years. Figure 2-5, Figure 2-6, Figure 2-7, Figure 2-8, Figure 2-9, and Figure 2-10 show the electrical symbols most commonly found on architectural and electrical plans. Because some items may have more

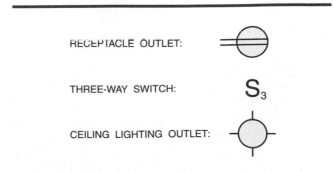

LIGHTING OUTLET

RECEPTACLE OUTLET:

THREE-WAY SWITCH: S_3

CEILING LIGHTING OUTLET:

Figure 2-3 When a luminaire (lighting fixture) is connected to the branch-circuit wires, the outlet is called a *lighting outlet*.

Figure 2-4 Some common electrical symbols.

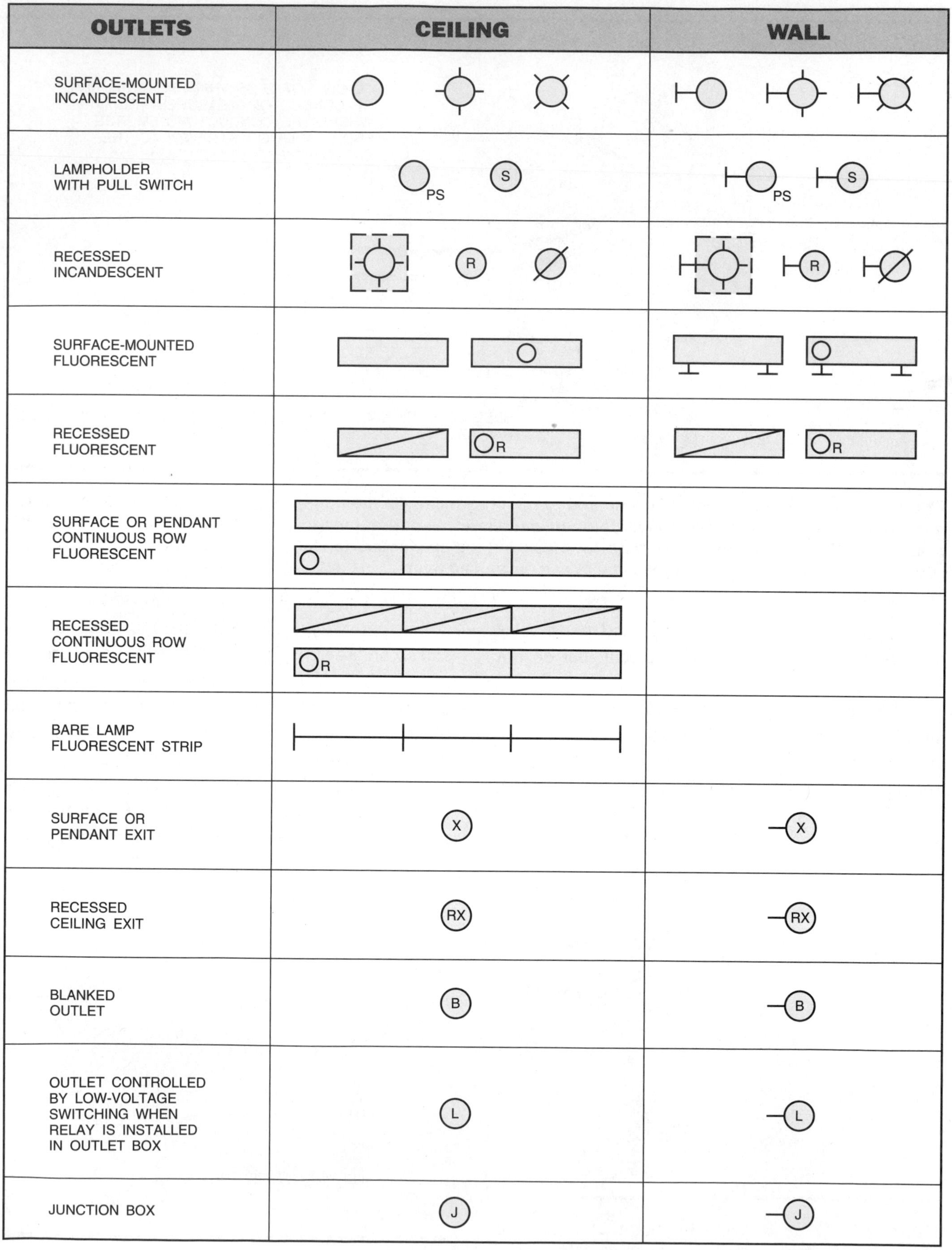

OUTLETS	CEILING	WALL
SURFACE-MOUNTED INCANDESCENT		
LAMPHOLDER WITH PULL SWITCH		
RECESSED INCANDESCENT		
SURFACE-MOUNTED FLUORESCENT		
RECESSED FLUORESCENT		
SURFACE OR PENDANT CONTINUOUS ROW FLUORESCENT		
RECESSED CONTINUOUS ROW FLUORESCENT		
BARE LAMP FLUORESCENT STRIP		
SURFACE OR PENDANT EXIT	Ⓧ	─Ⓧ
RECESSED CEILING EXIT	Ⓡⓧ	─Ⓡⓧ
BLANKED OUTLET	Ⓑ	─Ⓑ
OUTLET CONTROLLED BY LOW-VOLTAGE SWITCHING WHEN RELAY IS INSTALLED IN OUTLET BOX	Ⓛ	─Ⓛ
JUNCTION BOX	Ⓙ	─Ⓙ

Figure 2-5   Lighting outlet symbols.

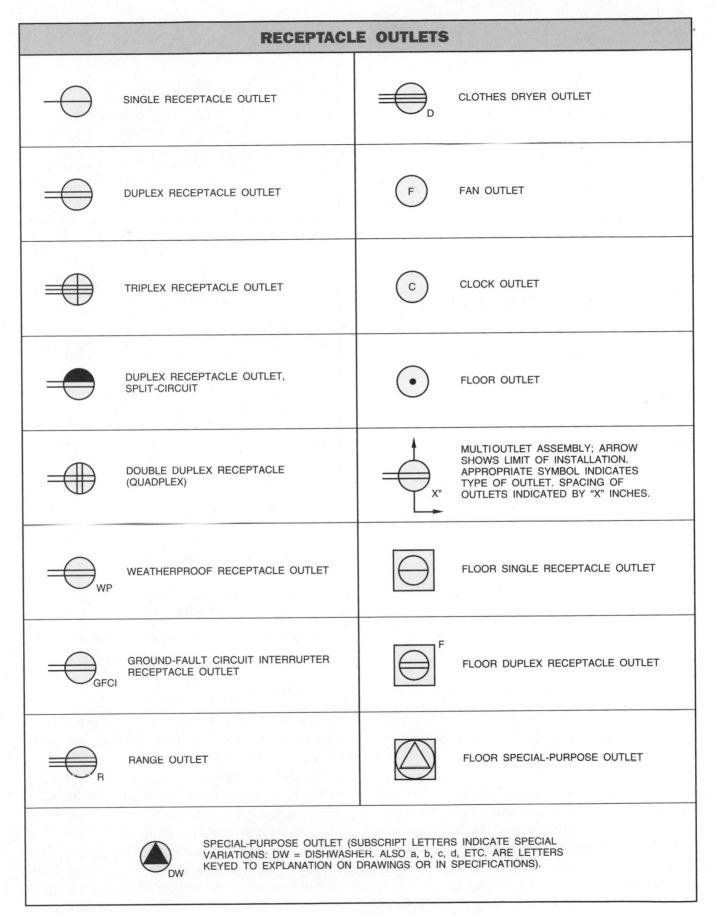

Figure 2-6   Receptacle outlet symbols.

SWITCH SYMBOLS

S	SINGLE-POLE SWITCH
S_2	DOUBLE-POLE SWITCH
S_3	THREE-WAY SWITCH
S_4	FOUR-WAY SWITCH
S_D	DOOR SWITCH
S_{DS}	DIMMER SWITCH
S_G	GLOW SWITCH TOGGLE— GLOWS IN OFF POSITION
S_K	KEY-OPERATED SWITCH
S_{KP}	KEY SWITCH WITH PILOT LIGHT
S_{LV}	LOW-VOLTAGE SWITCH
S_{LM}	LOW-VOLTAGE MASTER SWITCH
S_{MC}	MOMENTARY-CONTACT SWITCH
⟨M⟩	OCCUPANCY SENSOR—WALL-MOUNTED WITH "OFF-AUTO" OVERRIDE SWITCH
⟨M⟩ P	OCCUPANCY SENSOR—CEILING-MOUNTED "P" INDICATES MULTIPLE SWITCHES WIRE-IN PARALLEL
S_P	SWITCH WITH PILOT LIGHT ON WHEN SWITCH IS ON
S_T	TIMER SWITCH
S_R	VARIABLE-SPEED SWITCH
S_{WP}	WEATHERPROOF SWITCH

Figure 2-7 Switch symbols.

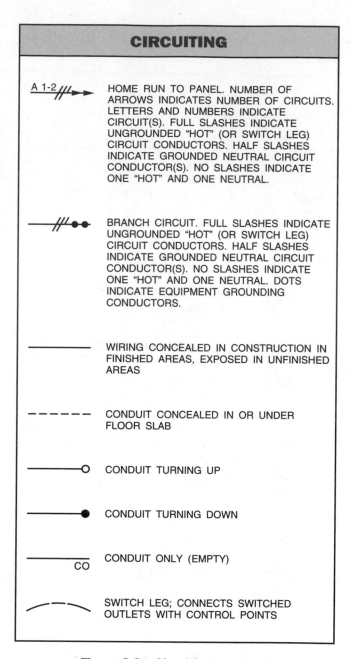

CIRCUITING

A 1-2 ⫻→	HOME RUN TO PANEL. NUMBER OF ARROWS INDICATES NUMBER OF CIRCUITS. LETTERS AND NUMBERS INDICATE CIRCUIT(S). FULL SLASHES INDICATE UNGROUNDED "HOT" (OR SWITCH LEG) CIRCUIT CONDUCTORS. HALF SLASHES INDICATE GROUNDED NEUTRAL CIRCUIT CONDUCTOR(S). NO SLASHES INDICATE ONE "HOT" AND ONE NEUTRAL.
—⫻••	BRANCH CIRCUIT. FULL SLASHES INDICATE UNGROUNDED "HOT" (OR SWITCH LEG) CIRCUIT CONDUCTORS. HALF SLASHES INDICATE GROUNDED NEUTRAL CIRCUIT CONDUCTOR(S). NO SLASHES INDICATE ONE "HOT" AND ONE NEUTRAL. DOTS INDICATE EQUIPMENT GROUNDING CONDUCTORS.
———	WIRING CONCEALED IN CONSTRUCTION IN FINISHED AREAS, EXPOSED IN UNFINISHED AREAS
- - - - -	CONDUIT CONCEALED IN OR UNDER FLOOR SLAB
—o	CONDUIT TURNING UP
—•	CONDUIT TURNING DOWN
—— CO	CONDUIT ONLY (EMPTY)
⌒	SWITCH LEG; CONNECTS SWITCHED OUTLETS WITH CONTROL POINTS

Figure 2-8 Circuiting symbols.

than one symbol, it is important to check the plans and specifications of any job you are working on for a symbol schedule to make sure you have interpreted the symbols correctly.

The dashed lines in Figure 2-1 run from the outlet to the switch or switches that control the outlet. These lines are usually curved so that they cannot be mistaken for invisible edge lines. Outlets shown on the plan without curved dashed lines are independent outlets and have no switch control.

A study of the plans for the single-family dwell-ing shows that many different electrical symbols are used to represent the electrical devices and equipment used in the building.

Figure 2-5, Figure 2-6, Figure 2-7, Figure 2-8, Figure 2-9, and Figure 2-10 list the standard approved electrical symbols and their meanings. Many of these symbols can be found on the accompanying plans of the residence. Note that in these figures several symbols have the same shape. However, differences in the interior presentation indicate that the meanings of the symbols are different. For

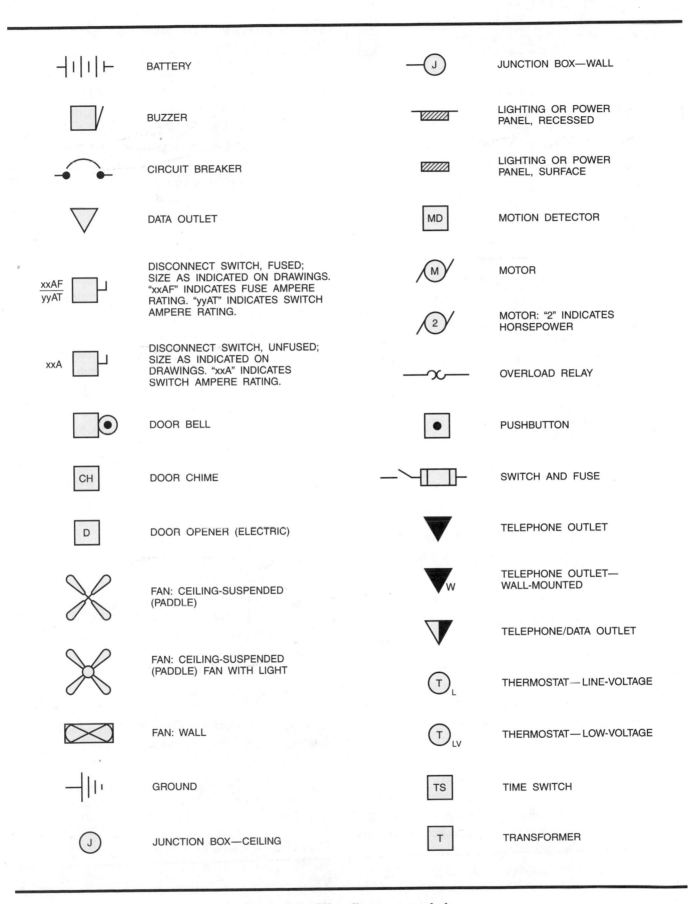

BATTERY		JUNCTION BOX—WALL
BUZZER		LIGHTING OR POWER PANEL, RECESSED
CIRCUIT BREAKER		LIGHTING OR POWER PANEL, SURFACE
DATA OUTLET		MOTION DETECTOR
xxAF/yyAT DISCONNECT SWITCH, FUSED; SIZE AS INDICATED ON DRAWINGS. "xxAF" INDICATES FUSE AMPERE RATING. "yyAT" INDICATES SWITCH AMPERE RATING.		MOTOR
		MOTOR: "2" INDICATES HORSEPOWER
xxA DISCONNECT SWITCH, UNFUSED; SIZE AS INDICATED ON DRAWINGS. "xxA" INDICATES SWITCH AMPERE RATING.		OVERLOAD RELAY
DOOR BELL		PUSHBUTTON
DOOR CHIME		SWITCH AND FUSE
DOOR OPENER (ELECTRIC)		TELEPHONE OUTLET
FAN: CEILING-SUSPENDED (PADDLE)		TELEPHONE OUTLET— WALL-MOUNTED
FAN: CEILING-SUSPENDED (PADDLE) FAN WITH LIGHT		TELEPHONE/DATA OUTLET
FAN: WALL		THERMOSTAT—LINE-VOLTAGE
GROUND		THERMOSTAT—LOW-VOLTAGE
JUNCTION BOX—CEILING		TIME SWITCH
		TRANSFORMER

Figure 2-9   Miscellaneous symbols.

SYMBOL	NOTATION
1	PLUGMOLD ENTIRE LENGTH OF WORKBENCH. OUTLETS 18 IN. (457 mm) O.C. INSTALL 48 IN. (1.2 m) TO CENTER FROM FLOOR. GFCI PROTECTED.
2	TRACK LIGHTING. PROVIDE 5 LAMPHOLDERS.
3	TWO 40-WATT RAPID START FLUORESCENT LAMPS IN VALANCE. CONTROL WITH DIMMER SWITCH.

Figure 2-10 **Example of how certain notations might be added to a symbol when a symbol itself does not fully explain its meaning. The architect or engineer has a choice of explaining fully the meaning directly on the plan if there is sufficient room; if insufficient room, then a notation could be used.**

SPLIT-CIRCUIT RECEPTACLE OUTLET

WEATHERPROOF RECEPTACLE OUTLET

Figure 2-11 **Variations in significance of outlet symbols.**

example, different meanings are shown in Figure 2-11 for the outlet symbol. A good practice to follow in studying symbols is to learn the basic forms first and then add the supplemental information to obtain different meanings.

In drawing electrical plans, most architects, designers, and electrical engineers use symbols approved by ANSI wherever possible. However, plans may contain symbols that are not found in these standards. The electrician must refer to a legend that describes these symbols. The legend may be included on the plans or in the specifications. In many instances, a notation on the plan will clarify the meaning of the symbol.

LUMINAIRES (FIXTURES) AND OUTLETS

The location of lighting outlets is determined by the amount and type of illumination required to provide the desired lighting effects. It is not the intent of this text to describe how proper and adequate lighting is determined. Rather, the text covers the proper methods of installing the circuits for such lighting.

Luminaire (fixture) manufacturers publish catalogs that provide a tremendous amount of information regarding recommendations for residential lighting. These publications are available at electrical distributors, lighting showrooms, and home centers. The Instructor's Guide lists some excellent publications relating to residential lighting that can be obtained by request.

Architects often include in the specifications a certain amount of money for the purchase of luminaires (lighting fixtures). The electrical contractor includes this amount in the bid, and the choice of luminaires (fixtures) is then left to the homeowner. If the owner selects luminaires (fixtures) whose total cost exceeds the luminaire (fixture) allowance, the owner is expected to pay the difference between the actual cost and the specification allowance. If the luminaires (fixtures) are not selected before the roughing-in stage of wiring the house, the electrician usually installs outlet boxes having standard luminaire (fixture) mounting studs.

Most modern surface-mount luminaires (lighting fixtures) can be fastened to a luminaire (fixture) stud in the box (Figure 2-12, top row, picture A) or an outlet box or plaster ring using appropriate No. 8-32 metal screws and mounting strap furnished with the luminaire (fixture) (Figure 2-12, third row, raised plaster cover; outlet boxes).

Symbols Used to Represent Lighting Outlets and Switches

Typical lighting outlet symbols are shown in Figure 2-12. Typical switch symbols are shown in Figure 2-13.

OUTLET, DEVICE, AND JUNCTION BOXES

Article 314 contains much detail about outlet boxes, device (switch) boxes, pull boxes, junction boxes, and conduit bodies (LB, LR, LL, C, etc.) all of which

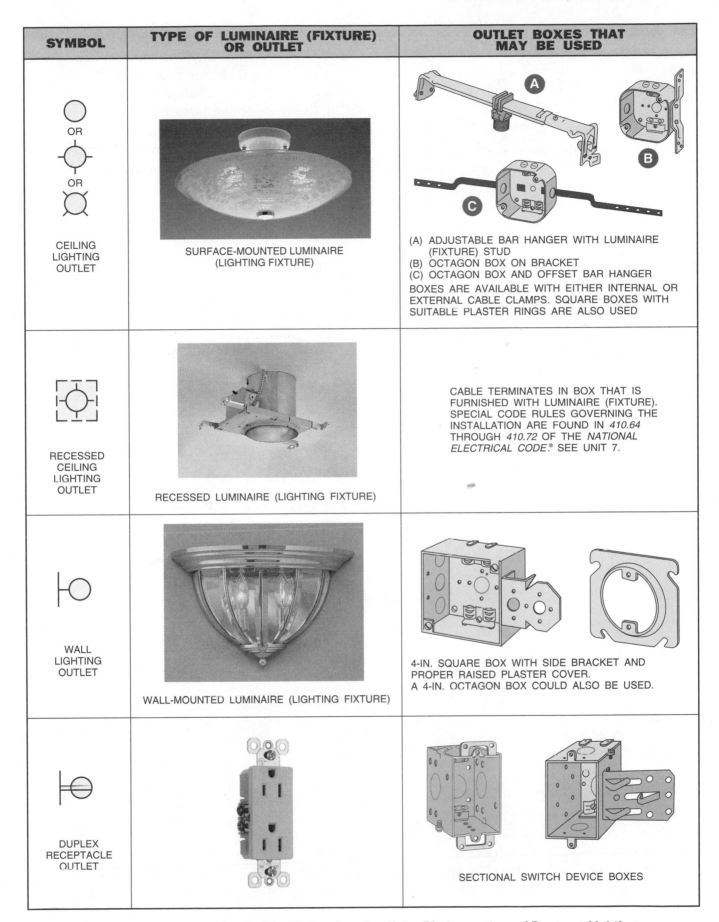

SYMBOL	TYPE OF LUMINAIRE (FIXTURE) OR OUTLET	OUTLET BOXES THAT MAY BE USED
CEILING LIGHTING OUTLET	SURFACE-MOUNTED LUMINAIRE (LIGHTING FIXTURE)	(A) ADJUSTABLE BAR HANGER WITH LUMINAIRE (FIXTURE) STUD (B) OCTAGON BOX ON BRACKET (C) OCTAGON BOX AND OFFSET BAR HANGER BOXES ARE AVAILABLE WITH EITHER INTERNAL OR EXTERNAL CABLE CLAMPS. SQUARE BOXES WITH SUITABLE PLASTER RINGS ARE ALSO USED
RECESSED CEILING LIGHTING OUTLET	RECESSED LUMINAIRE (LIGHTING FIXTURE)	CABLE TERMINATES IN BOX THAT IS FURNISHED WITH LUMINAIRE (FIXTURE). SPECIAL CODE RULES GOVERNING THE INSTALLATION ARE FOUND IN *410.64* THROUGH *410.72* OF THE *NATIONAL ELECTRICAL CODE.* SEE UNIT 7.
WALL LIGHTING OUTLET	WALL-MOUNTED LUMINAIRE (LIGHTING FIXTURE)	4-IN. SQUARE BOX WITH SIDE BRACKET AND PROPER RAISED PLASTER COVER. A 4-IN. OCTAGON BOX COULD ALSO BE USED.
DUPLEX RECEPTACLE OUTLET		SECTIONAL SWITCH DEVICE BOXES

Figure 2-12   Types of luminaires (fixtures) and outlets. *Photo courtesy* of Progress Lighting.

SYMBOL	FLUSH TOGGLE SWITCH	OPERATION	CONNECTIONS
S SINGLE-POLE		ON OFF	
S₂ DOUBLE-POLE		ON OFF	
S₃ THREE-WAY		POSITION POSITION 1 2	
S₄ FOUR-WAY		1 2	
Sₚ SWITCH AND PILOT LIGHT		FOR CONTROLLING LIGHTS FROM ONE POINT WITH PILOT LIGHT INDICATION	ALSO AVAILABLE IN THREE-WAY TYPE OF CONTROLLING LIGHT FROM TWO POINTS WITH PILOT LIGHT INDICATION

Figure 2-13 Standard switches and symbols.

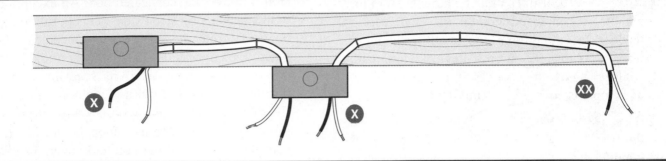

Figure 2-14   A box (or fitting) must be installed wherever there are splices, outlets, switches, or other junction points. Refer to the points marked X. A Code violation is shown at point XX, *300.15(A).*

are used to provide the necessary space for making electrical connections, contain wiring, and support luminaires (fixtures), and so on. *Article 314* is quite lengthy. For the moment, we will focus on some of the basic Code requirements for outlet boxes, device boxes, and junction boxes. Additional Code requirements are covered on an "as needed" basis in this chapter and in later chapters.

Some Basic Code Requirements

- *NEC® 300.11* and *314.23:* Boxes must be securely mounted and fastened in place.
- *NEC® 300.15*: Where conduit, electrical-metallic tubing, nonmetallic-sheathed cable, Type AC cable, or other cables are installed, a box or conduit body must be installed at each conductor splice connection point, outlet, switch point, junction point, or pull point. This requirement is illustrated in Figure 2-14. Exceptions to this are found in *300.15(A)* through *(M).*

- *NEC® 300.15(C):* Where cables enter or exit from conduit or tubing that is used to provide cable support or protection against physical damage, a fitting shall be provided on the end(s) of the conduit or tubing to protect the cable from abrasion. Figure 2-15 shows a transition from a cable to a raceway.

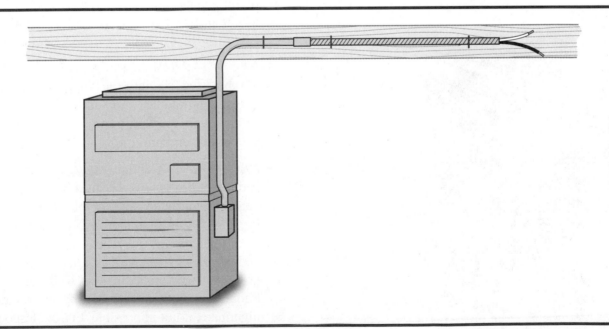

Figure 2-15   *NEC® 300.15(C)* permits a transition to be made from one wiring method to another wiring method. In this case, the armor of the Type AC cable is removed, allowing sufficient length of the conductors to be run through the conduit. A proper fitting must be used at the transition point, and the fitting must be accessible after installation.

- *NEC® 314.16:* Boxes must be large enough for all of the enclosed conductors and wiring devices. Installing boxes too small for the number of conductors, wiring devices, luminaire (fixture) studs, and splices is a real problem! This topic is covered in detail later.

- *NEC® 314.16(C):* The exception allows multiple cables to be run through a single knockout opening in a nonmetallic box. This permission is not given for metallic boxes. See Figure 2-16.

- *NEC® 314.20:* In walls or ceilings constructed of wood or other combustible material, install boxes so they are flush with the finished wall surface. In walls or ceilings constructed of concrete, tile, or other noncombustible material, it is acceptable to have the front edge of the box set back not more than ¼ in. (6 mm) from the finished surface. See the discussion of box mounting later in this chapter.

- *NEC® 314.22:* A surface extension from a box of a concealed wiring system is made by mounting and mechanically securing an exten-

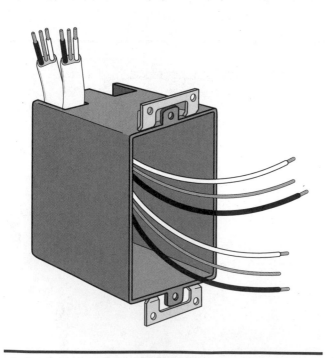

Figure 2-16 Multiple cables may run through a single knockout opening in a nonmetallic box, *314.17(C), Exception.*

sion ring over the concealed box. An exception to this is that a surface extension is permitted to be made from the cover of a concealed box where the cover is designed so it is unlikely to fall off, or be removed if its securing means becomes loose. The wiring method must be flexible and arranged so that any required grounding continuity is independent of the connection between the box and cover. This is shown in Figure 2-17.

- *NEC® 314.25:* In completed installations, all boxes shall have a cover, faceplate, or luminaire (fixture) canopy. Do not leave any electrical boxes uncovered.

- *NEC® 314.27(A):* Boxes used at lighting outlets must be designed or installed so that a luminaire (lighting fixture) may be attached to it.

- ►*NEC® 314.27(A)* and *(B):* Listed outlet boxes are suitable for hanging luminaires (lighting fixtures) that weigh 50 lb. (23 kg) or less. Luminaires (lighting fixtures) that weigh more than 50 lb. (23 kg) must be supported independently of the outlet box, or the outlet box must be "listed" for the weight to be supported. Wall-mounted luminaires (lighting fixtures) of less than 6 lb. (3 kg) may be supported by a box, or by a plaster ring secured to a box, by not less than two No. 6-32 or larger screws, as in Figure 2-18.◄

- *NEC® 314.27(D):* Standard outlet boxes must not be used as the sole support for ceiling-suspended (paddle) fans unless they are specifically listed for this purpose. See Unit 9 for detailed discussion of this issue.

- *NEC® 314.29:* Conduit bodies, junction, pull, and outlet boxes must be installed so the wiring contained in them is accessible without removing any part of the building. Never install outlet boxes, junction boxes, pull boxes, or conduit bodies where they will be inaccessible behind or above permanent finished walls or ceilings. Should anything ever go wrong, it would be a nightmare, if not impossible, to troubleshoot and locate hidden splices. Wiring located above "lay-in" ceilings is considered accessible because it is easy to remove a panel(s) to gain

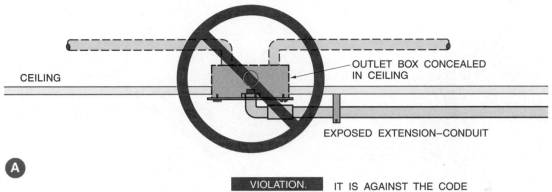

A

VIOLATION. IT IS AGAINST THE CODE
TO MAKE A RIGID EXTENSION FROM A COVER THAT IS ATTACHED TO
A CONCEALED OUTLET BOX, *314.22.* IT WOULD BE DIFFICULT TO GAIN
ACCESS TO THE CONNECTIONS INSIDE THE BOX.

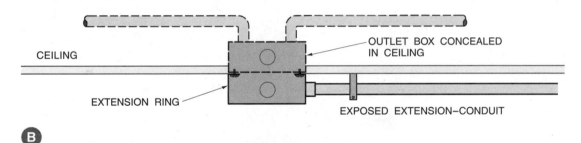

B

THIS MEETS CODE. AN EXTENSION RING
MUST BE MOUNTED OVER AND MECHANICALLY SECURED TO THE
CONCEALED OUTLET BOX, *314.22.*

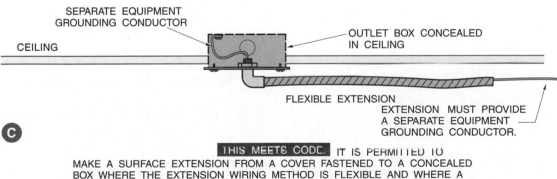

C

THIS MEETS CODE. IT IS PERMITTED TO
MAKE A SURFACE EXTENSION FROM A COVER FASTENED TO A CONCEALED
BOX WHERE THE EXTENSION WIRING METHOD IS FLEXIBLE AND WHERE A
SEPARATE GROUNDING CONDUCTOR IS PROVIDED SO THAT THE GROUNDING
PATH IS NOT DEPENDENT UPON THE SCREWS THAT ARE USED TO FASTEN
THE COVER TO THE BOX.

IT WOULD BE A CODE VIOLATION TO MAKE THE EXTENSION IF A
FLEXIBLE WIRING METHOD WAS USED, BUT HAD NO PROVISIONS FOR
A SEPARATE GROUNDING CONDUCTOR.

Figure 2-17   How to make an extension from a flush-mounted box.

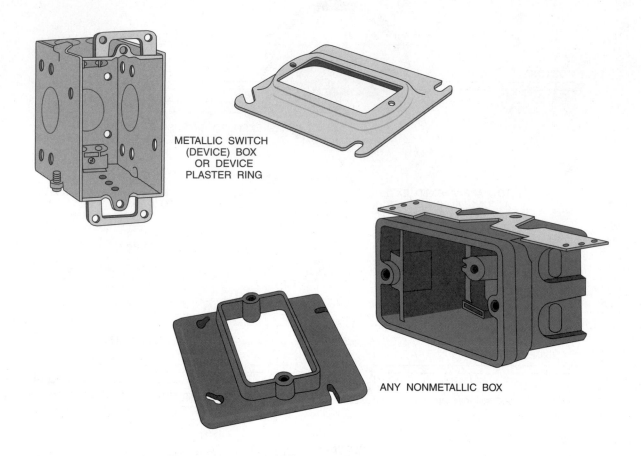

METALLIC SWITCH
(DEVICE) BOX
OR DEVICE
PLASTER RING

ANY NONMETALLIC BOX

Figure 2-18   ▶Wall mounted luminaires (lighting fixtures) that do not weigh more than 6 lb. (3 kg) are permitted to be secured to these types of boxes and plaster rings by not less than two No. 6-32 or larger screws, *314.27(A), Exception.* For standard outlet boxes, the maximum weight is 50 lb. (23 kg), *314.27(B).*◀

access to the space above the ceiling. The *NEC®* definition of *accessible* is "*Capable of being removed or exposed without damaging the building structure or finish, or not permanently closed in by the structure or finish of the building.*"

The same accessibility requirement is true for underground wiring. Boxes must be accessible without excavating sidewalks, paving, earth, or other material used to establish the finished grade.

• *NEC® 314.71:* Pull and junction boxes must provide adequate space and dimensions for the installation of conductors, and they shall comply with the specific requirements of this section.

• *NEC® 410.12:* This is another repeat of the requirement that boxes must be covered. The "cover" could be a blank cover, a luminaire (fixture) canopy, a lampholder, a switch or receptacle, or similar device with faceplate.

• *NEC® 410.16(A):* Here we find permission to use outlet boxes and fittings to support luminaires (lighting fixtures) provided the requirements found in *314.23* and *314.70* are met.

What to Do Where Recessed Luminaires (Lighting Fixtures) Will Be Installed. This topic is covered in detail in Unit 7.

NONMETALLIC OUTLET AND DEVICE BOXES

NEC® 314.3 permits nonmetallic outlet and device boxes to be installed where the wiring method is nonmetallic sheathed cable or nonmetallic raceway. *NEC® 314.3 (Exceptions)* allows metal raceways and metal-jacketed cables (BX) to be used with non-metallic boxes provided all metal raceways or cables entering the box are bonded together to maintain the integrity of the grounding path to other equipment in the installation.

The house wiring system usually is formed by a number of specific circuits. Each circuit consists of runs of cable from outlet to outlet or from box to box. The residence plans show many branch circuits for general lighting, appliances, electric heating, and other requirements. The specific Code rules for each of these circuits are covered in later units.

GANGED SWITCH (DEVICE) BOXES

A flush switch or receptacle for residential use fits into a standard 2 in. × 3 in. sectional switch box (sometimes called a *device box*). When two or more switches or receptacles are located at the same point, the switch boxes are ganged or fastened together to provide the required mounts, as in Figure 2-17.

Three switch boxes can be ganged together by removing and discarding one side from both the first and third switch boxes and both sides from the second (center) switch box. The boxes are then joined together as shown in Figure 2-19. After the switches are installed, the gang is trimmed with a gang plate having the required number of switch handle or receptacle openings. These plates are called two-gang wall plates, three-gang wall plates, and so on, depending upon the number of openings.

The dimensions of a standard sectional switch box (2 in. × 3 in.) are the dimensions of the opening of the box. The depth of the box may vary from 1½ in. to 3½ in., depending upon the requirements of the building construction and the number of conductors and devices to be installed. See Figure 2-20 for a complete listing of box dimensions.

Figure 2-19   Standard flush switches installed in ganged sectional device boxes.

QUIK-CHEK BOX SELECTION GUIDE
FOR METAL BOXES GENERALLY USED FOR RESIDENTIAL WIRING

DEVICE BOXES

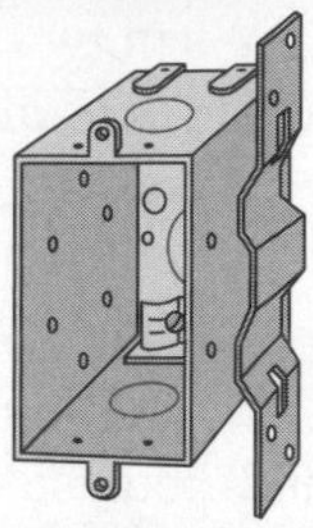

WIRE SIZE	3x2x1½ (7.5 in.³)	3x2x2 (10 in.³)	3x2x2¼ (10.5 in.³)	3x2x2½ (12.5 in.³)	3x2x2¾ (14 in.³)	3x2x3 (16 in.³)	3x2x3½ (18 in.³)
14 AWG	3	5	5	6	7	8	9
12 AWG	3	4	4	5	6	7	8

SQUARE BOXES

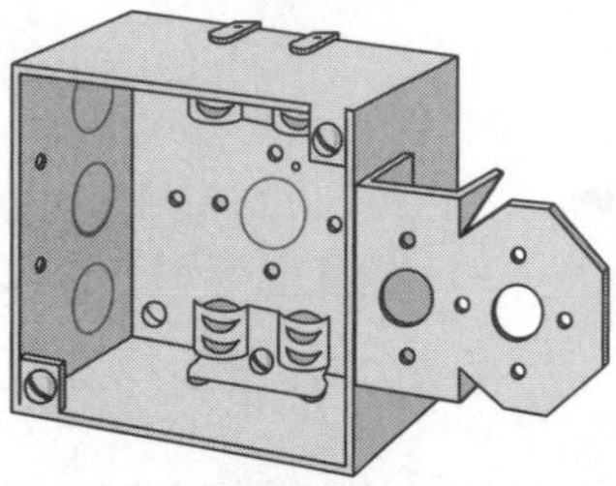

WIRE SIZE	4x4x1½ (21 in.³)	4x4x2⅛ (30.3 in.³)
14 AWG	10	15
12 AWG	9	13

OCTAGON BOXES

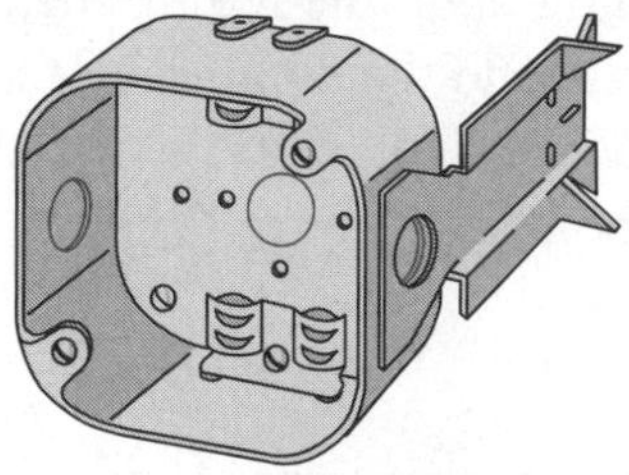

WIRE SIZE	4x1½ (15.5 in.³)	4x2⅛ (21.5 in.³)
14 AWG	7	10
12 AWG	6	9

HANDY BOXES

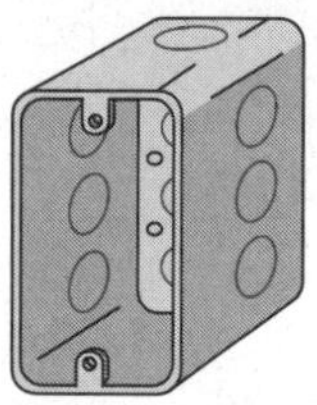

WIRE SIZE	4x2⅛x1½ (10.3 in.³)	4x2⅛x1⅞ (13 in.³)	4x2x2⅛ (14.5 in.³)
14 AWG	5	6	7
12 AWG	4	5	6

RAISED COVERS

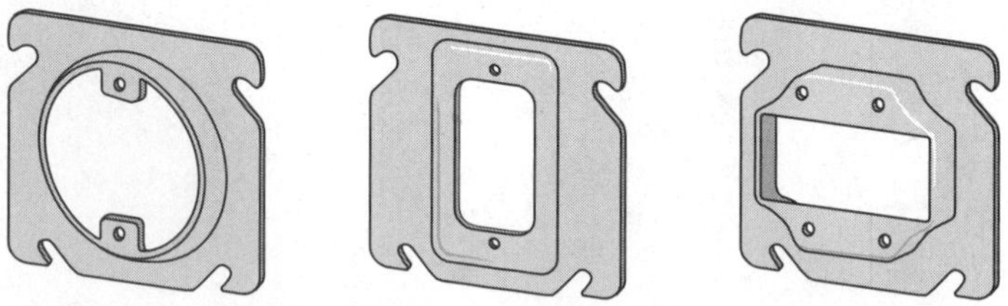

WHERE RAISED COVERS ARE MARKED WITH THEIR VOLUME IN CUBIC INCHES, THAT VOLUME MAY BE ADDED TO THE BOX VOLUME TO DETERMINE MAXIMUM NUMBER OF CONDUCTORS IN THE COMBINED BOX AND RAISED COVER.

NOTE: BE SURE TO MAKE DEDUCTIONS FROM THE ABOVE MAXIMUM NUMBER OF CONDUCTORS PERMITTED FOR WIRING DEVICES, CABLE CLAMPS, FIXTURE STUDS, AND GROUNDING CONDUCTORS. THE CUBIC INCH (IN.³) VOLUME IS TAKEN DIRECTLY FROM *TABLE 314.16(A)* OF THE *NEC*.® NONMETALLIC BOXES ARE MARKED WITH THEIR CUBIC INCH CAPACITY.

Figure 2-20   Quik-chek box selection guide.

BOX MOUNTING

NEC® 314.20 states that boxes must be mounted so that they will be set back not more than ¼ in. (6 mm) when the boxes are mounted in noncombustible walls or ceilings made of concrete, tile, or similar materials. When the wall or ceiling construction is of combustible material (wood), the box must be set flush with the surface, Figure 2-21. These requirements are meant to prevent the spread of fire if a short circuit occurs within the box.

Ganged sectional switch (device) boxes can be installed using a pair of metal mounting strips.

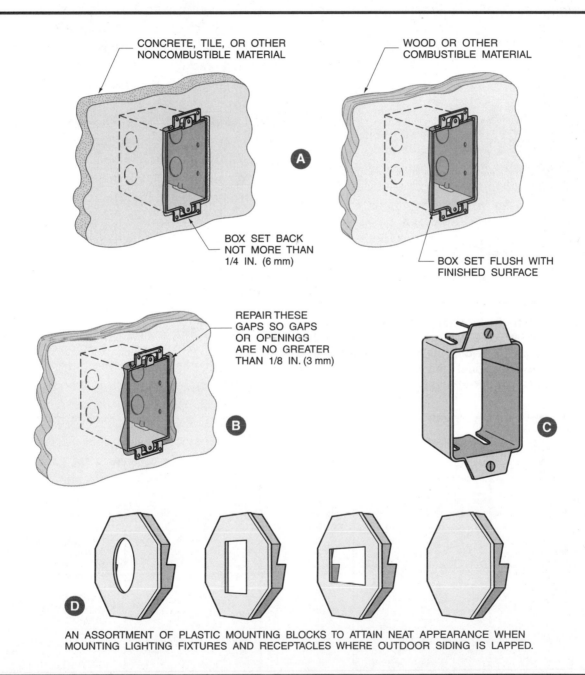

Figure 2-21   Box position requirements in walls and ceilings constructed of various materials. If the box is not flush with the finished surface (as a result of improper installation) or if paneling, drywall, tile, or a mirror is added, then install a box extender, sometimes referred to as a "goof ring." One type is shown in (C). Box extenders that are ¼ in., ⅜ in., ½ in., or ¾ in. deep are also available for one- and two-gang boxes. Refer to *314.20* and *314.21*.

These strips are also used to install a switch box between wall studs, Figure 2-22. The use of a bracket box in such an installation may result in an off-center receptacle outlet. When an outlet box is to be mounted at a specific location between joists, as for ceiling-mounted luminaires (fixtures), an offset bar hanger is used, as in Figure 2-12.

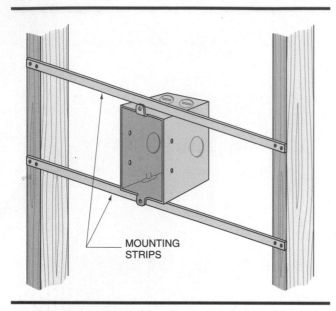

Figure 2-22   Switch (device) boxes installed between studs using metal switch box supports, often referred to as "Kruse strips." If wood strips are used, they must have a cross-sectional dimension of not less than 1 in. × 2 in. (25 mm × 50 mm), *314.23(B)(2)*.

▶The Code in *314.23(B)(1)* states that when a switch box or outlet box is mounted to a stud or ceiling joist with nails or screws through the box, the nails or screws must be not more than ¼ in. (6 mm) from the back or ends of the box, as in Figure 2-23.◀ This requirement ensures that when nails or screws pass through the box, they will not interfere with the wiring devices in the box.

Why bother with this Code requirement that takes away from the available wiring space in the box? It is much better to use boxes that have external mounting brackets.

Most residences are constructed with wood framing. However, metal studs and joists are being used more and more. Figure 2-24 shows types of boxes used for fast, easy attachment to metal framing members. These boxes "snap" onto the metal stud, and do not require drilling the metal studs and fastening the box with nuts and bolts. The first one is a 4-in. square box with ½-in. knockouts. The second shows a 4-in. square box with nonmetallic sheathed cable (Romex) clamps. The third shows a 4-in. square box with armored cable (BX) clamps. See Unit 4 for details about running cables through metal framing members.

Another type of device used for fastening electrical boxes to steel framing members is shown in Figure 2-25. The bracket is screwed to the steel framing members, then the box or boxes are attached to the bracket.

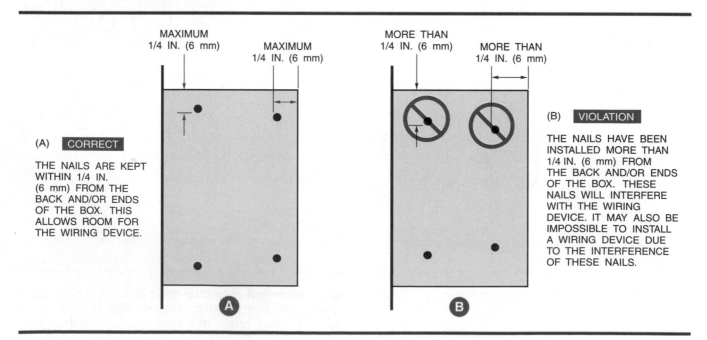

Figure 2-23   Using nails or screws to install a sectional switch box, *314.23(B)(1)*.

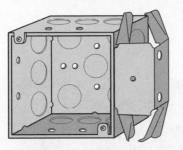
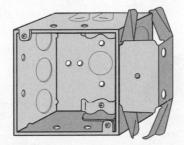
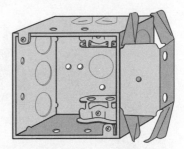

Figure 2-24 Types of boxes that can be used with steel framing construction. Note the side clamps that "snap" directly onto the steel framing members.

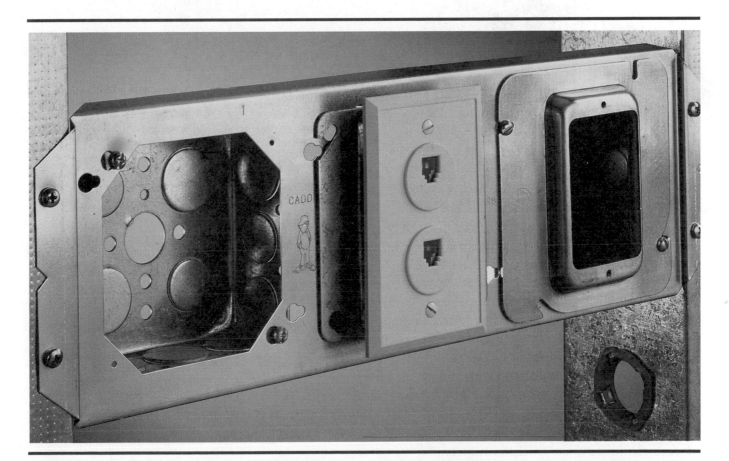

Figure 2-25 A bracket for use on steel framing members. The bracket spans the space between two studs. Note that more than one box can be mounted on this bracket. *Courtesy of* **ERICO,®** Inc.

Remodel (Old Work)

When installing switches, receptacles, and lighting outlets in remodel work where the walls and ceilings (paneling, drywall, sheetrock and plaster, or lath and plaster) are already in place, first make sure you will be able to run cables to where the switches, receptacles, and lighting outlets are to be installed. After making sure you can get the cables through the concealed spaces in the walls and ceilings, you can then proceed to cut openings the size of the specific type of box to be installed at each outlet location. Cables are then "fished" through the stud and joist spaces behind or above the finished walls and ceilings to where you have cut the openings. Then, boxes having plaster ears and snap-in brackets, Figure 2-26, can be inserted from the front,

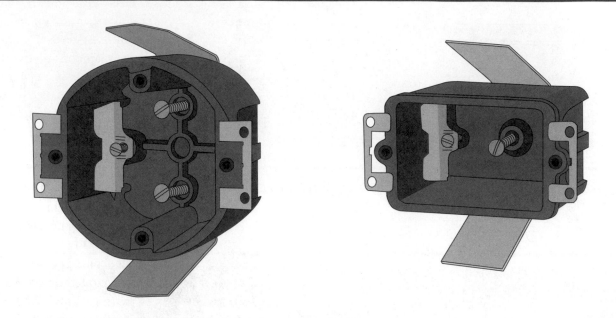

Figure 2-26 Nonmetallic boxes are used for remodel work so that when they are inserted into an opening cut into a wall or ceiling, they will "snap" into place, thus conforming to the *NEC®* requirements for the adequate support of boxes.

through the holes already cut at the locations where the wall or ceiling boxes are to be installed. After the boxes are snapped into place through the hole from the front, they become securely locked into place.

Another popular method of fastening wall and ceiling boxes in existing walls and ceilings is to use boxes that have plaster ears. Insert the box through the hole from the front. Then position a metal support into the hole on each side of the box. Be careful not to let the metal support fall into the hole. Next, bend the metal support over and into the box, (see Figure 2-27). Two metal supports are used. These metal switch box supports are known as Madison Hold-Its™* or "Madison Hangers"

Also available for old work are boxes that have a screw-type support on each side of the box, as in Figure 2-28. This type of box has plaster ears. It is inserted through the hole cut in the wall or ceiling, then the screws are tightened. This pulls up a metal support tightly behind the wall or ceiling, firmly holding the box in place. The action is very similar to that of Molly screw anchors.

---

*Hold-It™ is a brand name trademark owned by Madison Equipment, Inc., Cleveland, OH.

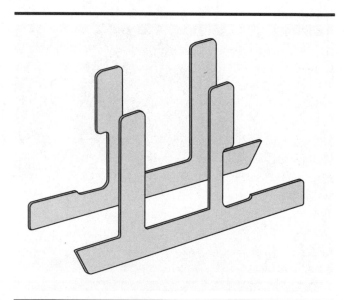

Figure 2-27 Metal supports referred to as Madison Hold-Its™ or "Madison Hangers."

Because these types of remodel boxes are supported by the drywall or plaster, as opposed to new work where the boxes are supported by the framing members, be careful not to hang heavy luminaires (lighting fixtures) from them. Be sure that there is proper support if a ceiling fan is to be installed. This is covered in Unit 9.

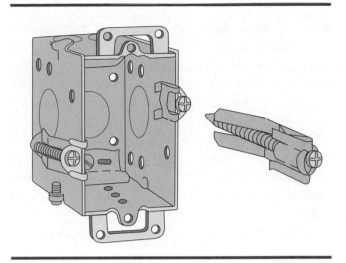

Figure 2-28 These boxes are commonly used in remodel work. The box is installed from the front, then the screws are tightened. The box is adequately supported.

Spread of Fire

To protect lives and property, fire must be contained and not be allowed to spread.

In addition to the *National Electrical Code*, you must also become familiar with building codes.

Building codes such as the International Code Council (ICC) *International Residential Code* generally do not require fire-resistance-rated walls and ceilings in one- and two-family buildings. Some communities have amendments to the basic building codes, such as requiring ⅝-in. (16-mm) gypsum wallboard [instead of the normal ½-in. (13-mm)] on the ceilings and on the walls between the garage and living areas and requiring *draft stopping* where ducts, cables, and piping run through bottom (sole) and top plates. Draft stopping is generally done by the insulation contractor. Townhouses, condominiums, and other multi-occupancy buildings do have fire-resistance rating requirements. To avoid costly mistakes, check with your local building officials to determine exactly what the requirements are!

In walls, partitions, and ceilings that are fire-resistance-rated, special consideration must be given when installing wall and ceiling electrical boxes "back-to-back" or in the same stud or joist space of common walls or ceilings, as might be found in multifamily buildings. Electrical wall boxes installed back-to-back or installed in the same stud or joist space defeat the fire-resistance rating of the wall or ceiling.

To be truly fire-resistance-rated, the construction and materials must conform to very rigid requirements, as set forth in the UL Standards discussed below.

Ratings of fire-resistance materials are expressed in hours. Terms such a 1-hour fire resistance rating, 2-hour fire resistance rating, and so on, are used.

UL Standard 263 *Fire Tests of Building Construction and Materials* and NFPA 251 cover fire resistance ratings. In UL Standard 263, and in all building codes, you will find requirements such as: "the surface area of individual metallic outlet or switch boxes shall not exceed 16 square inches" (i.e., a 4-in. square box is 4 in. × 4 in. = 16 in.²), "the aggregate surface area of the boxes shall not exceed 100 in.² per 100 ft² of wall surface," "boxes located on opposite sides of walls or partitions shall be separated by a minimum horizontal distance of 24 in. (600 mm)," "the metallic outlet or switch boxes shall be securely fastened to the studs and the opening in the wallboard facing shall be cut so that the clearance between the box and the wallboard does not exceed ⅛ in. (3 mm)," and "the boxes shall be installed in compliance with the *National Electrical Code*."

The UL *Fire Resistance Directory* covers fire-resistant materials and assemblies, and contains a rather detailed listing of *Outlet Boxes and Fittings Classified for Fire Resistance (CEYY)*, showing all manufacturers' product part numbers for *nonmetallic boxes* to be installed in walls, partitions, and ceilings that meet the fire-resistance standard. This category also covers special-purpose boxes for installation in floors.

When using nonmetallic boxes in fire-resistance-rated walls, the restrictions are more stringent than for metallic boxes. For example, the UL *Fire Resistance Directory* shows that one manufacturer of nonmetallic boxes limits the product to be installed so that no opening exceeds 10 in.² (64.5 cm²). Another manufacturer limits the product to be installed so that no opening exceeds 25 in.² (161 cm²).

►Recently, certain molded fiberglass outlet and device boxes have become available and are listed for use in fire-resistance-rated partitions without the use of putty pads, mineral wool batts, or fiberglass batts, provided a minimum horizontal distance of 3 in. (75 mm) is maintained between boxes and the boxes are not back-to-back. See Figure 2-31B. If you are dealing with fire-resistance-rated construction, be sure to check the manufacturers' instructions.◄

For both metallic and nonmetallic electric outlet boxes, the *minimum* distance between boxes on opposite sides of a fire-rated wall or partition is 24 in. (600 mm). There is an exception, which is discussed later.

For both metallic and nonmetallic electric outlet boxes, the maximum gap between the box and the wall material is ⅛ in. (3 mm). This holds true for fire-resistance-rated walls and non-fire-resistance-rated walls.

To maintain the hourly fire rating of a fire-rated wall that contains electrical outlet and switch boxes, the UL *Fire Resistance Directory* lists an *intumescent* (expands when heated) fire-resistant material that comes in "pads."

These pads are moldable, and can be used to wrap metal electrical boxes or inserted on the inside back wall of metal electrical boxes. When exposed to fire, the material expands and chars, sealing off the opening to prevent the spread of flames and limit the heat transfer from the fire-resistance-rated side of the wall to the non-fire-resistance-rated side of the wall. It is not to be used on nonmetallic boxes.

The ⅛ in. (3 mm) thickness provides a 1-hour fire-resistance rating. A 2-hour rating is obtained with ¼ in. (6 mm) thickness.

When this material is properly installed, the 24-in. (600-mm) separation between boxes is not required. However, the electrical outlet boxes *must not* be installed back-to-back. Using this material meets the requirements of *NEC® 300.21* of the *NEC®* and other building codes.

The International Code Council (ICC) *International Residential Code* states that "fireblocking" be provided to cut off all concealed draft openings (both vertical and horizontal) and to form an effective fire barrier between stories, and between a top story and the roof space. This includes openings around vents, pipes, ducts, chimneys, and fireplaces at ceiling and floor level. Oxygen supports fire. Firestopping cuts off or significantly reduces the flow of oxygen. Most communities that have adopted the ICC code require firestopping around electrical conduits. In residential work, this is generally done by the insulation installer, before the drywall installer closes up the walls.

The previous discussion makes it clear that the electrician must become familiar with certain building codes, in addition to the *NEC®*. Knowing the *NEC®* is not enough!

It is extremely important to check with the electrical inspector and/or building official for complete clarification on the matter of fire-resistance requirements before you find yourself cited with serious violations and lawsuits regarding noncompliance with fire codes. Changes can be costly, and will delay completion of the project. Improper wiring is hazardous. The *NEC®* addresses this issue in *300.21*.

Figure 2-29, Figure 2-30, Figure 2-31, and Figure 2-32 illustrate the basics for installing electrical outlet and switch boxes in fire-resistance-rated partitions. This is an extremely important issue. If in doubt, check with your local electrical inspector and/or building official.

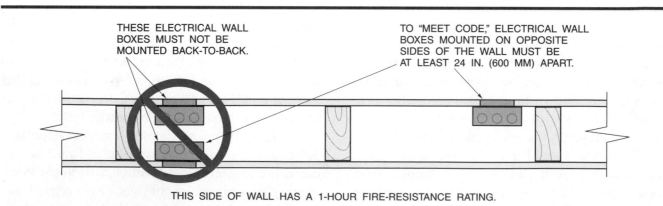

THESE ELECTRICAL WALL BOXES MUST NOT BE MOUNTED BACK-TO-BACK.

TO "MEET CODE," ELECTRICAL WALL BOXES MOUNTED ON OPPOSITE SIDES OF THE WALL MUST BE AT LEAST 24 IN. (600 MM) APART.

THIS SIDE OF WALL HAS A 1-HOUR FIRE-RESISTANCE RATING.

Figure 2-29 Building codes prohibit electrical wall boxes to be mounted back-to-back in fire-resistance-rated walls unless specific installation procedures, as illustrated in Figure 2-30, Figure 2-31, and Figure 2-32 are followed. One of these requirements is that electrical wall boxes must be kept at least 24 in. (600 mm) apart. This is to maintain the integrity of the fire-resistance rating of the wall.

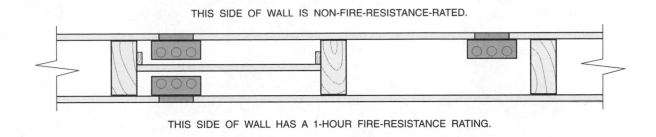

THIS SIDE OF WALL IS NON-FIRE-RESISTANCE-RATED.

THIS SIDE OF WALL HAS A 1-HOUR FIRE-RESISTANCE RATING.

Figure 2-30   Building codes permit boxes to be mounted back-to-back or in the same wall cavity when the fire-resistance rating is maintained. In this detail, gypsum board has been installed in the wall cavity between the boxes, creating a new fire resistance barrier. Verify if this is acceptable to the building authority in your area.

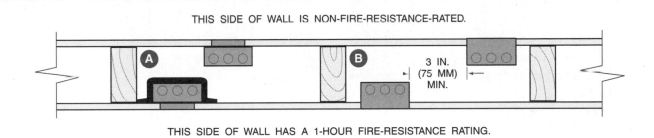

THIS SIDE OF WALL IS NON-FIRE-RESISTANCE-RATED.

3 IN.
(75 MM)
MIN.

THIS SIDE OF WALL HAS A 1-HOUR FIRE-RESISTANCE RATING.

Figure 2-31   Building codes permit electrical outlet and device boxes to be mounted in the same stud space under certain conditions. (A) Here we see "Wall Opening Protective Material" packed around one outlet box. (B) Here we see molded fiberglass boxes in the same stud space with a minimum horizontal clearance of 3 in. (75 mm) These boxes are UL-listed for this application. In both (A) and (B), the boxes are not permitted to be back-to-back.

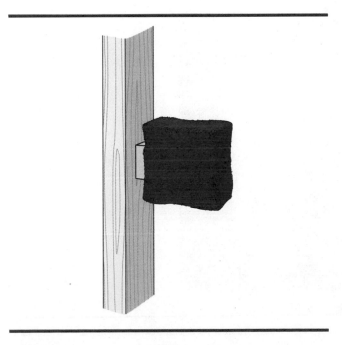

Figure 2-32   This illustration shows an electrical outlet box that has been covered with fire-resistance "putty" to retain the wall's fire-resistance rating per building codes.

BOXES FOR CONDUIT WIRING

Some local electrical ordinances require conduit rather than cable wiring. Conduit wiring is discussed in Unit 18. Examples of conduit fill (how many wires are permitted in a given size conduit) are presented.

When conduit is installed in a residence, it is quite common to use 4-in. square boxes trimmed with suitable plaster covers, Figure 2-33(A). There are sufficient knockouts in the top, bottom, sides, and back of the box to permit a number of conduits to run to the box. Plenty of room is available for the conductors and wiring devices. Note how easily these 4-in. square outlet boxes can be mounted back-to-back by installing a small fitting between the boxes. This is illustrated in Figure 2-33(B) and Figure 2-33(C).

Four-inch square outlet boxes can be trimmed with one-gang or two-gang plaster rings where wiring devices will be installed. Where luminaires (lighting fixtures) will be installed, a plaster ring having a round opening should be installed.

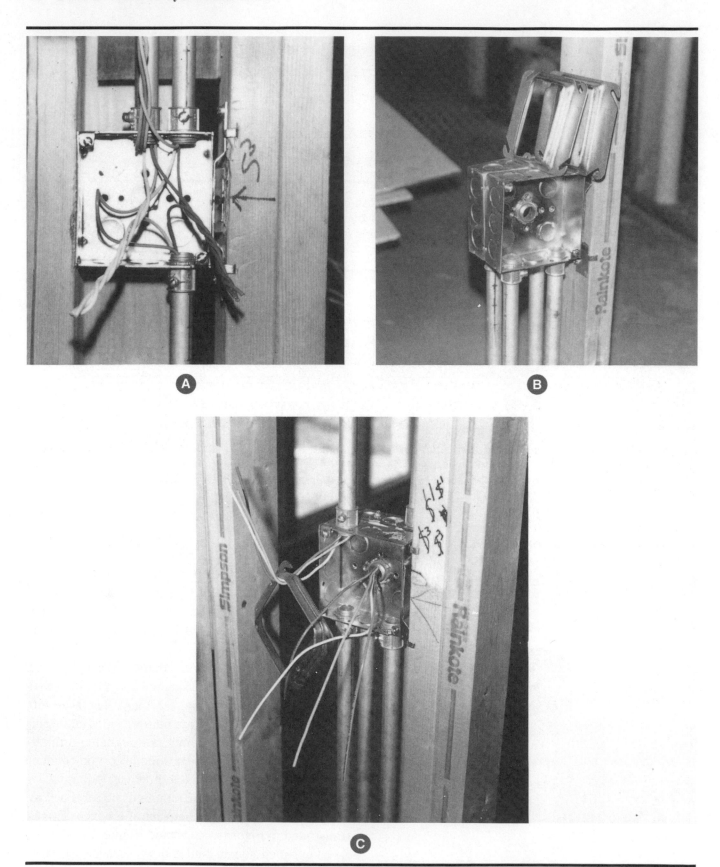

Figure 2-33   **(A) shows electrical metallic tubing run to a 4-in. square box. (B) and (C) show 4-in. square boxes attached together for "back-to-back" installations. When mounting boxes back-to-back, be sure to consider possible transfer of sound between the rooms. IMPORTANT: See text ("Spread of Fire") for details of fire-resistance ratings and the restrictions related to installing electrical boxes back-to-back.**

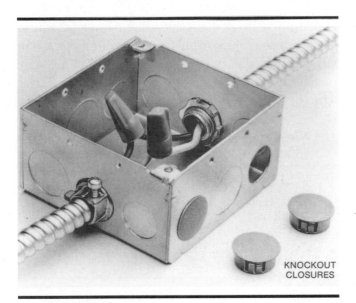

KNOCKOUT
CLOSURES

Figure 2-34 Unused openings in boxes must be closed according to *110.12(A)*. This is done to contain electrical short circuit problems inside the box or panel and to keep rodents out. Note that the conductors *do not* meet the 3-in. (75-mm) and 6-in. (150-mm) conductor length requirements of *300.14*. This is a good example of a Code violation. See Figure 2-2 for proper installation regarding conductor length at boxes.

▶Any unused cable or raceway openings in electrical boxes, cabinets, panelboards, meter socket enclosures, equipment cases, and similar equipment shall be effectively closed to afford protection substantially equivalent to the wall of the equipment. See *110.12(A)* and Figure 2-34. ◀

Close the Gap Around Boxes!

NEC® *314.21* requires that openings around electrical boxes shall be repaired so that "there will be no gaps or open space greater than ⅛ in. (3 mm) at the edge of the box or fitting." This is to minimize the spread of fire. The UL *Fire Resistance Directory* mirrors the *NEC®* rule by stating that "the outlet or switch boxes shall be securely fastened to the studs and the opening in the wallboard facing shall be cut so that the clearance between the box and the wallboard does not exceed ⅛ inch." See Figure 2-21(B).

This is easier said than done. The electrician installs the wiring (rough-in), followed by an inspection by the electrical inspector. Next comes the drywall/plaster/panel installer, who many times cuts out the box openings much larger than the outlet or

switch box. In most cases, the walls and ceilings are painted or wallpapered before the electrician returns to install the receptacles, switches, and luminaires (lighting fixtures). For gaps or openings greater than ⅛ in. (3 mm) around the electrical boxes, should the electrician repair the gap with patching plaster, possibly damaging or marring the finished wall? Whose responsibility is it? Is it the electrician's? Is it the drywall/plaster/panel installer's?

The electrician should check with the drywall/plaster/panel installer to clarify who is to be responsible for seeing to it that gaps or openings around electrical outlet and switch boxes do not exceed ⅛ in. (3 mm). *314.21* puts the responsibility on the electrician.

YOKE

We use the term "yoke." The *National Electrical Code®* uses the term "yoke." Yet, the *NEC®* does not define a "yoke." What is a "yoke"? It is the metal or polycarbonate mounting strap on which a wiring device or devices are attached. The yoke in turn is attached to a device box or device cover with No. 6-32 screws. One yoke may have one, two, or three wiring devices attached to it.

Visit an electrical distributor or home center. You will find many combination wiring devices on a single yoke or strap. Selecting one of these combination devices may eliminate the need to use combinations of interchangeable wiring devices.

SPECIAL-PURPOSE OUTLETS

Special-purpose outlets are usually indicated on the plans. These outlets are described by a notation and are usually detailed in the specifications. The plans indicate special-purpose outlets by a triangle inside a circle with subscript letters. In some cases, a subscript number is added to the letter.

When a special-purpose outlet is indicated on the plans or in the specifications, the electrician must check for special requirements. Such a requirement may be a separate circuit, a special 240-volt circuit, a special grounding or polarized receptacle, or other preparation.

A set of plans and specifications for this residence is found at the end of this text. The specifications include a *Schedule of Special Purpose Outlets*.

NUMBER OF CONDUCTORS IN BOX

NEC® 314.16 dictates that outlet boxes, switch boxes, and device boxes should be large enough to provide ample room for the wires in that box, without having to jam or crowd the wires into the box. Jamming the wires into the box can not only damage the insulation on the wires, but can result in heat buildup in the box that can further damage the insulation on the wires. The Code specifies the maximum number of conductors allowed in standard metal outlet, device, and junction boxes; see *Tables 314.16(A)* and *(B)*. Nonmetallic boxes are marked by the manufacturer with their cubic-inch capacity.

When conductors are the same size, the proper

Table 314.16(A) Metal Boxes

Box Trade Size			Minimum Volume		Maximum Number of Conductors*						
mm	in.		cm³	in.³	18	16	14	12	10	8	6
100 × 32	(4 × 1¼)	round/octagonal	205	12.5	8	7	6	5	5	5	2
100 × 38	(4 × 1½)	round/octagonal	254	15.5	10	8	7	6	6	5	3
100 × 54	(4 × 2⅛)	round/octagonal	353	21.5	14	12	10	9	8	7	4
100 × 32	(4 × 1¼)	square	295	18.0	12	10	9	8	7	6	3
100 × 38	(4 × 1½)	square	344	21.0	14	12	10	9	8	7	4
100 × 54	(4 × 2⅛)	square	497	30.3	20	17	15	13	12	10	6
120 × 32	(4¹¹⁄₁₆ × 1¼)	square	418	25.5	17	14	12	11	10	8	5
120 × 38	(4¹¹⁄₁₆ × 1½)	square	484	29.5	19	16	14	13	11	9	5
120 × 54	(4¹¹⁄₁₆ × 2⅛)	square	689	42.0	28	24	21	18	16	14	8
75 × 50 × 38	(3 × 2 × 1½)	device	123	7.5	5	4	3	3	3	2	1
75 × 50 × 50	(3 × 2 × 2)	device	164	10.0	6	5	5	4	4	3	2
75 × 50 × 57	(3 × 2 × 2¼)	device	172	10.5	7	6	5	4	4	3	2
75 × 50 × 65	(3 × 2 × 2½)	device	205	12.5	8	7	6	5	5	4	2
75 × 50 × 70	(3 × 2 × 2¾)	device	230	14.0	9	8	7	6	5	4	2
75 × 50 × 90	(3 × 2 × 3½)	device	295	18.0	12	10	9	8	7	6	3
100 × 54 × 38	(4 × 2⅛ × 1½)	device	169	10.3	6	5	5	4	4	3	2
100 × 54 × 48	(4 × 2⅛ × 1⅞)	device	213	13.0	8	7	6	5	5	4	2
100 × 54 × 54	(4 × 2⅛ × 2⅛)	device	238	14.5	9	8	7	6	5	4	2
95 × 50 × 65	(3¾ × 2 × 2½)	masonry box/gang	230	14.0	9	8	7	6	5	4	2
95 × 50 × 90	(3¾ × 2 × 3½)	masonry box/gang	344	21.0	14	12	10	9	8	7	2
min. 44.5 depth	FS — single cover/gang (1¾)		221	13.5	9	7	6	6	5	4	2
min. 60.3 depth	FD — single cover/gang (2⅜)		295	18.0	12	10	9	8	7	6	3
min. 44.5 depth	FS — multiple cover/gang (1¾)		295	18.0	12	10	9	8	7	6	3
min. 60.3 depth	FD — multiple cover/gang (2⅜)		395	24.0	16	13	12	10	9	8	4

*Where no volume allowances are required by 314.16(B)(2) through 314.16(B)(5).

Table 314.16(B) Volume Allowance Required per Conductor

Size of Conductor (AWG)	Free Space Within Box for Each Conductor	
	cm³	in.³
18	24.6	1.50
16	28.7	1.75
14	32.8	2.00
12	36.9	2.25
10	41.0	2.50
8	49.2	3.00
6	81.9	5.00

Reprinted with permission from NFPA 70-2002.

metal box size can be selected by referring to *Table 314.16(A)*. When conductors are of different sizes and nonmetallic boxes are installed, refer to *Table 314.16(B)* and use the cubic-inch volume for the particular size of wire being used.

Table 314.16(A) and *Table 314.16(B)* do not take into consideration the space taken by luminaire (fixture) studs, cable clamps, hickeys, switches, pilot lights, or receptacles that may be in the box. These require additional space. Table 2-1 shows the additional allowances that must be considered for different situations.

SELECTING THE CORRECT SIZE BOX

Selecting the correct size box depends on the mix of conductors and wiring devices that will be contained in the particular box. Here are some examples.

When using a nonmetallic box where all conductors are the same size. A nonmetallic box is marked as having a volume of 22.8 in.^{3} The box contains no luminaire (fixture) stud or cable clamps. How many 14 AWG conductors are permitted in this box?

From *Table 314.16(B)*, the volume requirement for a 14 AWG conductor is 2.00 in.^{3} Therefore, the maximum number of 14 AWG conductors permitted in this box is:

$$\frac{22.8}{2} = 11.4 \text{ (Round down to 11 conductors)}$$

When using a metal box where all conductors are the same size. A box contains one luminaire (fixture) stud and two internal cable clamps. Four 12-AWG conductors enter the box.

four 12 AWG conductors	4
one luminaire (fixture) stud	1
two cable clamps (only count one)	1
Total	6

Referring to *Table 314.16(A)* and *Table 314.16(B)*, a 4 in. × 1½ in. octagon box is suitable for this example.

When using a metal or nonmetallic box where the conductors are different sizes. What is the minimum cubic-inch volume required for a box that will contain one internal cable clamp, one switch, and one receptacle, all mounted on one yoke? Two 14 AWG and two 12 AWG conductors enter the box. The wiring method is armored cable.

two 14 AWG conductors @ 2 in.^{3} per wire	= 4.00 in.^{3}
two 12 AWG conductors @ 2.25 in.^{3} per wire	= 4.50 in.^{3}
one cable clamp @ 2.25 in.^{3}	= 2.25 in.^{3}
one switch and one receptacle on one yoke @ 2.25 in.^{3} × 2	= 4.50 in.^{3}
Total	15.25 in.^{3}

• If box contains no fittings, devices, fixture studs, cable clamps, hickeys, switches, receptacles, or equipment grounding conductors:	• Refer directly to *Table 314.16(A)* or *Table 314.16(B)*.
• **Clamps.** If box contains one or more internal cable clamps:	• Add a single-volume based on the largest conductor in the box.
• **Support Fittings.** If box contains one or more luminaire (fixture) studs or hickeys:	• Add a single-volume for each type based on the largest conductor in the box.
• **Device or Equipment.** If box contains one or more wiring devices on a yoke:	• Add a double-volume for each yoke based on the largest conductor connected to a device on that yoke.
• **Equipment Grounding Conductors.** If a box contains one or more equipment grounding conductors:	• Add a single-volume based on the largest equipment grounding conductor in the box.
• **Isolated Equipment Grounding Conductor.** If a box contains one or more additional "isolated" (insulated) equipment grounding conductors as permitted by *250.146(D)* for "noise" reduction:	• Add a single-volume based on the largest equipment grounding conductor in the box.
• For conductors running through the box without being spliced:	• Add a single-volume for each conductor that runs through the box.
• For conductors that originated outside of the box and terminate inside the box:	• Add a single-volume for each conductor that originates outside the box and terminates inside the box.
• If no part of the conductor leaves the box, as with a "jumper" wire used to connect three wiring devices on one yoke, or pigtails:	• Don't count this (these). No additional volume required.
• For small equipment grounding conductors or not more than four conductors smaller than 14 AWG that originate from a luminaire (fixture) canopy or similar canopy (like a fan) and terminate in the box:	• Don't count this (these). No additional volume required.
• For small fittings, such as locknuts and bushings:	• Don't count this (these). No additional volume required.

Table 2-1 Quick checklist for possibilities to be considered when determining proper size boxes. See *314.16(B)(1)* through *314.16(B)(5)*.

Select a box having a minimum volume of 15.25 in.³ of space. The cubic-inch volume may be marked on the box; otherwise refer to the second column of *Table 314.16(A)* entitled "Minimum Volume Capacity."

The Code in *314.16(A)(2)* requires that all boxes *other* than those listed in *Table 314.16(A)* be durably and legibly marked by the manufacturer with their cubic-inch capacity. Nonmetallic boxes have their cubic-inch volume marked on the box. When sectional boxes are ganged together, the calculated volume is the total cubic-inch volume of the assembled boxes.

When installing a box that will have a raised cover attached to it. When marked with their cubic-inch volume, the *additional* space provided by plaster rings, domed covers, raised covers, and extension rings is permitted to be used when determining the overall volume. This is illustrated in Figure 2-35.

How many 12 AWG conductors are permitted in this box and raised plaster ring? Refer to Figure 2-36 and *Table 314.16(A)* and *Table 314.16(B)*.

$$\frac{25.5 \text{ in.}^3 \text{ of total space}}{2.25 \text{ in.}^3 \text{ per 12 AWG conductor}} = \frac{\text{Eleven 12 AWG}}{\text{conductors}}$$

This calculation actually resulted in 11.33. The conductor fill numbers in *Table 314.16(A)* were created by dropping any excess above the whole number after the calculations were made. Following the same procedure, this conductor fill calculation dropped off the excess.

Electricians and electrical inspectors have become very aware of the fact that GFCI receptacles, dimmers, and timers take up a lot more space than regular wiring devices. Therefore, it is a good practice to install boxes that will provide plenty of room for the wires, instead of pushing, jamming, and crowding the wires into the box.

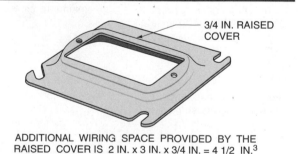

ADDITIONAL WIRING SPACE PROVIDED BY THE RAISED COVER IS 2 IN. x 3 IN. x 3/4 IN. = 4 1/2 IN.³

Figure 2-35   Raised cover. Raised covers are sometimes called plaster rings.

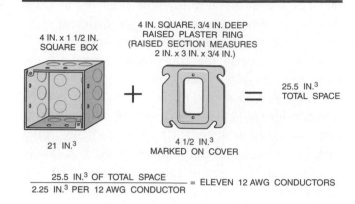

4 IN. x 1 1/2 IN. SQUARE BOX

4 IN. SQUARE, 3/4 IN. DEEP RAISED PLASTER RING (RAISED SECTION MEASURES 2 IN. x 3 IN. x 3/4 IN.)

25.5 IN.³ TOTAL SPACE

21 IN.³

4 1/2 IN.³ MARKED ON COVER

$$\frac{25.5 \text{ IN.}^3 \text{ OF TOTAL SPACE}}{2.25 \text{ IN.}^3 \text{ PER 12 AWG CONDUCTOR}} = \text{ELEVEN 12 AWG CONDUCTORS}$$

Figure 2-36   When marked with their cubic-inch volume, the volume of the cover may be added to the volume of the box to determine the total cubic-inch volume.

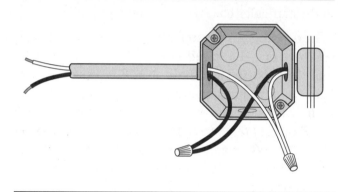

Figure 2-37   Transformer leads 18 AWG or larger must be counted when selecting the proper size box, *314.16(B)(1)*. In the example shown, the box contains four conductors.

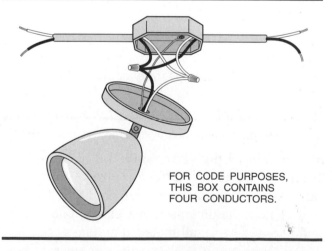

FOR CODE PURPOSES, THIS BOX CONTAINS FOUR CONDUCTORS.

Figure 2-38   Four or fewer luminaire (fixture) conductors, smaller than 14 AWG, and/or small equipment grounding conductors that originate in the luminaire (fixture) and terminate in the box, need not be counted when calculating box fill, *314.16(B)(1)*, Exception.

Figure 2-37 shows that transformer leads 18 AWG or larger must be counted when selecting a proper size box.

The minimum required size of equipment grounding conductors is shown in *Table 250.122*. The equipment grounding conductors are the same size as the circuit conductors in cables having 14-AWG, 12-AWG, and 10-AWG circuit conductors. Thus, for normal house wiring, box sizes can be calculated using *Table 314.16(A)*. Refer to Figure 2-38.

Figure 2-20, a quik-chek box selection guide, shows some of the most popular types of boxes used in residential wiring. This guide also shows the box's cubic-inch volume.

Figure 2-39 illustrates typical wiring using cable and conduit. Note in this figure how the conductor count is determined.

Using electrical metallic tubing (EMT) makes it possible to "loop" conductors through the box, which counts as one conductor only. Looping conductors through a box is not possible when using cable as the wiring method.

The basic rules for box fill are summarized in Table 2-1. Boxes that are intended to support ceiling fans are discussed in detail in Unit 9.

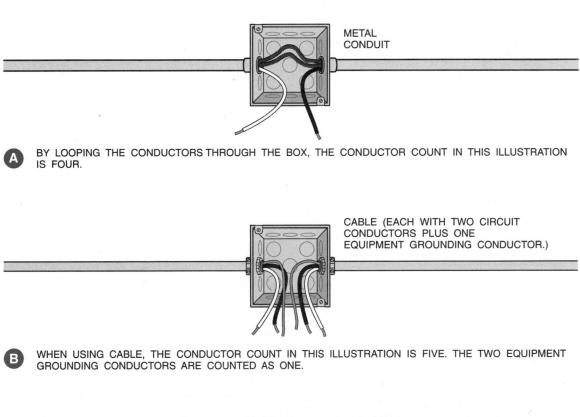

A BY LOOPING THE CONDUCTORS THROUGH THE BOX, THE CONDUCTOR COUNT IN THIS ILLUSTRATION IS FOUR.

B WHEN USING CABLE, THE CONDUCTOR COUNT IN THIS ILLUSTRATION IS FIVE. THE TWO EQUIPMENT GROUNDING CONDUCTORS ARE COUNTED AS ONE.

C WHEN USING CABLE AND A BOX THAT HAS INTERNAL CABLE CLAMPS, THE CONDUCTOR COUNT IN THIS ILLUSTRATION IS SIX. THE TWO EQUIPMENT GROUNDING CONDUCTORS ARE COUNTED AS ONE. THE TWO CABLE CLAMPS ARE COUNTED AS ONE.

Figure 2-39 Example of conductor count for both metal conduit and cable installations.

Means for Connecting Equipment Grounding Conductors in a Box

NEC® 314.40(D) requires that "a means shall be provided in each metal box for the connection of **equipment grounding conductors**." Generally, these are tapped No. 10-32 holes in the box that are marked GR, GRN, GRND, or similar identification. The screws used for terminating the equipment grounding conductor must be hexagonal and must be green. Electrical distributors sell green, hexagonal No. 10-32 screws for this purpose.

HEIGHT OF RECEPTACLE OUTLETS

There are no hard-and-fast rules for locating most outlets. A number of conditions determine the proper height for a switch box. For example, the height of the kitchen counter backsplash determines where the switches and receptacle outlets are located between the kitchen countertop and the cabinets.

The residence featured in this text is electrically heated and is discussed in Unit 23. The type of electric heat could be:

- electric furnace (as in this text).
- electric resistance heating buried in ceiling plaster or "sandwiched" between two layers of drywall material.
- electric baseboard heaters.
- heat pump

Let us consider the electric baseboard heaters. In most cases, the height of these electric baseboard units from the top of the unit to the finished floor seldom exceeds 6 in. (150 mm). The important issue here is that the manufacturer's receptacle accessories may have to be used to conform to the receptacle spacing requirements, as covered in *210.52*.

Electrical receptacle outlets are not permitted to be located above an electric baseboard heating unit. Refer to the section "Location of Electric Baseboard Heaters in Relation to Receptacle Outlets," in Unit 23.

It is common practice among electricians to consult the plans and specifications to determine the proper heights and clearances for the installation of electrical devices. The electrician then has these dimensions verified by the architect, electrical engineer, designer, or homeowner. This practice avoids unnecessary and costly changes in the locations of outlets and switches as the building progresses.

POSITIONING OF RECEPTACLES

Although no actual Code rules exist on positioning receptacles, there is a popular concept in the electrical industry that there is always the possibility of a metal wall plate coming loose and falling downward onto the blades of an attachment plug cap that is loosely plugged into the receptacle, thereby creating a potential shock and fire hazard. The options are shown in Figure 2-40, Figure 2-41, Figure 2-42, and Figure 2-43.

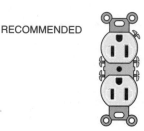

Figure 2-40   Grounding hole to the top. A loose metal plate could fall onto the grounding blade of the attachment plug cap, but no sparks would fly.

Figure 2-41   Ground*ing* hole to the bottom. A loose metal plate could fall onto both the grounded neutral and "hot" blades. Sparks would fly.

Figure 2-42   Ground*ed* neutral blades on top. A loose metal plate could fall onto these grounded neutral blades. No sparks would fly.

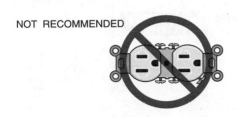

NOT RECOMMENDED

Figure 2-43 "Hot" terminal on top. A loose metal plate could fall onto these live blades. If this were a split-circuit receptacle fed by a three-wire, 120/240-volt circuit, the short would be across the 240-volt line. Sparks would fly.

TYPICAL HEIGHTS FOR SWITCHES AND OUTLETS

Typical heights for switches and outlets are shown in Table 2-2. These dimensions usually are satisfactory. However, the electrician must check the blueprints, specifications, and details for measurements that may affect the location of a particular outlet or switch. The cabinet spacing, available space between the countertop and the cabinet, and the tile height may influence the location of the outlet or switch. For example, if the top of the wall tile is exactly 4 ft (1.2 m) from the finished floor line, a wall switch should not be mounted 4 ft (1.2 m) to

SWITCHES	
Regular	46 in. (1.15 m). Some homeowners want wall switches mounted 30 in.–36 in. (750 mm–900 mm) so as to be easy for children to reach and easy for adults to reach when carrying something. The choice is yours!
Between counter and kitchen cabinets. Depends on backsplash.	44 in.–46 in. (1 m–1.15 m)
RECEPTACLE OUTLETS	
Regular	12 in. (300 mm)
Between counter and kitchen cabinets. Depends on backsplash.	44 in.–46 in. (1 m–1.15 m)
In garages	46 in. (1.15 m) Minimum 18 in. (450 mm)
In unfinished basements	46 in. (1.15 m)
In finished basements	12 in. (300 mm)
Outdoors (above grade or deck)	18 in. (450 mm)
WALL LUMINAIRE (FIXTURE) OUTLETS	
Outside entrances. Depends on luminaire (fixture).	66 in. (1.7 m). If luminaire (fixture) is "upward" from box on luminaire (fixture), mount wall box lower. If luminaire (fixture) is "downward" from box on luminaire (fixture), mount wall box higher.
Inside wall brackets	5 ft (1.5 m)
Side of medicine cabinet or mirror Above medicine cabinet or mirror	You need to know the measurement of the medicine cabinet. Check the rough-in opening of medicine cabinet measurement of mirror. Mount electrical wall box approx. 6 in. (150 mm) to center above rough-in opening or mirror. Medicine cabinets that come complee with luminaires (fixtures) have a wiring compartment with a conduit knockout(s) in which the supply cable or conduit is secured using the appropriate fitting. Many strip luminaires (fixtures) have a backplate with a conduit knockout in which the supply cable or conduit is secured using the appropriate fitting. Where to bring in the cable or conduit takes careful planning.

Note: All dimensions are from finished floor to center of the electrical box. If possible, try to mount wall boxes for luminaires (lighting fixtures) based upon the type of luminaire (fixture) to be installed. Verify all dimensions before "roughing in." If wiring for physically handicapped, the above heights may need to be lowered in the case of switches, and raised in the case of receptacles.

Table 2-2 Typical heights for switches, receptacles, and wall luminaire (lighting fixture) outlets. Local height preferences prevail.

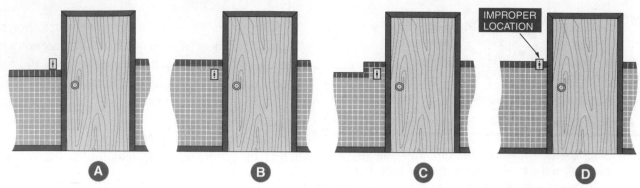

WHERE A WALL IS PARTIALLY TILED, A SWITCH OR CONVENIENCE OUTLET MUST BE LOCATED ENTIRELY OUT OF THE TILE AREA (A) OR ENTIRELY WITHIN THE TILE AREA (B), (C).

THE FACEPLATE IN (D) DOES NOT "HUG" THE WALL PROPERLY. THIS INSTALLATION IS CONSIDERED UNACCEPTABLE BY MOST ELECTRICIANS, AND IS A VIOLATION OF THE LAST SENTENCE OF *NEC® 404.9(A)*.

Figure 2-44 Locating an outlet on a tiled wall.

center. This is considered poor workmanship. The switch should be located entirely within the tile area or entirely out of the tile area, as in Figure 2-44. This situation requires the full cooperation of all craftspersons involved in the construction job.

Faceplates

Faceplates for switches and receptacles shall be installed so as to completely cover the wall opening and seat against the wall surface, *404.9(A)* and *406.5*.

Faceplates should be level! No matter how good the wiring inside the wall is, no matter how good the connections and splices are, no matter how proper the equipment grounding is, the only thing the homeowner will see is the wiring devices and the faceplates. That establishes the impression the homeowner has of the overall wiring installation. Because of the elongated slots in the yoke of a wiring device, it is rather easy to level a wiring device and faceplate on a single-gang box even if the box is not level. This is not true for multi-gang boxes. It is extremely important to make sure multi-gang boxes are level. If a multi-gang box is not level, it will be virtually impossible to make the wiring devices and faceplate level.

Another nice detail that can improve the appearance is to align the slots in all of the faceplate mounting screws in the same direction, as in Figure 2-45.

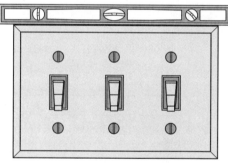

ALIGN SCREW—SLOTS
LIKE THIS,

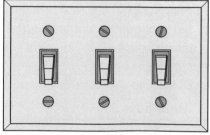

NOT LIKE THIS.

Figure 2-45 Aligning the slots of the faceplate. Mounting screws in the same direction makes the installation look neater. Make sure the faceplate is level.

REVIEW

Note: For assistance in finding the answers to these review questions, refer to the *NEC*,® the plans in the back of this text, and the specifications found in the Appendix.

PART 1—ELECTRICAL FEATURES

1. What does a plan show about electrical outlets? _____

   _____

   _____

   _____

2. What is an outlet? _____

   _____

   _____

3. Match the following switch types with the proper symbol.

   a. single-pole S_p

   b. three-way S_4

   c. four-way S

   d. single-pole with pilot light S_3

4. The plans show dashed lines running between switches and various outlets. What do these dashed lines indicate? _____

   _____

5. Why are dashed lines usually curved? _____

6. a. What are junction boxes used for? _____

   b. Are junction boxes normally used in wiring the first floor? Explain. _____

   _____

   _____

   c. Are junction boxes normally used to wire exposed portions of the basement? Explain. _____

   _____

7. How are standard sectional switch (device) boxes mounted? _____

   _____

   _____

   _____

8. a. What is an offset bar hanger? _____

   _____

   _____

   b. What types of boxes may be used with offset bar hangers? _____

   _____

9. What methods may be used to mount luminaires (lighting fixtures) to an outlet box fastened to an offset bar hanger? _____

_____

_____

10. What advantage does a 4-in. octagon box have over a 3¼-in. octagon box?

_____

_____

11. What is the size of the opening of a switch (device) box for a single device? _____

_____

_____

12. The space between a door casing and a window casing is 3½ in. (88.9 mm). Two switches are to be installed at this location. What type of switches will be used?

_____

_____

13. Three switches are mounted in a three-gang switch (device) box. The wall plate for this assembly is called a _____ plate.

14. The mounting holes in a device (switch) box are tapped for No. 6-32 screws. The mounting holes in an outlet box are tapped for No. 8-32 screws. The mounting holes in metal boxes for attaching equipment grounding conductors are tapped for _____ screws.

15. a. How high above the finished floor in the living room are switches located?

_____

b. How high above the garage floor are switches located?

_____

16. a. How high above the finished floor in the living room are receptacles located?

_____

b. How high above the garage floor are receptacles located?

_____

17. Outdoor receptacle outlets in this dwelling are located _____ _____ in. above grade.

18. In the spaces provided, draw the correct symbol for each of the descriptions listed in (a) through (r).

a. _____ Lighting panel

b. _____ Clock outlet

c. _____ Duplex outlet

d. _____ Outside telephone

e. _____ Single-pole switch

f. _____ Motor

g. _____ Duplex outlet, split-circuit

h. _____ Lampholder with pull switch

i. _____ Weatherproof outlet

j. _____ Special-purpose outlet

k. _____ Fan outlet

l. _____ Range outlet

m. _____ Power panel

n. _____ Three-way switch

o. _____ Push button

p. _____ Thermostat

q. _____ Electric door opener

r. _____ Multi-outlet assembly

19. The front edge of a box installed in a combustible wall must be _____ with the finished surface.

20. List the maximum number of 12 AWG conductors permitted in a

   a. 4 in. × 1½ in. octagon box. _____

   b. 4¹¹⁄₁₆ in. × 1½ in. square box. _____

   c. 3 in. × 2 in. × 3½ in. device box. _____

21. When a switch (device) box is nailed to a stud, and the nail runs through the box, the nails must not interfere with the wiring space. To accomplish this, keep the nail

   _____ (Select a, b, or c.)

   a. halfway between the front and rear of the box.

   b. a maximum of ¼ in. (6 mm) from the front edge of the box.

   c. a maximum of ¼ in. (6 mm) from the rear of the box.

22. Hanging a ceiling luminaire (fixture) directly from a plastic outlet box is permitted only if _____

   _____

   _____

   _____

23. It is necessary to count luminaire (fixture) wires when counting the permitted number of conductors in a box according to *314.16*.

   (True)   (False)   Underline or circle the correct answer.

24. *Table 314.16(A)* allows a maximum of ten wires in a certain box. However, the box will have two cable clamps and one fixture stud in it. What is the maximum number of wires allowed in this box? _____

   _____

25. When laying out a job, the electrician will usually make a layout of the circuit, taking into consideration the best way to run the cables and/or conduits and how to make up the electrical connections. Doing this ahead of time, the electrician determines exactly how many conductors will be fed into each box. With experience, the electrician will probably select two or three sizes and types of boxes that will provide adequate space to "meet Code." *Table 314.16(A)* of the Code shows the maximum number of conductors permitted in a given size box. In addition to counting the number of conductors that will be in the box, what is the additional volume that must be provided for the following items? Enter *single* or *double* volume allowance in the blank provided.

   a. one or more internal cable clamps: _____ volume allowance.

   b. for a fixture stud: _____ volume allowance.

   c. for one or more wiring devices on one yoke: _____ volume allowance.

   d. for one or more equipment grounding conductors: _____ volume allowance.

26. Is it permissible to install a receptacle outlet above an electric baseboard heater?

   _____

   _____

27. What is the maximum weight of a luminaire (lighting fixture) permitted to be hung directly from an outlet box in a ceiling? _____ pounds.

28. Two 12/2 and two 14/2 nonmetallic-sheathed AWG cables enter a box. Each cable has an equipment grounding conductor. The 12 AWG conductors are connected to a receptacle. Two of the 14 AWG conductors are connected to a toggle switch. The other two 14 AWG conductors are spliced together since they serve as a switch loop. The box contains two cable clamps. Calculate the minimum cubic-inch volume required for the box.

$$G = 2.25$$

$$2 \quad 12 = 9.00$$

$$4 \quad 14 = 8.00$$

$$2.28$$
$$\overline{21.50}$$

$$2 1.50$$
$$9.00$$
$$\overline{30.50}$$

29. Using the same number and size of conductors as in question 28, but using electrical metallic tubing, calculate the minimum cubic inch volume required for the box. There will be no separate equipment grounding conductors, nor will there be any clamps in the box.

30. To allow for adequate conductor length at electrical outlet and device boxes to make up connections, *300.14* requires that not less than [3 in. (75 mm), 6 in. (150 mm), 9 in. (225 mm)] of conductor length be provided. This length is measured from where the conductor emerges from the cable or raceway to the end of the conductor. For box openings having any dimension less than 8 in. (200 mm), the minimum length of conductor measured from the box opening in the wall to the end of the conductor is [3 in. (75 mm), 6 in. (150 mm), 9 in. (225 mm)]. Circle the correct answers.

31. When wiring a residence, what must be considered when installing wall boxes on both sides of a common partition that separates the garage and a habitable room? _____

_____

_____

32. Does the *NEC*® allow metal raceways to be used with nonmetallic boxes?

Yes _____ No _____ , *NEC*® _____

PART 2—STRUCTURAL FEATURES

1. To what scale is the basement plan drawn? _____

2. What is the size of the footing for the steel Lally columns in the basement? _____
   _____

3. To what kind of material will the front porch lighting bracket luminaire (fixture) be attached? _____

4. Give the size, spacing, and direction of the ceiling joists in the workshop. _____
   _____

5. What is the size of the lot on which this residence is located? _____
   _____

6. The front of the house is facing which compass direction? _____

7. How far is the front garage wall from the curb? _____

8. How far is the side garage wall from the property lot line? _____

9. How many steel Lally columns are in the basement, and what size are they? _____
   _____

10. What is the purpose of the I-beams that rest on top of the steel Lally columns? _____
    _____

11. To make sure that switch boxes and outlet boxes are set properly in the garage walls and ceilings, we need to know the thickness of the gypsum wallboard used in these locations. In the garage, what is the thickness and type of the gypsum wallboard?

    a) on the "warm walls" of the garage? _____

    b) on the "cold walls" of the garage? _____

    c) on the ceiling of the garage? _____

12. Where is access to the attic provided? _____

13. Give the thickness of the outer basement walls. _____

14. What material is indicated for the foundation walls? _____

15. Where are the smoke detectors located in the basement? _____
    _____

16. What is the ceiling height in the basement workshop from bottom of the joists to the floor?
    _____

17. Give the size and type of the front door.
    _____

18. What is the stud size for the partitions between the bathrooms in the bedroom area where substantial plumbing is to be installed? _____

19. Who is to furnish the range hood? _____
    _____

20. Who is to install the range hood? _____

Determining the Required Number and Location of Lighting and Small Appliance Circuits

OBJECTIVES

After studying this unit, the student should be able to

- understand the fundamental Code requirements for calculating branch-circuit sizing and loading.
- understand "volt-amperes per square foot" for arriving at total calculated general lighting loading.
- calculate the occupied floor area of a residence.
- determine the minimum number of lighting branch-circuits required.
- determine the minimum number of small appliance branch-circuits required.
- know where receptacle outlets must be installed in homes.
- know where GFCI receptacles must be installed in homes.
- know where switched and nonswitched luminaire (fixture) outlets must be installed in homes.

It is standard practice in the design and planning of dwelling units to permit the electrician to plan the circuits. Thus, the residence plans do not include layouts for the various branch-circuits. The electrician may follow the general guidelines established by the architect. However, any wiring systems designed and installed by the electrician must conform to the *NEC,* as well as local and state code requirements.

This unit focuses on lighting branch-circuits and small appliance branch-circuits. The circuits supplying the electric range, oven, clothes dryer, and other specific circuitry not considered to be a lighting branch-circuit or a small appliance branch-circuit are covered later in this text. Refer to the index for the specific circuit being examined.

BASICS OF WIRE SIZING AND LOADING

The *NEC®* establishes some very important fundamentals that weave their way through the decision-making process for an electrical installation. They are presented here in brief form, and are covered in detail as required throughout this text.

The *NEC®* defines, a **branch-circuit** as "the circuit conductors between the final overcurrent device protecting the circuit and the outlet(s)." See

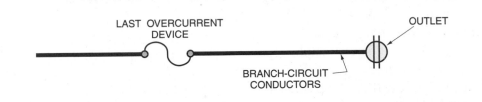

Figure 3-1 The branch-circuit is that part of the wiring that runs from the final overcurrent device to the outlet. The rating of the overcurrent device, not the conductor size, determines the rating of the branch-circuit.

Figure 3-1. In the residence discussed in this text, the wiring to wall outlets, the dryer, the range, and so on, are all examples of a branch-circuit.

The *NEC®* defines a **feeder** as "All circuit conductors between the service equipment, the source of a separately derived system, or other power supply source and the final branch-circuit overcurrent device." In the residence discussed in this text, the wiring between Main Panel A and Subpanel B is a feeder.

The **ampacity** (current-carrying capacity) of a conductor must not be less than the rating of the **overcurrent device** protecting that conductor, *210.19* and *210.20*. A common exception to this is a motor branch-circuit, where it is quite common to have overcurrent devices (fuses or breakers) sized larger than the ampacity of the conductor. Motors and motor circuits are covered specifically in *Article 430*. The **ampere** rating of the branch-circuit overcurrent protective device (fuse or circuit breaker) determines the rating of the branch-circuit. For example, if a 20-ampere conductor is protected by a 15-ampere fuse, the circuit is considered to be a 15-ampere branch-circuit, *210.3*.

Standard branch-circuits that serve more than one receptacle outlet or more than one lighting outlet are rated 15, 20, 30, 40, and 50 amperes. A branch-circuit that supplies an individual load can be of any ampere rating, *210.3*.

Where the ampacity of the conductor does not match up with a standard rating of a fuse or breaker, the next higher standard size overcurrent device may be used, provided the overcurrent device does not exceed 800 amperes, *240.4(B)*. This deviation is not permitted, however, if the circuit supplies receptacles where "plug-connected" appliances, and so on, could

be used, because too many "plug-in" loads could result in an overload condition, *210.19(B)* and *240.4(B)(1)*. You may go to the next standard size overcurrent device only when the circuit supplies other than receptacles for cord and plug-connected portable loads.

For instance, when a conductor having an allowable ampacity of 25 amperes (see *Table 310.16*) 12 AWG Type TW is derated to 70%:

$$25 \times 0.70 = 17.5 \text{ amperes}$$

It is permitted to use a 20-ampere overcurrent device if the circuit supplies *only* fixed lighting outlets or other fixed loads.

If the previous example were to supply receptacle outlets, then the rating of the overcurrent device would have to be dropped to 15 amperes; otherwise it is possible to overload the conductors by plugging in more load than the conductors can safely carry.

The allowable ampacity of conductors commonly used in residential occupancies is found in *Table 310.16*. The ampacities in *Table 310.16* are subject to **correction factors** that must be applied if high ambient temperatures are encountered—for example, in attics. See the bottom of *Table 310.16*.

Conductor ampacities are also subject to a **derating factor** if more than three current-carrying conductors are installed in a single raceway or cable. See *Table 310.15(B)(2)*.

For loads that are *continuous*, the load shall not exceed 80% of the rating of the branch-circuit. The *NEC®* defines *continuous* as "a load where the maximum current is expected to continue for three hours or more." The lighting loads in residences are not considered to be "continuous" loads like commercial lighting loads, such as store lighting. The *NEC®* permits 100% loading on an overcurrent device only if that overcurrent device is *listed* for 100% loading. At the time of this writing, Underwriters Laboratories (UL) has *no* listing of circuit breakers of the molded-case type used in residential installations that are suitable for 100% loading.

This is another judgment call on the part of the electrician and/or the electrical inspector. Will a branch-circuit supplying a heating cable in a driveway likely be on for 3 hours or more? The answer is "probably." Good design practice and experience tell us "never load a circuit conductor or overcurrent protective device to more than 80% of its rating."

The minimum lighting load for dwellings is 3 **volt-amperes** per square foot. See *220.3(A)* and *Table 220.3(A)*.

Most receptacle outlets in a residence are considered by the Code to be part of "general illumination," and additional load calculations are *not* necessary. See *220.3(B)(10)*. Appliance receptacle outlets in the kitchen, dining room, laundry, and workshop are *not* to be considered part of general illumination. Additional loads are figured in for these receptacle outlets and are discussed in detail later in this text.

VOLTAGE

All calculations throughout this text use **voltages** of 120 volts and 240 volts. This is in conformance to *220.2(A)*, which states that "unless other voltages are specified, for purposes of computing branch-circuits and feeder loads, nominal system voltages of 120, 120/240, 208Y/120, 240, 480Y/277, 480, and 600 volts shall be used." This is repeated in the second paragraph of *Annex D* of the *NEC*,® which states that "for uniform application of *Article 210*, *215*, and *220*, a nominal voltage of 120, 120/240, 240, and 208Y/120 shall be used in computing the ampere load on the conductor."

The word *nominal* means "in name only, not a fact." For example, in our calculations, we use 120 volts even though the actual voltage might be 110, 115, or 117 volts. This provides us with a uniformity that, without it, would lead to misleading and confusing results in our computations.

Most public service commissions require that the voltage for residential services be held to ±5 percent of the nominal voltage as measured at the service point. (See Definitions in the *NEC*.®)

COMPUTING LOADS

When wiring a house, it is all but impossible to know which appliances, lighting, heating, and other loads will be turned on at the same time. Different families lead different lifestyles. There is tremendous diversity. There is a big difference between "connected load" and "actual load." Who knows what will be plugged into a wall receptacle, now or in the future? It's a guess at best. Over the years, the *National Electrical Code*® has developed procedures for computing (calculating) loads in typical one- and two-family homes.

The rules for doing the computations (calculations) are found in *Article 220*. For lighting and receptacles, the computations are based on volt-amperes per square foot. For the small-appliance circuits in kitchens and dining rooms, the basis is 1500 volt-amperes per circuit. For large appliances such as dryers, electric ranges, ovens, cooktops, water heaters, air-conditioners, heat pumps, and so on, which are not all used continuously or at the same time, there are demand factors to be used in the computation. Following the requirements in the *NEC*,® the various computations roll together in steps that result in the proper sizing of branch-circuits, feeders, and service equipment.

As you work your way through this text, examples are provided for just about every kind and type of load found in a typical residence.

Inch-pounds vs. Metrics When Computing Loads

As discussed in Unit 1, converting inch-pound measurements to metric measurements and vice versa results in odd fractional results. Adding further to this problem are the values rounded off when the *NEC*® Code Making Panels did the metric conversions. When square feet is converted to square meters and the unit loads are computed for each, the end results are different; close, but nevertheless different. To show both computations would be confusing as well as space consuming. Many of the measurements in this text are shown in both inch-pound units and metric units. The computations throughout this text will use inch-pound values only, which is in agreement with the *Examples* given in *Annex D* of the 2002 *NEC*.®

CALCULATING FLOOR AREA

The general lighting load for a dwelling is based on the square footage of the dwelling. Here is how this is done.

First Floor Area

To estimate the total load for a dwelling, the occupied floor area of the dwelling must be calculated. Note in the residence plans that the first floor area has an irregular shape. In this case, the simplest method of calculating the occupied floor area is to determine the total floor area using the outside dimensions of the dwelling. Then, the areas of the following spaces are subtracted from the total area: open porches, garages, or other unfinished or unused spaces if they are not adaptable for future use, *220.3(A)*.

Many open porches, terraces, patios, and similar areas are commonly used as recreation and enter-tainment areas. Therefore, adequate lighting and receptacle outlets should be provided.

For practicality, the author has chosen to round up dimensions for the determining of total square footage and to round down dimensions for those areas (garage, porch, and portions of the inset at the front of the house) not to be included in the computation of the general lighting load. This produces a slightly larger result as opposed to being on the conservative side.

Figure 3-2 shows the procedure for calculating the total square footage of this residence.

Basement Area

The basement area of a home is not generally included in the calculating of general lighting loads because it could be classified as an "unfinished space not adaptable for future use," *220.3(A)*. Yet in this residence, more than half of the total basement

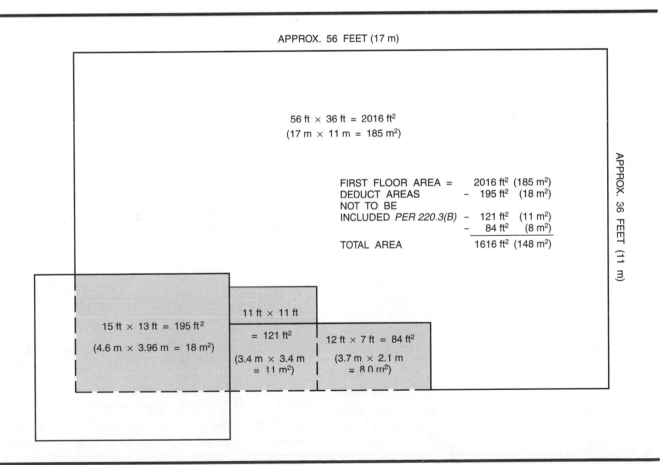

APPROX. 56 FEET (17 m)

56 ft × 36 ft = 2016 ft²
(17 m × 11 m = 185 m²)

FIRST FLOOR AREA =   2016 ft² (185 m²)
DEDUCT AREAS   –   195 ft² (18 m²)
NOT TO BE
INCLUDED *PER 220.3(B)* –   121 ft² (11 m²)
                    –   84 ft²   (8 m²)

TOTAL AREA   1616 ft² (148 m²)

APPROX. 36 FEET (11 m)

15 ft × 13 ft = 195 ft²

(4.6 m × 3.96 m = 18 m²)

11 ft × 11 ft

= 121 ft²

(3.4 m × 3.4 m
= 11 m²)

12 ft × 7 ft = 84 ft²

(3.7 m × 2.1 m
= 8.0 m²)

Figure 3-2   Determining the first floor square footage area. The basement area is the same as the first floor. See text for explanation of how the dimensions of the residence were rounded off to make the calculations much simpler and much more practical for the electrician and electrical inspector to compute.

area is finished off as a recreation room, which certainly is considered living area. The workshop area also is intended to be used.

To simplify the calculation for this residence, we will consider the entire basement as useable space. The basement square footage area is the same as the area of the first floor.

The combined occupied area of the dwelling is found by adding the first floor and basement areas together:

First Floor	1616 ft²	(149 m²)
Basement	1616 ft²	(149 m²)
Total	3232 ft²	(298 m²)

DETERMINING THE MINIMUM NUMBER OF LIGHTING BRANCH-CIRCUITS

To determine the minimum number of 15-ampere lighting branch-circuits required in a residence:

Step 1: $\dfrac{3 \text{ volt-amperes} \times \text{square feet}}{120 \text{ volts}} = \text{amperes}$

Step 2: $\dfrac{\text{amperes}}{15} = $ minimum number of 15-ampere circuits required

This equates to a *minimum* of:

- one 15-ampere branch-circuit for every 600 square feet (55.8 m²).

Although 20-ampere lighting branch-circuits are rarely used in residential installations, using the same procedure as indicated in Steps 1 and 2 would result in:

- one 20-ampere lighting branch-circuit for every 800 square feet (74.4 m²).

Load calculations for electric water heaters, ranges, heat pumps, air-conditioning units, dryers, motors, electric heat, small appliance branch-circuits, and other specific loads are in addition to the general lighting load. These are discussed throughout this text as they appear.

Table 220.3(A) shows that the minimum load required for dwelling units is 3 volt-amperes (VA) per square foot (0.093 m²) of occupied area. The calculated load for the total occupied area, 3232 ft² (298 m²), is:

$$1616 + 1616 = 3232 \text{ ft}^2$$

$$3232 \text{ (ft}^2) \times 3 \text{ (VA/ft}^2) = 9696 \text{ volt-amperes}$$

The total required amperage is determined as follows:

$$\text{Amperes} = \frac{\text{volt-amperes}}{\text{volts}}$$

$$= \frac{9696}{120} = 80.8 \text{ amperes}$$

The minimum number of 15-ampere lighting circuits required is equal to the total amperage divided by the maximum amperes of each circuit.

$$\frac{80.8 \text{ amperes}}{15 \text{ amperes}} = 5.39 \text{ circuits}$$

Therefore, a minimum of six lighting branch-circuits is required.

▶ Although it might seem obvious, the *NEC®* in *210.11(B)* states that "the loads shall be evenly proportioned in so far as is practical."◀

The *NEC®* in *Annex D* provides a number of examples that show calculations for services, feeders, neutrals, motor circuits, and other important issues.

In *Annex D* we also find that "except where the computations result in a major fraction of an ampere (0.5 or larger) such fractions may be dropped."

Table 3-1 is the schedule of the 15-ampere lighting and 20-ampere small appliance branch-circuits in this residence.

240.4(D) tells us that the maximum overcurrent protection is:

15 amperes for 14 AWG copper conductors

20 amperes for 12 AWG copper conductors

30 amperes for 10 AWG copper conductors

In conformance to *210.3*, a branch-circuit is rated according to the rating or setting of the overcurrent device. For example, a lighting branch circuit is rated 15 amperes even if 12 AWG copper conductors are installed.

TRACK LIGHTING LOADS

Track lighting loads are considered to be part of the general lighting load for residential installations, based upon the 3 volt-amperes per square foot as previously discussed. There is no need to add more wattage (volt-amperes) to the load calculations for track lighting in homes.

As required by *410.101(B)*, the connected load on a lighting track shall not exceed the rating of the

SUMMARY OF 15-AMPERE LIGHTING AND 20-AMPERE RECEPTACLE CIRCUITS

15-Ampere Circuits	20-Ampere Circuits
A14 Bathrooms, Hall Ltg.	A18 Workbench Receptacles
A15 Front Entry/Porch	A20 Workshop Receptacles, Window Wall
A16 Front Bedroom Ltg. & Outdoor Receptacle	A22 Master Bath Receptacle
A17 Workshop Ltg.	A23 Hall Bath Receptacle
A19 Master Bedroom Ltg. & Outdoor Receptacle	
A21 Study/Bedroom Ltg.	
B7 Kitchen Ltg.	B13 Kitchen North & East Wall Receptacles
B9 Wet Bar Ltg. & Receptacles	B15 Kitchen Receptacles, West Countertop
B10 Laundry, Rear Entry, Powder Room, Attic Ltg.	B16 Kitchen Receptacles, South Countertop, Refrigerator, & Island
B11 Recreation Room Receptacles	B18 Washer Receptacle
B12 Recreation Room Ltg.	B20 Laundry Room Receptacles & Outdoor Receptacle
B14 Garage, Post Light	B21 Powder Room Receptacle
B17 Living Room & Outdoor Receptacle	

Note: Panel "A" is the main panel located in the workshop. Panel "B" is the subpanel located in the corner of the recreation room.

Table 3-1

track. Also, be sure that a lighting track is supplied by a branch-circuit having a rating not more than that of the track. There is more on track lighting in Units 12 and 13.

DETERMINING THE NUMBER OF SMALL APPLIANCE BRANCH-CIRCUITS

NEC® 220.16(A) and *(B)* state that an additional load of not less than 1500 volt-amperes shall be included for each two-wire small appliance circuit and each laundry circuit.

The receptacle for the refrigerator in the kitchen is permitted to be supplied by a separate 15- or 20-ampere branch-circuit rather than connecting it to one of the two required 20-ampere small appliance circuits. This permission is found in *210.52(B)(1), Exception No. 2.* Because this separate branch-circuit diverts the refrigerator load from the receptacles that serve the countertop areas, it is not considered to be a small appliance branch-circuit. An additional 1500 volt-amperes *does not* have to be included in the load calculations, *220.16(A), Exception.* There is nothing in the Code that prohibits adding another 1500 volt-amperes to the load calculations for this additional branch-circuit, as this would result in more capacity in the electrical system.

NEC® 210.11(C)(1) requires that two or more 20-ampere small appliance circuits must be provided for supplying receptacle outlets, as specified in *210.52(B).*

NEC® 210.52(B)(1) states that the 20-ampere appliance circuits shall feed only the receptacle outlets in the kitchen, pantry, dining room, breakfast room, or similar rooms of dwelling units. *NEC® 210.52(B)(2)* also requires that these 20-ampere small appliance circuits shall not supply other outlets.

Exceptions in *210.52(B)(2)* allow clock receptacles, and receptacles that are installed to plug in the cords of gas-fired ranges, ovens, or counter-mounted cooking units in order to power up their timers and ignition systems, Figure 3-3.

NEC® 210.52(B)(3) requires that not less than two 20-ampere small appliance circuits must supply the receptacle outlets serving the countertop areas in kitchens.

Food waste disposers and/or dishwashers shall not be connected to these required 20-ampere small appliance circuits. They must be supplied by other circuits, as is discussed later. Generally, 12 AWG conductors are used in small appliance branch-circuits. The use of this larger conductor, rather than a 14 AWG conductor, helps improve appliance performance and lessens the danger of overloading the circuits.

Automatic washers draw a large amount of current during certain portions of their operating cycles. Thus, *210.11(C)(2)* requires that at least one additional 20-ampere branch-circuit must be provided to supply the laundry receptacle outlet(s) as required by *210.52(F).*

A complete list of the branch-circuits for this residence is found in Unit 27. The circuit directories for Main Panel A and Subpanel B show clearly the number of lighting, appliance, and special-purpose circuits provided.

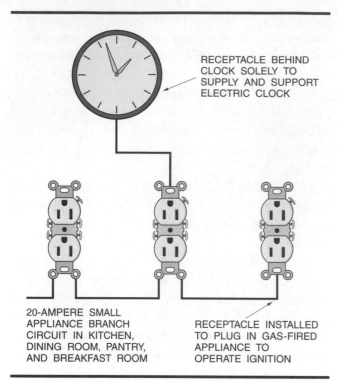

RECEPTACLE BEHIND CLOCK SOLELY TO SUPPLY AND SUPPORT ELECTRIC CLOCK

20-AMPERE SMALL APPLIANCE BRANCH CIRCUIT IN KITCHEN, DINING ROOM, PANTRY, AND BREAKFAST ROOM

RECEPTACLE INSTALLED TO PLUG IN GAS-FIRED APPLIANCE TO OPERATE IGNITION

Figure 3-3 A receptacle installed solely to supply and support an electric clock, or a receptacle that supplies power for clocks, clock timers, and electric ignition for gas ranges, ovens, and cooktops may be connected to a 20-ampere small appliance circuit or to a general-purpose lighting circuit, *210.52(B)(2), Exceptions.*

RECEPTACLE OUTLET BRANCH-CIRCUIT RATINGS

The receptacles in the kitchen, breakfast area, workshop, powder room, bathrooms, and laundry room are connected to 20-ampere branch-circuits. All other receptacles are connected to 15-ampere branch-circuits.

SUMMARY OF WHERE RECEPTACLE AND LIGHTING OUTLETS MUST BE INSTALLED IN RESIDENCES

The following is a brief explanation of where receptacle outlets and lighting outlets must be installed in residential occupancies. Throughout this text, each room is discussed in detail regarding outlet location. The following recaps all of the *NEC*® rules that relate to receptacle and lighting outlet requirements.

Bear in mind that the Code provides *minimum* requirements for the placement of receptacles.

A duplex receptacle located where there will be a concentration of such things as computers, printers, stereos, VCRs, CD players, television sets, radios, fax machines, answering machines, typewriters, adding machines, or lamps will probably not be adequate to serve these appliances without the use of a cord set having multiple receptacles. An electrician will do well to quiz the homeowner where this concentration might occur, then offer to install four receptacles at these locations. As an alternative, a quadplex receptacle (four receptacles) can replace a standard duplex receptacle in a single-gang device box. This same quadplex receptacle can be mounted on a 4-in. square or octagon outlet box, on a 1- or 2-gang device box, or on a $4^{11}/_{16}$-in. square box using a special adapter plate.

See Unit 6 for a discussion of surge protection.

Receptacle Outlets (125-Volts, Single-Phase, 15- or 20-Ampere)

The *NEC*® is very specific about the required number and location of receptacle outlets in homes. The following is a recap of these requirements.

Ground-Fault Circuit Interrupters. As we discuss the locations where receptacle outlets are required, **GFCI** stands for **Ground-Fault Circuit Interrupter**. This means that the receptacle must be of the GFCI type, or the receptacle must be protected with a GFCI circuit breaker. GFCIs are covered in detail in Unit 6.

- ►Wall receptacles must be placed so that no point measured horizontally along the floor line in any wall space is more than 6 ft (1.8 m) from a receptacle outlet. Fixed room dividers and railings are considered to be walls for the purpose of this requirement. See *210.52(A)(1)*.◄

- Any wall space 24 in. (600 mm) or more in width must have a receptacle outlet, *210.52(A)(2)(a)*.

- Nonsliding fixed glass panels on exterior walls are considered to be wall space and are to be figured in when applying the rule that no point along the floor line is more than 6 ft (1.8 m) from a receptacle. Sliding panels on exterior walls are not considered to be a wall, *210.52(A)(2)*.

- Receptacle outlets located in the floor more than 18 in. (450 mm) from the wall are not to be counted as meeting the required number of wall receptacle outlets, *210.52(A)(3)*. An example of this would be a floor receptacle installed under a dining room table for plugging in warming trays, coffee pots, or similar appliances.

- Hallways 10 ft (3.0 m) or longer in homes must have at least one receptacle outlet, *210.52(H)*. See Figure 3-4.

- Circuits supplying wall receptacle outlets for lighting loads are generally 15-ampere circuits, but are permitted by the Code to be 20 amperes.

- ▶A receptacle outlet shall be installed in an accessible location within 25 ft (7.5 m) of heating, air-conditioning, and refrigeration equipment. If the equipment is located on the side of the house, the receptacles required on the front and rear of the house might not be within 25 ft (7.5 m) of the equipment, in which case an additional receptacle is required. Do *not* connect this receptacle outlet to the load side of the equipment disconnecting means. See *210.63*.◀

- ▶A receptacle is *not* required on the rooftop of one- and two-family dwellings to serve heating, air-conditioning, and refrigeration equipment if the required outside receptacles are within 25 ft (7.5 m) of the equipment. ◀

Suggestion: In addition to installing receptacles per the *NEC®* spacing requirements, give serious thought to installing a receptacle below the wall switch in bedrooms, living rooms, family rooms, and similar rooms. In most cases, the required receptacles will be behind furniture, making it difficult to reach for plugging in a vacuum cleaner or other temporarily used appliance. A receptacle below the wall switch will be very easy to get to. A real plus!

See Figure 3-5 for an explanation of **receptacle** and **receptacle outlet**.

Basements

- At least one receptacle outlet must be installed in addition to the laundry outlet, *210.52(G)*.

- Connect to a 15- or 20-ampere circuit.

- GFCI protection is *not* required for receptacles that are not readily accessible, or for a single receptacle or duplex receptacle for two appliances that are not easily moved, are cord- and plug-connected, and are located in a dedicated space, *210.8(A)(5), Exceptions No. 1* and *No. 2*. Examples of this include the receptacle for laundry equipment, a sump pump, a water softener, a freezer, or a refrigerator.

- In finished basements, the required number, spacing, and location of receptacles, lighting outlets, and switches shall be in conformance to *210.52(A)* and *210.70*.

- In finished basements, GFCI protection for receptacles is *not* required unless the receptacles are located within 6 ft (1.8 m) from the outer edge of a wet bar sink and are intended to serve the countertop areas around the wet bar sink.

- Unfinished basements, or unfinished portions of basements that are not generally "lived in," such as work areas and storage areas, are not subject to the spacing requirements for receptacles as specified in *210.52(A)*.

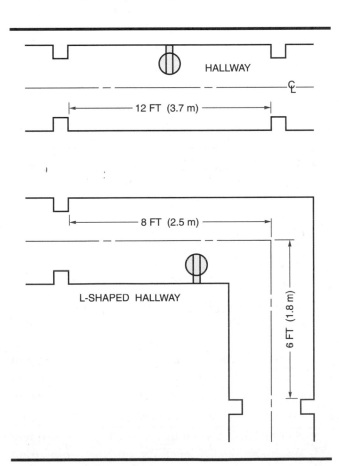

HALLWAY

12 FT (3.7 m)

8 FT (2.5 m)

L-SHAPED HALLWAY

6 FT (1.8 m)

Figure 3-4 Determination of hallway length. Hallways 10 ft (3.0 m) or longer must have at least one receptacle outlet, *210.52(H)*.

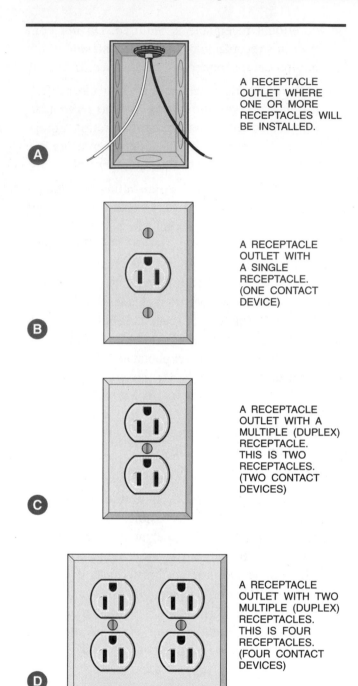

A RECEPTACLE OUTLET WHERE ONE OR MORE RECEPTACLES WILL BE INSTALLED.

A

A RECEPTACLE OUTLET WITH A SINGLE RECEPTACLE. (ONE CONTACT DEVICE)

B

A RECEPTACLE OUTLET WITH A MULTIPLE (DUPLEX) RECEPTACLE. THIS IS TWO RECEPTACLES. (TWO CONTACT DEVICES)

C

A RECEPTACLE OUTLET WITH TWO MULTIPLE (DUPLEX) RECEPTACLES. THIS IS FOUR RECEPTACLES. (FOUR CONTACT DEVICES)

D

Figure 3-5 The branch-circuit wiring and the box where one or more receptacles will be installed is defined by the Code as a *receptacle outlet*. The receptacle itself, whether a single or multiple device (one, two, or three receptacles) on one strap (yoke) is defined as a *receptacle*. For house wiring, receptacle loads are considered to be part of the general lighting load calculations, *220.3(B)(10)*.

- At least one receptacle must be installed for the laundry, *210.52(F)*. In this residence, one separate Circuit B18 is for the washer receptacle, and Circuit B20 serves a receptacle that is easily accessible for ironing.

- A separate 20-ampere branch circuit shall supply the laundry receptacle. This circuit shall not supply other loads, *210.11(C)(2)*. Allow 1500 volt-amperes for this circuit when doing the load calculations for determining the size of the service-entrance equipment, *220.16(B)*.

- ▶In each separate unfinished basement area that is not intended to be "lived in," such as storage areas or work areas, at least one 125-volt, single-phase, 15- or 20-ampere receptacle must be installed, *210.52(G)*. This receptacle(s) must be GFCI protected, *210.8(A)(5)*. This includes receptacles that are an integral part of porcelain or plastic pull-chain or keyless lampholders. ◀

- Refer to *210.8(A), 210.11(C)(2), 210.52(F), 210.52(G), 210.70,* and *220.16(B)*.

Small Appliance Receptacle Outlets

- At least two 20-ampere small appliance circuits must supply the receptacles that serve the countertop surfaces in the kitchen, *210.11(C)(1)* and *210.52(B)(3)*. Two small-appliance branch circuits is the minimum. It would be advisable to run additional 20-ampere branch circuits to these areas because of the heavy concentration of appliances.

- These two 20-ampere small-appliance circuits may also supply the receptacles in dining areas, pantries, or breakfast rooms, *210.52(B)(1)*.

- Space the receptacles in kitchens according to *210.52(C)*.

- GFCI protection must be provided for all receptacles in kitchens that serve countertop areas, *210.8(A)(6)*. GFCI protection may be GFCI receptacles or GFCI circuit breakers, as discussed in detail in Unit 6.

- These two or more 20-ampere small-appliance circuits shall not supply other outlets, *210.52(B)(2)*. Exceptions to this are: 1) a clock receptacle and 2) a receptacle(s) for plugging in a gas-fired range, oven, or counter-mounted cooktop which are permitted to be supplied by one of the 20-ampere small-appliance circuits.

- Allow 1500 volt-amperes for each 20-ampere small-appliance branch-circuit when computing loads for determining the size of the service-entrance equipment, *220.16(A)*.

- Refer to *210.8(A)(6), 210.11(C)(1), 210.52(B)(1), (2),* and *(3), 210.52(C),* and *220.16(A)*.

Countertops in Kitchens (See Figure 3-6)

- Receptacles that serve countertops in kitchens must be GFCI protected. This includes a receptacle located inside an "appliance garage."

- Receptacles are not permitted to be installed "face-up" in countertops.

- Receptacles must be installed above all countertop spaces 12 in. (300 mm) or wider.

- ►Receptacles must be installed not more than 20 in. (500 mm) above a countertop.◄

- Receptacles above countertop spaces shall be positioned so that no point along the wall line is more than 24 in. (600 mm), measured horizontally, from a receptacle outlet in that space. Example: A section of a countertop measures 5 ft 3 in. (1.6 m) along the wall line. The minimum number of receptacles is 5 ft 3 in. ÷ 4 ft = 1+. Therefore, install two receptacles in this space.

- Receptacles may be installed below countertops only:

  - where the construction is for the physically impaired, or
  - on islands and peninsulas where there is no way to mount a receptacle within 20 in. (500 mm) above the countertop.

  For either of these two conditions:
  - Receptacles shall be mounted not more than 12 in. (300 mm) below the countertop.
  - ►No receptacles shall be installed below a countertop that extends more than 6 in. (150 mm) beyond its base cabinet. The logic to this rule is the popular use of stools where the countertop extends beyond the base cabinet, increasing the possibility of someone pulling coffee pots and similar appliances off of the countertop.◄

- Peninsulas and islands that have a long dimension of 24 in. (600 mm) or greater and a short dimension of 12 in. (300 mm) or greater, must have at least one receptacle. Measure the peninsula from the edge where the peninsula connects to the wall base cabinet.

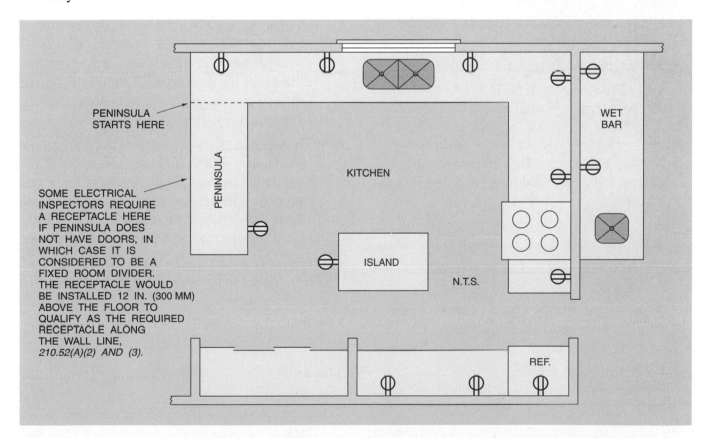

Figure 3-6   Floor plan showing placement of receptacles that serve countertop surfaces in kitchens and wetbar areas in conformance to _210.52(C)_. All 120-volt receptacles that serve countertops must be GFCI protected, _210.8(A)(6)_ and _(7)_. Receptacles that are located behind the refrigerator, under the sink for plugging in a food waste disposer, or inside the cabinets for plugging in a microwave oven do not have to be GFCI protected because they do not serve the countertop surfaces. Drawing is not to scale. See text for detailed explanation.

- If a countertop, peninsula, or island has a range, refrigerator, or sink that creates separate countertop spaces on both sides of these or similar items, each space is considered a separate space. Receptacles shall be installed to comply with the required number of receptacles in these spaces.

- Receptacle outlets not readily accessible, such as those behind a refrigerator, or behind an appliance that is fastened in place, or inside of an "appliance garage," are not considered part of the receptacle outlet requirement above countertops. Since receptacles behind these appliances are virtually impossible to use once the appliance is in place, these receptacles are not required to be GFCI protected.

- Receptacles that serve countertops must be supplied by not less than two 20-ampere small appliance branch-circuits. These circuits are permitted to supply other receptacles in the kitchen, dining room, pantry, breakfast room, or similar locations.

- Wet bar sinks: Receptacles installed within 6 ft (1.8 m) from the outside edge of a wet bar sink must be GFCI protected.

- See Unit 6 for details on GFCI protection.

- ►See *210.8(A)(6)* and *(7)*, *210.11(C)*, *210.52(C)*, *220.16(A)*, and *406.4(E)*.◄

Laundry

- At least one receptacle is required within 6 ft (1.8 m) of the intended location of the washing machine.

- A separate 20-ampere branch-circuit must be provided. This circuit shall not supply other outlets.

- A single receptacle does not require GFCI protection.

- A duplex receptacle requires GFCI protection because someone could plug something else into the second receptacle, which could present a potential shock hazard.

- The laundry receptacle and branch-circuit are in addition to other receptacles and branch-circuits.

- A single receptacle must have an ampere rating not less than that of the branch-circuit. Since the laundry circuit must be rated 20-amperes, the receptacle would have to be rated 20-amperes.

- Allow 1500 volt-amperes for this laundry branch-circuit when computing loads for determining the size of the service-entrance equipment.

- Refer to *210.11(C)(2)*, *210.21(B)(1)*, *210.8(A)(5)*, *210.50(C)*, *210.52(F)*, and *220.16(B)*.

Bathrooms

- ►At least one receptacle shall be installed within 900 mm (3 ft) from the outside edge of each basin, and shall be located on a wall or partition that is adjacent to the basin or basin countertop. One receptacle located between two basins might be acceptable if the above criteria are met.◄

- Receptacles in bathrooms must be GFCI-protected.

- A receptacle on a luminaire (fixture) or medicine cabinet does not qualify for the required receptacle.

- Do not install the receptacle(s) face up in the countertop.

- Connect the receptacle(s) to at least one separate 20-ampere branch-circuit. Although this separate branch-circuit is permitted to supply more than one bathroom, installing a separate 20-ampere branch-circuit to each bathroom is recommended because of the high-wattage appliances commonly used in bathrooms, such as electric hair dryers.

- The separate 20-ampere branch-circuit must not supply other outlets. An exception is that if this separate branch-circuit is supplying only one bathroom, then other equipment not exceeding 10 amperes (50% of 20) is permitted to be connected to the separate 20-ampere branch-circuit.

- For service and feeder load calculations, no additional load needs to be included for this separate circuit.

- ►See *210.8(A)(1)*, *210.11(C)(3)*, *210.23(A)(2)*, *210.52(D)*, *220.3(B)(10)(1)*, and *406.4(E)*.◄

Outdoors

- At least one receptacle outlet must be installed at grade level in the front and another in the back of all one-family dwellings and each unit of a two-family dwelling. These receptacles must be **accessible** and not more than 6½ ft (2.0 m) above grade level, Figure 3-7. A recep-

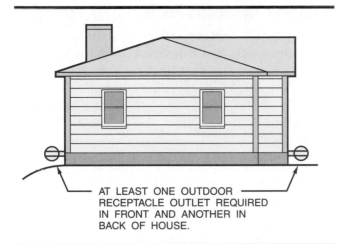

AT LEAST ONE OUTDOOR
RECEPTACLE OUTLET REQUIRED
IN FRONT AND ANOTHER IN
BACK OF HOUSE.

Figure 3-7 Outdoor receptacle requirements for residences, *210.52(E)*, require that for one-family dwellings, and each unit of a two-family dwelling that is at grade level, at least one receptacle outlet must be installed in the front and another in the back. These receptacles must not be mounted more than 6½ ft (2.0 m) above grade. Additional receptacles may be installed higher, such as receptacles used for plugging in ice and snow melting heating cables and/or decorative outdoor lighting. ▶If the required 15- or 20-ampere 120-volt receptacles on the front and rear of the house are not within 25 ft (7.5 m) of an air conditioner or heat pump that might be located on the side of the house, then another GFCI protected receptacle must be provided so as to be within 25 ft (7.5 m) of the air conditioner or heat pump, *210.63*. ◀

tacle on a porch or deck is acceptable, provided there is easy access onto the porch from grade level and to the receptacle. A screened-in porch or deck, or a porch or deck that requires access by climbing up a number of steps would not meet the intent of the Code. This is a judgment call. One or two steps from grade level onto the porch is acceptable. More steps might make it difficult to access the receptacle.

- Connect to a 15- or 20-ampere circuit.

- *All* outdoor receptacles must be GFCI protected. An exception is that personnel GFCI protection is *not* required for an outdoor receptacle(s) installed for snow and ice-melting equipment (i.e., heating cables) that is not **readily accessible**, and is supplied by a separate branch-circuit. GFCI devices for *personnel protection* trip in the range of 4 to 6 milliamperes.

- Ground-fault protection of equipment (GFPE) must be provided for "fixed" outdoor de-icing

and snow melting equipment. GFPE can be a circuit breaker, a receptacle, or be self-contained in the equipment. GFPE devices for *equipment protection* trip at approximately 30 milliamperes. When used for de-icing and snow melting, mineral insulated metal-sheathed cable embedded in non-combustible material such as concrete does not require gound-fault protection.

- See *210.8(A)(3)*, *210.52(E)*, and *426.28*.

Garages

- At least one receptacle outlet must be provided.

- Connect to a 15- or 20-ampere circuit.

- All receptacles must be GFCI protected. An exception to this is receptacles that are not readily accessible, such as a receptacle mounted on the ceiling for plugging in a garage door opener. Also exempt from the GFCI requirement are single receptacles or duplex receptacles intended for cord- and plug-connected appliances that are not easily moved and are located in dedicated spaces. Examples are a freezer, refrigerator, or central vacuum equipment.

- If the garage is detached, the Code does not require electric power, but if electric power is provided, then at least one GFCI receptacle outlet must be installed.

  Receptacles are not required in a storage shed (accessory building), but if installed, they must be GFCI protected if the floor is at or below grade level.

- See *210.8(A)(2)* and *210.52(G)*.

Receptacles in Other Locations. All of the required receptacles discussed above are *in addition* to any receptacles that are part of luminaires (lighting fixtures) or appliances, are located inside of cabinets, or are located more than 5½ ft (1.7 m) above the floor, *210.52*.

GFCI Exemption Recap

- Under certain conditions, GFCI protection is *not* required for receptacle outlets in garages, in crawl spaces at or below ground level, or in unfinished basements if the receptacles are:

  - not readily accessible.

  - single-type receptacles or duplex-type receptacles located in dedicated spaces to be used for cord- and plug-connected appliances that are not easily moved.

Examples of these would be such things as overhead door openers, refrigerators, freezers, water softeners, central vacuum equipment or sump pumps. These would be the receptacles that are dedicated to specific plug-in appliances. The homeowner would normally not unplug the appliance to plug in an extension cord. These receptacles are exempt from the GFCI requirement because they are not intended to be used for plugging in extension cords for portable electric tools or portable appliances. These GFCI-exempt receptacles are *in addition* to the receptacles required by *210.52(G)* for garages and basements. See *210.8(A)*.

Specific Loads. Appliances such as electric ranges, ovens, air conditioners, electric heat, heat pumps, water heaters, and similar loads are discussed later on in this text. See index for location.

Receptacle Ratings. The *NEC®* in *Article 100* defines the following:

- receptacle: A receptacle is a contact device installed at the outlet for the connection of an attachment plug.

- single receptacle: A single receptacle is a single contact device with no other contact device on the same yoke.

- multiple receptacle: A multiple receptacle is two or more contact devices on the same yoke.

Table 3-2 shows the requirements of *210.21(B)* and *Table 210.21(B)(3)* for receptacles installed in homes.

Replacing Existing Receptacles. See Unit 6 for detailed discussion and illustrations about replacing existing receptacles in older homes.

Lighting Outlets *(Section 210.70)*

The *NEC®* contains minimum requirements for providing lighting for dwellings.

Switched Lighting Outlets. Per *210.70*, lighting outlets in dwellings must be installed as shown in Figure 3-8. Here is a summary of these requirements.

Habitable Rooms. In every habitable room and in every bathroom, at least one wall-switch-controlled lighting outlet must be installed. Switched receptacles meet the intent of this requirement except in kitchens and bathrooms. However, if plug-in swag lighting as illustrated in Figure 3-9 is used in a dining room, a switched receptacle connected to a general lighting branch-circuit must be installed. This receptacle must be in addition to the receptacles supplied by the required 20-ampere small appliance branch-circuits, *210.52(B)(1)*, *Exception No. 1*, Figure 3-10.

Occupancy motion sensors may be used if: (1) they are in addition to the required wall switch, or (2) they have a manual override and are located where the wall switch would normally be located.

Additional Locations. ▶At least one wall-switch-controlled lighting outlet must be installed in hallways, stairways, attached garages, and detached garages with electric power, *210.70(A)(2)(a).*◀

The *NEC®* does *not* require that a detached garage have electric power, but if it does have electric power, then it must have at least one GFCI receptacle, and it must have wall switch-controlled lighting.

More specifically, the *International Residential Code* requires that lighting outlets be provided in the immediate vicinity of each landing at the top and bottom of interior stairs so as to illuminate the stairs, landings, and treads. An exception to this is when artificial lighting is provided directly over each stair section.

BRANCH-CIRCUIT RATING	RECEPTACLE RATING
15-ampere lighting branch-circuit: circuit has two or more receptacles or outlets. Receptacles may be single, duplex, or triplex.	15-ampere maximum
20-ampere small appliance branch-circuit: circuit has two or more receptacles or outlets. Receptacles may be single, duplex, or triplex.	15 or 20 amperes
Individual 20-ampere branch-circuit: circuit has only a single receptacle connected. An example of this is a dedicated branch circuit for a microwave oven.	20 amperes

Table 3-2 This table shows the ampere rating for receptacles used on 15- and 20-ampere branch circuits.

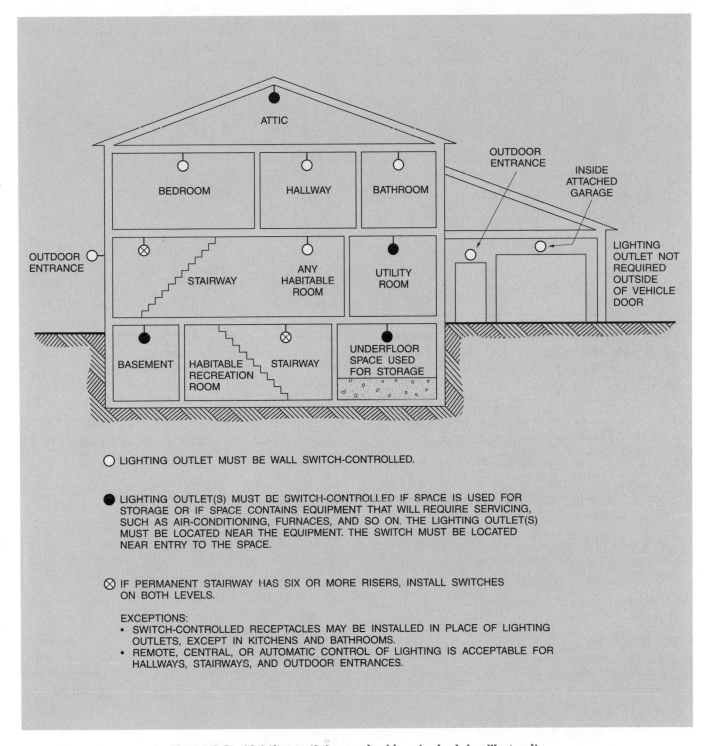

Figure 3-8 Lighting outlets required in a typical dwelling unit.

Some electrical inspectors consider a tool shed or other similar accessory building to be a storage or utility room that comes under the requirements of *210.70(A)(3)*, thus requiring a wall switch-controlled lighting outlet.

▶At least one wall-switch-controlled lighting outlet must be installed to provide illumination on the exterior side of outdoor entrances or exits that have grade-level access to dwelling units, attached garages, and detached garages if they have electric power. A sliding glass door is considered to be an outdoor entrance. A vehicle door is not considered to be an outdoor entrance. See *210.70(A)(2)(b)*.◀

Although not necessarily good design practice,

• One switch could control the required outdoor lighting at more than one entrance.

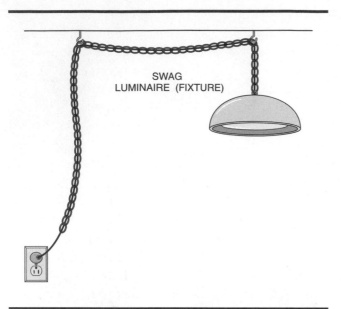

Figure 3-9   A receptacle controlled by a wall switch is acceptable as the required lighting outlet in rooms other than a kitchen or bathroom, *210.70(A)(1), Exception*.

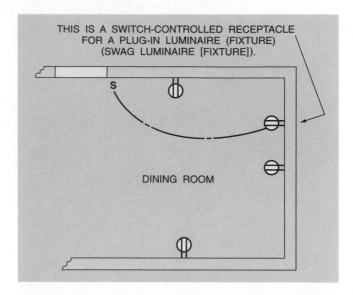

Figure 3-10   When a switch-controlled receptacle outlet is installed in rooms where the receptacle outlets normally would be 20-ampere small-appliance circuits (breakfast room, dining room, and so on.), it must be in addition to the required small appliance receptacle outlets, *210.52(B)(1), Exception No. 1*.

- One switch could control both the required outdoor and indoor lighting of a garage.

- One outdoor luminaire (fixture), properly located, could meet the required outdoor lighting for both the house entrance and the garage entrance.

►Where there are six stair risers or more on an interior stairway, the *NEC®* requires that wall-switch control of stairway lighting must be provided on each floor level and any landing level that has an entryway, *270.70(A)(2)(c)*.◄ *The International Residential Code, 303.4.1*, is more stringent in that it requires control "*at the top and bottom of each stairway without traversing any step of the stair.*"

This switching requirement includes permanent stairways to basements and attics, but does not include fold-down storable ladders commonly installed for access to residential attics.

Remote control, central control, motion sensors, photocells, and other automatic devices may be installed instead of wall switches for the control of lighting in hallways, stairways, and at outdoor entrances. For example, an outdoor post light controlled by a photocell does not require a separate switch.

►This is permitted by the *Exception* to *210.70(A)(2)(a), (b)*, and *(c)*.◄

The *NEC®* does not require lighting outlets in clothes closets, but some building codes do. You should check this out in your locality.

Storage or Equipment Spaces. At least one switch-controlled lighting outlet must be installed in spaces that are used for storage or contain equipment that requires servicing, such as attics, under-floor spaces, utility rooms, and basements. Examples of equipment that may need servicing are furnaces, air conditioners, heat pumps, sump pumps, and so on. The control is permitted to be a wall switch or a lighting outlet that has an integral switch such as a pull chain switch. The lighting outlet must be at or near the equipment that may require servicing. The control must be at the usual point of entrance to the space.

The Number of Branch-Circuits Required, and How to Determine the Number of Lighting and Receptacle Outlets to Be Connected to One Branch-Circuit

This unit has explained the *NEC®* rules on where receptacle and lighting outlets are to be installed. The question of *how many* receptacle outlets and lighting outlets should be connected to each branch-circuit is covered in detail in Unit 8, where we begin the actual layout of the wiring for the residence discussed throughout this text.

Common Areas in Two-Family or Multifamily Dwellings

NEC® 210.25 contains specific requirements for circuits that serve common areas of two-family or multifamily dwellings. Be careful when wiring a two-family or multifamily dwelling. For areas such as a common entrance, common basement, or common laundry area, the branch-circuits for lighting, central fire/smoke/security alarms, communication systems, or any other electrical requirements that serve the common area *shall not* be served by circuits from any individual dwelling unit's power service. Common areas will require separate circuits, associated panels, and metering equipment. The reasoning for this requirement is that power could be lost in one occupant's electrical system, resulting in no lights or alarms for the other tenant(s) using the common shared area.

For example, a two-family dwelling would probably require two separate stairways to the basement.

If only one stairway is provided to serve the basement of a two-family dwelling, then lighting would have to be provided and served from a branch-circuit originating from each tenant's electrical system, so that each tenant would have control of the lighting. For a common entrance, two luminaires (lighting fixtures) would be installed and connected to each tenant's electrical system.

Likewise, fire/smoke/security systems in a two-family dwelling shall be installed and connected to branch-circuits originating from each tenant's electrical panel. This becomes an architectural design issue. This is not generally a problem in two-family dwellings, but can become a problem in dwellings having more than two tenants. Rather than having to install a third watt-hour meter and associated electrical equipment on a two-family dwelling, the design of the structure can be such that there are no common areas, Figure 3-11. A four-plex will probably require a fifth watt-hour meter, four watt-hour meters for the tenants and one watt-hour meter to serve common area lighting and the fire/smoke/security systems.

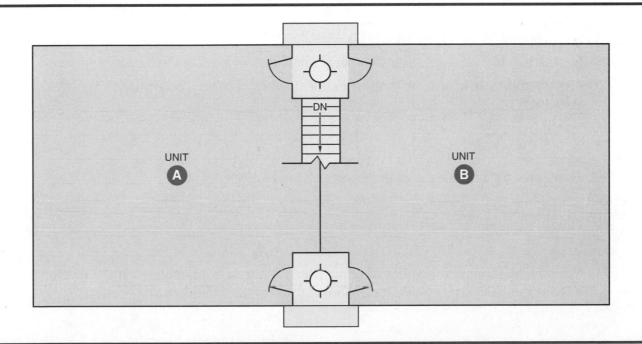

Figure 3-11 This outline of a two-family dwelling shows a common rear entry, a common stairway to the basement, and a common front entry. Note that each entry has one luminaire (lighting fixture). These common area luminaires (lighting fixtures) are not permitted to be connected to either of the tenant's electrical systems in the individual dwelling units, *210.25*. The choices are: (1) architect/designer must redo the layout of the building to eliminate the common entrances, (2) provide two luminaires (lighting fixtures) in each entry, with each luminaire (fixture) connected to the respective individual tenant's electrical circuit, or (3) install a third meter and panel to supply the electrical loads in the common areas.

REVIEW

Note: In the following questions the student may refer to the *NEC®* to obtain the answers. To help the student find these answers, references to Code sections are included after some of the questions. These references are not given in later reviews, where it is assumed that the student is familiar with the Code.

1. What is the meaning of computed load? _____

    _____

2. How are branch-circuits rated? See *NEC® 210.3* _____

    _____

3. How is the rating of the branch-circuit protective device affected when the conductors used are of a larger size than called for by the Code? See *NEC® 210.3.* _____

    _____

4. What dimensions are used when measuring the area of a building? See *NEC® 220.3(A)*

    _____

5. What spaces are not included in the floor area when computing the load in volt-amperes per square foot? See *NEC® 220.3(A)* _____

    _____

6. What is the unit load per square foot for dwelling units? See *Table 220.3(A)* _____

    _____

7. According to *NEC® 210.50(C)*, a laundry equipment outlet must be placed within _____ feet (_____ meters) of the intended location of the laundry equipment.

8. How is the total load in volt-amperes for lighting purposes determined? See *NEC® 220.3(A)* _____

    _____

    _____

9. How is the total lighting load in amperes determined? _____

    _____

10. How is the required number of branch-circuits determined? _____

    _____

11. What is the minimum number of 15-ampere lighting circuits required if the dwelling has an occupied area of 4000 ft² (368 m²)? Show all calculations.

12. How many lighting branch-circuits are provided in this dwelling? _____

13. a. What is the minimum load allowance for small appliance circuits for dwellings? See *NEC® 220.16(A)* _____

    b. A separate 15-ampere branch-circuit is run to the receptacle outlet behind the refrigerator instead of connecting it to one of the two 20-ampere small appliance branch-circuits that are required in kitchens. For this separate circuit, an additional 1500 volt-amperes (shall) (does not have to) be added to the load calculations for dwellings. Circle the correct answer.

14. What is the smallest size wire that can be used in a branch-circuit rated at 20 amperes?

    _____

15. How is the load determined for outlets supplying specific appliances? See *NEC® 220.3(B)* _____

    _____

16. What type of circuits must be provided for receptacle outlets in the kitchen, pantry, dining room, and breakfast room? See *NEC® 210.11(C)(1)*. _____

    _____

17. How is the minimum number of receptacle outlets determined for most occupied rooms? See *NEC® 210.52(A)*. _____

    _____

    _____

    _____

    _____

    _____

    _____

    _____

    _____

18. In a single-family dwelling, what types of overload protection for circuits are used? See *NEC® 240.6*. _____

    _____

    _____

    _____

    _____

19. The *NEC®* in *Article 100* defines a continuous load as "a load where the maximum current is expected to continue for three hours or more." According to *210.19(A)* and *210.20(A)*, the branch-circuit conductors and overcurrent protection for a continuous load shall be sized at not less than (100%) (125%) (150%) of the continuous load. Circle the correct answer.

20. The minimum number of outdoor receptacles for a residence is _____,

    NEC® _____. State the location. _____

21. The Code indicates the rooms in a dwelling that are required to have switched lighting outlets or switched receptacles. Write yes (switch required) or no (switch not required) for the following areas: See *NEC® 210.70(A)*.

   a. attic _____

   b. stairway _____

   c. crawl space (where used for storage) _____

   d. hallway _____

   e. bathroom _____

   f. closet _____

22. Is a receptacle required in a bedroom on a 3 ft (900 mm) wall space behind the door? The door is normally left open._____

23. The Code requires that at least one 20-ampere circuit feeding a receptacle outlet must be provided for the laundry. May this circuit supply other outlets?_____

24. In a basement, at least one receptacle outlet must be installed in addition to the receptacle outlet installed for the laundry equipment. This additional receptacle outlet and any other receptacle outlets in unfinished areas must be _____ protected.

25. *NEC® 210.8(A)(2)* requires that *all* receptacles installed in garages must have GFCI protection, but there are certain exceptions. What are these exceptions?

   _____

   _____

   _____

   _____

   _____

26. Define a branch-circuit. _____

   _____

27. Although the Code contains many exceptions to the basic overcurrent protection requirements for conductors, in general, the rating of the branch-circuit overcurrent device must (not be less than) (not be more than) the ampacity of a conductor. Circle the correct answer.

28. The rating of a branch-circuit is based upon (circle the correct answer)
   a. the rating of the overcurrent device
   b. the length of the circuit
   c. the branch-circuit wire size

29. a. A 25-ampere branch-circuit conductor is derated to 70%. It is important to provide proper overcurrent protection for these conductors. The derated conductor ampacity is _____

   _____

   b. If the connected load is a "fixed" nonmotor, noncontinuous load, the branch-circuit overcurrent device may be sized at (20) (25) amperes. Circle correct answer.

c. If the above circuits supply receptacle outlets, the branch-circuit overcurrent device must be sized at not over (15) (20) (25) amperes. Circle correct answer.

30. Small appliance receptacle outlets are (included) (not included) in the 3 volt-amperes per square foot calculations. Circle correct answer.

31. If a homeowner wishes to have a switched receptacle outlet for a swag lamp in a dining room, may this switched receptacle outlet be considered to be one of the receptacle outlets required for the 20-ampere small appliance circuits? _____

32. How many receptacle outlets are required on a 13 ft (4.0 m) wall space between two doors? Refer to *NEC® 210.52(H)*. Draw the outlets in on the diagram.

33. Is a receptacle required in a hallway in a home when the hallway is 8 ft (2.5 m) long?

_____

34. Is a receptacle required in a hallway in a home when the hallway is 13 ft (4.0 m) long?

_____

35. A sliding glass door is installed in a family room that leads to an outdoor deck. The sliding door has one 4 ft (1.22 m) sliding section and one 4 ft (1.22 m) permanently mounted glass section. Inside the recreation room, what does the Code say about wall receptacle outlets relative to the receptacle's position near the sliding glass door?

_____

_____

_____

_____

_____

_____

_____

36. The Code permits certain other receptacles to be connected to a 20-ampere small appliance circuit. What are they? Which section of the *NEC®* supports your answer?

_____

_____

_____

_____

_____

_____

37. What is the minimum number of 20-ampere small appliance circuits required by the Code to supply the receptacle outlets that serve the countertop surfaces in the kitchen? In which section of the Code is this requirement found? _____

38. Does the Code allow a 120-volt receptacle located behind a gas range to be supplied by a small appliance circuit? (Yes) (No) Circle the correct answer.

39. When determining the location and number of receptacles required, fixed room dividers and railings (shall be) (need not be) considered to be wall space. Circle correct answer.

40. No point along the wall line shall be more than 6 ft (1.8 m) from a receptacle outlet. This requirement (does apply) (does not apply) to unfinished residential basements. Circle correct answer.

41. Receptacle outlets that serve countertop surfaces in kitchens are required to be _____ protected, and shall be installed so that no point along the wall line above the base cabinets is more than [18 in. (450 mm)] [24 in. (600 mm)] [36 in. (900 mm)] from a receptacle. Circle the correct answer.

42. In kitchens, the Code requires that at least (one) (two) receptacle(s) be installed to serve the countertop surfaces on an island or peninsula. Circle the correct answer. Where in the Code is this requirement found? *NEC®* _____

43. In the past, it was common practice to connect the lighting and receptacle(s) in a bathroom to the same circuit. Because of overloads caused by high wattage grooming appliances, the Code now requires that these receptacle(s) be supplied by a separate 20-ampere branch-circuit. Where in the Code is this requirement found? *NEC®* _____

44. If a residence has two bathrooms, the Code states:

   a. The receptacles in each bathroom must be connected to a separate 20-ampere branch-circuit. For two bathrooms, this would mean two circuits. (True) (False) Circle the correct answer.

   b. The receptacles in both bathrooms are permitted to be served by the same 20-ampere branch-circuit. (True) (False) Circle the correct answer.

45. In your own words, explain the GFCI exemptions for receptacle outlets installed in unfinished basements and garages. Give some examples.

   _____

   _____

   _____

   _____

46. *NEC®* _____ of the Code prohibits connecting lighting and/or fire/smoke/ security systems to an individual tenant's electrical source where common (shared) areas are present, in which case the lighting provided for both tenants is connected on one tenant's electrical source.

47. When installing weatherproof outdoor receptacles high up under the eaves of a house intended to be used for plugging in de-icing and snow melting cable for the rain gutters, must these receptacles be GFCI protected for personnel protection? (Yes) (No) Circle the correct answer.

Conductor Sizes and Types, Wiring Methods, Wire Connections, Voltage Drop, Neutral Sizing for Services

OBJECTIVES

After studying this unit, the student should be able to

- define the terms used to size and rate conductors.
- understand the current-carrying capacity of flexible cords.
- understand overcurrent protection for conductors and the maximum loading on branch-circuits.
- understand aluminum conductors and understand the possible fire hazards associated with them.
- understand all of the Code rules for all types of cables, rigid raceways, and flexible raceways, used for wiring houses.
- understand the Code rules for service-entrance cables.
- understand the special ampacities for residential service-entrance conductors.
- understand the "weakest link" of an electrical system.
- size neutral conductors.
- understand why old knob and tube wiring must not be buried in thermal insulation.
- make voltage drop calculations.
- understand a new way to bring nonmetallic-sheathed cables into the top of a main panel.
- understand the markings on the terminals of wiring devices and wire connectors.

CONDUCTORS

Throughout this text, all references to conductors are for copper conductors, unless otherwise stated.

Wire Size

The copper wire used in electrical installations is graded for size according to the American Wire

Gauge (AWG) Standard. The wire diameter in the AWG standard is expressed as a whole number. The higher the AWG number, the smaller the wire. AWG sizes vary from fine, hairlike wire used in coils and small transformers to very large diameter wire required in industrial wiring to handle heavy loads.

The wire may be a single strand (solid conductor), or it may consist of many strands. Each strand of wire acts as a separate conducting unit. The wire size used for a circuit depends on the maximum current to be carried. The *NEC®* in *Table 210.24* shows that the minimum conductor size for branch-circuit wiring is 14 AWG. Smaller size conductors are permitted for bell wiring, thermostat wiring, communications wiring, intercom wiring, luminaire (fixture) wires, and similar low-energy circuits.

Don't Get Confused!

▶In the past, conductor size was shown as, for example, "No. 12 AWG." This same conductor now appears in the *NEC®* as "12 AWG." The "12" is a size, not a quantity.◀

Table 4-1 shows typical applications for different size conductors.

Solid and Stranded Conductors

For residential wiring, sizes 14, 12, and 10 AWG conductors are generally solid when the wiring method is nonmetallic-sheathed cable or armored cable.

In a raceway, the preference is to use stranded 10 AWG conductors because of the flexibility and ease of handling and pulling in the conductors.

NEC® *310.3* requires that conductors 8 AWG and larger must be stranded where installed in a raceway.

Ampacity

Ampacity means "the current, in amperes, a conductor can carry continuously under the conditions of use without exceeding its temperature rating." This value depends upon the conductor's cross-sectional area, whether the conductor is copper or aluminum, and the type of insulation around the conductor. Ampacity values, also referred to as current-carrying capacity, are found in *Article 310*. The most commonly used conductor ampacities are found in *Table 310.16*.

CONDUCTOR APPLICATIONS CHART		
Conductor Size	**Overcurrent Protection**	**Typical Applications** (Check wattage and/or ampere rating of load to select the correct size conductors based on *Table 310.16*.)
20 AWG	Class 2 circuit transformers provide overcurrent protection.	Telephone wiring is usually 20 or 22 AWG.
18 AWG	7 amperes. See *Table 430.72(B)* and *725.23*. Class 2 circuit transformers provide overcurrent protection.	Low-voltage wiring for thermostats, chimes, security, remote control, home automation systems, etc. For these types of installations, 18 or 20 AWG conductors can be used depending on the connected load and length of circuit.
16 AWG	10 amperes. See *Table 430.72(B)* and *725-23*. Class 2 circuit transformers provide overcurrent protection.	Same applications as above. Good for long runs to minimize voltage drop.
14 AWG	15 amperes	Typical lighting branch-circuits.
12 AWG	20 amperes	Small appliance branch-circuits for the receptacles in kitchens and dining rooms. Also laundry receptacles and workshop receptacles. Often used as the "home run" for lighting branch circuits. Some water heaters.
10 AWG	30 amperes	Most clothes dryers, built-in ovens, cooktops, central air conditioners, some water heaters, heat pumps.
8 AWG	40 amperes	Ranges, ovens, heat pumps, some large clothes dryers, large central air conditioners, heat pumps.
6 AWG	50 amperes	Electric furnaces, heat pumps.
4 AWG	70 amperes	Electric furnaces, feeders to subpanels.
3 AWG and larger	100 amperes	Main service entrance conductors, feeders to sub-panels, electric furnaces.

Table 4-1 Conductor Applications Chart

The allowable ampacity values in the tables are valid where there are no more than three current-carrying conductors in a raceway or cable, and where the temperature does not exceed 86°F (30°C).

When there are more than three current-carrying conductors in a raceway or cable, the allowable ampacity values are adjusted according to the factors in *Table 310.15(B)(2)(a)*.

When the ambient temperature exceeds 86°F (30°C), the allowable ampacity values are corrected according to the factors found below the tables.

If both conditions are present, then both penalties (adjustment and correction factors) must be applied. More on conductor ampacities, derating, adjusting, and correction factors is presented in Unit 18.

Conductors must have an ampacity not less than the maximum load that they are supplying, as shown in Figure 4-1. All conductors of a specific branch-circuit must have an ampacity of the branch-circuit's rating, as shown in Figure 4-2. There are exceptions to this rule, such as taps for electric ranges (Unit 20.)

Ampacity of Flexible Cords

Table 4-2 shows the allowable ampacities for some sizes of flexible cords. Refer to *Table 400.5(A)* in the *NEC*® for other specific types and sizes of flexible cords.

Conductor Sizing

The diameter of wire is usually given in a unit called a "mil." A *mil* is defined as one-thousandth of an inch (0.001 inch). Mils squared are known as "circular mils."

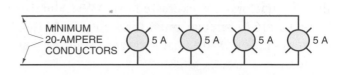

Figure 4-1 Branch-circuit conductors shall have an ampacity not less than the maximum load to be served, *210.19(A)*. See *210.23* for permissible loads.

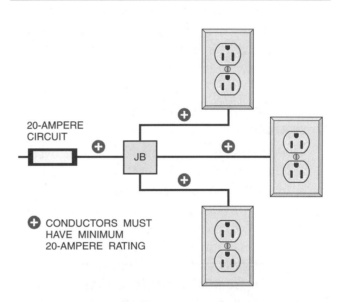

Figure 4-2 All conductors in this circuit supplying readily accessible receptacles shall have an ampacity of not less than the rating of the branch-circuit. In this example, 20-ampere conductors must be used, *210.19(A)* and *Table 210.24*.

ALLOWABLE AMPACITIES FOR FLEXIBLE CORDS

Conductor Size AWG Copper	Cords in Which Three Conductors Carry Current Example: black, white, red, plus equipment grounding conductor	Cords in Which Two Conductors Carry Current Example: black, white, plus equipment grounding conductor
14	15	18
12	20	25
10	25	30
8	35	40
6	45	55
4	60	70

Table 4-2 Allowable ampacities for some of the more common flexible cords used for residential applications. These values were taken from *Table 400.5(A), NEC*® Refer to the *NEC*® table for other specific types and sizes.

Table 8 in *Chapter 9* of the *NEC* clearly shows that conductors are expressed in AWG numbers from 18 (1620 circular mils) through 4/0 (211,600 circular mils). Wire sizes larger than 4/0 are expressed in circular mils.

Large conductors, such as 500,000 circular mils, are generally expressed as 500 kcmils. Since the letter *k* designates 1000, the term *kcmils* means "thousand circular mils." This is much easier to express in both written and verbal terms.

Older texts used the term MCM, which also means "thousand circular mils." The first letter *M* refers to the Roman numeral that represents 1000. Thus, 500 MCM means the same as 500 kcmils. Roman numerals are no longer used in the electrical industry for expressing conductor sizes.

Overcurrent Protection for Conductors

Conductors must be protected against overcurrent by fuses or circuit breakers rated not more than the ampacity of the conductors, *Section 240.4*.

NEC® 240.4(B) permits the use of the next higher standard overcurrent device rating as shown in *240.6(A)*. This permission is granted *only* when the overcurrent device is rated 800 amperes or less. For example, from *Table 310.16*, a 6 Type TW 140°F (60°C) conductor has an allowable ampacity of 55 amperes. In *240.6(A)*, we find that the next higher standard size overcurrent device is 60 amperes. Nonstandard ampere ratings of fuses and breakers are also permitted.

NEC® 240.4(D) spells out the maximum overcurrent protection for small branch-circuit conductors. Table 4-3 illustrates the requirements of *240.4(D)*.

There are exceptions for motor branch-circuits and HVAC equipment. Motor circuits are discussed

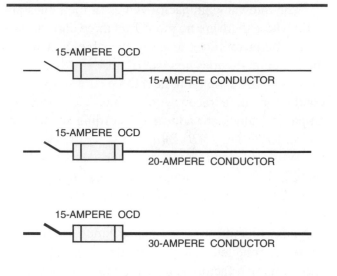

Figure 4-3   All three of the above branch-circuits are rated as 15-ampere circuits, even though larger conductors were used for some other reason, such as solving a voltage drop problem. The rating of the overcurrent device (OCD) determines the rating of a branch circuit, *210.3*.

in Unit 19 and Unit 22. HVAC circuits are discussed in Unit 23.

Branch-Circuit Rating

The rating of the overcurrent device (OCD) determines the rating of a branch-circuit, as shown in Figure 4-3.

PERMISSIBLE LOADS ON BRANCH-CIRCUITS (210.23)

The *NEC* is very specific about the loads permitted on branch-circuits. Here is a recap of these requirements.

- The load shall not exceed the branch-circuit rating.

- The branch-circuit must be rated 15, 20, 30, 40, or 50 amperes when serving two or more outlets.

- An individual branch-circuit can supply any size load.

- 15- and 20-ampere branch-circuits:
  a. May supply lighting, other equipment, or both types of loads.

▶b. Cord- and plug-connected equipment shall not exceed 80 percent of the branch-circuit rating. ◀

CONDUCTOR SIZE	MAXIMUM OVERCURRENT PROTECTION
14 AWG copper	15 amperes
12 AWG copper	20 amperes
10 AWG copper	30 amperes

Table 4-3   Maximum overcurrent protection for 14, 12, and 10 AWG copper conductors.

c. Equipment fastened in place shall not exceed 50 percent of the branch-circuit rating if the branch-circuit also supplies lighting, other cord- and plug-connected equipment, or both types of loads.

- The 20-ampere small appliance circuits in homes shall not supply other loads, *210.11(C)(1)*.

- 30-ampere branch-circuits may supply equipment such as dryers, cooktops, etc. Cord- and plug-connected equipment shall not exceed 80 percent of the branch-circuit rating.

- 40- and 50-ampere branch-circuits may supply cooking equipment that is fastened in place, such as an electric range.

- Over 50-ampere-rated branch-circuits are for electric furnaces, large heat pumps, air-conditioning equipment, large double-ovens, and similar large loads.

Aluminum Conductors

The conductivity of aluminum is not as great as that of copper for a given size. For example, checking *Table 310.16*, an 8 AWG Type THHN copper conductor has an allowable ampacity of 55 amperes. An 8 AWG Type THHN aluminum or copper-clad aluminum conductor has an ampacity of 45 amperes.

In *240.4(D)*, the maximum overcurrent protection for a 12 AWG copper conductor is 20 amperes but only 15 amperes for a 12 AWG aluminum or copper-clad aluminum conductor.

Aluminum conductors have a higher resistance compared to a copper conductor of the same size. When considering voltage drop, a conductor's resistance is a key ingredient. Voltage drop calculations are discussed later on in this unit.

Common Connection Problems

Some common problems associated with aluminum conductors when not properly connected may be summarized as follows:

- A corrosive action is set up when dissimilar wires come in contact with one another when moisture is present.

- The surface of aluminum oxidizes as soon as it is exposed to air. If this oxidized surface is not penetrated, a poor connection results. When installing aluminum conductors, particularly in large sizes, an inhibitor (anti-oxidant) is brushed onto the aluminum conductor, then the conductor is scraped with a stiff brush where the connection is to be made. The process of scraping the conductor breaks through the oxidation, and the inhibitor keeps the air from coming into contact with the conductor. Thus, further oxidation is prevented. Aluminum connectors of the compression type usually have an inhibitor paste already factory-installed inside of the connector.

- Aluminum wire expands and contracts to a greater degree than does copper wire for an equal load. This factor is another possible cause of a poor connection. Crimp connectors for aluminum conductors are usually longer than those for comparable copper conductors, thus resulting in greater contact surface of the conductor in the connector.

Proper Installation Procedures

Proper, trouble-free connections for aluminum conductors require terminals, lugs, and/or connectors that are suitable for the type of conductor being installed.

Terminals on receptacles and switches must be suitable for the conductors being attached. Table 4-4 shows how the electrician can identify these terminals. Listed connectors provide proper connection when properly installed.

See *110.14, 404.14(C)*, and *406.2(C)*.

Electrical Fires Caused by Improper Connections of Aluminum Conductors

Many electrical fires have been caused by improper connections and terminations of aluminum conductors. The problem is not the aluminum conductors. The problem is with the connections and terminations.

Many of these problems occurred in the meter base and at the main panel where aluminum service entrance conductors were installed years ago. In the course of time, the connections failed. Troubles also happened where aluminum conductors were terminated on receptacles and switches.

TYPE OF DEVICE	MARKING ON TERMINAL OR CONDUCTOR	CONNECTOR PERMITTED
15- or 20-ampere receptacles and switches	CO/ALR	aluminum, copper, copper-clad aluminum
15- and 20-ampere receptacles and switches	NONE	copper, copper-clad aluminum
30-ampere and greater receptacles and switches	AL/CU	aluminum, copper, copper-clad aluminum
30-ampere and greater receptacles and switches	NONE	copper only
Screwless pressure terminal connectors of the push-in type	NONE	copper or copper-clad aluminum
Wire connectors	AL	aluminum
Wire connectors	AL/CU or CU/AL	aluminum, copper, copper-clad aluminum
Wire connectors	CC	copper-clad aluminum only
Wire connectors	CC/CU or CU/CC	copper or copper-clad aluminum
Wire connectors	CU or CC/CU	copper only
Any of the above devices	COPPER OR CU ONLY	copper only

Table 4-4 Terminal identification markings and the acceptable types of conductors permitted to be connected to a specific type of terminal.

The Consumer Product Safety Commission (CPSC) (Washington, D.C. 20207) publishes an excellent free booklet (#516) entitled *Repairing Aluminum Wiring* that explains what you can do with connections and terminations for the small aluminum wires that were installed years ago for the branch-circuit wiring of receptacles and switches.

The CPSC states that older homes wired with aluminum conductors are 55 times more likely to have one or more connections that will reach "Fire Hazard Conditions" than are homes wired with copper.

Some local electrical codes do not allow aluminum conductors to be installed. Check this out before using aluminum conductors.

Some of the signs of potential problems might be:

- Wall switches and receptacles that feel warm to the touch.
- Smell of burning plastic or rubber at switches, receptacle, main panel, or meter.
- Flickering lights.
- Lights getting bright or dim.
- Circuits that don't work; some circuits are on, others are off.
- TV shifts to another channel for no apparent reason.
- Appliances with electronic controls shut off or change settings for no apparent reason.
- Security system sounds off for no apparent reason.

Wire Connections

Wire connectors are known in the trade by such names as *screw terminal, pressure terminal connector, wire connector, Wing-Nut,® Wire-Nut,® Scotchlok,® split-bolt connector, pressure cable connector, solderless lug, soldering lug, solder lug*, and others.

Solder-type lugs and connectors are rarely, if ever, used today. In fact, connections that depend upon solder are *not* permitted for connecting service-entrance conductors to service equipment, *230.81.* Nor is solder permitted for grounding and bonding connections, *250.8.* The labor costs and time spent make the use of the solder-type connections prohibitive.

Solderless connectors designed to establish Code-compliant electrical connections are shown in Figure 4-4.

As with the terminals on wiring devices (switches and receptacles), wire connectors must be marked AL when they are to be used with aluminum conductors. This marking is found on the connector itself, or appears on or in the shipping carton.

Connectors marked **AL/CU** are suitable for use with aluminum, copper, or copper-clad aluminum conductors. This marking is found on the connector itself, or it appears on or in the shipping carton.

Connectors not marked *AL* or *AL/CU* are for use with copper conductors only.

Unless specially stated on or in the shipping carton, or on the connector itself, conductors made of copper, aluminum, or copper-clad aluminum *may not* be used in combination in the same connector.

When combinations are permitted, the connector will be identified for the purpose and for the conditions when and where they may be used. The conditions usually are limited to dry locations only.

CRIMP CONNECTORS USED TO SPLICE AND TERMINATE 20 AWG TO 500 KCMILS ALUMINUM-TO-ALUMINUM, ALUMINUM-TO-COPPER, OR COPPER-TO-COPPER CONDUCTORS.

PROPERLY CRIMP
THEN TAPE

CONNECTORS USED TO CONNECT WIRES TOGETHER ON COMBINATIONS OF 18 AWG THROUGH 6 AWG CONDUCTORS. THEY ARE TWIST-ON, SOLDERLESS, AND TAPELESS.

*WIRE-NUT® and WING-NUT® are registered trademarks of IDEAL INDUSTRIES, INC. Scotchlok® is a registered trademark of 3M.

WIRE CONNECTORS VARIOUSLY KNOWN AS
WIRE-NUT,® WING-NUT,® AND SCOTCHLOK.®

CONNECTORS USED TO CONNECT WIRES TOGETHER IN COMBINATIONS OF 16, 14, AND 12 AWG CONDUCTORS. THEY ARE CRIMPED ON WITH A SPECIAL TOOL, THEN COVERED WITH A SNAP-ON INSULATING CAP.

CRIMP-TYPE WIRE CONNECTOR
AND INSULATING CAP

SOLDERLESS CONNECTORS ARE AVAILABLE IN SIZES 14 AWG THROUGH 500 KCMIL CONDUCTORS. THEY ARE USED FOR ONE SOLID OR ONE STRANDED CONDUCTOR ONLY, UNLESS OTHERWISE NOTED ON THE CONNECTOR OR ON ITS SHIPPING CARTON. THE SCREW MAY BE OF THE STANDARD SCREWDRIVER SLOT TYPE, OR IT MAY BE FOR USE WITH AN ALLEN WRENCH OR SOCKET WRENCH.

SOLDERLESS CONNECTORS

COMPRESSION CONNECTORS ARE USED FOR 8 AWG THROUGH 1000 KCMIL CONDUCTORS. THE WIRE IS INSERTED INTO THE END OF THE CONNECTOR, THEN CRIMPED ON WITH A SPECIAL COMPRESSION TOOL.

COMPRESSION CONNECTOR

SPLIT-BOLT CONNECTORS ARE USED FOR CONNECTING TWO CONDUCTORS TOGETHER, OR FOR TAPPING ONE CONDUCTOR TO ANOTHER. THEY ARE AVAILABLE IN SIZES 10 AWG THROUGH 1000 KCMIL. THEY ARE USED FOR TWO SOLID AND/OR TWO STRANDED CONDUCTORS ONLY, UNLESS OTHERWISE NOTED ON THE CONNECTOR OR ON ITS SHIPPING CARTON.

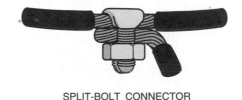

SPLIT-BOLT CONNECTOR

Figure 4-4   Types of wire connectors.

There are some "twist-on" wire connectors that have recently been listed for use with combinations of copper and aluminum conductors.

The preceding data is found in *110.14* and in the Underwriters Laboratories (UL) Standards. ▶When terminating large conductors such as service-entrance conductors, be sure to read any instructions that might be included with the equipment where these conduc-

tors will be terminated. Terminations for all electrical connections to devices and equipment shall be torqued as required by the manufacturer of such electrical device or equipment, *110.12(D).*◀

Conductor Insulation

The *NEC®* requires that all conductors be insulated, *310.2(A).* There are a few exceptions, such as the permission to use a bare neutral conductor for services, and bare equipment grounding conductors.

Thermoplastic insulation is the most common. It is like ice cream. It will soften and melt if heated above its rated temperature. It can be heated, melted, and reshaped. Thermoplastic insulation will stiffen at temperatures colder than 14°F (minus 10°C). Typical examples of thermoplastic insulation are Types THHN, THHW, THW, THWN, and TW.

Thermoset insulation can withstand higher and lower temperatures. It is like baking a cake. Once the ingredients have been mixed, heated, and formed, it can never be reheated and reshaped. If heated above its rated temperature, it will char and crack. Typical examples of thermoset insulation are Types RHH, RHW, XHH, and XHHW.

Table 310.13 lists the various conductor insulations and their applications.

The allowable ampacities of copper conductors are given in *Table 310.16* for various types of insulation. *Table 310.15(B)(6)* is a table of conductor ampacities that are greater than the ampacities found in *Table 310.16*. This table is permitted to be used *only* for dwelling unit 120/240-volt, three-wire, single-phase service-entrance conduc-tors, service lateral conductors, and feeder conduc-tors that serve as the main power feeder to a dwelling unit. This table is based upon the tremen-dous known load diversity found in homes. These special ampacities must not be used for feeder con-ductors that do not serve as the main power feeder to a dwelling. An example of this would be the feeder from the Main Panel A to Panel B in the res-idence discussed later on in this text.

Table 4-5 shows the special ampacities dis-cussed above. This table is a mirror image of *Table 310.15(B)(6)* in the *NEC.*®

The insulation covering wires and cables used in house wiring is usually rated at 600 volts or less. Exceptions to this statement are low-voltage wiring and luminaire (fixture) wiring.

WET, DAMP, DRY, AND SUNLIGHT LOCATIONS

Conductors are listed for specific locations. Be sure that the conductor you use is suitable for the loca-tion. Table 4-6 shows some of the more typical con-ductors used in house wiring. Here are some definitions taken directly from the *NEC.*®

SPECIAL AMPACITY RATINGS FOR RESIDENTIAL 120/240-VOLT, 3-WIRE, SINGLE-PHASE SERVICE-ENTRANCE CONDUCTORS, SERVICE LATERAL CONDUCTORS, AND FEEDER CONDUCTORS THAT SERVE AS THE MAIN POWER FEEDER TO A DWELLING UNIT.
► **CONDUCTOR TYPES RHH, RHW, RHW-2, THHN, THHW, THW, THW-2, THWN, THWN-2, XHHW, XHHW-2, SE, USE, USE-2.** ◄

Copper Conductor AWG for Insulation of RH-RHH-RHW-THW-THWN-THHW-THHN-XHHW-USE	Aluminum or Copper-Clad Aluminum Conductors (AWG)	Allowable Special Ampacity
4	2	100
3	1	110
2	1/0	125
1	2/0	150
1/0	3/0	175
2/0	4/0	200
3/0	250 kcmil	225
4/0	300 kcmil	250
250 kcmil	350 kcmil	300
350 kcmil	500 kcmil	350
400 kcmil	600 kcmil	400

Note: Where the conductors will be exposed to temperatures in excess of 86°F (30°C), apply the correction factors found at the bottom of *Table 310.16.* **Note:** Conductors having a suffix "2," such as THW-2, are permitted for continuous use at 194°F (90°C) in wet or dry locations.

Table 4-5

TRADE NAME	TYPE LETTER	MAXIMUM OPERATING TEMP.	APPLICATION PROVISIONS	INSULATION	AWG	OUTER COVERING
Heat-resistant thermoplastic	THHN	194°F (90°C)	Dry and damp locations	Flame-retardant, heat-resistant thermoplastic	14 through 1000 kcmil	Nylon jacket or equivalent
Moisture- and heat-resistant thermoplastic	THHW	167°F (75°C) 194°F (90°C)	Wet location Dry location	Flame-retardant, moisture- and heat-resistant thermoplastic	14 through 1000 kcmil	None
Moisture- and heat-resistant thermoplastic	THWN Note: If marked THWN-2, OK for 194°F (90°C) in dry or wet locations	167°F (75°C)	Dry and wet locations	Flame-retardant, moisture- and heat-resistant thermoplastic	14 through 1000 kcmil	Nylon jacket or equivalent
Moisture- and heat-resistant thermoplastic	THW Note: If marked THW-2, OK for 194°F (90°C) in dry or wet locations	167°F (75°C)	Dry and wet locations	Flame-retardant, moisture- and heat-resistant thermoplastic	14 through 2000 kcmil	None
Underground feeder and branch-circuit single conductor or multi-conductor cable. See *Article 339*	UF	140°F (60°C) 167°F (75°C)	See *Article 339* See *Article 339*	Moisture-resistant Moisture-resistant	14 through 4/0	Integral with insulation

Table 4-6 Typical conductors used for residential wiring.

- **Damp Location.** ▶Locations protected from weather and not subject to saturation with water or other liquids but subject to moderate degrees of moisture. Examples of such locations include partially protected locations under canopies, marquees, roofed open porches, and like locations; and interior locations subject to moderate degrees of moisture, such as some basements, some barns, and some cold-storage warehouses. ◀

- **Dry Location.** A location not normally subject to dampness or wetness. A location classified as dry may be temporarily subject to dampness or wetness, as in the case of a building under construction.

- **Wet Location.** Installations underground or in concrete slabs or masonry in direct contact with the earth; locations subject to saturation with water or other liquids, such as vehicle washing areas; and locations exposed to weather and unprotected.

- **Locations Exposed to Direct Sunlight.** Insulated conductors and cables used where exposed to direct rays of the sun shall be of a type listed or marked "sunlight resistant."

See *310.8* of the *NEC*® for more details.

Conductor Temperature Ratings

Conductors are also rated as to the temperature they can withstand. For example:

	Celsius	Fahrenheit
Type TW	60°	140°
Type THW	75°	167°
Type THHN	90°	194°

Without question, conductors with Type THHN insulation are the most popular and most commonly used, particularly in the smaller sizes, because of their small diameter (easy to handle; more conductors of a given size permitted in a given size raceway) and their suitability for installation where high temperatures are encountered, such as attics, buried in insulation, and supplying recessed fixtures.

The temperature rating of nonmetallic-sheathed cable is 194°F (90°C). However, *334.80* states that the allowable ampacity is that of the 140°F (60°C) column in *Table 310.16*.

The Weakest Link of the Chain

A conductor has two ends!

Selecting a conductor ampacity based solely on the allowable ampacity values found in *Table 310.16* is a very common mistake that can prove costly.

In any given circuit, we have an assortment of electrical components that have different maximum temperature ratings. Our job is to find out which component is the "weakest link" in the system and do our circuit design based on the lowest rated component in the circuit. Maximum temperature ratings are found in the *NEC®* and in the UL Standards.

- **Circuit breakers:** UL 489. Generally marked "75°C Only" or "60°C/75°C" (140°F/167°F).

- **Conductors:** *Table 310.13*, *Table 310.16*, and UL 83.

- **Disconnect switches:** UL 98. Generally marked "75°C Only" or "60°C/75°C" (140°F/167°F).

- **Receptacles and attachment plug caps:** UL 498. Most 15- and 20-ampere branch-circuits have receptacles that have a maximum temperature rating of 140°F (60°C). Some 30-ampere receptacles are rated 140°F (60°C) and some are rated for 167°F (75°C). Larger 50-ampere receptacles are rated 167°F (75°C).

- **Panelboards:** UL 67. Panelboards are generally marked 167°F (75°C). The temperature rating of a panelboard has been established as an assembly with circuit breakers in place. Do not use the circuit breakers' temperature marking by itself. It is the panelboard temperature marking that counts.

- **Snap switches:** UL 20. Most 15- and 20-ampere branch-circuits have snap switches with a maximum temperature rating of 140°F (60°C).

- **Wire splicing devices and connectors:** Wire-Nuts,® Wing-Nuts,® Twisters,® and Scotchloks® for copper conductors (UL 486) and for aluminum conductors (UL 486B). These types of connectors are rated 221°F (105°C).

Table 4-7 indicates the correct applications for conductors and terminations.

Another key *NEC®* requirement is found in *110.14(C)*, "Termination Limitations," where we find the following:

For circuits rated 100 amperes or less, or marked for conductor sizes 14 through 1 AWG, unless otherwise identified, the terminals on wiring devices, switches, breakers, motor controllers, and other electrical equipment are based upon the ampacity of 140°F (60°C) conductors. It is acceptable to install conductors having a higher temperature rating, such as 90°C (194°F) THHN, but you must use the 140°F (60°C) ampacity values.

For circuits rated over 100 amperes, or marked for conductors larger than 1 AWG, the conductors installed must have a minimum 167°F (75°C) rating. The ampacity of the conductors is based on the 167°F (75°C) values. It is acceptable to install conductors having a higher temperature rating, such as 194°F (90°C) THHN, but you must use the 167°F (75°C) ampacity values.

When using high-temperature insulated conductors where adjustment factors are to be applied, such as *derating* for more than three current-carrying conductors in a raceway or cable, or *correcting* where high temperatures are encountered, the final results of these adjustments (in amperes) must comply with

TERMINATION RATING	CONDUCTOR INSULATION RATING		
	60°C	75°C	90°C
60°C	OK	OK (at 60°C ampacity)	OK (at 60°C ampacity)
75°C	NO	OK	OK (at 75°C ampacity)
60/75°C	OK	OK (at 60°C or 75°C ampacity)	OK (at 60°C or 75°C ampacity)
90°C	NO	NO	OK (Only if equipment has a 90°C rating.)

Table 4-7 Types of conductor temperature ratings permitted for various terminal temperature ratings in conformance to the *NEC®* and the UL Standards.

the requirements of *110.14(C)*. Similar requirements are found in the UL Standards.

Examples of applying *derating factors* and *correcting factors* are found in Unit 18.

Temperature ratings of conductors and cables are discussed later in this unit.

In *240.3(D)*, there is a built-in code compliance limitation for applying small branch-circuit conductors. Here we find that the maximum overcurrent protection is 15 amperes for 14 AWG, 20 amperes for 12 AWG, and 30 amperes for 10 AWG.

From *Table 310.16* we find that:

- 14 AWG copper conductors' allowable ampacity is:

  20 amperes in the 60°C (140°F) column

  20 amperes in the 75°C (167°F) column

  25 amperes in the 90°C (194°F) column

- 12 AWG copper conductors' allowable ampacity is:

  25 amperes in the 60°C (140°F) column

  25 amperes in the 75°C (167°F) column

  30 amperes in the 90°C (194°F) column

- 10 AWG copper conductors' allowable ampacity is:

  30 amperes in the 60°C (140°F) column

  35 amperes in the 75°C (167°F) column

  40 amperes in the 90°C (194°F) column

There are exceptions to this maximum overcurrent protection rule for motor, air-conditioner, and heat pump branch-circuits. These exceptions are covered elsewhere in this text.

In summary, we cannot arbitrarily use the ampacity values for conductors as found in *Table 310.16*. Conductor ampacity is not a "stand-alone" issue. We must also consider the temperature limitations of the equipment, such as panelboards, receptacles, snap switches, receptacles, connectors, and so on. The lowest maximum temperature-rated device in an electrical system is the weakest link, and it is the weakest link that we must base our ultimate conductor ampacity decision on.

Knob-and-Tube Wiring

Older homes were wired with open, individual, insulated conductors supported by porcelain *knobs* and *tubes*. In its day, knob-and-tube wiring served its purpose well. This text does not cover knob-and-tube wiring because this method is no longer used in new construction. On occasion, it is necessary to make some modifications to knob-and-tube wiring when doing remodel work. The *NEC®* covers this wiring method in *Article 394*.

Many fires have occurred because thermal insulation has been blown into the attics of older homes, completely burying the knob-and-tube wiring. The conductors become overheated. *NEC® 394.4* in part states that "concealed knob-and-tube wiring shall not be used in hollow spaces of walls, ceilings, and attics where such spaces are insulated by loose, rolled, or foamed-in-place insulating material that envelops the conductors."

Keep an eye out for this potential fire hazard when working in older homes that have knob-and-tube wiring.

Neutral Conductor Size

Figure 4-5 shows how *line-to-line* loads and *line-to-neutral* loads are connected. The grounded neutral (white) conductor Ⓝ for residential services and feeders is permitted to be smaller than the "hot" ungrounded phase (black, red) conductors Ⓓ only when the neutral conductor is properly and adequately sized to carry the maximum unbalance loads computed according to *215.2, 220.22,* and *230.42*.

NEC® 215.2 relates to feeders and refers us to *Article 220* for computation requirements.

NEC® 230.42 relates to services, and refers us to *Article 220* for computation requirements.

Focusing on the neutral conductor, *220.22* states that "the feeder or service neutral load shall be the maximum unbalance of the load determined by this Article."

NEC® 220.22 further states that "the maximum unbalanced load shall be the maximum net computed load between the neutral and any one ungrounded conductor."

In this residence we find a number of loads that carry little or no neutral currents, such as the electric water heater, electric clothes dryer, electric oven and range, electric furnace, and air conditioner. These appliances for the most part are drawing their biggest current through the "hot" conductors because they are fed with 240-volt Ⓒ or 120/240-volt circuits Ⓔ. Loads Ⓑ are connected line to neutral only.

Thus we find logic in the *NEC®* which permits reducing the neutral size on services and feeders where the computations prove that the neutral will be carrying less current than the "hot" phase conductors.

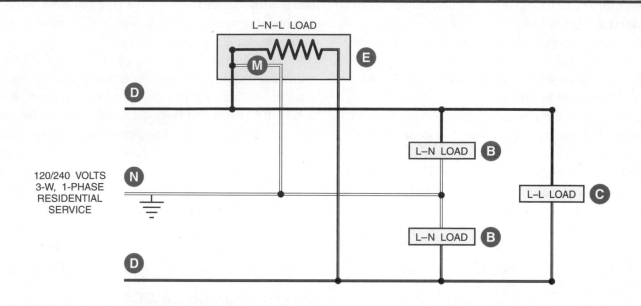

Figure 4-5 **Diagram showing line-to-line and line-to-neutral loads.**

See Unit 20 for sizing neutral conductors for electric range branch-circuits and Unit 29 for sizing neutral service-entrance conductors.

VOLTAGE DROP

Low voltage can cause lights to dim, some television pictures to "shrink," motors to run hot, electric heaters to not produce their rated heat output, and appliances to not operate properly.

Low voltage in a home can be caused by:

- wire that is too small for the load being served
- a circuit that is too long
- poor connections at the terminals
- conductors operating at high temperatures having higher resistance than when operating at lower temperatures

A simple formula for calculating voltage drop on single-phase systems considers only the dc resistance of the conductors and the temperature of the conductor. See *Table 8, Chapter 9* in the *NEC®* for the dc resistance values. The more accurate formulas consider ac resistance, reactance, temperature, spacing, in metal conduit and in nonmetallic conduit, *NEC® Table 9, Chapter 9*. Voltage drop is covered in great detail in *Electrical Wiring—Commercial* (Delmar, a division of Thomson Learning). The simple voltage drop formula is more accurate with smaller conductors, and gets less and less accurate as conductor size increases. It is sufficiently accurate for voltage drop calculations necessary for residential wiring.

To Find Voltage Drop in a Single-Phase Circuit:

$$E_d = \frac{K \times I \times L \times 2}{CMA}$$

To Find Conductor Size for a Single-Phase Circuit:

$$CMA = \frac{K \times I \times L \times 2}{E_d}$$

In the above formulas:

E_d = allowable voltage drop in volts.

K = approximate resistance in ohms per circular-mil foot at 167°F (75°C).
- for uncoated copper wire, use 12 ohms.*
- for aluminum wire, use 20 ohms.*

I = current in amperes flowing through the conductors.

L = length in feet from beginning of circuit to the load.

CMA = cross-sectional area of the conductors in circular mils.**

---

*Derived values from *Table 8, Chapter 9, NEC®.*
**From Table 4-8 in this text or from *Table 8, Chapter 9, NEC®.*

We use the factor of 2 for single-phase circuits because there is voltage drop in both conductors, to and from the connected load.

To Find Voltage Drop and Conductor Size for a Three-Phase Circuit

Although residential wiring does not use three-phase systems, commercial and industrial systems do. To make voltage drop and conductor sizing calculations for a three-phase system, substitute the factor of 1.73 in place of the factor 2 in the voltage drop formulas.

Code References to Voltage Drop

For branch-circuits, refer to *210.19(A)*, *FPN No. 4*. The recommended maximum voltage drop is 3 percent, as shown in Figure 4-6.

For feeders, refer to *215.2(D)*, *FPN No. 2*. The recommended maximum voltage drop is 3 percent. See Figure 4-7.

When both branch-circuits and feeders are involved, the total voltage drop should not exceed 5 percent. This is shown in Figure 4-7.

According to *90.5(C)*, *FPNs* are informational only, and are not to be enforced as Code requirement.

Examples of Voltage Drop Calculations

EXAMPLE 1: What is the approximate voltage drop on a 120-volt, single-phase circuit consisting of 14 AWG copper conductors where the load is 11 amperes and the distance of the circuit from the panel to actual load is 85 ft (25.91 m)?

SOLUTION:

$$E_d = \frac{K \times I \times L \times 2}{CMA}$$

$$E_d = \frac{12 \times 11 \times 85 \times 2}{4110}$$

$$E_d = 5.46 \text{ volts drop}$$

Note: Refer to Table 4-8 for the CMA value.

This exceeds the voltage drop permitted by the Code, which is:

$$3\% \text{ of } 120 \text{ volts } = 3.6 \text{ volts}$$

Let's try it again using 12 AWG conductors:

$$E_d = \frac{12 \times 11 \times 85 \times 2}{6530}$$

$$E_d = 3.44 \text{ volts drop}$$

This "meets Code."

EXAMPLE 2: Find the wire size needed to keep the voltage drop to no more than 3 percent on a single-phase, 240-volt, air-conditioner circuit. The nameplate reads: "Minimum Circuit Ampacity 40 Amperes." The circuit originates at the main panel located approximately 65 ft (19.81 m) from the air-conditioner unit. No neutral is required.

SOLUTION: Checking *Table 310.16*, the conductors could be 6 AWG Type TW copper or 8 AWG Type THHN copper.

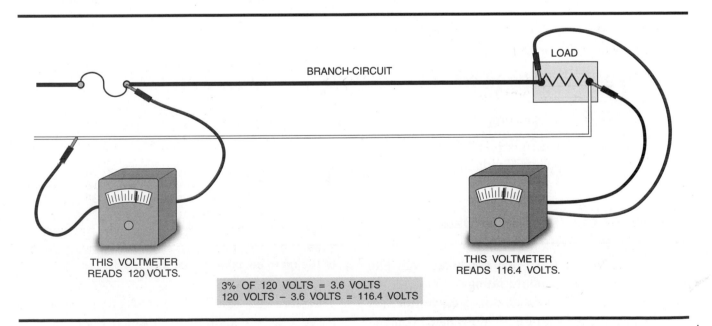

THIS VOLTMETER READS 120 VOLTS.

THIS VOLTMETER READS 116.4 VOLTS.

LOAD

BRANCH-CIRCUIT

3% OF 120 VOLTS = 3.6 VOLTS
120 VOLTS – 3.6 VOLTS = 116.4 VOLTS

Figure 4-6   Maximum recommended voltage drop on a branch-circuit is 3%, *210.19(A)*, *FPN No. 4*.

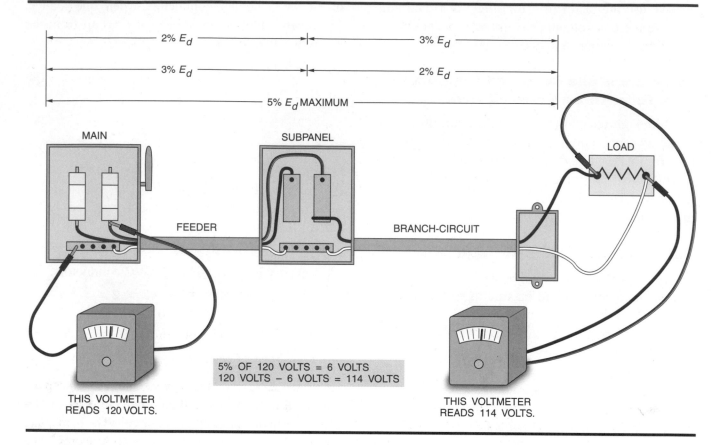

Figure 4-7 *210.19(A), FPN No. 4* and *215.2(D) FPN No. 2* of the Code state that the total voltage drop from the beginning of a feeder to the farthest outlet on a branch-circuit should not exceed 5%. In this figure, if the voltage drop in the feeder is 3%, then do not exceed 2% voltage drop in the branch-circuit. If the voltage drop in the feeder is 2%, then do not exceed 3% voltage drop in the branch-circuit.

The permitted voltage drop is:

$$E_d = 240 \times 0.03 = 7.2 \text{ volts}$$

Let's see what voltage drop we might experience if we installed the 8 AWG Type THHN conductors.

$$E_d = \frac{12 \times 40 \times 65 \times 2}{16,510}$$

$$= 3.78 \text{ volts drop}$$

This is well below the permissible 7.2 volts drop and would make an acceptable installation.

A Word of Caution When Using High-Temperature Wire

Note in *Table 310.16* that conductors fall into three classes of temperature ratings—60°C, 75°C, and 90°C. For a given conductor size, we find that the allowable ampacity of a 90°C insulated conductor is greater than that of a 60°C insulated conductor.

CONDUCTOR SIZE, AWG	CROSS-SECTIONAL AREA IN CIRCULAR MILS
18	1620
16	2580
14	4110
12	6530
10	10,380
8	16,510
6	26,240
4	41,740
3	52,620
2	66,360
1	83,690
0	105,600
00	133,100
000	167,800
0000	211,600

Table 4-8 Circular mil cross-sectional area for many of the more common conductors.

Therefore, be careful when selecting conductors based on their ability to withstand high temperatures. For example:

Table 310.16 Allowable Ampacities of Insulated Conductors Rated 0 Through 2000 Volts, 60°C Through 90°C (140°F Through 194°F), Not More Than Three Current-Carrying Conductors in Raceway, Cable, or Earth (Directly Buried), Based on Ambient Temperature of 30°C (86°F)

Size AWG or kcmil	Temperature Rating of Conductor (See Table 310.13.)						Size AWG or kcmil
	60°C (140°F)	75°C (167°F)	90°C (194°F)	60°C (140°F)	75°C (167°F)	90°C (194°F)	
	Types TW, UF	Types RHW, THHW, THW, THWN, XHHW, USE, ZW	Types TBS, SA, SIS, FEP, FEPB, MI, RHH, RHW-2, THHN, THHW, THW-2, THWN-2, USE-2, XHH, XHHW, XHHW-2, ZW-2	Types TW, UF	Types RHW, THHW, THW, THWN, XHHW, USE	Types TBS, SA, SIS, THHN, THHW, THW-2, THWN-2, RHH, RHW-2, USE-2, XHH, XHHW, XHHW-2, ZW-2	
	COPPER			ALUMINUM OR COPPER-CLAD ALUMINUM			
18	—	—	14	—	—	—	—
16	—	—	18	—	—	—	—
14*	20	20	25	—	—	—	—
12*	25	25	30	20	20	25	12*
10*	30	35	40	25	30	35	10*
8	40	50	55	30	40	45	8
6	55	65	75	40	50	60	6
4	70	85	95	55	65	75	4
3	85	100	110	65	75	85	3
2	95	115	130	75	90	100	2
1	110	130	150	85	100	115	1
1/0	125	150	170	100	120	135	1/0
2/0	145	175	195	115	135	150	2/0
3/0	165	200	225	130	155	175	3/0
4/0	195	230	260	150	180	205	4/0
250	215	255	290	170	205	230	250
300	240	285	320	190	230	255	300
350	260	310	350	210	250	280	350
400	280	335	380	225	270	305	400
500	320	380	430	260	310	350	500
600	355	420	475	285	340	385	600
700	385	460	520	310	375	420	700
750	400	475	535	320	385	435	750
800	410	490	555	330	395	450	800
900	435	520	585	355	425	480	900
1000	455	545	615	375	445.	500	1000
1250	495	590	665	405	485	545	1250
1500	520	625	705	435	520	585	1500
1750	545	650	735	455	545	615	1750
2000	560	665	750	470	560	630	2000

CORRECTION FACTORS							
Ambient Temp. (°C)	For ambient temperatures other than 30°C (86°F), multiply the allowable ampacities shown above by the appropriate factor shown below.						Ambient Temp. (°F)
21–25	1.08	1.05	1.04	1.08	1.05	1.04	70–77
26–30	1.00	1.00	1.00	1.00	1.00	1.00	78–86
31–35	0.91	0.94	0.96	0.91	0.94	0.96	87–95
36–40	0.82	0.88	0.91	0.82	0.88	0.91	96–104
41–45	0.71	0.82	0.87	0.71	0.82	0.87	105–113
46–50	0.58	0.75	0.82	0.58	0.75	0.82	114–122
51–55	0.41	0.67	0.76	0.41	0.67	0.76	123–131
56–60	—	0.58	0.71	—	0.58	0.71	132–140
61–70	—	0.33	0.58	—	0.33	0.58	141–158
71–80	—	—	0.41	—	—	0.41	159–176

* See 240.4(D).

Reprinted with permission from NFPA 70-2002.

- 8 AWG THHN (90°C) copper has an ampacity of 55 amperes.

- 8 AWG THW (75°C) copper has an ampacity of 50 amperes.

- 8 AWG TW (60°C) copper has an ampacity of 40 amperes.

We learned earlier in this unit that according to *110.14(C)*, we use the 60°C column of *Table 310.16* for circuits rated 100 amperes or less, and the 75°C column for circuits rated over 100 amperes. However, when the equipment is marked as being suitable for 75°C, we can take advantage of the higher ampacity of 75°C conductors. This often-times results in a smaller size conductor and a correspondingly smaller raceway.

But installing smaller conductors might result in excessive voltage drop.

Therefore, after selecting the proper size conductor for a given load, it is always a good idea to run a voltage-drop calculation to make sure the voltage drop is not excessive.

Effect of Voltage Variation

Unit 19 has more information on the effect voltage differences have on wattage output of appliances and on electric motors.

APPROXIMATE CONDUCTOR SIZE RELATIONSHIP

There is a definite relationship between the circular mil area and the resistance of conductors. The following rules explain this relationship.

Rule One: For wire sizes up through 0000, every third size doubles or halves in circular mil area.

Thus, a 1 AWG conductor is 2 times larger than a 4 AWG conductor (83,690 versus 41,470). Thus, a "0" wire is one-half the size of "0000" wire (105,600 versus 211,600).

Rule Two: For wire sizes up through 0000, every consecutive wire size is approximately 1.26 times larger or smaller than the preceding wire size.

Thus, a 3 AWG conductor is approximately 1.26 times larger than a 4 AWG conductor (41,740 × 1.26 = 52,592). A 2 AWG conductor is approximately 1.26 times smaller than a 1 AWG conductor (83,690 ÷ 1.26 = 66,420).

Try to fix in your mind that a 10 AWG conductor has a cross-sectional area of 10,380 circular mils, and that it has a resistance of 1.2 ohms per 1000 ft (300 m). The resistance of aluminum wire is approximately 2 ohms per 1000 ft (300 m). By remembering these numbers, you will be able to perform voltage drop calculations without having the wire tables readily available.

EXAMPLE: What is the approximate cross-sectional area, in circular mils, and resistance of a 6 AWG copper conductor?

SOLUTION:

Note in Table 4-9 that when the CMA of a wire is doubled, then its resistance is cut in half. Inversely, when a given wire size is reduced to one-half, its resistance doubles.

NONMETALLIC-SHEATHED CABLE
(ARTICLE 334)

The *NEC®* and UL Standard 719 describe the construction details of nonmetallic-sheathed cable. The following is a summary of these details.

WIRE SIZE (AWG)	CMA (In Circular Mils)	OHMS PER 1000 FT
10	10,380	1.2
9		
8		
7	20,760	0.6
6	26,158	0.0476

Table 4-9 "Every third size" and the 1.26 relationship between the circular-mil area (CMA) and the resistance in ohms per 1000 ft of copper conductors sizes 10 AWG through 6 AWG.

Description

Most electricians still refer to nonmetallic-sheathed cable as *Romex*, a name chosen years ago by the Rome Wire and Cable Company.

Nonmetallic-sheathed cable is a factory assembly of two or more insulated conductors having an outer sheath of moisture-resistant, flame-retardant, nonmetallic material. This cable is available with two or three current-carrying conductors. The conductors range in size from 14 AWG through 2 AWG for copper conductors, and from 12 AWG through 2 AWG for aluminum or copper-clad aluminum conductors. Two-wire cables contain one black conductor, one white conductor, and one bare equipment grounding conductor. Three-wire cables contain one black, one white, one red, and one bare equipment grounding conductor. Equipment grounding conductors are permitted to have green insulation, but bare equipment grounding conductors are the most common.

Figure 4-8 clearly shows a bare equipment grounding conductor. Sometimes the bare equipment grounding conductor is wrapped with paper or fiberglass, which acts as a filler. The equipment

Figure 4-8 A nonmetallic-sheathed cable showing (A) black "ungrounded" (hot) conductor, (B) bare equipment "grounding" conductor, and (C) white "grounded" conductor. *Courtesy of* Southwire Company.

grounding conductor is *not* permitted to be used as a current-carrying conductor.

UL lists three types of nonmetallic-sheathed cables:

- Type NM-B cable has a flame-retardant, moisture-resistant, nonmetallic outer jacket. The conductors are rated 194°F (90°C). The ampacity is based on 140°F (60°C).

- Type NMC-B cable has a flame-retardant, moisture-resistant, fungus-resistant, and corrosion-resistant nonmetallic outer jacket. The conductors are rated 194°F (90°C). The ampacity is based on 140°F (60°C).

- Type NMS-B cable contains the conventional insulated power conductors as well as telephone, coaxial, home entertainment, and signaling conductors all in one cable. This type of cable is used for home automation systems using the latest in digital technology. *Article 780* recognizes this wiring method for closed loop and programmed power distribution. Type NMS-B cable has a moisture-resistant, flame-retardant, nonmetallic outer jacket.

Although somewhat confusing, the UL standards use the suffix B whereas the *NEC®* refers to nonmetallic-sheathed cables as Type NM, NMC, and NMS. The conductors in these cables are rated 194°F (90°C). Types NM, NMC, and NMS cable identified by the markings NM-B, NMC-B, and NMS-B meet this 194°F (90°C) requirement. See *334.112*.

Why the suffix *B*?

Older nonmetallic-sheathed cable contained 140°F (60°C) rated conductors. There were many problems with the insulation becoming brittle and breaking off due to the extremely high temperatures associated with recessed and surface-mounted luminaires (fixtures): fires resulted. In hot attics, after applying correction factors, the adjusted allowable ampacity of the conductors could very well be zero. To solve this dilemma, since December 17, 1984 UL has required that the conductors be rated 194°F (90°C), and that the cable be identified with the suffix *B*. This suffix makes it easy to differentiate the newer 194°F (90°C) from the old 140°F (60°C) cable.

The ampacity of the conductors in nonmetallic-sheathed cable is based on that of 140°F (60°C) conductors. The 194°F (90°C) ampacity is permitted to be used for derating purposes, such as might be necessary in attics. The final derated ampacity must not exceed the ampacity for 140°F (60°C) conductors. More specifics in *334.80*.

Table 310.16 shows the allowable ampacities for conductors.

Overcurrent protection for small copper conductors, in conformance to *240.4(D)*, is:

- 14 AWG   15 amperes
- 12 AWG   20 amperes
- 10 AWG   30 amperes

A typical nonmetallic-sheathed cable is illustrated in Figure 4-8.

Where Nonmetallic-Sheathed Cable May Be Used

Table 4-10 shows the uses permitted for Types NM-B, NMC-B and NMS-B.

Equipment Grounding Conductors

Equipment grounding conductors are selected based upon the rating or setting of the overcurrent device protecting a given branch-circuit or feeder. See *Table 250.122*. The equipment grounding conductor in a nonmetallic-sheathed cable is sized according to this in Table 4-11.

Installation

Nonmetallic-sheathed cable is probably the least expensive of the various wiring methods. It is relatively lightweight and easy to install. It is widely used for dwelling unit installations on circuits of 600 volts or less. Figure 4-9 shows an example of a stripper used for stripping nonmetallic-sheathed cable. A knife also works fine with care. The installation of nonmetallic-sheathed cable must conform to the requirements of *Article 334*. Refer to Figure 4-10, Figure 4-11, Figure 4-12, Figure 4-13, Figure 4-14, Figure 4-15, and Figure 4-16.

- The cable must be strapped or stapled not more than 12 in. (300 mm) from a box or fitting.

- Do not staple flat nonmetallic-sheathed cables on edge, Figure 4-15.

	TYPE NM-B	TYPE NMC-B	TYPE NMS-B
May be used on circuits of 600 volts or less.	Yes	Yes	Yes
May be run exposed or concealed in dry locations.	Yes	Yes	Yes
May be installed exposed or concealed in damp and moist locations.	No	Yes	No
Has flame-retardant and moisture-resistant outer covering.	Yes	Yes	Yes
Has fungus-resistant and corrosion-resistant outer covering.	No	Yes	No
May be used to wire one- and two-family dwellings, or multifamily dwellings.	Yes	Yes	Yes
May be embedded in masonry, concrete, plaster, adobe, fill.	No	No	No
May be installed or fished in the hollow voids of masonry blocks or tile walls where not exposed to excessive moisture or dampness.	Yes	Yes	Yes
May be installed or fished in the hollow voids of masonry blocks or tile walls where exposed to excessive moisture.	No	Yes	No
In outside walls of masonry block or tile.	No	Yes	No
In inside wall of masonry block or tile.	Yes	Yes	Yes
May be used as service-entrance cable.	No	No	No
Must be protected against physical damage.	Yes	Yes	Yes
May be run in shallow chase of masonry, concrete, or adobe if protected by a steel plate at least 1/16 in. (1.6 mm) thick, then covered with plaster, adobe, or similar finish.	No	Yes	No

Table 4-10   Uses permitted for nonmetallic-sheathed cable.

- The intervals between straps or staples must not exceed 4½ ft (1.4 m).

- The cable must be protected against physical damage where necessary.

- The 4½ ft (1.4 m) securing requirement is not needed where nonmetallic-sheathed cable is run *horizontally* through holes in wood or metal framing members (studs, joists, rafters, etc.). The cable is considered to be adequately supported by the framing members, *334.30*.

- The inner edge of the bend shall have a minimum radius not less than five times the cable diameter, *334.24*.

Rating or Setting of Overcurrent Device	MINIMUM SIZE (AWG) EQUIPMENT GROUNDING CONDUCTOR	
	Copper	Aluminum or Copper-Clad Aluminum
15	14	12
20	12	10
30	10	8
40	10	8
60	10	8
100	8	6
200	6	4

Table 4-11   Minimum size equipment grounding conductors is based on the rating or setting of the overcurrent device protecting that particular branch circuit. This table is based on *Table 250.122* of the *NEC*.

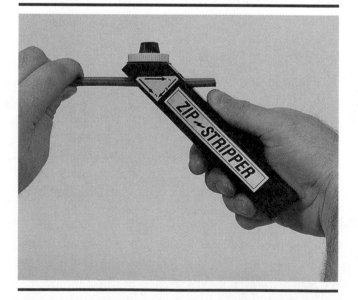

Figure 4-9   This stripper is used to remove the outer jacket from nonmetallic-sheathed cable. *Courtesy of Seatek Co., Inc.*

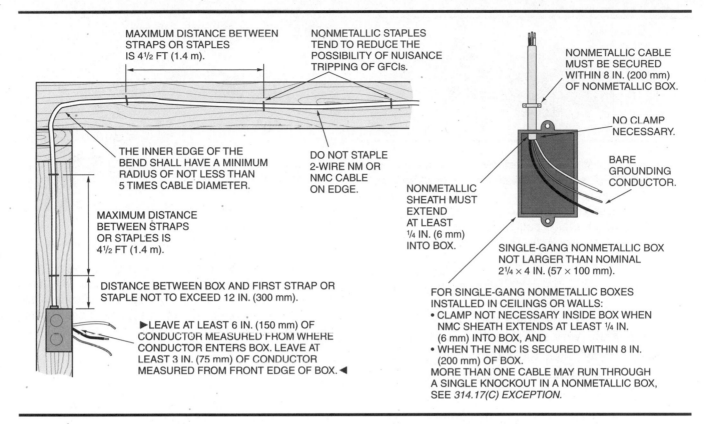

MAXIMUM DISTANCE BETWEEN STRAPS OR STAPLES IS 4½ FT (1.4 m).

NONMETALLIC STAPLES TEND TO REDUCE THE POSSIBILITY OF NUISANCE TRIPPING OF GFCIs.

THE INNER EDGE OF THE BEND SHALL HAVE A MINIMUM RADIUS OF NOT LESS THAN 5 TIMES CABLE DIAMETER.

DO NOT STAPLE 2-WIRE NM OR NMC CABLE ON EDGE.

MAXIMUM DISTANCE BETWEEN STRAPS OR STAPLES IS 4½ FT (1.4 m).

DISTANCE BETWEEN BOX AND FIRST STRAP OR STAPLE NOT TO EXCEED 12 IN. (300 mm).

▶LEAVE AT LEAST 6 IN. (150 mm) OF CONDUCTOR MEASURED FROM WHERE CONDUCTOR ENTERS BOX. LEAVE AT LEAST 3 IN. (75 mm) OF CONDUCTOR MEASURED FROM FRONT EDGE OF BOX. ◀

NONMETALLIC CABLE MUST BE SECURED WITHIN 8 IN. (200 mm) OF NONMETALLIC BOX.

NO CLAMP NECESSARY.

BARE GROUNDING CONDUCTOR.

NONMETALLIC SHEATH MUST EXTEND AT LEAST ¼ IN. (6 mm) INTO BOX.

SINGLE-GANG NONMETALLIC BOX NOT LARGER THAN NOMINAL 2¼ × 4 IN. (57 × 100 mm).

FOR SINGLE-GANG NONMETALLIC BOXES INSTALLED IN CEILINGS OR WALLS:
• CLAMP NOT NECESSARY INSIDE BOX WHEN NMC SHEATH EXTENDS AT LEAST ¼ IN. (6 mm) INTO BOX, AND
• WHEN THE NMC IS SECURED WITHIN 8 IN. (200 mm) OF BOX.
MORE THAN ONE CABLE MAY RUN THROUGH A SINGLE KNOCKOUT IN A NONMETALLIC BOX, SEE *314.17(C) EXCEPTION.*

Figure 4-10   Installation of nonmetallic-sheathed cable.

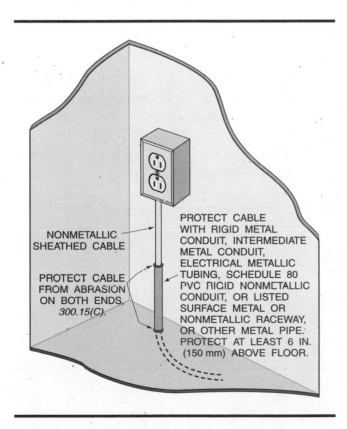

NONMETALLIC SHEATHED CABLE

PROTECT CABLE FROM ABRASION ON BOTH ENDS, *300.15(C).*

PROTECT CABLE WITH RIGID METAL CONDUIT, INTERMEDIATE METAL CONDUIT, ELECTRICAL METALLIC TUBING, SCHEDULE 80 PVC RIGID NONMETALLIC CONDUIT, OR LISTED SURFACE METAL OR NONMETALLIC RACEWAY, OR OTHER METAL PIPE. PROTECT AT LEAST 6 IN. (150 mm) ABOVE FLOOR.

Figure 4-11   Installation of exposed nonmetallic-sheathed cable where passing through a floor. See *334.15(B).*

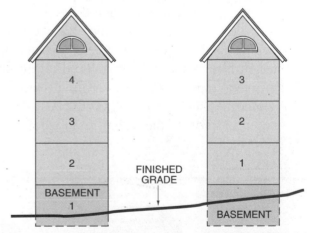

THIS IS A 4-STORY BUILDING   THIS IS A 3-STORY BUILDING

STORY ABOVE GRADE: ANY STORY HAVING ITS FINISHED FLOOR SURFACE ENTIRELY ABOVE GRADE EXCEPT THAT A BASEMENT SHALL BE CONSIDERED AS A STORY ABOVE GRADE WHEN THE FINISHED SURFACE OF THE FLOOR ABOVE THE BASEMENT IS:

1. MORE THAN 6 FEET (1.8 M) ABOVE GRADE PLANE; OR

2. MORE THAN 6 FEET (1.8 M) ABOVE THE FINISHED GROUND LEVEL FOR MORE THAN 50 PERCENT OF THE TOTAL BUILDING PERIMETER; OR

3. MORE THAN 12 FEET (3.7 M) ABOVE THE FINISHED GROUND LEVEL AT ANY POINT.

Figure 4-12   Illustration of how a story above grade is defined by building Codes.

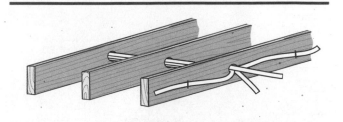

Figure 4-13 ▶When nonmetallic-sheathed cables are "bundled" or "stacked" for distances of more than 24 in. (600 mm), their ampacities must be reduced (derated, adjusted). Type AC (armored cable) and Type MC (metal clad) cables are not subject to this rule. See text and *310.15(B)(2)* for details.◀

- The cable must not be used in circuits of more than 600 volts.

- *Caution*: Be very careful when installing staples. Driving them too hard can squeeze and damage the insulation on the conductors, causing short circuits and/or ground faults.

- See Figure 4-10 for special conditions when using single-gang nonmetallic device boxes.

- Nonmetallic-sheathed cable must be protected where passing through a floor by at least 6 in. (150 mm) of rigid metal conduit, intermediate metal conduit, electrical metallic tubing, Schedule 80 PVC rigid nonmetallic conduit, listed surface metal or nonmetallic raceway, or other metal pipe, *334.15(B)*. A fitting (bushing or connector) must be used at both ends of the conduit to protect the cable from abrasion, Figure 4-11.

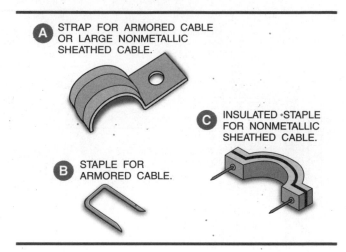

Figure 4-14 (A) One-hole strap used for conduit, EMT, large nonmetallic-sheathed cable and armored cable. (B) Metal staple for armored cable and nonmetallic-sheathed cable. (C) Insulated staple for nonmetallic-sheathed cable. Certain types of insulated and noninsulated staples can be applied with a staple gun.

- ▶When nonmetallic-sheathed cables are "bundled" or "stacked" for distances of more than 24 in. (600 mm), their ampacities must be reduced (adjusted, derated) according to *Table 310.15(B)(2)(a)*. When cables are bunched together, the heat generated by the conductors cannot easily dissipate. See Figure 4-13.◀

- ▶When Type AC and Type MC cables are "bundled" or "stacked," derating is not necessary if the following is true:
  - the conductors are 12 AWG copper
  - if the bundling or stacking is not more than 24 in. (600 mm)

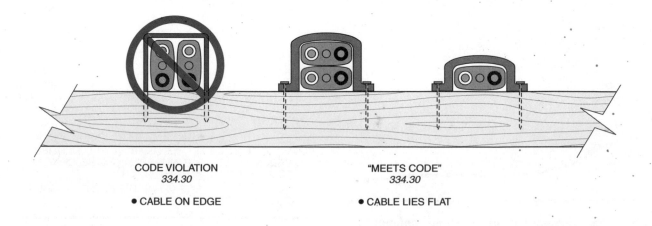

CODE VIOLATION
334.30

"MEETS CODE"
334.30

- CABLE ON EDGE

- CABLE LIES FLAT

Figure 4-15 It is a Code violation to staple flat nonmetallic-sheathed cables on edge, *334.30*.

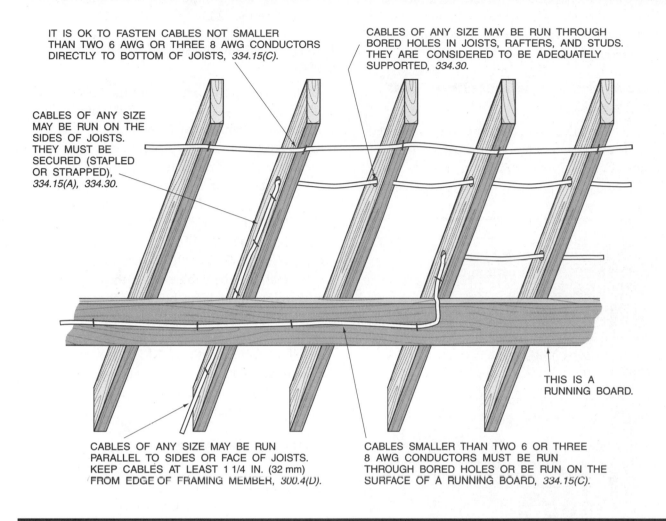

IT IS OK TO FASTEN CABLES NOT SMALLER THAN TWO 6 AWG OR THREE 8 AWG CONDUCTORS DIRECTLY TO BOTTOM OF JOISTS, *334.15(C).*

CABLES OF ANY SIZE MAY BE RUN THROUGH BORED HOLES IN JOISTS, RAFTERS, AND STUDS. THEY ARE CONSIDERED TO BE ADEQUATELY SUPPORTED, *334.30.*

CABLES OF ANY SIZE MAY BE RUN ON THE SIDES OF JOISTS. THEY MUST BE SECURED (STAPLED OR STRAPPED), *334.15(A), 334.30.*

THIS IS A RUNNING BOARD.

CABLES OF ANY SIZE MAY BE RUN PARALLEL TO SIDES OR FACE OF JOISTS. KEEP CABLES AT LEAST 1 1/4 IN. (32 mm) FROM EDGE OF FRAMING MEMBER, *300.4(D).*

CABLES SMALLER THAN TWO 6 OR THREE 8 AWG CONDUCTORS MUST BE RUN THROUGH BORED HOLES OR BE RUN ON THE SURFACE OF A RUNNING BOARD, *334.15(C).*

Figure 4-16   The requirements for running nonmetallic-sheathed cable in unfinished basements.

– there are no more than three current-carrying conductors in each cable
– there are no more than 20 current-carrying conductors in the bundle
– When there are more than 20 current-carrying conductors in the bundle, a 60% adjustment factor shall be applied. See *310.15(B)(2), Exception No. 5.*◄

• ►The cable may be run in unsupported lengths of not more than 4½ ft (1.4 m) between the last point of support of the cable and a luminaire (fixture) or other equipment within an accessible ceiling. For this last 4½ ft (1.4 m), the 12-in. (300-mm) and 4½-ft (1.4-m) securing requirements mentioned above are not necessary if the nonmetallic-sheathed cable is used as a luminaire (fixture) whip to connect luminaires (fixtures) of other equipment. See *334.30.*

Luminaire (fixture) whips are discussed in Unit 7 and Unit 17.◄

Figure 4-16 shows the installation of nonmetallic-sheathed cable in unfinished basements, *334.15(C).*

See Unit 15 and Figure 15-13 for additional text and diagrams covering the installation of cables in attics.

Figure 4-10 shows the Code requirements for securing nonmetallic-sheathed cable when using nonmetallic boxes.

The Code in *314.40(D)* requires that all metal boxes have provisions for the attachment of an equipment grounding conductor. Figure 4-17 shows a gang-type switch (device) box that is tapped for a screw by which the grounding conductor may be connected underneath. Figure 4-18 shows an outlet box, also with provisions for attaching grounding conductors. Figure 4-19 illustrates the use of a small grounding clip.

NEC® 410.20 requires that there be a provision whereby the equipment grounding conductor can be attached to the exposed metal parts of luminaires (lighting fixtures).

Alternate Method of Installing NMC Above Panel

Nonmetallic-sheathed cables are usually secured to a panel, box, or cabinet, with a cable clamp or connector. For surface-mounted enclosures only, *312.5(C), Exception* provides an alternate method

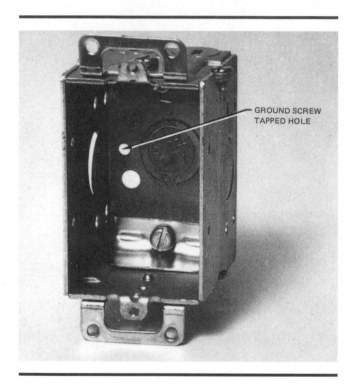

Figure 4-17   Gang-type switch (device) box.

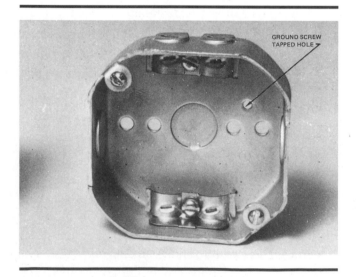

Figure 4-18   Outlet box.

for running nonmetallic-sheathed cables as illustrated in Figure 4-20.

Identifying Nonmetallic-Sheathed Cables

One nice thing about nonmetallic-sheathed cable, particularly that with a white outer jacket, is that at panelboards, load centers, and device and outlet boxes, you can use a permanent marking pen to mark the outer jacket with a few words indicating what the cable is for. Examples: ***FEED TO ATTIC; FEED TO L.R; FEED TO DINING ROOM RECEPTACLES; FEED TO PANEL; BC#6; TO OTHER 3-WAY SWITCH; SWITCH LEG TO CEILING LIGHT; TO OUTSIDE LIGHT; ETC.***

By marking the cables when they are installed, confusion when making the connections is reduced considerably.

Another great idea is to make a sleeve by cutting off a short 2-in. length of NM cable, pull out the conductors, then mark an identifying statement on the sleeve. Slide the sleeve over the specific conductors inside the panel or box.

Multistory Buildings

▶Nonmetallic-sheathed cable is permitted to be installed in one- and two-family dwellings, and in multifamily dwellings where the building construction is Type III, Type IV, or Type V. Refer to *NEC® 334.10(1)* and *(2).*◀ Construction types are defined in building Codes. These definitions are found in the glossary of this text.

▶For years, the *NEC®* limited nonmetallic-sheathed cable to a maximum of three floors above

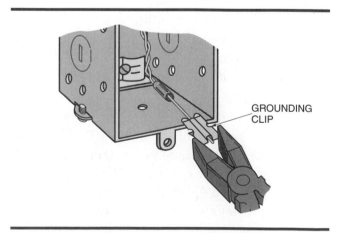

Figure 4-19   Method of attaching ground clip to a metal switch (device) box. See *250.148.*

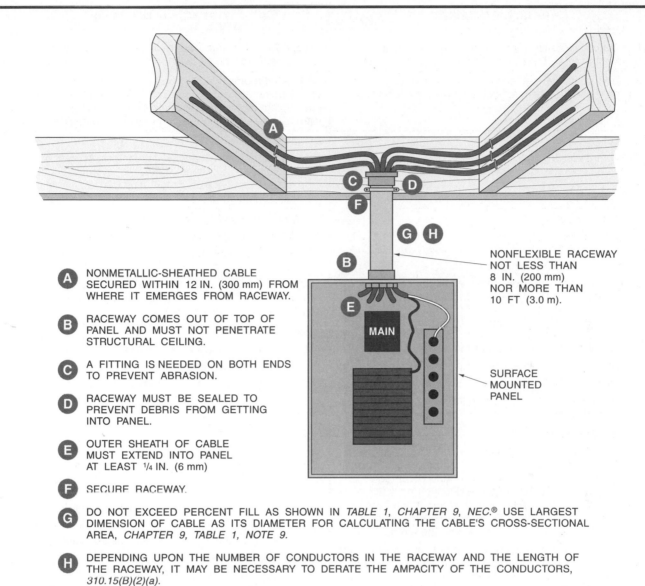

(A) NONMETALLIC-SHEATHED CABLE SECURED WITHIN 12 IN. (300 mm) FROM WHERE IT EMERGES FROM RACEWAY.

(B) RACEWAY COMES OUT OF TOP OF PANEL AND MUST NOT PENETRATE STRUCTURAL CEILING.

(C) A FITTING IS NEEDED ON BOTH ENDS TO PREVENT ABRASION.

(D) RACEWAY MUST BE SEALED TO PREVENT DEBRIS FROM GETTING INTO PANEL.

(E) OUTER SHEATH OF CABLE MUST EXTEND INTO PANEL AT LEAST 1/4 IN. (6 mm)

(F) SECURE RACEWAY.

(G) DO NOT EXCEED PERCENT FILL AS SHOWN IN *TABLE 1, CHAPTER 9, NEC.*® USE LARGEST DIMENSION OF CABLE AS ITS DIAMETER FOR CALCULATING THE CABLE'S CROSS-SECTIONAL AREA, *CHAPTER 9, TABLE 1, NOTE 9.*

(H) DEPENDING UPON THE NUMBER OF CONDUCTORS IN THE RACEWAY AND THE LENGTH OF THE RACEWAY, IT MAY BE NECESSARY TO DERATE THE AMPACITY OF THE CONDUCTORS, *310.15(B)(2)(a).*

NONFLEXIBLE RACEWAY NOT LESS THAN 8 IN. (200 mm) NOR MORE THAN 10 FT (3.0 m).

SURFACE MOUNTED PANEL

MAIN

Figure 4-20   An alternate method for running nonmetallic-sheathed cables into a surface-mounted panel. This is permitted for surface-mounted panels only, *312.50(C), Exception*.

finished grade. The *2002 NEC*® no longer limits the use of nonmetallic-sheathed cable based on the number of floors above finished grade for multifamily dwellings. See Figure 4-12. ◀

In previous editions, the *NEC*® has defined the first floor of a building as "that floor that has 50 percent or more of the exterior wall surface area level with or above finished grade."

ARMORED CABLE (TYPE AC) AND METAL-CLAD CABLE (TYPE MC)

Type AC and Type MC cables look very much the same (see Figure 4-21), yet there are significant differences. Table 4-12 compares some of these differences.

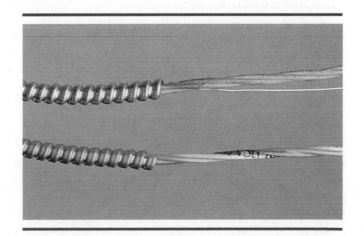

Figure 4-21   Flexible armored cable. *Courtesy of* AFC Cable Systems, Inc.

TYPE AC (*ARTICLE 320*)	TYPE MC (*ARTICLE 330*)
Article 320: This article contains all of the installation requirements for the installation of Type AC armored cable.	*Article 330:* This article contains all of the installation requirements for the installation of Type MC metal-clad cable.
320-1. **Definition:** Type AC cable is a fabricated assembly of insulated conductors in a flexible metallic enclosure.	*330.1.* **Definition:** Type MC cable is a factory assembly of one or more insulated circuit conductors with or without optical fiber members and enclosed in an armor of interlocking metal tape or a smooth or corrugated metallic sheath.
Number of Conductors: 2 to 4 current-carrying conductors plus bonding wire. It may also have a separate equipment grounding conductor.	**Number of Conductors:** Any number of current-carrying conductors. At least one manufacturer has a "Home Run Cable" that has conductors for more than one branch circuit, plus the equipment grounding conductors.
Conductor Size and Type: 14 AWG thru 1 AWG for copper conductors. 12 AWG thru 1 AWG for aluminum conductors will be marked "AL" on a tag on the carton or reel. 12 AWG thru 1 AWG for copper-clad aluminum conductors will be marked "*AL (CU-CLAD)*" or "*Cu-Clad Al*" on a tag on the carton or reel.	**Conductor Size and Type:** 18 AWG thru 2000 kcmils for copper conductors. 12 AWG thru 2000 kcmils for aluminum or copper-clad aluminum conductors. 12 AWG thru 2000 kcmils for copper-clad aluminum conductors will be marked "*AL (CU-CLAD)*" or "*Cu-Clad Al*" on a tag on the carton or reel.
Color Coding: Two-conductor: one black, one white. Three-conductor: one black, one white, one red. Four-conductor: one black, one white, one red, one blue. These are in addition to any equipment grounding and bonding conductors.	**Color Coding:** Two-conductor: one black, one white. Three-conductor: one black, one white, one red. Four-conductor: one black, one white, one red, one blue. These are in addition to any equipment grounding and bonding conductors.
Bonding & Grounding: Has a 16 AWG bare bonding wire. Bonding wire is in continuous direct contact with the metal armor. The bonding wire and armor act together to serve as the acceptable equipment ground. The bonding wire does not have to be terminated, just folded back over the armor. It can even be cut off where it comes out of the armor. Figure 4-22 shows how the bonding strip is folded back over the armor. Never use the bonding strip as a grounded neutral conductor or as a separate equipment grounding conductor. It is an internal bonding strip—nothing more!	**Bonding & Grounding:** Has a separate green insulated equipment grounding conductor. May have two equipment grounding conductors for isolated ground requirements, such as for computer wiring. Unless specifically listed for grounding, the jacket of MC cable is not to be used as an equipment grounding conductor. The sheath of smooth or corrugated tube Type MC is listed as acceptable as the required equipment grounding conductor. Has no bare bond wire as does Type AC.
Insulation: Type ACTHH has 194°F (90°C) thermoplastic insulation, by far the most common. Type ACTH has 167°F (75°C) thermoplastic insulation. Type AC has thermoset insulation. Type ACHH has 194°F (90°C) thermoset insulation.	**Insulation:** Type MC has 194°F (90°C) thermoplastic insulation. Some Type MC has thermoset insulation.
Conductors are individually wrapped with a flame-retardant fibrous cover: light brown Kraft paper. *Author's comment:* This is an easy way to distinguish Type AC from Type MC.	No individual wrap. Has a polyester (Mylar) tape over all conductors. Has no bare bond wire. *Author's comment:* These are easy ways to distinguish Type AC from Type MC.
Armor: Galvanized steel or aluminum. Armored cable with aluminum armor is marked *Aluminum Armor*.	**Armor:** Galvanized steel, aluminum, or bronze. Three types: *Interlocked* (*most common*). Requires a separate equipment grounding conductor. The armor by itself is not an acceptable equipment grounding conductor. *Smooth tube.* Does not require a separate equipment grounding conductor. *Corrugated tube.* Does not require a separate equipment grounding conductor.

Table 4-12   Comparisons of Type AC and Type MC cables.

TYPE AC (*ARTICLE 320*)	TYPE MC (*ARTICLE 330*)
Covering over armor: None	**Covering over armor:** Available with PVC outer covering.
Locations: OK for dry locations. Not OK for damp or wet locations. Not OK for direct burial.	**Locations:** OK for wet locations if so listed. OK for direct burial is so listed. OK to be installed in a raceway.
Insulating bushings: Required to protect conductor insulation from damage. Referred to as "anti-shorts," these keep any sharp metal edges of the armor from cutting the conductor insulation. Usually supplied in a bag with cable. Figure 4-22(A) shows an anti-short being inserted. Figure 4-22(B) shows the recommended way of bending the bonding strip over the antishort to hold the antishort in place.	**Insulating bushings:** Recommended but not required to protect conductor insulation from damage. Usually in a bag supplied with cable.
Minimum radius of bends: 5× diameter of cable.	**Minimum radius of bends:** Smooth Sheath: • 10× for sizes not over ¾ in. (19 mm) in diameter. • 12× for sizes over ¾ in. (19 mm) and not over 1½ in. (38 mm) in diameter. • 15× for sizes over 1½ in. (38 mm) in diameter. Corrugated or interlocked: • 7× diameter of cable.
Not available with "super neutral."	Available with "super neutral" when required for computer branch circuit wiring. For example, three 12 AWG and one 10 AWG neutral.
Not available with fiber optic cables.	Available with power conductors and fiber optic cables in one cable.
Support: Not more than 4½ ft (1.4 m) apart. Not more than 12 in. (300 mm) from box or fitting. Support not needed where cable is fished through walls and ceilings. Cables run horizontally through holes in studs, joists, rafters, etc., are considered supported by the framing members. ►Support not needed when used as a fixture whip within an accessible ceiling for lengths not over 6 ft (1.8 m) from last point of support. ◄	**Support:** Not more than 6 ft (1.8 m) apart. Not more than 12 in. (300 mm) from box or fitting for cables having four or fewer 10 AWG or smaller conductors. Support not needed where cable is fished through walls and ceilings. Cables run horizontally through holes in studs, joists, rafters, etc., are considered supported by the framing members. ►Support not needed when used as a fixture whip within an accessible ceiling for lengths not over 6 ft (1.8 m) from last point of support. ◄
Voltage: 600 volts or less	**Voltage:** 600 volts or less. Some MC has a voltage rating not to exceed 2000 volts.
Ampacity: Use the 60°C column of *Table 310.16*.	**Ampacity:** Use the 60°C column of *Table 310.16*. For 1/0 and larger conductors, OK to use the 75°C.
Thermal insulation: If buried in thermal insulation, the conductors must be rated 194°F (90°C).	**Thermal insulation:** No mention in NEC. Type MC has 194°F (90°C) thermoplastic insulation.
Connectors: Set screw-type connectors not permitted with aluminum armor. Listed connectors are suitable for grounding purposes.	**Connectors:** Set screw-type connectors not permitted with aluminum armor. Listed connectors are suitable for grounding purposes.
UL Standard 4. Because UL standards have evolved in numerical order, it is apparent that the armored cable standard is one of the first standards to be developed.	**UL Standard 1569.**

Table 4-12 (*continued*)

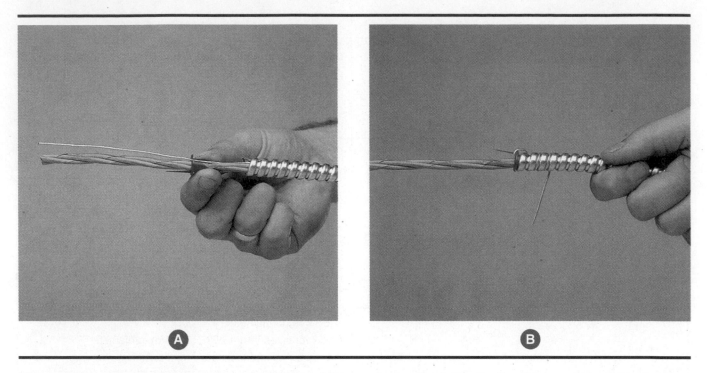

Ⓐ   Ⓑ

Figure 4-22 Antishort bushing prevents cutting of the conductor insulation by the sharp metal armor. *Courtesy of* **AFC Cable Systems, Inc.**

To remove the outer metal armor, a tool like that shown in Figure 4-23 may be used.

If a hacksaw is used, make an angle cut on one of the armor's convolutions as long as the conductors need to be, perhaps 10 to 12 in. Be careful not to cut too deep or you might cut into the conductor insulation. Bend the armor at the cut and snap it off. Then slide the 10–12 in. of armor off the end of the cable. See Figure 4-24.

Most electricians, when using armored cable, will make three cuts with their hacksaw. The middle cut is where the cable and conductors will be cut off

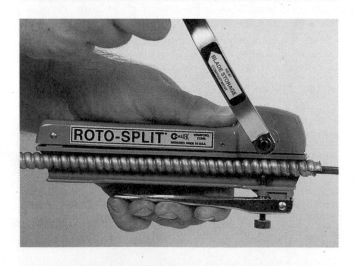

Figure 4-23 This is an example of a tool that precisely cuts the outer armor of armored cable, making it easy to remove the armor with a few turns of the handle. *Courtesy of* **Seatek Co., Inc.**

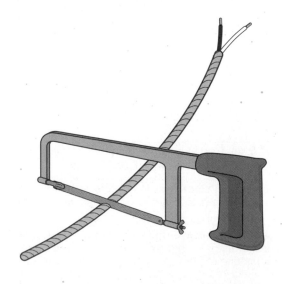

Figure 4-24 Using a hacksaw to cut through a raised convolution of the cable armor.

USES & INSTALLATION IN DWELLINGS	Type AC	Type MC
a) Branch-circuits & feeders	Yes	Yes
b) Services	No	Yes
c) Exposed & concealed work in:		
Dry locations	Yes	Yes
Wet locations	No	Yes (See *300.10(A)(12)* for specifics
d) Where not subject to physical damage	Yes	Yes
e) May be embedded in plaster finish on masonry walls or run through the hollow spaces of such walls if these locations are not considered damp or wet.	Yes	Yes
f) Run or fished in air voids of masonry block or tile walls where not exposed to excessive moisture or dampness. Masonry in direct contact with earth is considered to be a wet location.	Yes	Yes
g) In any raceway	No	Yes
h) Direct burial	No	Yes, if identified for such use.
i) In accessible attics or roof spaces, must be protected by guard strips at least as high as the cable when run across top of floor joists, or within 7 ft (2.1 m) of the floor or floor joists when the cable is run across the face of studs or rafters. If there is no permanent stairway or ladder, protection is needed only within 6 ft (1.8 m). from edge of scuttle hole or entrance to attic. No guard strips or running boards are necessary for cables run parallel to the sides of joists, rafters, or studs. See Unit 15 for additional discussion and diagrams regarding physical protection of cables in attics.	Yes	Yes

Table 4-13 Uses and installation in dwellings (Type AC and Type MC).

completely. The first and third cuts are where the armor will be snapped off. This saves time because it actually prepares two ends of the cable.

Most electricians still refer to armored cable as BX, supposedly derived from an abbreviation of the Bronx in New York City, where armored cable was once manufactured. BX was a trademark owned by the General Electric Company, but over the years it has become a generic term.

Figure 4-21 shows armored cable with the armor removed to show the conductors that are wrapped with paper and the bare bonding wire.

Table 4-13 shows the uses and installation requirements for Type AC and Type MC cables for residential wiring.

INSTALLING CABLES THROUGH WOOD AND METAL FRAMING MEMBERS *(300.4)*

To complete the wiring of a residence, cables must be run through studs, joists, and rafters. One method of installing cables in these locations is to run the cables through holes drilled at the approximate centers of wood building members, or at least 1¼ in. (32 mm) from the nearest edge. Holes bored through the centers of standard 2 × 4s meet the requirements of *300.4(A)(1)*. Refer to Figure 4-25.

If the 1¼-in. (32-mm) distance cannot be maintained, or if the cables are to be laid in a notch, a metal plate at least ¹⁄₁₆ in. (1.6 mm) thick must cover the notch to protect the cable from nails, *300.4(A)(1)*

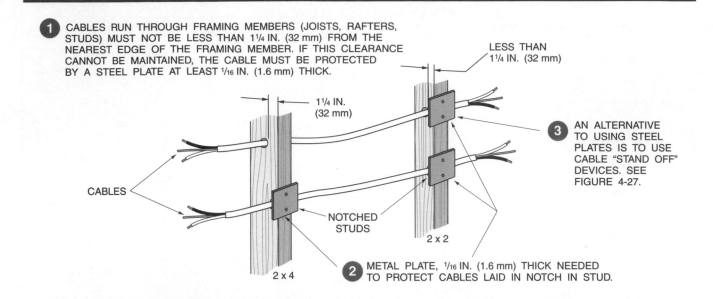

1 CABLES RUN THROUGH FRAMING MEMBERS (JOISTS, RAFTERS, STUDS) MUST NOT BE LESS THAN 1¼ IN. (32 mm) FROM THE NEAREST EDGE OF THE FRAMING MEMBER. IF THIS CLEARANCE CANNOT BE MAINTAINED, THE CABLE MUST BE PROTECTED BY A STEEL PLATE AT LEAST ¹⁄₁₆ IN. (1.6 mm) THICK.

LESS THAN 1¼ IN. (32 mm)

1¼ IN. (32 mm)

CABLES

3 AN ALTERNATIVE TO USING STEEL PLATES IS TO USE CABLE "STAND OFF" DEVICES. SEE FIGURE 4-27.

NOTCHED STUDS

2 x 2

2 x 4

2 METAL PLATE, ¹⁄₁₆ IN. (1.6 mm) THICK NEEDED TO PROTECT CABLES LAID IN NOTCH IN STUD.

Figure 4-25 **Methods of protecting nonmetallic and armored cables so nails and screws that "miss" the studs will not damage the cables, *300.4(A)(1)* and *(2)*. Because of their strength, intermediate and rigid metal conduit, rigid nonmetallic conduit, and electrical metallic tubing are exempt from this rule per the *Exceptions* found in *300.4*. See Figure 4-27 for alternative methods of protecting cables.**

and *(2)*. Metal plates for this purpose are commercially available at electrical supply houses, home centers, and hardware stores. Refer to Figure 4-25.

Metal studs are being used more and more. They don't burn and termites don't eat them. They are stronger than wood, and can be set 20 in. (500 mm) on center instead of the usual 16-in. (400-mm) spacing. Metal studs are 10 times more conductive than wood, so treatment of insulation must be addressed.

▶Where nonmetallic-sheathed cables pass through holes or slots in metal framing members, the cable must be protected by a listed bushing or listed grommet securely fastened in place prior to installing the cable. The bushing or grommet shall:

• remain in place during the wall finishing process,
• cover the complete opening, and
• be listed for the purpose of cable protection.

This information is found in *300.4(B)(1)* and *334.17*.◀ See Figure 4-26.

Bushings that snap into a hole sized for the specific size bushing (½-in., ¾-in., etc.), or a two-piece snap-in bushing that fits just about any size precut hole in the metal framing members are acceptable to use, as shown in Figure 4-26. These types of bushings (insulators) are not required in metal framing

members when the wiring method is armored cable, EMT, FMC, or electrical nonmetallic tubing.

To reduce the possibility of nails being driven into electrical cables and certain types of raceways, Figure 4-27 illustrates some devices that can be used to *stand off* the cables from a framing member.

When running nonmetallic-sheathed cable or electrical nonmetallic tubing through metal framing members where there is a likelihood that nails or screws could be driven into the cables or tubing, a steel plate, steel sleeve, or steel clip at least ¹⁄₁₆-in. (1.6 mm) thick must be installed to protect the cables and/or tubing. This is spelled out in *300.4(B)(2)*. See Figure 4-26 and Figure 4-28.

Where cables and certain types of raceways are run concealed or exposed, parallel to a stud, joist, or rafter (any building framing member), the cables and raceways must be supported and installed in such a manner that there is not less than 1¼ in. (32 mm) from the edge of the stud or joist (framing member) to the cable or raceway. If the 1¼-in. (32-mm) clearance cannot be maintained, then a steel plate, sleeve, or equivalent that is at least ¹⁄₁₆ in. (1.6 mm) thick must be provided to protect the cable or raceway from damage that can occur when nails or screws are driven into the wall or ceiling, *300.4(A)(1)*, see Figure 4-25 and Figure 4-28.

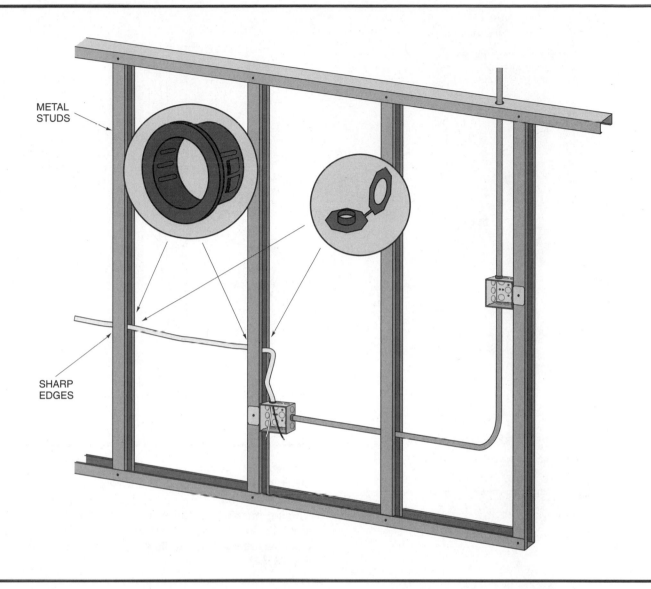

Figure 4-26 *NEC® 300.4(B)* and *334.17* **requires protection of nonmetallic-sheathed cable where run through holes in metal framing members. Shown are two types of listed bushings (grommets) that can be installed to protect nonmetallic-sheathed cable from abrasion. Electricians generally avoid using nonmetallic-sheathed cable through metal studs and joists. They use electrical metallic tubing, flexible metal conduit, armored cable (Type AC or MC), or electrical nonmetallic tubing. Where there is a likelihood that nails or screws might penetrate the nonmetallic-sheathed cable, ¹⁄₁₆-in. (1.6 mm) thick steel plates, sleeves, or clips must be installed to protect the cable or electrical nonmetallic tubing.**

NEC® 300.4(E) addresses the wiring problems associated with cutting grooves into building material because there is no hollow space for the wiring. Nonmetallic-sheathed cable and some raceways must be protected when laid in a groove. Protection must be provided with a ¹⁄₁₆-in. (1.6-mm) steel plate, sleeve, or equivalent, or the groove must be deep enough to allow not less than 1¹⁄₄-in. (32-mm) free space for the entire length of the groove. This situation can be encountered where styrofoam insulation

building blocks are grooved to receive the electrical cables, then covered with wallboard, wood paneling, or other finished wall material. Another typical example is where solid wood planking is installed on top of wood beams. The top side of the planking is covered with roofing material, and the bottom side is exposed and serves as the finished ceiling. The only way to install wiring for ceiling luminaires (fixtures) and fans is to groove the planking in some manner, as shown in Figure 4-29.

Figure 4-27 Devices that can be used to meet the Code requirement to maintain a 1¼-in. (32 mm) clearance from framing members.

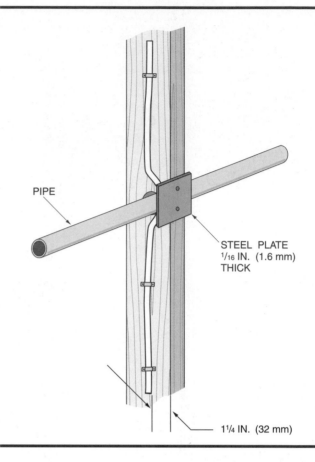

PIPE

STEEL PLATE
1/16 IN. (1.6 mm)
THICK

1¼ IN. (32 mm)

Figure 4-28 Where it is necessary to cross pipes or other similar obstructions, making it impossible to maintain at least 1¼-in. (32-mm) clearance from the framing member to the cable, install a steel plate at least 1/16 in. (1.6 mm) thick to protect the cable from possible damage from nails or screws. See *300.4(A)(1)* and *(A)(2)*.

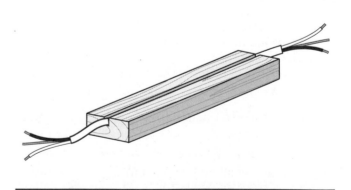

Figure 4-29 Example of how a solid wood framing member or styrofoam insulating block might be grooved to receive the nonmetallic-sheathed cable. After the cable is laid in the groove, a steel plate not less than 1/16 in. (1.6 mm) thick must be installed over the groove to protect the cable from damage, such as from driven nails or screws. See *300.4(E)*.

This additional protection is not required when the raceway is intermediate metal conduit *(Article 342)*, rigid metal conduit *(Article 344)*, rigid non-metallic conduit *(Article 352)*, or electrical metallic tubing *(Article 358)* (see Figure 4-26).

Additional protection obviously cannot be installed inside the walls or ceilings of an existing building where the walls and ceilings are already closed in.

NEC® 334.30(B)(1), and *300.4(D), Exception 2* allow cables to be "fished" between outlet boxes or other access points without additional protection.

In the recreation room of the residence plans, precautions must be taken so that there is adequate protection for the wiring. The walls in the recreation room are to be paneled. Therefore, if the carpenter uses 1×2 or 2×2 furring strips, as in Figure 4-30, nonmetallic-sheathed cable would require the addi-

tional 1/16-in. (1.6 mm) steel plate protection or the use of *stand off* devices similar to those illustrated in Figure 4-27.

Some building contractors attach 1-in. insulation to the walls, then construct a 2×4 wall in front of the basement structural foundation wall, leaving 1-in. spacing between the insulation and the back side of the 2×4 studs. This makes it easy to run cables or conduits behind the 2×4 studs, and eliminates the need for the additional mechanical protection, as shown in Figure 4-31.

Watch out for any wall partitions where 2×4 studs are installed "flat," as in Figure 4-32. In this case, the choice is to provide the required mechanical protection as required by *300.4*, or to install the wiring in EMT.

See Unit 15 concerning the installation of cables in attics.

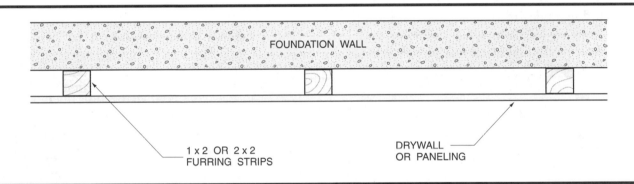

Figure 4-30   When 1×2 or 2×2 furring strips are installed on the surface of a basement wall, additional protection, as shown in Figure 4-25, Figure 4-26, Figure 4-27, and Figure 4-29, must be provided.

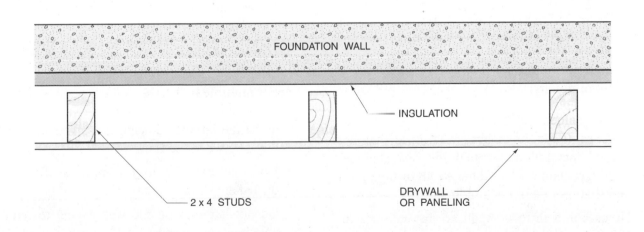

Figure 4-31   This sketch shows how to insulate a foundation wall and how to construct a wall that provides space to run cables and/or raceways that will not require the additional protection as shown in Figure 4-25, Figure 4-26, Figure 4-27, and Figure 4-29.

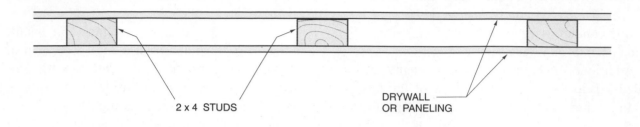

Figure 4-32 When 2 × 4 studs are installed "flat," such as might be found in nonbearing partitions, cables running parallel to or through the studs must be protected against the possibility of having nails or screws driven through the cables. This means protecting the cable with 1/16-in. (1.6-mm) steel plates or equivalent for the entire length of the cable, or installing the wiring using intermediate metal conduit, rigid metal conduit, rigid nonmetallic conduit, or EMT. See Figure 4-25, Figure 4-26, Figure 4-27, and Figure 4-28.

Building Codes

When complying with the *National Electrical Code,*® be sure to also comply with the building code. Building codes are concerned with the weakening of framing members when drilled or notched. Meeting the requirements of the *NEC*® can be in conflict with building codes and vice versa. Here is a recap of some of the key building code requirements found in the ICC *International Residential Code.*

Studs:

* Cuts or notches in exterior or bearing partitions shall not exceed 25 percent of the stud's width.

* Cuts or notches in nonbearing partitions shall not exceed 40 percent of the stud's width.

* Bored or drilled holes in any stud:
  - shall not be greater than 40 percent of the stud's width, and
  - shall not be closer than 5/16 in. (16 mm) to the edge of the stud, and
  - shall not be located in the same section as a cut or notch.

Joists:

* Notches on the top or bottom of joists shall not exceed one-sixth the depth of the joist, shall not be longer than one-third the width of the joists, and shall not be in the middle third of the joist.

* Holes bored in joists shall not be within 2 in. (50 mm) of the top or bottom of the joist or any other hole, and the diameter of any such hole shall not exceed one-third the depth of the joist. If the joist is also notched, the hole shall not be closer than 2 in. (50 mm) to the notch.

Engineered Wood Products

Today, we find many homes constructed with engineered wood I-joists and beams. The strength of these factory-manufactured framing members conform to various building codes. It is critical that certain precautions be taken relative to drilling and cutting these products so as not to weaken them. In general,

* Do not cut, notch, or drill holes in trusses, laminated veneer lumber, glue-laminated members, or I-joists unless specifically permitted by the manufacturer.

* Do not cut, drill, or notch flanges. Flanges are the top and bottom pieces.

* Do not cut holes too close to supports or to other holes.

* For multiple holes, the amount of web to be left between holes must be at least twice the diameter of the largest adjacent hole. The web is the piece between the flanges.

* Holes not over 1½ in. (38 mm) usually can be drilled anywhere in the web.

* Do not hammer on the web except to remove knockout prescored holes.

* Check and follow the manufacturer's instructions!

INSTALLATION OF CABLES THROUGH DUCTS

NEC® 300.22 is extremely strict as to what types of wiring methods are permitted for installation of cables through ducts or plenum chambers. These stringent rules are for fire safety.

The Code requirements are somewhat relaxed by *300.22(C), Exception*. In this exception, permission is given to run nonmetallic-sheathed and armored cables in joist and stud spaces (i.e., cold air returns), but only if they are run perpendicular to the long dimensions of such spaces, as illustrated in Figure 4-33.

CONNECTORS FOR INSTALLING NONMETALLIC-SHEATHED AND ARMORED CABLE

The connectors shown in Figure 4-34 are used to fasten nonmetallic-sheathed cable and armored cable to the boxes and panels in which they terminate. These connectors clamp the cable securely to each outlet box. Many boxes have built-in clamps and do not require separate connectors.

The question continues to be asked: May more than one cable be inserted in one connector? Unless the UL listing of a specific cable connector indicates that the connector has been tested for use with more

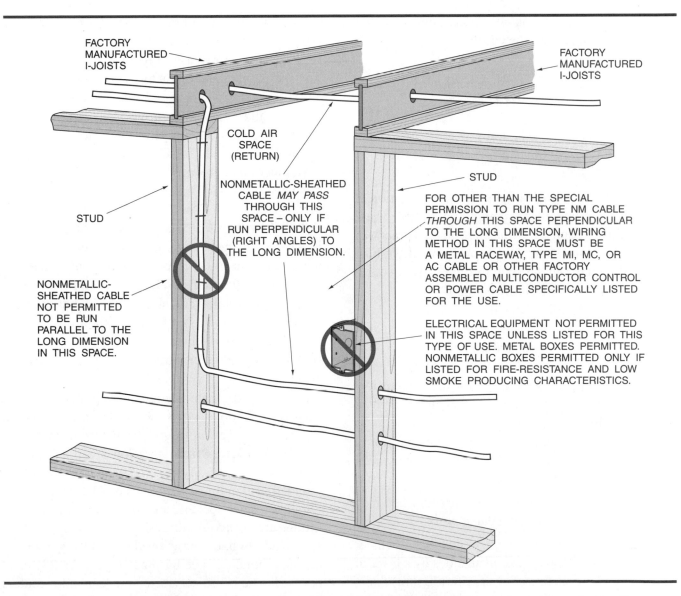

Figure 4-33 Nonmetallic-sheathed cable may pass through cold air return, joist, or stud spaces in dwellings, but only if it is run at right angles to the long dimension of the space, *300.22(C), Exception.*

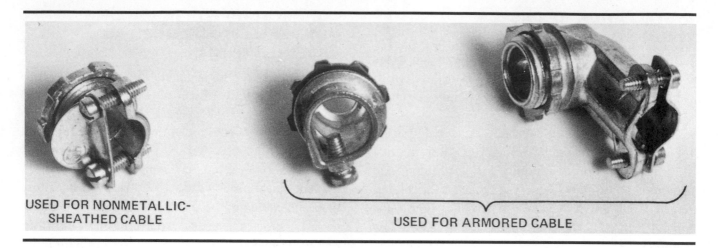

USED FOR NONMETALLIC-SHEATHED CABLE

USED FOR ARMORED CABLE

Figure 4-34 Cable connectors.

than one cable, the rule is *one cable, one connector*. This will be marked on the carton.

INTERMEDIATE METAL CONDUIT *(ARTICLE 342)*, RIGID METAL CONDUIT *(ARTICLE 344)*, RIGID NONMETALLIC CONDUIT *(ARTICLE 352)*, AND ELECTRICAL METALLIC TUBING *(ARTICLE 358)*

Some communities do not permit the installation of cable of any type in residential buildings. These communities require the installation of a raceway system of wiring, such as electrical metallic tubing (thinwall), intermediate metal conduit, rigid metal conduit, or rigid nonmetallic conduit.

Where the building construction is cement block, cinderblock, or poured concrete, it will be necessary to make the electrical installation in conduit.

Sufficient data is given in this text for both cable and conduit wiring methods. Thus, the student will be able to complete the type of installation permitted or required by local codes.

According to installation requirements for a raceway system of wiring, EMT, intermediate metal conduit, and rigid metal conduit

- may be buried in concrete or masonry and may be used for open or concealed work.

- may *not* be installed in cinder concrete or cinder fill unless (1) protected on all sides by a layer of noncinder concrete at least 2 in. (50 mm) thick, or (2) the conduit is at least 18 in. (450 mm) below the fill.

In general, conduit must be supported within 3 ft (900 mm) of each box or fitting and at 10-ft (3.0-m) intervals along runs, as in Figure 4-35.

The number of conductors permitted in conduit and tubing is found in *Chapter 9* of the Code. Heavy-wall rigid conduit provides greater protection against mechanical injury to conductors than does either EMT or intermediate metal conduit, which have much thinner walls.

The residence specifications indicate that a meter pedestal is to be furnished by the utility and installed by the electrical contractor. Main Panel A is located in the workshop. The meter pedestal is located on the outside of the house near the air-conditioning unit. The electrical contractor must furnish and install trade size 1½ EMT between the meter pedestal and Main Panel A.

NEC® *300.4(F)* and *312.6(C)* require that smoothly rounded insulating fittings or equivalent be used where 4 AWG or larger conductors enter a raceway, box, cabinet, auxiliary gutter, or other enclosure. Various types of fittings are shown in Figure 4-36. Examples of insulated fittings that protect conductors from abrasion are illustrated in Figure 27-28 and Figure 27-29. Where smoothly rounded or flared threaded hubs and bosses, such as found on meter sockets, are encountered, insulating fittings are not required, as shown in Figure 27-16.

In addition to these insulating fitting requirements, a bonding type bushing (Figure 27-28) must be used where the service-entrance conduit enters Main Panel A. Unit 27 covers grounding and bonding of service equipment.

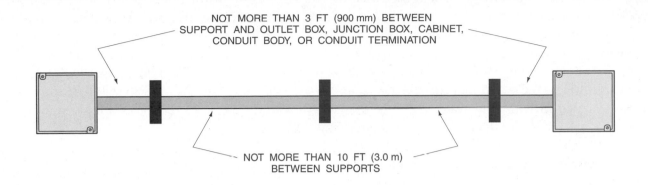

NOT MORE THAN 3 FT (900 mm) BETWEEN
SUPPORT AND OUTLET BOX, JUNCTION BOX, CABINET,
CONDUIT BODY, OR CONDUIT TERMINATION

NOT MORE THAN 10 FT (3.0 m)
BETWEEN SUPPORTS

Figure 4-35 Code requirements for securing and supporting raceways are found in *342.30(A)* and *(B)* (intermediate metal conduit), *344.30(A)* and *(B)* (rigid metal conduit), *352.30(A)* and *(B)* (rigid nonmetallic conduit), and *358.30(A)* and *(B)* (electrical metallic tubing). There are exceptions to the measurements indicated in the above illustration where raceways are "fished" through finished or concealed partitions or where structural members do not readily permit fastening according.

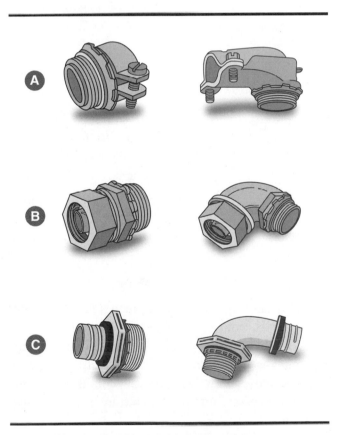

Figure 4-36 Various fittings for (A) flexible metal conduit, (B) liquidtight flexible metal conduit, and (C) liquidtight flexible nonmetallic conduit.

The plans also indicate that EMT is to be used in the workshop. Because the recreation room and basement stairwell have finished ceilings and walls, they will be wired with cable.

Many of the *NEC*® rules are applicable to both cable and conduit installations. Some Code requirements are applicable only to cable installations. For example, there are special requirements for the attachment of the grounding conductors and when protection from physical damage must be provided. Many of the Code requirements governing the installation of cables do not apply to a raceway system. Properly installed EMT provides good continuity for equipment grounding, a means of withdrawing or pulling in additional conductors, and excellent protection against physical damage to the conductors.

RIGID NONMETALLIC CONDUIT (RNC) *(ARTICLE 352)*

Rigid nonmetallic conduit (RNC) generally is made of polyvinyl chloride (PVC), and is covered in *Article 352* of the *NEC*.®

RNC comes in 10-ft (3.0-m) lengths. It comes in two types:

- Schedule 40 (heavy wall thickness)

- Schedule 80 (extra-heavy wall thickness)

The outside diameter is the same for both Schedule 40 and Schedule 80. The inside diameter of Schedule 40 is greater than that of Schedule 80.

Schedule 40 has a thinner wall than Schedule 80 and is used where it is not subject to physical damage. Schedule 80 is stronger and is used where exposed to physical damage.

The fittings used with RNC are attached with a solvent-type cement, much the same as what plumbers use. Plumbing PVC pipe may look the same as electrical PVC—but they are not the same. One is designed to withstand water pressure from the inside—the other is designed to withstand forces from the outside.

The applications where RNC are permitted are found in *Article 352*. For residential use, here are some of the key requirements for RNC.

- RNC may be used concealed in walls, ceilings, and floors.

- RNC may be used in wet locations, such as outdoors.

- RNC may be used underground.

- RNC may not be used to support luminaires (lighting fixtures).

- RNC may be used to support nonmetallic conduit bodies.

- If used to support nonmetallic conduit bodies, the conduit body must not contain wiring devices or support luminaires (lighting fixtures).

- RNC is nonmetallic, therefore, an equipment grounding conductor may be needed.

- RNC shall not have more than the equivalent of four 90° bends between pull points. Simply stated, the sum of all bends and offsets must not exceed 360°.

- RNC must be supported within 3 ft (900 mm) of each outlet box, junction box, device box, conduit body, or other termination.

- RNC must be supported at intervals as shown in Table 4-14.

- RNC must have expansion fittings on long runs,

352.44. Instructions furnished with the RNC will provide the necessary details. *Table 10* in *Chapter 9* of the *NEC®* has dimensional requirements for anticipated expansion. Long runs of RNC in direct sunlight will pull out of the hubs of boxes or will pull the boxes off the wall if proper attention is not given to thermal expansion.

- When all conductors are the same size and type, the number of conductors permitted in RNC Schedule 80 is found in *Table C9* in *Annex C* of the *NEC.*®

- When conductors are different sizes, refer to Chapter 9 in the *NEC.*® The number of conductors permitted is determined using the percentage fill shown in *Table 1*, the dimensions of RNC found in *Table 4*, and the dimensions of the conductors as shown in *Table 5*, *Table 5A*, and *Table 8*.

- Conductors emerging from RNC must be protected from abrasion, either by the rounded edges of the box, fitting, or other enclosure.

- *NEC®* 352.24 refers back to *Table 344.24* for the minimum radius for bends on RNC.

- Bends can be made with a proper bender that electrically heats the PVC to just the right temperature for bending. Manufactured 90°, 45°, and 30° elbows are available.

FLEXIBLE CONNECTIONS

The installation of certain equipment requires flexible connections, both to simplify the installation and to stop the transfer of vibrations. In residential wiring, flexible connections are used to wire attic fans, food waste disposers, dishwashers, air conditioners, heat pumps, recessed luminaires (fixtures), and similar equipment.

Figure 4-36 shows many of the types of connectors used with flexible metal conduit, flexible liquidtight metal conduit, and flexible liquidtight nonmetallic conduit. The three types of flexible conduit used for these connections are: flexible metal conduit, flexible liquidtight metal conduit, and flexible liquidtight nonmetallic conduit, as shown in Figure 4-37.

RIGID NONMETALLIC CONDUIT TRADE SIZE	MAXIMUM SPACING BETWEEN SUPPORTS
½, ¾, 1	3 ft (900 mm)
1¼, 1½, 2	5 ft (1.5 m)

Table 4-14   Maximum distance between supports for rigid nonmetallic conduit.

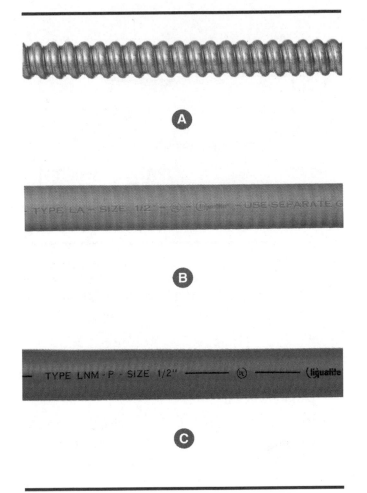

Figure 4-37 (A) Flexible metal conduit. (B) Flexible liquid-tight metal conduit. (C) Flexible liquidtight non-metallic conduit. *Courtesy* **of Electro-Flex Co.**

FLEXIBLE METAL CONDUIT (FMC)
(ARTICLE 348)

Article 348 of the Code covers the use and installation of flexible metal conduit (FMC). This wiring method is similar to armored cable, except that the conductors are installed by the electrician. For armored cable, the cable armor is wrapped around the conductors at the factory to form a complete cable assembly. FMC is often referred to as *Greenfield*, named after Harry Greenfield, the individual who submitted the product for listing back in 1902.

Common installations using FMC are shown in Figure 4-38. Note that the flexibility required to make the installation is provided by the FMC. Figure 4-38 calls attention to the *NEC*® and UL restrictions on the use of flexible metal conduit with

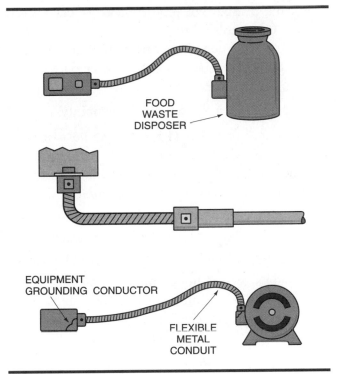

Figure 4-38 Some of the more common places where flexible metal conduit (FMC) may be used.

regard to relying on the metal armor as a grounding means.

The following are some of the common uses, limitations, and applications for FMC, combining the material found in *Article 348* of the Code and in the UL White Book:

• Do not install in wet locations unless the conductors are suitable for use in wet locations (TW, THW, THWN), and the installation is made so water will not enter the enclosure or other raceways to which the flex is connected.

• Do not bury in concrete.

• Do not bury underground.

• Do not use in locations subject to corrosive conditions.

• FMC is acceptable as an equipment grounding conductor if:
 – the FMC is listed for grounding.
 – the fittings are listed for grounding.

• if the FMC is not listed for grounding, then all of the following conditions must be met:
 – the fittings must be listed for grounding.

– the overcurrent protection must not exceed 20 amperes.
– the total length of the return ground path through the flex must not exceed 6 ft (1.8 m). See Figure 4-39.

• the FMC is not installed for flexibility.

• If longer than 6 ft (1.8 m), regardless of trade size, FMC is not acceptable as grounding means. You must install a separate equipment grounding conductor sized per *Table 250.122*.

• Fittings larger than trade size ¾ that have been listed as acceptable for grounding means will be marked "GRND" or some equivalent marking.

• Trade size ⅜ is not permitted to be longer than 6 ft (1.8 m). Luminaire (fixture) whips are a good example. See "Recreation Room," Unit 17, for a discussion of luminaire (fixture) whips.

• FMC must be supported every 4½ ft (1.4 m).

• FMC must be supported within 12 in. (300 mm) of box, cabinet, or fitting.

Note: The 4½ ft (1.4 m) and the 12 in. (300 mm) requirements are waived when the FMC is fished through walls and ceilings, where flexibility is needed and the length of the flex is not over 3 ft (900 mm), or for luminaire (fixture) whips where the length of the flex is not over 6 ft (1.8 m).

• Horizontal runs through holes, supported by framing members not more than 4½ ft (1.4 m) apart and secured within 12 in. (300 mm) of a box, cabinet, fitting, or enclosure are considered to be adequately supported.

• If flexibility is needed, support within 3 ft (900 mm) of termination.

• If the FMC is not acceptable as a grounding means, then install a bonding jumper inside or outside the FMC. If run outside of the FMC, the bonding jumper must not be longer than 6 ft (1.8 m) and it must be routed with FMC, *250.102(E)*. Size the bonding jumper per *Table 250.122*. For mechanical protection and to keep the impedance of the circuit to a minimum, it is best to run the bonding jumper inside the FMC.

• Do not have more than four quarter bends (360°) between boxes.

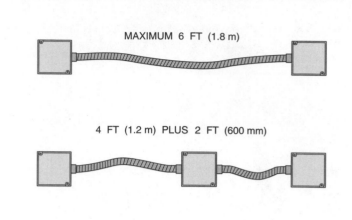

MAXIMUM 6 FT (1.8 m)

4 FT (1.2 m) PLUS 2 FT (600 mm)

Figure 4-39   If a flexible metal conduit is serving as an equipment ground return path, the total length shall not exceed 6 ft (1.8 m). Refer to *250.118(6)c, 250.118(7)d, 250.118(8)b*.

• Do not conceal angle-type fittings because of the difficulty that would present itself when pulling wires in.

• If flexibility is needed, always install a separate equipment grounding conductor, regardless of trade size. Size equipment grounding conductors per *Table 250.122*.

• FMC may be used for exposed or concealed installations.

• Calculate conductor fill same as regular conduit. See *Chapter 9, NEC,® Table 1* and *Table 4*.

• Calculate conductor fill for trade size ⅜, *Table 348.12*.

• Fittings used with FMC must be listed.

• Be sure to remove rough edges at "end-cuts" unless fittings are used that thread into the convolutions of the flex.

See Figure 4-40 when the installation has combined lengths of flex.

NEC® 250.118(5), (6), and *(7)* also discuss flexible metal conduit installations.

LIQUIDTIGHT FLEXIBLE METAL CONDUIT (LFMC) *(ARTICLE 350)*

The use and installation of liquidtight flexible metal conduit (LFMC) are described in *Article 350* of the Code. LFMC has a "tighter" fit of its spiral turns as compared to standard flexible metal conduit. It also

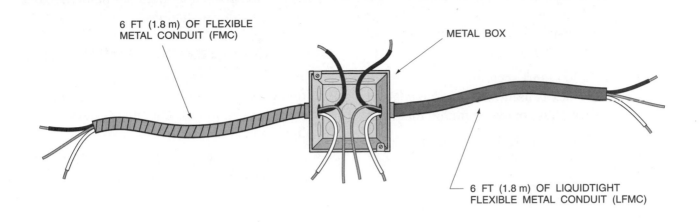

6 FT (1.8 m) OF FLEXIBLE
METAL CONDUIT (FMC)

METAL BOX

6 FT (1.8 m) OF LIQUIDTIGHT
FLEXIBLE METAL CONDUIT (LFMC)

Figure 4-40 The "combined" length of flexible metal conduit, liquidtight flexible metal conduit, and flexible metal tubing shall not exceed 6 ft (1.8 m) if it is to serve as the same ground return path, *250.118(6)(c), (7)(d),* and *(8)(b).* In the above illustration, a separate equipment grounding conductor would have to be provided.

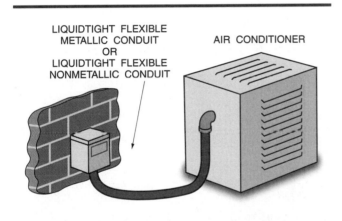

LIQUIDTIGHT FLEXIBLE
METALLIC CONDUIT
OR
LIQUIDTIGHT FLEXIBLE
NONMETALLIC CONDUIT

AIR CONDITIONER

Figure 4-41 Use of liquidtight flexible metal conduit or flexible liquidtight nonmetallic conduit.

has a thermoplastic sunlight-resistant outer jacket that is liquidtight. Liquidtight flexible metal conduit is commonly used as the flexible connection to a central air-conditioning unit located outdoors, Figure 4-41.

Figure 4-42 shows the Code rules for the use of liquidtight flexible metal conduit as a grounding means.

Listed here are some of the common uses, limitations, and applications for LFMC, combining *NEC®* rules and UL White Book requirements:

- LFMC may be used for exposed and concealed installations.

- LFMC may be buried directly in the ground if so listed and marked.

- LFMC may not be used where subject to physical damages.

- LFMC may not be used if ambient temperature and heat from conductors will exceed the temperature limitation of the nonmetallic outer jacket. UL listing will indicate maximum temperature.

- LFMC must not be used in trade sizes smaller than ½ inch except that ⅜-inch trade size is acceptable for enclosing the leads to a motor, or as a luminaire (fixture) whip.

- For conductor fill, see *Chapter 9, NEC® Table 1* and *Table 4.*

- To calculate conductor fill for trade size ⅜, see *Table 348.12.*

- The LFMC flex must be listed.

- All fittings must be listed.

- LFMC must be supported every 4½ ft (1.4 m).

- LFMC must be supported within 12 in. (300 mm) of a box, cabinet, fitting, or enclosure.

- Horizontal runs through holes, supported by framing members not more than 4½ ft (1.4 m) apart and secured within 12 in. (300 mm) of a box, cabinet, fitting, or enclosure are considered to be adequately supported.

- The 4½-ft (1.4-m) and 12-in. (300-mm) securing requirements are waived if the LFMC is used as a luminaire (fixture) whip whose length is not over 6 ft (1.8 m).

- The 12-in. (300-mm) securing requirement is waived if the flex is used for a flexible connection whose length is not over 3 ft (900 mm).

- The 4½-ft (1.4-m) and 12-in. (300-mm) securing requirements are waived if the LFMC is fished.

- If flexibility is needed, support within 3 ft (900 mm) of termination.

- See Figure 4-40 and Figure 4-42 for requirements if using LFMC for grounding purposes.

- If flexibility is needed, always install a separate equipment grounding conductor regardless of the trade size of the LFMC. Size equipment grounding conductor per *Table 250.122*.

- Do not conceal angle-type fittings.

- Do not have more than four quarter bends (360°) between boxes.

NEC® 250.118(7) also discusses liquidtight flexible metal conduit installations.

LIQUIDTIGHT FLEXIBLE NONMETALLIC CONDUIT (*ARTICLE 356*)

The following are some of the important Code and UL rules pertaining to liquidtight flexible nonmetallic conduit (LFNC):

- Do not use in direct sunlight unless specifically marked for use in direct sunlight.

- LFNC may be used in exposed or concealed installations.

- LFNC can become brittle in extremely cold applications.

- LFNC may be used outdoors when listed and marked as suitable for this application.

- LFNC may be buried directly in the earth when listed and marked as suitable for this application.

- LFNC may not be used where subject to physical abuse.

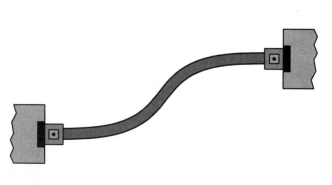

A PERMITTED AS A GROUNDING MEANS IF IT IS NOT OVER TRADE SIZE 1¼, IS NOT OVER 6 FT (1.8 m) LONG, AND IS CONNECTED BY FITTINGS LISTED FOR GROUNDING. THE 6 FT (1.8 m) LENGTH INCLUDES THE TOTAL LENGTH OF ANY AND ALL FLEXIBLE CONNECTIONS IN THE RUN. SEE FIGURE 4-40.

B MINIMUM TRADE SIZE ½. TRADE SIZE ⅜ PERMITTED AS A LUMINAIRE (FIXTURE) WHIP NOT OVER 6 FT (1.8 m) LONG.

C WHEN USED AS THE GROUNDING MEANS, THE MAXIMUM OVERCURRENT DEVICE IS 20 AMPERES FOR TRADE SIZES ⅜ AND ½, AND 60 AMPERES FOR TRADE SIZES ¾, 1, AND 1¼.

D NOT SUITABLE AS A GROUNDING MEANS UNDER ANY OF THE FOLLOWING CONDITIONS:
- TRADE SIZES 1½ AND LARGER
- TRADE SIZES ⅜ AND ½ WHEN OVERCURRENT DEVICE IS GREATER THAN 20 AMPERES
- TRADE SIZES ⅜ AND ½ WHEN LONGER THAN 6 FT (1.8 m)
- TRADE SIZES ¾, 1 AND 1¼ WHEN OVERCURRENT DEVICE IS GREATER THAN 60 AMPERES
- TRADE SIZES ¾, 1 AND 1¼ WHEN LONGER THAN 6 FT (1.8 m)

FOR THESE CONDITIONS, INSTALL A SEPARATE EQUIPMENT GROUNDING CONDUCTOR SIZED PER *TABLE 250.122*.

Figure 4-42   Grounding capabilities of liquidtight flexible metal conduit (LFMC), See *Article 350*.

- LFNC may not be used if ambient temperature and heat from the conductors will exceed the temperature limitation of the nonmetallic material. UL listing will indicate this.

- Do not have more than four quarter bends (360°) between boxes.

- LFNC must not be used in sizes smaller than trade size ½ except that trade size ⅜ is acceptable for enclosing the leads to a motor, or as a luminaire (fixture) whip.

- LFNC must be secured every 3 ft (900 mm) and within 12 in. (300 mm) of a box, cabinet, fitting, or enclosure.

- Horizontal runs through holes, supported by framing members not more than 3 ft (900 mm) apart and secured within 12 in. (300 mm) of a box, cabinet, fitting, or enclosure are considered to be adequately supported.

- The 3-ft (900-mm) and 12-in. (300 mm) securing requirements are waived if the LFNC is used as a luminaire (fixture) whip whose length is not over 6 ft (1.8 m).

- The 12-in. (300-mm) securing requirement is waived if the LFNC is used for a flexible connection whose length is not over 3 ft (900 mm).

- The 3-ft (900-mm) and 12-in. (300 mm) securing requirements are waived if the LFNC is fished.

- Calculate conductor fill same as for regular conduit. See *Table 1, Chapter 9, NEC.*®

- Calculate conductor fill for trade size ⅜ LFNC. See *Table 348.12.*

- Do not conceal angle-type fittings.

- The LFNC must be listed.

- LFNC must be used with listed fittings.

- If an equipment grounding conductor is needed, the equipment grounding conductor may be run inside or outside of the LFNC. The equipment grounding conductor shall not be longer than 6 ft (1.8 m) if run on the outside of the LFNC. Size the equipment grounding conductor per *Table 250.122.*

- LFNC must not be used in lengths greater than 6 ft (1.8 m) unless absolutely necessary for flexibility, such as when connecting a motor.

- LFNC must not be used in lengths longer than 6 ft (1.8 m) unless it is of the integral reinforcement type and is secured at intervals not over 3 ft (900 mm) and within 12 in. (300 mm) of a box or other similar enclosure.

ELECTRICAL NONMETALLIC TUBING (ENT) *(ARTICLE 362)*

Electrical nonmetallic tubing (ENT), is covered in *Article 362.* It is a pliable corrugated raceway made of PVC and requires special fittings made specifically for the ENT.

ENT is not often used in house wiring, but is mentioned here to emphasize that it is a recognized wiring method that could be used.

SERVICE-ENTRANCE CABLE *(ARTICLE 338)*

Service-entrance cable is covered in *Article 338* of the *National Electrical Code.*®

Service-entrance cable is defined as *"a single conductor or multi-conductor assembly provided with or without an overall covering, primarily used for services."*

Where permitted by local electrical codes, service-entrance cable is most often run from the utility's **service point** to the meter base, then from the meter base to the main panel.

It is also permitted as a branch-circuit to connect major electrical appliances and as a feeder to a sub panel. For these situations, the choice of whether to use Type SE cable or nonmetallic-sheathed cable becomes a matter of choice, the major issue being the cost of the cable. Some Type SE cable with aluminum conductors may cost less than that for nonmetallic-sheathed cable with copper conductors. The labor for the installation of either type is the same.

There are differences in service-entrance cables. Table 4-15 shows the key differences.

SERVICE-ENTRANCE CABLES TYPE SE, SER, SEU, AND USE COMPARISONS

TYPE	COVERING	INSULATION	USES PERMITTED	BENDS	AMPACITY
SE Service-Entrance Cable Conforms to UL Standard 854: Service Entrance Cables. See Figure 4-43.	Flame-retardant, moisture- and sunlight-resistant covering.	Type RHW, XHHW, or THWN insulation. Sizes 12 AWG and larger for copper. Sizes 10 AWG and larger for aluminum or copper-clad aluminum. If aluminum conductors, jacket will be marked "AL." If copper-clad aluminum conductors, jacket will be marked "AL (CU-CLAD)" or "Cu-Clad AL". If letter "H" on insulation, temperature rating 167°F (75°C), wet or dry locations. If letter "HH" on insulation, temperature rating 167°F (75°C) wet/194°F (90°C) dry locations. If marked with a "-2", suitable for wet or dry locations as high as 194°F (90°C). May have one conductor uninsulated. May have reduced neutral as permitted by 220.22 and 310.15(B)(6).	600 volts or less. Above ground. Suitable for use where exposed to sun. ▲ Permitted for interior wiring (feeder to a sub-panel, ranges, dryers, cooktops, ovens) where all conductors are insulated. Must be installed according to the requirement in *Article 334* (Nonmetallic-sheathed cable). ▼ ▲ A bare, uninsulated conductor permitted only for use as an equipment grounding conductor. ▼ For existing installations only, may have a bare neutral, 250.140. Not permitted for new work. Grounding frames of electric ranges and dryers to the neutral was permitted prior to the 1996 *NEC*. No longer permitted for new work. Use "sill plate" where cable is bent to pass through a wall. See Figure 27-3.	Internal radius of bend not to exceed 5 times the diameter of the cable.	See *Table 310.16*. Refer to the 75°C and 90°C columns. For single-phase dwelling services, see *310.15(B)(6)*.
SER Conforms to UL Standard 854: Service-Entrance Cables. Subject to the same rules as Type SE.	Same as above.	Same as above. Available with two insulated phase conductors and one insulated neutral. Also available with two insulated phase conductors, one insulated neutral, and one bare equipment grounding conductor. Available with reduced neutral as permitted by 220.22 and 310.15(B)(6). Type ZHHW-2 conductors. Rated 194°F (90°C), wet or dry.	Same as above.	Same as above.	Refer to the 75°C and 90°C columns. For single-phase dwelling services, see *310.15(B)(6)*.

Table 4-15

SERVICE-ENTRANCE CABLES TYPE SE, SER, SEU, AND USE COMPARISONS (*continued*)

TYPE	COVERING	INSULATION	USES PERMITTED	BENDS	AMPACITY
SEU Conforms to UL Standard 854: Service-Entrance Cables. Subject to the same rules as Type SE.	Same as above.	Type XHHW-2 conductors. Rated 194°F (90°C), wet or dry. Type THHN or THWN Easily identified by the bare wraparound neutral. Available with two and three insulated conductors plus the bare wraparound neutral. Available with reduced neutral as permitted by 220.22 and 310.15(B)(6).	Same as above.	Same as above.	Refer to the 75°C and 90°C columns. For single-phase dwelling services, see 310.15(B)(6).
USE Service-Entrance Conductor. Individual conductors may be bundled in an assembly. Conforms to UL Standard 854: Service-Entrance Cables. USE cable is shown in Figure 4-44.	Moisture-resistant covering. Not required to have flame-retardant covering.	Insulation equivalent to Type RHW or XHHW. The temperature rating of the cable is the same as the temperature rating of the individual conductors. If this is not marked on the outer jacket, then the cable is rated 167°F (75°C). If Type USE-2, insulation equivalent to Type RHW-2 or XHHW-2. Rated 194°F (90°C), wet or dry. Cabled single conductor may have bare copper conductor with the assembly. May have reduced neutral if multiple conductor assembly, as permitted by 220.22 and 310.15(B)(6).	Same as above except USE is also permitted for underground, direct burial. Permitted to emerge above ground in meter bases and similar enclosures. If run to a termination box inside of building after emerging from underground, shall not exceed 1.8 m (6 ft) in length inside building.	Same as above.	Refer to the 75°C and 90°C columns. For single-phase dwelling services, see 310.15(B)(6).

Table 4-15 *continued*

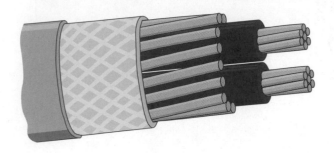

Figure 4-43 ▶A Type SE service-entrance cable with two insulated conductors and one wrap-around bare neutral conductor. This cable has a reinforcement tape plus a final sunlight-resistant outer jacket. Type SE cable with a bare neutral is permitted only when used for services. For interior wiring, or as a feeder to a second building, all current-carrying conductors must be insulated. A bare equipment grounding conductor in the cable is permitted.◀

Figure 4-44   A Type USE service-entrance conductor. The insulation is abrasion-, moisture-, heat-, and sunlight-resistant suitable for direct burial. Utility companies generally use Type USE conductors that are prebundled cable assemblies having the proper number of conductors.

REVIEW

Note: Refer to the Code or the plans where necessary.

1. The largest size solid conductor permitted to be installed in a raceway is (10 AWG) (8 AWG) (6 AWG). Circle the correct answer.

2. What is the minimum branch-circuit wire size that may be installed in a dwelling?
   _____

3. What exceptions, if any, are there to the answer for question 2? _____
   _____
   _____
   _____
   _____
   _____

4. What determines the ampacity of a wire? _____
   _____
   _____
   _____
   _____

5. What unit of measurement is used for the diameter of wires? _____

6. What unit of measurement is used for the cross-sectional area of wires? _____
   _____

7. What is the voltage rating of the conductors in Type NMC cable? _____

8. Indicate the allowable ampacity of the following Type THHN (copper) conductors. Refer to *Table 310.16*.

   a. 14 AWG _____ amperes   d. 8 AWG _____ amperes

   b. 12 AWG _____ amperes   e. 6 AWG _____ amperes

   c. 10 AWG _____ amperes   f. 4 AWG _____ amperes

9. What is the maximum operating temperature of the following conductors? Give the answer in degrees Fahrenheit and in degrees Celsius.

   a. Type XHHW _____   c. Type THHN _____

   b. Type RH _____   d. Type TW _____

10. What are the colors of the conductors in nonmetallic-sheathed cable for

   a. two-wire cable? _____ , _____

   b. three-wire cable? _____ , _____ , _____

11. For nonmetallic-sheathed cable, may the uninsulated conductor be used for purposes other than grounding? _____

12. What size equipment grounding conductor is used with the following sizes of non-metallic-sheathed cable?

   a. 14 AWG _____   c. 10 AWG _____

   b. 12 AWG _____   d. 8 AWG _____

13. Under what condition may nonmetallic-sheathed cable (Type NM) be fished in the hollow voids of masonry block walls? _____

   _____

   _____

14. a. What is the maximum distance permitted between straps on a cable installation?

   _____

   b. What is the maximum distance permitted between a box and the first strap in a cable installation? _____

   c. Does the *NEC®* permit two-wire Romex stapled on edge? _____

   d. When Type NMC cable is run through holes in studs and joists, must additional support be provided? _____

15. What is the difference between Type AC and Type ACT armored cables? _____

   _____

   _____

   _____

   _____

   _____

   _____

16. Type ACT armored cable may be bent to a radius of not less than _____ times the diameter of the cable.

17. When armored cable is used, what protection is provided at the cable ends?

_____

_____

_____

_____

18. What protection must be provided when installing a cable in a notched stud or joist, or when a cable is run through bored holes in a stud or joist where the distance is less than 1¼ in. (32 mm) from the edge of the framing member to the cable, or where the cable is run parallel to a stud or joist and the distance is less than 1¼ in. (32 mm) from the edge of the framing member to the cable? _____

_____

_____

_____

_____

19. For installing directly in a concrete slab, (armored cable) (nonmetallic-sheathed cable) (conduit) may be used. Circle the correct method of installation.

20. Circle the correct answer to the following statements:

    a. Type SE service-entrance cable is for aboveground use.   True   False

    b. Type USE service-entrance cable is suitable for direct burial in the ground.   True   False

    c. Type SE cable with an uninsulated neutral is permitted for hooking up an electric range.   True   False

    d. Type SE cable with an insulated neutral is permitted for hooking up an electric range.   True   False

    e. Are Types SE and USE service-entrance cables used in your area?   Yes   No

21. When running NMC through a bored hole in a stud, the nearest edge of the bored hole shall not be less than _____ in. from the face of the stud unless the cable is protected by a ¹⁄₁₆ in. (1.6 mm) metal steel plate.

22. Where is the main service entrance panel located in this residence? _____

_____

23. a. Is nonmetallic-sheathed cable permitted in your area for residential wiring?

_____

    b. From what source is this information obtained? _____

24. Is it permitted to use flexible metal conduit over 6 ft (1.8 m) in length as a grounding means? (Yes) (No)   Circle the correct answer.

25. Liquidtight flexible metal conduit may serve as a grounding means in sizes up to and including _____ in. where used with listed fittings.

26. The allowable current-carrying capacity (ampacity) of aluminum wire, or the maximum overcurrent protection in the case of 14 AWG, 12 AWG, and 10 AWG conductors, is less than that of copper wire for a given size, insulation, and temperature of 86°F (30°C). Refer to *Table 310.16* and *240.4(D)* and complete the following table. Important: where the ampacity of the conductor does not match the rating of a standard fuse or circuit breaker as listed in *240.6(A)*. *240.3(B)* permits the selection of the next "higher" standard rating of fuse or circuit breaker if the next higher standard rating does not exceed 800 amperes. Nonstandard ampere ratings are also permitted.

WIRE (AWG-KCMIL)	COPPER		ALUMINUM	
	Ampacity	Overcurrent Protection	Ampacity	Overcurrent Protection
12 AWG THHN				
10 AWG THHN				
3 AWG THW				
0000 THWN				
500 KCMIL THWN				

27. It is permissible for an electrician to connect aluminum, copper, or copper-clad aluminum conductors together in the same connector. (True) (False). Circle the correct answer.

28. Terminals of switches and receptacles marked CO/ALR are suitable for use with _____ , _____ , and _____ conductors.

29. Wire connectors marked AL/CU are suitable for use with _____ , _____ , and _____ conductors.

30. A wire connector bearing no marking or reference to AL, CU, or ALR is suitable for use with (copper) (aluminum) conductors only. Circle the correct answer.

31. When Type NM or NMC cable is run through a floor, it must be protected by at least _____ in. (_____ mm) of _____
_____
_____

32. When nonmetallic-sheathed cables are bunched or bundled together for distances longer than 24 in. (600 mm), what happens to their current-carrying ability?
_____

33. In diagrams A and B, nonmetallic-sheathed cable is run through the cold air return. Which diagram "meets Code"? A _____ B _____ Check one.

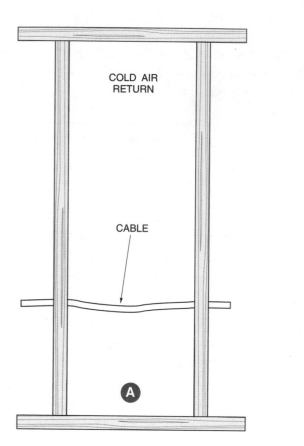

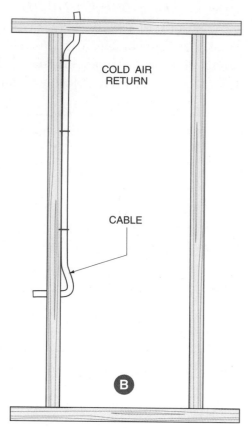

34. The marking on the outer jacket of a nonmetallic-sheathed cable indicates the letters NMC-B. What does the letter **B** signify?_____

_____

_____

35. A 120-volt branch-circuit supplies a resistive heating load of 10 amperes. The distance from the panel to the heater is approximately 140 ft. Calculate the voltage drop using (a) 14 AWG, (b) 12 AWG, (c) 10 AWG, (d) 8 AWG copper conductors. See *210.19(A), FPN No. 4* and *215.2(D), FPN No. 2*.

36. In question 35, it is desired to keep the voltage drop to 3% maximum. What minimum size wire would be installed to accomplish this 3% maximum voltage drop? See *210.19(A), FPN No. 4* and *215.2(D), FPN No. 2.*

_____

_____

_____

_____

_____

37. *NEC® 220.22* and *310.15(B)(6)* states that the neutral conductor of a residential service or feeder can be smaller than the phase conductors, but only:

_____

_____

_____

38. The allowable ampacity of a 4 AWG THHN from *Table 310.16* is 95 amperes. What is this conductor's ampacity if connected to a terminal listed for use with 60°C wire?

_____

39. If, because of some obstruction in a wall space, it is impossible to keep an NMC cable at least 1¼ in. (32 mm) from the edge of the stud, then it shall be protected by a metal plate at least _____ in. thick.

40. The recessed fluorescent luminaires (fixtures) installed in the ceiling of the recreation room of this residence are connected with trade size ⅜ flexible metal conduit. These flexible connections are commonly referred to as fixture whips. Does the flexible metal conduit provide adequate grounding for the luminaires (fixtures), or must a separate equipment grounding conductor be installed? _____

_____

_____

_____

_____

41. Flexible liquidtight metal conduit will be used to connect the air-conditioner unit. It will be trade size ¾. Must a separate equipment grounding conductor be installed in this FLMC to ground the air conditioner properly? _____

_____

_____

_____

_____

42. What size overcurrent device protects the air-conditioning unit? _____

_____

43. May the 20-ampere small appliance branch-circuits in the kitchen of a residence also supply the lighting above the kitchen sink and under-cabinet lighting? Check the correct answer. Yes _____ No _____

44. A 30-ampere branch-circuit is installed for an electric clothes dryer. What is the maximum ampere rating of a dryer permitted by the *NEC*® to be cord-and-plug connected to this circuit? Check the correct answer.

    a._____ 30 amperes

    b._____ 24 amperes

    c._____ 15 amperes

45. In many areas, metal framing members are being used in residential construction. When using nonmetallic-sheathed cable, what must be used where the cable is run through the metal framing members?_____

46. Are set-screw-type connectors permitted to be used with armored cable that has an aluminum armor? _____

47. Most armored cable today has 90°C conductors. What is the correct designation for this type of armored cable?_____

Switch Control of Lighting Circuits, Receptacle Bonding, and Induction Heating Resulting from Unusual Switch Connections

OBJECTIVES

After studying this unit, the student should be able to

- identify the grounded and ungrounded conductors in cable or conduit (color coding).
- identify the various types of toggle switches for lighting circuit control.
- select a switch with the proper rating for the specific installation conditions.
- describe the operation that each type of toggle switch performs in typical lighting circuit installations.
- demonstrate the correct wiring connections for each type of switch per Code requirements.
- analyze some unusual three-way switch connections.
- understand the various ways to bond wiring devices to the outlet box.
- understand how to design circuits to avoid heating by induction.

The electrician installs and connects various types of lighting switches. To do this, both the operation and method of connection of each type of switch must be known. In addition, the electrician must understand the meanings of the current and voltage ratings marked on lighting switches, as well as the *NEC®* requirements for the installation of these switches.

CONDUCTOR IDENTIFICATION
(ARTICLES 200 AND 210)

Before making any wiring connections to devices, the electrician must be familiar with the ways in which conductors are identified. For alternating-current circuits, the *NEC®* requires that the grounded (identified) conductor have an outer finish that is either continuous white or gray, or have an outer finish (not green) that has three continuous white stripes along the conductor's entire length. In multiwire circuits, the grounded conductor is also called a *neutral* conductor.

An ungrounded conductor must have an outer finish that is a color other than green, white, natural gray, or three continuous white stripes. This conductor generally is called the "hot" conductor. A shock is felt if this "hot" conductor and a grounded conductor or grounded surface such as a metal water pipe are touched at the same time.

The hazards associated with electric shock are discussed in detail in Unit 6.

Grounded Neutral Conductor

For residential wiring, the white grounded conductor is commonly referred to as the *neutral*.

For branch circuits and feeders, the neutral conductor must be insulated, *310.2(A)*.

For services, the neutral conductor is permitted to be a bare conductor, *230.41*.

For residential wiring, the 120/240-volt electrical system is grounded by the electric utility at their transformer, and again by the electrician at the main service. This is in accordance with *250.20(B)(1)* requiring that the system be grounded so the maximum voltage to ground on the ungrounded conductors does not exceed 150 volts. *250.26(2)* requires that the neutral conductor be grounded for single-phase, three-wire systems.

Electricians often use the term "neutral" whenever they refer to a white grounded circuit conductor. Proposals have been made to include a definition of a neutral in the *NEC®*—with no success. Even electrical engineering textbooks are evasive as to whether the definition of a neutral refers only to the actual source (transformer), or down into the system such as a two-wire branch-circuit.

A neutral conductor is:

- the conductor that carries only the unbalanced current from the other conductors, as in the case of multiwire circuits of three or more conductors, and

- the conductor where the voltage from every other conductor to it is equal under normal operating conditions.

By these definitions, the white grounded conductor in a two-wire circuit is not truly a neutral conductor. In a multiwire three-wire branch-circuit, the white grounded conductor is a neutral conductor. See Figure 5-1.

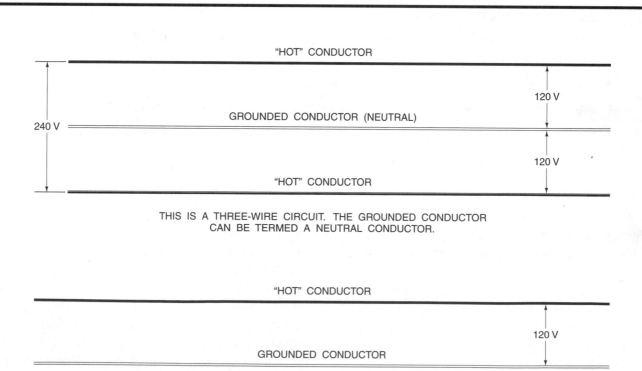

"HOT" CONDUCTOR

120 V

GROUNDED CONDUCTOR (NEUTRAL)

240 V

120 V

"HOT" CONDUCTOR

THIS IS A THREE-WIRE CIRCUIT. THE GROUNDED CONDUCTOR
CAN BE TERMED A NEUTRAL CONDUCTOR.

"HOT" CONDUCTOR

120 V

GROUNDED CONDUCTOR

THIS IS A TWO-WIRE CIRCUIT. THE GROUNDED CONDUCTOR
IS NOT TRULY A NEUTRAL CONDUCTOR.

Figure 5-1 Definition of a true neutral conductor.

For residential wiring, if you are comfortable calling the white grounded conductor a neutral, continue to do so. Do not get hung up on technicalities.

For those of you who will broaden your training into commercial and industrial wiring, you will find that there are electrical systems such as a 3-phase Delta corner grounded system where the grounded conductor is a phase conductor and is not a neutral conductor. This is covered in detail in the *Electrical Wiring—Commercial* text.

A neutral conductor in residential wiring is always a grounded conductor—but a grounded conductor is not always a neutral conductor.

Multiwire Branch-Circuit

The Code defines a multiwire branch-circuit as ▶"*A branch circuit consisting of two or more ungrounded conductors having a voltage between them, and a grounded conductor having equal voltage between it and each ungrounded conductor of the circuit and that is connected to the neutral or grounded conductor of the system.*"◀

For additional information relative to the hazards involved with multiwire circuits, refer to Figure 17-5, Figure 17-6, Figure 17-7, Figure 17-8, and Figure 17-9 in Unit 17.

Color Coding (Cable Wiring)

The conductors in nonmetallic-sheathed cable (Romex) are color coded as follows.

Two-wire: one black ("hot" phase conductor)
one white (grounded "identified" conductor)
one bare (equipment grounding conductor)

Three-wire: one black ("hot" phase conductor)
one white (grounded "identified" conductor)
one red ("hot" phase conductor)
one bare (equipment grounding conductor)

Four-wire: one black ("hot" phase conductor)
one white (grounded "identified" conductor)
one red ("hot" phase conductor)
one blue ("hot" phase conductor)
one bare (equipment grounding conductor)

Four-wire nonmetallic-sheathed cable and armored cable are not commonly used for house wiring. In fact, it may be impossible to find this type of cable in stock at electrical supply houses, hardware stores, or home centers. It may need to be ordered special. The cost is very high when compared to two-wire and three-wire cables. It is generally easier and more economical to lay out the wiring using two-conductor and three-conductor cables.

Color Coding (Raceway Wiring)

When the wiring method is a raceway such as EMT, the choice of colors for the ungrounded conductors is virtually unlimited, with the following restrictions:

Green: reserved for use as an equipment grounding conductor only. See *250.119* and *310.12(B)*.

White, Gray, or an insulated conductor (not green) with three continuous white stripes: reserved for use as a grounded circuit conductor (neutral) only. The striping is done by the manufacturer—not by the electrician on the job site. See *200.6(A)* and *(B)*, *200.7(A)*, *(B)*, and *(C)*, and *310.12(B)*.

▶For years, conductors with gray insulation were used in commercial and industrial work as both grounded conductors and ungrounded conductors. Past editions of the *NEC*® used the term *natural gray*, but no one really knew what *natural gray* was. Was it "dirty white?" Was it "light gray?" Was it "dark gray"? Confusion was the end result. The 2002 edition of the *NEC*® now recognizes a conductor that is insulated with gray insulation as a grounded conductor. Gray is no longer permitted to be used on an ungrounded "hot" conductor. All references to the term *natural gray* have been deleted in the 2002 edition of the *NEC*.® See *200.6(A)* and *(B)*.◀

Changing Colors When Conductors Are in a Cable

The basic rule in *200.7* is that grounded conductors must be white, gray, or have three continuous white stripes on other than green insulation.

For cable wiring such as nonmetallic-sheathed cable or armored cable, *200.7(C)(1)* and *(2)* permits

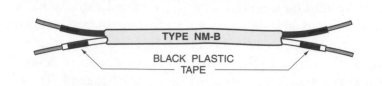

Figure 5-2 A white wire may be wrapped with a piece of black plastic tape so that everyone will know that this wire is *not* a grounded conductor. Use red tape if it is desired to mark the conductor red.

the white or gray conductor to be used for single-pole, three-way, or four-way switch loops. These sections require that when used for a switch "loop," the conductor that is white, gray, or marked with three continuous white stripes is to be used for the supply to the switch, and not as the return conductor from the switch to the switched outlet. Also, the conductor must be permanently reidentified to indicate its use by painting or other effective means, at its terminations, and at each location where the conductor is visible and accessible. See Figure 5-2.

In the past, merely attaching a white conductor to the terminal of a switch was acceptable as proper identification of an ungrounded conductor at the switch location. This practice is no longer permitted.

Much confusion resulted and inadvertent electric shocks happened because we always think of a white conductor as being a grounded conductor. Now when wiring with cable, we must reidentify a white conductor that is used as an ungrounded conductor wherever it is visible and accessible, and whenever it is used as a switch leg. The intent of the Code is that we do not want to see white conductor insulation when, in fact, the conductor is really a "hot" ungrounded conductor.

The wiring diagrams that follow are good examples of where permanent reidentification of the white conductor is necessary.

Changing Colors When Conductors Are in a Raceway

Occasionally, it may be necessary to change the actual color of the conductor's insulation. Table 5-1 provides guidelines for doing this.

CONNECTING WIRING DEVICES

Terminating and splicing conductors is a skill. The connections must be tight so as not to overheat under load.

▶*NEC® 110.3(B)* requires us to follow the manufacturers instructions regarding the listing and labeling of all electrical products. *NEC® 110.14* provides clear requirements for making electrical connections◀

Table 4-4 shows in detail the terminal identification markings found on wiring devices (switches and receptacles) indicating the types of conductors permitted to be connected to them.

Most commonly used wiring devices have screw terminals.

Some wiring devices have back-wiring holes only. For these, the conductor insulation is stripped off for the desired length, inserted into the hole, and held in place by the device's internal spring pressure mechanism. These wiring devices are referred to as having *push-in* terminals.

For convenience, some wiring devices have both screw terminals and back-wiring holes, as illustrated in Figure 5-3 and Figure 5-4. These can be connected or hooked up in different ways. One way is if the conductor insulation is stripped off for the desired length and inserted into the back hole, then the screw terminal is properly tightened. In another way, the conductor insulation is stripped off for the desired length and terminated under the screw terminal, in which case the back-wiring holes are not used.

Using both the back-wiring terminals and the screw terminals, it would be possible to terminate as many as four conductors to the white terminal, and four conductors to the brass terminal. But this would

TO CHANGE FROM THIS	TO THIS	DO THIS
red, black, blue, etc.	an *equipment grounding* conductor	**The Basic Rule:** No reidentification permitted for conductors 6 AWG and smaller. An equipment grounding conductor must be bare, covered, or insulated. If covered or insulated, the outer finish must be green or green with yellow stripes. The color must run the entire length of the conductor. **Conductors larger than 6 AWG:** At time of installation, at both ends and at every point where the conductor is accessible, do any of the following for the entire exposed length: • strip off the covering or insulation • paint the covering or insulation green • use green tape, adhesive labels, or green heat shrink tubing. See *210.5(B)*, *250.119*, and *310.12(B)*.
red, black, blue, etc.	a *grounded* conductor	**Conductors 6 AWG or smaller:** For the conductor's entire length, the insulation must be white, gray, or have three continuous white stripes on other than green insulation. Reidentification in the field is not permitted when the wiring method is a raceway. **Conductors larger than 6 AWG:** For the conductor's entire length, the insulation must be white, gray, or have three continuous white stripes on other than green insulation. At terminations, reidentification may be white paint, white tape, or white heat shrink tubing. This marking must encircle the conductor or insulation. Do this marking at time of installation. See *200.2*, *200.6(A)* and *(B)*, *200.7(A)*, *(B)*, and *(C)*, *210.5(A)*, and *310.12(A)*.
White, gray	an *ungrounded* (hot) conductor	Insulated conductors that are white, gray, or have three continuous white stripes are to be used as grounded conductors only. There is no provision in the *NEC®* for reidentifying white or gray conductors for use as ungrounded conductors in raceways. The only exception to use white or gray insulated conductors as ungrounded conductors is for cable wiring. See *200.7(A)*, *(B)*, and *(C)*, and *310.12(C)*.

Table 5-1.

be considered poor workmanship. So many conductors on the wiring device would make it extremely difficult to work with. A much better way is shown in Figure 5-5 where all of the necessary splicing is done independent of the receptacle. Short lengths of wire called *pigtails* are included in the splice, and these pigtails in turn connect to the receptacle. Removal of the receptacle would not have any effect on the splice.

Receptacle Configuration

Figure 5-6 illustrates some of the common receptacle configurations used in homes in conformance to *Table 210.21(B)(3)*. Range and dryer receptacles are discussed in Unit 20.

PUSH-IN TERMINATIONS

Be careful when using screwless push-in terminals.

Screwless push-in terminals on receptacles are "listed" by Underwriters Laboratories (UL) for use *only* with solid 14 AWG copper conductors. They *are not* to be used with:

• aluminum or copper-clad aluminum conductors.

• stranded conductors.

• 12 AWG conductors. By design, the holes are large enough to take only a 14 AWG solid conductor.

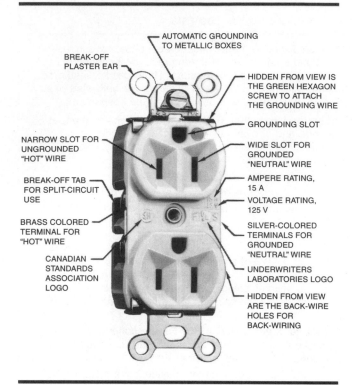

Figure 5-3 Grounding-type receptacle detailing various parts of the receptacle.

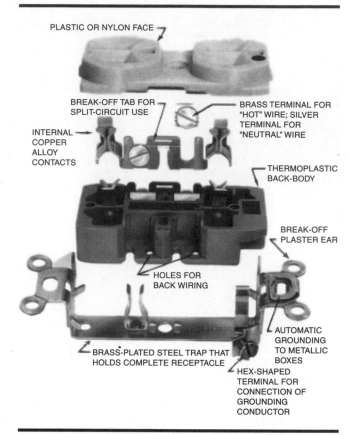

Figure 5-4 Exploded view of a grounding-type receptacle, showing all internal parts. *200.10(B)* requires that terminals on receptacles, plugs, and connectors that are intended for the ground circuit conductor be substantially white (usually silver colored), or marked with the word "white" or the letter "W" adjacent to the terminal. The terminal for the ungrounded "hot" conductor is usually brass colored.

Prior to January 1, 1995, screwless push-in terminals on receptacles accepted both 14 and 12 AWG conductors. Manufacturers are no longer permitted to produce these.

Push-in terminals for 12 AWG solid copper conductors are still permitted on snap switches.

TOGGLE SWITCHES *(ARTICLE 404)*

UL refers to them as snap switches. Manufacturers and electricians refer to them as toggle switches. They are one and the same.

The most frequently used switch in lighting circuits is the toggle flush switch, Figure 5-7. When mounted in a flush switch box, the switch is concealed in the wall with only the insulated handle or toggle protruding.

Figure 5-8 shows how switches are weatherproofed.

Toggle Switch Ratings

UL lists toggle switches used for lighting circuits as *general-use snap switches*. The UL requirements are the same as *404.14(A) and (B)* of the *NEC.*®

AC/DC General-Use Snap Switches *[404.14(B)]*

The requirements for general-use snap switches include the following:

- alternating-current (ac) or direct-current (dc) circuits.

- resistive loads not to exceed the ampere rating of the switch at rated voltage.

- inductive loads not to exceed one-half the ampere rating of the switch at rated voltage unless otherwise marked.

- tungsten filament lamp loads not to exceed the ampere rating of the switch at 125 volts when marked with the letter **T**.

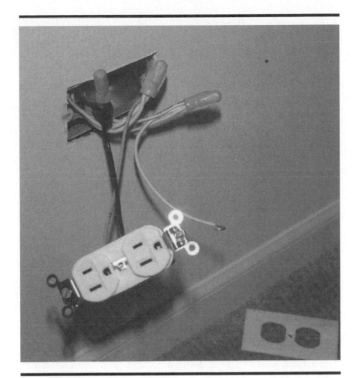

Figure 5-5 A receptacle connected with pigtails. This particular installation has one outlet switched and one outlet on continuously. Six conductors enter the box because the wiring runs from receptacle to receptacle around the room. A 4-in. square outlet box trimmed with a single-gang plaster (dry-wall) cover is used. Removal of the receptacle does not affect the circuit to the other receptacles.

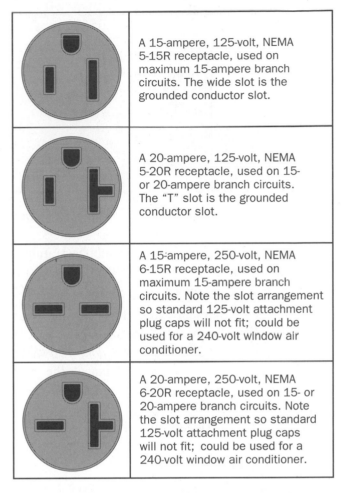

A 15-ampere, 125-volt, NEMA 5-15R receptacle, used on maximum 15-ampere branch circuits. The wide slot is the grounded conductor slot.

A 20-ampere, 125-volt, NEMA 5-20R receptacle, used on 15- or 20-ampere branch circuits. The "T" slot is the grounded conductor slot.

A 15-ampere, 250-volt, NEMA 6-15R receptacle, used on maximum 15-ampere branch circuits. Note the slot arrangement so standard 125-volt attachment plug caps will not fit; could be used for a 240-volt window air conditioner.

A 20-ampere, 250-volt, NEMA 6-20R receptacle, used on 15- or 20-ampere branch circuits. Note the slot arrangement so standard 125-volt attachment plug caps will not fit; could be used for a 240-volt window air conditioner.

Figure 5-6 Slot configuration of receptacles.

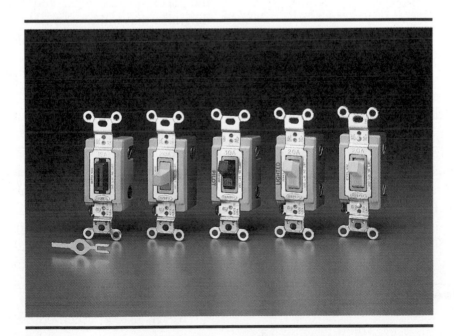

Figure 5-7 Toggle flush switches. (*Courtesy* Hubbell, Incorporated).

Figure 5-8 Switch is protected by a weatherproof cover.

- for switches marked with a horsepower rating, a motor load shall not exceed the rating of the switch at rated voltage.

A tungsten filament lamp draws a very high momentary inrush current at the instant the circuit is energized. This is because the *cold resistance* of tungsten is very low. For instance, the cold resistance of a typical 100-watt lamp is approximately 9.5 ohms. This same lamp has a *hot resistance* of 144 ohms when operating at 100% of its rated voltage.

Normal operating current would be

$$I = \frac{E}{R} = \frac{120}{144} = 0.83 \text{ ampere}$$

But, *maximum* instantaneous inrush current could be as high as

$$I = \frac{E}{R} = \frac{170 \text{ (peak voltage)}}{9.5} = 17.9 \text{ amperes}$$

This instantaneous inrush current drops off to normal operating current in about 6 cycles (0.10 second). The contacts of T-rated switches are designed to handle these momentary high inrush currents. See Unit 13 for more information pertaining to inrush currents.

The ac/dc general-use snap switch normally is not marked ac/dc. However, it is always marked with the current and voltage rating, such as 10A-125V or 5A-250V-T.

AC General-Use Snap Switches
[404.14(A)]

Alternating-current general-use snap switches are marked "ac only," in addition to identifying their current and voltage ratings. A typical switch mark-

ing is 15A, 120-277V ac. The 277-volt rating is required on 277/480-volt systems. Other requirements include:

- ac only.
- resistive and inductive loads not to exceed the ampere rating of the switch at the voltage involved.
- tungsten-filament lamp loads not to exceed the ampere rating of the switch at 120 volts.
- motor loads not to exceed 80% of the ampere rating of the switch at rated voltage. A UL requirement is that the load shall not exceed 2 horsepower.

Conductor Color Coding for Switch Connections

The insulated conductors in nonmetallic-sheathed cable or armored cable are *black-white*, *black-white-red*, or, in some cases, *black-white-red-blue*. We usually think of the black insulated conductor as the feed to a switch, and a red insulated conductor as the switch leg. However, when wiring with nonmetallic-sheathed cable or armored cable, we are "stuck" with the colors found in the cables. We do not want to use a white conductor as the return conductor from a switch, *200.7(C)(2)*. A color-coding scheme needs to be established for the feeds, switch legs, and travelers. The wiring diagrams that follow show the recommended color schemes when wiring with cable.

When the wiring method is a raceway or electrical metallic tubing (EMT), the choices of insulation colors are virtually unlimited.

Always connect a white wire to the white (silverish) terminal or to the white wire of a lampholder or receptacle, *200.9* and *200.10*. Always connect the black switch leg conductor (or red in some cases) to the black wire (or dark brassy terminal) of a lampholder or receptacle.

In cables, always reidentify white conductors when they are used as ungrounded (hot) conductors. This reidentification must be done wherever the conductors are visible and accessible. These requirements are found in *200.7(C)(2)*.

Never use a green-colored insulation for a grounded or ungrounded conductor. Green is reserved for equipment grounding conductors. See *210.5* and *250.119*.

Connecting Switches, Receptacles, and Lighting Outlets

A conductor carrying an alternating current produces a magnetic field (flux) around the conductor. This magnetic field extends outside of the conductor. The greater the current, the stronger the magnetic field. In a 60-Hz (pronounced 60-Hertz) circuit, the current and magnetic field reverses 120 times each cycle. If this conductor is run through a steel raceway, steel jacketed cable, or through a knockout in a steel box, the alternating magnetic field "induces" heat in the steel. Heat in the steel, in turn, can harm the insulation on the conductor. When all conductors of the same circuit are run through the same raceway, the magnetic fields around the conductors are equal and opposite, thereby canceling one another out.

To prevent induction heating and to keep the impedance of a circuit as low as possible where conductors are run in metal raceways, metal jacketed cables, and in metal enclosures, the installation must follow certain *NEC®* requirements.

▶*NEC® 300.3(B)* requires that *"all conductors of the same circuit and, where used, the grounded conductor, all equipment grounding conductors, and bonding conductors shall be contained within the same raceway, auxiliary gutter, cable tray, trench, cable, or cord, unless otherwise permitted."*◀

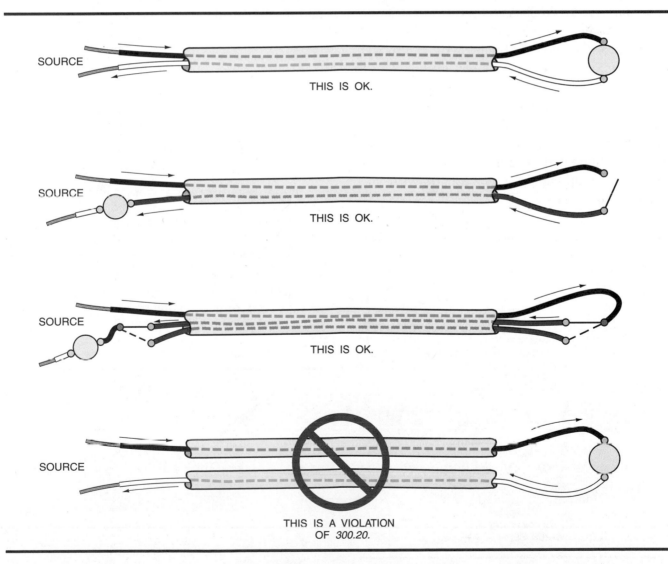

THIS IS OK.

THIS IS OK.

THIS IS OK.

THIS IS A VIOLATION OF *300.20.*

Figure 5-9 A few examples of arranging circuitry to avoid induction heating according to *300.20* when metal raceway or metal cables (BX) are used. Always ask yourself, "Is the same amount of current flowing in both directions in the metal raceway?" If the answer is "no," there will be induction heating of the metal that can damage the insulation on the conductors.

There are four exceptions: for paralleled installations, for grounding conductors, for nonferrous wiring methods, and certain enclosures.

NEC® 300.20(A) requires that "*Where conductors carrying alternating current are installed in metal enclosures or metal raceways, they shall be so arranged as to avoid heating the surrounding metal by induction. To accomplish this, all phase conductors and, where used, the grounded conductor and all equipment grounding conductors shall be grouped together.*"

▶*NEC® 404.2(A)* requires that "*Three-way and four-way switches shall be so wired that all switching is done only in the ungrounded circuit conductor. Where in metal raceways or metal-armored cables, wiring between switches and outlets shall be in accordance with 300.20(A).*" Switch loops do not require a grounded conductor. ◀

Figure 5-8, Figure 5-9, Figure 5-10, Figure 5-11, and Figure 5-12 illustrate the intent of the *NEC®* to prevent induction heating.

Unusual switch connections may or may not conform to the *NEC,®* depending upon how the circuitry is run. Some electricians have found that they can save substantial lengths of three-wire cable when hooking up three-way and four-way switches by using two-wire cable.

Connections as shown in Figure 5-13 are in violation of *300.3(B)*, *300.20(A)*, and *404.2(A)* when metal boxes and/or steel jacketed Types AC and MC are used. If nonmetallic-sheathed cable and plastic boxes are used, this connection is not a Code violation because there would be no induction heating. If the framing members are steel, we have a Code violation because of the induction heating issue. The circuitry in Figure 5-13 results in greater impedance and greater voltage drop than if wired in a conventional manner. High impedance will cause the tripping time of a breaker or opening time of a fuse to increase. Another possible problem would be that of picking up "noise" from the magnetic field surrounding the conductor. This could affect the operation of a computer or telephone located in the vicinity, particularly if non-twisted telephone cables are

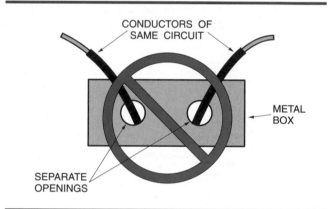

Figure 5-11 Another typical example of how induction heating can occur is to run conductors of the same circuit through different openings of a metal box.

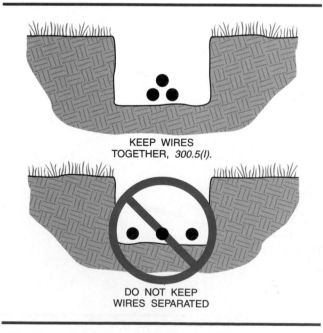

Figure 5-10 Arrangement of conductors in trenches. Refer to Unit 16 for complete information pertaining to underground wiring.

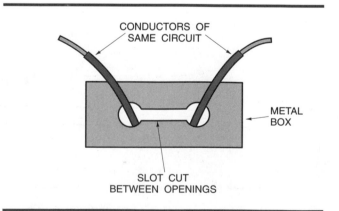

Figure 5-12 Inductive effects can be minimized by cutting slots between the openings.

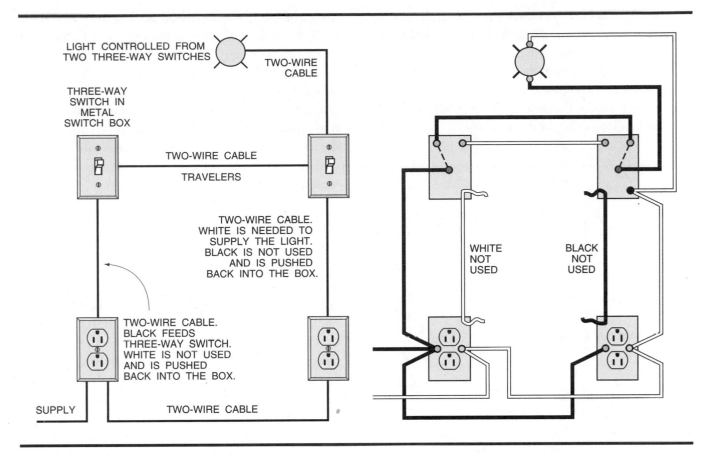

Figure 5-13 This connection is a violation of *300.3(B)*, *300.20(A)*, and *404.2(A)* when using steel raceways, cables, or boxes. See text for full explanation.

used. The circuitry is very confusing and is certainly an odd way to do a simple job. Because of all of the possible adverse affects, the circuitry in Figure 5-13 is strongly discouraged.

Switch Types

Switches are available in single-pole, three-way, four-way, double-pole, dimmer, and speed control types. They are available in snap (toggle), push, slider, rotary, electronic, and with or without pilot lights. Faceplates are commonly available in white, brown, and ivory plastic, stainless steel, brass, chrome, and wood.

In all of the following wiring diagrams, the equipment grounding conductors are not shown as this is a separate issue covered elsewhere in this text.

Single-Pole Switches. A single-pole switch is marked with "ON" and "OFF" positions. Used when the load is to be controlled from one switching point, a single-pole switch is connected in series with the ungrounded (hot) conductor.

Single-Pole Switch, Feed at Switch. In Figure 5-14, the 120-volt source enters the switch box where the white grounded conductor is spliced directly through to the lamp. The black conductor from the supply circuit connects to one terminal on

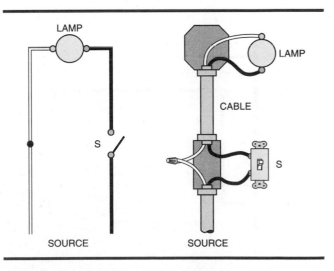

Figure 5-14 Single-pole switch in circuit with feed at switch.

the switch. The black conductor in the cable coming from the outlet box at the lamp is connected to the other terminal on the switch. Because a single-pole switch connection is simply a series circuit, it makes no difference which terminal is the supply and which terminal is the load. This layout needs no reidentification of the white conductor.

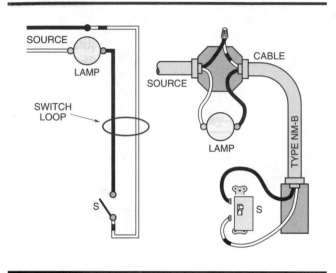

Figure 5-15   Single-pole switch in circuit with feed at light.

Single-Pole Switch, Feed at Light.

In Figure 5-15, the 120-volt source enters the outlet box at the lamp. The switch loop is a two-wire cable. The white wire of a cable is permitted in *200.7(C)(2)* to be used as a switch loop (switch leg). When so used, *200.7(C)(2)* requires that the white conductor be reidentified in the ceiling outlet box and in the device box. Color plastic tape is most often used to reidentify a conductor. For this diagram, black tape would be suitable.

Single-Pole Switch, Feed at Switch, Receptacle "On" Continuously.

Figure 5-16 shows the 120-volt source entering the switch box where the switch is located. The switch controls the lamp. The receptacle remains "hot" at all times. The white conductor in the switch box is spliced straight through to the outlet box. The black conductor in the switch box is spliced through to the outlet box and also feeds the switch. The red conductor in the three-wire cable is the switch leg, connecting to the switch and to the lamp. A two-wire cable from the outlet box carries the circuit to the receptacle. The circuitry is such that no reidentification of the white conductor is necessary.

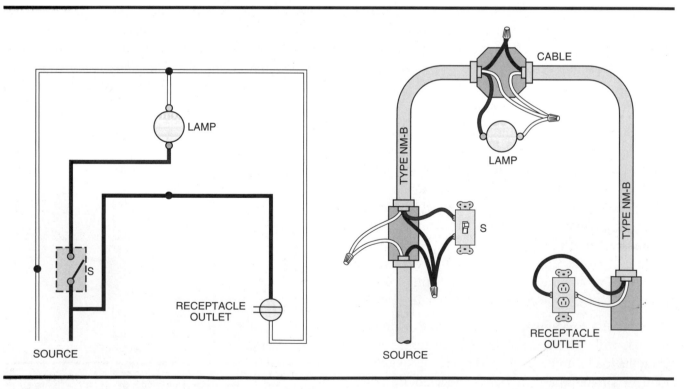

Figure 5-16   Ceiling outlet controlled by single-pole switch with live receptacle outlet and feed at switch.

Three-Way Switches. Three-way switches are used when a load is to be controlled from two locations. The term "three-way" is very misleading. Three-way switches have three terminals. Maybe they should have been named "three-terminal" switches. One terminal is called the "common" and is darker in color than the other two terminals. The other two terminals are called "traveler terminals."

Figure 5-17 shows the two positions of a three-way switch. Actually, a three-way switch is a single-pole, double-throw switch.

There is no "ON/OFF" marking on a three-way switch, as can be seen in Figure 5-18.

Three-Way Switch Control, Feed at Switch. Figure 5-19 shows the 120-volt source entering at the first switch location. A three-wire cable is run between the two switches. A two-wire cable is run from the second switch to the lamp. The white conductors in both switch boxes are spliced directly through to the lamp. The red and black conductors in the three-wire cable are the "travelers." The black conductor in the three-wire cable is reidentified with red tape, consistent with keeping both "travelers" red in color. The black conductor from the source is connected to the "common" terminal of the first switch. The black conductor in the cable coming from the outlet box at the lamp is connected to the "common" terminal of the second switch. The circuitry is such that no reidentification of the white conductor is necessary.

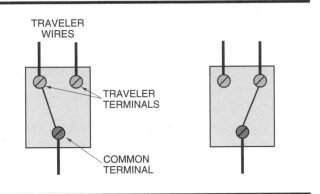

Figure 5-17 Two positions of a three-way switch.

Figure 5-18 Toggle switch: three-way flush switch. *Courtesy of* **Pass & Seymour.**

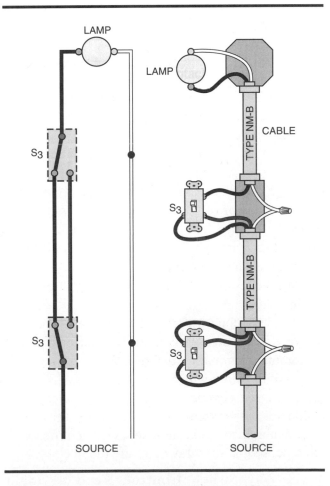

Figure 5-19 Circuit with three-way switch control. The feed is at the first switch. The load is connected to the second switch. The black and red conductors are used for the travelers. The black conductor has been reidentified with red tape.

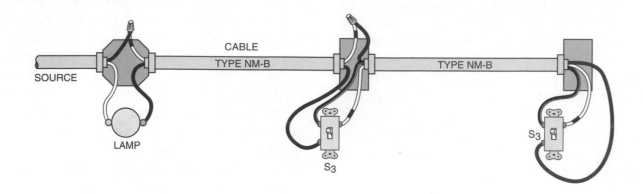

Figure 5-20 Circuit with three-way switch control and feed at light.

Three-Way Switch Control, Feed at Light.
Figure 5-20 shows the 120-volt source feeding into the outlet box at the lamp. A two-wire cable is run from the outlet box to the first switch location. A three-wire cable is run from the first switch to the second switch. In all locations where the white wire is to be used as an ungrounded "hot" conductor, it must be reidentified, *200.7(C)(2)*. In this diagram, reidentification of the white conductor in the outlet box is with black tape.

The black conductor of the two-wire cable is connected to the "common" terminal of the first switch. At the first switch, the white conductor in the two-wire cable is reidentified with black tape. The white conductor of the three-wire cable is reidentified with red tape. The red conductor and the reidentified white conductor in the three-wire cable are the "travelers." The white conductor is reidentified with red tape, consistent with keeping both "travelers" red in color.

At the second switch location, the black conductor of the three-wire cable is connected to the "common" terminal of the switch. The white conductor of the three-wire cable is reidentified with red tape. The red conductor and the reidentified white conductor in the three-wire cable are the "travelers." Here again, the white conductor is re-identified with red tape, so that the "travelers" are red in color.

Three-Way Switch Control, Feed at Light (Alternate Connection). Figure 5-21 shows the 120-volt source entering at the outlet box. The white conductor from the 120-volt source at the outlet box goes directly to the lampholder. In all locations

where the white wire is to be used as an ungrounded "hot" conductor, it must be reidentified, *200.7(C)(2)*. From the outlet box to the switch on the left, the black conductor in the three-wire cable serves as the switch return to the lamp, and the red conductor and the reidentified white conductor are the "travelers." From the outlet box to the switch on the right, the black conductor serves as the feed to the three-way switch, and the red conductor and the reidentified white conductor are the "travelers." For uniformity, reidentification of the white conductors used for "travelers" through the outlet box and in the switch boxes is with red tape.

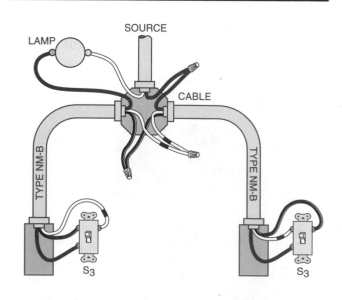

Figure 5-21 Alternative circuit with three-way switch control and feed at light.

Four-Way Switches. Four-way switches are used where switch control is needed from more than two locations. The term "four-way" is misleading. Four-way switches have four terminals. Maybe four-way switches should have been called "four-terminal" switches. A four-way switch does not have "ON" and "OFF" markings. Four-way switches are connected to the "travelers" *between* two three-way switches. One three-way switch is fed by the source—the other three-way switch is connected to the load. Make sure that the "travelers" are connected to the proper "pairs" of terminals on a four-way switch; otherwise the switching will not work.

Figure 5-22 shows the internal switching of a four-way switch.

Figure 5-23 shows a lamp controlled from three switching points. Because the white conductor from the source is connected straight through all of the switch boxes and on to the lamp, no reidentification of the white wire is necessary. The black conductor from the source is connected to the common terminal on the first three-way switch. The black conductor at the third switch is connected to the common terminal on the three-way switch. The black and red conductors in the three-wire cables are the "travelers" and are connected to the proper "traveler"

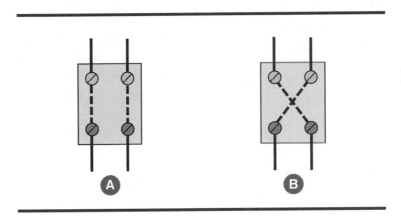

Figure 5-22   Two positions of a four-way switch.

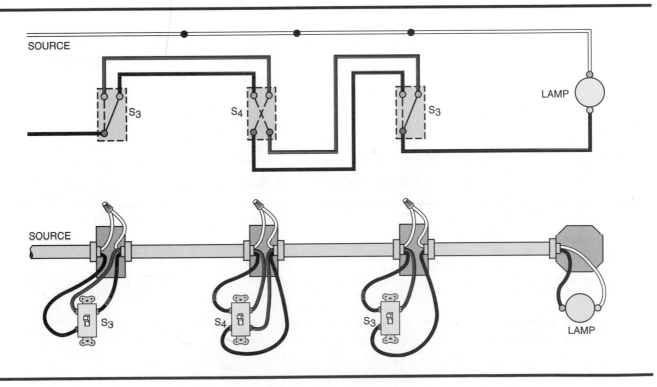

Figure 5-23   Circuit with switch control at three different locations.

terminals of the three-way and four-way switches. The black conductors in the three-wire cables have been reidentified with red tape, consistent with keeping both "travelers" red in color. At the lamp, the white conductor of the two-wire cable connects to the white conductor (or silverish terminal) of the lampholder, and the black conductor connects to the black conductor (or dark brassy terminal) of the lampholder.

Double-Pole Switches. Double-pole switches are not common in residential work. They can be used when two separate circuits must be controlled with one switch, as in Figure 5-24. They are also used to control 240-volt loads, such as electric heat, motors, electric clothes dryers, and similar 240-volt loads, as in Figure 5-25. When line-voltage thermostats for electric heat are double-pole, they switch both ungrounded "hot" conductors, and are marked with "ON" and "OFF" positions.

Switches with Pilot Lights.   There are instances when a pilot light is desired at the switch location. Figure 5-26 shows a "locator" type of switch where the lamp in the toggle glows when the switch is in the "OFF" position. *This is sometimes referred to as a "glow" or "lit-handle" switch.* A resistor inside the switch is connected in series with the pilot light. Be careful when working on this type of switch because when the switch is in the "OFF" position, the load and the pilot light are actually connected in "series." Even though these switches are marked with an "OFF" position, they are not truly off. The circuit's path through the switch and connected load is still complete. There is a voltage drop (almost full open circuit voltage) across the pilot light/resistor, and an infinitesimal voltage drop across the connected load. Although there is not enough voltage at the light to light it, there may be enough voltage and current present to cause a person to "flinch," which might result in the person jerking back and falling off a ladder.

Figure 5-27 shows a true pilot light that is "on" when the light is on.

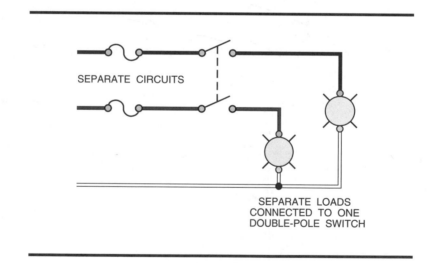

Figure 5-24   Application of a double-pole switch.

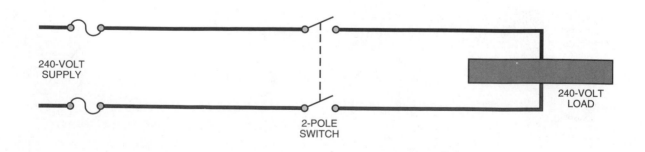

Figure 5-25   Double-pole (2-pole) disconnect switch.

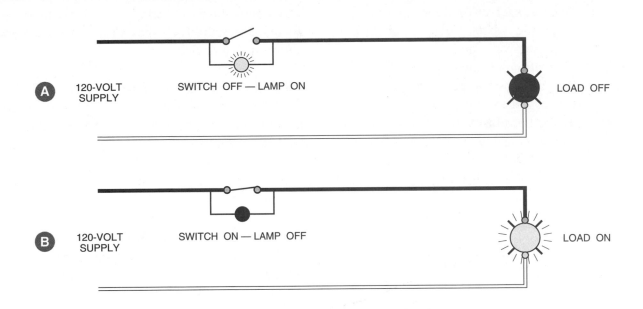

Figure 5-26 In (A) the switch is in the "OFF" position. The neon lamp in the handle of the switch glows and the load is off. This is a series circuit. The neon lamp has extremely high resistance. In a series circuit voltage divides proportionately to the resistance. Therefore, the neon lamp "sees" line voltage (120 volts) for all practical purposes and the load "sees" zero voltage. The neon lamp glows.

In (B) the switch is in the "ON" position. The neon lamp is shunted out and therefore has zero voltage across it. Thus, the neon lamp does not glow. Full voltage is supplied to the load. This type of switch might be referred to as a *locator* because it glows when the load is off and does not glow when the load is on.

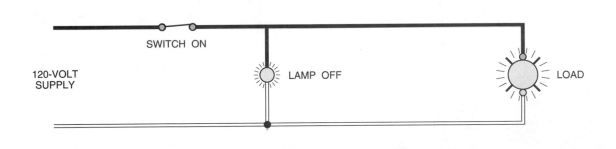

Figure 5-27 A true pilot light. The pilot lamp is an integral part of the switch. When the load is turned on, the pilot light is also on. When the load is turned off, the pilot light is also off. This switch has three terminals because it requires a grounded neutral circuit conductor.

COMBINATION AND INTERCHANGEABLE WIRING DEVICES

In the space of one standard wiring device, it is possible to install one, two, or three wiring devices using combination or interchangeable wiring devices. These are available in single-pole, three-way, four-way, pilot lights, grounding-type receptacles, and GFCI receptacles. Using these types of wiring devices can offer a neater appearance than having two-gang or three-gang wallplates at a given location. This arrangement is also used where the wall space is limited, such as between a door casing and a window casing where there is not enough space for a multi-gang wallplate.

When combination or interchangeable wiring devices are installed, standard sectional switch boxes generally are not used because of the lack of wiring space in the box. Instead, a 4-in. square box with a raised single-gang cover is used. Another possibility is to install 3-in. or 3½-in. deep device boxes. The maximum number of conductors and wiring devices permitted in a box is critical. Crowding of conductors in a box can only lead to problems such as short-circuits and ground-faults. Box fill calculations were discussed in Unit 2 and elsewhere in this text.

Figure 5-28(A) shows two switches and a pilot light of the interchangeable type. Figure 5-28(B) shows an interchangeable grounding-type receptacle, a three-hole mounting bar, and a single-pole switch. Figure 5-28(C) shows three types of combination wiring devices.

Miscellaneous Connections.   Figures 5-29 and 5-30 are diagrams that somewhat combine parts of the previous wiring diagrams. These connections become rather complicated using cable because of the limitation on the available insulation colors. These connections are more suited to raceway wiring.

Do You Need the Grounded Circuit Conductor at the Switch?   In most instances, the answer is no, *404.2(A), Exception*. However, a grounded conductor is needed when a true pilot light is connected at the switch location.

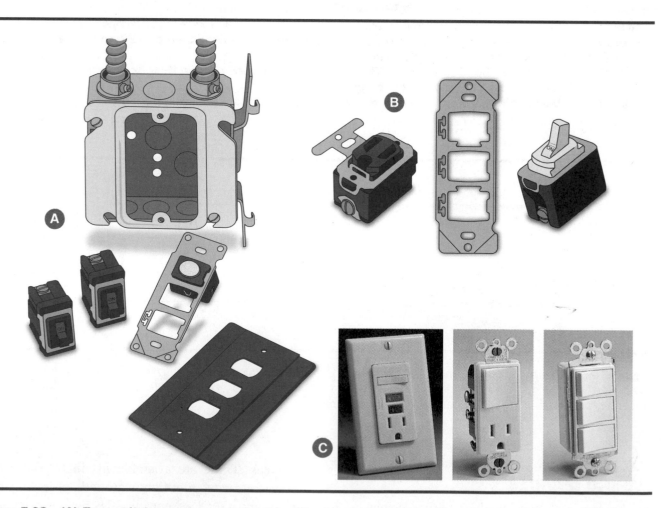

Figure 5-28   (A) Two switches and one pilot light of the interchangeable type. (B) Interchangeable type single grounding receptacle and single-pole switch and a three-hole mounting bar. (C) Three types of combination wiring devices. *Photos courtesy* of Leviton Manufacturing Co., Inc.

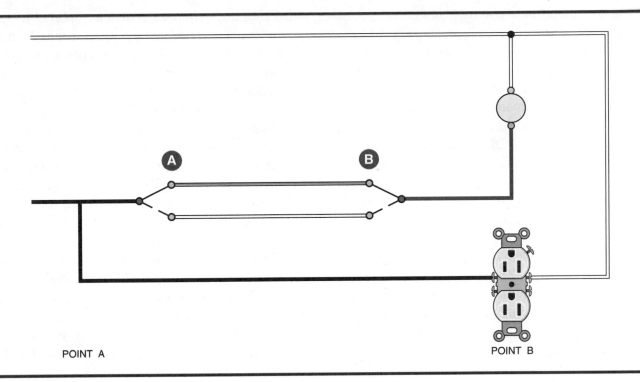

POINT A   POINT B

Figure 5-29   A lighting outlet controlled by two three-way switches. The receptacle is "live" at all times. Four conductors are run between Point A and Point B. A typical example of this might be between a house and a detached garage.

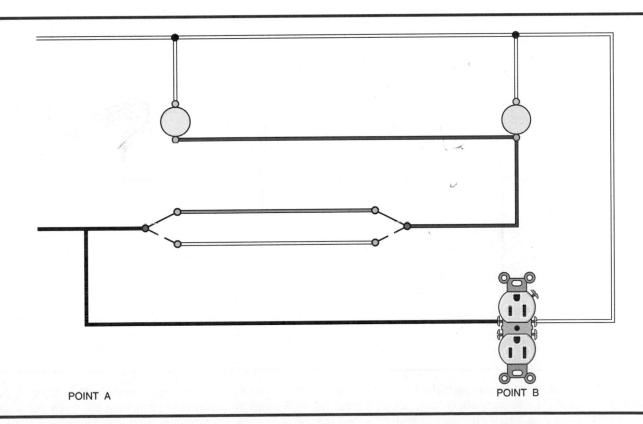

POINT A   POINT B

Figure 5-30   Two three-way switches controlling two lighting outlets. The receptacle outlet is "live" at all times. Five conductors are needed between Point A and Point B. An example of this hookup might be between a house and a detached garage, where one lighting outlet is on the house and another lighting outlet is on the garage.

Certain electronic controls, such as motion (occupancy) sensors, need a small voltage drop across the device to operate. In the motion sensing mode, when the lamp is not burning, there is still an ever-so-slight amount of voltage present at the load. The device is not in a full "OFF" position. This can be hazardous to someone working on the luminaire (fixture), thinking it is "OFF." For a typical human, a current flow of ½ milliampere cannot be felt; a current of 1 milliampere will cause a slight tingling sensation. The surprise of an electrical shock might cause the person to fall off the ladder.

►When a switching device has a marked "OFF" position, it must completely disconnect the ungrounded conductor(s) to the load it serves when turned to the "OFF" position. See *404.15(B)*. Switches of the "lit-handle" type shown in Figure 5-26 can be a problem since they do not totally break the "hot" conductor.◄

For the most common house wiring, the grounded branch circuit conductor is not present at a switch location. Never use the equipment grounding conductor as a grounded conductor in order to hook up an electronic control that really needs both the grounded and ungrounded circuit conductor present in order to operate properly. Using the equipment grounding conductor as a substitute for the circuit's grounded conductor could be very hazardous.

Proposals to change the *NEC*® have been made that would require a branch-circuit's grounded conductor to be brought to every switch location. To date, the Code Making Panel (CMP) has agreed that this extra cost burden is unrealistic since most switching locations do not actually need the grounded circuit conductor. Issues that were considered by the CMP were; what do you do with the conductor when it is not used; the extra conductor will take up more room in the box; someone in the future might use this "spare" unused conductor to add an additional circuit. If you anticipate using electronic switching devices that require a grounded circuit conductor, then definitely run the full circuit (grounded and ungrounded conductors) to the switching location.

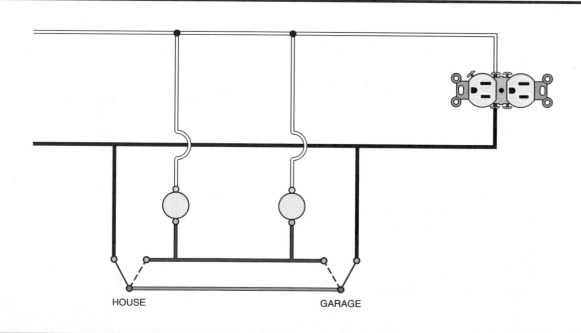

Figure 5-31   A set of three-way switches controlling one light on the side of the house, and another light on the side of the garage. The receptacle is "hot" at all times. The lights can be connected anywhere between the white and red wire. Normally, five wires are needed, whereas this hookup gets the job done with only four wires. Insofar as the Code is concerned, there are no Code violations. This is a very confusing hookup and is not a recommended way to make the connections.

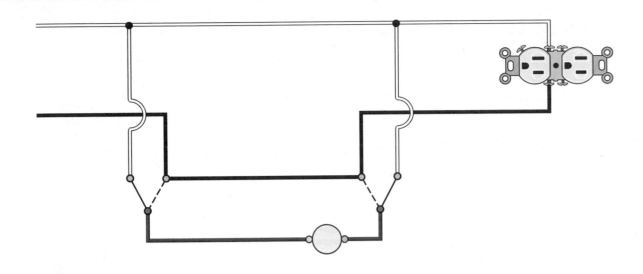

Figure 5-32   This hookup is *not* in conformance to the Code. Here we have a set of three-way switches controlling one light. The receptacle is "hot" at all times. The switches as shown have the light turned off. If you were to snap any one of the three-way switches to its other position, the light would go on. But note that the reason the light went on is that you are now connecting both terminals of the light with the same circuit conductor. Snapping the other switch to its other position turns the light off again. Each time a three-way switch is snapped, the wires on the light socket reverse. Sometimes the "shell" of the lampholder is connected to the grounded conductor as it should be, and other times the shell is connected to the "hot" circuit conductor. This is a violation of *200.10(C)* and *200.11* of the Code. This is a very unsafe hookup. Here are the four possible three-way switch positions and the respective polarity of the lampholder:

	SHELL	CENTER CONTACT	LAMP
POSITION 1	grounded	"live"	on
POSITION 2	"live"	grounded	on
POSITION 3	grounded	grounded	off
POSITION 4	"live"	"live"	off

Refer to Unit 6 for additional information relating to the effects of an electrical shock.

Unusual Switch Connections.   Over the years, "creative" electricians have come up with some very unusual ways to connect three-way switches. Figure 5-31, Figure 5-32, and Figure 5-33 are shown here only as a challenge for the student. These unusual connections are difficult to troubleshoot, are not recommended, and in some cases are not in conformance to the *NEC.*® But if you ever run across one of these hookups, you now have some wiring diagrams that might help you out of a jam.

Bonding and Grounding at Receptacles and Switches

A metal box is considered to be adequately grounded when the wiring method is armored cable, nonmetallic-sheathed cable with ground, or a metal raceway such as EMT, *250.118.* A separate equipment grounding conductor can also provide the required grounding, *250.134(B).* Equipment grounding conductors may be bare or green.

Figure 5-34 illustrates a switch that has an equipment grounding conductor terminal. Using a switch of this type provides an easy means for

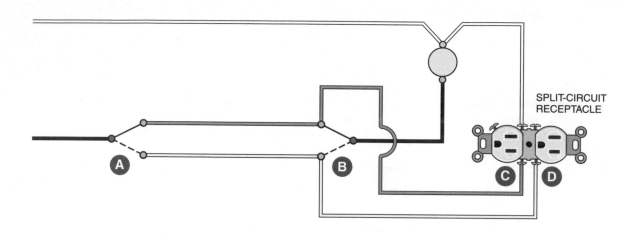

Figure 5-33 This wiring diagram is shown for informational purposes only. It should not be used for new installations. A pair of three-way switches control a lighting outlet. A receptacle is also installed, a "split circuit" type. The white grounded conductor is common to both outlets on the receptacle. The "hot" blue and yellow conductors connect to different "hot" terminals on the receptacle. Depending upon the position of the three-way switch (A), one or the other outlet is "hot." Both outlets cannot be "live" at the same time. In the diagram, the left outlet (C) on the receptacle is "live." Flip the three-way switch (A) to its other position, and the right outlet (D) on the receptacle becomes "live." Three-way switch (B) has no effect on the switch control of the receptacle.

providing the required grounding of a metal faceplate when using nonmetallic boxes.

Grounding and bonding of the equipment grounding conductor to a metal box, switch, or receptacle is important. Figure 5-35, Figure 5-36, and Figure 5-37 show how an equipment grounding conductor of a nonmetallic-sheathed cable can be attached to a metal box and to the equipment grounding screw of a receptacle. Most metal boxes have a No. 10-32 tapped hole for securing a green hexagon-shaped equipment grounding screw.

Other than for special "isolated ground" receptacles, the metal yoke of a receptacle is integrally bonded to the equipment grounding terminal of the receptacle. Isolated ground receptacles are discussed in Unit 6.

Figure 5-3 and Figure 5-4 show a receptacle that has a special "self-grounding" feature for the No. 6-32 mounting screws that results in effective grounding of the metal yoke and equipment grounding terminal without the use of a separate equipment grounding conductor.

To ensure the continuity of the equipment grounding conductor path, *250.148* requires that where more than one equipment grounding conductor enters a box, they shall be spliced with devices "suitable for the use." Splices shall not depend on solder. The splicing must be done so that if a receptacle or other wiring device is removed, the continuity of the equipment grounding path shall not be interrupted. This is clearly shown in Figure 5-36. Splicing must be done in accordance with *110.14(B)*.

Grounding Metal Faceplates and Switches

Assuming that an equipment grounding conductor has been carried to and properly connected at all metallic and nonmetallic boxes, we now need to be concerned with carrying that equipment grounding means to receptacles, switches, and metal faceplates.

Metal faceplates for receptacles shall be grounded, *406.5(B)*. This is accomplished through the No. 6-32 screw that secures the faceplate to the metal yoke on the receptacle.

A similar requirement for the grounding of metal faceplates is found in *314.25(A)*.

Snap switches and dimmer switches must provide a means to ground metal faceplates, *404.9(B)*. This is a requirement whether metallic or nonmetallic face plates are installed. The actual grounding of

Figure 5-34 Single-pole toggle switch. Note that this switch has a terminal for the connection of an equipment ground conductor. Thus, when a metal faceplate is installed, it will be grounded. See *404.9(A)* and *(B)*. *Courtesy of* Pass & Seymour.

the faceplate is through the No. 6-32 screws that secure the faceplate to the metal yoke on the switch.

In *UL Standard 20*, we find in *14.1* the requirement that flush-type switches intended for mounting in a flush-device box shall provide a grounding means. In most cases this is a green hexagon-shaped screw connected to the metal yoke of the switch. This can be seen on the switches illustrated in Figure 5-7, Figure 5-18, and Figure 5-34. The grounding means could also be a special self-grounding clip on the metal yoke, as illustrated in Figure 5-3 and Figure 5-4, a bare copper wire, a copper wire with green insulation, or a copper wire with green insulation having one or more yellow stripes.

If the equipment grounding path is accomplished through a self-grounding clip on the metal yoke of a switch or receptacle, you do not have to connect an equipment grounding conductor to the green hexagon-shaped grounding screw on the switch or receptacle.

Many wiring devices have small pieces of cardboard holding the No. 6-32 mounting screws from falling out. Remove these small pieces of cardboard

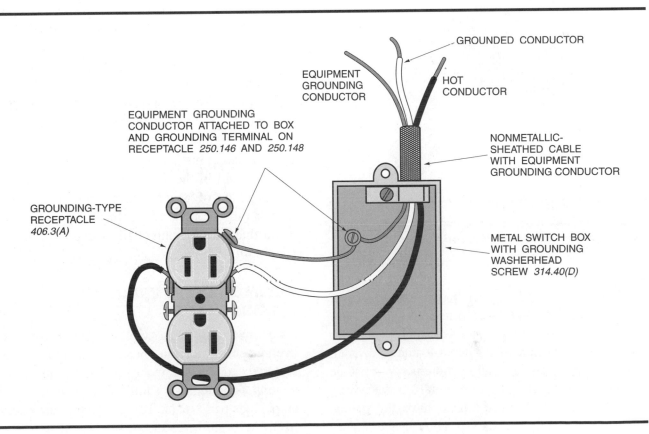

Figure 5-35 Connections for grounding-type receptacles

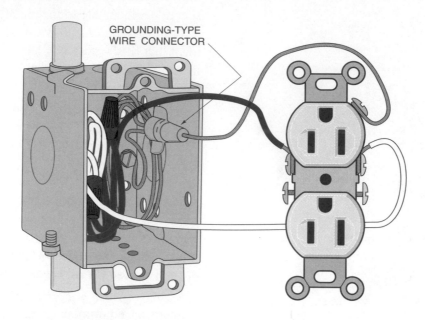

Figure 5-36 An approved method of connecting the grounding conductor using a special grounding-type wire connector. See *250.146* and *250.148*.

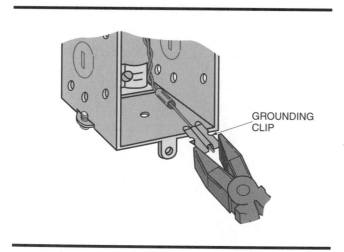

Figure 5-37 Method of attaching grounding clip to switch (device) box. See *250.146* and *250.148*.

when metal-to-metal contact between the yoke of the wiring device and the box is needed, *250.146(A)*. An example of this would be a receptacle secured to a handy box. If there is no metal-to-metal contact, then an equipment grounding conductor must be connected to the grounding terminal on the wiring device or the wiring device must have the special self-grounding clips discussed above. See Figure 5-35 and Figure 5-36.

▶For surface-mounted boxes where cover-mounted receptacles are used, the metal-to-metal requirement mentioned above is not acceptable. Some other "listed" means for providing satisfactory ground continuity must be used, *250.146(A)*. This section makes it clear that there must be direct metal-to-metal contact between the metal box and the device yoke to be considered an acceptable means of grounding the receptacle equipment grounding terminal and the metal cover.◀

In existing installations where there is no equipment ground in the switch box, and where located within reach of a conductive floor (i.e., concrete, tile) or other conductive surfaces (i.e., metal pipes, metal ducts), use nonmetallic faceplates, *404.9(B), Exception*.

COMMON CODE VIOLATION— MAKING "TAPS"

As defined in *240.2*, a "tap" is a conductor that has overcurrent protection ahead of its point of supply that exceeds the ampacity of the conductor. In general, conductors shall be protected at their ampacities per *310.15*. Tap conductors are excluded from the basic rule, and this is found in *240.4(E)*. Some examples of "taps" are small luminaire (fixture) wires. In some case, taps can be cost-effective for hooking up electric ranges, counter-mounted cooktops, and wall-mounted ovens. This is discussed in Unit 20.

Table 5-2 shows the maximum overcurrent protection for small conductors as permitted by *240.4(D)*.

Figure 5-38 illustrates a code violation. The switch leg (loop) conductors are part of the branch-circuit wiring and are not to be considered a "tap." Although the connected load may be well within the ampacity of the switch leg, the switch leg must also be capable of handling short-circuits and ground-faults. The switch leg shown in Figure 5-38 must be the same size as the branch-circuit conductors,

which are 12 AWG 20-ampere conductors protected by a 20-ampere circuit breaker.

Connecting Receptacles to 20-Ampere Branch Circuits

Another very common Code violation is using short 14 AWG pigtails to connect receptacles to 20-ampere small appliance branch-circuit conductors in a kitchen or dining room. These short pigtails must be rated 20 amperes because they are an *extension* of the 20-ampere branch-circuit wiring ... they are not "taps." Figure 5-5 illustrates this type of connection.

Consider Using Larger Size Conductors for Home Runs

To minimize voltage drop, sometimes 12 AWG conductors are used for long "home runs" (50 ft or longer) even though the branch-circuit overcurrent device is rated 15 amperes. A "home run" is the branch-circuit wiring from the panel to the first outlet or junction box in that circuit. This is permitted by *210.3*, which states, *"where conductors of higher*

CONDUCTOR SIZE	MAXIMUM OVERCURRENT PROTECTION
14 AWG copper	15 amperes
12 AWG copper	20 amperes
10 AWG copper	30 amperes

Table 5-2

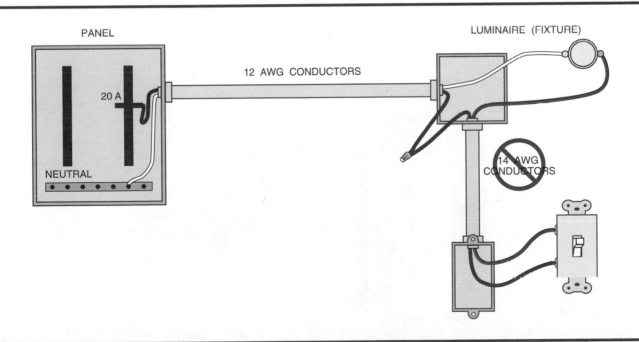

Figure 5-38 *Violation:* The electrician has run a 20-ampere branch-circuit consisting of 12 AWG conductors to the outlet box where a lumiaire (fixture) will be installed. 14 AWG conductors are run as the switch legs between the lighting outlet and the switch. The switch legs are part of the branch-circuit and must have an ampacity equal to that of the branch-circuit ampere rating, 20 amperes. Switch legs are not "taps." For the same reasons, it is a Code violation to use short "pigtails" of 14 AWG wire to connect receptacles to the 12 AWG conductors fed from a 20-ampere branch-circuit.

ampacity are used for any reason, the ampere rating or setting of the specified overcurrent device shall determine the circuit rating." In this case, 14 AWG conductors are permitted for the switch legs. But this leads to confusion later when someone removes the cover from the panel and wonders why 12 AWG branch-circuit conductors are connected to 15-ampere circuit breakers. For house wiring, it is best to use the same size conductor throughout a given branch-circuit.

Specifications for commercial and industrial installations oftentimes will call for a "home run" of 50 ft or more to be 10 AWG conductors, even though the branch-circuit rating is 20 amperes.

Table 210.24 is a summary of branch-circuit requirements. This table lists branch-circuit ratings, branch-circuit conductor sizes, tap conductor sizes, overcurrent protection, outlet (lampholder and receptacle) ratings, maximum loads, and permissible loads.

TIMERS

Timers are unique in that they provide automatic control of electrical loads. Timers are also referred to as time clocks.

Timers are used where a load is to be controlled for specific "ON/OFF" times of the day or night. The capabilities of timers are endless.

Figure 5-39 shows two types of spring-loaded timers that are connected in series with the load—the same as a standard wall switch. A typical application is in a bathroom for the control of an exhaust fan or an electric heater. These timers install in a standard single-gang device box. Because they take up quite a bit of space, make sure the wall box is large enough to meet the box-fill requirements of *314.16*.

Figure 5-40 is a 24-hour time clock commonly used to control security lighting, decorative lighting, or energy management. This particular time clock has four adjustable and removable "pins" that control two "ON" and two "OFF" operations (additional "pins" can be added). There is an "override" switch for manual control.

Figure 5-41 is an electronic 24-hour, 7-day time clock. It can be programmed to skip certain days, and up to sixteen events can be programmed. There is an "override" switch for manual control.

Figure 5-39 Spring-loaded timers. *Courtesy* Paragon Electric Company, Inc.

Figure 5-40   A 24-hour time clock. *Courtesy* Paragon Electric Company, Inc.

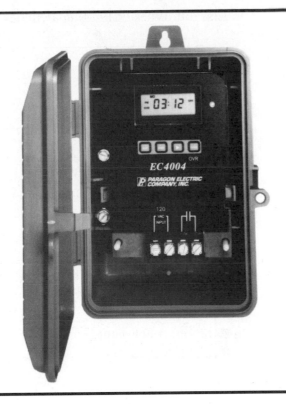

Figure 5-41   An electronic time clock. *Courtesy* Paragon Electric Company, Inc.

REVIEW

Note: Refer to the Code or the plans where necessary.

1. The identified grounded circuit conductor must be _____ or _____ in color.

2. Explain how lighting switches are rated. _____
_____
_____
_____
_____

3. A T-rated switch may be used to its _____ *full* _____ current capacity when controlling an incandescent lighting load.

4. What switch type and rating is required to control five 300-watt tungsten filament lamps on a 120-volt circuit? Show calculations. _____
_____
_____
_____
_____

5. List four types of lighting switches.

   a. _____   c. _____

   b. _____   d. _____

6. To control a lighting load from one control point, what type of switch would be used?

_____

7. Single-pole switches are always connected to the _____ wire.

8. Complete the connections in the following arrangement so that both ceiling light outlets are controlled from the single-pole switch. Assume the installation is in cable.

9. Complete the connections for the diagram. Installation is cable.

10. A three-way switch may be compared to a _____ switch.

11. What type of switch is installed to control a luminaire (fixture) from two different control points? How many switches are needed and what type are they? _____

_____

12. Complete the connections in the following arrangement so that the lamp may be controlled from either three-way switch.

13. When connecting four-way switches, care must be taken to connect the "travelers" to the _____ terminals.

14. Show the connections for a ceiling outlet that is to be controlled from any one of three switch locations. The 120-volt feed is at the light. Label the conductor colors. Assume the installation is in cable.

15. Match the following switch types with the correct number of terminals for each.

Three-way switch	Two terminals
Single-pole switch	Four terminals
Four-way switch	Three terminals

16. When connecting single-pole, three-way, and four-way switches, they must be wired so that all switching is done in the _____ circuit conductor.

17. What section of the Code emphasizes the fact that all circuiting must be done so as to avoid the damaging effects of induction heating? _____

18. If you had to install an underground three-wire feeder to a remote building using three individual conductors, which of the following installations "meets Code"? Circle the correct installation.

19. Is it always necessary to attach the bare equipment grounding conductor of a nonmetallic-sheathed cable to the green hexagon-shaped grounding screw on a receptacle? Explain.

_____

_____

_____

_____

20. List the methods by which an equipment grounding conductor is connected to a device box.

_____

_____

_____

_____

_____

21. When two nonmetallic-sheathed cables (Romex) enter a box, is it permitted to bring both of the bare equipment grounding conductors directly to the grounding terminal of a receptacle, using the terminal as a splice point? _____

22. This installation is in electrical metallic tubing. Because receptacle outlet "B" is only a few feet from switch "C" and switch "D" is only a few feet from receptacle outlet "A," the electrician saved a considerable amount of wire by picking up the feed for switch "C" from receptacle "B." The switch leg return is at switch "D." Please comment on this installation.

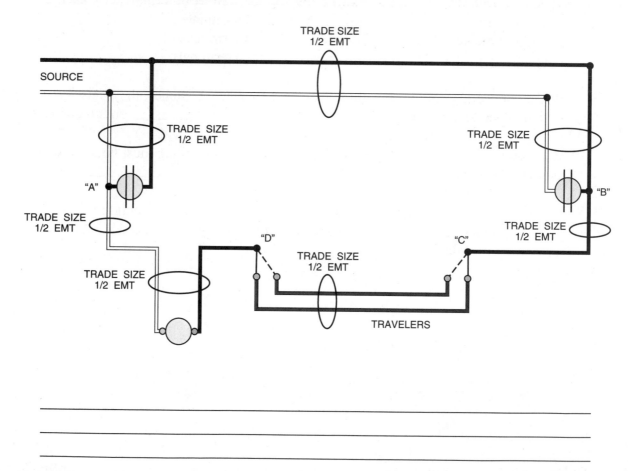

_____

_____

_____

_____

23. Define an equipment grounding conductor. _____

_____

_____

_____

24. When metal toggle switchplates are used with nonmetallic boxes, the faceplate must be grounded according to *404.9(B)*. How is this accomplished?

_____

25. Does the Code permit the ampacity of switch legs to be less than the ampere rating of the branch-circuit? _____

26. Does the Code permit connecting receptacles to a 20-ampere branch-circuit using short lengths of 14 AWG conductors as the "pigtail" between the receptacle terminals and the 12 AWG branch-circuit conductors? _____

27. Do receptacles that have back-wired "push-in" terminals accept 12 AWG conductors? Give a brief explanation for your answer. _____

_____

_____

_____

Ground-Fault Circuit Interrupters, Arc-Fault Circuit Interrupters, Transient Voltage Surge Suppressors, and Immersion Detection Circuit Interrupters

OBJECTIVES

After studying this unit, the student should be able to

- understand the theory of operation of ground-fault circuit-interrupting (GFCI) devices.

- explain the operation and connection of GFCIs.

- discuss the dos and don'ts relative to the use of GFCI devices.

- explain why GFCIs are required.

- discuss locations where GFCIs must be installed in homes.

- understand the proper way to install and utilize feedthrough GFCI receptacles.

- understand testing, recording, and identification of GFCI receptacles.

- discuss the Code rules relating to the replacement of existing receptacles with GFCI receptacles.

- understand the Code requirements for GFCI protection on construction sites.

- understand the logic of the exemptions to GFCI mandatory requirements for receptacles in certain locations in kitchens, garages, and basements.

- understand the operation of and Code requirements for arc-fault circuit interrupters (AFCI).

- discuss and understand the basics of transient voltage surge suppressors.

- discuss and understand the principles of *immersion detection circuit interrupters*.

ELECTRICAL HAZARDS

There can be no compromise where human life is involved!

Many injuries have occurred and many lives have been lost because of electrical shock. Coming in contact with live wires or with an appliance or other equipment that is "hot" spells danger. Problems arise in equipment when there is a breakdown of insulation because of wear and tear, defective construction, or misuse of the equipment. Insulation failure can result when the "hot" ungrounded conductor in the appliance comes in contact with the metal frame of the appliance. If the equipment is not properly grounded, the potential for an electrical shock is present.

The shock hazard exists whenever the user can touch both the defective equipment and grounded surfaces such as water pipes, stainless steel sinks, metal faucets, grounded metal luminaires (lighting fixtures), earth, concrete in contact with the earth, or water.

A severe shock can cause considerably more damage to the human body than is visible. A person may suffer internal hemorrhages, destruction of tissues, nerves, and muscles. Further injury can result from a fall, cuts, burns, or broken bones.

The effect of an electric current passing through a human body is determined by *how much current is flowing* and *the length of time the current will flow.* Figure 6-1 shows a time/current curve that indicates the amount of current that a normal healthy adult can stand for a certain time. Table 6-1 shows expected sensations at various current levels in milliamperes.

Experts say that the resistance of the human body varies from a few hundred ohms to many thousand ohms, but generally agree on values of 800 to 1000 ohms as typical. Body internal resistance, body contact resistance, moisture, duration, voltage, as well as the path of current, all have an effect on the severity of the shock. The current could flow through finger to finger, hand to hand, a hand holding a "hot" wire, a hand holding a pair of pliers, a hand holding an electric drill, a hand grasping a grounded metal pipe, two hands grasping a grounded metal pipe, a hand in water, standing in water, hand to foot, foot to foot, and so on.

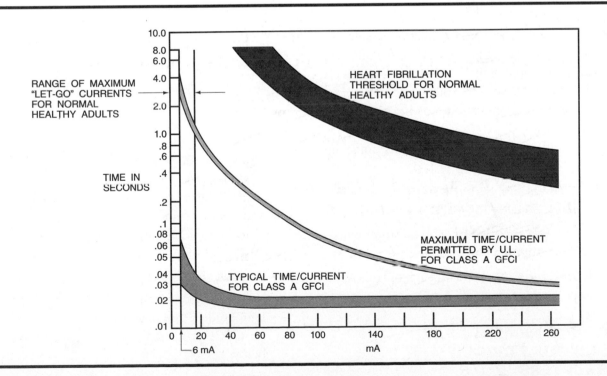

Figure 6-1   The time/current curve shows the tripping characteristics of a typical Class A GFCI. Note that if you follow the 6-mA line vertically to the crosshatched typical time/current curve, you will find that the GFCI will open in from approximately 0.035 second to just less than 0.1 second. One electrical cycle is 1/60 of a second (0.0167 second). An air bag in an automobile inflates in approximately 1/20 of a second (0.05 seconds).

EFFECT OF ELECTRIC SHOCK		
	Current in milliamperes @ 60 hertz	
	Men	Women
• cannot be felt	0.4	0.3
• a little tingling—mild sensation	1.1	0.7
• shock—not painful—can still let go	1.8	1.2
• shock—painful—can still let go	9.0	6.0
• shock—painful—just about to point where you can't let go—called "threshold"—you may be thrown clear	16.0	10.5
• shock—painful—severe—can't let go—muscles immobilize—breathing stops	23.0	15.0
• ventricular fibrillation (usually fatal) length of time . . . 0.03 sec. length of time . . . 3.0 sec.	1000 100	1000 100
• heart stops for the time current is flowing—heart may start again if time of current flow is short	4 amperes	4 amperes
• burning of skin—generally not fatal unless heart or other vital organs burned	5 or more amperes	5 or more amperes
• cardiac arrest, severe burns, and probable death	10,000	10,000

Table 6-1

Generally, voltages of less than 40 volts to ground are considered safe. Voltages 50 volts and greater are considered lethal. All ac systems (with few exceptions) of 50 volts to 1000 volts that supply premise wiring are required to be grounded, *250.20(B)*.

Some circuits of less than 50 volts must be grounded according to *250.20(A)*:

- if the circuit is supplied by a transformer where the primary exceeds 150 volts to ground

- if the circuit is supplied by a transformer where the primary supply is ungrounded

- if the circuit is overhead wiring outside of a building

CODE REQUIREMENTS FOR GROUND-FAULT CIRCUIT INTERRUPTERS *(210.8)*

To protect against electric shocks, *210.8(A)* requires that personnel ground-fault protection be provided for *all* 125-volt, 15- and 20-ampere receptacle outlets in dwellings as follows:

- in bathrooms. See Figure 10-7 for the definition of a bathroom.
- outdoors.
- in kitchens, for all receptacles that serve the countertop surfaces. This includes receptacles installed on islands and peninsulas, as in Figure 6-2.
  GFCI protection is *not* required for:
  – a receptacle installed solely for a clock.
  – a receptacle installed inside of an upper kitchen cabinet for plugging in a microwave oven that is fastened to the underside of the cabinet.
  – a receptacle installed below the kitchen sink for plugging in a food-waste disposer or dishwasher.
- within 6 feet (1.8 m) from the edge of a wet-bar sink, as in Figure 6-2.
- in attached or detached garages.
  GFCI protection is *not* required for:
  – receptacles that are not readily accessible, such as a receptacle installed on the ceiling for an overhead garage door opener.
  – a single receptacle, or a duplex receptacle installed for two cord- and plug-connected appliances in a dedicated space where the appliances are not easily moved from one location to another, such as a freezer or a refrigerator.
- in other buildings (sheds, workshops, or storage buildings).
  GFCI protection is *not* required for:
  – receptacles that are not readily accessible.
  – a single receptacle, or a duplex receptacle installed for two cord- and plug-connected appliances in a dedicated space where the appliances are not easily moved from one location to another, such as a freezer or a refrigerator.
- in crawl spaces that are at or below grade.
- in unfinished basements. Porcelain or plastic lampholders that have a plug on them also must be GFCI protected. See the exceptions listed below.

Exceptions:
– GFCI protection is *not* required for receptacles that are not readily accessible. The Code defines "readily accessible" as able to be reached without the use of ladders or chairs.

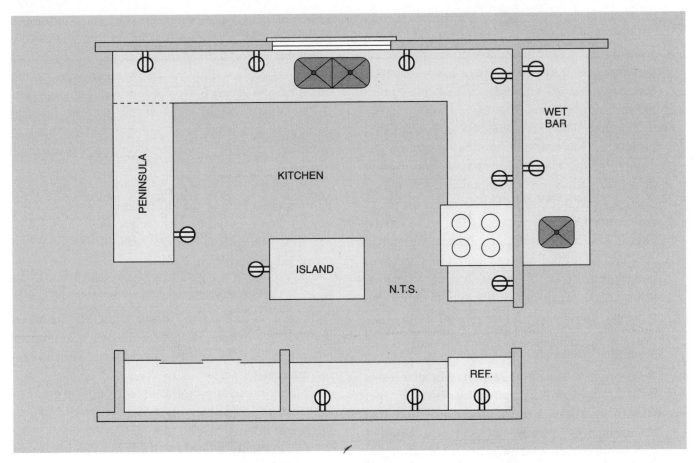

Figure 6-2 In kitchens, *all* 125-volt, single-phase, 15- and 20-ampere receptacles that serve countertop surfaces (walls, islands, or peninsulas) must be GFCI protected. The key words are "that serve countertop surfaces." GFCI personnel protection is *not* required for the refrigerator receptacle, a receptacle under the sink for plugging-in a food waste disposer, or a receptacle installed inside cabinets above the countertops for plugging in a microwave oven attached to the underside of the cabinets. Where wet-bar sinks are installed outside of the kitchen area, *all* 125-volt, single-phase, 15- and 20-ampere receptacles that serve the countertops must be GFCI protected if they are within 6 ft (1.8 m) measured from the outer edge of the wet-bar sink. See *210.8(A)(6)* and *(7)*. Drawing is not to scale.

— GFCI protection is *not* required for a single receptacle, or a duplex receptacle installed for cord- and plug-connected appliances in a dedicated space where the appliance is not easily moved from one location to another. Examples are a freezer, refrigerator, water softener, and a gas-fired electric-controlled water heater. Do not plug sump pumps into a GFCI receptacle because of the possibility of the GFCI tripping off, unknown to the homeowner. Result—a flooded basement. Since a single receptacle is exempt from the GFCI requirement, it is better to install a single receptacle for a sump pump. Also, plan on installing a single receptacle for the laundry receptacle that is required by *210.52(F)*. The laundry receptacle must be supplied by a sep-arate 20-ampere branch-circuit, and must not supply any other loads, *210.11(C)(2)*.

— ▶GFCI protection is not required for a receptacle that supplies only a permanently installed fire alarm or burglar alarm system. ◀

Some individuals try to get around the requirement for GFCI personnel protection in basements by calling the basement "finished," even though it is clearly evident that it is not "finished." If the basement is intended to be "finished," then the Code requirements for the spacing of receptacles, required lighting outlets, and switch controls would have to be followed. This is covered in Unit 3.

• there is a special rule for residential underground branch-circuit wiring that can result in a labor cost savings when the underground branch-circuit wiring is GFCI protected. *Table 300.5* of the *NEC®* indicates that for residential underground 120-volt or less branch-circuits that are rated 20 amperes or less, the underground wiring need be buried only 12 in. (300 mm) below grade when the circuit is GFCI protected. An example of this would be the Type UF cable supplying the post light in front of the residence. Type UF cable must be buried at least 24 in. (600 mm) below grade if the circuit is not GFCI protected. If the underground circuit is installed in rigid or intermediate metal conduit, a depth of not less than 6 in. (150 mm) is required. Refer to Unit 16 for more detailed information regarding residential underground wiring.

- for swimming pools, see Unit 30.

- ▶for the branch-circuit supplying heated floors in bathrooms, and hydromassage bathtub, spa, and hot tub locations, *424.44(G)*. This GFCI requirement is regardless of the type of flooring or heating cables used. ◀

The Code requirements for ground-fault circuit protection can be met in many ways. Figure 6-3

illustrates a GFCI circuit breaker installed on a branch-circuit. A ground fault in the range of 4 to 6 milliamperes or more shuts off the entire circuit. If for example, a GFCI circuit breaker were to be installed on Circuit B14, a ground-fault at any point in the circuit would shut off the garage lighting, garage receptacles, rear garage door outside bracket luminaire (fixture), post light, and overhead door opener.

When a GFCI receptacle is installed, then only that receptacle is shut off when a ground fault greater than 6 milliamperes occurs, as in Figure 6-4.

GFCI receptacles break both the ungrounded (hot) and grounded conductors.

GFCI circuit breakers break only the ungrounded (hot) conductor.

Figure 6-5 shows the effect of a GFCI installed as part of a feeder supplying 15- and 20-ampere receptacle branch-circuits. This is rarely used in residential wiring.

Figure 6-6 is a pictorial view of how a ground-fault circuit interrupter operates. Figure 6-7 shows the actual internal wiring of a GFCI.

Occasionally a GFCI will trip for seemingly no reason. This could be "leakage" in extremely long runs of cable from a GFCI circuit breaker in a panel to the protected branch-circuit wiring. One manu-

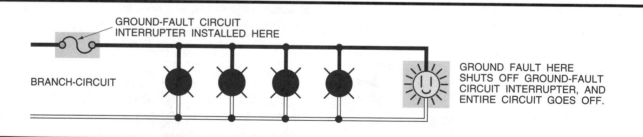

Figure 6-3 Ground-fault circuit interrupter as a part of the branch-circuit overcurrent device.

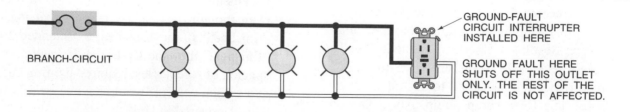

Figure 6-4 Ground-fault circuit interrupter as an integral part of a receptacle outlet.

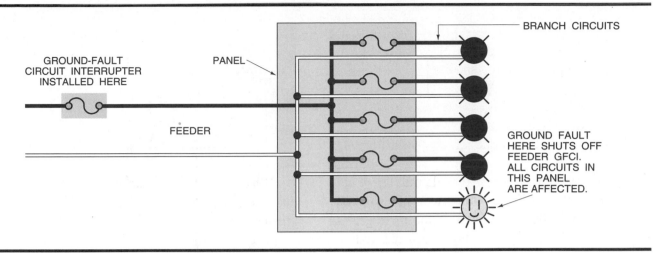

Figure 6-5 Ground-fault circuit interrupter as part of the feeder supplying a panel that serves a number of 15- and 20-ampere branch-circuits. This is permitted by *215.9*, but is rarely—if ever—used in residential wiring.

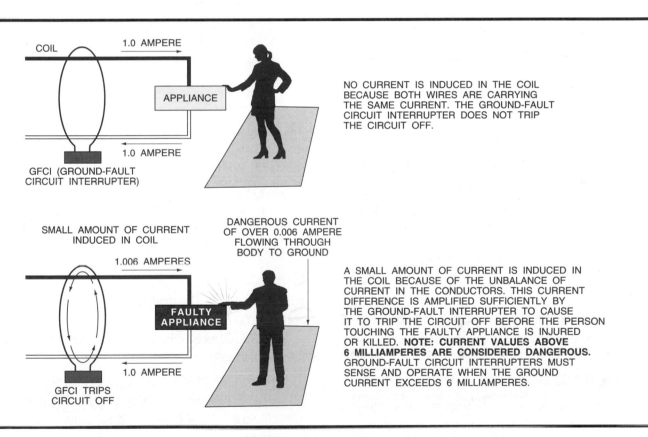

Figure 6-6 Basic principle of how a ground-fault circuit interrupter operates.

facturer's literature specifies a maximum one-way length of 250 feet for the branch-circuit.

The allowable "leakage" for UL-listed appliances is so small that the appliances should not contribute to nuisance tripping of GFCIs. However, if nuisance tripping can be traced to an appliance, the

appliance should be checked by a qualified appliance repairman.

Nuisance tripping occasionally can be traced to moisture somewhere in the circuit wiring, receptacles, or lighting outlets.

Some electricians are of the opinion that using

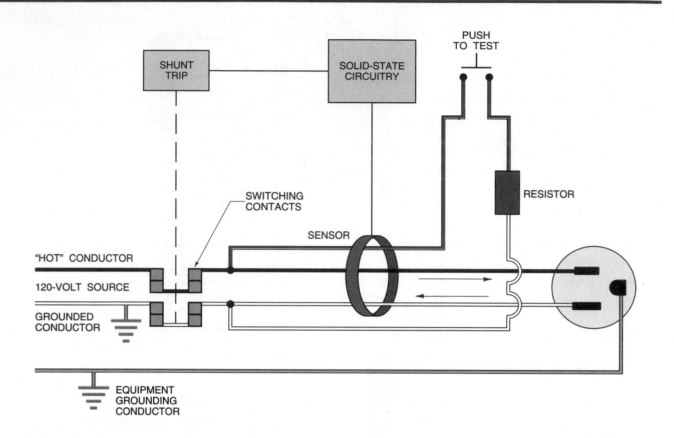

Figure 6-7 Ground-fault circuit interrupter internal components and connections. Receptacle-type GFCIs switch both the phase (hot) and grounded conductors. Note that when the test button is pushed, the test current passes through the test button, the sensor, then back around (bypasses, outside of) the sensor, then back to the opposite circuit conductor. This is how the "unbalance" is created, then monitored by the solid-state circuitry to signal the GFCI's contacts to open. Note that because both normal "load" currents pass through the sensor, no unbalance is present.

nonmetallic staples (see Figure 4-15(C)) will reduce the possibility of nuisance tripping. No actual testing has been done to prove or disprove this opinion.

What a GFCI Does

A GFCI monitors the current balance between the ungrounded "hot" conductor and the grounded conductor. As soon as the current flowing through the "hot" conductor is in the range of 4 to 6 milliamperes more than the current flowing in the "return" grounded conductor, the GFCI senses this unbalance and trips (opens) the circuit off. The unbalance indicates that part of the current flowing in the circuit is being diverted to some path other than the normal return path along the grounded return conductor. If the "other" path is through a

human body, as illustrated in Figure 6-6, the outcome could be fatal.

UL Standard No. 943 covers ground-fault circuit interrupters.

- Class "A" GFCI devices are designed to trip when a ground fault in the range of 4 to 6 milliamperes ($4/1000$ to $6/1000$ of an ampere) or more occurs.

- Class "B" GFCI devices are pretty much obsolete. They were designed to trip on ground faults of 20 milliamperes ($20/1000$ of an ampere) or more. They were used only for underwater swimming pool lighting installed before the adoption of the 1965 *NEC.* For this application, Class "A" devices were too sensitive and would nuisance trip.

What a GFCI Does *Not* Do

- It *does not* protect against electrical shock when a person touches both circuit conductors at the same time (two "hot" wires, or one "hot" wire and one grounded neutral wire) because the current flowing in both conductors is the same. Thus, there is no unbalance of current for the GFCI to sense and trip.

- It *does not* limit the magnitude of ground-fault current. It *does* limit the length of time that a ground fault will flow. In other words, you will still receive a severe shock during the time it takes the GFCI device to trip "off." See Figure 6-1.

- It *does not* sense solid short-circuits between the "hot" conductor and the grounded "neutral" conductor. The branch-circuit fuse or circuit breaker provides this protection.

- It *does not* sense solid short-circuits between two "hot" conductors. The branch-circuit fuse or circuit breaker provides this protection.

- It *does not* sense and protect against the damaging effects of arcing faults, such as would occur with frayed extension cords. This protection is provided by an arc-fault circuit interrupter (AFCI) discussed later in this unit.

- A GFCI receptacle *does not* provide overload protection for the branch-circuit wiring. It provides *ground-fault protection only*.

GROUND-FAULT CIRCUIT INTERRUPTERS IN RESIDENCE CIRCUITS

In this residence, GFCI personnel protection must be provided for receptacle outlets installed outdoors, in the bathrooms, in the garage, in the workshop, specific receptacles in the basement area, in the kitchen, and adjacent to the wet-bar in the recreation room. The GFCI protection can be provided by GFCI circuit-breakers or by GFCI receptacles of the type shown in Figure 6-8. This is a design issue that confronts the electrician. The electrician must decide how to provide the GFCI personnel protection required by *210.8*. These receptacle outlets are connected to the circuits listed in Table 6-2.

Swimming pools also have special requirements for GFCI protection. These requirements are covered in Unit 30.

GFCIs operate properly only on grounded electrical systems, as is the case in all residential, condo, apartment, commercial, and industrial wiring. The GFCI will operate on a two-wire circuit even though an equipment grounding conductor is not included with the circuit conductor. In this residence, equipment grounding conductors are in the cables. In the case of metal conduit, the equipment grounding conductor is the metal conduit.

Never ground a system neutral conductor except at the service equipment. A GFCI would be inoperative.

Never connect the neutral of one circuit to the neutral of another circuit.

(A) GROUND-FAULT CIRCUIT INTERRUPTER AS AN INTEGRAL PART OF A DUPLEX GROUNDING-TYPE CONVENIENCE RECEPTACLE.

(B) A GROUND-FAULT CIRCUIT-INTERRUPTER CIRCUIT BREAKER.

Figure 6-8   (A) A ground-fault circuit-interrupter receptacle, *courtesy of* Pass & Seymour, and (B) a ground-fault circuit-interrupter circuit breaker. The switching mechanism of a GFCI receptacle opens both the ungrounded "hot" conductor and the grounded conductor. The switching mechanism of a GFCI circuit breaker opens the ungrounded "hot" conductor only. *Courtesy of* Square D Company.

CIRCUITS	GFCI RECEPTACLE LOCATION
A15	Front porch receptacle
A16	Outdoor receptacle on rear of residence outside of Front Bedroom
A18	Workbench receptacles
A19	Outdoor receptacle on rear of residence outside of Master Bedroom
A20	Workshop receptacles on window wall
A22	Master Bath receptacle
A23	Bath (off hall) receptacle
B9	Wet-bar (Recreation Room) receptacles
B13	Kitchen receptacles
B14	Garage receptacles
B15	Kitchen receptacles
B16	Kitchen receptacles
B17	Outdoor receptacle outside of Living Room next to sliding door
B20	Outdoor receptacle outside of laundry
B21	Powder Room receptacle

Table 6-2 Circuit numbers and general locations of GFCI receptacles in this residence.

When a GFCI feeds an isolation transformer (separate primary winding and separate secondary winding), as might be used for swimming pool underwater luminaires (fixtures), the GFCI *will not* detect any ground faults on the secondary of the transformer.

Ground-fault circuit interrupters may be installed on other circuits, in other locations, and even when rewiring existing installations where the Code does not specifically call for GFCI protection,

Do not connect fire and smoke detector systems to a circuit that is protected by a GFCI breaker. A nuisance tripping of the GFCI would render the fire alarm and smoke detectors inoperative. This requirement is found in *NFPA 72, The National Fire Alarm Code.*

FEEDTHROUGH GROUND-FAULT CIRCUIT INTERRUPTER

The decision to use more GFCIs rather than trying to protect many receptacles through one GFCI becomes one of economy and practicality. GFCI receptacles are more expensive than regular receptacles. Here is where knowledge of material and labor costs comes into play. The decision must be made separately for each installation, keeping in mind that

GFCI protection is a safety issue recognized and clearly stated in the Code. However, the actual circuit layout is left up to the electrician.

Figure 6-9 illustrates how a feedthrough GFCI receptacle supplies many other receptacles. Should a ground fault occur anywhere on this circuit, all 11 receptacles lose power—not a good circuit layout. Attempting to locate the ground-fault problem, unless it's obvious, can be very time-consuming. The use of more GFCI receptacles is generally the more practical approach.

Figure 6-10 shows how a feedthrough receptacle is connected into a circuit. Figure 6-11 shows a feedthrough GFCI receptacle connected midway into a circuit. In both diagrams, the feedthrough GFCI receptacle and all downstream outlets are ground-fault protected.

In UL Standard 943, we find that a GFCI receptacle may have black-and-white line wire leads, red-and-gray load wire leads, and a green lead for the equipment grounding conductor as illustrated in Figure 6-10. GFCI receptacles may have ordinary screw terminals: brass color for the "hot" conductors, silver color for the grounded conductors, and a green hexagon-shaped screw for the equipment grounding conductor. The line and load connections are clearly marked.

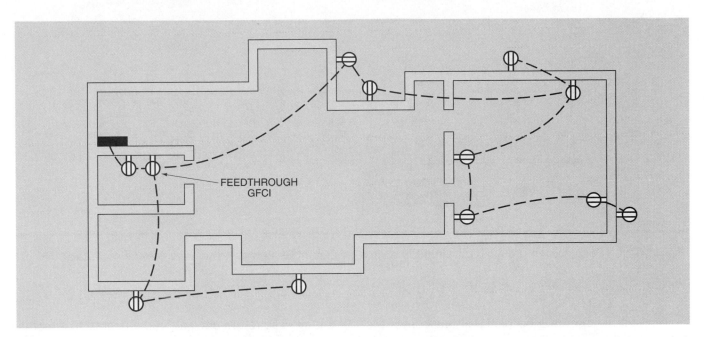

Figure 6-9   Illustration showing one GFCI feedthrough receptacle protecting ten other receptacles. Although this might be an economical way to protect many receptacle outlets with only one GFCI feedthrough receptacle, there is great likelihood that the GFCI will nuisance trip because of leakage currents. Use common sense and the manufacturer's recommendations as to how many devices should be protected through one GFCI.

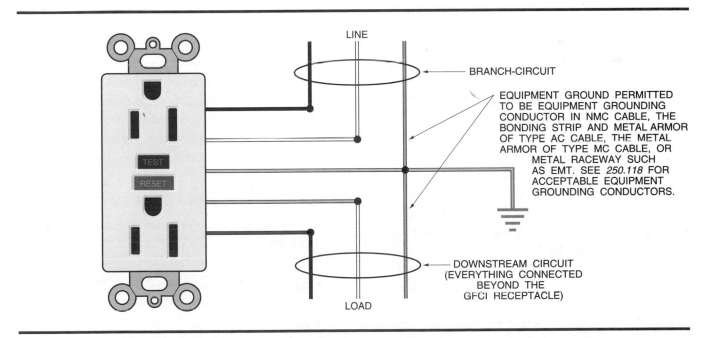

Figure 6-10   Connecting feedthrough ground-fault circuit interrupter into circuit.

Where the wiring method is a grounded method such as electrical metallic tubing, flexible metal conduit, or armored cable, the grounding terminal of the GFCI receptacle is considered adequately grounded by the metallic grounded wiring method, and a separate connection to the green lead or to the green hexagon-shaped screw is not necessary. This was discussed and illustrated in Unit 5.

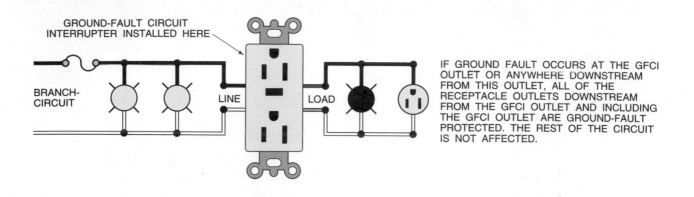

GROUND-FAULT CIRCUIT INTERRUPTER INSTALLED HERE

BRANCH-CIRCUIT

LINE   LOAD

IF GROUND FAULT OCCURS AT THE GFCI OUTLET OR ANYWHERE DOWNSTREAM FROM THIS OUTLET, ALL OF THE RECEPTACLE OUTLETS DOWNSTREAM FROM THE GFCI OUTLET AND INCLUDING THE GFCI OUTLET ARE GROUND-FAULT PROTECTED. THE REST OF THE CIRCUIT IS NOT AFFECTED.

Figure 6-11 Feedthrough ground-fault circuit interrupter as an integral part of the convenience receptacle outlet.

A "safety yellow" adhesive label must cover the load terminals, or be wrapped around the load wire leads. The label must be marked with wording like this: *"Attention! The load terminals under this label are for feeding additional receptacles. Miswiring can leave this outlet without ground-fault protection. Read instructions prior to wiring."*

The most important factors to be considered are the continuity of electrical power and the economy of the installation. The decision must be made separately for each installation.

Warning: Do not reverse the line and load connections on a feedthrough GFCI receptacle. This would result in the feedthrough GFCI receptacle itself still being "live" even though the GFCI mechanism has tripped "off" and has shut off the circuit downstream from the feedthrough GFCI receptacle. It is easy to tell if a GFCI receptacle has the line and load leads reversed. Push the test button. The GFCI will trip. If the GFCI receptacle is still "hot," it has been wired incorrectly.

Warning: When hooking up GFCIs or AFCIs, *never share a neutral* such as when using a multiwire branch-circuit. A GFCI OR AFCI device will not work! Be careful when making connections! Follow the manufacturer's instructions. GFCI and AFCI circuit breakers and receptacles have both the ungrounded conductor and the grounded conductor going in and coming out of them. This allows the sensor inside of the GFCI or AFCI to detect any unbalance current. Think of it this way: "what goes in must come out—we hope." In a multiwire circuit, the neutral carries the unbalanced current of the

circuit. The return current on the neutral may not be the same as the current flowing in the ungrounded conductor. Figure 6-12(A) is a three-wire, multiwire circuit. Trace the current flow and note that the GFCI or AFCI will sense a current unbalance of 5 amperes and trip "OFF." In fact, as long as the unbalance exists, it will never be able to be turned "ON."

Figure 6-12(B) and Figure 6-12(C) show a two-wire circuit. Figure 6-12(B) is an improper connection of a feedthrough GFCI receptacle. Note how the GFCI detects a current unbalance of 1 ampere and will trip "OFF." As long as the unbalance exists, the GFCI receptacle will continue to trip "OFF" until the misconnection is corrected. Figure 6-12(C) illustrates the proper connection of a GFCI receptacle.

Other Features for GFCI Receptacles

Look for these features when purchasing GFCI receptacles.
Some GFCI receptacles

- have a lockout feature so the GFCI cannot be reset if the internal GFCI circuit is not functioning properly.

- have a built-in line/load reversal feature that prevents the GFCI from being reset if the line and load connections are reversed.

- have an indicator light that glows when the test button is pushed if the line/load connections are properly connected. This same indicator light glows when the GFCI has tripped.

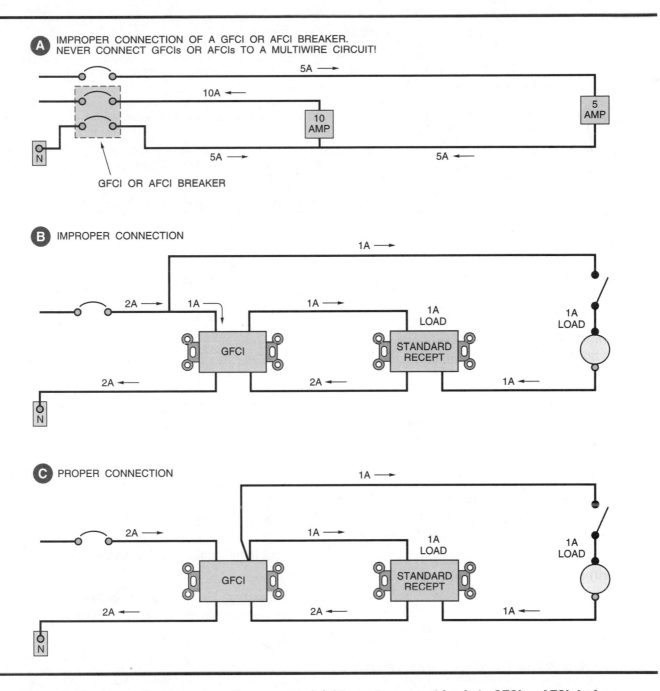

Figure 6-12   These diagrams show the wrong and right way to connect loads to GFCI or AFCI devices.

TESTING AND RECORDING OF TEST DATA FOR GFCI RECEPTACLES

Refer to Figure 6-8 and note that the GFCI receptacle has "T" (Test) and "R" (Reset) buttons. Pushing the test button places a small ground fault on the circuit. If operating properly, the GFCI receptacle should trip to the "OFF" position. Pushing the reset button will restore power. The operation is similar for the GFCI circuit breakers.

Detailed information and testing instructions are furnished with GFCI receptacles and GFCI circuit breakers. Monthly testing is recommended to ensure that the GFCI mechanism will operate properly when called upon to do so.

Because of the fragile nature of the electronic circuitry in GFCIs, test them often, particularly in high lightning strike areas of the country. GFCIs and other delicate electronic circuitry are susceptible to transient voltage surges.

OCCUPANT'S TEST RECORD

TO TEST, depress the "TEST" button, the "RESET" button should extend. Should the "RESET" button not extend, the GFCI will not protect against electrical shock. Call qualified electrician.

TO RESET, depress the "RESET" button firmly into the GFCI unit until an audible click is heard. If reset properly, the "RESET" button will be flush with the surface of the "TEST" button.

This label should be retained and placed in a conspicuous location to remind the occupants that for maximum protection against electrical shock, each GFCI should be tested monthly.

Year	Jan	Feb	Mar	Apr	May	Jun	Jul	Aug	Sep	Oct	Nov	Dec

Figure 6-13   Homeowner's testing chart for recording GFCI testing dates.

Underwriters Laboratories requires that detailed installation and testing instructions be included in the packaging for GFCI receptacle or circuit breaker. Instructions are not only for the electrician but also for homeowners, and must be left in a conspicuous place so homeowners can familiarize themselves with the receptacle, its operation, and its need for testing. Figure 6-13 shows a chart homeowners can use to record monthly GFCI testing.

REPLACING EXISTING RECEPTACLES

The *NEC*® is very specific on the type of receptacle permitted to be used as a replacement for an existing receptacle. These Code rules are found in *406.3(D)*.

In *406.3(D)(2)* we find a retroactive ruling requiring that when an existing receptacle is replaced in *any* location where the present code requires GFCI protection (bathrooms, kitchens, outdoors, unfinished basements, garages, etc.), the replacement receptacle must be GFCI protected. This could be a GFCI-type receptacle, or the branch-circuit could be protected by a GFCI-type circuit breaker.

It is important to note that on existing wiring systems, such as "knob and tube," or older non-

metallic-sheathed cable wiring that did not have an equipment grounding conductor, that a GFCI receptacle will still properly function. It does not need an equipment grounding conductor. Thus, protection against lethal shock is still provided. See Figure 6-7.

Replacing Existing Two-Wire Receptacles Where Grounding Means Does Exist

When replacing an existing receptacle (Figure 6-14(A)) where the wall box is properly grounded (Figure 6-14(E)) or where the branch-circuit wiring contains an equipment grounding conductor (Figure 6-14(D)), at least two easy choices are possible for the replacement receptacle:

1. The replacement receptacle must be of the grounding type (Figure 6-14(B)) unless . . .

2. The replacement receptacle is of the GFCI type (Figure 6-14(C)).

If the enclosure is not grounded, *250.130(C)* permits an equipment grounding conductor to be run from the enclosure to be grounded, or from the green hexagon grounding screw of a grounding-type

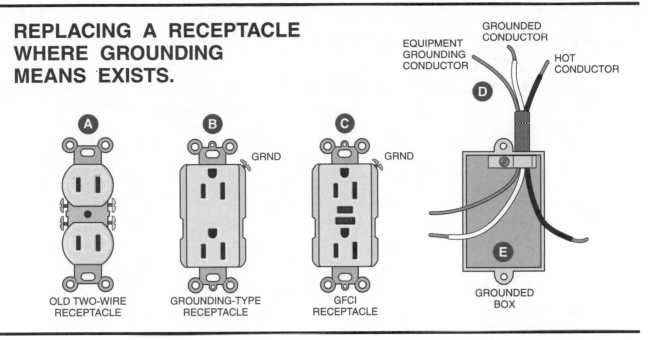

REPLACING A RECEPTACLE WHERE GROUNDING MEANS EXISTS.

A — OLD TWO-WIRE RECEPTACLE

B — GROUNDING-TYPE RECEPTACLE — GRND

C — GFCI RECEPTACLE — GRND

D — EQUIPMENT GROUNDING CONDUCTOR / GROUNDED CONDUCTOR / HOT CONDUCTOR

E — GROUNDED BOX

Figure 6-14   If a wall box (E) is properly grounded by any of the methods specified in *250.118*, or if the branch-circuit wiring contains an equipment grounding conductor (D), an existing receptacle of the type shown in (A) *must* be replaced with a grounding-type receptacle (B) or with a GFCI receptacle (C). See *406.3(D)(1)*.

receptacle, to any accessible point on the grounding electrode system or to any accessible point on the grounding electrode conductor. Note however, that this permission is only for existing installations. See Figure 6-17.

Be sure that the equipment grounding conductor of the circuit is connected to the receptacle's green hexagon-shaped grounding terminal.

Do not connect the white grounded circuit conductor to the green hexagon-shaped grounding terminal of the receptacle.

Do not connect the white grounded circuit conductor to a metal box.

Replacing Existing Two-Wire Receptacles Where Grounding Means Does Not Exist

When replacing an existing two-wire non-grounding-type receptacle (Figure 6-15(A)), where the box is not grounded (Figure 6-15(E)), or where an equipment grounding conductor has not been run with the circuit conductors (Figure 6-15(D)), four choices are possible for selecting the replacement receptacle:

1. The replacement receptacle may be a non-grounding type (Figure 6-15(A)).

2. The replacement receptacle may be a GFCI type (Figure 6-15(C)).

   • The GFCI replacement receptacle must be marked "*No Equipment Ground*," as in Figure 6-16.

   • The green hexagonal grounding terminal of the GFCI replacement receptacle does not have to be connected to any grounding means. It can be left "unconnected." The GFCI's trip mechanism will operate properly when ground faults occur anywhere on the load side of the GFCI replacement receptacle. Ground-fault protection is still there. Refer to Figure 6-7.

   • Do not connect an equipment grounding conductor from the green hexagonal grounding terminal of a replacement GFCI receptacle (Figure 6-15(C)) to any other downstream receptacles that are fed through the replacement GFCI receptacle.

   The reason this is not permitted is that if at a later date someone saw the conductor connected to the green hexagonal grounding terminal of the downstream receptacle, there would be an immediate assumption that the

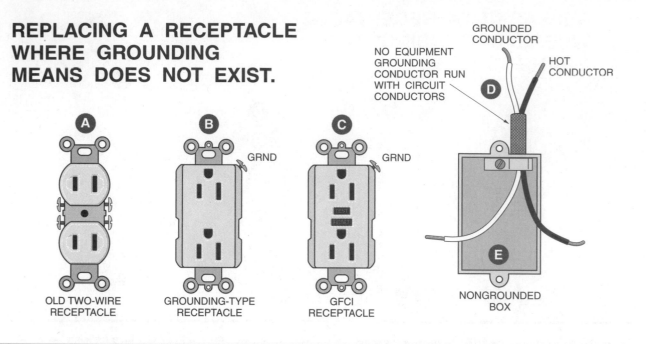

REPLACING A RECEPTACLE WHERE GROUNDING MEANS DOES NOT EXIST.

A OLD TWO-WIRE RECEPTACLE

B GROUNDING-TYPE RECEPTACLE   GRND

C GFCI RECEPTACLE   GRND

D NO EQUIPMENT GROUNDING CONDUCTOR RUN WITH CIRCUIT CONDUCTORS   GROUNDED CONDUCTOR   HOT CONDUCTOR

E NONGROUNDED BOX

Figure 6-15 If a box (E) is *not* grounded, or if an equipment grounding conductor has *not* been run with the circuit conductors (D), an existing nongrounding-type receptacle (A) may be replaced with a nongrounding-type receptacle (A), a GFCI-type receptacle (C), a grounding-type receptacle (B) if supplied through a GFCI receptacle (C), or a grounding-type receptacle (B) if a separate equipment grounding conductor is run from the receptacle to any accessible point on the grounding electrode system, *250.130(C)*. Most electricians would replace an existing nongrounding-type receptacle with a GFCI type rather than install an equipment grounding conductor, which could be very labor-intensive. See *406.3(D)(3)*.

other end of that conductor had been properly connected to an acceptable grounding point of the electrical system. The fact is that the so-called equipment grounding conductor had been connected to the replacement GFCI receptacle's green hexagonal grounding terminal that was not grounded in the first place. We now have a false sense of security, a real shock hazard.

3. The replacement receptacle may be a grounding type (Figure 6-15(B)) if it is supplied through a GFCI-type receptacle (Figure 6-15(C)).

 • The replacement receptacle must be marked "*GFCI Protected*" and "*No Equipment Ground*," as in Figure 6-16.

 • The green hexagonal equipment grounding terminal of the replacement grounding-type receptacle (Figure 6-15(B)) need not be connected to any grounding means. It may be left "unconnected." The upstream

feedthrough GFCI receptacle (Figure 6-15(C)) trip mechanism will work properly when ground faults occur anywhere on its load-side. Ground-fault protection is still there. Refer to Figure 6-7.

 • Do not connect an equipment grounding conductor from the green hexagonal grounding terminal of a replacement GFCI receptacle (Figure 6-15(C)) to any other downstream receptacles that are fed through the replacement GFCI receptacle.

4. The replacement receptacle may be a grounding type (Figure 6-15(B)) if:

 • an equipment grounding conductor, sized per *Table 250.122*, is run from the replacement receptacle's green hexagonal grounding screw and properly connected to a listed ground clamp at some point on the first 5 ft (1.5 m) of the metal water piping as it enters the building. You are *not* per-

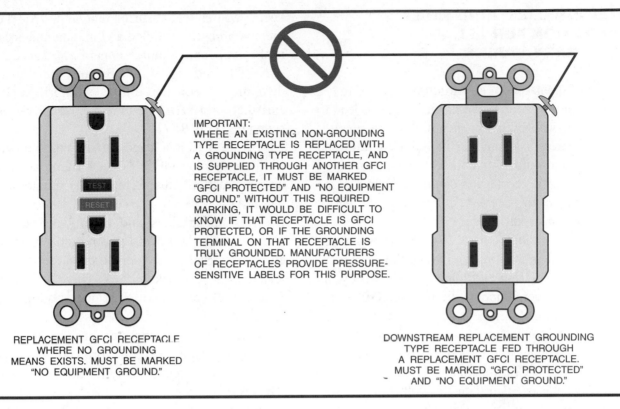

IMPORTANT:
WHERE AN EXISTING NON-GROUNDING TYPE RECEPTACLE IS REPLACED WITH A GROUNDING TYPE RECEPTACLE, AND IS SUPPLIED THROUGH ANOTHER GFCI RECEPTACLE, IT MUST BE MARKED "GFCI PROTECTED" AND "NO EQUIPMENT GROUND." WITHOUT THIS REQUIRED MARKING, IT WOULD BE DIFFICULT TO KNOW IF THAT RECEPTACLE IS GFCI PROTECTED, OR IF THE GROUNDING TERMINAL ON THAT RECEPTACLE IS TRULY GROUNDED. MANUFACTURERS OF RECEPTACLES PROVIDE PRESSURE-SENSITIVE LABELS FOR THIS PURPOSE.

REPLACEMENT GFCI RECEPTACLE WHERE NO GROUNDING MEANS EXISTS. MUST BE MARKED "NO EQUIPMENT GROUND."

DOWNSTREAM REPLACEMENT GROUNDING TYPE RECEPTACLE FED THROUGH A REPLACEMENT GFCI RECEPTACLE. MUST BE MARKED "GFCI PROTECTED" AND "NO EQUIPMENT GROUND."

Figure 6-16   Where a grounding means *does not* exist, there are certain marking requirements as presented in this diagram. *Do not* connect an equipment grounding conductor between these receptacles. See *406.3(D)(3)*.

mitted to make the equipment ground connection to the interior metal water piping anywhere beyond the first 5 ft (1.5 m). See *250.130(C)* and the *FPN*. Also read *250.50*

for the explanation of a "grounding electrode system." This permission is only for replacing a receptacle in existing installations. See Figure 6-17.

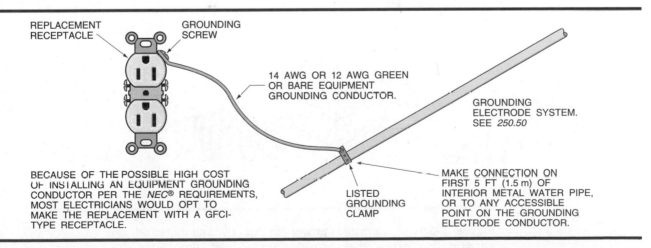

REPLACEMENT RECEPTACLE

GROUNDING SCREW

14 AWG OR 12 AWG GREEN OR BARE EQUIPMENT GROUNDING CONDUCTOR.

GROUNDING ELECTRODE SYSTEM. SEE *250.50*

BECAUSE OF THE POSSIBLE HIGH COST OF INSTALLING AN EQUIPMENT GROUNDING CONDUCTOR PER THE *NEC®* REQUIREMENTS, MOST ELECTRICIANS WOULD OPT TO MAKE THE REPLACEMENT WITH A GFCI-TYPE RECEPTACLE.

LISTED GROUNDING CLAMP

MAKE CONNECTION ON FIRST 5 FT (1.5 m) OF INTERIOR METAL WATER PIPE, OR TO ANY ACCESSIBLE POINT ON THE GROUNDING ELECTRODE CONDUCTOR.

Figure 6-17   In existing installations that do not have an equipment grounding conductor as part of the branch-circuit wiring, *406.3(D)* permits replacing an old-style nongrounding-type receptacle with a grounding-type receptacle. The grounding terminal on the receptacle must be properly grounded. One of the acceptable ways to do this is to run an equipment grounding conductor to any accessible point on the grounding electrode system. Another way is to run an equipment grounding conductor to any accessible point on the grounding electrode conductor. See *250.130(C)*.

PERSONNEL GROUND-FAULT PROTECTION FOR ALL TEMPORARY WIRING

Because of the nature of construction sites, there is a continual presence of shock hazard that can lead to serious personal injury or death through electrocution. Workers are standing in water, standing on damp or wet ground, or in contact with steel framing members. Electric cords and cables are lying on the ground, subject to severe mechanical abuse. All of these conditions spell danger.

All 125-volt, single-phase, 15-, 20- and 30-ampere receptacle outlets that are not part of the permanent wiring of the building and that will be used by the workers on the construction site must be GFCI protected, *527.6(A)*. Receptacle outlets that are part of the actual permanent wiring of a building and are used by personnel for temporary power are also required to be GFCI protected.

For receptacles *other* than 125-volt, single-phase, 15-, 20-, and 30-ampere receptacles, there are two options for providing personnel protection. In *527.6(B)(1)*, GFCI protection is one of the choices. In *527.6(B)(2)*, a written *Assured Equipment Grounding Conductor Program* is the second choice. This program shall be continuously enforced on the construction site. A designated person must keep a written log, ensuring that all electrical equipment is properly installed and maintained according to the applicable requirements of *250.114, 250.138, 406.3(C)*, and *527.4(D)*. Because this option is difficult to enforce, many Authorities Having Jurisdiction (AHJs) require GFCI protection as stated in *527.6(B)(1)*.

Figure 6-18 is a listed manufactured power outlet for providing GFCI protected temporary power on construction sites. It meets the requirements of *527.6(A)*.

Most electricians and other construction site workers carry their own listed portable ground-fault circuit-interrupter cord sets to be sure they are protected against shock hazard. These are shown in Figure 6-19 and also meet the requirements of *527.6(A)*.

Portable GFCI devices are easy to carry around. They have "open neutral" protection should the neutral in the circuit supplying the GFCI open for whatever reason.

Portable GFCI devices are available with manual reset, which is advantageous should a power outage occur or if the GFCI is unplugged, so that equipment (drills, saws, etc.) will not start up again when the power is restored. This could cause injury to anyone using the equipment when the power is restored.

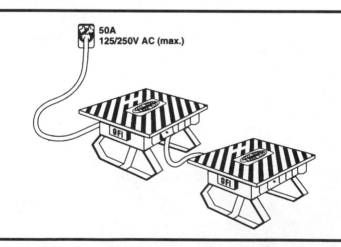

50A
125/250V AC (max.)

Figure 6-18 A portable power outlet commonly used on construction sites that provide ground-fault protection for workers. This particular unit is connected with a 50-ampere 125/250-volt power cord. Each of the six 20-ampere receptacles in the unit is connected to individual GFCI circuit breakers in the unit that provide both ground-fault and overload protection for anything plugged into one of the receptacles. *Courtesy of* Hubbell, Incorporated, Wiring Device Division.

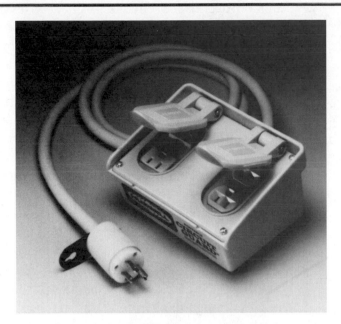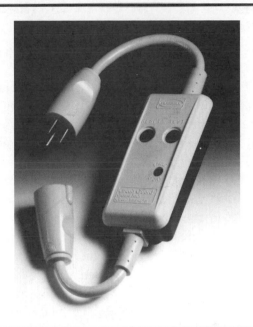

Figure 6-19   Plug- and cord-connected "portable" GFCI devices are easy to carry around. They can be used anywhere when working with electrical tools and extension cords. These devices operate independently and are not dependent upon whether or not the branch-circuit is GFCI protected. These portable GFCI in-line cords are relatively inexpensive and should always be used when using portable electric tools. *Courtesy of* Hubbell Incorporated, Wiring Device Division.

Portable devices are also available that reset automatically. These should be used on lighting, engine heaters, water (sump) pumps and similar equipment where it would be advantageous to have the equipment start up as soon as power is restored.

IMMERSION DETECTION CIRCUIT INTERRUPTERS (IDCIs)

You will recognize an **Immersion Detection Circuit Interrupter (IDCI)** by the large attachment plug cap on hand-held grooming appliances such as electric hair dryers, combs, and curlers. The major culprit for electrical shock was found to be personal grooming appliances because they are most commonly used at a vanity having a wash basin with water in it, or even worse, next to a water-filled bathtub with a person (child or adult) in it. Accidentally dropping or pulling the appliance into the tub when reaching for a towel can be lethal. Electricity and water do not mix!

An IDCI is usually an integral part of the attachment plug cap. An IDCI is required to open the circuit regardless of whether the appliance switch is in the "ON" or "OFF" position.

The operation of an IDCI depends upon a "third" wire *probe* in the cord, which is connected to a sensor within the appliance. This probe will detect leakage current flow when conductive liquid enters the appliance and makes contact with any live part inside the appliance and the internal "sensor." When this happens, the IDCI trips "OFF." Never replace the special attachment plug cap on these appliances with a standard plug-cap because that would make the appliance absolutely unsafe. The UL requirement is that the plug cap shall be marked: "Warning: To Reduce The Risk Of Electric Shock, Do Not Remove, Modify, Or Immerse This Plug."

Like a GFCI, an IDCI must open the circuit when the sensor detects a leakage current to ground in the range of 4 to 6 milliamperes. If you plug an IDCI protected appliance into a GFCI receptacle, then you have double protection. An IDCI on a hair dryer or similar appliance must be manually reset. Protection against electrocution for free-standing hydromassage appliances and hand-held hair dryers is a requirement in *422.41*. This is intended to include heated stylers, heated air combs, heated air curlers, and curling iron–hair dryer combinations.

ARC-FAULT CIRCUIT INTERRUPTERS (AFCIs)

Arc-fault circuit interrupters (AFCIs) are the "new kid on the street."

▶The *NEC*® requires that effective January 1, 2002, arc-fault circuit interrupter (AFCI) protection must be provided for all 125-volt, single-phase, 15- and 20- ampere outlets in dwelling bedrooms. An AFCI circuit breaker installed in the panel provides arc-fault protection for the entire branch circuit. A feedthrough AFCI receptacle installed at the first receptacle in the branch circuit provides arc-fault protection for that receptacle and all downstream outlets connected to that receptacle. AFCI receptacles are available that have the technology to "look" upstream as well as downstream to sense arcing on the circuit. The *NEC*® requires that AFCIs be listed by a nationally recognized testing laboratory (NRTL). See *210.12.*◀

As presently worded, this requirement includes all receptacle outlets, lighting/fan outlets, and 120-volt smoke detector outlets in bedrooms. It does not include closet lighting. In most cases, the closet lighting is probably connected to the same branch circuit as the bedroom lighting and receptacles, and would be protected by an AFCI circuit breaker supplying the circuit.

To meet the requirement of *210.12*, Circuits A16, A19, and A21 are protected by AFCI circuit breakers installed in Panel A. All lighting, fan, and receptacle outlets on these circuits are AFCI protected.

Why AFCI Protection?

Electrical arcing is one of the leading causes of electrical fires in homes. The Consumer Product Safety Commission recently cited 150,000 residential electrical fires annually in the United States, 850 deaths, 6000 injuries, and more than $1.5 billion in property loss. These statistics are pretty convincing of the need for greater electrical protection. The heat of an arc is extremely high, and the hot particles of metal expelled by the arc are enough to ignite most surrounding combustible materials. The temperature of an arc can reach 10,000°F or more. Arcing faults can be line-to-line, line-to-neutral, or line-to-ground. Arcing is considered an "early" event in the history of a typical electrical fire. If arcing can be detected early on, many electrical fires can be prevented.

We have all been guilty of blaming electrical fires on overloaded circuits, overloaded extension cords, and too many extension cords plugged into a receptacle. The heat of an overloaded cord *by itself* is generally not hot enough to ignite surrounding combustible materials. An overload condition can cause the breakdown of the insulation on the conductors, allowing the conductors to touch one another. This sets the stage for an arcing condition between the conductors, or between the "hot" conductor and ground. We now have the ingredients of an electrical fire.

Conventional circuit breakers and fuses cannot detect these low-current level arcing conditions.

Let's say that the cord on an electric iron becomes frayed and the conductor(s) break. The iron draws 10 amperes. The arcing of 10 amperes trying to bridge the gap in the broken conductor(s) develops the tremendous heat capable of igniting surrounding combustible material, yet 10 amperes flowing in the circuit *will not* trip a 15-ampere circuit breaker or open a 15-ampere fuse. This condition is not an overload, and initially is not a short circuit. These overcurrent devices were not designed to detect and open under current levels below their trip setting. What is needed is a device that can detect the characteristics of the arcing condition.

Why Bedrooms?

Consumer Product Safety Commission studies indicate that a high percentage of residential electrical fires occur in bedrooms. The Code Making Panels saw the benefit of arc-fault protection and introduced the AFCI requirement for residential bedrooms. As time passes and more is learned, we can expect AFCI protection requirements to extend to other rooms of a home, in the same manner that ground-fault circuit interrupter (GFCI) protection spread to more locations over a number of Code cycles. Over time, the cost of AFCI/GFCI combination circuit breakers will become so low that it will become cost effective to have AFCI/GFCI protection for all lighting and appliance branch circuits in homes. Although the 2002 edition of the *NEC*® specifies AFCI protection for all outlets in bedrooms, there already have been Code Proposals to extend the AFCI requirement beyond bedrooms. This will probably happen as more field experience is compiled on the effectiveness and cost/benefit comparison of AFCI protection.

How Does an AFCI Work?

A new protective device was needed. *Voila!* The industry responded with an *Arc-Fault Circuit Interrupter (AFCI).*

When a conductor that is carrying current breaks, or when a loose connection occurs, it is a "series" arcing fault because the current is trying to travel in its intended path. A broken or frayed cord could result in a "series" or "parallel" arc. Sputtering is another term to define an arc. Arc faults can go undetected for long periods of time without AFCI protection.

When arcing occurs between the black ("hot") and white grounded (neutral) circuit conductors, or between the black ("hot") conductor and ground, it is referred to as a "parallel" arc. These conditions might be caused by a nail being driven through a cable, a staple driven too tightly, or a clamp or connector squeezing the cable too tightly. The term "ground" refers to any equipment ground such as a grounded metal raceway or cable, a grounded metal box or cabinet, metal water or gas piping, sheet metal ducts, or similar grounded metal items.

UL Standard 1699 spells out the sensing and tripping characteristics of an AFCI. This UL Standard shows AFCIs in the following types:

1. **Branch/feeder AFCI:** Installed at the panel. Protects the branch-circuit wiring. Provides somewhat limited protection for extension cords.

2. **Outlet AFCI:** This is an AFCI receptacle. Provides protection for cord-sets that are plugged into the receptacle. Receptacle can be of the "feedthrough" type. Some AFCI receptacles are available that can sense upstream as well as downstream arcing in the branch circuit. Be sure to check the manufacturer's installation instructions.

3. Combination of a branch/feeder AFCI and an outlet AFCI.

4. **Portable AFCI:** A plug-in device with one or more receptacles. Provides AFCI protection for cords that are plugged into the device. Similar to the portable GFCI devices.

5. **Cord AFCI:** A plug-in device that provides AFCI protection to the power cord connected to it. It has no additional outlets.

An AFCI is designed to sense the rapid fluctuations of current flow typical of an arcing condition. It will recognize the unique characteristic sometimes referred to as the "signature" of a dangerous arc. It will de-energize the circuit when dangerous arcing is detected. AFCI devices do not qualify as GFCI protection unless also listed to UL 943, the GFCI standard. Some manufacturers have AFCI circuit breakers that do meet both AFCI and GFCI requirements.

The design of the electronic circuitry of an AFCI is such that it will not nuisance trip because of the operational arcing that occurs when a switching device is turned on or off, when something is plugged into or unplugged from a receptacle, when a light bulb burns out, brush contact of universal motors, and similar harmless arcs. In other words, an AFCI can distinguish between dangerous arcs and harmless arcs.

Types of AFCI Devices

The most common AFCIs that electricians install are the circuit-breaker type and receptacle type. Circuit breaker AFCIs are the same width as a standard circuit breaker so they will fit into a standard electrical panel. AFCI receptacles fit into standard device boxes.

When installing a 125-volt, single-pole AFCI circuit breaker, the white lead from the AFCI circuit breaker is connected to the neutral bar in the panel. The black branch-circuit conductor is connected to the brass-colored load side terminal on the breaker. The white branch-circuit conductor is connected to the silver-colored load side terminal on the breaker. This is the same as for GFCI circuit breakers discussed earlier in this unit.

Never share the neutral of a multiwire branch-circuit when using GFCIs and AFCIs.

The mandatory requirement in *210.12* for AFCI protection is for new work only. It is not retroactive. It is easy to provide arc-fault protection in an older home. Just replace existing circuit breaker(s) that feed bedrooms with an AFCI circuit breaker(s). This added protection might be a good idea in older homes where the insulation on the original wiring is rated 140°F (60°C). In and above a luminaire (fixture), the heat from the luminaire (fixture) literally bakes and chars the insulation on the conductors. AFCI protection could "catch" an arcing problem in its early stages. Today, conductors used in house wiring are rated 194°F (90°C), so the deterioration of conductor insulation because of heat has pretty well become a thing of the past.

Other Types of "People Protectors"

Most older homes do not have GFCI protection. There are other devices that manufacturers of appliances are required to design into certain electrical products in conformance to various UL standards. In addition to GFCIs and IDCIs, other devices currently being used are *appliance leakage circuit interrupters (ALCI)* that trip at 6 milliamperes, *equipment leakage circuit interrupters (ELCI)* that trip at 50 milliamperes, and *leakage monitoring receptacles (LMR)* that trip at 50 milliamperes.

TRANSIENT VOLTAGE SURGE SUPPRESSORS (TVSSs)

▶Transient voltage surge suppressor (TVSS) Code rules that are permanently installed are found in *Article 285* in the *NEC.*®◀ Always install TVSS devices that are listed under the requirements of UL Standard 1449. Carefully read and follow the instructions furnished with the TVSS.

In today's homes, we find many electronic appliances (television, stereo, personal computer, fax equipment, word processor, VCR, CD player, digital stereo equipment, microwave oven), all of which contain many sensitive, delicate electronic components (printed circuit boards, chips, microprocessors, transistors, etc.).

Voltage transients, called *surges* or *spikes*, can stress, degrade, and/or destroy these components. They can cause loss of memory in the equipment or "lock up" the microprocessor.

The increased complexity of electronic integrated circuits makes this equipment an easy target for "dirty" power that can and will affect the performance of the equipment.

Voltage transients cause abnormal current to flow through the sensitive electronic components. This energy is measured in joules. A *joule* is the unit of energy when one ampere passes through a one-ohm resistance for one second (like wattage, only on a much smaller scale).

Line surges can be line-to-neutral, line-to-ground, and line-to-line.

Transients

Transients are generally grouped into two categories:

- **Ring wave:** These transients originate within the building and can be caused by copiers, computers, printers, HVAC cycling, spark igniters on furnaces, water heaters, dryers, gas ranges, ovens, motors, and other inductive loads.
- **Impulse:** These transients originate outside of the building and are caused by utility company switching, lightning, and so on.

To minimize the damaging results of these transient line surges, service equipment, panels, load centers, feeders, branch-circuits, and individual receptacle outlets can be protected with a TVSS, as shown in Figure 6-20.

A TVSS contains one or more metal-oxide varistors (MOV) that clamp the transients by absorbing the major portion of the energy (joules) created by the surge, allowing only a small, safe amount of energy to enter the actual connected load.

The MOV clamps the transient in times of less than 1 nanosecond, which is one billionth of a second, and keeps the voltage spike passed through to the connected load to a maximum range of 400 to 500 peak volts.

Typical TVSS devices for homes are available as an integral part of receptacle outlets that mount in the same wall boxes as regular receptacles. They look the same as a normal receptacle and may have an audible alarm that sounds when an MOV has failed. The TVSS device may also have a visual indication, such as a light-emitting diode (LED) that glows continuously until an MOV fails. Then the LED starts to flash on and off, as in Figure 6-21.

TVSS devices are available in plug-in strips (Figure 6-22) and as part of desktop computer hardware power direction stands commonly placed under a monitor for plugging in a monitor, printer, scanner, and other peripheral equipment. These minimize the accumulation of the multitude of cords under and behind the equipment.

A TVSS on a branch-circuit will provide surge suppression for all of the receptacles on the same circuit. These surges are classified as Category A low-level surges.

"Whole-house" surge protectors are available that offer surge protection for the entire house.

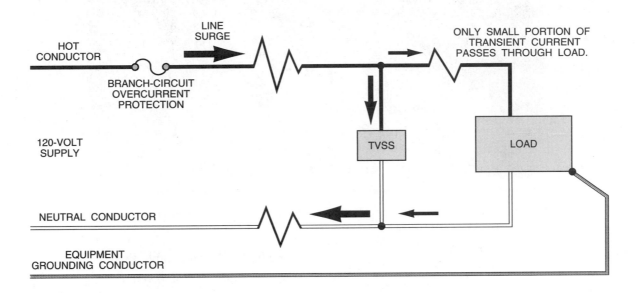

Figure 6-20   A transient voltage surge suppressor absorbs and bypasses (shunts) transient currents around the load. The MOV dissipates the surge in the form of heat.

Figure 6-21   Surge suppressor protector. *Courtesy of Hubbell, Incorporated.*

These are generally available at electrical supply houses and are installed at the main electrical panel of a home.

When a whole-house surge protector is installed, it is still a good idea to install surge protectors at the plug-in locations to more closely protect against low level surges at the computer or other delicate electronic equipment location.

Noise

"Noise" is recognized as snow on a TV screen or as static on a radio or telephone. Low-level nondamaging transients can also be present. These can be caused by fluorescent lamps and ballasts, other electronic equipment in the area, x-ray equipment, motors running or being switched on and off, improper grounding, and so forth. Although not physically damaging to the electronic equipment, computers can lose memory or perform wrong calculations. Their intended programming can malfunction.

"Noise" comes from *electromagnetic interference* (EMI) and *radio frequency interference* (RFI). EMI is usually caused by ground currents of very low values from motors, utility switching loads, lightning, and so forth, and is transmitted through metal conduits. RFI is noise, like the buzzing heard on a car radio when driving under a high-voltage transmission line. This interference "radiates" through the air from the source and is picked up by the grounding system of the building.

This undesirable noise can be reduced by reducing the number of ground reference points on a system. This can be accomplished by installing an **isolated ground receptacle**, but they are not too common in residential wiring.

Figure 6-22 Surge suppressors. *Courtesy of* Hubbell, Incorporated.

Here are some related UL Standards:

- UL 498 Attachment Plugs and Receptacles

- UL 943 Ground-Fault Circuit Interrupters

- UL 1283 Electromagnetic Interference Filters

- UL 1363 Relocatable Power Taps

- UL 1449 Transient Voltage Surge Suppressors

- UL 1644 Immersion Detection Circuit Interrupters

- UL 1699 Arc-Fault Circuit Interrupters

REVIEW

1. Explain the operation of a ground-fault circuit interrupter. Why are GFCI devices used? Where are GFCI receptacles required? _____

_____

_____

_____

_____

2. Residential GFCI devices are set to trip a ground-fault current above _____ milliamperes.

3. Where must GFCI receptacles be installed in residential garages? _____

_____

_____

_____

_____

4. The Code requires GFCI protection for certain receptacles in the kitchen. Explain where these are required. _____

_____

_____

5. Is it a Code requirement to install GFCI receptacles in a fully carpeted, finished recreation room in the basement? _____

6. A homeowner calls in an electrical contractor to install a separate circuit in the basement (unfinished) for a freezer. Is a GFCI receptacle required? _____

7. GFCI protection is available as (a) branch-circuit breaker GFCI, (b) feeder circuit breaker GFCI, (c) individual GFCI receptacles, (d) feedthrough GFCI receptacles. In your opinion, for residential use, what type would you install? _____

8. Extremely long circuit runs connected to a GFCI branch-circuit breaker might result in
    a. nuisance tripping of the GFCI
    b. loss of protection
    c. the need to reduce the load on the circuit
    Circle the correct answer.

9. If a person comes in contact with the "hot" and grounded conductors of a two-wire branch-circuit that is protected by a GFCI, will the GFCI trip "OFF"? Why?

    _____

    _____

    _____

    _____

    _____

10. What might happen if the line and load connections of a feedthrough GFCI receptacle were reversed? _____

    _____

    _____

    _____

    _____

    _____

11. May a GFCI receptacle be installed as a replacement in an old installation where the 2-wire circuit has no equipment grounding conductor? _____

12. What two types of receptacles may be used to replace a defective receptacle in an older home that is wired with knob-and-tube wiring where no equipment grounding means exists in the box? _____

    _____

13. You are asked to replace a receptacle. Upon checking the wiring, you find that the wiring method is conduit and that the wall box is properly grounded. The receptacle is of the older style two-wire type that does not have a grounding terminal. You remove the old receptacle and replace it with (Circle the letter of the correct answer.)
    a. the same type of receptacle as the type being removed.
    b. a receptacle that is of the grounding type.
    c. a GFCI receptacle.

14. What color are the terminals of a standard grounding-type receptacle? _____

    _____

15. What special shape and color are the grounding terminals of receptacle outlets and other devices? _____

    _____

16. Construction sites can be dangerous because of the manner in which extension cords, portable electrical tools, and other electrical equipment are used and abused. To reduce this hazard, *527.6(A)* of the *NEC*® requires that all 125-volt, single-phase, 15-, 20-, and 30-ampere receptacle outlets used for temporary wiring during construction must be _____ protected.

17. In your own words, explain why the Code does not require certain receptacle outlets in kitchens, garages, and basements to be GFCI protected. _____

    _____

    _____

    _____

    _____

    _____

    _____

    _____

    _____

    _____

    _____

18. Circuit A1 supplies GFCI receptacles ① and ②. Circuit A2 supplies GFCI receptacles ③ and ④. Receptacles ① and ③ are feedthrough type. Using colored pencils or marking pens, complete all connections.

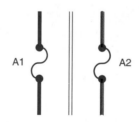

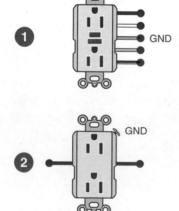

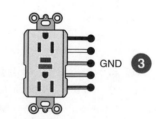

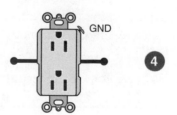

19. The term TVSS is becoming quite common. What do the letters stand for? _____

    _____

    _____

20. Transients (surges) on a line can cause spikes or surges of energy that can damage delicate electronic components. A TVSS device contains one or more _____ _____ _____ that bypass and absorb the energy of the transient.

21. Undesirable "noise" on a circuit can cause computers to lock up, lose their memory, and/or cause erratic performance of the computer. This "noise" does not damage the equipment. The two types of this "noise" are EMI and RFI. What do these letters mean?

    _____

    _____

    _____

    _____

22. Can TVSS receptacles be installed in standard device boxes? _____

23. Some line transients are not damaging to electronic equipment, but can cause the equipment to operate improperly. The effects of these transients can be minimized by installing _____ _____ _____ .

24. Briefly explain the operation of an immersion detection circuit interrupter (IDCI).

    _____

    _____

    _____

    _____

    _____

    _____

25. What range of leakage current must trip an IDCI? _____

    _____

    _____

26. What amount of current flowing through a male human being will cause muscle contractions that will keep him from letting go of the live wire? _____ milliamperes

27. Other than for a few exceptions, the *NEC*® requires that all systems having a voltage of over _____ volts shall be grounded.

28. In an old house, an existing nongrounding non-GFCI-type receptacle in a bathroom needs to be replaced. According to *406.3(D)*, this replacement receptacle shall be (circle the correct letter):

    a. a nongrounding-type receptacle.

    b. a GFCI-type receptacle or be GFCI protected in some other way, such as changing the breaker that supplies the receptacle to a GFCI type.

29. Do all receptacle outlets in kitchens have to be GFCI protected? Explain. _____

_____

_____

_____

30. What special device is required for all 120-volt outlets in bedrooms? _____

_____

31. When installing GFCI and AFCI circuit breakers in a panel, it is permissible to connect them to a multiwire branch circuit. (True) (False). Circle the correct answer.

Luminaires (Fixtures), Ballasts, and Lamps

OBJECTIVES

After studying this unit, the student should be able to

- understand luminaire (fixture) terminology, such as Type IC, Type NON IC, suspended ceiling luminaires (fixtures), recessed luminaires (fixtures), and surface-mounted luminaires (fixtures).
- connect recessed luminaires (fixtures), both prewired and nonprewired types, according to Code requirements.
- specify insulation clearance requirements.
- discuss the importance of temperature effects while planning recessed luminaire (fixture) installations.
- describe thermal protection for recessed luminaires (fixtures).
- understand the cautions to be observed when installing outdoor lighting.
- understand Class P and other types of fluorescent ballasts.

WHAT IS A LUMINAIRE (FIXTURE)?

▶The *FPN* in *410.1* defines a luminaire (fixture) as "*a complete lighting unit consisting of a lamp or lamps together with the parts designed to distribute the light, to position and protect the lamps and ballast (where applicable), and to connect the lamps to a power supply.*" "Luminaire" is the international term for "lighting fixture" and is used throughout the *NEC.*® In the *NEC,*® the term is followed by the word "fixture."◀

TYPES OF LUMINAIRES (FIXTURES)

There are literally thousands of different types of luminaires (fixtures) from which to choose to satisfy certain needs, wants, desires, space requirements, and last but not least, price considerations.

Note in Table 7-1, that whether the luminaire (fixture) is incandescent or fluorescent, the basic categories are surface mounted, recessed mounted, and suspended ceiling mounted.

FLUORESCENT	INCANDESCENT
• Surface	• Surface
• Recessed	• Recessed
• Suspended Ceiling	• Suspended Ceiling

Table 7-1

The Code Requirements

Article 410 sets forth the requirements for installing luminaires (fixtures). The electrician must "meet Code" with regard to mounting, supporting, grounding, live-parts exposure, insulation clearances, supply conductor types, maximum lamp wattages, and so forth.

Probably the two biggest contributing factors to fires caused by luminaires (fixtures) are installing lamp wattages that exceed that for which the luminaire (fixture) has been designed, and burying recessed luminaires (fixtures) under thermal insulation when the luminaire (fixture) has not been designed for such an installation.

UL tests, lists, and labels luminaires (fixtures) that are in conformance to their standards. Always install luminaires (fixtures) that bear the UL label.

In addition to the *NEC*, the UL *Electrical Construction Materials Directory* (Green Book) and the UL *General Information for Electrical Equipment* (White Book), and manufacturers' catalogs and literature are excellent sources of information about luminaires (fixtures).

Read the Label

Always carefully read the label on a luminaire (fixture). Many Code requirements can be met by simply doing what the label says. It will provide such information as:

- For wall mount only.
- Ceiling mount only.
- Maximum lamp wattage.
- Type of lamp.
- Access from above ceiling required.
- Access from behind wall required.
- Suitable for air handling only.
- For chain or hook suspension only.
- Suitable for operation in ambient temperature not exceeding _____°F (°C).
- Suitable for installation in poured concrete.
- For line volt-amperes, multiply lamp wattage by 1.25.
- Suitable for use in suspended ceilings.
- Suitable for use in noninsulated ceilings.
- Suitable for use in insulated ceilings.
- Suitable for damp locations.
- Suitable for wet locations.
- Suitable for use as a raceway.
- Suitable for mounting on low-density cellulose fiberboard.
- For supply connections, use wire rated for at least _____°F (°C).
- Not for use in dwellings.
- Thermally protected.
- Type-IC.
- Type Non-IC.
- Inherently protected.

Surface-Mounted Luminaires (Fixtures)

These are easy to install. It is simply a matter of following the manufacturers' instructions furnished with the luminaires (fixtures). The luminaire (fixture) is attached to a ceiling outlet box or wall outlet box using luminaire (fixture) studs, hickeys, bar straps, or luminaire (fixture) extensions. Do not exceed the maximum lamp wattage marked on the luminaire (fixture).

Special attention must be given when installing surface-mounted luminaires (fixtures) on low-density cellulose fiberboard, Figure 7-1. Because of the potential fire hazard, *410.76(B)* states that fluorescent luminaires (fixtures) that are surface mounted on this material must be spaced at least 1½ in. (38 mm) from the fiberboard surface unless the fixture is marked "Suitable for Surface Mounting on Low-Density Cellulose Fiberboard."

The *Fine Print Note* to *410.76(B)* explains combustible low-density cellulose fiberboard as sheets, panels, and tiles that have a density of 20 lb/ft³ (320 kg/m³) or less that are formed of bonded plant fiber material. It does not include fiberboard that has a density of over 20 lb/ft³ (320 kg/m³) or material that has been integrally treated with fire-retarding chemicals to meet specific standards. Solid or laminated wood does not come under the definition of "combustible low-density cellulose fiberboard."

To obtain fire-resistance ratings, most acoustical ceiling panels today are made from mineral wool fibers or fiberglass.

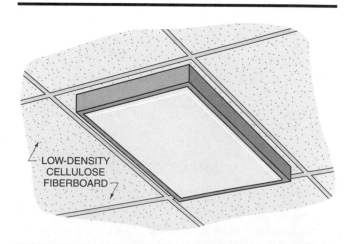

LOW-DENSITY CELLULOSE FIBERBOARD

Figure 7-1   If they are to be mounted on *low-density cellulose fiberboard*, surface-mounted fluorescent luminaires (fixtures) must be marked "Suitable for Surface Mounting on Low-Density Cellulose Fiberboard."

Recessed Luminaires (Fixtures)

Installing recessed luminaires (fixtures) requires more attention than simple surface mounted luminaires (fixtures). The Code requirements for the installation of recessed luminaires (fixtures) are found in *410.64* through *410.72*. UL Standard 1571 covers recessed incandescent luminaires (fixtures).

Recessed luminaires (fixtures) as shown in Figure 7-2 have an inherent heat problem. They

must be suitable for the application and must be installed properly.

It is absolutely essential for the electrician to know early in the roughing-in stages of wiring a house what types of luminaires (fixtures) are to be installed. This is particularly true for recessed-type luminaires (fixtures). To ensure that such factors as location, proper and adequate framing, and possible obstructions have been taken into consideration, the electrician must work closely with the general building contractor, carpenter, plumber, heating contractor, the thermal insulation installer, and the other building trades.

Figure 7-3(A) shows a Type Non-IC (insulated ceiling) recessed luminaire (fixture) where a minimum 3-in. (75-mm) clearance between the sides of the luminaire (fixture) and thermal insulation is needed. Above the luminaire (fixture), thermal insulation must be installed so as not to entrap heat in the cavity. How far should the thermal insulation be above the luminaire? It is a judgment call: 3 in. (75 mm) should suffice. Check the manufacturer's instructions.

Figure 7-2   Typical recessed luminaire (fixture).

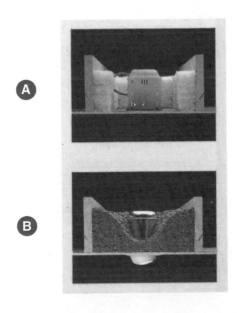

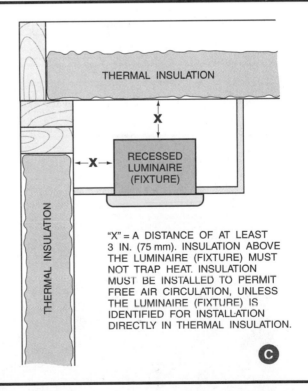

"X" = A DISTANCE OF AT LEAST 3 IN. (75 mm). INSULATION ABOVE THE LUMINAIRE (FIXTURE) MUST NOT TRAP HEAT. INSULATION MUST BE INSTALLED TO PERMIT FREE AIR CIRCULATION, UNLESS THE LUMINAIRE (FIXTURE) IS IDENTIFIED FOR INSTALLATION DIRECTLY IN THERMAL INSULATION.

Figure 7-3   (A) is a Type Non-IC recessed luminaire (fixture) that requires clearance between the luminaire (fixture) and the thermal insulation. (B) is a Type IC recessed luminaire (fixture) that may be completely covered with thermal insulation. (C) shows the required clearances from thermal insulation for a Type Non-IC luminaire (fixture). See *410.66*.

Figure 7-3(B) shows a Type IC recessed luminaire (fixture) that may be completely buried in thermal insulation.

Figure 7-3(C) is a sketch that summarizes the minimum clearances for a Type Non-IC recessed luminaire (fixture) that is not listed for direct burial in thermal insulation, *410.66(B)*.

Recessed luminaires (fixtures) are available for both new work and remodel work. Remodel work recessed luminaires (fixtures) can be installed from below an existing ceiling by cutting a hole in the ceiling, bringing power to the luminaire (fixture), making the electrical connections in the integral junction box on the luminaire (fixture), then installing the housing through the hole.

Energy Efficient Housings

Some recessed luminaires (fixtures) have energy efficient housings that restrict the airflow (leakage)

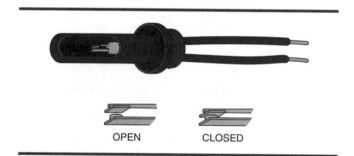

Figure 7-4   One type of recessed luminaire (fixture) thermal cutout.

through the luminaire (fixture) to less than 2 ft³ of air per minute. These luminaires (fixtures) meet energy codes that have been adopted by some states. You should check with your local electrical inspector and/or building official to see if such laws are applicable in your area for one- and two-family dwellings.

Thermal Protection

To protect against the hazards of overheating, UL and the *NEC®* in *410.65(C)* require that recessed luminaires (fixtures) be equipped with an integral thermal protector, as in Figure 7-4. These devices will cycle on and off repeatedly until the heat problem has been resolved. There are two exceptions to this requirement. Thermal protection is not required for recessed incandescent luminaires (fixtures) designed for installation directly in a concrete pour, or for recessed incandescent luminaires (fixtures) that are constructed so as to not exceed "temperature performance characteristics" equivalent to thermally protected luminaires (fixtures). These luminaires (fixtures) are so identified.

What the Junction Box on the Recessed Luminaire (Fixture) Housing Is for

Most recessed luminaires (fixtures) come equipped with a junction box that is an integral part of the luminaire (fixture) housing. This is clearly shown in Figure 7-2 and Figure 7-5. For most resi-

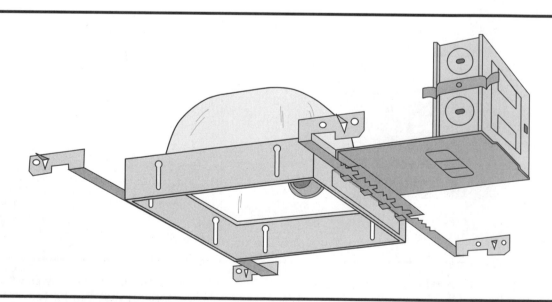

Figure 7-5   Roughing-in box of a recessed luminaire (fixture) with mounting brackets and junction box.

dential-type recessed luminaire (fixture) installations, nonmetallic-sheathed cable or armored cable can be run directly into this junction box. These cables have conductors rated 194°F (90°C). Check the marking on the luminaire (fixture) for supply conductor temperature limitations.

Figure 7-6 shows the clearances normally needed between recessed luminaires (fixtures) and combustible framing members. This figure also summarizes the *NEC®* requirements for connecting the supply conductors to a recessed luminaire (fixture). Note in one case that the branch-circuit wiring is brought directly into the integral junction box on the luminaire (fixture). In the other case, a "fixture whip" is run from an adjacent junction box to the luminaire (fixture). In some instances, the luminaire (fixture) comes with a flexible metal "fixture whip" and in other situations, the "fixture whip" is supplied by the electrician.

The short length of flex between an outlet box and a luminaire (fixture) is referred to as a "fixture whip" by electricians across the country. The term is a trade expression as common as apple pie. Until the term "luminaire" becomes the term used in the trade, let's continue to use the term "fixture whip."

In *410.11*, we find that *"Branch-circuit wiring, other than 2-wire or multiwire branch-circuits supplying power to luminaires (fixtures) connected together, shall not be passed through an outlet box that is an integral part of a luminaire (fixture) unless the luminaire (fixture) is identified for through-wiring."*

In *410.31*, we find that *"luminaires (fixtures) shall not be used as a raceway for circuit conductors unless listed and identified for use as a raceway."*

Some recessed luminaires (fixtures) are marked "identified for through-wiring," in which case branch-circuit conductors in addition to the conductors that supply the luminaire (fixture) are permitted to be run through the outlet box or wiring compartment on the luminaire (fixture). A typical label will look like Figure 7-7:

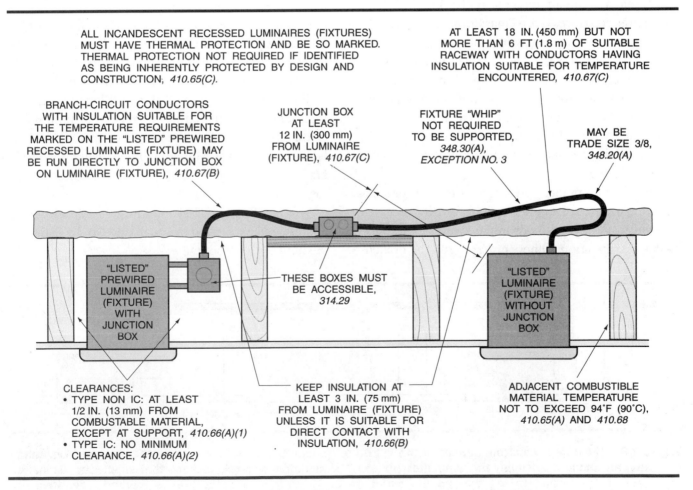

ALL INCANDESCENT RECESSED LUMINAIRES (FIXTURES) MUST HAVE THERMAL PROTECTION AND BE SO MARKED. THERMAL PROTECTION NOT REQUIRED IF IDENTIFIED AS BEING INHERENTLY PROTECTED BY DESIGN AND CONSTRUCTION; *410.65(C)*.

AT LEAST 18 IN. (450 mm) BUT NOT MORE THAN 6 FT (1.8 m) OF SUITABLE RACEWAY WITH CONDUCTORS HAVING INSULATION SUITABLE FOR TEMPERATURE ENCOUNTERED, *410.67(C)*

BRANCH-CIRCUIT CONDUCTORS WITH INSULATION SUITABLE FOR THE TEMPERATURE REQUIREMENTS MARKED ON THE "LISTED" PREWIRED RECESSED LUMINAIRE (FIXTURE) MAY BE RUN DIRECTLY TO JUNCTION BOX ON LUMINAIRE (FIXTURE), *410.67(B)*

JUNCTION BOX AT LEAST 12 IN. (300 mm) FROM LUMINAIRE (FIXTURE), *410.67(C)*

FIXTURE "WHIP" NOT REQUIRED TO BE SUPPORTED, *348.30(A), EXCEPTION NO. 3*

MAY BE TRADE SIZE 3/8, *348.20(A)*

"LISTED" PREWIRED LUMINAIRE (FIXTURE) WITH JUNCTION BOX

THESE BOXES MUST BE ACCESSIBLE, *314.29*

"LISTED" LUMINAIRE (FIXTURE) WITHOUT JUNCTION BOX

CLEARANCES:
• TYPE NON IC: AT LEAST 1/2 IN. (13 mm) FROM COMBUSTABLE MATERIAL, EXCEPT AT SUPPORT, *410.66(A)(1)*
• TYPE IC: NO MINIMUM CLEARANCE, *410.66(A)(2)*

KEEP INSULATION AT LEAST 3 IN. (75 mm) FROM LUMINAIRE (FIXTURE) UNLESS IT IS SUITABLE FOR DIRECT CONTACT WITH INSULATION, *410.66(B)*

ADJACENT COMBUSTIBLE MATERIAL TEMPERATURE NOT TO EXCEED 94°F (90°C), *410.65(A)* AND *410.68*

Figure 7-6   Requirements for installing recessed luminaires (fixtures).

Maximum of ___ ___ AWG through branch-circuit conductors suitable for at least ___°C (___°F) permitted in a junction box (___ in ___ out).

Figure 7-7 A typical label found on a recessed luminaire (fixture) specifying the maximum number and size of conductors permitted to be run through the wiring compartment on the luminaire (fixture).

The manufacturer of the luminaire (fixture), following UL requirements, will determine the maximum number, size, and temperature rating for the conductors entering and leaving the outlet box or wiring compartment, and will so mark the label.

Only those luminaires (fixtures) that have been tested for the extra heat generated by the additional branch-circuit conductors will bear the label marking shown in Figure 7-7.

These fixtures have been tested for the added heat generated by the additional branch-circuit conductors.

Figure 7-8 shows three recessed luminaires (fixtures), each with an integral outlet box. If these luminaires (fixtures) are marked "Identified for through-wiring," then it would be permitted to run conductors through the outlet boxes *in addition* to the conductors that supply the luminaires (fixtures). If the "Identified for through-wiring" marking is not found on the luminaire (fixture), then it is a Code violation to run any conductors through the box other than the conductors that supply the luminaire (fixture) itself.

Some electricians prefer to run the branch-circuit wiring to the junction box on the recessed luminaire (fixture), then drop the switch loop to the switch box location. Others prefer to run the branch-circuit to the switch location, then to the junction box on the luminaire (fixture). Either way is acceptable.

Recessed Luminaire (Fixture) Trims

Recessed incandescent luminaires (fixtures) are listed by Underwriters Laboratories with specific trims. Luminaire (fixture)/trim combinations are marked on the label. Do not use trims that are not listed for use with the particular recessed luminaire (fixture). Installing a trim not listed for use with a specific recessed luminaire (fixture) can result in overheating and possible on-off cycling of the luminaire's (fixture's) thermal protective device. Mismatching luminaires (fixtures) and trims is a violation of *110.3(B)* of the *National Electrical Code,®* which states that *"Listed or labeled equipment shall be installed or used in accordance with any instructions included in the listing or labeling."*

Sloped Ceilings

►Be very careful when selecting recessed luminaires (fixtures). Conventional recessed luminaires (fixtures) are listed for installation in flat ceilings, not in sloped ceilings! Recessed luminaires (fixtures) intended for installation in a sloped ceiling are submitted to UL specifically for testing in a sloped ceiling. Look for the marking "Sloped Ceiling." Instructions furnished with recessed luminaires (fixtures) provide the necessary information, such as being suitable for sloped ceilings having a 2/12, 6/12, or 12/12 pitch. Check the manufacturer's catalog and the UL listing.◄

Remember, *110.3(B)* requires that all electrical equipment be installed in accordance to the listing or labeling of the product.

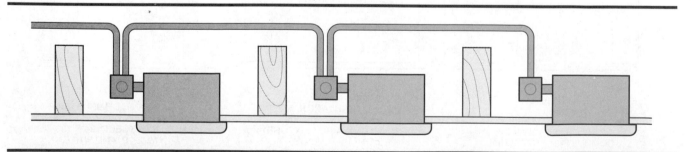

Figure 7-8 The only conductors permitted to run into or through the junction boxes on the recessed luminaires (fixtures) are those that supply the luminaires (fixtures). Conductors other than those that supply the luminaires (fixtures) are permitted to run through the outlet boxes if the luminaire (fixture) is marked "Identified For Through-Wiring." See *410.11* and *410.31*.

Suspended Ceiling Lay-In Luminaires (Fixtures)

Figure 7-9 illustrates luminaires (fixtures) installed in a typical suspended ceiling.

The recreation room of this residence has a "dropped" suspended acoustical paneled ceiling. Luminaires (fixtures) installed in a "dropped" ceiling must bear a label stating "Suspended Ceiling Luminaire (Fixture)." These types of luminaircs (fixtures) are not classified as recessed luminaires (fixtures) because in most cases, there is a great deal of open space above and around them.

Support of Suspended Ceiling Luminaires

In *410.16(C)* we find that when framing members of a suspended ceiling grid are used to support luminaires (fixtures), all framing members must be securely fastened together and to the building structure. It is the responsibility of the installer of the ceiling grid to do this.

All lay-in suspended ceiling luminaires (fixtures) must be securely fastened to the ceiling grid members with bolts, screws, or rivets. Listed clips supplied by the manufacturer of the luminaire (fixture) are also permitted to be used. Clips that meet this criteria are shown in Figure 17-1.

The logic behind all of the "securely fastening" requirements is that in the event of a major problem such as an earthquake or fire, the luminaires (fixtures) will not fall down and injure someone.

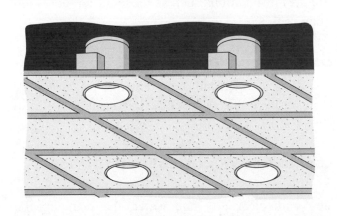

Figure 7-9 Suspended ceiling luminaires (fixtures).

Connecting Suspended Ceiling Luminaires (Fixtures)

The most common way to connect suspended ceiling lay-in luminaires (fixtures) is to complete all of the wiring above the suspended ceiling using standard wiring methods such as electrical metallic tubing, nonmetallic-sheathed cable, or armored cable.

Then, from outlet boxes strategically placed above the ceiling near the intended location of the lay-in luminaires (fixtures), a flexible connection is made between the outlet boxes and the luminaires (fixtures).

In the electrical trade, these flexible connections are referred to as "fixture whips." A fixture whip must be at least 18 in. (450 mm) but not more than 6 ft (1.8 m) in length, *410.67(C)*. Figure 7-10 shows a typical installation using a fixture whip between an outlet box and the luminaire (fixture).

A fixture whip might be:

- flexible metal conduit (FMC).

- armored cable (AC).

- metal clad cable (MC).

- nonmetallic-sheathed cable (NM and NMC).

- liquidtight flexible nonmetallic conduit (LFNC), trade size ⅜.

- liquidtight flexible metallic conduit (LFMC), trade size ⅜.

- liquidtight flexible nonmetallic conduit (LFNC), trade size ⅜.

- electrical nonmetallic-tubing (ENT).

All of the Code rules pertaining to these wiring methods were discussed in Unit 4. As a reminder when installing fixture whips, the 4½ ft (1.4 m) and 12 in. (300 mm) securing and supporting requirements are waived.

Grounding Luminaires (Fixtures)

All luminaires (fixtures) must be grounded, *410.18(A)*. Assuming that the wiring is a metallic wiring method or nonmetallic-sheathed cable with ground, you have an acceptable equipment ground at the lighting outlet. Grounding of a surface-mounted luminaire (fixture) is easily accomplished

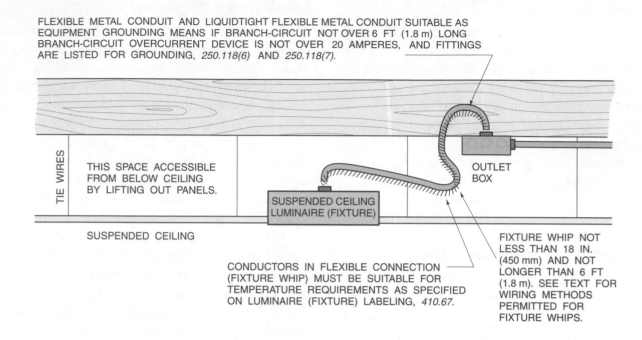

FLEXIBLE METAL CONDUIT AND LIQUIDTIGHT FLEXIBLE METAL CONDUIT SUITABLE AS EQUIPMENT GROUNDING MEANS IF BRANCH-CIRCUIT NOT OVER 6 FT (1.8 m) LONG BRANCH-CIRCUIT OVERCURRENT DEVICE IS NOT OVER 20 AMPERES, AND FITTINGS ARE LISTED FOR GROUNDING, *250.118(6)* AND *250.118(7)*.

TIE WIRES

THIS SPACE ACCESSIBLE FROM BELOW CEILING BY LIFTING OUT PANELS.

SUSPENDED CEILING LUMINAIRE (FIXTURE)

SUSPENDED CEILING

OUTLET BOX

FIXTURE WHIP NOT LESS THAN 18 IN. (450 mm) AND NOT LONGER THAN 6 FT (1.8 m). SEE TEXT FOR WIRING METHODS PERMITTED FOR FIXTURE WHIPS.

CONDUCTORS IN FLEXIBLE CONNECTION (FIXTURE WHIP) MUST BE SUITABLE FOR TEMPERATURE REQUIREMENTS AS SPECIFIED ON LUMINAIRE (FIXTURE) LABELING, *410.67*.

Figure 7-10   A suspended ceiling luminaire (fixture) connected with a flexible fixture whip between an outlet box and the luminaire (fixture).

by securing the luminaire (fixture) to the properly grounded metal outlet box with the hardware provided with the luminaire (fixture). Metal outlet boxes have a tapped No. 10-32 hole in which to insert a green hexagonal grounding screw. The bare copper equipment grounding conductor in the nonmetallic-sheathed cable is usually terminated under this screw. Grounding and bonding was discussed in Unit 5. If the outlet box is nonmetallic, the small bare equipment grounding conductor from the luminaire (fixture) is connected to the equipment grounding conductor in the outlet box. For suspended ceiling luminaires (fixtures), grounding of the luminaire (fixture) is accomplished by using metallic fixtures whips or nonmetallic-sheathed cable with ground between the outlet box and the luminaire (fixture). Always follow the manufacturers' instructions for making the electrical connections.

▶What about replacing a luminaire (fixture) in an older home? If there is no equipment grounding means in the outlet box, then a luminaire (fixture) that is made of insulating material and has no exposed conductive parts must be used, *410.18(B)*. The exception to *410.18(B)* allows a metallic

replacement luminaire (fixture) to be installed and connected to a system that does not have an equipment grounding conductor. In this case, a separate equipment grounding conductor must be installed in conformance with *250.130(C)*. This requirement was discussed in Unit 6 in the topic of replacing receptacles, so it will not be repeated here. Acceptable types of equipment grounding conductors are listed in *250.118*.◀

In Unit 17, Figure 17-1 and Figure 17-2 have additional information regarding recessed luminaire (fixture) installations such as those in the recreation room of this residence.

Luminaires (Fixtures) in Closets

There are special requirements for installing luminaires (fixtures) in closets. This is covered in Unit 8.

Luminaires (Fixtures) in Bathrooms

Because of the electrical shock hazards associated with electricity and water, there are special restrictions for luminaires (fixtures) installed in bathrooms. This is covered in Unit 10.

FLUORESCENT BALLASTS AND LAMPS, INCANDESCENT LAMPS

The following text introduces the basics of fluorescent ballasts and lamps. This topic is extremely complex. The student is urged to send for the technical brochures, bulletins, and booklets mentioned in the Instructor's Guide. These brochures contain a wealth of information on ballasts, lamps, wiring diagrams, troubleshooting hints, application suggestions, and so on.

Fluorescent Ballasts

An incandescent lamp contains a filament that has a specific "hot" resistance value when operating at rated voltage. The current flowing through the lamp filament is governed by Ohm's law, where Amperes = Volts/Ohms. When energized, the lamp will provide the light output for which the lamp was designed.

Fluorescent lamps do not have a filament running from end-to-end. Instead, they have a filament at each end. The filaments are connected to a ballast, as in Figure 7-11. A ballast is needed to control the energy, control voltage to heat the filaments, control voltage across the lamp to start the arc within the tube (end-to-end) needed to ionize the gas and vaporize the droplets of mercury inside the lamp, and to "limit" the current flowing through the lamp. Without a ballast, the fluorescent lamp would probably not start. But, if it did, then the internal resistance (impedance) would get lower and lower, and the current flowing through the lamp would get higher and higher, and in a very short time, the lamp would be destroyed.

Ballast Types (Circuitry)

Preheat. Preheat ballasts are connected in a simple series circuit configuration. They are easily identified because the luminaire (fixture) will have a "starter." One type of starter is automatic, and looks like a small roll of life-savers with two "buttons" on one end. When the luminaire (fixture) is turned on, the starter is in series with the lamp's filaments and the ballast. The lamp filaments glow for a few seconds. The starter's bimetal element heats up and "warps," causing the starter contacts to automatically open the series circuit. At the instant the starter contacts open, an arc is established inside the lamp, end-to-end, and the lamp starts.

Another type of starter is a manual "ON/OFF" switch that has a momentary "make" position just beyond the "ON" position. When you push the switch on, and hold it there for a few seconds, the lamp filaments glow. When the switch is released, the start contacts open, an arc is initiated within the lamp—and the lamp lights up.

The fluorescent lamps for use with preheat ballasts have two pins (bi-pin) on each end of the lamp.

Preheat lamps and ballasts are not used for dimming applications.

Rapid Start. Rapid start lamps do not require a starter. This is probably the most common type used today. The lamps start in less than 1 second. The filaments remain energized with a low-voltage circuit from the ballast. The lamps start up much faster than the preheat type. To ensure proper starting, ballast manufacturers recommend that there be a grounded reflector within ½ in. (12.7 mm) of the lamp and

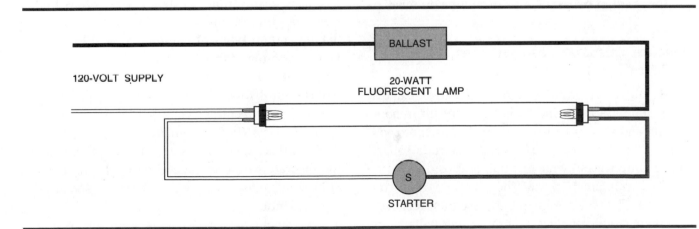

Figure 7-11 Simple series circuit.

running the full length of the lamp, that the ballast be grounded, and that the supply circuit originate from a grounded system. T5 lamps do not require that a grounded surface be present for starting. Fluorescent luminaires (fixtures) can be nonmetallic. Residential electrical systems in the United States are grounded systems and operate at 240 volts between phases and 120 volts to ground from either phase. Commercial installations might be 208/120 volts or 480/277 volts.

Fluorescent lamps for use with rapid start ballast have two pins (bi-pin) on each end of the lamp.

A *modified* rapid start ballast operates just like a standard rapid start ballast, but has a delay feature that shuts off or reduces the filament heat after the lamps have started.

Another version of a rapid start ballast is the *trigger start* ballast. Trigger start ballasts are used with preheat fluorescent lamps up to 32 watts. The trigger start ballast was the forerunner of the rapid start ballast.

Rapid start lamps can be dimmed using a special dimming ballast. See Unit 13.

Instant Start. Instant start lamps do not require a starter. These ballasts provide a high voltage "kick" to start the lamp instantly. They require special fluorescent lamps that do not require preheating of the lamp filaments. Because instant start fluorescent lamps are started by brute force, they have a shorter life (as much as 40% less) than rapid start lamps. Most instant start fluorescent lamps have a single pin on each end of the lamp, although some have two pins (bi-pin), in which case a short-circuiting arrangement is designed into the lampholder.

Instant start ballasts/lamps cannot be dimmed.

Mismatching of Fluorescent Lamps and Ballasts

Be sure to use the proper lamp in a given ballast. Mismatching a lamp and ballast may result in poor starting and poor performance, as well as shortened lamp and/or ballast life. The manufacturer's ballast and/or lamp warranty may be null and void.

T8 lamps are bi-pin as they have two-pins on each end. They are designed to be used interchangeably on magnetic or electronic rapid start ballasts, or electronic instant start ballasts. Lamp life is reduced slightly when used with an instant start ballast.

Ballast Types (Applications)

The three types of ballast discussed previously are available in:

High Efficiency. High efficiency ballasts are highly recommended because they save energy. Energy-saving ballasts can be of the *electronic type* or *magnetic type* (core & coil). If they are of the magnetic type, then they will probably have steel laminations and copper wire instead of iron laminations and aluminum wire. High efficiency, energy-saving ballasts have a high power factor rating. Power factor is the ratio of watts to volt-amperes.

For example, a 40-watt fluorescent lamp is installed in a luminaire (fixture) that has a ballast marked 42 VA. The power factor of the complete luminaire (fixture) is:

$$\text{Power Factor} = \frac{\text{Watts}}{\text{Volt-amperes}} = \frac{40}{42} = 0.95 \ (95\%)$$

Energy-saving ballasts cost more to buy initially than low power factor ballasts, but over time, the savings in energy consumption far exceed the higher initial costs. A standard fluorescent ballast might consume 14 to 16 watts. An energy-efficient ballast consumes 8 to 10 watts. Combined with an energy-efficient fluorescent lamp that uses 32 or 34 watts instead of 40 watts, there is a considerable energy savings. You are buying lighting—not heat. Always install luminaires (fixtures) that have a high power factor rating. This data is found on the nameplate of the ballast. Today, electronic ballasts are the most energy efficient when compared to magnetic ballasts containing coils, capacitors, and transformers.

The federal government's National Appliance Energy Conservation Amendment of 1988, Public Law 100-357, prohibited manufacturers from producing certain types of commonly used low power factor (less than 90%) ballasts as of January 1990. Among these were ballasts for single and two-lamp ballasts for 4-foot T12 rapid start lamps. luminaire (fixture) manufacturers had to start using high power factor ballasts (90% or greater) as of April 1991. Ballasts that "meet or exceed" the requirements of the federal law are marked with an "E" in a circle. Dimming ballasts and ballasts designed specifically for residential use were exempted.

Following the ballast legislation, the National Energy Policy Act of 1992 enacted restrictions on lamps. Lamp manufacturers could no longer produce certain 8-foot fluorescent lamps as of April 1994. In October 1995, the common 4-foot T12 linear and U-tube fluorescent lamps, and the R30, R40, and PAR38 incandescent lamps were eliminated. These were replaced by energy-efficient lamps that are direct replacements for the discontinued lamps. For example, the typical PAR38 reflector flood lamp could be replaced with a 60PAR/HIR incandescent lamp. An F40T12WW fluorescent lamp could be replaced with an F40SP30RS/WM fluorescent lamp that consumes 34 watts instead of 40 watts for just about the same amount of light (lumen output).

It has been estimated that the total electric bill savings across the country will exceed $250 billion over the next 15 years. Table 7-2 clearly illustrates how much energy saving is possible when using high-efficiency (high power factor) ballasts. It compares a standard F40T12 fluorescent lamp when used with different types of ballasts.

Watts versus Amperes

It is possible to overload a branch-circuit if too many low power factor ballasts are connected to the circuit. In this residence, the use of low cost, low power factor ballasts for the recessed fluorescent luminaires (fixtures) in the recreation room could result in the need for two branch-circuits instead of one—as well as higher energy (bigger light bills) consumption. High-efficiency, high power factor ballasts should always be used.

Fluorescent loads are a good example of why volt-amperes, rather than watts, must be considered in order to find the current draw on the circuit.

Let's make a comparison of possibilities for the recreation room lighting where there are six recessed fluorescent luminaires (fixtures), each containing two 2-lamp ballasts. The nameplate on each ballast indicates a line current rating of 0.70 ampere at 120 volts, which is 84 volt-amperes. The lamps are marked 40-watts.

The total current in amperes is

$$12 \times 0.70 = 8.4 \text{ amperes}$$

The total power in volt-amperes is

$$8.4 \times 120 = 1008 \text{ volt-amperes}$$

The total lamp wattage is

$$40 \times 24 = 960 \text{ watts}$$

If we had calculated the load for these luminaires (fixtures) based upon wattage, the result would have been

$$\frac{W}{E} = \frac{960}{120} = 8 \text{ amperes}$$

This result is not much different than the 8.4 amperes using data from the ballast nameplate.

Had we used low-cost, low-efficiency ballasts like No. 4 shown in Table 7-2, the result would be entirely different. The ballasts in Table 7-2 are single-lamp ballasts. We double the current values for a good approximation of a similar 2-lamp ballast. Using the low-power factor ballast, we find that the total current in amperes is:

$$12 \times 0.85 \times 2 = 20.4 \text{ amperes}$$

The total power in volt-amperes is

$$20.4 \times 120 = 2,448 \text{ volt-amperes}$$

or

$$12 \times 102 \times 2 = 2,448 \text{ volt-amperes}$$

The total lamp wattage is

$$24 \times 40 = 960 \text{ watts}$$

BALLAST	LINE CURRENT	LINE VOLTAGE	LINE VOLT/AMPERES	LAMP WATTAGE	LINE POWER FACTOR
No. 1	0.35	120	42	40	0.95 (95%)
No. 2	0.45	120	54	40	0.74 (74%)
No. 3	0.55	120	66	40	0.61 (61%)
No. 4	0.85	120	102	40	0.39 (39%)

Table 7-2 Various line currents, volt-amperes, wattages, and overall power factor for various single-lamp fluorescent ballasts. These values were taken from actual ballast manufacturers' data. For two-lamp ballasts, double the values in this table.

As previously shown, if we use lamp wattage to calculate the current draw:

$$I = \frac{W}{E} = \frac{960}{120} = 8 \text{ amperes}$$

But we really have a current draw of 20.4 amperes, which would overload the 15-ampere branch-circuit B12. In fact, a load of 20.4 amperes is too much for a 20-ampere branch-circuit. Two 15-ampere branch-circuits would have been needed to hook-up the recreation room recessed fluorescent luminaires (fixtures). Remember, *do not* load any circuit to more than 80% of the branch-circuit's rating. That is 16 amperes for a 20-ampere branch-circuit and 12 amperes for a 15-ampere branch-circuit. And that is the maximum.

It is apparent that the *installed* cost using cheap luminaires (fixtures) is greater and more complicated than if high-quality luminaires (fixtures) using energy-efficient ballasts and lamps had been used. However, after the initial installation, the energy savings add up significantly.

Dimming Ballast. Dimming ballasts are discussed in Unit 13. They are used where the dimming of rapid start fluorescent lamps is desired. Instant start lamps and ballasts cannot be used for dimming applications. Dimming ballasts are more costly and are less energy efficient than standard ballasts.

Noise Levels. All ballasts will hum, but some do so more than others. Ballasts containing transformers and choke coils hum when the metal laminations vibrate because of the alternating current reversals. This hum can be magnified by the luminaire (fixture) and/or the surface the luminaire (fixture) is mounted on. Electronic ballasts have little if any hum.

Do not insert spacers, washers, or shims between a ballast and the luminaire (fixture) in attempts to make the ballast more quiet. This will cause the ballast to run much hotter, and could result in shortened ballast life, and possible fire hazard. Instead, replace the noisy ballast with a quiet, sound-rated ballast. Sometimes checking and tightening the many nuts, bolts, and screws of the luminaire (fixture) will solve the problem.

Ballasts are sound rated. Again, cost is a consideration. Ballasts are marked with letters *A* through *F*: *A* is the quietest, and *F* is the noisiest.

A—homes, churches, libraries, hospital rooms.

B—classrooms, private offices.

C—open office areas.

D—retail stores.

E & F—outdoors and in factories.

Lamp Diameters. There is a simple way to determine the diameter of a lamp (light bulb or tube) at its widest measurement. The industry uses an "eighths of an inch rule." For example, a T5 fluorescent lamp is ⅝ in. in diameter. A T8 fluorescent lamp is ⅝ in., or 1 in. in diameter. A T12 lamp is ¹²⁄₈ in., or 1½ in. in diameter. Incandescent lamps follow the same system. A PAR38 lamp is ³⁸⁄₈ in., or 4¾ in. in diameter. An R30 lamp is ³⁰⁄₈ in., or 3¾ in. in diameter.

Other Considerations

- Because heat trapped by insulation around and on top of a luminaire (fixture) can shorten the life of a ballast, always follow the manufacturer's installation requirements and the requirements found in *Article 410*, discussed earlier in this unit. A 21°F (10°C) temperature rise (*Note:* This is a temperature difference, not an absolute temperature measurement), above the ballast's rated temperature of 194°F (90°C) can reduce the ballast's life by one-half. This "half-life rule" is also true for conductors, motors, transformers, and other electrical equipment.

- Fluorescent lamps that are intensely blackened on both ends should be replaced. Operating a two-lamp ballast with only one lamp working will cause the ballast to run hot, and will shorten the life of the ballast. Severe blackening of one end of the lamp can also ruin the ballast. A flickering lamp should be replaced.

- Poor starting of a fluorescent lamp can be caused by poor contact in the lamp holder, poor grounding, excessive moisture on the outside of the tube, or cold temperature (approximately 50°F (10°C)), as well as dirt, dust, and grime on the lamp.

- Most ballasts will operate satisfactorily within a range of +5% to −7% of their rated voltage. The higher quality CBM-certified ballasts will operate satisfactorily within a range of ±10%.

- Most T12 fluorescent lamps sold today are of the energy-efficient type. Generally, energy-efficient fluorescent lamps should not be used with old-style magnetic ballasts. To do so could result in poor starting, reduced lamp life, flickering, or spiraling. Home centers carry the most common T12 fluorescent lamps. Electrical supply houses usually carry different types of fluorescent lamps that can be matched with different types of ballasts. Check the markings on the ballast and the lamp carton, as well as the instructional literature available at the point-of-sale of the ballast or lamp, and/or in the manufacturer's catalogs.

- Lamp life generally is rated in "X" number of hours of operation based on 3 hours per start. Frequent switching results in shortened expected lamp life. Inversely, leaving the lamps on for long periods of time extends the expected lamp life. Vibration, rough handling, cleaning, and so on, shortens lamp life.

- Be careful what type of conductors you use to connect fluorescent luminaires (fixtures). Branch-circuit conductors within 3 in. (75 mm) of a ballast must have an insulation temperature rating of not less than 194°F (90°C), *410.33*. Type THHN conductors, the conductors in non-metallic-sheathed cable and in Type ACTHH armored cable are rated 194°F (90°C).

Voltage Limitations

The maximum voltage allowed for residential lighting is 120 volts between conductors, per *210.6(A)*.

In or on a home, lighting equipment that operates with an open-circuit voltage over 1000 volts is not permitted, *410.80(B)*. This pretty much eliminates most neon lighting systems for decorative lighting purposes, as shown in Figure 7-12.

Incandescent Lamp Life at Different Voltages

Operating an incandescent lamp at other than rated voltage will result in longer—or shorter—lamp life. The following formula can predict the approximate expected lamp life at different voltages. For example, assume that a gas-filled 120-volt incandescent lamp has a published lamp life of 1000 hours. The calculations show the expected lamp life of this lamp when operated at 130 volts, and the expected lamp life when operated at 110 volts.

- At 130 volts:

$$\text{Expected Life} = \left(\frac{\text{Rated Volts}}{\text{Actual Volts}}\right)^{13.1} \times \text{Rated Life}$$

$$= \left(\frac{120}{130}\right)^{13.1} \times 1000$$

$$= 0.350 \times 1000$$

$$= 350 \text{ hours}$$

- At 110 volts:

$$\text{Expected Life} = \left(\frac{\text{Rated Volts}}{\text{Actual Volts}}\right)^{13.1} \times \text{Rated Life}$$

$$= \left(\frac{120}{110}\right)^{13.1} \times 1000$$

$$= 3.126 \times 1000$$

$$= 3126 \text{ hours}$$

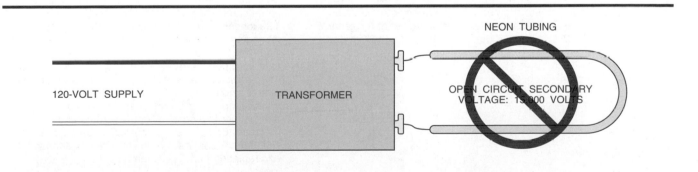

120-VOLT SUPPLY   TRANSFORMER   NEON TUBING   OPEN CIRCUIT SECONDARY VOLTAGE: 15,000 VOLTS

Figure 7-12 It is *not* permitted to install lighting in or on residences where the open-circuit voltage is over 1000 volts. This is a violation of *410.80(B)*.

Incandescent Lamp Lumen Output at Different Voltages

Notice that operating an incandescent lamp at a voltage lower than the lamp's voltage rating will result in longer lamp life. A strong case might be made to install 130-volt lamps, particularly where they are hard to reach, such as flood lights located high up. However, the lumen output is reduced. For example, calculate the approximate lumen output of a 100-watt, 130-volt incandescent lamp that has an initial lumen output of 1750 lumens at rated voltage. The lamp is to be operated at 120 volts.

$$\frac{\text{Expected}}{\text{Lumens}} = \left(\frac{\text{Actual Volts}}{\text{Rated Volts}}\right)^{3.38} \times \text{Rated Lumens}$$

$$= \left(\frac{120}{130}\right)^{3.38} \times 1750$$

$$= 0.763 \times 1750$$

$$= 1335 \text{ Lumens}$$

Formulas such as these are found in lamp manufacturer's catalogs. These formulas are useful in determining the effect of applied voltage on lamp wattage, line current, lumen output, lumens per watt, and lamp life. A calculator that has a y^x power function is needed to solve these equations.

LAMP EFFICIENCY

Lighting efficiency is measured in *lumens per watt*. A *lumen* is a unit of light measurement. One lumen falling on one square foot of surface produces one *foot-candle*. The following is a very brief comparison of different types of lamps used in residences.

The more lumens per watt, the greater the efficiency of the lamp, as shown in Table 7-3.

The smaller the diameter of a fluorescent lamp, the less power is needed to operate it, and the more lumens per watt it produces.

	LUMENS/ WATT	DIMMING	COLOR AND APPLICATION	LIFE (Approx. Hours)	TYPICAL SHAPES
Incandescent	18	Yes	Warm and Natural. Great for general lighting.	500, 1000, 1500; Depends on type of lamp.	Standard, spots, floods, decorative, tubes, globes, PAR (similar to standard spots and floods but much stronger). Use rough service where there is vibration, like garage door openers and ceiling fans.
Halogen	22	Yes	Brilliant white. Excellent for accent and task lighting. Are filled with halogen gas and have an inner lamp allowing the filament to run hotter (whiter).	2000 to 4000	PAR spots and floods, mini-reflector spots and floods, double and single end quartz.
Fluorescent	T12 84 T8 92 T5 104	Yes, but only 40-watt rapid start lamps using special dimming ballast. See Unit 13.	Warm and deluxe warm white, cool and deluxe cool white plus many other shades of white. Great for general lighting, like the Recreation Room in this residence.	6000 to 24000. Compact fluorescents can last 9× to 13× longer than incandescent lamps. Compact fluorescents are complete units having both ballast and lamp. When lamp burns out, ballast is also replaced. Compacts with screw-in bases can replace most incandescents.	Straight tubular, U-shaped tubular, circular tubular, compacts. Compact screw-in can have twin, double twin, and triple twin tubes. This refers to the number of "loops" in the lamp. Compacts can save $20 to $30 over its lifetime when compared to an equivalent incandescent lamp. Compacts last longer than incandescent lamps.

Table 7-3 Comparisons of various lamps' characteristics.

Figure 7-13 An electronic Class P ballast, thermally protected as required in *410.73(E)*. Electronic ballasts provide improved performance in fluorescent lighting installations. Electronic ballast lighting systems are 25–40% more energy efficient than conventional magnetic (core & coil) ballast fluorescent systems. *Courtesy of* Motorola Lighting.

Class P Ballasts

In *410.73(E)*, we find that all fluorescent ballasts installed indoors (except simple reactance-type ballasts), both for new and replacement installations, must have thermal protection built into the ballast by the manufacturer of the ballast, as shown in Figure 7-13. Ballasts provided with built-in thermal protection are listed by the UL as Class P ballasts. Under normal conditions, the Class P ballast has a case temperature not exceeding 194°F (90°C). The thermal protector must open within two hours when the case temperature reaches 230°F (110°C).

Some Class P ballasts also have a nonresetting fuse integral with the capacitor to protect against capacitor leakage and violent rupture. The Class P ballast's internal thermal protector will disconnect the ballast from the circuit in the event of excessive temperature. Excessive temperatures can be caused by abnormal voltage and improper installation, such as being covered with insulation.

The reason for thermal protection is to reduce the hazards of possible fire due to an overheated ballast when the ballast becomes shorted, grounded, covered with insulation, lacking in air circulation, and so on. Ballast failure has been a common cause of electrical fires.

Additional backup protection can be provided by an in-line fuseholder connected in series with the black line lead and the black ballast lead, Figure 7-14. The in-line fuseholder acts as a disconnect for the ballast. Should a ballast fail (short out), the indi-

vidual fuse opens, thus isolating the faulty ballast but not affecting the rest of the circuit. When the ballast is to be replaced, the fuse is removed to disconnect the ballast, but the entire circuit need not be turned off. In general, the ballast manufacturer will recommend the correct type and ampere rating of the fuse to be used for the particular ballast.

See Figure 7-15 for some important requirements for fluorescent and incandescent luminaires (fixtures).

Incandescent Tungsten Filament Lamp Load Inrush Currents

On rare occasions, a branch-circuit breaker may nuisance trip where a single dimmer controls a substantial incandescent tungsten filament lamp load. Why?

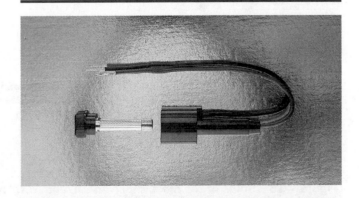

Figure 7-14 In-line fuseholder for ballast added protection. *Photo courtesy of* Cooper Bussmann, Inc.

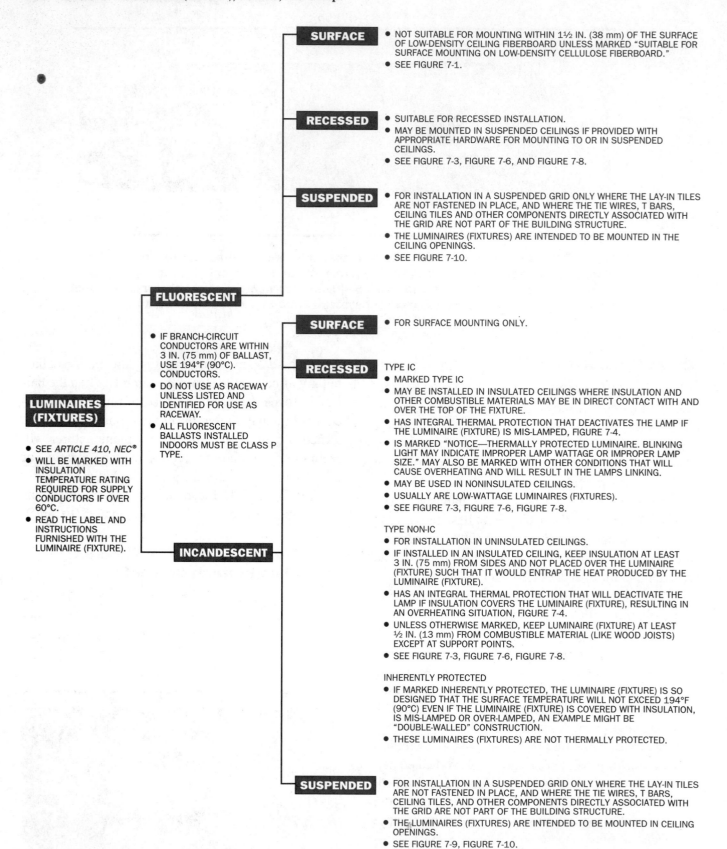

Figure 7-15   Some of the most important UL and *NEC*® requirements for fluorescent and incandescent luminaires (fixtures). Always refer to the UL Standards, the *NEC*,® and the label and/or instructions furnished with the luminaire (fixture).

The cold resistance of a tungsten filament is very low. When the lamp is connected to the proper voltage, the filament resistance increases very quickly—within $\frac{1}{240}$ second or one-quarter of an electrical cycle after the circuit is energized. This momentary inrush current might be 10 to 20 times greater than the normal operating current of the lamp.

T-rated switches have this momentary high inrush current taken into consideration. T-rated switches were discussed in Unit 5. Branch-circuit conductors are not affected by this momentary high surge of current. However, a circuit breaker might nuisance trip as a result of this momentary high inrush of current.

In cases where a dimmer is set at a low (dim) setting and the circuit is turned on with a separate or integral switch, the lower voltage will heat up the lamp filament slower, and the inrush current will last slightly longer.

Nuisance tripping problems will generally not occur in residential installations because the lamp loads controlled by a dimmer are usually small. But nuisance tripping can happen in commercial dimming applications where the connected lamp load controlled by a single dimmer is quite large. Manufacturers can furnish high magnetic trip circuit breakers if the problem happens repeatedly.

Branch-circuit fuses are not affected by this momentary inrush current phenomena because they have sufficient time-delay to override the inrush.

Outdoor Lighting

Before installing outdoor lighting, check with the local electrical inspector and/or building official to find out if there are any restrictions regarding outdoor lighting.

A virtually unlimited array of outdoor luminaires (fixtures) is available that provide uplighting, downlighting, diffused lighting, moonlighting, shadow and texture lighting, accent lighting, silhouette lighting, and bounce lighting. After the luminaires (fixtures) are selected, the type and color of the lamp is then selected. Some luminaires (fixtures) have specific light "cut-off" data that is useful in determining whether the emitting light will spill over onto the neighbor's property. The method of control must also be considered. Switch control, timer control, dusk-to-dawn control, and motion sensors are ways of turning outdoor lighting on and off.

In recent years, more and more complaints are coming from neighbors claiming that they are being bothered by glare, brightness, and light spillover from their neighbor's outdoor luminaires (fixtures). Security lighting, yard flood lights, driveway lighting, "moonlighting" in trees, and shrub lighting are examples of sources of light that might cross over the property line and be a "nuisance" to the next door neighbor.

Outdoor "lightscape" lighting considered by a homeowner to be aesthetically wonderful might be offensive to the neighbor. Nuisance lighting is also referred to as *light pollution, trespassing, intrusion, glare, spill-over,* and *brightness.* Some quiet residential neighborhoods are beginning to look like commercial areas, used car lots, and airport runways because no restrictions govern outdoor lighting methods. This is not a safety issue, and is not addressed in the *NEC.* However, the issue might be found in local building codes.

Many communities are being forced to legislate strict outdoor lighting laws, specifying various restrictions on the location, type, size, wattage, and/or footcandles for outdoor luminaires (fixtures). Checking building codes in your area might reveal requirements such as:

Light Source: The source of light (the lamp) must not be seen directly.

Glare: Glare, whether direct or reflected, such as from floodlights, and as differentiated from general illuminations, shall not be visible at any property line.

Exterior Lighting: Any lights used for exterior illuminations shall direct light away from adjoining properties.

Before installing exterior luminaires (fixtures), be sure to check with the building authorities in your community to see if there are any requirements or restrictions for outdoor lighting.

The International Dark-Sky Association has a lot of information regarding light pollution issues on their web site: www.darksky.org.

Replacing a Luminaire (Lighting Fixture) in an Older Home: Will There Be Problems?

Very possible!

Older homes were wired with conductors that were insulated with rubber (Type R) or thermoplastic (Type T). Both of these older style insulations were rated for maximum 60°C. The heat generated by the lamps over the years literally baked the conductor insulation to a hard, brittle material. If you aren't careful, pulling on or moving these conductors can easily break off the conductor insulation, which will result in a hazardous short circuit or ground fault situation.

Handle with care! Being very careful in handling the conductors while making up the splices and mounting the new luminaire—possibly sliding some readily available tubing-like electrical insulation—or taping over the conductor insulation with plastic electrical tape might be all that is needed.

Worst case. You may need to do some rewiring using the latest in nonmetallic-sheathed cables Type NM-B, Type NMC-B, or Type NMS-B. If the conductors are in a raceway, replacing the conductors with Type THHN might be a good idea. All of these conductors have a maximum temperature rating of 90°C.

More on this subject is found in Unit 4 in the nonmetallic-sheathed cable section.

REVIEW

1. Is it permissible to install a recessed luminaire (fixture) directly against wood ceiling joists when the label on the luminaire (fixture) does not indicate that the luminaire (fixture) is suitable for insulation to be in direct contact with the luminaire (fixture)? This is a Type Non-IC fixture.

   _____

2. If a recessed luminaire (fixture) without an integral junction box is installed, what extra wiring must be provided? _____

   _____

   _____

   _____

   _____

   _____

   _____

   _____

3. Thermal insulation shall not be installed within _____ in. (mm) of the top or _____ in. (mm) of the side of a recessed luminaire (fixture) unless the luminaire (fixture) is identified for use in direct contact with thermal insulation. This is a Type Non-IC fixture.

4. Recessed luminaires (fixtures) are available for installation in direct contact with thermal insulation. These luminaires (fixtures) bear the UL mark "Type _____."

5. Unless specifically designed, all recessed incandescent luminaires (fixtures) must be provided with factory-installed _____ .

6. Plans require the installation of surface-mounted fluorescent luminaires (fixtures) on the ceiling of a recreation room that is finished with low-density ceiling fiberboard. What sort of mark would you look for on the label of the luminaire (fixure)? _____

_____

_____

7. A recessed luminaire (fixture) bears no marking indicating that it is "Identified for Through-Wiring." Is it permitted to run branch-circuit conductors *other* than the conductors that supply the luminaire (fixture) through the integral junction box on the luminaire (fixture)? _____

8. Fluorescent ballasts for all indoor applications must be _____ type. These ballasts contain internal _____ protection to protect against overheating.

9. Additional backup protection for ballasts can be provided by connecting a(an) _____ with the proper size fuse as recommended by the ballast manufacturer.

10. You are called upon to install a number of luminaires (fixtures) in a suspended ceiling. The ceiling will be dropped approximately 8 in. (200 mm) from the ceiling joists. Briefly explain how you might go about wiring these luminaires (fixtures). _____

_____

_____

_____

_____

_____

_____

_____

_____

_____

_____

11. The Code places a maximum open-circuit voltage on lighting equipment in or on homes. This maximum voltage is (600) (750) (1000). (Circle the correct answer.) Where in the *NEC*® is this voltage maximum referenced?

_____

_____

12. The letter **E** in a circle on a ballast nameplate indicates that the ballast _____

_____

_____

13. A 2-lamp fluorescent ballast for two 40-watt lamps is marked 85 volt-amperes. What is the power factor of the ballast? _____

_____

_____

14. Can an incandescent lamp dimmer be used to control a fluorescent lamp load? _____

15. A good "rule-of-thumb" to estimate the expected life of a motor, a ballast, or other electrical equipment is that for every _____°C above rated temperature, the expected life will be cut in _____ .

16. A post light has a 120-volt, 60-watt lamp installed. The lamp has an expected lamp life of 1000 hours. The homeowner installed a dimmer ahead of the post light, and leaves the dimmer set so the output voltage is 100 volts. The lamp burns slightly dimmer when operated at 100 volts, but this is not a problem. What is the expected lamp life when operated at 100 volts? _____

_____

_____

_____

17. Does your community have exterior outdoor lighting restrictions? _____

If yes, what are they? _____

_____

_____

18. a. What is meant by incandescent lamp load inrush? _____

_____

_____

b. What type of snap switch is required to control an incandescent load? _____

_____

c. What problems could occur when a single dimmer controls a large incandescent tungsten filament lamp load? _____

_____

_____

19. Circle the correct answer (True) or (False) for the following statements.

a. Always match a fluorescent lamp's wattage and type designation to the type of ballast in the luminaire (fixture). (True) (False)

b. It really makes no difference what type of fluorescent lamp is used, just so the wattage is the same as marked on the ballast. (True) (False)

20. Circle the correct answer (True) or (False) for the following statements.

a. When selecting a trim for a recessed incandescent luminaire (fixture), select any trim that physically fits and can be attached to the luminaire (fixture). (True) (False)

b. When selecting a trim for a recessed incandescent luminaire (fixture), select a trim that the manufacturer indicates may be used with that luminaire (fixture). (True) (False)

Lighting Branch-Circuit for Front Bedroom

OBJECTIVES

After studying this unit, the student should be able to

- **explain the factors that influence the grouping of outlets into circuits.**
- **understand the meaning of general, accent, task, and security lighting.**
- **estimate loads for the outlets of a circuit.**
- **draw a cable layout and a wiring diagram based on information given in the residence plans, the specifications, and Code requirements.**
- **understand where and how to provide arc-fault circuit-interrupter protection for all 120-volt outlets in bedrooms.**
- **select the proper wall box for a particular installation.**
- **explain how wall boxes can be grounded.**
- **list the requirements for the installation of luminaires (fixtures) in clothes closets.**

This unit is the first exposure to real-life installations. Most issues confronting the electrician for a typical residential installation are covered in this unit. Repetition in later units is kept to a minimum.

RESIDENTIAL LIGHTING

Residential lighting is a personal thing. The homeowner, builder, and electrical contractor must meet to decide on what types of luminaires (lighting fixtures) are to be installed in the residence. Many variables (cost, personal preference, construction obstacles, etc.) must be taken into consideration.

Residential lighting can be segmented into four groups.

General Lighting

General lighting provides overall illumination for a given area, such as the front hall, the bedroom hall, the workshop, or the garage lighting in this residence. General lighting can be very basic, which means just getting the job done by providing adequate lighting for the area involved. Or, it can take the form of decorative lighting, such as a chandelier over a dining room table or other decorative luminaires (fixtures) that can be attached to ceiling paddle fans.

Accent Lighting

Accent lighting provides focus and attention to an object or area in the home. Examples are the recessed "spots" over the fireplace. Another example is the track lighting in the living room of this residence, which can accent a picture, photo, painting, or sculpture that might be hung on the wall. To be effective, accent lighting should be at least five times that of the surrounding general lighting.

Task Lighting

This is sometimes referred to as "activity" lighting. Task lighting provides proper lighting where "tasks" or "activities" are performed. The fluorescent luminaires (fixtures) above the workbench, the kitchen range hood, and the recessed luminaire (fixture) over the kitchen sink are all examples of "task" lighting. To avoid too much contrast, task lighting should not exceed three times that of the surrounding general lighting.

Security Lighting

Security lighting generally includes outdoor lighting, such as post lamps, wall luminaires (fixtures), walkway lighting, and all other lighting that serves the purpose of providing lighting for security and safety reasons. Security lighting in many instances is provided by the normal types of luminaires (lighting fixtures) found in the typical residence. In the residence discussed in this text, the outdoor bracket luminaires (fixtures) in front of the garage and next to the entry doors, as well as the post light, might be considered to be security lighting even though they add to the beauty of the residence. Security lighting can be controlled manually with regular wall switches, timers, light sensors, or motion detectors.

An organization called the American Lighting Association offers excellent reading materials with residential lighting suggestions and other materials relating to residential lighting. This organization comprises the many manufacturers of luminaires (lighting fixtures), as well as electrical distributors that maintain extensive lighting showrooms across the country. A visit to one of these showrooms offers the prospective buyer of luminaires (lighting fixtures) an opportunity to select from hundreds (in some cases thousands) of luminaires (lighting fixtures). Most of the finer lighting showrooms are staffed with individuals highly qualified to make recommendations to homeowners so that they will get the most value for their money.

The lighting provided throughout this residence conforms to the residential lighting recommendations of the American Lighting Association. Certainly many other variations are possible. For the purposes of studying an entire wiring installation for this residence, all of the load calculations, wiring diagrams, and so on, have been accomplished using the luminaire (fixture) selection as indicated on the plans and in the specifications.

GROUPING LIGHTING AND RECEPTACLE OUTLETS

The grouping of outlets into circuits must conform to *NEC*® standards and good wiring practices. There are many possible combinations or groupings of outlets. In most residential installations, circuit planning is usually done by the electrician, who must ensure that the circuits conform to Code requirements. In larger, more costly residences, the circuit layout may be completed by the architect and included on the plans.

Because many circuit arrangements are possible, there are few guidelines for selecting outlets for a particular circuit. An electrician plans circuits that are economical without sacrificing the quality of the installation.

For example, some electricians prefer to have more than one circuit feed a room. Should one circuit have a problem, the second circuit would continue to supply power to the other outlets in that room, as shown in Figure 8-1.

Some electricians keep receptacles on different branch circuits than the lighting outlets. Again, it is a matter of choice.

Another excellent way to provide good wiring while economizing on wiring materials is to connect

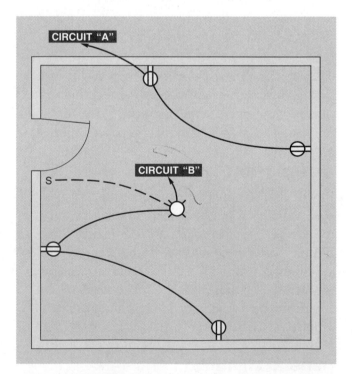

Figure 8-1   Wiring layout showing one room that is fed by two different circuits. If one of the circuits goes out, the other circuit will still provide electricity to the room.

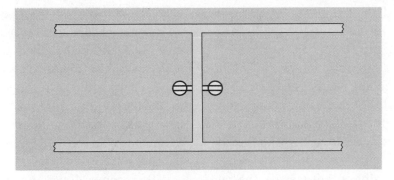

Figure 8-2 Receptacle outlets connected back-to-back. This can reduce the cost of the installation because of the short distance between the outlets. Do not do this in fire-rated walls unless special installation procedures are followed as illustrated in Figure 2-29, Figure 2-30, and Figure 2-31.

receptacle outlets back-to-back, as illustrated in Figure 8-2.

Although not a Code issue, it is poor practice to include outlets on different floors on the same circuit. Some local building codes limit this type of installation to lights at the head and foot of a stairway.

Cable Runs

A student studying the great number of wiring diagrams throughout this text will find many different ways to run the cables and make up the circuit connections. Sizing of boxes that will conform to the Code for the number of wires and devices is covered in great detail in this text. When a circuit is run "through" a recessed luminaire (fixture) like the type shown in Figure 7-8, that recessed luminaire (fixture) must be identified for "through-wiring." This subject is discussed in detail in Unit 7. When in doubt as to whether the luminaire (fixture) is suitable for running wires to and beyond it to another luminaire (fixture) or part of the circuit, it is recommended that the circuitry be designed so as to end up at the recessed luminaire (fixture) with only the two wires that will connect to that luminaire (fixture).

ESTIMATING LOADS FOR OUTLETS

Although this text is not intended to be a basic electrical theory text, it should be mentioned that, when calculating loads,

$$\text{Volts} \times \text{Amperes} = \text{Volt-amperes}$$

Yet, many times we say that

$$\text{Volts} \times \text{Amperes} = \text{Watts}$$

What we really mean to say is that

$$\text{Volts} \times \text{Amperes} \times \text{Power factor} = \text{Watts}$$

In a pure resistive load such as a simple light bulb, a toaster, an iron, or a resistance electric heating element, the power factor is 100 percent. Then

$$\text{Volts} \times \text{Amperes} \times 1 = \text{Watts}$$

With transformers, motors, ballasts, and other "inductive" loads, wattage is not necessarily the same as volt-amperes.

EXAMPLE: Calculate the wattage and volt-amperes of a 120-volt, 10-ampere resistive load.

SOLUTIONS:
a. $120 \times 10 \times 1 = 1200$ watts
b. $120 \times 10 \qquad = 1200$ volt-amperes

EXAMPLE: Calculate the wattage and volt-amperes of a 120-volt, 10-ampere motor load at 50% power factor.

SOLUTIONS:
a. $120 \times 10 \times 0.5 = 600$ watts
b. $120 \times 10 \qquad = 1200$ volt-amperes

Therefore, to be sure that adequate ampacity is provided for in branch-circuit wiring, feeder sizing, and service-entrance calculations, the Code requires that we use the term *volt-amperes*. This allows us to ignore power factor and address the *true* current draw that will enable us to determine the correct ratings of electrical equipment.

However, in some instances the terms *watts* and *volt-amperes* can be used interchangeably without creating any problems. For instance, *220.19* recognizes that for electric ranges and other cooking equipment, the kVA and kW ratings shall be considered to be equivalent for the purpose of branch-circuit and feeder calculations.

The load calculation examples in *Annex D* of the *NEC®* show kW values in the list of connected loads, but the actual calculations are done using volt-ampere values. Throughout this text, *wattage* and *volt-amperes* are used when calculating and/or estimating loads.

Building plans typically do not specify the ratings in watts for the outlets shown. When planning circuits, the electrician must consider the types of luminaires (fixtures) that may be used at the various outlets. To do this, the electrician must know the general uses of the receptacle outlets in the typical dwelling.

Unit 3 shows that the general lighting load of a residence is determined by allowing a load of 3 volt-amperes per ft² (33 volt-amperes per m²) of floor area, *220.3(A)*. For the residence in the plans, it is shown that six lighting circuits meet the minimum standards set by the Code. However, to provide sufficient capacity, 13 lighting circuits are to be installed in this residence.

Number of Outlets Permitted on One Branch-Circuit

The *NEC®* does not specify the maximum number of receptacle outlets or lighting outlets that may be connected to one 120-volt lighting or small appliance branch-circuit in a residence. It may seem to be ridiculous and illogical that 10, 20, or more receptacle outlets and lighting outlets can be connected to one branch-circuit and not be in violation of the Code. Consider the fact, however, that having many "convenience" receptacles is "safer" because many receptacle outlets will virtually eliminate the use of extension cords, one of the most frequent causes of electrical fires. Rarely, if ever, would all receptacle outlets and lighting outlets be fully loaded at the same time. There is much diversity of load in residential occupancies.

Circuit Loading Guidelines

Recessed and surface-mounted incandescent luminaires (lighting fixtures) are marked with their maximum lamp wattage. Fluorescent luminaires (fixtures) are marked with lamp wattage and the bal-

last is marked in volt-amperes. But for new construction, usually the exact type of luminaire (fixture) that will be installed at each lighting outlet is unknown. Also unknown is the exact load that will be connected to the receptacle outlets. Checking a luminaire (lighting fixture) manufacturer's catalog provides excellent guidelines for lamp wattages recommended for various types of luminaires (lighting fixtures).

So it comes down to *approximating* and *estimating* what the connected loads might be.

In Unit 3, we discussed how typical 15-ampere lighting branch-circuits in homes can be figured at one 15-ampere branch-circuit for every 600 ft² (55.8 m²). Although not often used in residential wiring, if you choose to install 20-ampere lighting branch-circuits, figure one 20-ampere branch-circuit for every 800 ft² (74.4 m²). This results in a good approximation for the minimum number of lighting branch-circuits needed.

Another guideline to follow in residential wiring is *never load the circuit to more than 80 percent of its rating*. Although this is a Code requirement for *continuous* loads—*210.19(A), 210.20(A)*, and *215.2(A), 215.3, 230.42(A)*—it makes good sense to follow this guideline for most residential loads.

A 15-ampere, 120-volt lighting branch-circuit would be calculated as:

$$15 \times 0.80 = 12 \text{ amperes}$$

or

$$12 \text{ amperes} \times 120 \text{ volts} = 1440 \text{ watts}$$
$$\text{(volt-amperes)}$$

A 20-ampere, 120-volt lighting branch-circuit would be calculated as:

$$20 \times 0.80 = 16 \text{ amperes}$$

or

$$16 \text{ amperes} \times 120 \text{ volts} = 1920 \text{ watts}$$
$$\text{(volt-amperes)}$$

Estimating Number of Outlets by Assigning an Amperage Value to Each

One method of determining the number of lighting and receptacle outlets to be included on one circuit is to assign a value of 1 to 1½ amperes to each outlet to a total of 15 amperes. Thus, a total of 10 to 15 outlets can be included in a 15-ampere circuit.

As stated previously, the *NEC®* does not limit the number of outlets on one circuit. However, many local building codes do specify the maximum number of outlets per circuit. Before planning any circuits, the electrician must check the local building code requirements.

All outlets will not be required to deliver 1½ amperes. For example, closet lights, night lights, and clocks will use only a small portion of the allowable current. A 60-watt closet light will draw less than 1 ampere.

$$I = \frac{W}{E} = \frac{60}{120} = 0.5 \text{ ampere}$$

If low-wattage luminaires (fixtures) are connected to a circuit, it is quite possible that 15 or more lighting outlets (many times referred to as "openings") could be connected to the circuit without a problem. On the other hand, if the load consists of high-wattage lamps, the number of outlets would be less. In this residence, estimated loads were figured at 120 volt-amperes (1 ampere) for those receptacle outlets obviously intended for general use. Receptacle outlets in the garage and workshop were figured at 180 volt-amperes (1½ amperes) because of the use these outlets will be supplying (i.e., tools, drills, table saws, etc.).

Receptacle outlets in the kitchen and laundry have been included in the small appliance circuit calculations.

If an electrician were to *estimate* the number of 15-ampere lighting branch-circuits desired for a new residence where the total count of lighting and receptacle outlets is, for example, 80, then

$$\frac{80}{10} = 8 \text{ (eight 15-ampere lighting circuits)}$$

If the circuit consists of low-wattage loads, then

$$\frac{80}{15} = 5.3 \text{ (six 15-ampere lighting circuits)}$$

In residential occupancies there is great diversity in the loading of lighting branch-circuits.

CAUTION: The foregoing procedure for estimating the number of lighting and receptacle outlets is for branch-circuits only. Obviously, where the small appliance circuits in the kitchen, laundry area, and other areas supply heavy concentrations of plug-in appliances, the 1½ amperes per receptacle outlet would not be applicable. These small appliance circuits are discussed later on in this text.

Review of How to Determine the Minimum Number of Lighting Branch-Circuits in a Residence

To determine the minimum number of lighting circuits:

1. Use 3 volt-amperes per ft² (33 volt-amperes per m²) of floor area.

2. Figure a minimum of one 15-ampere branch-circuit for every 600 ft² (55.8 m²) of floor area.

3. Figure a minimum of one 20-ampere branch-circuit for every 800 ft² (55.8 m²) of floor area.

4. Estimate the probable loading for each lighting outlet and receptacle outlet (do not include the small appliance receptacle outlets). Try to have a total VA not over 1440 for a 15-ampere lighting circuit, and not over 1920 for a 20-ampere lighting circuit.

5. Connect approximately 10 to 15 outlets per circuit—more than 10 if the probable connected load is small, less than 10 if the probable connected load is large. For instance, the wattage of a 120-volt smoke detector is approximately ½ watt.

Divide Loads Evenly

It is mandatory that loads be divided evenly among the various circuits, *210.11(B)*. The obvious reason is to avoid overload conditions on some circuits when other circuits might be lightly loaded. This is common sense. At the Main Service, don't connect the branch-circuits and feeders to result in, for instance, 120 amperes on phase A and 40 amperes on phase B. Look at the probable and/or calculated loads to attain as close a balance as possible, such as 80 amperes on each A and B phase.

SYMBOLS

The symbols used on the cable layouts in this text are the same as those found on the actual electrical plans for the residence. Refer to Figure 2-5, Figure 2-6, Figure 2-7, Figure 2-8, and Figure 2-9.

Pictorial illustrations are used on all wiring diagrams in this text to make it easy for the reader to complete the wiring diagrams, as in Figure 8-3.

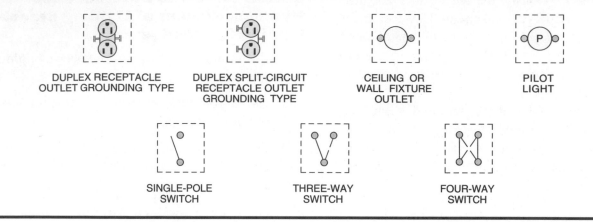

DUPLEX RECEPTACLE OUTLET GROUNDING TYPE	DUPLEX SPLIT-CIRCUIT RECEPTACLE OUTLET GROUNDING TYPE	CEILING OR WALL FIXTURE OUTLET	PILOT LIGHT
SINGLE-POLE SWITCH	THREE-WAY SWITCH	FOUR-WAY SWITCH	

Figure 8-3   Pictorial illustrations used on wiring diagrams in this text.

DRAWING THE WIRING DIAGRAM OF A LIGHTING CIRCUIT

The electrician must take information from the building plans and convert it to the forms that will be most useful in planning an installation that is economical, conforms to all Code regulations, and follows good wiring practice. To do this, the electrician first makes a cable layout. Then a wiring diagram is prepared to show clearly each wire and all connections in the circuit.

A skilled electrician, after many years of experience, will not prepare wiring diagrams for most residential circuitry because he has the ability to "think" the connections through mentally. An unskilled individual should make detailed wiring diagrams.

The following steps will guide the student in the preparation of wiring diagrams. The student is required to draw wiring diagrams in later units of this text. This is an exercise in how to make a color-coded cable layout and wiring diagram. There are other ways to lay out the cable runs for this particular "typical" room. Conductor fill in the boxes must always be considered, and this was covered in Unit 2.

1. Refer to plans and make a cable layout of all lighting and receptacle outlets, as in Figure 8-4.

2. Draw a wiring diagram showing the traveler conductors for all three-way switches, if any, as in Figure 8-5.

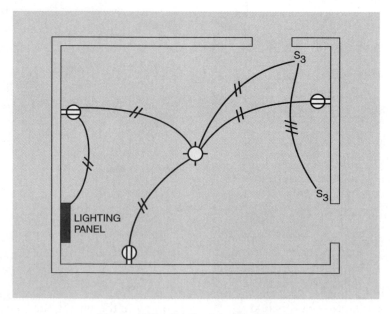

Figure 8-4   Typical cable layout.

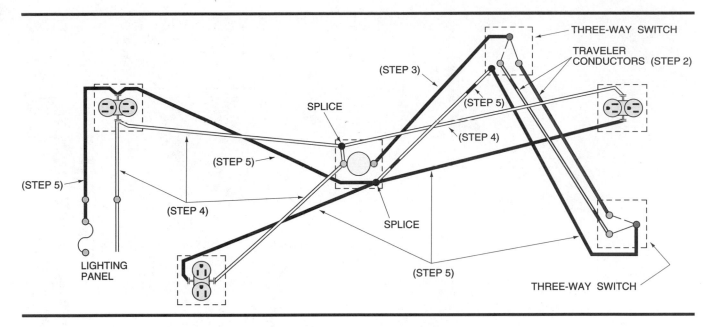

Figure 8-5 Wiring diagram of circuit shown in Figure 8-4.

3. Draw a line between each switch and the outlet or outlets it controls. For three-way switches, do this for one switch only. This is the "switch leg."

4. Draw a line from the grounded terminal in the lighting panel to each current-consuming outlet. This line may pass through switch boxes, but must not be connected to any switches. This is the white grounded circuit conductor, often called the "neutral" conductor.

   Note: An exception to step 4 may be made for double-pole switches. For these switches, all conductors of the circuit are opened simultaneously. They are rarely used in residential wiring.

5. Draw a line from the ungrounded "hot" terminal in the lighting panel to each switch and to each unswitched outlet. For three-way switches, do for one switch only. This is the "feed." The "switch leg" (switch loop, switch return) is covered in Step 3.

6. Show splices as small dots where the various wires are to be connected together. In the wiring diagram, the terminal of a switch or outlet may be used for the junction point of wires. In actual wiring practice, however, the Code does not permit more than one wire to be connected to a terminal unless the terminal is a type identified for use with more than one conductor. The standard screw-type terminal is not acceptable for more than one wire, *110.14.*

7. The final step in preparing the wiring diagram is to mark the color of the conductors, as in Figure 8-5. Note that the colors selected—black (B), white (W), and red (R)—are the colors of two- and three-conductor cables (refer to Unit 5). It is suggested that the reader use colored pencils or markers for different conductors when drawing the wiring diagram.

   Use a fine line for the white conductor.

LIGHTING BRANCH-CIRCUIT A16 FOR FRONT BEDROOM

Circuit A16 is a 15-ampere branch-circuit. ►Overcurrent protection for this circuit is provided by a 15-ampere AFCI circuit breaker in Panel A. AFCIs are discussed in detail in Unit 6. ◄ As we look at this circuit, we find five split-circuit receptacle outlets. These will provide the general lighting through the use of table or other lamps that will be plugged into the switched receptacles, as shown in Figure 8-6.

Radio, television, clocks, and other electrical and electronic items not intended to be controlled by the wall switch will be plugged into the "hot continuously" receptacle.

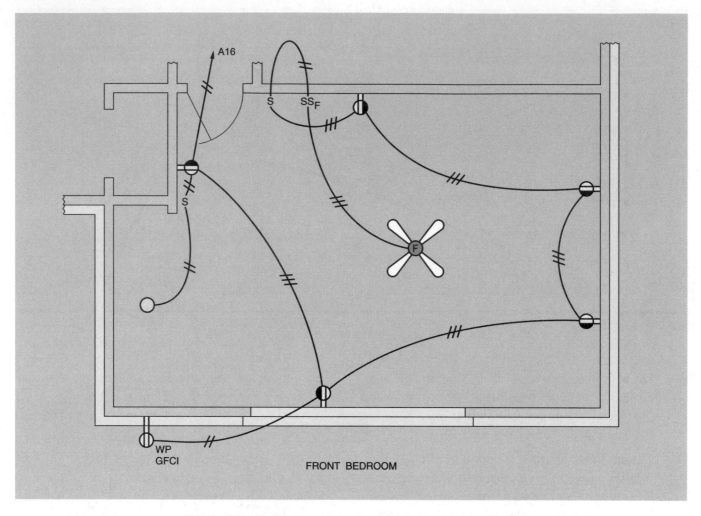

Figure 8-6   Cable layout for front bedroom circuit, A16.

Next to the wall switch we find the control for the ceiling fan, which also has a luminaire (lighting fixture) as an integral part of the fan, as shown in Figure 8-7. A single-pole switch controls the "ON/OFF" and speed of the fan motor.

The closet light is controlled by a single-pole switch outside and to the right of the closet door.

Note that the outside weatherproof receptacle connected to this circuit is required to have GFCI protection, *210.8(A)(3)*. For this particular installation, there are two ways to accomplish GFCI protection for this receptacle:

1. Install a GFCI receptacle.

2. Install a combination GFCI/AFCI circuit breaker in the main panel. Be careful when selecting the circuit breaker to be sure the nameplate indicates that it will provide both GFCI and AFCI protection.

If you cannot locate a combination GFCI/AFCI

circuit breaker, then install a GFCI outdoor receptacle and an AFCI circuit breaker in the panel.

GFCI and AFCI protection are discussed in Unit 6.

Checking the actual electrical plans, we find one television outlet and one telephone outlet are to be installed in the front bedroom. Television and telephones are discussed in Unit 25. Table 8-1 summarizes the outlets in the front bedroom and the estimated load.

DETERMINING THE SIZE OF OUTLET BOXES, DEVICE BOXES, JUNCTION BOXES, AND CONDUIT BODIES

Box fill was discussed in detail in Unit 2. Here is more on the subject. One of the most common problems found in residential wiring are wires jammed into electrical boxes. Factors that need to be consid-

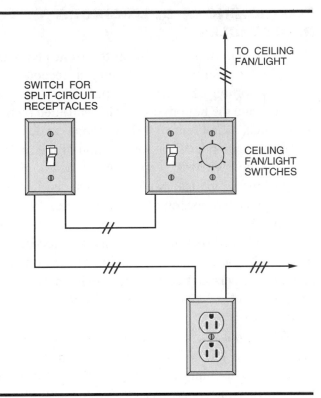

SWITCH FOR
SPLIT-CIRCUIT
RECEPTACLES

TO CEILING
FAN/LIGHT

CEILING
FAN/LIGHT
SWITCHES

Figure 8-7   Conceptual view of how the front bedroom switch arrangement is to be accomplished.

DESCRIPTION	QUANTITY	WATTS	VOLT-AMPERES
Receptacles @ 120 watts each	5	600	600
Weatherproof receptacle	1	120	120
Closet fixture	1	75	75
Ceiling fan/light	1		
3 - 50-W lamps		150	150
Fan motor (0.75 A @ 120 V)		80	90
TOTALS	8	1025	1035

Table 8-1   Front bedroom: outlet count and estimated load (Circuit A16).

ered are the size and number of conductors, the number of wiring devices, cable clamps, luminaire (fixture) studs, hickeys, and splices. *NEC® 314.16* provides the rules for sizing boxes.

Experience has shown that *minimum* does not always make for easy working with the splices, conductors, wiring devices, and carefully putting them all into the box without crowding. Do not skimp on box sizing. The small additional cost of larger boxes

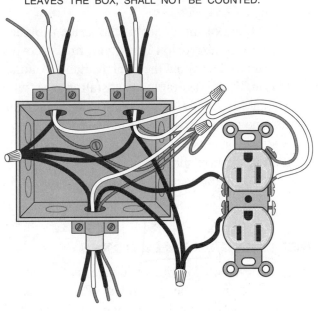

THE "PIGTAILS" CONNECTED TO THE RECEPTACLE NEED NOT BE COUNTED WHEN DETERMINING THE CORRECT BOX SIZE. THE CODE IN *314.16(B)(1)* STATES, "A CONDUCTOR, NO PART OF WHICH LEAVES THE BOX, SHALL NOT BE COUNTED."

Figure 8-8   Determining size of box according to number of conductors.

is worth much more than the grief and added time it will take to trim out the box using *minimum* size boxes. You will be much happier with boxes that have more than the *minimum* size. When it comes to box sizing, think big—not small.

The following example shows how to calculate the *minimum* size box for a particular installation, as in Figure 8-8.

1. Add the 14 AWG circuit conductors:
   2 + 3 + 3 =   8

2. Add equipment grounding wires (count one only)   1

3. Add two conductors for the receptacle   2

   Total   11   Conductors

Note that the pigtails connected to the receptacle in Figure 8-8 need not be counted when determining the correct box size. *NEC® 314.16(B)(1)* states, "*A conductor, no part of which leaves the box, shall not be counted.*"

Once the total number of conductors, plus the volume count required for wiring devices, luminaire (fixture) studs, hickeys, and clamps is known, refer to *Table 314.16(A)* and *(B)* and Figure 2-20 to select a box that has sufficient space. Also refer to Figure 2-34 for a summary of box fill requirements. For example, a 4 × 2⅛ in. square box with a suitable plaster ring can be used.

The volume of the box plus the space provided by plaster rings, extension rings, and raised covers may be used to determine the total available volume. In addition, it is desirable to install boxes with external cable clamps. Remember that if the box contains one or more devices, such as cable clamps, fixture studs, or hickeys, the number of conductors permitted in the box shall be one less than shown in *Table 314.16(A)* for *each type* of device contained in the box.

GROUNDING OF WALL BOXES

The specifications for the residence state that *all* metal boxes are to be grounded. The means of grounding is armored cable or nonmetallic-sheathed cable containing an extra equipment grounding conductor. This conductor is used only to ground the metal box. This bare equipment grounding conductor must *not* be used as a current-carrying conductor because severe shocks can result.

According to *406.3(A)*, grounding-type receptacles must be installed on 15-ampere and 20-ampere branch-circuits. The methods of attaching the equipment grounding conductor to the proper terminal on the receptacle are covered in Unit 5.

POSITIONING OF SPLIT-CIRCUIT RECEPTACLES

The receptacle outlets shown in the front bedroom are called two-circuit or split-circuit receptacles. The top portion of such a receptacle is "hot" at all times, and the bottom portion is controlled by the wall switch, as in Figure 8-9. It is recommended that the electrician wire the bottom section of the receptacle as the switched section. As a result, when theattachment plug cap of a lamp is inserted into the bottom switched portion of the receptacle, the cord does not hang in front of the unswitched section. This unswitched section can be used as a receptacle for a clock, vacuum cleaner, radio, stereo, compact disc player, personal computer, television, or other appliances where a switch control is not necessary or desirable.

When split-circuit receptacles are horizontally mounted, which is common when 4-in. square boxes are used because the plaster ring is easily fastened to the box in either a vertical or horizontal position, locate the switched portion to the right.

POSITIONING OF RECEPTACLES NEAR ELECTRIC BASEBOARD HEATING

Electric heating for a home can be accomplished with an electric furnace, electric baseboard heating units, heat pump, or resistance heating cables embedded in plastered ceilings or sandwiched between two layers of drywall sheets on the ceiling.

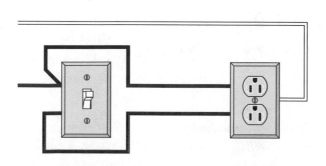

Figure 8-9 Split-circuit wiring for receptacles. The top receptacle is "hot" at all times. The bottom receptacle is controlled by the switch.

The important thing to remember at this point is that some baseboard electric heaters may not be permitted to be installed below a receptacle outlet, *210.52 FPN*.

According to *210.52*, receptacles that are part of an electric baseboard heating unit may be counted as one of the required number of receptacles in a given space, but only if the receptacles are not connected to the electric baseboard heating branch-circuit.

See Unit 23 for a detailed discussion of installing electric baseboard heating units and their relative positions below receptacle outlets.

The front bedroom does have a ceiling fan/light on the ceiling. Ceiling fans are discussed in Unit 9.

LUMINAIRES (FIXTURES) IN CLOTHES CLOSETS

The *NEC®* does not require luminaires (lighting fixtures) in clothes closets, *210.70*. However, some local codes and some building codes do. When luminaires (lighting fixtures) are installed in clothes closets, they must be of the correct type and must be installed properly.

Clothing, boxes, and other material normally stored in clothes closets are a potential fire hazard. These items may ignite on contact with the hot surface of exposed lamps. Incandescent lamps have a hotter surface temperature than fluorescent lamps.

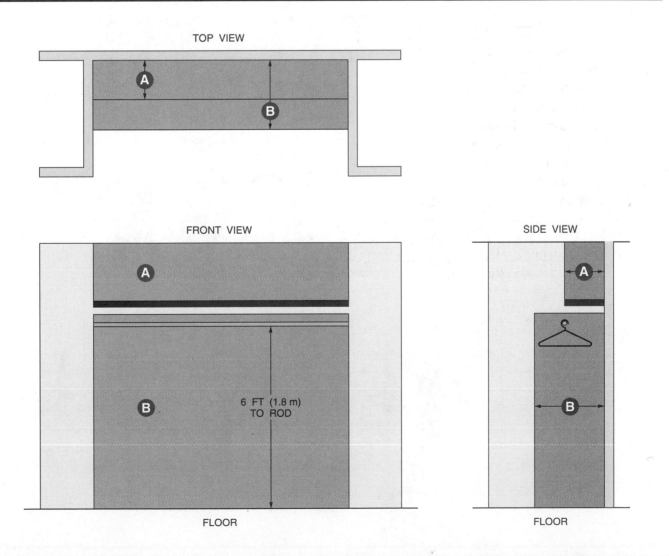

Figure 8-10   Typical closet with one shelf and one rod. The shaded area defines "storage space." Dimension A is width of shelf or 12 in. (300 mm) from wall, whichever is greater. Dimension B is below rod, 24 in. (600 mm) from wall. See *410.8(A)*.

In *410.8*, there are very specific rules for the location and types of luminaires (lighting fixtures) permitted to be installed in clothes closets. Figure 8-10, Figure 8-11 Figure 8-12, Figure 8-13, and Figure 8-14 illustrate the requirements of *410.8*.

In this residence, the closets in the bedrooms and front hall are 24 in. (600 mm) deep. The storage closet in the recreation room is 30 in. (750 mm) deep. The shelving is approximately 12 in. (300 mm) wide. Because of the required clearances between luminaires (lighting fixtures) and the storage space, locating luminaires (lighting fixtures) in

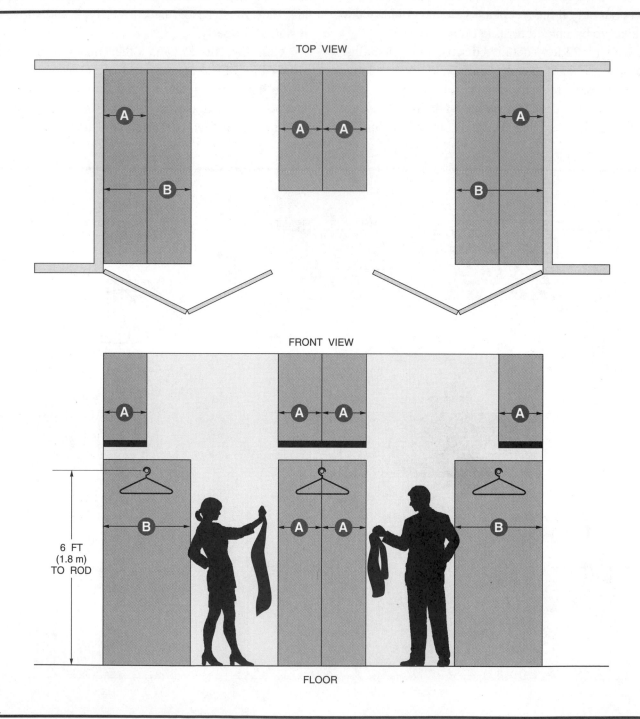

Figure 8-11 Large walk-in closet where there is access to the center rod from both sides. The shaded area defines "storage space." Dimension A is width of shelf or 12 in. (300 mm), whichever is greater. Dimension B is 24 in. (600 mm) from wall. See *410.8(A)*.

closets can be difficult. In fact, some clothes closets are so small that luminaires (lighting fixtures) of any kind are not permitted.

Figure 8-13 shows the possibilities: recessed incandescents, recessed fluorescents, surface-mounted incandescents, and surface-mounted fluorescents.

These can be on the ceiling, or on the wall space above the closet door. Whichever type is chosen, be sure that the minimum clearances are provided as required by *410.8*.

Figure 8-15 shows typical bedroom-type luminaires (lighting fixtures).

LUMINAIRES (FIXTURES) IN CLOTHES CLOSETS

PERMITTED BY
410.8(B)(1).

SURFACE-MOUNTED INCANDESCENT LUMINAIRE (FIXTURE) WITH COMPLETELY ENCLOSED LAMP(S)

RECESSED INCANDESCENT LUMINAIRE (FIXTURE) WITH COMPLETELY ENCLOSED LAMP(S)

PERMITTED BY
410.8(B)(2).

SURFACE-MOUNTED FLUORESCENT LUMINAIRE (FIXTURE)

RECESSED FLUORESCENT LUMINAIRE (FIXTURE)

NOT PERMITTED BY *410.8(C).*

INCANDESCENT OPEN OR PARTIALLY OPEN LAMPS IN LAMPHOLDERS

PENDANT LUMINAIRE (FIXTURES)

PENDANT LAMPHOLDERS

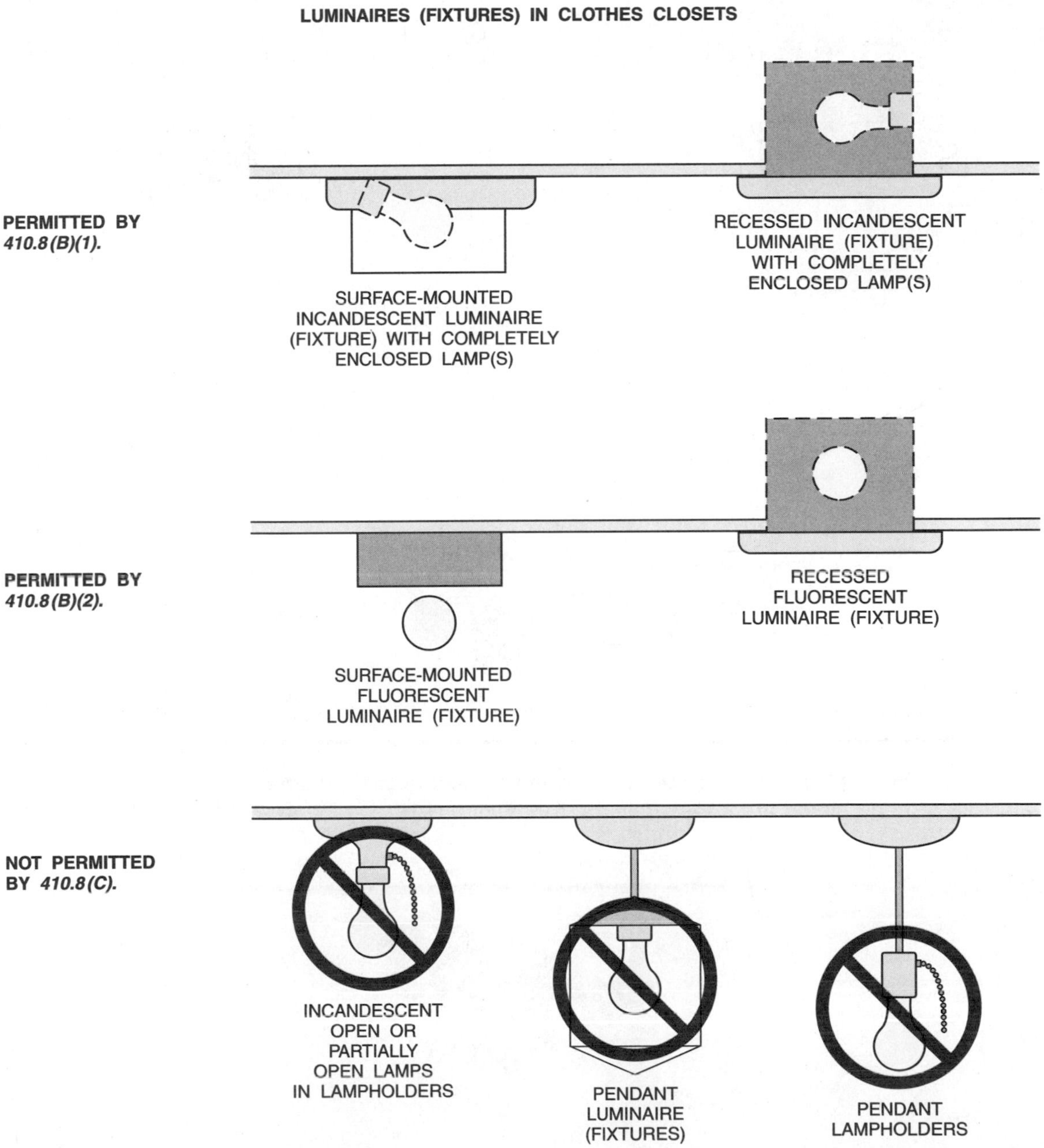

Figure 8-12 Illustrations of the types of luminaires (lighting fixtures) permitted in clothes closets. The Code does not permit bare incandescent lamps, pendant luminaires (fixtures), or pendant lampholders to be installed in clothes closets.

LOCATION OF LUMINAIRES (FIXTURES) IN CLOTHES CLOSETS

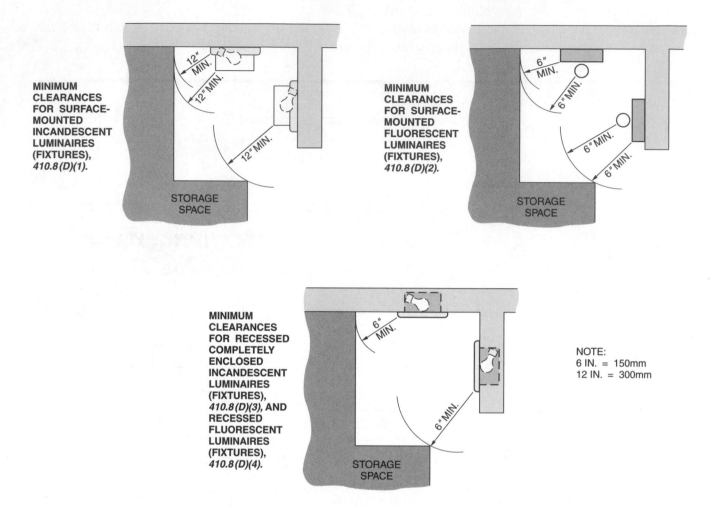

MINIMUM CLEARANCES FOR SURFACE-MOUNTED INCANDESCENT LUMINAIRES (FIXTURES), *410.8(D)(1).*

12" MIN.
12" MIN.
12" MIN.

STORAGE SPACE

MINIMUM CLEARANCES FOR SURFACE-MOUNTED FLUORESCENT LUMINAIRES (FIXTURES), *410.8(D)(2).*

6" MIN.
6" MIN.
6" MIN.
6" MIN.

STORAGE SPACE

MINIMUM CLEARANCES FOR RECESSED COMPLETELY ENCLOSED INCANDESCENT LUMINAIRES (FIXTURES), *410.8(D)(3),* **AND RECESSED FLUORESCENT LUMINAIRES (FIXTURES),** *410.8(D)(4).*

6" MIN.
6" MIN.

STORAGE SPACE

NOTE:
6 IN. = 150mm
12 IN. = 300mm

Figure 8-13 The above illustrations show the minimum clearances required between luminaires (lighting fixtures) and the storage space in clothes closets. See Figure 8-10 and Figure 8-11 for a definition of storage space.

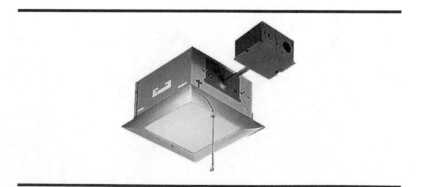

Figure 8-14 Recessed incandescent closet luminaire (lighting fixture) with pull-chain switch that may be used where a separate wall switch is not installed. *Courtesy of* Progress Lighting.

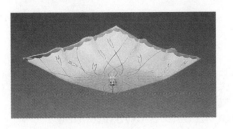

Figure 8-15   Typical bedroom-type luminaires (lighting fixtures). *Courtesy of* Progress Lighting.

REVIEW

Note: Refer to the Code or the plans where necessary.

1. Can the outlets in a circuit be arranged in different groupings to obtain the same result? Why? _____

_____

_____

_____

_____

_____

2. Is it good practice to have outlets on different floors on the same circuit? Why?

_____

_____

_____

_____

_____

3. A good rule to follow is to never load a circuit to more than _____ percent of the branch-circuit rating.

4. The *NEC*® does not limit the number of lighting and receptacle outlets permitted on one branch-circuit for residential installations. A guideline that is often used by electricians for house wiring is to allow _____ to _____ amperes per outlet. For a 15-ampere circuit, this results in _____ to _____ outlets on the branch-circuit.

5. For residential wiring, not less than one 15-ampere lighting branch-circuit should be provided for every_____ ft² of floor area. If 20-ampere lighting branch-circuits are used, provide one 20-ampere lighting branch-circuit for every _____ ft² of floor area.

6. For this residence, what are the estimated wattages used in determining the loading of branch-circuit A16?

Receptacles _____ watts (volt-amperes)

Closet fixture _____ watts (volt-amperes)

7. a. What is the ampere rating of circuit A16? _____

b. What is the conductor size of circuit A16? _____

8. The *NEC®* requires what type of unique type of protection for all outlets in residential bedrooms? In which section of the *NEC®* is this requirement found? _____

_____

_____

9. How many and what type of receptacles are connected to this circuit? _____

_____

_____

_____

10. What main factors influence the choice of wall boxes?_____

_____

_____

11. How is a wall box grounded? _____

_____

_____

12. What is a split-circuit receptacle? _____

_____

_____

_____

_____

_____

13. Is the switched portion of an outlet mounted toward the top or the bottom? Why?

_____

_____

_____

_____

14. The following questions pertain to luminaires (lighting fixtures) in clothes closets.

a. Does the Code permit bare incandescent lamps to be installed in porcelain keyless or porcelain pull-chain lampholders? _____

b. Does the Code permit bare fluorescent lamp luminaires (fixtures) to be installed? _____

_____

c. Does the Code permit pendant lampholders to be installed?_____

_____

d. What is the minimum clearance from the storage area to surface-mounted incandescent luminaires (fixtures)? _____

e. What is the minimum clearance from the storage area to surface-mounted fluorescent luminaires (fixtures)? _____

f. What is the minimum distance between recessed incandescent or recessed fluorescent luminaires (fixtures) and the storage area? _____

g. Define the "storage area."

_____

_____

_____

_____

_____

_____

h. If a clothes hanging rod is installed where there is access from both sides, such as might be found in a large walk-in closet, define the storage area under that rod.

_____

_____

_____

_____

15. How many switches are in the bedroom circuit, what type are they, and what do they control? _____

_____

16. The following is a layout of the lighting circuit for the front bedroom. Using the cable layout shown in Figure 8-7, make a complete wiring diagram of this circuit. Use colored pencils or pens.

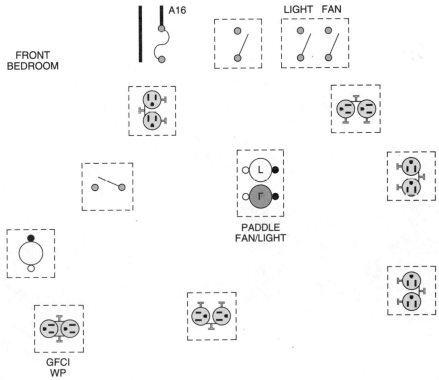

17. When planning circuits, what common practice is followed regarding the division of loads? _____

_____

18. The Code uses the terms *watts, volt-amperes, kW,* and *kVA.* Explain their significance in calculating loads. _____

_____

_____

_____

_____

19. How many 14 AWG conductors are permitted in a device box that measures 3 in. × 2 in. × 2¾ in.? _____

20. A 4 in. × 1½ in. octagon box has one cable clamp and one luminaire (fixture) stud. How many 14 AWG conductors are permitted? _____

_____

_____

_____

_____

Lighting Branch-Circuit for Master Bedroom

OBJECTIVES

After studying this unit, the student should be able to

- draw the wiring diagram of the cable layout for the master bedroom.
- understand where AFCIs are required for bedrooms.
- study Code requirements for the installation of ceiling fans.
- estimate the probable connected load for a room based on the number of luminaires (fixtures) and outlets included in the circuit supplying the room.
- gain more practice in determining box sizing based upon the number of conductors, devices, and clamps in the box.
- make the connections for three-way switches.

The discussion in Unit 3 and Unit 8 about grouping outlets, estimating loads, selecting wall box sizes, and drawing wiring diagrams can also be applied to the circuit for the master bedroom.

The residence panel schedules show that the master bedroom is supplied by Circuit A19. Circuit A19 is a 15-ampere branch-circuit. ▶Overcurrent protection for this circuit is provided by a 15-ampere AFCI circuit breaker in Panel A. AFCI protection is required for all 120-volt outlets in bedrooms, *210.12*. In Unit 8, we discussed how GFCI/AFCI protection could be attained in the Front Bedroom. The same principle applies to the Master Bedroom. AFCIs and GFCIs are discussed in detail in Unit 6.◀ Because Panel A is located in the basement below the wall switches next to the sliding doors, the home run for Circuit A19 is brought into the outdoor weatherproof receptacle. This results in six conductors in the outdoor receptacle box. The home run could have been brought into the corner receptacle outlet in the bedroom. Again, it is a matter of studying the circuit

to determine the best choice for conservation of cable or conduit in these runs and to economically select the correct size of wall boxes.

LIGHTING BRANCH-CIRCUIT A19 FOR MASTER BEDROOM

Figure 9-1, Table 9-1, and the electrical plans show that Circuit A19 has four split-circuit receptacle outlets in this bedroom, one outdoor weatherproof GFCI receptacle outlet, two closet luminaires (fixtures), each on a separate switch, plus a ceiling fan/luminaire (fixture), one telephone outlet, and one television outlet.

In addition, an outdoor bracket luminaire (fixture) is located adjacent to the sliding door and is controlled by a single-pole switch just inside the sliding door.

The split-circuit receptacle outlets are controlled by two three-way switches. One is located next to the sliding door. As in the front bedroom, living room, and study/bedroom, the use of split-circuit

DESCRIPTION	QUANTITY	WATTS	VOLT-AMPERES
Receptacles @ 120 watts each	4	480	480
Weatherproof receptacle	1	120	120
Outdoor bracket luminaire (fixture)	1	150	150
Closet luminaires (fixtures) One 75-W lamp each	2	150	150
Ceiling fan/light	1		
Three 50-W lamps		150	150
Fan motor (0.75 @ 120 V)		80	90
TOTALS	9	1130	1140

Table 9-1   Master bedroom outlet count and estimated load. (Circuit A19).

receptacles offers the advantage of having switch control of one of the receptacles at a given outlet, while the other receptacle remains "live" at all times. See Figure 3-5 and Figure 12-14 for definition of a receptacle.

Next to the bedroom door are a three-way switch and the ceiling fan/light control, which is installed in a separate two-gang box, as shown in Figure 9-2.

SLIDING GLASS DOORS AND FIXED GLASS PANELS

Receptacle outlets must be installed so that no point along the floor line in any wall space is more than 6 ft (1.8 m), measured horizontally, from an outlet in that space, including any wall space 24 in. (600 mm) or more in width. Wall space occupied by fixed glass

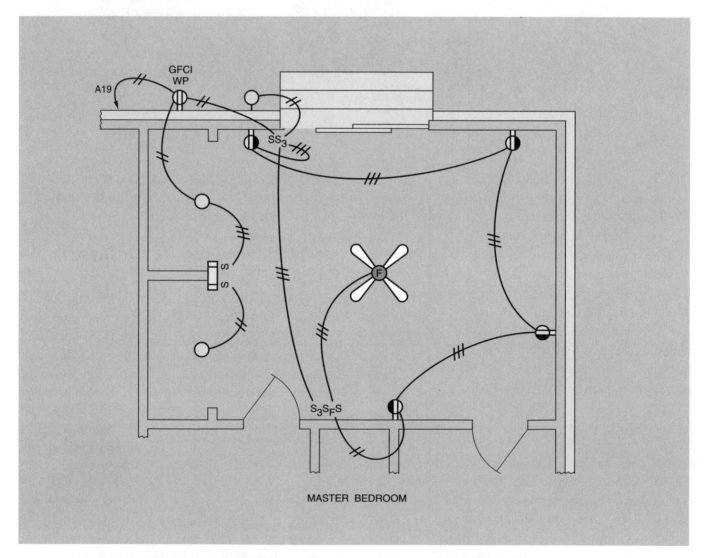

Figure 9-1   Cable layout for master bedroom.

panels in exterior walls are considered to be wall space. Sliding glass panels are not considered to be wall space, the same as any other interior or exterior door. This is covered in *210.52(A)(2)*.

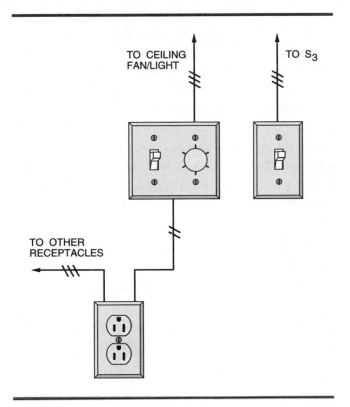

Figure 9-2 Conceptual view of how the switching arrangement is to be accomplished in the master bedroom Circuit A19.

Because fixed glass panels are considered to be wall space, in the master bedroom a receptacle must be located so as to be not more than 6 ft (1.8 m) from the left-hand edge of the fixed glass panel.

SELECTION OF BOXES

As discussed in Unit 2, the selection of outlet boxes and switch boxes is made by the electrician. These decisions are based on Code requirements, space allowances, good wiring practices, and common sense.

For example, in the master bedroom, the electrician may decide to install a 4-in. square box with a two-gang raised plaster cover at the box location next to the sliding door. Or two sectional switch boxes ganged together may be installed at that location, as shown in Figure 9-3 and Table 9-2.

The type of box to be installed depends upon the number of conductors entering the box. The suggested cable layout, Figure 9-1, shows that four cables enter this box for a total of 14 conductors (ten 14 AWG circuit conductors and four equipment grounding conductors).

The requirements for calculating box fill are found in *314.16*. These requirements are summarized in Figure 2-36. Also refer to *Table 314.16(A), 314.16(B)*, and Figure 2-20.

In this example, we have ten circuit conductors, four equipment grounding conductors (counted as

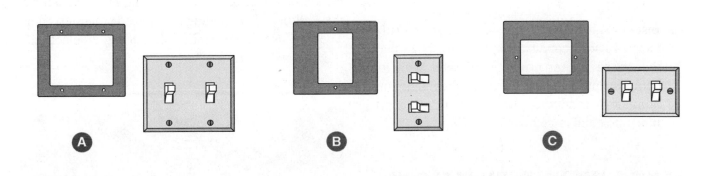

Figure 9-3 (A) shows two switches and wall plate that attach to a two-gang, 4 in.-square raised plaster cover or to two device boxes that have been ganged together. (B) and (C) show two interchangeable-type switches and wall plate that attach to a one-gang, 4 in.-square raised plaster cover or to a single device (switch) box. These types of wiring devices may be mounted vertically or horizontally.

NO. OF GANGS	HEIGHT	WIDTH
1	4½ in. (114.3 mm)	2¾ in. (69.85 mm)
2	4½ in. (114.3 mm)	4⁹⁄₁₆ in. (115.9 mm)
3	4½ in. (114.3 mm)	6⅜ in. (161.9 mm)
4	4½ in. (114.3 mm)	8³⁄₁₆ in. (207.9 mm)
5	4½ in. (114.3 mm)	10 in. (254 mm)
6	4½ in. (114.3 mm)	11¹³⁄₁₆ in. (300 mm)

Table 9-2   Height and width of standard wall plates.

one conductor), four cable clamps (counted as one conductor), one single-pole switch (counted as two conductors), and one three-way switch (counted as two conductors), for a total of 16 conductors.

1. Count the circuit conductors
   $2 + 2 + 3 + 3 =$ 10

2. Add one for one or more equipment grounding conductors   1

3. Add two for each switch $(2 + 2)$ 4

4. Add one for one or more cable clamps   1

                                 Total   16

Now look at the Quik-Chek Box Selection Guide (Figure 2-20), and select a box or a combination of gangable device boxes that are permitted to hold 16 conductors.

Possibilities are:

1. Gang two 3 in. × 2 in. × 3½ in. device boxes together ($9 × 2 =$ eighteen 14 AWG conductors allowed).

2. Install a 4 in. × 4 in. × 2⅛ in. square box (fifteen 14 AWG conductors allowed). Note that the raised plaster cover, if marked with its cubic-inch volume, can increase the maximum number of conductors permitted for the combined box and raised cover. See Figure 2-37.

CEILING-SUSPENDED (PADDLE) FANS

For appearance and added comfort, ceiling-suspended (paddle) fans are very popular. Figure 9-4 is a typical ceiling-suspended (paddle) fan. Ceiling-suspended (paddle) fans rotate slowly in the range of approximately 60 r/min to 250 r/min for residential

fans. Without forced air movement, warm air rises to the ceiling and cool air drops to the floor. Air movement results in increased evaporation from our skin, and we feel cooler. Fans that hug the ceiling are not quite as efficient in moving air as those that are suspended a few inches or more from the ceiling.

Most ceiling-suspended (paddle) fans have a reversing switch, a "HIGH/MEDIUM/LOW/OFF" speed control pull chain switch for the fan, and an "ON/OFF" pull chain switch for the light accessory kit. Upscale models feature remote "ON/OFF" infinite speed and light control, eliminating the need for "hard wiring" of wall switches. These remote controls are similar to those used with television, VCR, stereo, and cable channel selectors. Wiring for wall switches is discussed later in this unit.

What Direction Should a Ceiling-Suspended (Paddle) Fan Rotate?

Major manufacturers of ceiling-suspended (paddle) fans recommend directing the air downward (counter clockwise) in the summer and upward (clockwise) in the winter. Some people like air blowing on them; others do not. Upward or downward, the key is to get the stratified air moving. Try different speeds and direction, and do whatever feels best.

Supporting Ceiling-Suspended (Paddle) Fans

This is a very important issue. There have been many reports of injuries (contusions, concussions, lacerations, fractures, head trauma, etc.) caused by the falling of a ceiling-suspended (paddle) fan. There is also the possibility of starting an electrical

Figure 9-4   Typical home-type ceiling fan. *Courtesy of NuTone.*

fire when the wires in the outlet box and fan canopy "short-circuit" as the fan falls from the ceiling. These problems are usually traced back to an improper installation.

The safe supporting of a ceiling-suspended (paddle) fan involves three issues:

1. the actual weight of the fan

2. the twisting and turning motion when the fan is started

3. vibration caused by unbalanced fan blades

See Figure 9-5 for a typical mounting. Because of the hazards involved, the *NEC*® has strict requirements relating to ceiling-suspended (paddle) fans. Here are the Code rules.

Outlet boxes (metal or nonmetallic)

• shall be properly supported, *314.23*.

• ►shall not be used as the sole support of a ceiling-suspended (paddle) fan unless listed for the application and for the weight of the fan to be supported, *314.27(D)*.◄

• shall be permitted to support a ceiling-suspended (paddle) fan that weighs 35 lb. (16 kg) or less if identified for such use, *422.18(A)*. The weight limitation includes the combined weight of the fan and light kit.

• shall not support a ceiling-suspended (paddle) fan that weighs more than 35 lb. (16 kg), *422.18(B)*. The exception to this is that some outlet boxes might be marked as suitable to support a ceiling-suspended (paddle) fan that weighs 70 lb. (32 kg) or less. The weight limitation includes the combined weight of the fan and light kit.

For new work where ceiling-suspended (paddle) fans are likely to be installed, use outlet boxes that are marked "Acceptable For Fan Support" or "Suitable For Fan Support." This generally includes all habitable rooms. Ceiling boxes installed close to a wall, in hallways, in closets and similar non-habitable rooms, or boxes installed specifically for security systems, smoke and fire detectors, etc., do not require the fan support marking. More and more electrical contractors and electricians are tired of trying to guess what the homeowner might do at some later date. They install suitable fan-rated outlet boxes for all ceiling outlets, other than those that are clearly intended for security systems, etc.

When replacing an existing luminaire (fixture) with a ceiling-suspended (paddle) fan, make sure that the outlet box is suitable for fan support. Conventional outlet boxes that have No. 8-32 tapped holes do not meet this requirement, in which case the fan will have to be supported from the building structure! Carefully read the manufacturer's installation instructions.

Instead of using conventional outlet boxes, listed assemblies that include a hanger, an outlet box, and mounting hardware are available. Figure 9-6 shows one type of box/hanger assembly that can be used for new work or for existing installations.

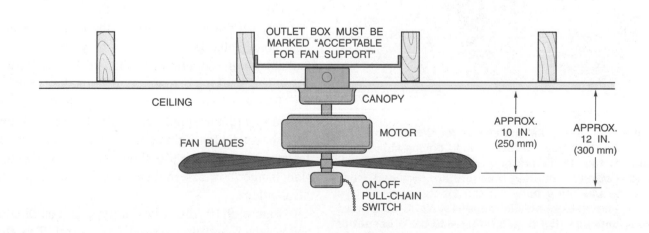

Figure 9-5 Typical mounting of a ceiling-suspended (paddle) fan.

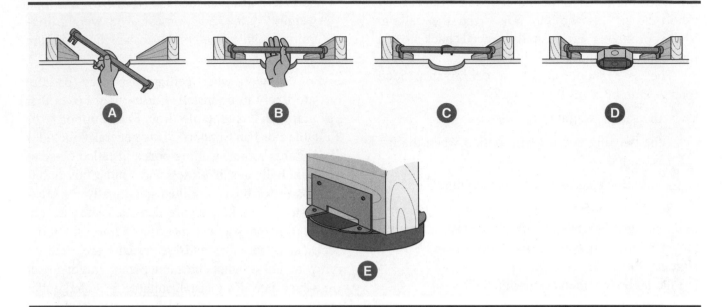

Figure 9-6   (A), (B), (C), and (D) show how a listed ceiling fan hanger and box assembly is installed through a properly sized and carefully cut hole in the ceiling. This is done for existing installations. Similar hanger/box assemblies are used for new work. The hanger adjusts for 16-in. (400-mm) and 24-in. (600 mm) joist spacing but can be cut shorter if necessary. (E) shows a type of box listed and identified for the purpose where the fan is supported from the joist, independent of the box, as required by *422.18* for fans that weigh more than 35 lb. (16 kg). *Courtesy of* Reiker Enterprises Inc.

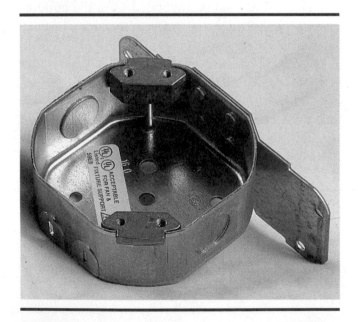

Figure 9-7   A special outlet box for the support of a luminaire (lighting fixture) or a ceiling-suspended (paddle) fan. There are no ears on the box. Instead, special strong mounting brackets are provided. Note that the mounting holes on the brackets are of two sizes—one set of holes is tapped for No. 8-32 screws [for a luminaire (lighting fixture)], and the other set of holes is tapped for No. 10-32 screws (for a ceiling-suspended [paddle] fan). *Photo courtesy of* Reiker Enterprises.

Figure 9-7 illustrates an outlet box that is "listed" for both luminaire (fixture) and fan support for fans of 35 lb. (16 kg) or less, including all accessories. This box has a set of No. 8-32 holes for luminaire (fixture) mounting, and a set of No. 10-32 holes for fan support.

Requirements relating to the support of ceiling-suspended (paddle) fans are also found in the UL White Book, and in the UL Standards 507 (electric fans), 514A (metallic boxes), and 514C (nonmetallic boxes).

Pay careful attention to the marking on metallic and nonmetallic boxes to see if they are "listed" as suitable for the support of ceiling-suspended (paddle) fans. This marking is required to be on the product.

Figure 9-8 shows the wiring for a fan/light combination with the supply at the fan/light unit. Figure 9-9 shows the fan/light combination with the supply at the switch. Figure 9-6 shows two types of ceiling fan hanger/box assemblies. Many other types are available.

Figure 9-10 illustrates a typical combination light switch and three-speed fan control. The electrician is required to install a deep 4-in. box with a two-gang raised plaster ring for this light/fan control.

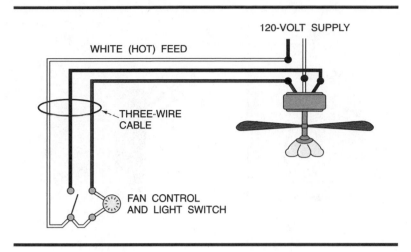

Figure 9-8 Fan/light combination with the supply at the fan/light unit.

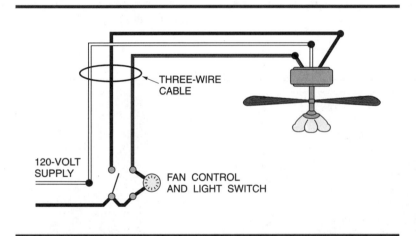

Figure 9-9 Fan/light combination with the supply at the switch.

Figure 9-11 illustrates a "slider" switch for speed control of a fan, and a "dual control" that provides four-speed fan control and two-level "hi/lo" dimming for the incandescent lamps. Figure 9-12 shows two other styles of fan/light controls. Dual controls are needed where the ceiling fan also has a luminaire (lighting fixture). These controls fit into a deep single-gang device box, or preferably into a 4-in. square box with single-gang plaster cover. Be careful of "box fill." Make sure the wall box is large enough. Dimmer and speed control devices are quite large when compared to regular wall switches. Do not jam the control into the wall box.

A typical home-type ceiling fan motor draws 50 to 100 watts (50 to 100 volt-amperes). This is approximately 0.4 to slightly over 0.8 ampere.

Ceiling fan/light combinations increase the load requirements somewhat. Some ceiling fan/light units have one lamp socket, whereas others have four or five lamp sockets.

For the ceiling fan/light unit in the master bedroom, 240 volt-amperes were included in the load calculations. The current draw is

$$I = \frac{VA}{V} = \frac{240}{120} = 2 \text{ amperes}$$

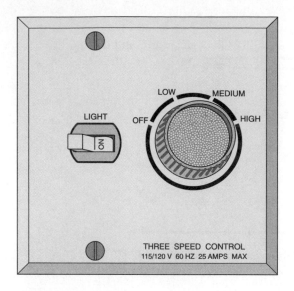

Figure 9-10 Combination light switch and three-speed fan switch.

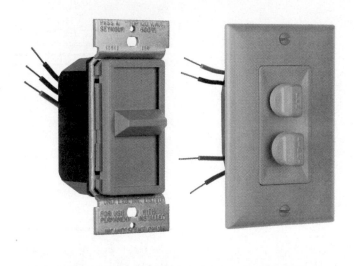

Figure 9-11 The "slider" switch is for speed control of a fan. The "dual-control" provides 4-speed fan control and a 2-level "hi/lo" dimming for an incandescent lamp. *Courtesy of* Pass & Seymour.

Figure 9-12 Two types of fan/light controls. One has a toggle switch for turning a ceiling fan off and on, plus a "slider" that provides 3-speed control of the fan. The other has a toggle switch for turning a light off and on, plus two "sliders," one for "ON/OFF" and 3-speed control of a ceiling fan, the other for full range dimming of lighting. *Courtesy of* Lutron.

For most fans, switching on and off may be accomplished by the pull-switch, an integral part of the fan ("OFF/HIGH/MED/LOW"), or with a solid-state switch having an infinite number of speeds. Some of the more expensive models offer a remote control similar to that used with television, VCRs, stereos, and cable channel selectors.

Most ceiling fan motors have an integral reversing switch to blow air downward or upward.

Do not use a standard incandescent lamp dimmer to control fan motors, fluorescent ballasts, transformers, low-voltage lighting systems (they use transformers to obtain the desired low voltage), motor-operated appliances, or other "inductive" loads. Serious overheating and damage to the motor can result. Fan speed controllers have internal circuitry that is engineered specifically for use on inductive load circuits. Fan speed controllers virtually eliminate any fan motor "hum." Read the label and the instructions furnished with a dimmer to be sure it is suitable for the application.

REVIEW

Note: Refer to the Code or the plans where necessary.

1. What circuit supplies the master bedroom? _____

2. What unique type of electrical protection is required for all 120-volt outlets in residential bedrooms? Where in the *NEC*® is this requirement found?_____

3. What type of receptacles are provided in this bedroom? How many receptacles are there? _____

4. How many ceiling outlets are included in this circuit? _____

5. What wattage for the closet luminaires (fixtures) was used for calculating their contribution to the circuit? _____

6. What is the current draw for the ceiling fan/light? _____

7. What is the estimated load in volt-amperes for the circuit supplying the master bedroom? _____

8. Which is installed first, the switch and outlet boxes or the cable runs? _____

9. How many conductors enter the ceiling fan/light wall box? _____

10. What type and size of box may be used for the ceiling fan/light wall box? _____

11. What type of covers are used with 4-in. square outlet boxes? _____

_____

_____

_____

12. a. Does the circuit for the master bedroom have a grounded circuit conductor?

    b. Does it have an equipment grounding conductor? _____

    c. Explain the difference between a "grounded" conductor and a "grounding" conductor. _____

_____

_____

_____

_____

_____

_____

_____

13. Approximately how many feet (meters) of two-wire cable and three-wire cable are needed to complete the circuit supplying the master bedroom? Two-wire cable _____ ft (_____ m). Three-wire cable _____ ft (_____ m).

14. If the cable is laid in notches in the corner studs, what protection for the cable must be provided? _____

_____

_____

_____

15. How high are the receptacles mounted above the finished floor in this bedroom?

_____

_____

16. Approximately how far from the bedroom door is the first receptacle mounted? (See the plans.) _____

17. What is the distance from the finished floor to the center of the wall switches in this bedroom? _____

18. The master bedroom features a sliding glass door. For the purpose of providing the proper receptacle outlets, answer the following statements true or false.

    • Sliding glass panels are considered to be wall space.   _____

    • Fixed panels of glass doors are considered to be wall space.   _____

19. What type of receptacle will be installed outdoors, just outside of the master bedroom?

_____

_____

_____

20. Show your calculations of how to select a proper wall box for the closet luminaire (fixture) switch. Keep in mind that the available space between the wood casings is small. See Figure 9-1. _____

_____

_____

_____

_____

_____

_____

_____

21. When an outlet box is to be installed to support a ceiling fan, how must it be marked?

_____

_____

22. The following is a layout of the lighting circuit for the master bedroom. Using the cable layout shown in Figure 9-1, make a complete wiring diagram of this circuit. Use colored pencils or pens.

23. Connect the three-wire cable, fan, light, fan speed control and light switch. Refer to *200.7* to review the permitted use of the white conductor in the cable.

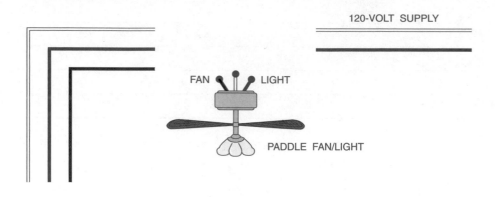

24. What is the width of a two-gang wall plate? _____

25. May a standard electronic dimmer be used to control the speed of a fan motor?

_____

26. Outlet boxes that pass the tests at Underwriters Laboratories for the support of ceiling-suspended (paddle) fan are listed and marked "Acceptable For Fan Support" or "Suitable For Fan Support." When roughing-in the electrical wiring for a new home, this type of outlet box must be installed wherever there is a likelihood of installing a ceiling-suspended (paddle) fan. Name those locations in homes that you consider likely to support a ceiling-suspended (paddle) fan.

_____

_____

_____

Lighting Branch-Circuit– Bathrooms, Hallway

OBJECTIVES

After studying this unit, the student should be able to

- list equipment grounding requirements for bathroom installations.
- draw a wiring diagram for the bathroom and hallway.
- understand Code requirements for receptacles installed in bathrooms.
- understand Code requirements for receptacle outlets in hallways.
- discuss fundamentals of proper lighting for bathrooms.

Circuit A14 is a 15-ampere branch-circuit that supplies the lighting outlets and receptacle outlet in the hallway, and the vanity lighting in both bathrooms. The receptacle outlets in the bathrooms are supplied by separate circuits A22 and A23. This is a Code requirement, discussed later in this unit and in Unit 15.

Table 10-1 summarizes outlets and estimated load for the bathrooms and bedroom hall.

DESCRIPTION	QUANTITY	WATTS	VOLT-AMPERES
Receptacles @ 120 W each	1	120	120
Vanity luminaires (fixtures) @ 200 W each	2	400	400
Hall luminaire (fixture)	1	100	100
TOTALS	4	620	620

Table 10-1 Bathroom and bedroom hall: outlet count and estimated load (Circuit A14). The receptacles in the master bathroom and hall bathroom are not included in this table as they are connected to separate 20-ampere branch-circuits A22 and A23.

Note that each bathroom shows a ceiling heater/light/fan that is connected to separate circuit Ⓐ J and Ⓐ K. These are discussed in detail in Unit 22.

A hydromassage tub is located in the bathroom serving the master bedroom. It is connected to a separate circuit Ⓐ A and is also discussed in Unit 22.

The attic exhaust fan in the hall is supplied by a separate circuit Ⓐ L, also covered in Unit 22.

LIGHTING BRANCH-CIRCUIT A14 FOR HALLWAY AND BATHROOMS

Figure 10-1 and the electrical plans for this area of the home show that each bathroom has a luminaire (fixture) above the vanity mirror. Some typical luminaires (fixtures) are shown in Figure 10-2. Of course, the homeowner might decide to purchase a medicine cabinet complete with a self-contained luminaire (lighting fixture). See Figure 10-3, which illustrates how to rough-in the wiring for each type. These luminaires (fixtures) are controlled by single-pole switches at the doors.

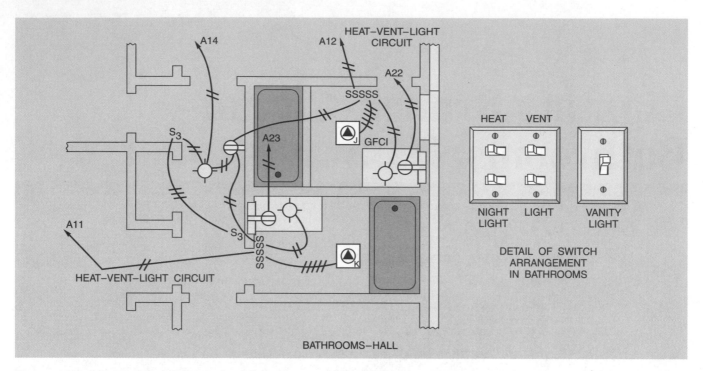

Figure 10-1   Cable layout for the master bathroom, hall bathroom, and hall. Layout includes lighting Circuit A14, two circuits (A11 and A12) for the heat/vent/lights, and two circuits (A22 and A23) for the receptacles in the bathrooms. The special purpose outlets for the hydromassage tub, attic exhaust fan, and smoke detector are covered elsewhere in this text.

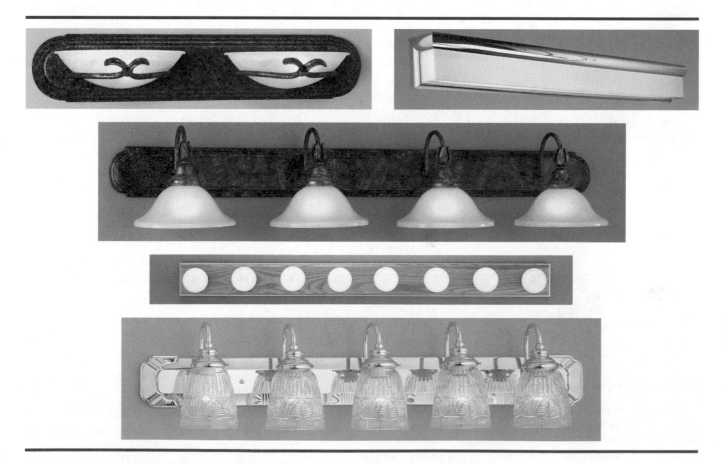

Figure 10-2   Typical vanity (bathroom) luminaires (lighting fixtures) of the side bracket and strip types. *Courtesy of* Progress Lighting.

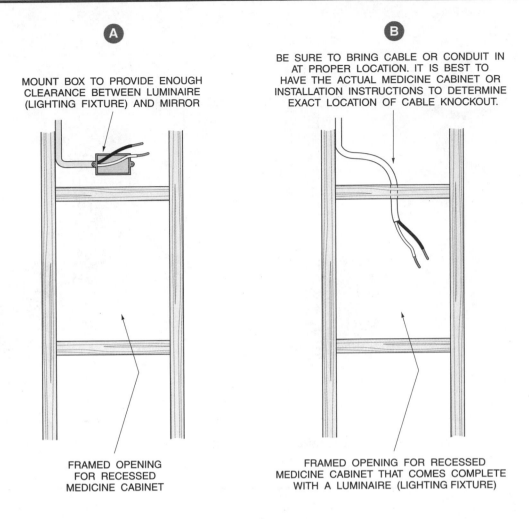

Figure 10-3   Two methods most commonly used for roughing-in the wiring for lighting above a vanity.

You must remember that a bathroom (powder room) should have proper lighting for shaving, combing hair, grooming, and so on. Mirror lighting can accomplish this, because a mirror will reflect what it "sees." If the face is poorly lit, with shadows on the face, that is precisely what will be reflected in the mirror.

A luminaire (lighting fixture) directly overhead will light the top of one's head, but will cause shadows on the face. Mirror lighting and/or adequate lighting above and forward of the standing position at the vanity can provide excellent lighting in the bathroom. Figure 10-4, Figure 10-5, and Figure 10-6 show pictorial as well as section views of typical soffit lighting above a bathroom vanity.

Bathroom Receptacles

The *NEC®* in *Article 100* defines a bathroom as "an area including a basin, with one or more of the following: a toilet, a tub, or a shower." See Figure 10-7.

Bathroom receptacles are required to be GFCI protected per *210.8(A)(1)* as discussed in Unit 6.

►At least one wall receptacle shall be installed within 3 ft (900 mm) of the outside edge of each basin, and must be located on a wall or partition adjacent to the basin, *210.52(D)*.◄

Receptacles shall not be installed face-up on counters or work surfaces, *406.4(E)*.

At least one 20-ampere branch circuit must be provided for receptacles in bathrooms, *210.11(C)(3)*. This separate 20-ampere branch-circuit shall not

WRONG

RIGHT

Figure 10-4 Positioning of bathroom luminaires (lighting fixtures). Note the wrong way and the right way to achieve proper lighting.

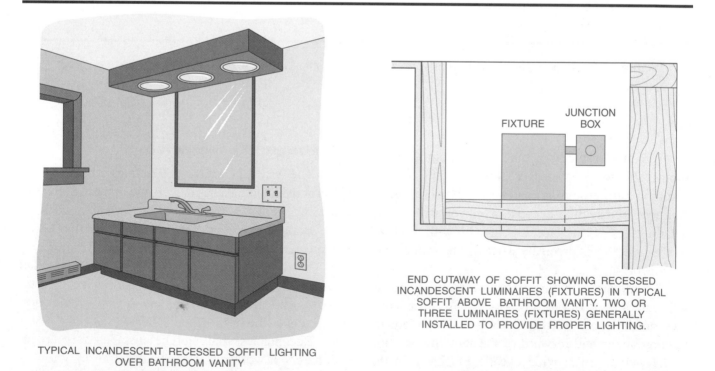

TYPICAL INCANDESCENT RECESSED SOFFIT LIGHTING OVER BATHROOM VANITY

FIXTURE

JUNCTION BOX

END CUTAWAY OF SOFFIT SHOWING RECESSED INCANDESCENT LUMINAIRES (FIXTURES) IN TYPICAL SOFFIT ABOVE BATHROOM VANITY. TWO OR THREE LUMINAIRES (FIXTURES) GENERALLY INSTALLED TO PROVIDE PROPER LIGHTING.

Figure 10-5 Incandescent soffit lighting. Refer to Unit 7 for minimum clearance Code requirements for installation of recessed luminaires (fixtures).

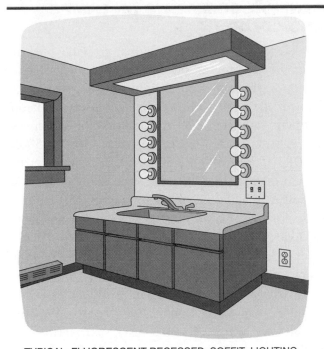

TYPICAL FLUORESCENT RECESSED SOFFIT LIGHTING
OVER BATHROOM VANITY. NOTE ADDITIONAL
INCANDESCENT "SIDE-OF-MIRROR" LIGHTING.

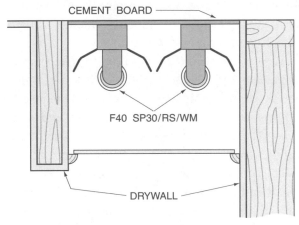

END CUTAWAY VIEW OF SOFFIT ABOVE
BATHROOM VANITY SHOWING RECESSED
FLUORESCENT LUMINAIRES (FIXTURES) CONCEALED
ABOVE TRANSLUCENT ACRYLIC LENS

Figure 10-6   Combination fluorescent and incandescent bathroom lighting.

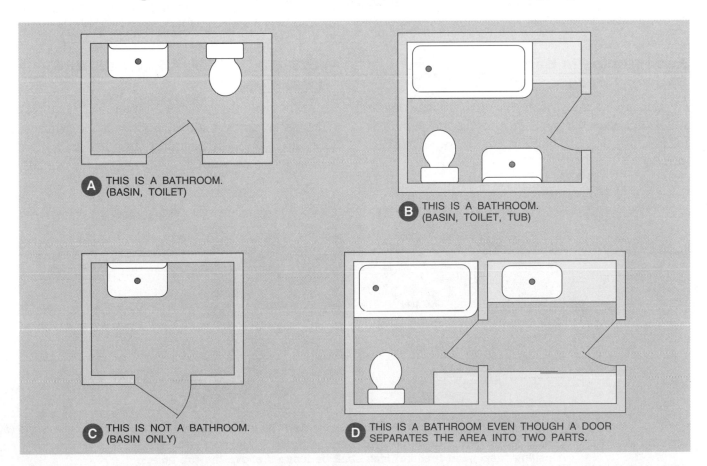

A THIS IS A BATHROOM.
(BASIN, TOILET)

B THIS IS A BATHROOM.
(BASIN, TOILET, TUB)

C THIS IS NOT A BATHROOM.
(BASIN ONLY)

D THIS IS A BATHROOM EVEN THOUGH A DOOR
SEPARATES THE AREA INTO TWO PARTS.

Figure 10-7   The Code in *Article 100* defines a *bathroom* as an area including a basin with one or more of the following: a toilet, a tub, or a shower.

supply any other outlets, except that if the circuit supplies a single bathroom, then that 20-ampere branch-circuit is permitted to supply other electrical equipment in the same bathroom. The other equipment shall not exceed 10 amperes which is 50 percent of the 20-ampere branch-circuit.

The intent of *210.11(C)(3)* is to take bathroom receptacles off of the lighting branch-circuits in homes. In many instances, overloads caused by plugging in high-wattage hair dryers and similar high-wattage appliances have resulted in total loss of power to that particular branch-circuit.

The present wording of *210.11(C)(3)* permits more than one bathroom to be connected to the one required separate 20-ampere branch-circuit.

In this text, we have chosen to run a separate 20-ampere branch-circuit A22 to the receptacle in the master bedroom bathroom, a separate 20-ampere branch-circuit A23 to the receptacle in the hall bathroom, and another separate 20-ampere branch-circuit B21 to the receptacle in the powder room located near the laundry. These separate circuits are included in the general lighting load calculations, so no additional load need be added.

Receptacles in Bathtub and Shower Spaces

Because of the obvious hazards associated with water and electricity, receptacles are *not* permitted to be installed in bathtub and shower spaces, *406.8(C)*.

GENERAL COMMENTS ON LAMPS AND COLOR

Incandescent lamps (light bulbs) provide pleasant color tones, bringing out the warm red flesh tones similar to those of natural light. This is particularly true for the "soft" white lamps. Fluorescent lamps available today provide a wide range of "coolness" to "warmth." Rated in degrees Kelvin (i.e., 2500K, 3000K, 3500K, 4000K, 5000K, etc.), the lower the Kelvin degrees, the "warmer" the color tone. Conversely, the higher the Kelvin degrees, the "cooler" the color tone. Warm fluorescent lamps bring out the red tones, whereas cool fluorescent lamps tend to give a person's skin a pale appearance. Thus, you will find many types of fluorescent lamps, each providing different color tones and different efficiencies. These might be marked daylight **D** *(very cool)*, cool white **CW** *(cool)*, white **W** *(moderate)*, and warm white **WW** *(warm)*. These categories break down further into a *deluxe* **X** series, *specification* **SP** series, and *specification deluxe* **SPX** series. Unit 7 of this text discusses some basics of lighting. The Instructor's Guide lists some lamp manufacturers' publications that cover the subject of lighting in greater detail.

HANGING LUMINAIRES (FIXTURES) IN BATHROOMS

No parts of cord-connected lumnaires (fixtures), hanging lumnaires (fixtures), lighting track, pendants, or ceiling-suspended (paddle) fans shall be located within a zone measuring 3 ft (900 mm) horizontally

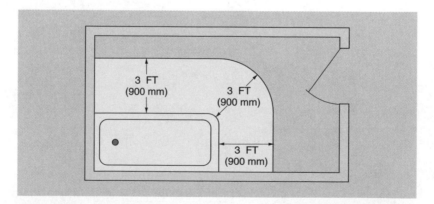

Figure 10-8 No parts of cord-connected luminaires (lighting fixtures), hanging luminaires (fixtures), lighting track, pendants, or ceiling-suspended (paddle) fans shall be located above the tub and the area within 3 ft (900 mm) measured horizontally from the top of the bathtub rim (top view), *410.4(D)*.

and 8 ft (2.5 m) vertically from the top of the bathtub rim, *410.4(D)*. Figure 10-8 projects the top view of the 3 ft (900 mm) restriction. Figure 10-9 shows both the allowable installation and Code violations of the given dimensions. Recessed or surface mounted lumnaires (lighting fixtures) and recessed exhaust fans may be located within the restricted zone.

HALLWAY LIGHTING

The hallway lighting is provided by one ceiling lumnaire (fixture) that is controlled with two three-way switches located at either end of the hall. The home

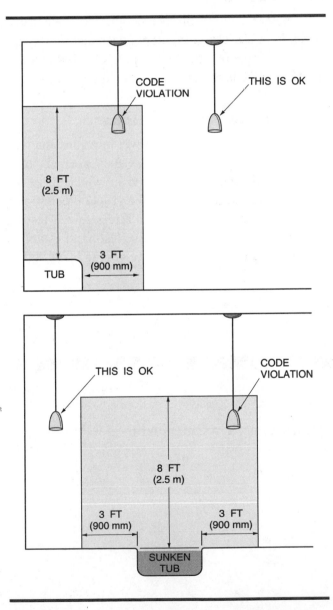

Figure 10-9   No parts of cord-connected luminaires (fixtures), hanging luminaires (fixtures), lighting track, pendants, or ceiling- suspended (paddle) fans shall be allowed in the shaded areas, *410.4(D)*.

run to Main Panel A has been brought into this ceiling outlet box.

RECEPTACLE OUTLETS IN HALLWAYS

One receptacle outlet has been provided in the hallway as required in *210.52(H)*, which states that "for hallways of 10 ft (3 m) or more in length, at least one receptacle outlet shall be required."

For the purpose of determining the length of a hallway, the measurement is taken down the centerline of the hall, turning corners if necessary, but not passing through a doorway, Figure 10-10.

EQUIPMENT GROUNDING

With few exceptions, equipment with exposed metal parts that could become energized shall be grounded. This includes such things as luminaire (fixtures), in-the-wall and baseboard electric heaters, metal faceplates, medicine cabinets that have lighting, and similar items. Equipment that must be grounded is itemized in *250.110*, *250.112* and *250.114*.

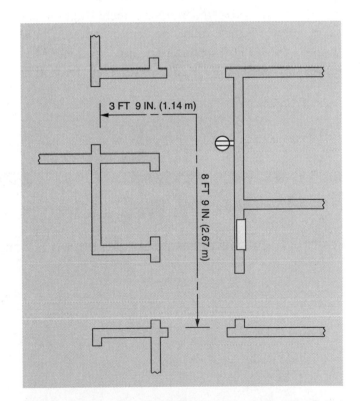

Figure 10-10   The centerline measurement of the bedroom hallway in this residence is 12 ft 6 in. (3.81 m), which requires at least one wall receptacle outlet, *210.52(H)*. This receptacle may be installed anywhere in the hall. The Code does not specify a location.

In general, according to *250.110*, all exposed noncurrent-carrying metal parts of electrical equipment must be grounded

- if they are within 8 ft (2.5 m) vertically or 5 ft (1.5 m) horizontally of the ground or other grounded metal objects.

- if the ground and the electrical equipment can be touched at the same time.

- if they are located in wet or damp locations, such as in bathrooms, showers, and outdoors.

- if they are in electrical contact with metal. This requirement includes such things as metal lath, aluminum foil insulation, and metal sidings.

Equipment is considered grounded when it is properly and permanently connected to metal raceway, the armor of armored cable or metal-clad cable, the equipment grounding conductor in nonmetallic-sheathed cable, or a separate equipment grounding conductor. Of course, the means of grounding (the metal raceway, the armored cable, or the equipment grounding conductor in the nonmetallic-sheathed cable) must itself be properly grounded.

A list of cord- and plug-connected appliances in residences that must be grounded is found in *250.114(3)*. The second paragraph of *250.114* is an exception to the main rule. This exception accepts "double insulation" or its equivalent in lieu of grounding. Many hand-held appliances, such as electric drills, electric razors, and electric toothbrushes, make use of the "double insulation" technique. Such appliances have a two-wire cord and two-wire plug cap instead of three-wire cord that has an equipment grounding conductor and a three-wire plug cap that has a ground prong.

Double-insulated appliances are clearly marked to indicate that they are double insulated.

Immersion Detection Circuit Interrupters

Another way to protect people from electrical shock is to use grooming appliances that are equipped with a special attachment plug-cap that is a liquid IDCI. IDCIs are discussed in Unit 6.

Switches in Wet Locations

Do *not* install switches in wet locations, such as in bathtubs or showers, unless they are part of a listed tub or shower assembly, in which case the manufacturer has taken all of the proper precautions, and submitted the assembly to a recognized testing laboratory to undergo exhaustive testing to establish the safety of the equipment. See *404.4*.

REVIEW

1. List the number and types of switches and receptacles used in Circuit A14.

2. There is a three-way switch in the bedroom hallway leading into the living room. Show your calculation of how to determine the box size for this switch. The box contains cable clamps.

3. What wattage was used for each vanity luminaire (fixture) to calculate the estimated load on Circuit A14? _____

4. What is the current draw for the answer given in question 3?

5. Exposed noncurrent-carrying metallic parts of electrical equipment must be grounded if installed within _____ ft (_____ m) vertically or _____ ft (_____ m) horizontally of bathtubs, plumbing fixtures, pipes, or other grounded metal work or grounded surfaces.

6. What color are the faceplates in the bathrooms? Refer to the specifications.

_____

7. Most appliances of the type commonly used in bathrooms, such as hair dryers, electric shavers, and curling irons, have two-wire cords. These appliances are _____ insulated or _____ _____ _____ _____ protected.

8. a. The *NEC*® in section _____ requires that all receptacles in bathrooms be _____ protected.

   b. The *NEC*® in section _____ requires that all receptacles in bathrooms be connected to one or more separate 20-ampere branch-circuits that serve no other outlets.

   c. The *NEC*® in section _____ permits the additional required 20-ampere branch-circuit for bathroom receptacles to supply more than one bathroom.

   d. The *NEC*® in section _____ permits other electrical equipment to be connected to the additional required 20-ampere branch-circuit for bathroom receptacles, but only if the branch-circuit supplies a single bathroom and the other equipment is located in that same bathroom.

   e. The *NEC*® in section _____ prohibits mounting receptacles in bathrooms face-up in the countertops and work surfaces near basins.

9. Hanging luminaires (lighting fixtures) must be kept at least _____ ft (_____ m) from the edge of the tub as measured horizontally. In bathrooms with high ceilings, where the hanging luminaire (fixture) is installed directly over the tub, it must be kept at least _____ ft (_____ m) above the edge of the tub.

10. The following is a layout of a lighting circuit for the bathroom and hallway. Using the cable layout shown in Figure 10-1, make a complete wiring diagram of this circuit. Use colored pencils or pens.

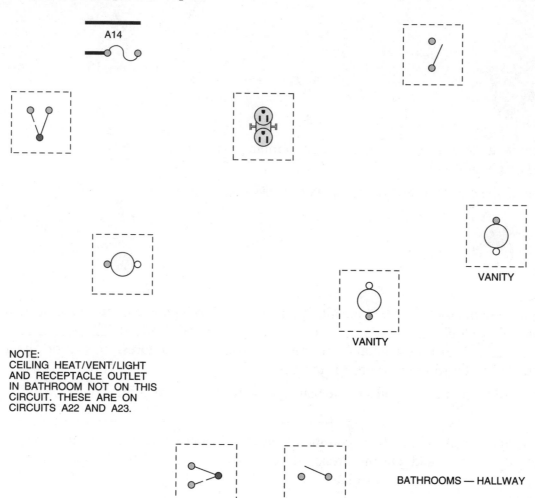

A14

NOTE:
CEILING HEAT/VENT/LIGHT
AND RECEPTACLE OUTLET
IN BATHROOM NOT ON THIS
CIRCUIT. THESE ARE ON
CIRCUITS A22 AND A23.

VANITY

VANITY

BATHROOMS — HALLWAY

11. Circle the correct answer as to whether a receptacle outlet is required in the following hallways.

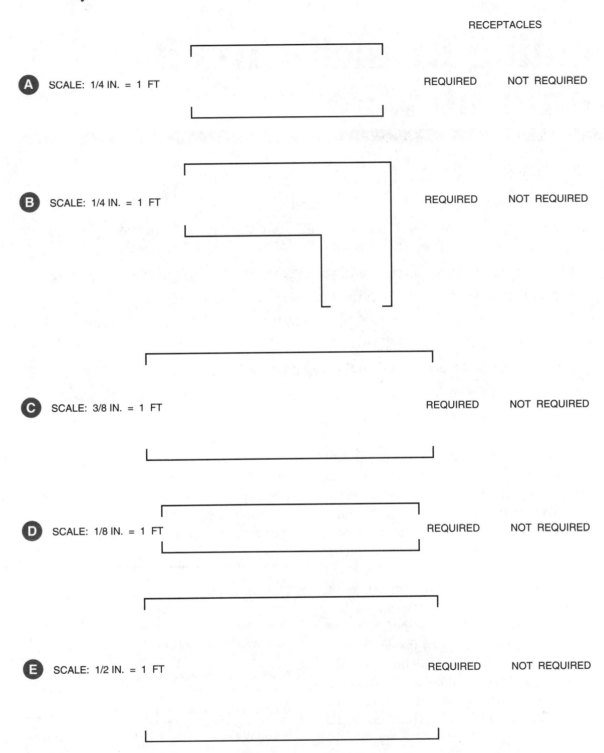

RECEPTACLES

A   SCALE: 1/4 IN. = 1 FT   REQUIRED   NOT REQUIRED

B   SCALE: 1/4 IN. = 1 FT   REQUIRED   NOT REQUIRED

C   SCALE: 3/8 IN. = 1 FT   REQUIRED   NOT REQUIRED

D   SCALE: 1/8 IN. = 1 FT   REQUIRED   NOT REQUIRED

E   SCALE: 1/2 IN. = 1 FT   REQUIRED   NOT REQUIRED

Lighting Branch-Circuit–Front Entry, Porch

OBJECTIVES

After studying this unit, the student should be able to

- understand how to install a switch in a doorjamb for automatic "ON/OFF" when the door is opened or closed.
- discuss types of luminaires (fixtures) recommended for porches and entries.
- complete the wiring diagram for the entry-porch circuit.
- discuss the advantages of switching outdoor receptacles from indoors.
- define wet and damp locations.
- understand box fill for sectional ganged device boxes.

The front entry-porch is connected to Circuit A15. The home run enters the ceiling box in the front entry. From this box, the circuit spreads out, feeding the closet light, the porch bracket luminaire (fixture), the two bracket luminaires (fixtures) on the front of the garage, one receptacle outlet in the entry, and one outdoor weatherproof GFCI receptacle on the porch, as shown in Figure 11-1. Table 11-1 summarizes the outlets and estimated load for the entry and porch.

Note that the receptacle on the porch is controlled by a single-pole switch just inside the front door. This allows the homeowner to plug in outdoor lighting, such as strings of ornamental Christmas lights or decorative lighting, and have convenient switch control of that receptacle from inside the house, a nice feature.

Typical ceiling luminaires (fixtures) commonly installed in front entryways, where it is desirable to make a good first impression on guests, are shown in Figure 11-2.

Typical outdoor wall-bracket-style porch and entrance luminaires (lighting fixtures) are shown in Figure 11-3.

A location exposed to the weather is a wet location. A partially protected area such as an open porch, or under a roof or canopy, is a damp location. For more precise definitions of dry, wet, and damp locations, see *Article 100* of the *NEC*.

DESCRIPTION	QUANTITY	WATTS	VOLT-AMPERES
Receptacles @ 120 W	1	120	120
Weatherproof receptacles @ 120 W	1	120	120
Outdoor porch bracket luminaire (fixture)	1	100	100
Outdoor garage bracket luminaires (fixtures) @ 100 W each	2	200	200
Ceiling luminaire (fixture)	1	150	150
Closet recessed luminaire (fixture)	1	75	75
TOTALS	7	765	765

Table 11-1 Entry and porch: outlet count and estimated load (Circuit A15).

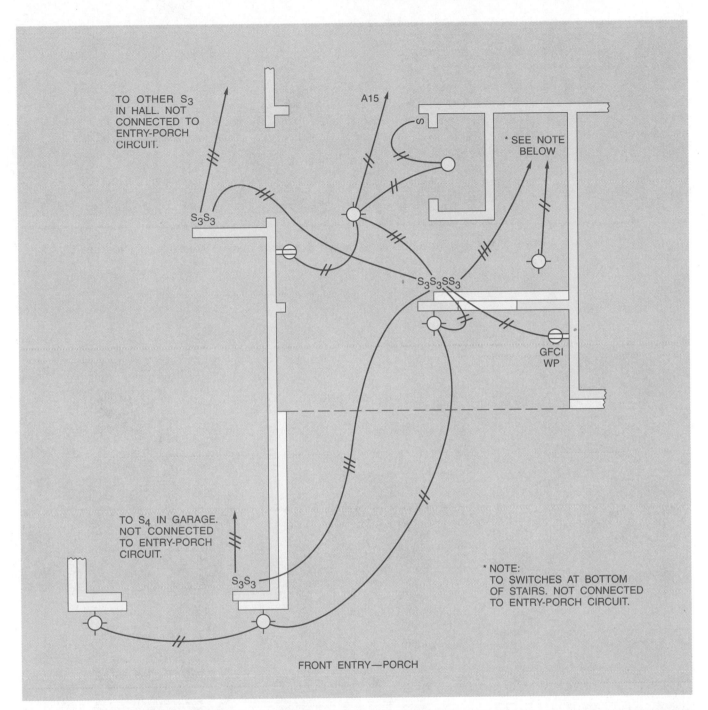

Figure 11-1 Cable layout of the front entry-porch circuit A15. Note that cables from some of the other circuits are shown so that you can get a better idea of exactly how many wires will be found at the various locations.

Luminaires (fixtures) installed in wet or damp locations must be installed so that water cannot enter or accumulate in wiring compartments, lampholders, or other electrical parts. Luminaires (fixtures) installed in wet locations shall be marked, "Suitable for Wet Locations." Luminaires (fixtures) installed in damp locations shall be marked, "Suitable for Wet Locations" or "Suitable for Damp Locations." This information is found in *410.4(A)*.

CIRCUIT A15

Circuit A15 is a rather simple circuit that has two sets of three-way switches: one set controlling the front entry ceiling luminaire (fixture), the other set controlling the porch bracket luminaire (fixture) and the two bracket luminaires (fixtures) on the front of the garage. Also, one single-pole switch controls the weatherproof porch receptacle outlet, and another

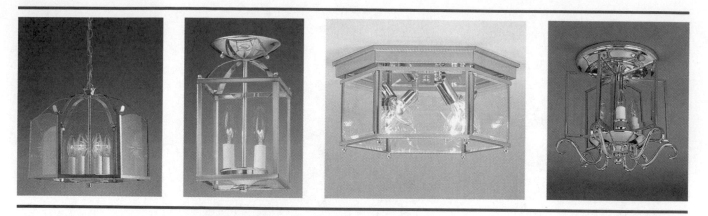

Figure 11-2   Hall luminaires (lighting fixtures): ceiling mount and chain mount. *Courtesy of* Progress Lighting.

Figure 11-3   Outdoor porch and entrance luminaires (lighting fixtures): wall bracket styles. *Courtesy of* Progress Lighting.

single-pole switch controls the closet light. In the Review questions you will be asked to follow the suggested cable layout for making up all of the circuit connections.

Probably the most difficult part of this circuit is planning and making up the connections at the switch location just inside the front door. Here we find four toggle switches: (1) a three-way switch for a ceiling luminaire (fixture); (2) a three-way switch for porch and garage luminaires (fixtures); (3) a single-pole switch for a weatherproof receptacle on the porch; and (4) a three-way switch for the luminaire (fixture) over the stairway leading to the basement (connected to the Recreation Room lighting circuit B12, not the entry-porch circuit A15).

This location is another challenge for determining the proper size wall box. Let's give it a try.

1. Add the circuit conductors
   $2 + 2 + 3 + 3 + 3 + 3 =$ 16

2. Add for equipment grounding wires
   (count 1 only)   1

3. Add eight for the four switches   8

4. Add for cable clamps (count 1 only)   <u>1</u>

   Total   26

To select a box that has adequate cubic-inch capacity for this example, refer to *Table 314.16(A)* in the *NEC*,® and Figure 2-20 and Table 2-1 in this text. One possibility would be to use four 3 in. × 2 in. × 3½ in. device boxes ganged together, providing $4 \times 9 = 36$ conductor count based on 14 AWG conductors.

When using nonmetallic boxes, refer to *Table 314.16(B)* and base the volume allowance on 2 in.^{3} for each 14 AWG conductor and for the volume allowance for the wiring devices, cable clamps, and equipment grounding conductors. In the previous example, we are looking for a cubic-in. volume of $26 \times 2 = 52$ in.^{3}

An interesting possibility presents itself for the front entry closet. Although the plans show that the closet luminaire (fixture) is turned on and off by a standard single-pole switch to the left as you face the closet, a doorjamb switch could have been installed. A doorjamb switch is usually mounted about 6 feet (1.8 m) above the floor, on the inside of the 2 × 4 framing for the closet door. The electrician will run the cable to this point, and let it hang out until the carpenter can cut the proper size opening into the doorjamb for the box. After the finished woodwork is completed, the electrician then finishes installing the switch. Figure 11-4 illustrates a door switch. Note that the plunger on the switch is pushed inward when the edge of the door pushes on it as the door is closed, shutting off the light. The plunger can be adjusted in or out to make the switch work properly. These doorjamb switches come complete with their own special wall box.

The integral box furnished with door switches is generally suitable for one cable only. Wiring space is very limited. Do not plan on using this box for splices other than those necessary to make up the connections to the switch.

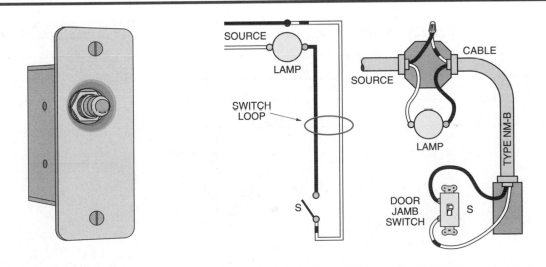

Figure 11-4   Door jamb switch. Feed at light. Switch loop run from ceiling box to door jamb box.

REVIEW

1. a. How many circuit wires enter the entry ceiling box? _____
    b. How many equipment grounding conductors enter the entry ceiling box? _____

2. Assuming that an outlet box for the entry ceiling has a fixture stud but has external cable clamps, what size box could be installed? Show calculations for both metal and nonmetallic boxes.

    _____

    _____

    _____

    _____

    _____

3. How many receptacle outlets and lighting outlets are supplied by Circuit A15?

    _____

4. Outdoor luminaires (fixtures) directly exposed to the weather must be marked as:

    a. suitable for damp locations

    b. suitable for dry locations only

    c. suitable for wet locations

    Circle the correct answer.

5. Make a material list of all types of switches and receptacles connected to Circuit A15.

    _____

    _____

    _____

6. From left to right, facing the switches, what do the switches next to the front door control?

    _____

    _____

    _____

    _____

7. Who is to select the entry ceiling luminaire (fixture)? _____

8. The following layout is for Lighting Circuit A15, the entry, porch, and front garage lights. Using the cable layout shown in Figure 11-1, make a complete wiring diagram of this circuit. Use colored pencils to indicate the color of the conductors' insulation.

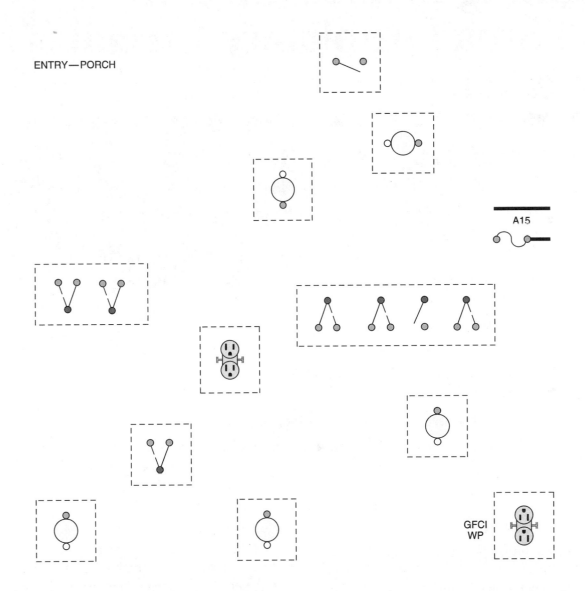

Lighting Branch-Circuit and Small Appliance Circuits for Kitchen

OBJECTIVES

After studying this unit, the student should be able to

- discuss the features of an exhaust fan for the removal of kitchen cooking odors and humidity.
- explain the Code requirements for small appliance circuits in kitchens.
- understand the Code requirements for ground-fault protection for receptacles that serve countertops.
- discuss split-circuit receptacles.
- discuss multiwire circuits.
- discuss exhaust fan noise ratings.
- understand the overall general concept of grounding electrical equipment.
- know the difference between a receptacle outlet and a lighting outlet.
- discuss typical kitchen, dining room, undercabinet, and swag lighting.

LIGHTING CIRCUIT B7

The lighting circuit supplying the kitchen originates at Panel B. The circuit number is B7. The home run is very short, leading from Panel B in the corner of the recreation room to the switch box to the right of the kitchen sink. Here, the circuit continues upward through the junction box on the recessed luminaire (fixture) above the sink, which is approved for feed-through wiring (see Unit 7 for a complete explanation of Code requirements for recessed luminaires [fixtures]) through the clock outlet box, and then across the kitchen ceiling connecting the ceiling luminaires (light fixtures), the lighting track, the outdoor bracket luminaire (fixture), and range hood exhaust fan/light unit, Figure 12-1.

Another way to run this circuit would be to run the branch-circuit home run to the clock outlet box, then over to the recessed luminaire (fixture). This is a matter of considering how many conductors enter the boxes, the ease in making up the connections, and whether the junction box on the recessed luminaire (fixture) is listed for through-wiring. Table 12-1 shows the outlets and estimated loads for the kitchen.

KITCHEN LIGHTING

The general lighting for the kitchen is provided by two four-lamp square fluorescent surface-mounted luminaires (lighting fixtures) shown in Figure 12-2.

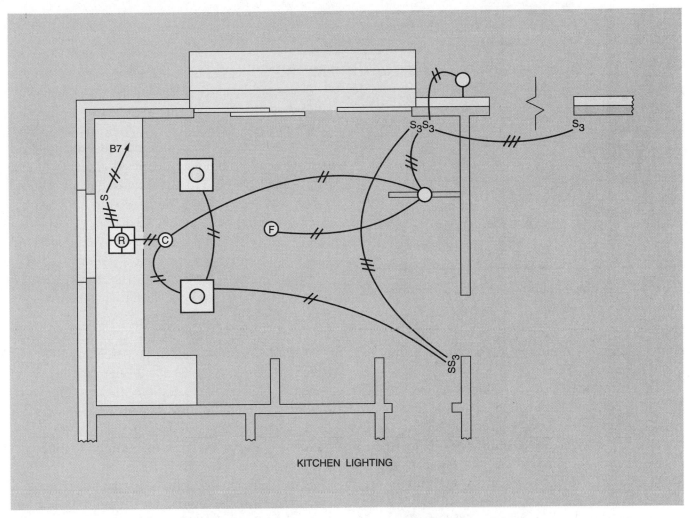

KITCHEN LIGHTING

Figure 12-1   Cable layout for the kitchen lighting. This is Circuit B7. The appliance circuit receptacles are connected to 20-ampere circuits B13, B15, and B16 and are not shown on this lighting circuit layout.

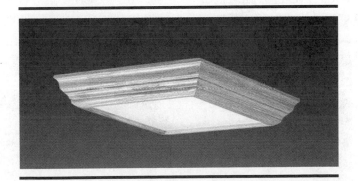

Figure 12-2   Typical ceiling luminaire (fixture) of the type installed in the kitchen of this residence. This type of luminaire (fixture) might have four 20-watt fluorescent lamps, or it might have two 40-watt U-shaped lamps. *Courtesy of* Progress Lighting.

DESCRIPTION	QUANTITY	WATTS	VOLT-AMPERES
Ceiling luminaires (fixtures) (eight 20-W FL lamps)	2	160	180
Recessed luminaire (fixture) over sink	1	100	100
Lighting track	1	100	100
Outdoor rear bracket luminaire (fixture)	1	100	100
Clock outlet	1	15	15
Range hood fan/light Two 60-W lamps		120	120
Fan motor (1⅓ A @ 120 V)		160	180
TOTALS	7	755	795

Table 12-1   Kitchen: outlet count and estimated load (Circuit B7).

These luminaires (fixtures) are controlled by a single-pole switch located next to the doorway leading to the living room.

Above the breakfast area table is a short section of track lighting controlled by two three-way switches: one located adjacent to the sliding door, the other at the doorway leading to the living room. Figure 12-3 illustrates just a few of the many types of luminaires (lighting fixtures) that might be secured to the track.

Figure 12-4 shows a variety of chain-suspended luminaires (lighting fixtures) of the type commonly used in dining rooms.

One goal of good lighting is to reduce shadows in work areas. This is particularly important in the kitchen. The two ceiling luminaires (fixtures), the luminaire (fixture) above the sink, plus the luminaire (fixture) within the range hood will provide excellent lighting in the kitchen area. Of course, the track

lighting in the eating area offers good lighting for that area.

Undercabinet Lighting

In some instances, particularly in homes with little outdoor exposure, the architect might specify strip fluorescent luminaires (lighting fixtures) under the kitchen cabinets. Several methods may be used to install undercabinet luminaires (lighting fixtures). They can be fastened to the wall just under the upper cabinets (Figure 12-5[A]), installed in a recess that is part of the upper cabinets so that they are hidden from view (Figure 12-5[B]), or fastened under and to the front of the upper cabinets (Figure 12-5[C]). All three possibilities require close coordination with the cabinet installer to be sure that the wiring is brought out of the wall at the proper location to connect to the undercabinet luminaires (fixtures).

Flat fluorescent and incandescent luminaires (lighting fixtures) are available that mount on the underside of upper cabinets. Some of these luminaires (lighting fixtures) are direct connected—others are cord- and plug-connected. These are pretty much out of sight, but provide excellent lighting for countertop work surfaces.

Luminaire (lighting fixture) manufacturers also offer shallow undercabinet track lighting systems in both 120 volts and 12 volts. These systems include track, track lighting heads for small halogen lamps, end fittings, and transformer with cord and plug. Some have two-level, high-low switching to vary the light intensity.

Visiting a lighting showroom, an electrical supply house, a home center, or a hardware store is an exciting experience. The choice of luminaires (lighting fixtures) is almost endless.

Figure 12-3   Fixtures that can be hung either single or on a track over dinette tables, such as in the eating area of this kitchen. *Courtesy of* Progress Lighting.

Figure 12-4   Types of luminaires (fixtures) used over dining room tables. *Courtesy of* Progress Lighting.

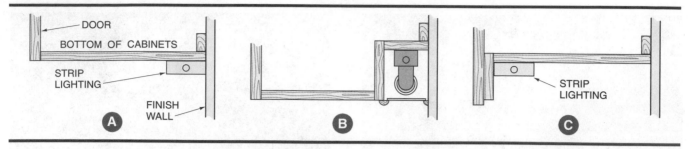

Figure 12-5   Methods of installing undercabinet fluorescent strip lighting.

Lamp Types

Good color rendition in the kitchen area is achieved by using incandescent lamps, or by using warm white deluxe (WWX), warm white (WW), or deluxe cool white (CWX) 3000-K lamps wherever fluorescent lamps are used. See Unit 7 for details regarding energy-saving ballasts and lamps.

FAN OUTLET

The fan outlet is connected to branch lighting circuit B7. Fans can be installed in a kitchen either to exhaust the air to the outside of the building or to filter and recirculate the air. Both types of fans can be used in a residence. The ductless hood fan does not exhaust air to the outside and does not remove humidity from the air.

Homes heated electrically with resistance-type units usually have excess humidity. The heating units neither add nor remove moisture from the air unless humidity control is provided. The proper use of weather stripping, storm doors, and storm windows, and a vapor barrier of polyethylene plastic film, tends to retain humidity.

Vapor barriers are installed to keep moisture out of the insulation. This protection normally is installed on the warm side of ceilings, walls, and floors, under concrete slabs poured directly on the ground, and as a ground cover in crawl spaces. Insulation must be kept dry to maintain its effectiveness. A 1% increase of moisture in insulating material can reduce its efficiency 5%. Humidity control using a humidistat on the attic exhaust fan is covered in Unit 22.

An exhaust fan simultaneously removes odors and moisture from the air and exhausts heated air, Figure 12-6. The choice of the type of fan to be installed is usually left to the owner. The electrician, builder, and/or architect may make suggestions to guide the owner in making the choice.

For the residence in the plans, a separate wall switch is not required because the speed switch, light, and light switch are integral parts of the fan.

Fan Noise

The fan motor and the air movement through an exhaust fan generate a certain amount of noise. The manufacturers of exhaust fans assign a sound-level rating so that it is possible to get some idea as to the noise level one might expect from the fan after it is installed.

The unit used to define fan noise is the *sone* (rhymes with tone). For simplicity, one sone is the noise level of an average refrigerator. The lower the sone rating, the quieter the fan. Technically, the sone is a unit of loudness equal to the loudness of a sound of one kilohertz at 40 decibels above the threshold of hearing of a given listener.

All manufacturers of exhaust fans provide this information in their descriptive literature.

Figure 12-6   Typical range hood exhaust fan. Speed control and light are integral parts of the unit. *Courtesy of* NuTone.

CLOCK OUTLETS

The clock outlet is connected to lighting circuit B7. The clock outlet in the face of the soffit may be installed in any sectional switch box or 4-in. square box with a single-gang raised plaster cover. A deep sectional box is recommended because a recessed clock outlet takes up considerable room in the box.

Some decorative clocks have the entire motor recessed so that only the hands and numbers are exposed on the surface of the soffit or wall. Some of these clocks require special outlet boxes, usually furnished with the clock. Other clocks require a standard 4-in. square outlet box. Accurate measurements must be taken to center the clock between the ceiling and the bottom edge of the soffit. If the clock is available while the electrician is roughing in the wiring, the dimensions of the clock can be checked to help in locating the clock outlet.

If an electrical clock outlet has not been provided, battery-operated clocks are available.

SMALL APPLIANCE BRANCH-CIRCUITS FOR RECEPTACLES IN KITCHEN

The Code requirements for small appliance circuits and the spacing of receptacles in kitchens are covered in *210.11(C)(1)*, *210.52(B)*, and *220.16*. This was discussed in detail in Unit 3. Important points are:

- At least two 20-ampere small appliance circuits shall be installed, as in Figure 12-7.

- Either (or both) of the two circuits required in the kitchen is permitted to supply receptacle outlets in other rooms, such as a dining room, breakfast room, or pantry, Figure 12-7.

- These small appliance circuits shall be assigned a load of 1500 volt-amperes each when calculating feeders and service-entrance requirements.

- The receptacle for the refrigerator in the kitchen is *permitted* (not required) to be supplied by a separate 15- or 20-ampere branch-circuit rather than connecting it to one of the two required 20-ampere small appliance branch-circuits. This permission is found in *210.52(B)(1)*, *Exception No. 2*. No additional load has actually been added. The refrigerator load would have been there no matter what circuit it was connected to. The advantage of doing this is that it diverts the refrigerator load from the small appliance branch circuits in the kitchen. An additional 1500 volt-amperes does not have to be added to the service-entrance calculations, *220.16(A)*, *Exception*.

- The clock outlet may be connected to an appliance circuit when this outlet is to supply and support the clock only (see Figure 3-3). In the residence in the plans, the clock outlet is connected to the lighting circuit.

- Countertop receptacles must be supplied by at least two 20-ampere small appliance circuits, as in Figure 12-7.

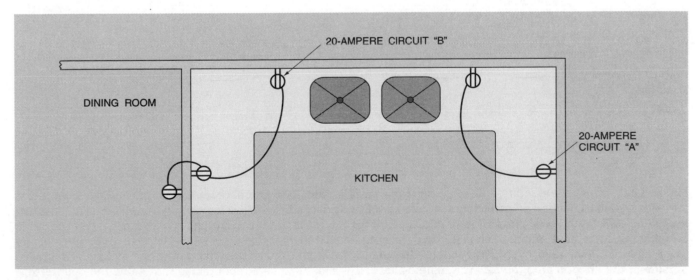

Figure 12-7 A receptacle outlet in a dining room may be connected to one of the two required small appliance circuits in the kitchen, *210.52(B)(1)*. Countertop receptacles must be supplied by at least two 20-ampere small appliance circuits, *210.52(B)(3)*.

- All 125-volt, single-phase, 15- and 20-ampere receptacles installed in kitchens to serve the countertops shall be GFCI protected, *210.8(A)(6),* Figure 12-8. A person can receive an electrical shock by touching a "live" wire and a grounded surface. The kitchen sink (plumbing, water) was always considered to be the hazard. The Code recognizes that there are other grounded "things" a person might touch, such as a range, oven, cooktop, refrigerator, microwave oven, or countertop food processor.

- A receptacle may be installed for plugging in a gas appliance to serve the ignition and/or clock devices on the appliance. See Figure 3-3.

- The small appliance circuits are provided to serve plug-in portable appliances. They are not permitted by the Code to serve lighting or appliances such as dishwashers, garbage disposals, or exhaust fans.

- When a receptacle is installed below the sink for easy plug-in connection of a food waste disposer, it is *not* to be connected to the required small appliance circuits, *210.52(B)(2),* and it is *not* required to be GFCI protected, *210.8(A)(6).* Refer to Unit 3 for spacing requirements for receptacles located in kitchens.

The 20-ampere small appliance circuits prevent

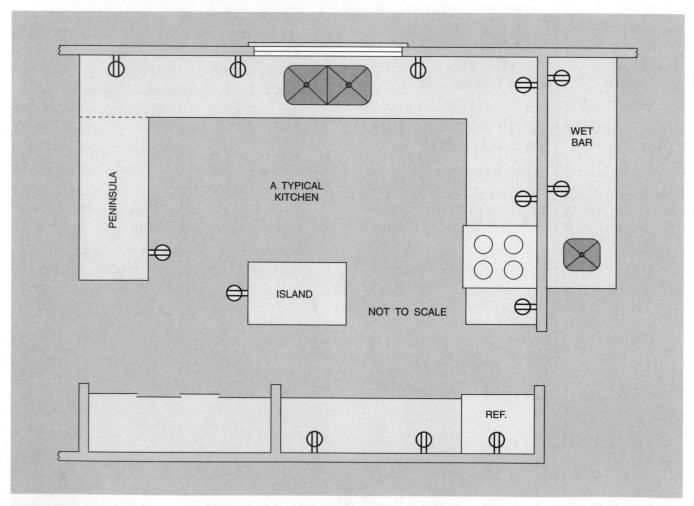

Figure 12-8 In kitchens, *all* 125-volt, single-phase, 15- and 20-ampere receptacles that serve countertop surfaces (walls, islands, or peninsulas) must be GFCI protected. The key words are "that serve countertop surfaces." GFCI personnel protection is *not* required for the refrigerator receptacle, a receptacle under the sink for plugging-in a food waste disposer, or a receptacle installed inside cabinets above the countertops for plugging-in a microwave oven attached to the underside of the cabinets. Where wet-bar sinks are installed outside of the kitchen area, *all* 125-volt, single-phase, 15- and 20-ampere receptacles that serve the countertops must be GFCI protected if they are within 6 ft (1.8 m) measured from the outer edge of the wet-bar sink. See *210.8(A)(6)* and *210.8(A)(7).* Drawing is not to scale.

circuit overloading in the kitchen because of the heavy concentration and use of electrical appliances. For example, one type of cord-connected microwave oven is rated 1500 watts at 120 volts. The current required by this appliance alone is:

$$I = \frac{W}{E} = \frac{1500}{120} = 2.5 \text{ amperes}$$

In 90.1(B) we find a statement that Code requirements are considered necessary for safety. According to the Code, compliance with these rules will not necessarily result in an efficient, convenient, or adequate installation for good service or future expansion of electrical use.

Figure 12-9 shows four 20-ampere branch-circuits feeding the receptacles in the kitchen area. These circuits are B13, B15, B16, and B22. The receptacle located behind the refrigerator is not required to be GFCI protected since it does not serve a counter surface area. By putting the refrigerator on a separate circuit, a ground-fault tripping of any of the GFCI-protected receptacles will not affect the refrigerator.

Figure 12-9 shows one way to arrange the 20-ampere branch-circuits in the kitchen area. Obviously, there are other ways to arrange these branch-circuits. The intent is to arrange the circuits so as to provide the best availability of electrical power for appliances that will be used in the heavily concentrated work areas in the kitchen.

All receptacles serving countertops in kitchens must be GFCI protected, *210.8(A)(6)*.

GFCI receptacles or GFCI circuit breakers can provide this protection.

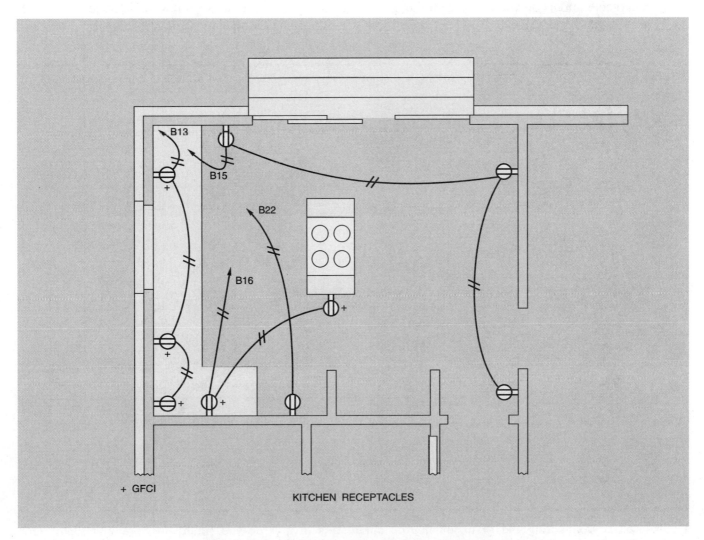

Figure 12-9   Cable layout for kitchen receptacles.

In Figure 12-9, a feedthrough GFCI receptacle is installed to the right of the sink. Standard grounding-type receptacles are installed for the other two receptacles on Circuit B13. For Circuit B16, a feedthrough GFCI receptacle is installed for the first receptacle in the run, then a standard grounding-type receptacle is installed on the side of the range cooktop island.

Circuit B15 does not require any GFCI receptacles because none of these receptacles serve countertops.

Circuit B22 supplies only the non-GFCI receptacle behind the refrigerator.

Refer to Unit 6 for detailed discussion of GFCI protection.

Sliding Glass Doors

See Units 3 and 9 for discussions on how to space receptacle outlets on walls where sliding glass doors are installed.

Small Appliance Branch-Circuit Load Calculations

The 20-ampere small appliance branch circuits are figured at 1500 volt-amperes for each two-wire circuit when calculating feeders and services. The separate 20-ampere branch-circuit for the refrigerator is excluded from this requirement, *220.16(A), Exception*. See Unit 29 for complete coverage of load calculations.

SPLIT-CIRCUIT RECEPTACLES AND MULTIWIRE CIRCUITS

Split-circuit receptacles may be installed where a heavy concentration of plug-in load is anticipated, in which case each receptacle is connected to a separate circuit, as in Figure 12-10 and Figure 12-11.

A popular use of split-circuit receptacles is to control one receptacle with a switch and leave one receptacle "hot" at all times, as is done in the bedrooms and living room of this residence.

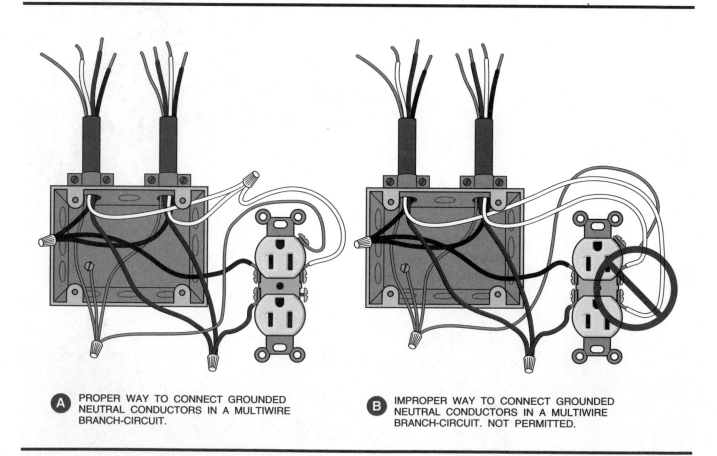

A PROPER WAY TO CONNECT GROUNDED NEUTRAL CONDUCTORS IN A MULTIWIRE BRANCH-CIRCUIT.

B IMPROPER WAY TO CONNECT GROUNDED NEUTRAL CONDUCTORS IN A MULTIWIRE BRANCH-CIRCUIT. NOT PERMITTED.

Figure 12-10 Connecting the grounded neutral conductors in a three-wire (multiwire) branch-circuit, NEC® 300.13(B).

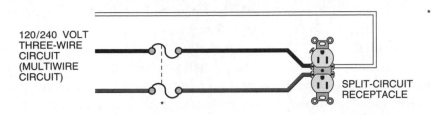

* *NEC® 210.4(B)* REQUIRES THAT A MEANS MUST BE PROVIDED TO SIMULTANEOUSLY DISCONNECT BOTH UNGROUNDED CONDUCTORS AT THE PANELBOARD WHERE THE BRANCH-CIRCUIT ORIGINATES. THIS COULD BE A 2-POLE SWITCH WITH FUSES, A 2-POLE CIRCUIT BREAKER, OR TWO 1-POLE CIRCUIT BREAKERS WITH A LISTED HANDLE TIE.

Figure 12-11   Split-circuit receptacle connected to multiwire circuit.

Split-circuit GFCI receptacles are not available. Refer to Unit 6, where GFCIs are discussed in great detail.

Therefore, the use of split-circuit receptacles is rather difficult to incorporate where GFCI receptacles are required. A multiwire branch-circuit could be run to a box where proper connections are made, with one circuit feeding one feedthrough GFCI receptacle and its downstream receptacles.

This would be used when it becomes more economical to run one three-wire circuit rather than installing two two-wire circuits. Again, it is a matter of knowing what to do, and when it makes sense to do it, as in Figure 12-12.

When multiwire circuits are installed, do not use the terminals of a receptacle or other wiring device to serve as the splice for the grounded neutral conductor. The *NEC®* in *300.13(B)* states that: "*In multiwire branch circuits, the continuity of a grounded conductor shall not be dependent upon device connections . . .* " Any splicing of the grounded conductor in a multiwire circuit must be made

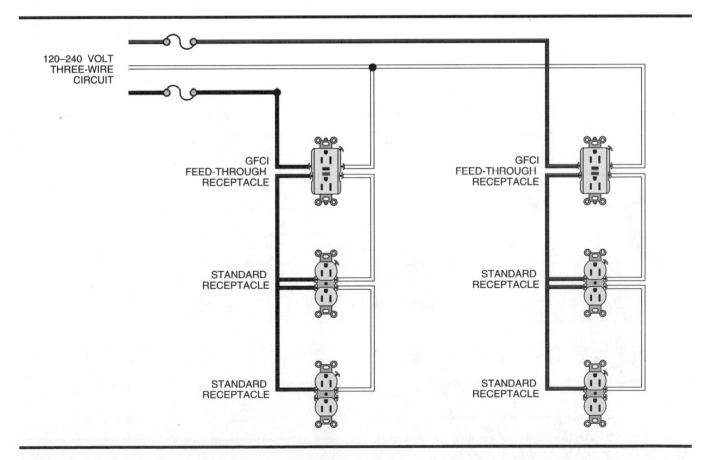

Figure 12-12   This diagram shows how a multiwire circuit can be used to carry two circuits to a box, then split the three-wire circuit into two two-wire circuits. This circuitry does not require simultaneous disconnect of the ungrounded conductors as shown in Figure 12-11.

independently of the receptacle or lampholder, or other wiring device. This requirement is illustrated in Figure 12-10. Note that the screw terminals of the receptacle *must* not be used to splice the neutral conductors. The hazards of an open neutral are discussed in Unit 17. Figure 12-13 shows two types of grounding receptacles commonly used for installations of this type.

When multiwire branch-circuits supply more than one receptacle on the same yoke, a means must be provided to disconnect simultaneously all of the "hot" conductors at the panelboard where the branch-circuit originates. Figure 12-11 shows that 240 volts are present on the wiring device; only 120 volts are connected to each receptacle on the wiring device. See *210.4(B)*.

The "simultaneous disconnect" rule applies to any receptacles, lampholders, or switches mounted on one yoke.

RECEPTACLES AND OUTLETS

Article 100 defines a receptacle and a receptacle outlet, as illustrated in Figure 3-5, Figure 12-14, and Figure 12-15.

Receptacles are selected for circuits according to the following guidelines. A single receptacle connected to a circuit must have a rating not less than the rating of the circuit, *210.21(B)*. In a residence, typical examples for this requirement

for single receptacles are the clothes dryer outlet (30 amperes), the range outlet (50 amperes), or the freezer outlet (15 amperes).

Circuits rated 15 amperes supplying two or more receptacles shall not contain receptacles rated over 15 amperes. For circuits rated 20 amperes supplying two or more receptacles, the receptacles connected to the circuit may be rated 15 or 20 amperes. See *Table 210.21(B)(3)*.

GENERAL GROUNDING CONSIDERATIONS

All electrical equipment located within reach of a grounded surface must be grounded. Metal water pipes, gas pipes, sinks, concrete floors in basements and garages, hot and cold air ducts, outdoor lighting, and similar scenarios leave little excuse for any electrical equipment not to be grounded. *NEC® 250.110* and *250.112* provide the details for the above statement.

NEC® 250.114 tells us that cord- and plug-connected appliances must be grounded unless they are designed with "double insulation." Appliances that are cord- and plug-connected with a three-wire cord are grounded by the equipment grounding conductor in the cord. Appliances that are cord- and plug-connected with a two-wire cord are "double-insulated." Double-insulated appliances provide two levels of protection against electric shock. Accepted levels of protection can be the insulation on a conductor, added insulating material, the plastic non-

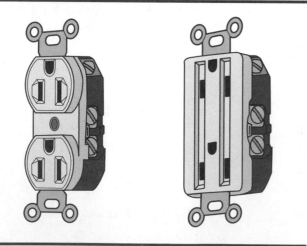

Figure 12-13 Grounding-type duplex receptacles, 15A–125V rating. Most duplex receptacles can be changed into "split-circuit" receptacles by breaking off the small metal tab on the terminals. The above photo clearly shows this feature.

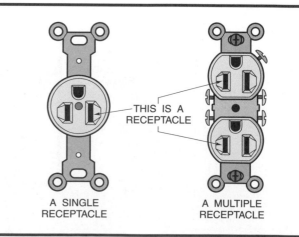

A SINGLE RECEPTACLE

THIS IS A RECEPTACLE

A MULTIPLE RECEPTACLE

Figure 12-14 Receptacles: See *NEC®* definitions, *Article 100*.

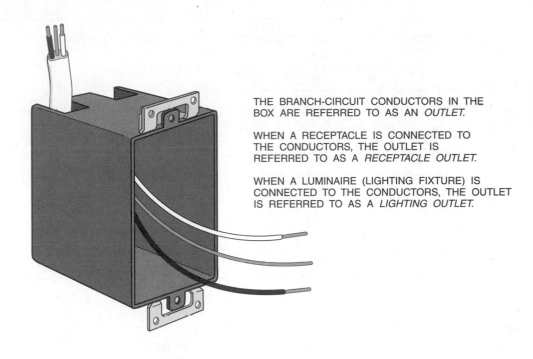

THE BRANCH-CIRCUIT CONDUCTORS IN THE BOX ARE REFERRED TO AS AN *OUTLET.*

WHEN A RECEPTACLE IS CONNECTED TO THE CONDUCTORS, THE OUTLET IS REFERRED TO AS A *RECEPTACLE OUTLET.*

WHEN A LUMINAIRE (LIGHTING FIXTURE) IS CONNECTED TO THE CONDUCTORS, THE OUTLET IS REFERRED TO AS A *LIGHTING OUTLET.*

Figure 12-15   An *outlet* **is defined in the** *NEC*® **as "A point on the wiring system at which current is taken to supply utilization equipment."**

metallic enclosure of the appliance itself, or even a specified air-gap at an exposed terminal. It would take two failures to create a shock hazard to the user. UL Standard 1097 *Double-Insulated Systems for Use in Electrical Equipment* is the standard that electrical appliance manufacturers conform to.

The equipment grounding conductor found in nonmetallic-sheathed cables, the metal armor of Type AC and Type MC cable, metal raceways such as EMT, and flexible metal conduit provide the required grounding means, *250.118.*

In the case of nonmetallic-sheathed cable, never use the bare *equipment grounding conductor* as a *grounded circuit conductor* and never use the *grounded circuit conductor* as an *equipment grounding conductor.* These two conductors come together only at the main service panel, never in branch-circuit wiring. Refer to *250.142.*

REVIEW

Note: Refer to the Code or plans where necessary.

1. If everything on Circuit B7 were turned on, what would be the total current draw?

_____

_____

2. From what panel does the kitchen lighting circuit originate? What size conductors are used? _____

_____

3. How many luminaires (lighting fixtures) are connected to the kitchen lighting circuit?

_____

4. What color fluorescent lamps are recommended for residential installations?

_____

_____

5. a. What is the minimum number of 20-ampere small appliance circuits required for a kitchen according to the Code? _____

   b. How many are there in this kitchen? _____

6. How many receptacle outlets are provided in the kitchen? _____

7. What is meant by the term two-circuit (split-circuit) receptacles? _____

_____

_____

_____

_____

_____

8. Circle the correct response. Duplex receptacles connected to the 20-ampere small appliance branch-circuits in kitchens and dining rooms:

   a. may be rated 15 amperes.   (True) (False)

   b. must be rated 15 amperes.   (True) (False)

   c. may be rated 20 amperes.   (True) (False)

   d. must be rated 20 amperes.   (True) (False)

9. In kitchens, a receptacle must be installed at each counter space _____ in. or wider.

10. A fundamental rule regarding the grounding of metal boxes, luminaires (fixtures), and so on, is that they must be grounded when "in reach of _____ ."

11. How many circuit conductors enter the box

    a. where the clock will be installed? _____

    b. in the ceiling box over which the track will be installed? _____

    c. at the switch location to the right of the sliding door? _____

12. How much space is there between the countertop and upper cabinets? _____

13. Where is the speed control for the range hood fan located? _____

14. Who is to furnish the range hood? _____

15. List the appliances in the kitchen that must be electrically connected. _____

_____

_____

_____

_____

_____

16. Complete the wiring diagram, connecting feedthrough GFCI 2 to also protect receptacle 1, both to be supplied by Circuit A1. Connect feedthrough GFCI 3 to also protect receptacle 4, both to be supplied by Circuit A2. Use colored pencils or markers to show proper color. Assume that the wiring method is EMT, where more freedom in the choice of insulation colors is possible.

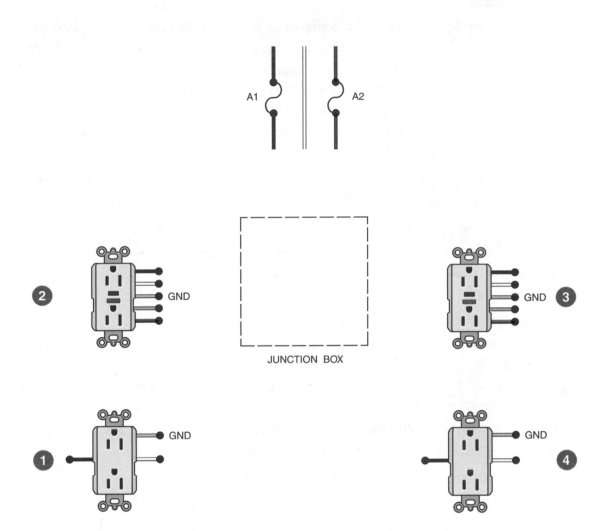

JUNCTION BOX

17. Each 20-ampere small appliance branch-circuit load demand shall be determined at _____ . Choose one.

    a. 2400 volt-amperes   b. 1500 volt-amperes   c. 1920 volt-amperes

18. Is it permitted to connect an outlet supplying a clock receptacle to a 20-ampere small appliance circuit? _____

   _____

   _____

   _____

19. a. The Code requires a minimum of two small appliance circuits in a kitchen. Is it permitted to connect a receptacle in a dining room to one of the kitchen small appliance circuits? _____

    What Code section applies to this situation? _____

    b. May outdoor weatherproof receptacles be connected to a 20-ampere small appliance circuit? _____

    c. Receptacles located above the countertops in the kitchen must be supplied by at least _____ 20-ampere small appliance circuits.

20. According to *210.52*, no point along the floor line shall be more than _____ ft (_____ m) from a receptacle outlet. A receptacle must be installed in any wall space _____ ft (_____ m) wide or greater.

21. The Code states that in multiwire circuits, the screw terminals of a receptacle must *not* be used to splice the neutral conductors. Why?

    _____

    _____

    _____

    _____

    _____

    _____

    _____

22. Electric fans produce a certain amount of noise. It is possible to compare the noise levels of different fans prior to installation by comparing their _____ ratings.

23. Is it permitted to connect the white grounded circuit conductor to the equipment grounding terminal of a receptacle? _____

24. What is unique about 120-volt appliances that are two-wire cord- and plug-connected instead of being three-wire cord- and plug-connected? _____

    _____

    _____

25. a. All 125-volt, single-phase 15- and 20-ampere receptacles installed in a kitchen that serve countertop surfaces must be GFCI protected. (True) (False) Circle the correct answer.

    b. All 125-volt, single-phase 15- and 20-ampere receptacles installed within 6 ft (1.8 m) of a wet-bar sink outside of the kitchen that serve countertop surfaces must be GFCI protected. (True) (False) Circle the correct answer.

26. The kitchen features a sliding glass door. For the purpose of providing the proper receptacle outlets, answer the following statements *true* or *false*.

    a. Sliding glass panels are considered to be wall space. _____

    b. Fixed panels of glass doors are considered to be wall space. _____

27. The following is a layout for the lighting circuit for the kitchen. Complete the wiring diagram using colored pens or pencils to show the conductors' insulation color.

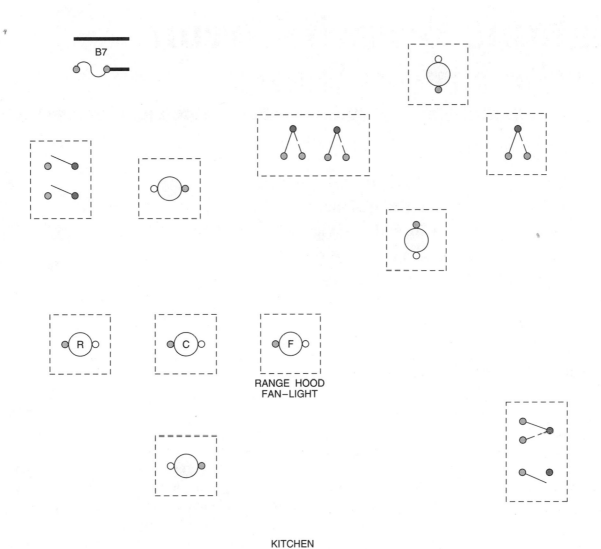

RANGE HOOD
FAN–LIGHT

KITCHEN

Lighting Branch-Circuit for the Living Room

OBJECTIVES

After studying this unit, the student should be able to

- understand various types of dimmer controls.
- connect dimmers to incandescent lamp loads and fluorescent lamp loads.
- understand the phenomenon of incandescent lamp inrush current.
- discuss Class P ballasts and ballast overcurrent protection.
- discuss the basics of track lighting.

LIGHTING CIRCUIT B17 OVERVIEW

The feed for the living room lighting branch-circuit is connected to Circuit B17, a 15-ampere circuit. The home run is brought from Panel B to the weatherproof receptacle outside of the living room next to the sliding door. The circuit is then run to the split-circuit receptacle just inside the sliding door, almost back-to-back with the outdoor receptacle, Figure 13-1.

A three-wire cable is then carried around the living room feeding in and out of the eight receptacles. This three-wire cable carries the black and white circuit conductors plus the red wire, which is the switched conductor. Remember, because these receptacles will provide most of the general lighting for the living room through the use of floor lamps and table lamps, it is advantageous to have some of these lamps controlled by the three switches—two three-way switches and one four-way switch.

The receptacle below the four-way switch in the corner of the living room is required because the Code requires that a receptacle be installed in any wall space 24 in. (600 mm) or more in width, *210.52(A)(2)(1)*. It is highly unlikely that a split-circuit receptacle connected for switch control, as are the other receptacles in the living room, will ever

be needed. Plus, the short distance to the study/bedroom receptacle makes it economically sensible to make up the connections as shown on the cable layout.

Accent lighting above the fireplace is in the form of two recessed luminaires (fixtures) controlled by a single-pole dimmer switch. See Unit 7 for installation data and Code requirements for recessed luminaires (fixtures).

Table 13-1 summarizes the outlets and estimated load for the living room circuit.

The spacing requirements for receptacle outlets and location requirements for lighting outlets are covered in Unit 3, as is the discussion on the spacing of receptacle outlets on exterior walls where sliding glass doors are involved, *210.52(A)(2)(2)*.

TRACK LIGHTING *(ARTICLE 410, PART XV)*

The plans show that track lighting is mounted on the ceiling of the living room on the wall opposite the fireplace. The lighting track is controlled by a single-pole dimmer switch.

Track lighting provides accent lighting for a fireplace, a painting on the wall, or some item (sculpture or collection) that the homeowner wishes to focus attention on, or it may be used to light work

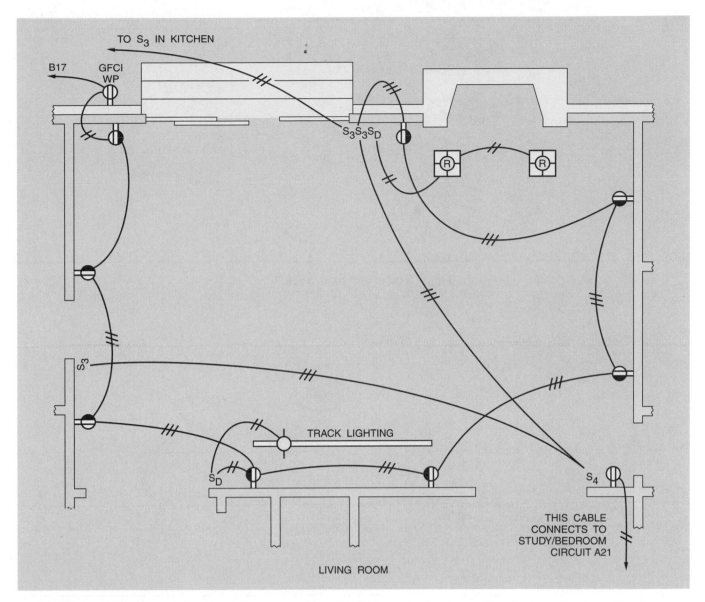

TO S₃ IN KITCHEN

B17

GFCI WP

S₃S₃S_D

R **R**

TRACK LIGHTING

S₃

S_D

S₄

THIS CABLE CONNECTS TO STUDY/BEDROOM CIRCUIT A21

LIVING ROOM

Figure 13-1 Cable layout for the living room. Note that the receptacle on the short wall where the four-way switch is located is fed from the study/bedroom circuit A21.

DESCRIPTION	QUANTITY	WATTS	VOLT-AMPERES
Receptacles @ 120 W Note: Receptacle under S₄ is connected to Study/Bedroom circuit.	8	960	960
Weatherproof receptacle	1	120	120
Track light five lamps @ 40 W each	1	200	200
Fireplace recessed luminaires (fixtures) @ 75 W each	2	150	150
TOTALS	12	1430	1430

Table 13-1 Living room outlet count and estimated load (Circuit B17).

areas such as counters, game tables, or tables in such areas as the kitchen eating area.

Differing from recessed luminaires (fixtures) and individual ceiling luminaires (fixtures) that occupy a very definite space, track lighting offers flexibility because the actual lampholders can be moved and relocated on the track as desired.

Lampholders from track lighting selected from hundreds of styles are inserted into the extruded aluminum or PVC track at any point (the circuit conductors are in the track), and the plug-in connector on the lampholder completes the connection. In addition to being fastened to the outlet box, the track is generally fastened to the ceiling with toggle bolts

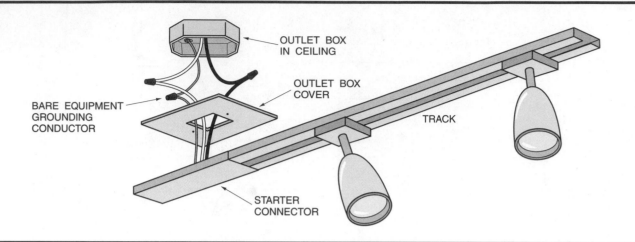

Figure 13-2   End-feed track light.

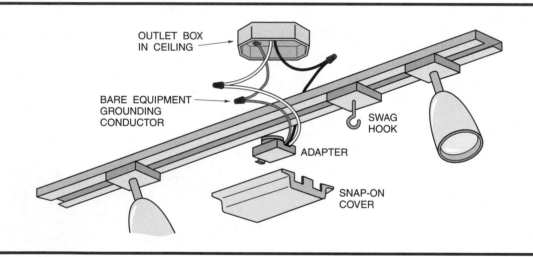

Figure 13-3   Center-feed track light.

or screws. Various track light installations are shown in Figure 13-2, Figure 13-3, and Figure 13-4.

Note on the plans for the residence that the living room ceiling has wood beams. This presents a challenge when installing track lighting that is longer than the space between the wood beams.

There are a number of things the electrician can do:

1. Install pendant kit assemblies that will allow the track to hang below the beam, as in Figure 13-5.

2. Install conduit conductor fittings on the ends of sections of the track, then drill a hole in the beam through which ½-in. EMT can be

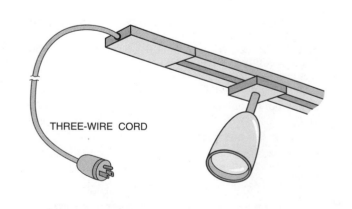

Figure 13-4   Plug-in track light.

installed and connected to the conduit connector fittings, as in Figure 13-6. Running track lighting through walls or partitions is prohibited by *410.101(C)(7)*.

3. Run the branch-circuit wiring concealed above the ceiling to outlet boxes located at the points where the track is to be fed by the branch-circuit wiring, as in Figure 13-7.

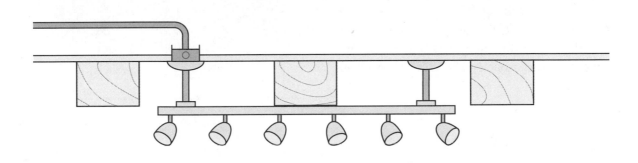

Figure 13-5 Track lighting mounted on the bottom of a wood beam. The supply wires to the track assembly are fed through the pendant.

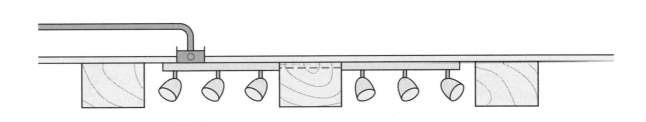

Figure 13-6 Track lighting mounted on the ceiling between the wood beams. The two track assembly sections are connected together by means of a conduit that has been installed through a hole drilled in the beam.

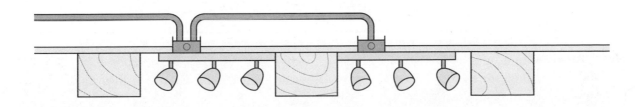

Figure 13-7 Track lighting mounted on the ceiling between the wood beams. Each of the two track assembly sections are connected to outlet boxes in the ceiling in the same manner that a regular luminaire (lighting fixture) is hung.

Where to Mount Lighting Track

Lighting track manufacturers' catalogs provide excellent recommendations for achieving the desired lighting results. For example, consider a lighting track to illuminate an oil painting centered 5½ ft (1.7 m), average eye level, above the floor. One manufacturers' catalog suggests an aiming angle of 60°. For a ceiling height of 9 ft (2.7 m), the track should be mounted approximately 24 in. (600 mm) from the wall. For a ceiling height of 8 ft (2.5 m), the track should be approximately 18 in. (450 mm) from the wall.

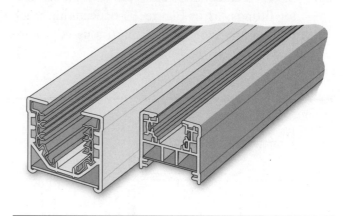

Figure 13-8   Tracks in cross-sectional views.

Lighting Track and the Code

Lighting track is listed under UL Standard 1574.

The code rules for lighting track are found in *410.100* through *410.105*.

Lighting track in residential installations:

- shall be listed by an NRTL.

- shall be installed according to the manufacturers' instructions, per *110.3(B)*.

- shall not be installed:
  - where subject to physical damage.
  - in damp or wet locations.
  - concealed.
  - through walls or partitions.
  - less than 5 ft (1.5 m) above finished floor unless protected against physical damage, or if the track is of the low-voltage type operating at less than 30 volts rms open-circuit voltage. See *410.11* for low-voltage system requirements.
  - In the zone 3 ft (900 mm) horizontally and 8 ft (2.5 m) vertically from the top of a bathtub rim. This is discussed in Unit 10.

- shall be permanently installed and permanently connected to a branch circuit.

- shall be supported twice if the track is not more than 4 ft (1.2 m), and shall have one additional support for each additional 4 ft (1.2 m) of track.

- shall be grounded. Grounding of equipment is covered in previous units.

- shall not be cut to length in the field unless permitted by the manufacturer. Instructions will specify where and how to make the cut, and how to close off the cut with a proper end cap.

- shall have lampholders designed specifically for the particular track.

- shall not have so many lampholders that the load would exceed the rating of the track.

Figure 13-8 shows cross-sectional views of a typical lighting track.

Figure 13-9 shows typical lighting track lampholders.

The track lighting in the living room of this residence is controlled by a single-pole dimmer switch.

Figure 13-9   Typical lamp holders that attach to track lighting systems. *Courtesy of* Progress Lighting.

Load Calculations for Track Lighting

For track lighting installed in homes, it is not necessary to add additional loads for the purpose of branch-circuit, feeder, or service-entrance calculations. Track lighting is just another form of general lighting. This is confirmed in *220.3(B)(10)* and *220.12(B)*.

In a commercial building, a loading factor of 150 VA for each 24 in. (600 mm) of lighting track must be added to the calculations for branch-circuit, feeder, and service-entrance computations, *220.12(B)*.

For further information regarding track lighting, refer to *Part XV* of *Article 410*.

Portable Lighting Tracks

Portable lighting tracks, per UL Standard 153 shall:

- not be longer than 8 ft (2.4 m).

- have a cord not less than 5 ft (1.5 m) permanently attached to the track by the manufacturer.

- have overcurrent protection (fuse or circuit breaker) built into the plug cap. Overcurrent protection in the plug cap is not required if the conductors in the cord are 12 AWG, rated 20 amperes.

- not have its length altered in the field. To do so would void the UL listing.

DIMMER CONTROLS FOR HOMES

Dimmers are used in homes to lower the level of light. There are two basic types.

Electronic Dimmers

Figure 13-10 shows two types (slide and rotate) of electronic dimmers. These are very popular. They "fit" in a standard switch (device) box. They can replace standard single-pole or three-way toggle switches. See Figure 13-11 for the typical hookup of a single-pole and a three-way dimmer. Because dimmer and fan speed controls are physically larger than most wiring devices (plugs and switches), make sure there is enough room in the wall box to accommodate the dimmer, taking into consideration the number of conductors that are in the switch box. To "jam" the dimmer in may cause trouble, such as short-circuits and ground faults. The most common type of electronic dimmer for residential use is rated 125 volts, 60-hertz alternating current, 600 watts. Also available are 1000-watt, 1500-watt, and 2000-watt dimmer switches. The 1500-watt and 2000-watt dimmers fit into a single-gang wall box, but require a two-gang plate because of the need to cover the heat sinking "fins." To prevent an overload, check the total wattage to be controlled with the rated capacity of the dimmer. The maximum wattage that can safely be connected is shown on the nameplate of the dimmer.

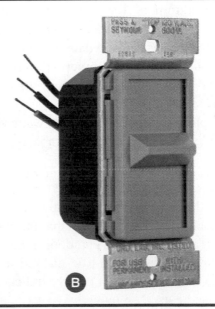

Figure 13-10   Some typical electronic dimmers. Both perform the same function of varying the voltage to the connected incandescent lamp load; (A) rotating knob, (B) slider knob. *Courtesy of* Pass & Seymour/Legrand.

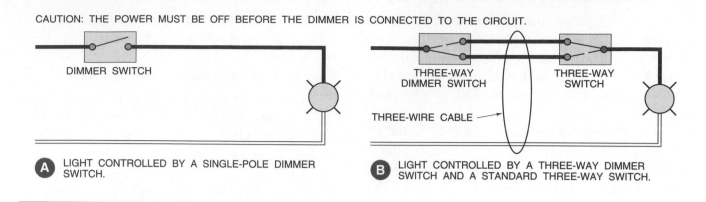

Figure 13-11 Use of single-pole and three-way electronic solid-state dimmer controls in circuits.

▶Dimmers that install in flush device boxes are intended to control only permanently installed incandescent lighting unless listed for other applications, *404.14(E)*.◀ The instructions will state something like "Do not use to control receptacle outlets." The reasons for this are:

1. Serious overloading could result, because it may not be possible to determine or limit what other loads might be plugged in the receptacle.

2. Dimmers reduce voltage. Reduced voltage supplying a television, radio, stereo components, personal computer, vacuum cleaner, and so on, can result in costly damage to the appliance.

Electronic dimmers for use *only* to control incandescent lamps are marked with a *wattage* rating. Serious overheating and damage to the transformer and to the electronic dimmer will result if an "incandescent only" dimmer is used to control inductive loads.

For example, a dimmer marked 600 watts @ 120 volts is capable of safely carrying 5 amperes.

$$\text{Amperes} = \frac{\text{Watts}}{\text{Volts}} = \frac{600}{120} = 5 \text{ amperes}$$

Dimmers that are suitable for control of "inductive" loads, such as lighting systems that incorporate a transformer (as in low voltage outdoor lighting and low voltage track lighting) will be marked in *volt-amperes*. The instructions will clearly indicate that the dimmer can be used on inductive loads.

See Unit 7 for a discussion of watts versus volt-amperes.

Do *not* hook up a dimmer with the circuit "hot." Not only is it dangerous because of the possibility of receiving an electrical shock, but the electronic solid-state dimmer can be destroyed when the splices are being made as a result of a number of "make and breaks" before the actual final connection is completed. The momentary inrush of current each time there is a "make and break" can cause the dimmer's internal electronic circuitry to heat up beyond its thermal capability. As one lead is being connected, the other lead could come in contact with "ground." Unless the dimmer has some sort of "built-in" short-circuit protection, the dimmer is destroyed. The manufacturer's warranty will probably be null and void if the dimmer is worked "hot." It is well worth the time to take a few minutes to de-energize the circuit.

Electronic dimmers use triacs to "chop" the ac sine wave, so some incandescent lamps might "hum" because the lamp filament vibrates. This hum can usually be eliminated or greatly reduced by turning the dimmer setting to a different brightness level, changing the lamp, installing a "rough service lamp," or as a last resort, changing the dimmer.

Autotransformer Dimmers

Figure 13-12 shows an autotransformer (one winding)-type dimmer. These are physically larger than the electronic dimmer, and require a special wall box. They are not ordinarily used for residential applications unless the load to be controlled is very large. They are generally used in commercial applications for the control of large loads. The lamp intensity is controlled by varying the voltage to the lamp load. When the knob is rotated, a brush contact moves over a bared portion of the transformer winding. As with electronic dimmers, they will not work on direct current, but that is not a problem because all homes are supplied with alternating current. The maximum wattage to be controlled is marked on the nameplate of the dimmer. Overcurrent protection is built into the dimmer control to prevent burnout due to an overload.

Figure 13-13 illustrates the connections for an autotransformer dimmer that controls incandescent lamps.

Dimming Fluorescent Lamps

To dim incandescent lamps, the voltage to the lamp filament is varied (raised or lowered) by the dimmer control. The lower the voltage, the less intense is the light. The dimmer is simply connected "in series" with the incandescent lamp load.

Fluorescent lamps are a different story. The connection is not a simple "series" circuit. To dim fluorescent lamps, special dimming ballasts and special dimmer switches are required. The dimmer control allows the dimming ballast to maintain a voltage to the cathodes that will maintain the cathodes' proper

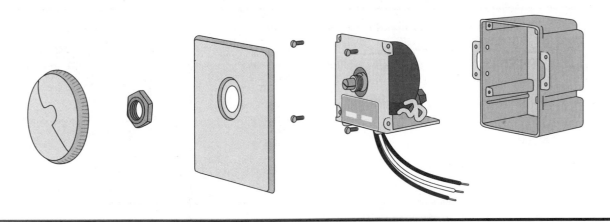

Figure 13-12   Dimmer control, autotransformer type.

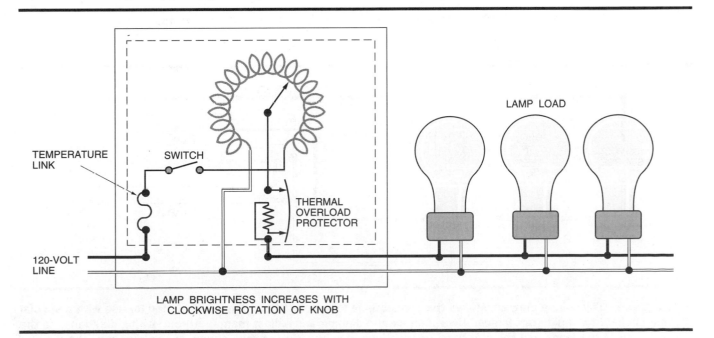

TEMPERATURE LINK

SWITCH

THERMAL OVERLOAD PROTECTOR

LAMP LOAD

120-VOLT LINE

LAMP BRIGHTNESS INCREASES WITH CLOCKWISE ROTATION OF KNOB

Figure 13-13   Dimmer control (autotransformer) wiring diagram for incandescent lamps.

operating temperature, and also allows the dimming ballast to vary the current flowing in the arc. This in turn varies the intensity of light coming from the fluorescent lamp.

There is an exception to the need to use special fluorescent dimming ballasts and special dimmer controls. Some compact fluorescent lamps are available that have an integrated circuit built right into the lamp, allowing the lamp to be controlled by a standard incandescent wall dimmer. These dimmable compact fluorescent lamps can be used to replace standard incandescent lamps of the medium screw shell base type. No additional wiring is neces-

sary. They are designed for use with dimmers, photocells, occupancy sensors, and electronic timers that are marked "Incandescent Only."

Figure 13-14 illustrates the connections for an autotransformer-type dimmer for the control of fluorescent lamps using a special dimming ballast. The fluorescent lamps are F40T12 rapid start lamps. With the newer electronic dimming ballasts, 32-watt, rapid start T8 lamps are generally used.

Figure 13-15 shows an electronic dimmer for the control of fluorescent lamps using a special dimming ballast.

Fluorescent dimming ballasts and controls are

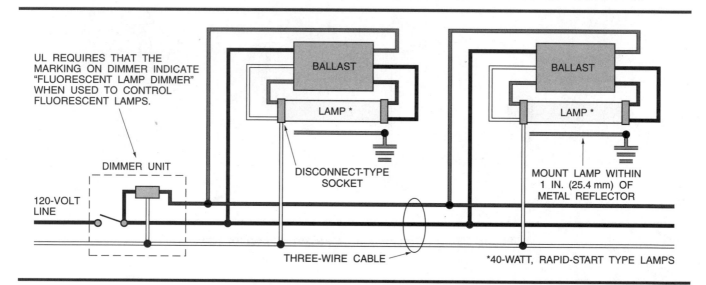

Figure 13-14   Dimmer control wiring diagram for fluorescent lamps.

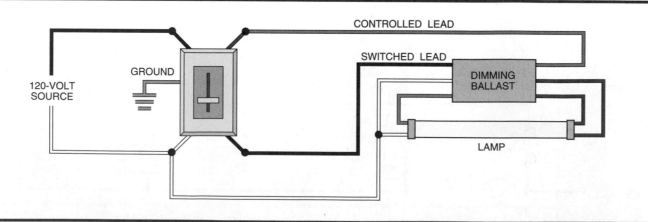

Figure 13-15   This wiring diagram shows the connections for an electronic dimmer designed for use with a special dimming ballast for rapid start lamps. Always check the dimmer and ballast manufacturers' wiring diagrams, as the color coding of the leads and the electrical connections may be different than shown in this diagram. The dimmer shown is a "slider" type, and offers wide-range (20% to 100%) adjustment of light output for the fluorescent lamp.

available in many types such as: simple "ON-OFF," 100% to 1% variable light output, 100% to 10% variable light output, and two-level 100%/50% light output. Wireless remote control is also an option.

Here are some hints to ensure proper dimming performance of fluorescent lamps:

- Make sure that the lamp reflector and the ballast case are solidly grounded.
- Do not mix ballasts or lamps from different manufacturers.
- Do not mix single- and two-lamp ballasts.
- "Age" the lamps for approximately 100 hours before dimming. This allows the gas cathodes in the lamps to stabilize. Without "aging" (sometimes referred to as "seasoning"), it is possible to have unstable dimming, lamp striation (stripes), and/or unbalanced dimming characteristics between lamps.

- Because incandescent lamps and fluorescent lamps have different characteristics, they cannot be controlled simultaneously with a single dimmer control.
- Do not control electronic ballasts and magnetic ballasts simultaneously with a single dimmer control.

Always read and follow the manufacturer's instructions furnished with dimmers and ballasts.

Without special dimming fluorescent ballasts and controls, a simple way to have two light levels is to use four-lamp fluorescent luminaires (fixtures), such as those installed in the Recreation Room. Then, control the ballast that supplies the center two lamps with one wall switch, and control the ballast that supplies the outer two lamps with a second wall switch. This is sometimes referred to as an "inboard/outboard" connection.

REVIEW

Note: Refer to the Code or the plans where necessary.

1. To what circuit is the living room connected? _____

2. How many receptacles are connected to the living room circuit? _____

3. a. How many wires enter the switch box at the four-way switch location? _____

   b. What type and size of box may be installed at this location? _____

   _____

   _____

   _____

4. How many wires must be run between an incandescent lamp and its dimmer control?

   _____

5. Complete the wiring diagram for the dimmer and lamp.

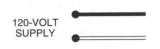
120-VOLT
SUPPLY

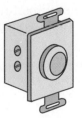

6. Is it possible to dim standard fluorescent ballasts? _____

7. a. How many wires must be run between a dimming-type fluorescent ballast and the dimmer control? _____

   b. Is a switch needed in addition to the dimmer control? _____

   _____

   _____

   _____

   _____

8. Explain why fluorescent lamps having the same wattage draw different current values.

   _____

   _____

   _____

   _____

   _____

9. What is the total current consumption of the track lighting and recessed luminaires (fixtures) above the fireplace? Show your calculations.

10. How many television outlets are provided in the living room? _____

11. Where is the telephone outlet located in the living room? _____

   _____

12. Solid-state electronic dimmers of the type sold for residential use (shall) (shall not) be used to control fluorescent luminaires (fixtures). These dimmers (are) (are not) intended for speed control of small motors. Circle the correct answers.

13. When calculating a dwelling lighting load on the "volt-ampere per ft^2" basis, the receptacle load used for floor lamps, table lamps, and so forth,
   a. must be added to the general lighting load at 1½ amperes per receptacle.
   b. must be added to the general lighting load at 1500 volt-amperes for every 10 receptacles.
   c. is already included in the calculations as part of the volt-amperes per ft^2.
   Circle the correct letter.

14. A layout of the outlets, switches, dimmers, track lighting, and recessed luminaires (fixtures) is shown in the following diagram. Using the cable layout of Figure 13-1, make a complete wiring diagram of this circuit. Use colored pencils or marking pens to indicate conductors.

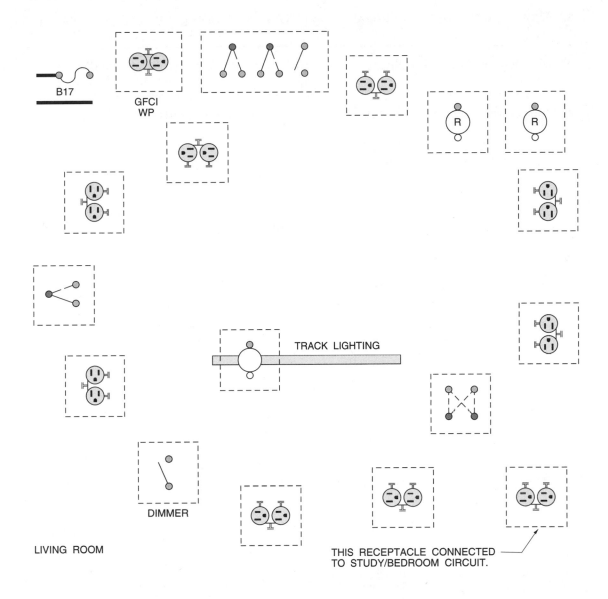

15. Prepare a list of television outlets, telephone outlets, and wiring devices shown on the plans and cable layout for the living room area. Include the number of each type present.

_____

_____

_____

_____

_____

_____

16. The living room features a sliding glass door. For the purpose of providing the proper receptacle outlets, answer the following statements *true* or *false*.

   a. Sliding glass panels are considered wall space. _____

   b. Fixed panels of glass doors are considered wall space. _____

17. a. May lengths of track lighting be added when the track is permanently connected?

   _____

   b. May lengths of track lighting be added when the track is cord-connected?

   _____

18. Must track lighting always be fed (connected) at one end of the track? _____

   _____

19. May a standard electronic dimmer be used to control low-voltage lighting that incorporates a transformer, or for controlling the speed of a ceiling fan motor?

   Yes _____   No _____ .

20. Does the Code permit cutting lengths of portable track lighting to change their length?

   Yes _____   No _____ .

21. Why should the power be turned off when hooking up a dimmer? _____

   _____

   _____

Lighting Branch-Circuit for the Study/Bedroom

OBJECTIVES

After studying this unit, the student should be able to

- discuss valance lighting.
- make all connections in the study/bedroom for the receptacles, switches, fan, and lighting.

CIRCUIT A21 OVERVIEW

The study/bedroom circuit is so named because it can be used as a study for the present, providing excellent space for home office, personal computer, or den. Should it become necessary to have a third bedroom, the change is easily made.

Circuit A21 originates at Main Panel A. Circuit A21 is a 15-ampere branch-circuit. ▶Overcurrent protection for this circuit is provided by a 15-ampere AFCI circuit breaker in Panel A. AFCIs are discussed in detail in Unit 6.◀ This circuit feeds the study/bedroom at the receptacle in the living room just outside of the study. From this point, the circuit feeds the split-circuit receptacle just inside the door of the study. A three-wire cable runs around the room to feed the other four split-circuit receptacles. Figure 14-1 shows the cable layout for Circuit A21.

The black and white conductors carry the "live" circuit, and the red conductor is the "switch return" for the control of one portion of the split-circuit receptacles.

There are four switched split-circuit receptacles in the study/bedroom controlled by two three-way switches. A fifth receptacle on the short wall leading to the bedroom hallway is not switched. The ready access to this receptacle makes it ideal for plugging in vacuum cleaners or similar appliances.

A ceiling fan/light is mounted on the ceiling. This installation is covered in Unit 9. Table 14-1 summarizes the outlets and estimated load for the study/bedroom circuit.

VALANCE LIGHTING

Figure 14-2 illustrates an interesting indirect fluorescent lighting treatment above the windows and behind the valance board.

If the fluorescent valance lighting is to have a dimmer control, then special dimming ballasts

DESCRIPTION	QUANTITY	WATTS	VOLT-AMPERES
Receptacles @ 120 W each	6	720	720
Closet luminaire (fixture)	1	75	75
Paddle fan/light	1		
Three 50-W lamps		150	150
Fan motor (0.75 A 120 V)		80	90
Valance lighting	1		
Two 40-W fluorescent lamps		80	90
TOTALS	9	1105	1125

Table 14-1 Study/bedroom outlet count and estimated load (Circuit A21).

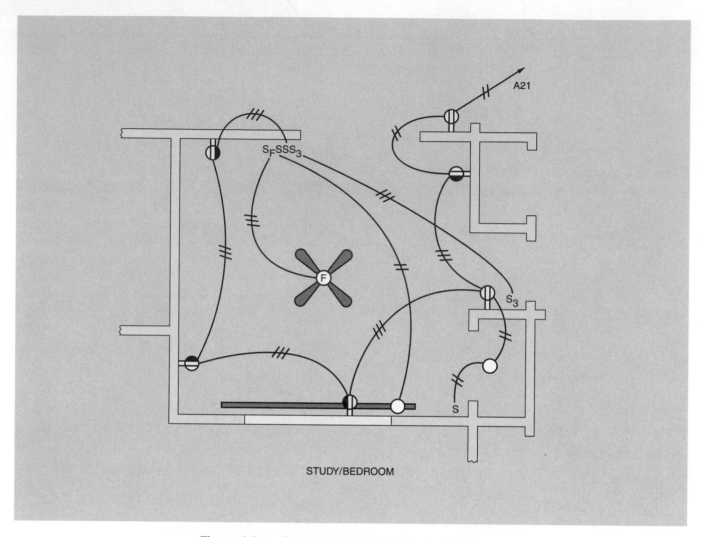

STUDY/BEDROOM

Figure 14-1   Cable layout for the study/bedroom.

would be required. See the discussion on fluorescent lamp dimming in Unit 13.

The study/bedroom wiring is rather simple. Most of the concepts are covered in previous units in this text.

Figure 14-3 suggests one way to arrange the wiring for the ceiling fan/light control, the valance lighting control, and the receptacle outlet control.

SURGE SUPPRESSORS

Because this room is intended initially to be utilized as a study (home office), it is very likely that personal computers (PCs) will be located here. Transient voltage surge suppressors, covered in Unit 6, minimize the possibility that voltage surges might damage PCs.

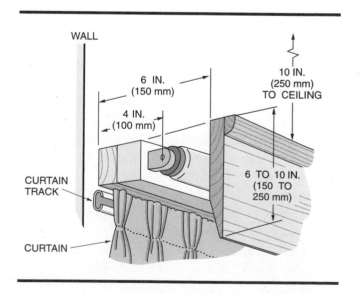

Figure 14-2   Fluorescent lighting behind valance.

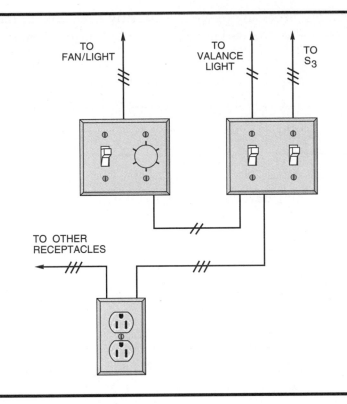

Figure 14-3   Conceptual view of how the study/bedroom switch arrangement is to be accomplished (Circuit A21).

REVIEW

Note: Refer to the Code or plans where necessary.

1. Based upon the total estimated load calculations, what is the current draw on the study/bedroom circuit? _____

2. a. The study/bedroom is connected to circuit _____ .

   b. The conductor size for this circuit is _____ .

3. What unique type of electrical protection is required for all outlets in residential bedrooms?

   _____

4. Why is it necessary to install a receptacle in the wall space leading to the bedroom hallway? _____

   _____

   _____

5. Show the calculations needed to select a properly sized box for the receptacle outlet mentioned in question 4. Nonmetallic-sheathed cable is the wiring method.

   _____

   _____

   _____

   _____

   _____

6. Show the calculations necessary to select a properly sized box for the receptacle outlet mentioned in question 4. The wiring method is EMT, usually referred to as "thin-wall conduit."

_____

_____

_____

_____

_____

7. Prepare a list of *all* wiring devices used in Circuit A21.

_____

_____

_____

_____

_____

8. Using the cable layout shown in Figure 14-1, make a complete wiring diagram of Circuit A21. Use colored pencils or pens.

A21

STUDY/BEDROOM

VALANCE
LIGHT

Dryer Outlet and Lighting Circuit for the Laundry, Powder Room, Rear Entry Hall, and Attic

OBJECTIVES

After studying this unit, the student should be able to

- understand the Code requirements for receptacles installed in bathrooms.
- make the proper wiring and grounding connections for large appliances, based on the type of wiring method used.
- make load calculations for electric dryers.
- understand how to connect clothes dryers with cord- and plug-connections.
- understand the Code requirements governing the receptacle outlet(s) for laundry areas.
- calculate proper conduit sizing for conductors of the same size and for conductors of different sizes in one conduit.
- discuss the subject of "reduced size neutrals."
- discuss the Code rules pertaining to wiring methods in attics.
- demonstrate the proper way to connect pilot lights and pilot light switches.

LIGHTING CIRCUIT B10

The estimated loads for the various receptacles and luminaires (fixtures) in the laundry, rear entry, powder room, and attic are shown in Table 15-1.

RECEPTACLE CIRCUIT B21

Here is a brief recap of the Code requirements for 120-volt receptacles in bathrooms. This subject is also discussed in Unit 10.

- A bathroom is defined as an area that has a basin, plus one or more of the following: a toilet, a tub, or a shower. Although the plans refer to this particular area as a *powder room*, for Code purposes, it is a bathroom.

DESCRIPTION	QUANTITY	WATTS	VOLT-AMPERES
Rear entry hall receptacle	1	120	120
Powder room vanity luminaire (fixture)	1	200	200
Rear entrance hall ceiling luminaires (fixtures) @ 100 W each	2	200	200
Laundry room fluorescent luminaire (fixture) Four 40-W lamps	1	160	180
Attic luminaires (fixtures) @ 75 W each	4	300	300
Laundry exhaust fan	1	80	90
Powder room exhaust fan	1	80	90
TOTALS	11	1140	1180

Table 15-1 Laundry-rear entrance-powder room-attic (Circuit B10).

- At least one receptacle must be installed in a bathroom, *210.52(D)*.

- The receptacle must be installed on the wall adjacent to the basin, and not more than 36 in. (914 mm) from the outside edge of the basin, *210.52(D)*. In large bathrooms, additional receptacles may be installed elsewhere in the bathroom.

- Receptacle(s) shall not be installed face up in the countertop, *406.4(E)*.

- Receptacles in bathrooms must be GFCI protected, *210.8(A)(1)*.

- The receptacle(s) in the bathroom(s) must be connected to a separate 20-ampere branch-circuit, *210.11(C)(3)*. More than one bathroom may be served by the separate 20-ampere branch-circuit. In this residence, a separate 20-ampere branch-circuit has been run to each bathroom. Circuit B21 supplies the receptacle in the powder room.

- If the required 20-ampere branch-circuit for the receptacle(s) supplies more than one bathroom, then the circuit shall not supply other electrical equipment.

- If a separate 20-ampere branch-circuit supplies the receptacle(s) in a single bathroom, then the circuit is permitted to supply other equipment in that bathroom, *210.11(C)(3)*. The other equipment shall not exceed 10 amperes, which is 50 percent of the 20-ampere branch-circuit.

CLOTHES DRYER CIRCUIT ▲D

The electric clothes dryer requires a separate 120/240-volt, single-phase, three-wire branch-circuit. In this residence, the electric clothes dryer is indicated on the Electrical Plans in the laundry by the symbol ▲D. It is supplied by Circuit B(1-3) located in Panel B.

Operation

Clothes dryers are available in many models. Electric clothes dryers have an electric heating element and a motor-operated drum that tumbles the clothes as the heat evaporates and expels the moisture. Dryers are generally vented to the outdoors to expel moisture-laden air. In this residence, a ceiling exhaust fan has also been installed in the laundry room.

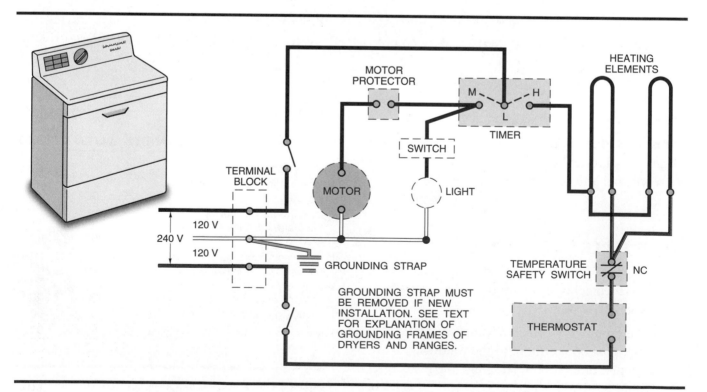

Figure 15-1 Clothes dryer: typical wiring and components.

A thermostat regulates the temperature and a timer regulates the length of time for the various drying cycles. Upscale electric clothes dryers have a moisture sensing feature that stops the drying process at a preselected level of dryness for various types of fabrics. Figure 15-1 shows the internal components and wiring of a typical electric clothes dryer.

The electrician is really interested in the dryer's wattage, ampere rating, and voltage rating so the proper size circuit can be determined and installed.

Dryer Connection Methods

Electric clothes dryers can be connected in several ways.

One method is to run a three-wire armored cable directly to a junction box on the dryer provided by the manufacturer, Figure 15-2. This flexible connection allows the dryer to be moved for servicing without disconnecting the wiring.

Another method is to run a conduit (EMT) to a point just behind the dryer. A combination coupling makes the transition from the EMT to a length of flexible metal conduit 24 in. (600 mm) to 3 ft

(900 mm) long, as in Figure 15-3. This flexible connection also allows the dryer to be moved for servicing without disconnecting the wiring.

Probably the most common method is to install a four-wire, 30-ampere receptacle on the wall behind the dryer. A four-wire, 30-ampere cord is then connected to the dryer. This arrangement is referred to as a "cord- and plug-connection" and is shown in Figure 15-4. In conformance to *210.23*, the connected load shall not exceed the branch-circuit rating.

Disconnecting Means

All electrical equipment is required to have a disconnecting means, *422.30*.

For permanently connected motor-driven appliances, *422.32* states that a disconnect switch or a circuit breaker in a load-center (panel) may serve as the disconnecting means, provided the switch or circuit breaker is within sight of the appliance, or is capable of being locked in the "OFF" position. Breaker manufacturers supply "lock-off" devices that fit over the circuit breaker handle. Disconnect switches have provisions for locking the switch "OFF."

The "unit switch" on an appliance is acceptable as a disconnect for an appliance, provided the unit switch has a marked "OFF" position, *422.34*. This certainly turns off the power within the appliance on the load side of the unit switch, but does not shut off the power to the appliance. To consider the unit switch as the appliance's disconnect, another disconnecting means must be provided, and this is permitted to be the service disconnect, *422.34(C)*. If this arrangement is used, the service disconnect serving as the "other" disconnect does not have to be within sight of the appliance, *422.32, Exception*.

Cord- and Plug-Connection. For cord- and plug-connected appliances, *422.33(A)* accepts the

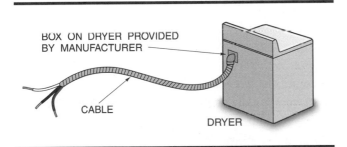

Figure 15-2   Dryer connected by armored cable.

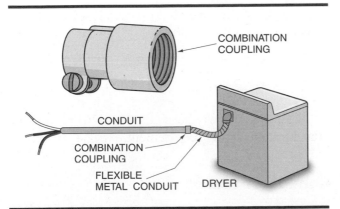

Figure 15-3   Dryer connected by EMT and flexible metal conduit.

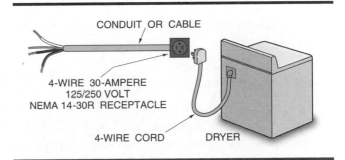

Figure 15-4   Dryer connection using a cord set.

SURFACE OUTLET   FLUSH OUTLET

Figure 15-5 **(A) shows a surface-mount four-wire receptacle. (B) shows a flush-mount four-wire receptacle. These are the types required for hooking up electric dryers and ranges. Prior to the adoption of the 1996 *NEC*, three-wire receptacles and cords were permitted. This is no longer permitted. It is mandatory that a separate equipment grounding conductor be installed to ground the frames of electric clothes dryers and electric ranges. Also see Figure 20-2.**

cord-and-plug arrangement as the disconnecting means.

Figure 15-4 shows a 30-ampere, four-wire receptacle configuration for an electric clothes dryer installation in conformance to the latest *NEC.* Figure 15-5 shows a surface-mount four-wire receptacle (A) and a flush mount four-wire receptacle (B).

For most residential electric clothes dryers, receptacles and cord sets are rated 30 amperes, 125/250 volts. Some large electric clothes dryers might require a 50-ampere, 125/250-volt receptacle and cord set.

Receptacles for dryers can be surface mounted or flush mounted. Flush-mounted receptacles are quite large, and require a large wall box. A 4-in. square, 1½-inch or 2⅛-in. deep outlet box with a suitable single-gang or two-gang plaster ring generally works fine. Correct box sizing is determined by referring to *Table 314.16(A)* or *Table 314.16(B)*, or to Figure 2-20 and Table 2-1.

Dryer and range receptacles and cord sets are identical. To avoid duplication of text, refer to Unit 20 for a complete description of ratings and blade and slot configurations for three-wire and four-wire dryer and range receptacles, and their associated cord sets.

Load Calculations

In this residence, the dryer is installed in the laundry room. The Schedule of Special Purpose Outlets in the specifications shows that the dryer is rated 5700 watts, 120/240 volts. The schematic wiring diagram in Figure 15-1 shows that the heating

element is connected across 240 volts. The motor and lamp are connected to the 120-volt terminals.

It is impossible for the electrician to know the ampere rating of the motor, the heating elements, timers, lights, relays, and so on, and in what sequence they actually operate. This dilemma is answered in the *NEC®* and UL listing requirements for all appliances.

The nameplate must show volts and amperes, or volts and watts, *422.60(A)*. The nameplate must also show the minimum supply circuit conductor's ampacity and the maximum overcurrent protection, *422.62(B)(1)*.

A minimum load of 5000 watts (volt-amperes) or the nameplate rating of the dryer, whichever is larger, is used when calculating feeder or service-entrance requirements, *220.18*.

The rating of an appliance branch-circuit must not be less than the marked rating on the appliance, *422.10(A)*.

The electric clothes dryer in this residence has a nameplate rating of 5700 watts. This calculates out to be:

$$I = \frac{W}{E} = \frac{5700}{240} = 23.75 \text{ amperes}$$

Rounded up, we have a minimum 30-ampere branch-circuit rating requirement.

Conductor Sizing

The minimum branch-circuit rating for the electric clothes dryer was found to be 30 amperes, and that is the minimum rating of the overcurrent protective device—fuses or circuit breaker. We need to find a conductor that has an ampacity of 30 amperes that will serve as the branch-circuit for the dryer. The conductors will be protected by the 30-ampere overcurrent device.

According to *Table 310.16*, the allowable ampacity for a 10 AWG copper conductor is 35 amperes. The maximum overcurrent protection for these conductors must not exceed 30 amperes, *240.4(D)*.

In this residence, Circuit B(1-3) is a three-wire, 30-ampere, 240-volt branch-circuit.

Nonmetallic-sheathed cable is run concealed in the walls from Panel B to a flush-mounted 4-in. square box behind the dryer location in the laundry room. The nonmetallic-sheathed cable contains three 10 AWG conductors plus a 10 AWG equipment grounding conductor.

Where EMT is used as the wiring method, *Table C2, Annex C* of the *NEC*® shows that three 10 AWG THHN conductors require trade size ½ EMT. The EMT serves as the equipment ground.

Overcurrent Protection for the Dryer

The motor of the dryer has integral thermal overload protection as required by *422.11(G)*. Note the motor protector in Figure 15-1. This protection is required by UL standards and prevents the motor from reaching dangerous temperatures as the result of an overload, bearing failure, or failure to start. Also note the temperature safety switch. This is a high-temperature cut-off that shuts off the heating element should the appliance's thermostat fail to operate. This high temperature cut-off is also a UL requirement.

Overcurrent Protection for Branch-Circuit

NEC® *422.62* requires the appliance to be marked with the minimum supply circuit conductor ampacity and the maximum size overcurrent device.

The rating of the branch-circuit must not be less than the rating marked on the appliance, *422.10(A)*.

In *422.11(A)* we find that the overcurrent protection for the branch-circuit conductors must be sized according to *240.4*, which refers us right back to *Article 422*—Appliances.

The maximum overcurrent protection for 10 AWG copper conductors is 30 amperes, *240.4(D)*.

Grounding Frames of Electric Dryers and Electric Ranges

For protection against electrical shock hazard, frames of electric clothes dryers and electric ranges must be grounded. Some key Code references are *250.134*, *250.138*, *250.140*, and *250.142(B)*.

There are now two methods for grounding the frames of electric dryers and electric ranges, namely *New Installations* and *Existing Installations*.

New Installations.   One accepted grounding means is to connect the appliance with a metal raceway such as EMT, flexible metal conduit (Greenfield), armored cable (BX), or other means listed in *250.118*. There are limitations on the use of flexible metal conduit as an equipment grounding conductor. See *250.118* and *348.22*. This topic is covered in Unit 4.

The separate equipment grounding conductor found in nonmetallic-sheathed cable is also an acceptable grounding means. The equipment grounding conductor in nonmetallic-sheathed cable is sized according to *Table 250.122*. See Figure 15-6.

Service-entrance cable is also permitted as a branch-circuit to an electric dryer or electric range. The Code rules governing service-entrance cable are covered in Unit 4.

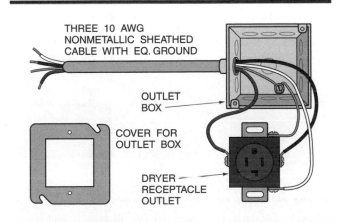

THREE 10 AWG NONMETALLIC SHEATHED CABLE WITH EQ. GROUND

OUTLET BOX

COVER FOR OUTLET BOX

DRYER RECEPTACLE OUTLET

Figure 15-6 New installations of branch-circuit wiring for electric ranges, wall-mounted ovens, surface cooking units, and clothes dryers require that all equipment grounding be done according to *250.134*, *250.138*, *250.140*, and *250.142(B)*. Receptacles and cord sets must be four-wire (three circuit conductors *plus* equipment grounding conductor).

When electric clothes dryers and electric ranges are cord- and plug-connected, the receptacle and the cord set must be four-wire, the fourth wire being the separate equipment grounding conductor.

The small copper bonding strap (link) that is furnished with an electric clothes dryer or electric range *must not* be connected between the neutral terminal and the metal frame of the appliance.

Existing Installations. Prior to the 1996 edition of the *NEC,* 250.60 permitted the frames of electric ranges, wall-mounted ovens, surface cooking units, and clothes dryers to be grounded to the neutral conductor, as in Figure 15-7. By way of *Tentative Interim Amendment No. 53*, this special permission was put into effect in July of 1942, and was supposedly an effort to conserve raw materials during World War II. In effect, this special permission allowed the neutral conductor to serve a dual purpose: (1) the neutral conductor and (2) the equipment grounding conductor. This special permission remained in effect until the 1996 *NEC.*

Since then, grounding equipment to a neutral conductor is not permitted!

Existing installations need not be changed, because they were in conformance to the *NEC* at the time.

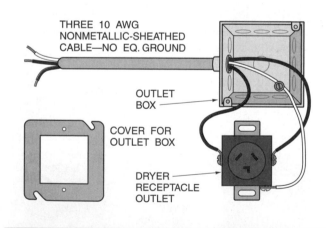

THREE 10 AWG NONMETALLIC-SHEATHED CABLE—NO EQ. GROUND

OUTLET BOX

COVER FOR OUTLET BOX

DRYER RECEPTACLE OUTLET

Figure 15-7 Prior to the 1996 *National Electrical Code,* it was permitted to ground the junction box to the neutral only if the box was part of the circuit for electric ranges, wall-mounted ovens, surface cooking units, and clothes dryers. The receptacle was permitted to be a three-wire type because the appliance grounding was accomplished by making a connection between the neutral conductor and the metal frame of the appliance. This is no longer permitted.

If an electric clothes dryer or electric range is to be installed in a residence where the dryer branch-circuit wiring had been installed according to pre-1996 *NEC* rules, the wiring *does not* have to be changed. Merely use the grounding strap furnished by the dryer manufacturer to make a connection between the neutral conductor and the frame of the dryer.

RECEPTACLE OUTLETS—LAUNDRY

The Code in *210.11(C)(2)* requires that a separate 20-ampere branch-circuit be provided for the laundry equipment and that this circuit "shall have no other outlets."

A receptacle outlet must be installed within 6 ft (1.8 m) of the intended location of the washer, *210.50(C)*.

At least one receptacle outlet must be installed for the laundry equipment, *210.52(F)*.

A load value of 1500 volt-amperes must be included for each two-wire branch-circuit serving laundry equipment, *220.16(B)*. This is for the purpose of computing the size of service-entrance and/or feeders. See the complete calculations in Unit 29.

In the laundry room in this residence, the 20-ampere Circuit B18 supplies the receptacle for the washer. Circuit B18 serves no other outlets. The receptacle is not required to be GFCI protected. Figure 15-8 shows the cable layout for the laundry room.

Circuit B20 is a 20-ampere branch-circuit. It supplies two receptacles in the laundry room and will serve an iron, sewing machine, or other small portable appliance. This circuit also serves the weatherproof receptacle on the outside wall of the laundry area. This branch-circuit is *in addition* to the required separate 20-ampere circuit (B18) for the laundry equipment.

Circuits B18 and B20 are included in the calculations for service-entrance and feeder conductor sizing at a computed load of 1500 volt-amperes per circuit, *220.16(B)*. This load is included with the general lighting load for the purpose of computing the service-entrance and feeder conductor sizing, and is subject to the demand factors that are applicable to these calculations. See the complete calculations in Unit 29.

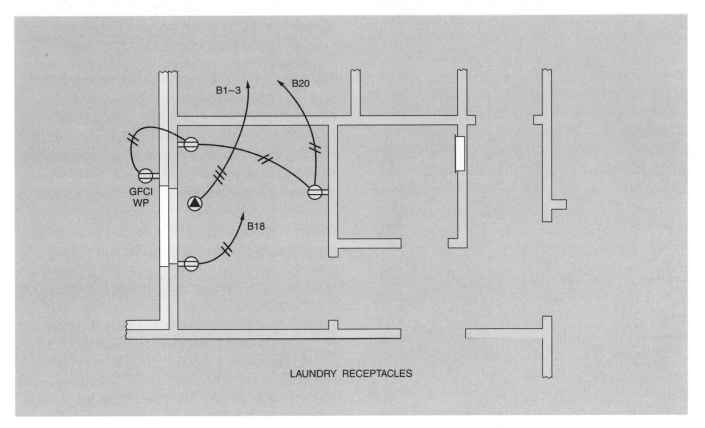

Figure 15-8   Cable layout for laundry room receptacles.

COMBINATION WASHER/DRYERS

Combination washer/dryers are often found in multi-family dwellings where space is an issue. Combination washer/dryers take up half the floor space of individual conventional washers and dryers. If you are involved in an electrical installation, verify and resolve the electrical requirements before you start the project. These combination appliances might require a 30-ampere branch-circuit. Installation manuals will provide this information. The electrical inspector might *not* require the traditional 20-ampere branch-circuit and receptacle that is required by *210.52(F)* and *210.11(C)(2)*.

LIGHTING CIRCUIT

The general lighting circuit for the laundry, powder room, and hall area is supplied by Circuit B10. Note on the cable layout, Figure 15-9, that in order to help balance loads evenly, the attic lights are also connected to Circuit B10.

The types of vanity and ceiling luminaires (fixtures) and wall receptacle outlets, as well as the circuitry and switching arrangements, are similar to types discussed in other units of this text and will not be repeated.

The receptacle in the powder room is supplied by Circuit B21, a separate 20-ampere branch-circuit as required by *210.52(D)*, discussed in Unit 10.

Exhaust Fans

Ceiling exhaust fans are installed in the laundry and in the powder room. These exhaust fans are connected to Circuit B10.

The exhaust fan in the laundry will remove excess moisture resulting from the use of the clothes washer.

Exhaust fans may be installed in walls or ceilings, as in Figure 15-10. The wall-mounted fan can be adjusted to fit the thickness of the wall. If a ceiling-mounted fan is used, sheet-metal duct must be installed between the fan unit and the outside of the house. The fan unit terminates in a metal hood or grille on the exterior of the house. The fan has a shutter that opens as the fan starts up and closes as the fan stops. The fan may have an integral

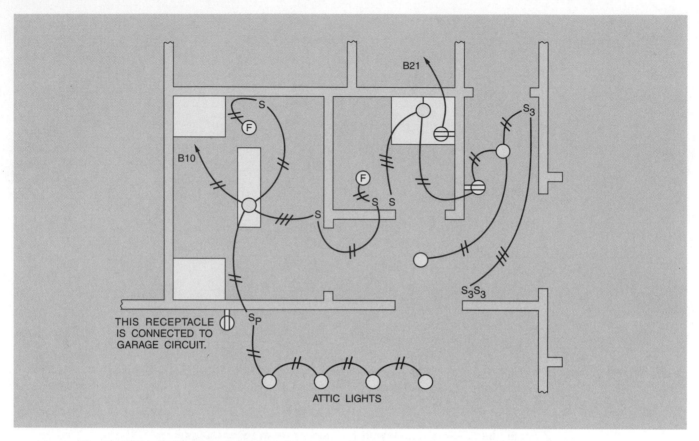

Figure 15-9 Cable layout for laundry, rear entrance, powder room, and attic (Circuit B10).

Figure 15-10 Exhaust fans. _Courtesy of_ NuTone.

pull-chain switch for starting and stopping, or it may be used with a separate wall switch. In either case, single-speed or multispeed control is available. The fan in use has a very small power demand, 90 volt-amperes.

To provide better humidity control, both ceiling-mounted and wall-mounted fans may be controlled with a humidistat. This device starts the fan when the humidity reaches a certain value. When the humidity drops to a preset level, the humidistat turns the fan off.

Some of the more expensive exhaust fans have a built-in sensor that will automatically turn on the exhaust fan at a predetermined humidity level. The fan will automatically turn off when the humidity has been lowered to a preset level. This eliminates the need for a separate wall-mounted humidity control (humidistat).

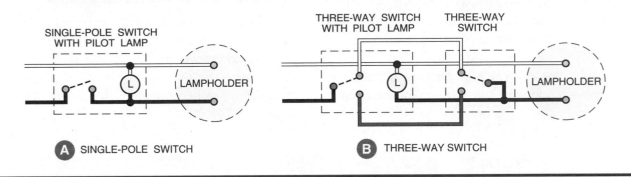

Figure 15-11   Pilot lamp connections.

ATTIC LIGHTING AND PILOT LIGHT SWITCHES

NEC® 210.70(A)(3) requires that at least one lighting outlet must be installed in an attic, and that this lighting outlet contain a switch such as a pull-chain lampholder, or be controlled by a switch located near the entry to the attic. This section of the Code addresses attics that are used for storage and attics that contain equipment that might need servicing, such as air-conditioning equipment. *NEC® 210.70(A)(3)* requires that the lighting outlet(s) be installed near this equipment.

A 125-volt, single-phase, 15- or 20-ampere receptacle outlet shall be installed in an accessible location within 25 ft (7.5 m) of air-conditioning or heating equipment in attics, in crawl spaces, or on the roof, *210.63*. Connecting this receptacle outlet to the load side of the equipment's disconnecting means is not permitted because this would mean that if you turned off the equipment, the power to the receptacle would also be off and thus would be useless for servicing the equipment.

The residence discussed in this text does not have air-conditioning, heating, or refrigeration equipment in the attic or on the roof.

The porcelain lampholders are available with a receptacle outlet for convenience in plugging in an extension cord. However, most electrical inspectors (authority having jurisdiction) would not accept the porcelain lampholder's receptacle in lieu of the required receptacle outlet as stated in *210.63*.

The four porcelain lampholders in the attic are turned on and off by a single-pole switch on the garage wall close to the attic storable ladder. Associated with this single-pole switch is a pilot light. The pilot light may be located in the handle of the switch or it may be separately mounted. Figure 15-11 shows how pilot lamps are connected in circuits containing either single-pole or three-way switches.

Because the attic in this residence is served by a folding storable ladder, switch control in the attic is not required. Where a permanent stairway of six or more risers is installed, *210.70(A)(2)(c)* requires switch control at both levels.

If a neon pilot lamp in the handle (toggle) of a switch does not have a separate grounded conductor connection, then it will glow only when the switch is in the "OFF" position, as the neon lamp will then be in series with the lamp load, Figure 15-12.

The voltage across the load lamp is virtually zero, so it does not burn, and the voltage across the

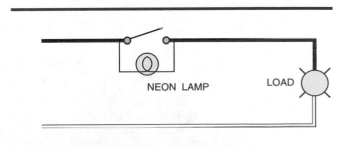

Figure 15-12   Example of neon pilot lamp in handle of toggle switch.

neon lamp is 120 volts, allowing it to glow. When the switch is turned on, the neon lamp is bypassed (shunted), causing it to turn off and the lamp load to turn on.

Use this type of switch when it is desirable to have a switch glow in the dark to make it easy to locate.

Installation of Cable in Attics

When nonmetallic-sheathed cable is installed in accessible attics, the installation must conform to the requirements of *334.23*. This section refers the reader directly to *320.23*, which describes how the cable is to be protected. See Figure 15-13 and Figure 15-14.

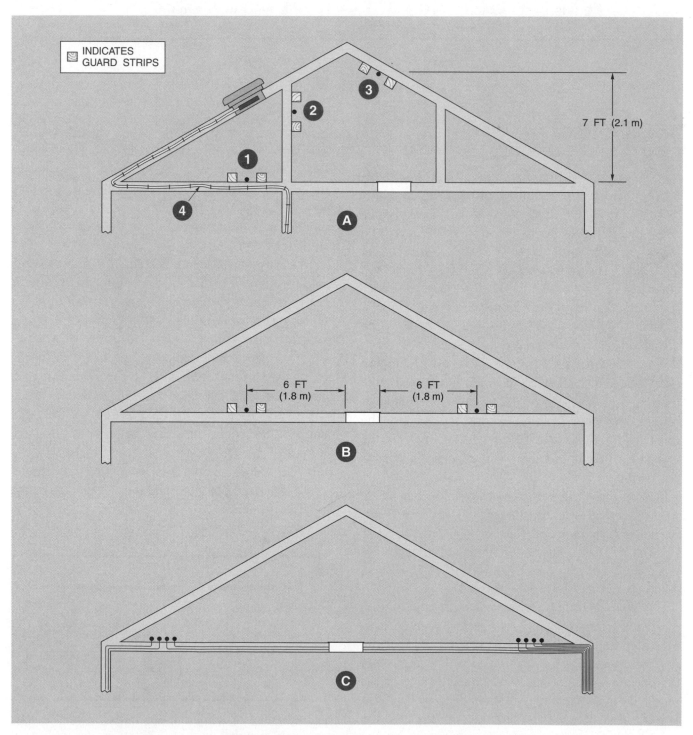

Figure 15-13 Protection of nonmetallic-sheathed cable or armored cable in an attic. Refer to *334.23* and *320.23*.

In accessible attics, cables must be protected by guard strips if:

- they are run across the top of floor joists ①.

- they are run across the face of studs ② or rafters ③ within 7 ft (2.1 m) of the floor or floor joists.

Guard strips are *not* required if the cable is run along the sides of rafters, studs, or floor joists ④.

In attics not accessible by permanent stairs or ladders, as in Figure 15-13(B), guard strips are required only within 6 ft (1.8 m) of the nearest edge of the scuttle hole or entrance.

Figure 15-13(C) illustrates a cable installation that most electrical inspectors consider to be safe. Because the cables are installed close to the point where the ceiling joists and the roof rafters meet, they are protected from physical damage. It would be very difficult for a person to crawl into this space, or store cartons in an area with such a low clearance. Although the plans for this residence show a 2-ft-wide catwalk in the attic, the owner may decide to install additional flooring in the attic to obtain more storage space. Because of the large number of cables required to complete the circuits, it would interfere with flooring to install guard strips wherever the cables run across the tops of the joists, as in Figure 15-14. However, the cables can be run through holes bored in the joists and along the sides of the joists and rafters. In this way, the cables do not interfere with the flooring.

When running cables parallel to framing members, be careful to maintain at least 1¼ in. (32 mm) between the cable and the edge of the framing member. This is a Code requirement referenced in *300.4(D)* to minimize the possibility of driving nails into the cable. All of *300.4* is devoted to the subject of "protection against physical damage."

See Unit 4 for more detailed discussion and illustrations relating to the mechanical protection of cables.

GUARD STRIPS REQUIRED WHEN CABLES ARE RUN ACROSS THE TOP OF JOISTS.

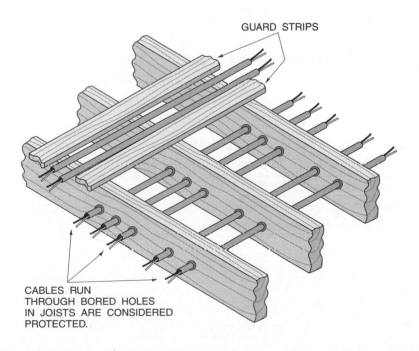

GUARD STRIPS

CABLES RUN
THROUGH BORED HOLES
IN JOISTS ARE CONSIDERED
PROTECTED.

Figure 15-14   Methods of protecting cable installations in accessible attics. See *320.23* and *340.23*.

REVIEW

Note: Refer to the Code or plans where necessary.

1. List the switches, receptacles, and other wiring devices that are connected to circuit B10. _____

   _____

   _____

   _____

   _____

   _____

   _____

   _____

2. a. Fill in the blank spaces. At least _____ 120-volt receptacle must be installed in a bathroom within _____ in. of the basin, and it must be on the wall _____ to the basin. The receptacle must not be installed _____ on the countertop. The receptacle must be _____ protected.

   b. Circle the correct answers. A separate (15) (20)-ampere branch-circuit must supply the receptacle(s) in bathrooms and (shall) (shall not) serve other loads. If a separate 20-ampere branch-circuit serves a single bathroom, then that circuit (is permitted) (is not permitted) to supply other loads in that bathroom.

3. What special type of switch is controlling the attic lights? _____

   _____

4. When installing cables in an accessible attic along the top of the floor joists _____ _____ must be installed to protect the cables.

5. The total estimated volt-amperes for Circuit B10 has been calculated to be 1300 volt-amperes. How many amperes is this at 120 volts? _____

6. If an attic is accessible through a scuttle hole, guard strips are installed to protect cables run across the top of the joists only within (6 ft [1.8 m]) (12 ft [3.7 m]) of the scuttle hole. Circle correct answer.

7. What section(s) of the Code refers (refer) to the receptacle required for the laundry equipment? State briefly the requirements.

   _____

   _____

   _____

   _____

   _____

   _____

   _____

   _____

8. What is the current draw of the exhaust fan in the laundry? _____

9. The following is a layout of the lighting circuit for the laundry, powder room, rear entry hall, and attic. Complete the wiring diagram using colored pens or pencils to indicate conductor insulation color. The GFCI receptacle in the powder room is not shown in the diagram because it is connected to a separate circuit, B21.

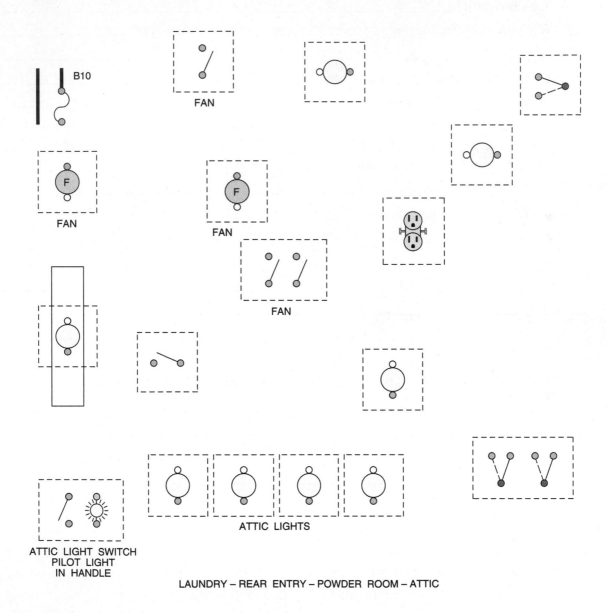

LAUNDRY – REAR ENTRY – POWDER ROOM – ATTIC

10. Laundry receptacle outlets are included in the residential load calculations at a value of _____ volt-amperes per circuit. Choose one: 1000, 1500, 2000 volt-amperes.

11. a. What regulates the temperature in the dryer? _____

   _____

   b. What regulates drying time? _____

   _____

12. List the various methods of connecting an electric clothes dryer. _____

   _____

13. a. What is the minimum power demand allowed by the Code for an electric dryer if no actual rating is available for the purpose of calculating feeder and service-entrance conductor sizing? _____

   b. What is the current draw? _____

14. What is the maximum permitted current rating of a portable appliance on a 30-ampere branch-circuit? _____

15. What provides motor running overcurrent protection for the dryer? _____
_____

16. a. Must the metal frame of an electric clothes dryer be grounded? _____

   b. May the metal frame of an electric clothes dryer be grounded to the neutral conductor? _____

17. An electric dryer is rated at 7.5 kW and 120/240 volts, three-wire, single-phase. The terminals on the dryer and panelboard are marked 75°C.

   a. What is the wattage rating? _____

   b. What is the current rating? _____

   c. What minimum size type THHN copper conductors are required? _____
_____
_____
_____
_____
_____

   d. If EMT is used, what minimum size is required? _____
_____
_____
_____
_____
_____

18. When a metal junction box is installed as part of the cable wiring to a clothes dryer or electric range, may this box be grounded to the circuit neutral? _____

19. A residential air-conditioning unit is installed in the attic.

   a. Is a lighting outlet required? _____

   b. What Code section? _____

   c. If a lighting outlet is required, how shall it be controlled? _____
_____

   d. What Code section? _____

   e. Is a receptacle outlet required? _____

   f. What Code section? _____

20. The Code states that a means must be provided to disconnect an appliance. In the case of an electric clothes dryer, the disconnecting means can be: Circle the correct statement(s).

    a. a separate disconnect switch located within sight of the appliance.

    b. a separate disconnect switch located out of sight from the appliance. The switch is capable of being locked in the "OFF" position.

    c. a cord- and plug-connected arrangement.

    d. a circuit breaker in a panel that is within sight of the dryer or capable of being locked in the "OFF" position.

Lighting Branch-Circuit for the Garage

OBJECTIVES

After studying this unit, the student should be able to

* understand the fundamentals of providing proper lighting in residential garages.
* understand the application of GFCI protection for receptacles in residential garages.
* understand the Code requirements for underground wiring, both conduit and underground cable.
* complete the garage circuit wiring diagram.
* discuss typical outdoor lighting and related Code requirements.
* describe how conduits and cables are brought through cement foundations to serve loads outside of the building structure.
* understand the application of GFCI protection for loads fed by underground wiring.
* make a proper installation for a residential overhead garage door opener.
* select the proper overload protective devices based on the ampere rating of the connected motor load.

LIGHTING BRANCH-CIRCUIT B14

Circuit B14 is a 15-ampere circuit that originates at Panel B in the recreation room, and is brought into the wall receptacle box on the front wall of the garage. Figure 16-1 shows the cable layout for Circuit B14. From this point, the circuit is then carried to the switch box adjacent to the side door of the garage. From there the circuit conductors and the switch leg are carried upward to the ceiling lampholders, and then to the receptacle outlets on the right-hand wall of the garage.

The garage ceiling's porcelain lampholders are controlled from switches located at all three entrances to the garage.

The post light in the front of the house is fed from the receptacle located just inside the overhead door. The post light turns on and off by a photocell, an integral part of the post light.

Note that the luminaires (fixtures) on the outside next to the overhead garage door are controlled by two three-way switches, one just inside the overhead door and the other in the front entry. These luminaires (fixtures) are not connected to the garage lighting circuit.

All of the wiring in the garage will be concealed.

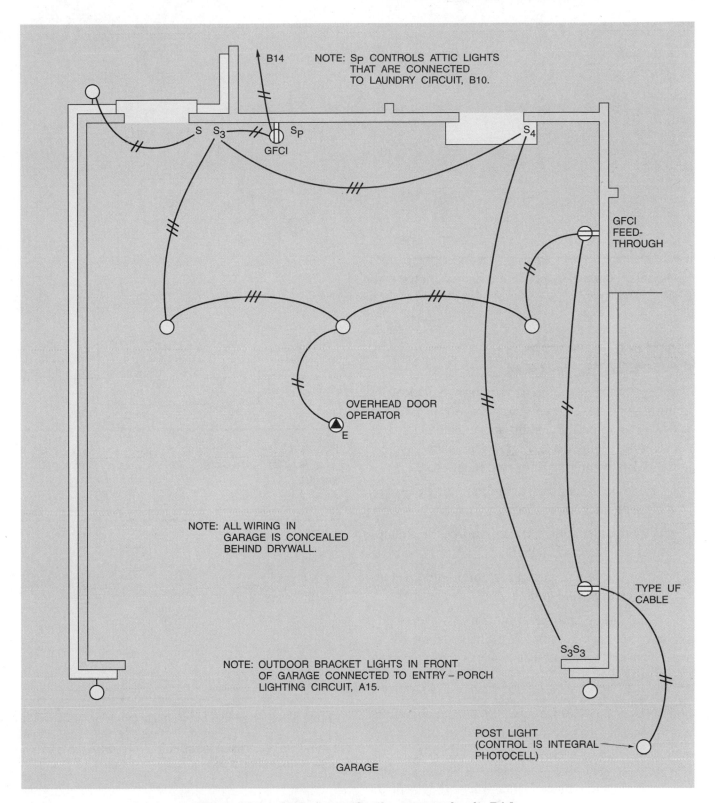

Figure 16-1 Cable layout for the garage circuit, B14.

Building codes for one- and two-family dwellings require that the walls separating habitable areas and a garage have a fire-resistance rating. In this residence, the walls and ceilings have a fire-resistance rating of 1 hour. Fire resistance rating requirements are discussed in Unit 2.

Table 16-1 summarizes the outlets and the estimated load for the garage circuit.

DESCRIPTION	QUANTITY	WATTS	VOLT-AMPERES
Receptacles @ 180 W each	3	540	540
Ceiling lampholders @ 100 W each	3	300	300
Outdoor garage bracket luminaire (fixture) (rear)	1	100	100
Post light	1	100	100
Overhead door opener Two 40-W lamps	1	80	80
Motor (½ A @ 120 V)		180	200
TOTALS	9	1300	1320

Table 16-1   Garage outlet count and estimated load (Circuit B14).

LIGHTING A TYPICAL RESIDENTIAL GARAGE

The recommended approach to provide adequate lighting in a residential garage is to install one 100-watt lamp on the ceiling above each side of an automobile, Figure 16-2. Figure 16-3 shows typical lampholders commonly mounted in garages.

- For a one-car garage, a minimum of two lampholders is recommended.

- For a two-car garage, a minimum of three lampholders is recommended.

- For a three-car garage, a minimum of four lampholders is recommended.

Lampholders arranged in this manner eliminate shadows between automobiles. Shadows are a hazard because they hide objects that can cause a person to trip or fall.

It is highly recommended that these ceiling lampholders be mounted toward the front end of the automobile as it is normally parked in the garage. This arrangement will provide better lighting where it is most needed when the owner is working under the hood. Do not place lights where they will be covered by the open overhead door.

Fluorescent luminaires (fixtures) are quite often used in the warmer parts (south and southwest) of the country. For example, instead of the three porcelain lampholders indicated on the electrical plans on the ceiling of the garage for the residence discussed in this text, three 8-ft fluorescent strip lights could

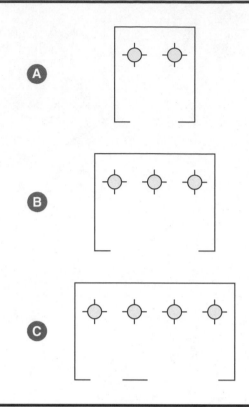

Figure 16-2   Positioning of lights in (A) a one-car garage, (B) a two-car garage, and (C) a three-car garage.

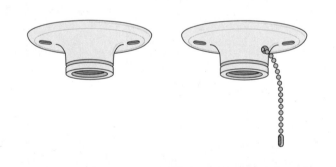

Figure 16-3   Typical porcelain lampholders that would be installed in garages, attics, basements, crawl spaces, and similar locations. Note that one has a pull-chain and one does not.

be installed in the direction of front to back of the garage. These 8-ft luminaires (fixtures) each have two 40-watt fluorescent lamps, for a total of six 40-watt fluorescent lamps. Fluorescent lamps installed in a cold garage (northern parts of the country) would be difficult if not impossible to start, thus the need to use incandescent lamps.

RECEPTACLE OUTLETS IN A GARAGE

At least one receptacle must be installed in an attached residential garage, *210.52(G)*. The garage in this residence has three receptacles. If a detached garage has no electricity provided, then obviously no rules are applicable, but just as soon as a detached garage is wired, then all pertinent Code rules become effective and must be followed. These would include rules relating to grounding, lighting outlets, receptacle outlets, GFCI protection, and so on.

NEC® 210.8(A)(2) requires that all 125-volt, single-phase, 15- or 20-ampere receptacles installed in garages shall have ground-fault circuit-interrupter protection. GFCIs are discussed in detail in Unit 6.

There are a few exceptions to *210.8(A)(2)* requiring GFCI protection of receptacles in residential garages. They are:

1. GFCI protection is *not* required if the receptacle is not readily accessible, such as the receptacle installed for the overhead door opener.

2. GFCI protection is *not* required for single or duplex receptacles intended for cord- and plug-connected appliances that are not easily moved, and are located in dedicated spaces. Examples include a freezer, refrigerator, or central vacuum equipment. The intent is that these dedicated receptacles will have the appliance cord plugged into it, and therefore would not be used to plug in portable electric tools and extension cords. This exception is difficult to enforce, because the electrician may not know during the rough-in that a certain receptacle is for the sole purpose of plugging in a specific appliance, say a large freezer. Such appliances are not at the construction site during rough-in.

Note that one GFCI receptacle is installed on the front wall of the garage, in the same box as the single-pole switch for the attic lighting. A second feedthrough GFCI receptacle is installed on the right-hand garage wall near the entry to the house. This GFCI receptacle also provides the required GFCI protection for the other receptacle on the same wall and for the protection of the underground wiring to the post light.

If we were to install one GFCI feedthrough receptacle at the point where Circuit B14 feeds the garage, then the entire circuit would be GFCI protected as required. But nuisance tripping of the GFCI might occur because of:

- the overall length of the circuit wiring.
- possible problems in the overhead door opener.
- possible problems in the underground run to the post light.
- problems in the post lamp itself.

All of these would result in a complete outage on Circuit B14. The electrician must consider the cost of GFCI receptacles and compare it with the cost of additional cable or conduit and wire needed. The possibility of a power outage must also be kept in mind when trying to get by with fewer GFCI receptacles.

All receptacles and switches in the garage are to be mounted 46 in. (1.15 m) to center according to a notation on the First Floor Electrical Plan.

LANDSCAPE LIGHTING

One of the nice things about residential wiring is the availability of a virtually unlimited variety of sizes, types, and designs of outdoor luminaires (lighting fixtures) for gardens, decks, patios, and for accent lighting. Luminaire (fixture) showrooms at electrical distributors and the electrical departments of home centers offer a wide range of outdoor luminaires (lighting fixtures), both 120-volt and low-voltage types. Figure 16-4 and Figure 16-5 illustrate landscape lighting mounted on the ground. See also the following section on outdoor wiring.

UL Standard 1838 covers *Low-Voltage Landscape Lighting Systems* generally used outdoors in wet locations.

UL Standard 2108 covers *Low-Voltage Lighting Systems* for use in dry locations.

Both of these UL standards require the manufacturer to include installation instructions with the product. It is extremely important to read these instructions to be sure the luminaires (lighting fixtures) and associated wiring are installed and used in a safe manner.

Low-voltage lighting products will bear markings such as *Outdoor Use Only*, *Indoor Use Only*, or *Indoor/Outdoor Use*. If marked for use in damp or wet locations, the maximum voltage rating is 15 volts

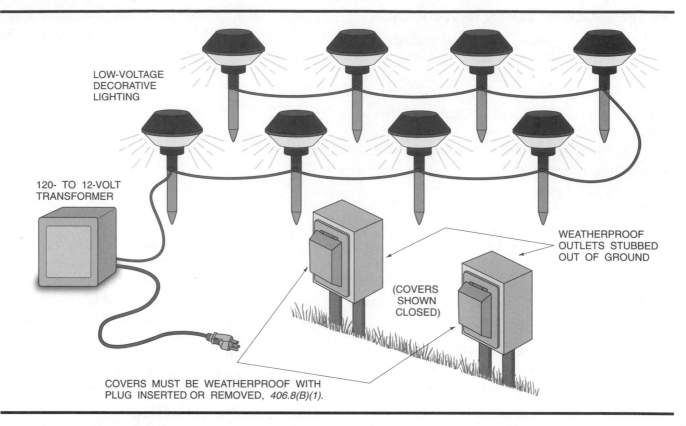

LOW-VOLTAGE
DECORATIVE
LIGHTING

120- TO 12-VOLT
TRANSFORMER

WEATHERPROOF
OUTLETS STUBBED
OUT OF GROUND

(COVERS
SHOWN
CLOSED)

COVERS MUST BE WEATHERPROOF WITH
PLUG INSERTED OR REMOVED, *406.8(B)(1)*.

Figure 16-4   Weatherproof receptacle outlets stubbed out of the ground; either low-voltage decorative landscape lighting or 120-volt PAR luminaires (lighting fixtures) may be plugged into these outlets.

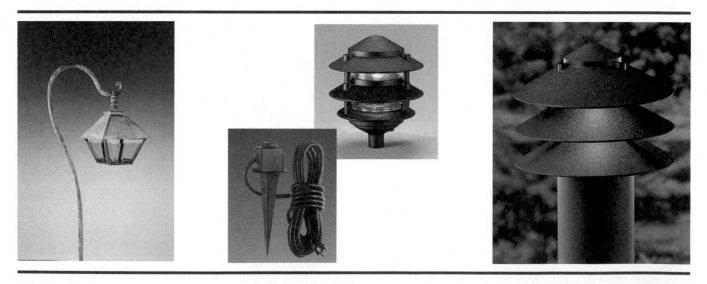

Figure 16-5   Luminaires (fixtures) used for decorative purposes outdoors, under shrubs and trees, in gardens, and for lighting paths and driveways. *Courtesy of* Progress Lighting.

unless live parts are made inaccessible to contact during normal use. Low-voltage luminaires (lighting fixtures) are intended for specific applications, such as surface mounting, suspended installations, or recessed installations, and are so marked. If of the recessed type, they must be installed according to the provisions of *410.66*, as discussed in Unit 7.

Low-voltage lighting systems are covered in *Article 411*. The basic requirements for these systems are that they shall:

- operate at 30 volts rms (42.4 volts peak) or less.

- have one or more secondary circuits, each limited to 25 amperes maximum.

- be supplied through an isolating transformer. A standard step-down transformer does not meet this criteria. Look carefully at the nameplate on the transformer.

- be listed for the purpose.

- not have the wiring concealed or extended through a building unless the wiring meets the requirements of *Chapter 3*, *NEC.*® An example of this is running Type NMC nonmetallic-sheathed cable for the low-voltage portion of the wiring that is concealed or extended through the structure of the building.

- not be installed within 10 ft (3.0 m) of pools, spas, or water fountains unless specifically permitted in *Article 680*.

- not have the secondary of the isolating transformer grounded.

- not be supplied by a branch-circuit that exceeds 20 amperes.

OUTDOOR WIRING

If the homeowner requests that 120-volt receptacles be installed away from the building structure, the electrician can provide weatherproof receptacle outlets as illustrated in Figure 16-4.

Outdoor receptacles must have GFCI protection. See Unit 6 for the complete discussion of GFCI protection.

These types of boxes have trade size ½ female threaded openings or hubs in which conduit fittings secure the conduit "stub-ups" to the box. Any unused openings are closed with trade size ½ plugs that are screwed tightly into the threads. Refer to Figure 16-6.

CONDUIT BODIES THAT *DO NOT* CONTAIN WIRING DEVICES OR SUPPORT LUMINAIRES (FIXTURES):
- SHALL HAVE THREADED ENTRIES OR HUBS.
- SHALL BE SUPPORTED BY AT LEAST TWO CONDUITS.
- MAY BE SUPPORTED BY ONE RMC, IMC, RNC, OR EMT, BUT ONLY IF THE ENTRY OR HUB ON THE CONDUIT BODY IS NOT LARGER THAN THE TRADE SIZE OF THE CONDUIT.

CONDUIT BODIES THAT *DO* CONTAIN WIRING DEVICES OR SUPPORT LUMINAIRES (FIXTURES):
- SHALL HAVE THREADED ENTRIES OR HUBS.
- SHALL BE SUPPORTED BY AT LEAST TWO CONDUITS.
- MUST BE SUPPORTED WITHIN 18 IN. (450 mm) OF THE ENCLOSURE.
- MAY BE SUPPORTED BY ONE RMC OR IMC, BUT ONLY IF THE ENTRY OR HUB ON THE CONDUIT BODY IS NOT LARGER THAN THE TRADE SIZE OF THE CONDUIT.

NOT A GOOD IDEA TO SUPPORT A CONDUIT BODY WITH ONE CONDUIT. THE CONDUIT BODY COULD EASILY TWIST, RESULTING IN CONDUCTOR INSULATION DAMAGE, AND A POOR GROUND CONNECTION BETWEEN THE CONDUIT AND THE CONDUIT BODY. SUGGEST ADDITIONAL SUPPORT WITH AN ANGLE IRON OR FLAT 1/4 IN. x 1 IN. STEEL ATTACHED TO THE ENCLOSURE AND DRIVEN INTO THE GROUND.

CONDUIT BODIES MAY CONTAIN SPLICES, TAPS, OR DEVICES. BUT ONLY IF MARKED WITH THEIR SPACE VOLUME. USE *TABLE 314.16(B)* FOR DETERMINING THE MAXIMUM NUMBER OF CONDUCTORS.

RMC = RIGID METAL CONDUIT (*ARTICLE 344*)
IMC = INTERMEDIATE METAL CONDUIT (*ARTICLE 342*)
RNC = RIGID NONMETALLIC CONDUIT (*ARTICLE 352*)
EMT = ELECTRICAL METALLIC TUBING (*ARTICLE 358*)

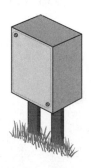

Figure 16-6   Supporting threaded conduit bodies, *314.23(E)* and *(F)*.

▶All receptacles installed outdoors or in wet locations must have an enclosure that is weatherproof, whether or not an attachment plug cap is inserted, *406.8(B)(1)*.◀ This requires a self-closing cover that is deep enough to shelter the attachment plug cap of the cord. See Figure 16-4 and Figure 16-7.

Where flush-mounted weatherproof receptacles are desired for appearance purposes, a recessed weatherproof receptacle, as illustrated in Figure 16-7, may be used. This type of installation takes up more depth than a standard receptacle and requires special mounting, as indicated in the "exploded" view. Careful thought must be given during the "roughing-in" stages to ensure that the proper depth "roughing-in" box is installed.

Wiring with Type UF Cable
(Article 340)

The plans show that Type UF underground cable, Figure 16-8, is used to connect the post light.

The current-carrying capacity (ampacity) of Type UF cable is found in *Table 310.16*, using the 60°C column. According to *Article 340*, Type UF cable:

- is marked underground feeder cable.

- is available in sizes from 14 AWG through 4/0 AWG (for copper conductors) and from 12 AWG through 4/0 AWG (for aluminum conductors).

- ▶may be used in direct exposure to the sun if the cable is of a type listed for sunlight resistance, or is listed and marked for sunlight resistance.◀

- may be used with nonmetallic-sheathed cable fittings.

- is flame-retardant.

- is moisture-, fungus-, and corrosion-resistant.

- may be buried directly in the earth.

- may be used for branch-circuit and feeder wiring.

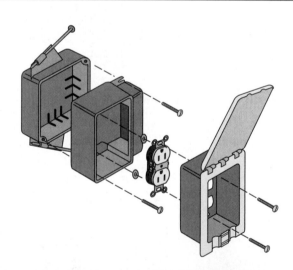

Figure 16-7 A recessed type of weatherproof receptacle that meets the requirements of *406.8(B)(1)* and *(B)(2)*, which requires that the enclosure be weatherproof while in use if unattended, where the cord will be plugged in for long periods of time. The "exploded" view illustrates one method of "roughing-in" this type of weatherproof receptacle. *Photos courtesy of* TayMac Corporation.

Figure 16-8 Type UF underground cable. *Courtesy of* Southwire Company.

- may be used in interior wiring for wet, dry, or corrosive installations.

- is installed by the same methods as non-metallic-sheathed cable *(Article 336)*.

- must not be used as service-entrance cable.

- must not be embedded in concrete, cement, or aggregate.

- must be buried in the same trench where single-conductor cables are installed.

- shall be installed according to *300.5*.

Equipment fed by Type UF cable is grounded by properly connecting the bare equipment grounding conductor found in the UF cable to the equipment to be grounded, Figure 16-9.

For underground wiring, it is significant to note that whereas the Code does permit running up the side of a tree with conduit or cable (when protected from physical damage), as in Figure 16-10, the Code does *not* permit supporting conductors between trees, as in Figure 16-11. Figure 16-12 shows one type of luminaire (lighting fixture) that is permitted for the installation shown in Figure 16-10. Splices are permitted to be made directly in the ground, but only if the connectors are listed by an NRTL for such use, *110.14(B)* and *300.5(E)*.

UNDERGROUND WIRING

Underground wiring is common in residential applications. Examples include decorative landscape lighting and wiring to post lamps and detached buildings such as garages or tool sheds.

Figure 16-13 illustrates a Code violation where the post lantern is grounded to the white grounded circuit conductor. This violation results in a parallel return path for a ground fault. First of all, *250.4(A)(5)* prohibits using the earth as the sole equipment grounding conductor, and secondly, *250.24(A)(5)* and *250.142(B)* prohibit making any connections of equipment to the grounded circuit conductor anywhere on the load side of the service disconnect.

Figure 16-10 This installation conforms to the Code. *NEC® 225.26* **does not permit spans of conductors to be run between live or dead trees, but it does not prohibit installing decorative lighting as shown in this figure.** *NEC® 410.16(H)* **permits outdoor luminaires (lighting fixtures) and associated equipment to be supported by trees. Recognized wiring methods to carry the conductors up the tree must be used.**

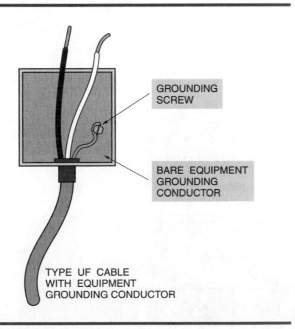

GROUNDING SCREW

BARE EQUIPMENT GROUNDING CONDUCTOR

TYPE UF CABLE WITH EQUIPMENT GROUNDING CONDUCTOR

Figure 16-9 Illustration showing how the bare equipment grounding conductor of a Type UF cable is used to ground the metal box.

Figure 16-11 *Violation:* Permanent overhead conductor spans are not permitted to be supported by live or dead vegetation, such as trees. Lumber and treated poles are not considered to be vegetation for this requirement. ►Temporary wiring is permitted to be supported by live or dead vegetation. See *225.26* and *527.4(J)*.◄

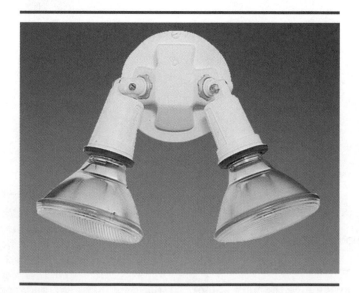

Figure 16-12 Outdoor luminaire (fixture) with lamps. *Courtesy of* Progress Lighting.

Table 300.5 shows the minimum depths required for the various types of wiring methods. Refer to Figure 16-14 and Figure 16-15.

Across the top of *Table 300.5* are listed wiring methods such as direct burial cables or conductors, rigid metal conduit, intermediate metal conduit, rigid nonmetallic conduit that is approved for direct burial, special considerations for residential branch-circuits rated 120 volts or less having GFCI protection and overcurrent protection not over 20 amperes, and low-voltage landscape lighting supplied with Type UF cable.

Down the left-hand side of the table we find the location of the wiring methods that are listed across the top of the table.

The notes to *Table 300.5* are very important.

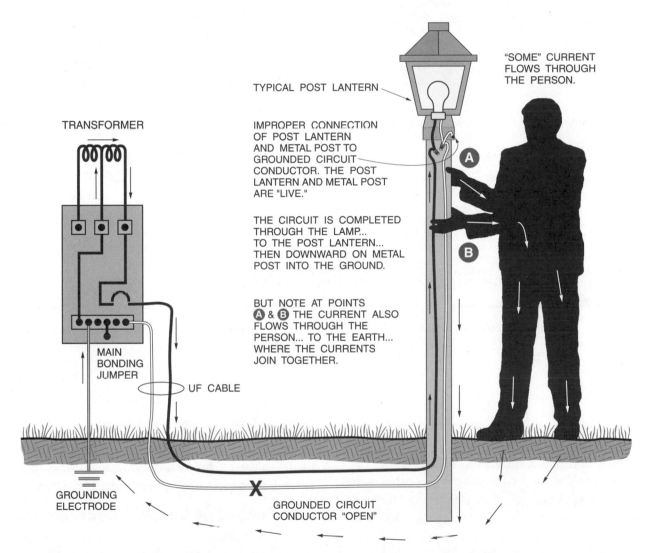

TYPICAL POST LANTERN

"SOME" CURRENT FLOWS THROUGH THE PERSON.

TRANSFORMER

IMPROPER CONNECTION OF POST LANTERN AND METAL POST TO GROUNDED CIRCUIT CONDUCTOR. THE POST LANTERN AND METAL POST ARE "LIVE."

THE CIRCUIT IS COMPLETED THROUGH THE LAMP... TO THE POST LANTERN... THEN DOWNWARD ON METAL POST INTO THE GROUND.

BUT NOTE AT POINTS Ⓐ & Ⓑ THE CURRENT ALSO FLOWS THROUGH THE PERSON... TO THE EARTH... WHERE THE CURRENTS JOIN TOGETHER.

MAIN BONDING JUMPER

UF CABLE

GROUNDING ELECTRODE

GROUNDED CIRCUIT CONDUCTOR "OPEN"

IF THE GROUNDED CIRCUIT CONDUCTOR "OPENS" FOR ANY REASON, THE EARTH PROVIDES AN INEFFECTIVE RETURN PATH FOR THE CURRENT. THE CURRENT TRAVELS THROUGH THE GROUND BACK TO THE GROUNDING ELECTRODE... TO THE PANEL GROUNDED NEUTRAL BUS... THROUGH THE SERVICE NEUTRAL CONDUCTOR... THROUGH THE TRANSFORMER... BACK TO THE SERVICE PANEL... THEN THROUGH THE "HOT" CONDUCTOR TO POST LANTERN AND METAL POST.

Figure 16-13   VIOLATION! This diagram clearly illustrates the electric shock hazard associated with the Code violation practice of grounding metal objects to the grounded circuit conductor. In this illustration, the post lantern and metal post have been improperly grounded (connected) to the grounded circuit conductor at (A). The grounded circuit conductor is "open." One path for the return current is through the metal post into the ground. The second path is through the person. This is a violation of *250.24(A)(5)* and *250.142(B)*.

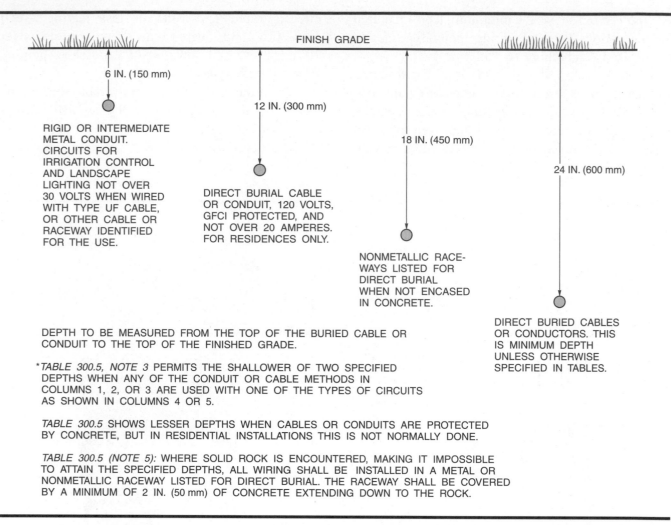

FINISH GRADE

6 IN. (150 mm)

RIGID OR INTERMEDIATE
METAL CONDUIT.
CIRCUITS FOR
IRRIGATION CONTROL
AND LANDSCAPE
LIGHTING NOT OVER
30 VOLTS WHEN WIRED
WITH TYPE UF CABLE,
OR OTHER CABLE OR
RACEWAY IDENTIFIED
FOR THE USE.

12 IN. (300 mm)

DIRECT BURIAL CABLE
OR CONDUIT, 120 VOLTS,
GFCI PROTECTED, AND
NOT OVER 20 AMPERES.
FOR RESIDENCES ONLY.

18 IN. (450 mm)

NONMETALLIC RACE-
WAYS LISTED FOR
DIRECT BURIAL
WHEN NOT ENCASED
IN CONCRETE.

24 IN. (600 mm)

DIRECT BURIED CABLES
OR CONDUCTORS. THIS
IS MINIMUM DEPTH
UNLESS OTHERWISE
SPECIFIED IN TABLES.

DEPTH TO BE MEASURED FROM THE TOP OF THE BURIED CABLE OR
CONDUIT TO THE TOP OF THE FINISHED GRADE.

*TABLE 300.5, NOTE 3 PERMITS THE SHALLOWER OF TWO SPECIFIED
DEPTHS WHEN ANY OF THE CONDUIT OR CABLE METHODS IN
COLUMNS 1, 2, OR 3 ARE USED WITH ONE OF THE TYPES OF CIRCUITS
AS SHOWN IN COLUMNS 4 OR 5.

TABLE 300.5 SHOWS LESSER DEPTHS WHEN CABLES OR CONDUITS ARE PROTECTED
BY CONCRETE, BUT IN RESIDENTIAL INSTALLATIONS THIS IS NOT NORMALLY DONE.

TABLE 300.5 (NOTE 5): WHERE SOLID ROCK IS ENCOUNTERED, MAKING IT IMPOSSIBLE
TO ATTAIN THE SPECIFIED DEPTHS, ALL WIRING SHALL BE INSTALLED IN A METAL OR
NONMETALLIC RACEWAY LISTED FOR DIRECT BURIAL. THE RACEWAY SHALL BE COVERED
BY A MINIMUM OF 2 IN. (50 mm) OF CONCRETE EXTENDING DOWN TO THE ROCK.

Figure 16-14   Minimum depths for cables and conduits installed underground. See *Table 300.5* for depths for other conditions.

Note 4 allows us to select the shallower of two depths when the wiring methods in Columns 1, 2, or 3 are combined with Columns 4 or 5. For example, the basic rule for direct buried cable is that it be covered by a minimum of 24 in. (600 mm). However, this minimum depth is reduced to 12 in. (300 mm) for residential installations when the branch-circuit is not over 120 volts, is GFCI protected, and the overcurrent protection is not over 20 amperes. This is for residential installations only.

The measurement for the depth requirement is from the top of the raceway or cable to the top of the finished grade, concrete, or other similar cover.

Do not back-fill a trench with rocks, debris, or similar coarse material, *300.5(F)*.

Protection From Damage

►Underground direct buried conductors and cables must be protected against physical damage according to the rules set forth in *300.5(D)*.◄

Installation of Conduit Underground

By definition, an underground installation is a wet location. Cables and insulated conductors installed in underground raceways shall be listed for use in wet locations. Conductors suitable for use in wet locations will have the letter "W" in their type designation. For example RHW, TW, THW, THHW, THWN, and XHHW. See *300.5(D)(5)* and *310.8(C)*.

All conduit installed underground must be protected against corrosion. The manufacturer of the

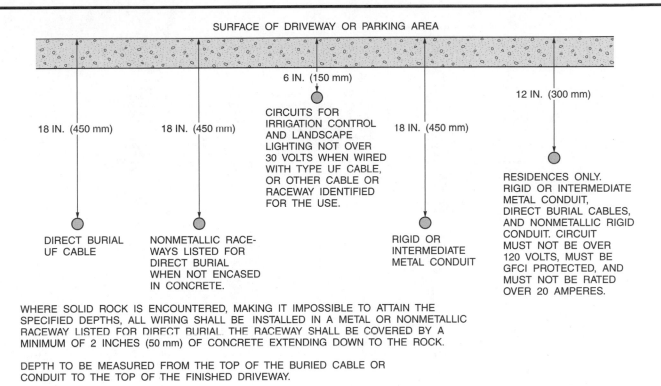

SURFACE OF DRIVEWAY OR PARKING AREA

6 IN. (150 mm)

12 IN. (300 mm)

18 IN. (450 mm)   18 IN. (450 mm)

CIRCUITS FOR
IRRIGATION CONTROL
AND LANDSCAPE
LIGHTING NOT OVER
30 VOLTS WHEN WIRED
WITH TYPE UF CABLE,
OR OTHER CABLE OR
RACEWAY IDENTIFIED
FOR THE USE.

18 IN. (450 mm)

RESIDENCES ONLY.
RIGID OR INTERMEDIATE
METAL CONDUIT,
DIRECT BURIAL CABLES,
AND NONMETALLIC RIGID
CONDUIT. CIRCUIT
MUST NOT BE OVER
120 VOLTS, MUST BE
GFCI PROTECTED, AND
MUST NOT BE RATED
OVER 20 AMPERES.

DIRECT BURIAL
UF CABLE

NONMETALLIC RACE-
WAYS LISTED FOR
DIRECT BURIAL
WHEN NOT ENCASED
IN CONCRETE.

RIGID OR
INTERMEDIATE
METAL CONDUIT

WHERE SOLID ROCK IS ENCOUNTERED, MAKING IT IMPOSSIBLE TO ATTAIN THE
SPECIFIED DEPTHS, ALL WIRING SHALL BE INSTALLED IN A METAL OR NONMETALLIC
RACEWAY LISTED FOR DIRECT BURIAL. THE RACEWAY SHALL BE COVERED BY A
MINIMUM OF 2 INCHES (50 mm) OF CONCRETE EXTENDING DOWN TO THE ROCK.

DEPTH TO BE MEASURED FROM THE TOP OF THE BURIED CABLE OR
CONDUIT TO THE TOP OF THE FINISHED DRIVEWAY.

Figure 16-15 Depth requirements under residential driveways and parking areas. These depths are permitted for one- and two-family residences. See *Table 300.5*.

conduit and UL Standards will furnish information as to whether the conduit is suitable for direct burial. Additional supplemental protection of metal conduit can be accomplished by "painting" the conduit with a nonmetallic coating. See Table 16-2.

When metal conduit is used as the wiring method, the metal boxes, post lantern, and so forth that are properly connected to the metal raceways are considered to be grounded because the conduit serves as the equipment ground, *250.118* (Figure 16-16). See *Article 410, Part V* for details on grounding requirements for luminaires (fixtures).

When nonmetallic raceways are used, then a separate equipment grounding conductor, either green or bare, must be installed to accomplish the adequate grounding of the equipment served. See Figure 16-17. This equipment grounding conductor must be sized according to *Table 250.122, NEC.*®

Depending upon the type of soil and moisture content of the soil, the earth may or may not be a good ground. The Code recognizes that the earth by itself is not considered to be an adequate ground.

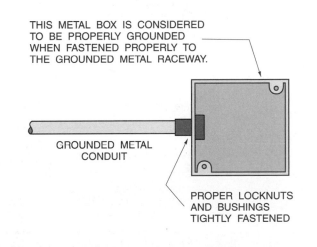

THIS METAL BOX IS CONSIDERED
TO BE PROPERLY GROUNDED
WHEN FASTENED PROPERLY TO
THE GROUNDED METAL RACEWAY.

GROUNDED METAL
CONDUIT

PROPER LOCKNUTS
AND BUSHINGS
TIGHTLY FASTENED

Figure 16-16 This illustration shows that a box is grounded when properly fastened to the grounded metal raceway. This is acceptable in *250.118*. More and more Code-enforcing authorities and consulting engineers are requiring that a separate equipment grounding conductor be installed in all raceways. This ensures effective grounding of the installation.

Table 300.5 Minimum Cover Requirements, 0 to 600 Volts, Nominal, Burial in Millimeters (Inches)

Location of Wiring Method or Circuit	Column 1 Direct Burial Cables or Conductors		Column 2 Rigid Metal Conduit or Intermediate Metal Conduit		Column 3 Nonmetallic Raceways Listed for Direct Burial Without Concrete Encasement or Other Approved Raceways		Column 4 Residential Branch Circuits Rated 120 Volts or Less with GFCI Protection and Maximum Overcurrent Protection of 20 Amperes		Column 5 Circuits for Control of Irrigation and Landscape Lighting Limited to Not More Than 30 Volts and Installed with Type UF or in Other Identified Cable or Raceway	
	mm	in.	mm	in.	mm	in.	mm	in.	mm	in.
All locations not specified below	600	24	150	6	450	18	300	12	150	6
In trench below 50-mm (2-in.) thick concrete or equivalent	450	18	150	6	300	12	50	6	50	6
Under a building	0 (in raceway only)	0	0	0	0	0	0 (in raceway only)	0	0 (in raceway only)	0
Under minimum of 102-mm (4-in.) thick concrete exterior slab with no vehicular traffic and the slab extending not less than 152 mm (6 in.) beyond the underground installation	450	18	100	4	100	4	150 (direct burial) 100 (in raceway)	6 4	150	6
Under streets, highways, roads, alleys, driveways, and parking lots	600	24	600	24	600	24	600	24	600	24
One- and two-family dwelling driveways and outdoor parking areas, and used only for dwelling-related purposes	450	18	450	18	450	18	300	12	450	18
In or under airport runways, including adjacent areas where trespassing prohibited	450	18	450	18	450	18	450	18	450	18

Notes:

1. Cover is defined as the shortest distance in millimeters (inches) measured between a point on the top surface of any direct-buried conductor, cable, conduit, or other raceway and the top surface of finished grade, concrete, or similar cover.

2. Raceways approved for burial only where concrete encased shall require concrete envelope not less than 50 mm (2 in.) thick.

3. Lesser depths shall be permitted where cables and conductors rise for terminations or splices or where access is otherwise required.

4. Where one of the wiring method types listed in Columns 1–3 is used for one of the circuit types in Columns 4 and 5, the shallower depth of burial shall be permitted.

5. Where solid rock prevents compliance with the cover depths specified in this table, the wiring shall be installed in metal or nonmetallic raceway permitted for direct burial. The raceways shall be covered by a minimum of 50 mm (2 in.) of concrete extending down to rock.

Reprinted with permission from NFPA 70-2002.

IS SUPPLEMENTAL CORROSION PROTECTION REQUIRED?

	In Concrete Above Grade?	In Concrete Below Grade?	In Direct Contact with Soil?
Rigid Conduit[1]	No	No	No[2]
Intermediate Conduit[1]	No	No	No[2]
Electrical Metallic Tubing[1]	No	Yes	Yes[3]

[1] Severe corrosion can be expected where ferrous metal conduits come out of concrete and enter the soil. Here again, some electrical inspectors and consulting engineers might specify the application of some sort of supplemental nonmetallic corrosion protection.

[2] Unless subject to severe corrosive effects. Different soils have different corrosive characteristics.

[3] In most instances, electrical metallic tubing is not permitted to be installed underground in direct contact with the soil because of corrosion problems.

Table 16-2

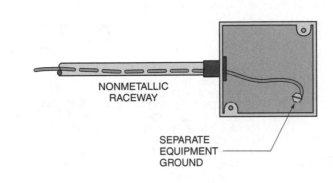

NONMETALLIC
RACEWAY

SEPARATE
EQUIPMENT
GROUND

Figure 16-17   Grounding a metal box with a separate equipment grounding conductor.

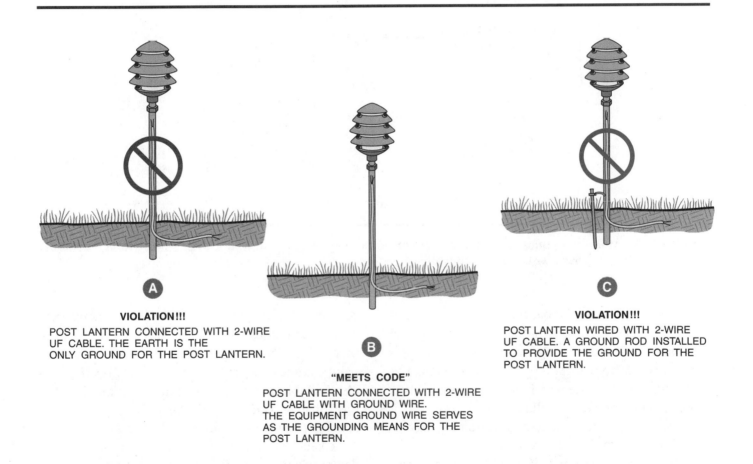

A

VIOLATION!!!
POST LANTERN CONNECTED WITH 2-WIRE
UF CABLE. THE EARTH IS THE
ONLY GROUND FOR THE POST LANTERN.

B

"MEETS CODE"
POST LANTERN CONNECTED WITH 2-WIRE
UF CABLE WITH GROUND WIRE.
THE EQUIPMENT GROUND WIRE SERVES
AS THE GROUNDING MEANS FOR THE
POST LANTERN.

C

VIOLATION!!!
POST LANTERN WIRED WITH 2-WIRE
UF CABLE. A GROUND ROD INSTALLED
TO PROVIDE THE GROUND FOR THE
POST LANTERN.

Figure 16-18   ▶These drawings illustrate the intent of the last sentence of *250.4(A)(5)*, which states, "The earth shall not be used as the sole equipment grounding conductor or effective ground fault current path."◀

The last sentence of *250.4(A)(5)* states that: ▶"The earth shall not be used as the sole equipment grounding conductor or effective ground fault current path."◀ See Figure 16-18.

Figure 16-19 shows two methods of bringing the conduit into the basement: (1) it can be run below ground level and then can be brought through the basement wall, or (2) it can be run up the side of the

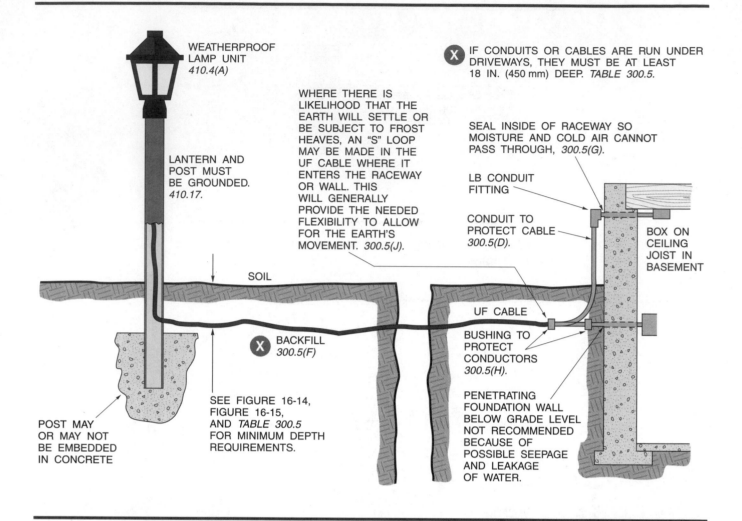

WEATHERPROOF
LAMP UNIT
410.4(A)

IF CONDUITS OR CABLES ARE RUN UNDER
DRIVEWAYS, THEY MUST BE AT LEAST
18 IN. (450 mm) DEEP. *TABLE 300.5.*

WHERE THERE IS
LIKELIHOOD THAT THE
EARTH WILL SETTLE OR
BE SUBJECT TO FROST
HEAVES, AN "S" LOOP
MAY BE MADE IN THE
UF CABLE WHERE IT
ENTERS THE RACEWAY
OR WALL. THIS
WILL GENERALLY
PROVIDE THE NEEDED
FLEXIBILITY TO ALLOW
FOR THE EARTH'S
MOVEMENT. *300.5(J).*

SEAL INSIDE OF RACEWAY SO
MOISTURE AND COLD AIR CANNOT
PASS THROUGH, *300.5(G).*

LB CONDUIT
FITTING

CONDUIT TO
PROTECT CABLE
300.5(D).

BOX ON
CEILING
JOIST IN
BASEMENT

LANTERN AND
POST MUST
BE GROUNDED.
410.17.

SOIL

UF CABLE

BACKFILL
300.5(F)

BUSHING TO
PROTECT
CONDUCTORS
300.5(H).

POST MAY
OR MAY NOT
BE EMBEDDED
IN CONCRETE

SEE FIGURE 16-14,
FIGURE 16-15,
AND *TABLE 300.5*
FOR MINIMUM DEPTH
REQUIREMENTS.

PENETRATING
FOUNDATION WALL
BELOW GRADE LEVEL
NOT RECOMMENDED
BECAUSE OF
POSSIBLE SEEPAGE
AND LEAKAGE
OF WATER.

Figure 16-19   Methods of bringing cable and/or conduit through a concrete wall and/or upward from a concrete wall into the hollow space within a framed wall.

building and through the basement wall at the ceiling joist level. When the conduit is run through the basement wall, the opening must be sealed to prevent moisture from seeping into the basement. The electrician must decide which of the two methods is more suitable for each installation.

According to the Code, any location exposed to the weather is considered a wet location.

Luminaires (fixtures) installed in wet locations shall be marked "Suitable for Wet Locations." These luminaires (fixtures) are constructed so that water cannot enter or accumulate in lampholders, wiring compartments, or other electrical parts, *410.4(A).*

A damp location is defined as being protected from the weather but subject to moderate degrees of moisture. Partially protected areas such as roofed

open porches or areas under canopies are damp locations. Luminaires (fixtures) in these locations must be marked "*Suitable for Damp Locations.*" A luminaire (fixture) that is "*Suitable for Wet Locations*" is also suitable for damp locations. A luminaire (fixture) that is "*Suitable for Damp Locations*" is not suitable for wet locations.

Check the UL label on a luminaire (fixture) to determine the suitability of the (luminaire) fixture for a wet or damp location.

The post in Figure 16-19 may or may not be embedded in concrete, depending on the consistency of the soil, the height of the post, and the size of the post lantern. Most electricians prefer to embed the base of the post in concrete to prevent rotting of wood posts and rusting of metal posts.

Figure 16-20 illustrates some typical post lights.

Figure 16-20   Post-type lanterns. *Courtesy of* Progress Lighting.

OVERHEAD GARAGE DOOR OPERATOR Ⓐₑ

The overhead garage door operator in this residence is plugged into the receptacle provided for this purpose on the garage ceiling. This particular overhead door operator is rated 5.8 amperes at 120 volts. The actual installation of the overhead door will usually be done by a specially trained overhead door installer.

Principles of Operation

The overhead door operator contains a motor, gear reduction unit, motor reversing switch, and electric clutch. This unit is preassembled and wired by the manufacturer.

Almost all residential overhead door operators use split-phase, capacitor-start motors in sizes from ¼ to ½ horsepower. The motor size selected depends upon the size (weight) of the door to be raised and lowered. By using springs in addition to the gear reduction unit to counterbalance the weight of the door, a small motor can lift a fairly heavy door.

There are two ways to change the direction of a split-phase motor: (1) reverse the starting winding with respect to the running winding; (2) reverse the running winding with respect to the starting winding. The motor can be run in either direction by properly connecting the starting and running winding leads of the motor to the reversing switch.

When the motor shaft is connected to a chain drive or screw drive, the door can be raised or lowered according to the direction in which the motor is rotating.

Wiring of Garage Door Operators

The wiring of overhead door operators for residential use is quite simple, because these units are completely prewired by the manufacturer. All residential-type overhead door operators come equipped with a cord and plug. This is shown in Figure 16-21. A 120-volt receptacle outlet is

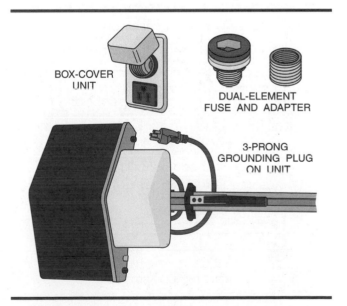

BOX-COVER UNIT

DUAL-ELEMENT FUSE AND ADAPTER

3-PRONG GROUNDING PLUG ON UNIT

Figure 16-21   Overhead door operator with three-wire cord connector.

installed on the ceiling near the operator. Because a receptacle on a ceiling is not *readily accessible* as defined by the *NEC*,® it does not have to be GFCI protected, *210.8(A)(2), Exception No. 1.*

Some overhead door operators can be "hard-wired" as shown in Figure 16-22. This is not often done in residential installations.

The current draw of a typical residential overhead door operator is usually so small that it could be connected to the garage lighting branch-circuit. In the residence discussed in this text, the overhead door operator is served by a separate 15-ampere branch-circuit.

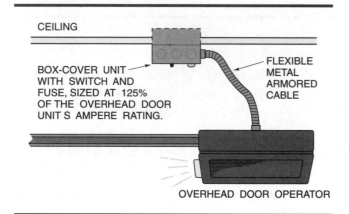

Figure 16-22   Connections for overhead door operator.

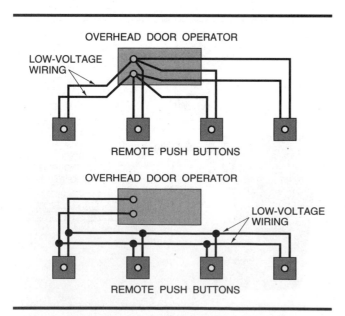

Figure 16-23   Push-button wiring for overhead door operator.

Push Buttons

In addition to remote controllers, push buttons should be installed next to each door leading to the garage, including the overhead door. Key-operated weatherproof switches can also be installed outside next to the overhead door. The wiring to each push button or key-operated switch can be a low-voltage type similar to that used for chime wiring. Figure 16-23 shows how push buttons are connected. Because the operating relay in an overhead door operator works on the momentary contact principle, a two-wire cable is all that is needed between the operator and the push buttons or key-operated switch.

Quite often, electricians will install the push-button wiring during the "rough-in" stage. An easy way to do this is to nail a single-gang, 4-in. square plaster "mud" ring to the stud at the desired locations, then run a two-wire, 18 or 16 AWG low-voltage cable to a point where the overhead door operator will be mounted. When the house is "trimmed out," a single-gang wallplate with the pushbutton is installed. See Figure 16-24.

Box-Cover Unit

The overhead door operator unit has overcurrent protection built into it according to UL standards.

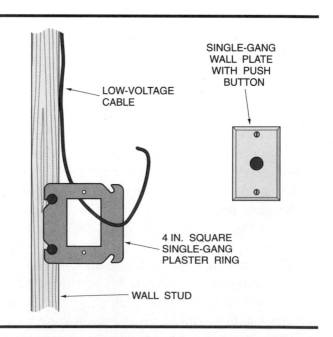

Figure 16-24   A sketch of how a 4-in. square cover can be used to provide an opening for low-voltage wiring.

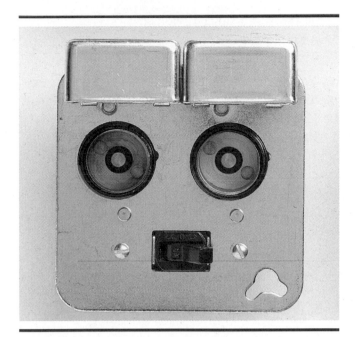

Figure 16-25 Box-cover unit that provides disconnecting means plus individual overcurrent protection. *Courtesy of* Bussmann Division, Cooper Industries.

Additional "back-up" protection can be installed in the field. Here's how.

Instead of a standard receptacle on the ceiling near the operator, a box-cover unit can be installed. Box-cover units are available with *switch/fuse* or *receptacle/fuse* combinations, as in Figure 16-21, Figure 16-22, Figure 16-25, and Figure 16-26. Box-cover units provide a convenient disconnecting

means in addition to having a fuse holder in which to install a dual-element, time-delay Type S fuse.

Size the fuse at approximately 125% of the overhead door operator's ampere rating.

For example, let us say the overhead door operator is marked 5.8 amperes. Then:

$$5.8 \times 1.25 = 7.25 \text{ amperes}$$

The next standard size dual-element time-delay fuse is 8 amperes. An 8-ampere fuse will open faster than the 15- or 20-ampere branch-circuit breaker or fuse, should a short-circuit occur within the operator. This isolates the door operator, and does not shut off the power to the rest of the branch-circuit.

VOLTAGE DROP PROBLEMS WITH LOW-VOLTAGE LIGHTING

Low voltage lighting systems that operate at 30 volts or less are covered in *Article 411*. These systems are supplied through an isolating (two-winding) transformer that is connected to a 120-volt, 20-ampere-or-less branch-circuit.

Careful consideration must be given to the length of the low-voltage wiring. When the low-voltage wiring is long, larger conductors are necessary to compensate for voltage drop. If voltage drop is ignored, the lamps will not get enough voltage to burn properly; the lamps will burn dim. Voltage drop calculations are discussed in Unit 4.

You must also be concerned with the current draw of a low-voltage lighting system.

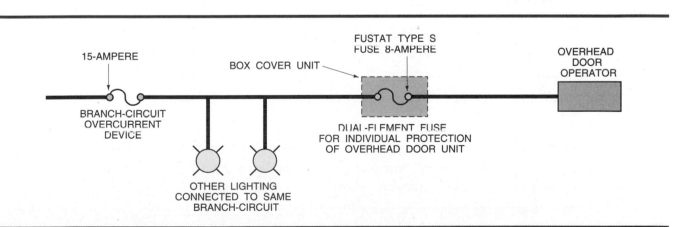

Figure 16-26 This diagram shows the benefits of installing a small-ampere-rated dual element time-delay fuse that is sized to provide overload protection for the motor. If an electrical problem occurs at the overhead door operator, only the 8-ampere fuse will open. The 15-ampere branch-circuit is not affected. All other loads connected to the 15-ampere circuit remain energized.

For the same wattage, the current draw of a low-voltage lamp will be much greater than that of a 120-volt lamp. If not taken into consideration, the higher current can cause the conductors to run at dangerous temperatures; that could cause a fire!

EXAMPLE:

A 480-watt load at 120 volts draws:
480 ÷ 120 = 4 amperes
This is easily carried by a 14 AWG conductor.

A 480-watt load at 12 volts draws:
480 ÷ 12 = 40 amperes
This requires an 8 AWG conductor.

You can readily see that for this 480-watt example, the conductors must be considerably larger for the 12-volt application to safely carry the current.

Prewired low-voltage lighting assemblies that bear the UL label have the correct size and length of conductors. Adding conductors in the field must be done according to the manufacturers' instructions.

Voltage drop calculations are covered in Unit 4.

REVIEW

Note: Refer to the Code or plans where necessary.

1. What circuit supplies the garage? _____

2. What is the circuit rating? _____

3. How many GFCI-protected receptacle outlets are connected to the garage circuit?

   _____

4. a. How many cables enter the wall switch box located next to the side garage door?

   _____

   b. How many circuit conductors enter this box? _____

   c. How many equipment grounding conductors enter this box? _____

5. Show calculations on how to select a proper size box for question 4. What kind of box would you use? _____

   _____

   _____

   _____

   _____

   _____

6. a. How many lights are recommended for a one-car garage? _____ for a two-car garage? _____ for a three-car garage? _____

   b. Where are these lights to be located? _____

   _____

7. From how many points in the garage of this residence are the ceiling lights controlled? _____

8. GFCI breakers or GFCI receptacles are relatively expensive. How would *you* arrange the wiring of the garage circuit to make as economical an installation as possible that complies with the Code? _____

_____

_____

_____

_____

_____

9. The total estimated volt-ampere load of the garage circuit draws how many amperes? Show your calculations. _____

_____

_____

_____

_____

_____

10. How high from the floor are the switches and receptacles to be mounted? _____

11. What type of cable feeds the post light? _____

12. All raceways, cables, and direct burial-type conductors require a "cover." Explain the term "cover." _____

_____

_____

_____

_____

_____

13. In the spaces provided, fill in the cover (depth) for the following residential underground installations.
    a. Type UF. No other protection. _____ in.
    b. Type UF below driveway. _____ in.
    c. Rigid metal conduit under lawn. _____ in.
    d. Rigid metal conduit under driveway. _____ in.
    e. Electrical metallic tubing between house and detached garage. _____ in.
    f. UF cable under lawn. Circuit is 120 volts, 20 amperes, GFCI protected. _____ in.
    g. Rigid nonmetallic conduit approved for direct burial. No other protection. _____ in.
    h. Rigid nonmetallic conduit. Circuit is 120 volts, 20 amperes, GFCI protected. _____ in.
    i. Rigid conduit passing over solid rock and covered by 2 in. (50 mm) of concrete extending down to rock. _____ in.

14. What section of the Code prohibits embedding Type UF cable in concrete? _____
_____

15. The following is a layout of the garage circuit. Using the suggested cable layout, make a complete wiring diagram of this circuit, using colored pens or pencils to indicate conductor insulation colors.

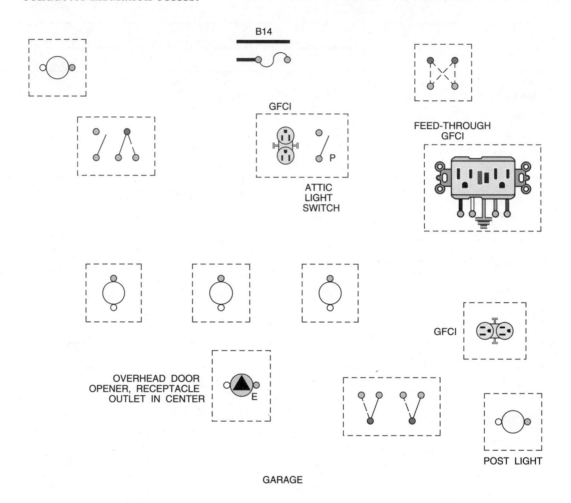

GARAGE

16. What types of motors are generally used for garage door openers? _____

17. How is the direction of a split-phase motor reversed? _____
_____
_____
_____

18. When a box-cover unit is installed for an overhead door operator, a time-delay dual-element fuse ampere rating would generally be sized at _____% of the unit's name-plate current rating.

19. When outdoor receptacles are installed where a cord will be plugged in permanently, such as in the case of decorative lighting, the cover must provide weatherproof integrity when the cord is plugged in. True or false? _____ NEC® _____ .

20. A homeowner purchased some single-conductor cable that was marked as suitable for direct burial in the ground to wire a post light. A separate equipment grounding conductor to the post light was not carried. Instead, the bottom of the metal post buried two ft into the ground was used as a ground for the post light. The inspector "red tagged" (turned down) the installation. Who is right? The homeowner or the electrical inspector? Explain.

_____

_____

_____

_____

21. What concerns are there when installing a low-voltage lighting system?

_____

_____

_____

UNIT 17

Recreation Room

OBJECTIVES

After studying this unit, the student should be able to

- understand three-wire (multiwire) branch-circuits.
- understand how to install lay-in luminaires (fixtures).
- calculate watts loss and voltage drop in two-wire and three-wire circuits.
- understand the term *fixture whips*.
- understand the advantages of installing multiwire branch-circuits.
- understand problems that can be encountered on multiwire branch-circuits as a result of open neutrals.

RECREATION ROOM LIGHTING (B9, 11, 12)

The recreation room is well-lighted through the use of six lay-in 2 ft × 4 ft fluorescent luminaires (fixtures). These fixtures are exactly the same size as two 2 ft × 2 ft ceiling tiles. The luminaires (fixtures) rest on top of the ceiling tee bars, as in Figure 17-1.

Junction boxes are mounted above a dropped (suspended) ceiling usually within 12 to 24 in. (300 to 600 mm) of the intended luminaire (fixture) location. A flexible conduit referred to as a *fixture whip*, 4 to 6 ft (1.2 to 1.8 m) long, is installed between the junction box and the luminaire (fixture). These fixture whips contain the correct type and size of conductor suitable for the temperature ratings and load required by the Code for recessed luminaires (fixtures). See Unit 7 for a complete explanation of Code regulations for the installation of recessed luminaires (fixtures), as in Figure 17-2.

Fluorescent luminaires (fixtures) of the type shown in Figure 17-1 might bear a label stating "Recessed Fluorescent Luminaire (Fixture)." The label might also state "Suspended Ceiling Fluorescent Luminaire (Fixture)." Although they might look the same, there is a difference.

Recessed fluorescent luminaires (fixtures) are intended for installation in cavities in ceilings and walls and are to be wired according to *410.64*. These luminaires (fixtures) may also be installed in suspended ceilings if they have the necessary mounting hardware.

Suspended fluorescent luminaires (fixtures) are intended only for installation in suspended ceilings where the acoustical tiles, lay-in panels, and suspended grid are not part of the actual building structure.

Underwriters Laboratories (UL) Standard 1570 covers recessed and suspended ceiling luminaires (fixtures) in detail.

The six recessed "lay-in" fluorescent luminaires (fixtures) in the recreation room are four-lamp luminaires (fixtures) with warm white energy-saving F40SPX30/RS lamps installed in them. The ballasts in these luminaires (fixtures) are high power factor, energy-saving ballasts. Five of the luminaires (fixtures) are controlled by a single-pole switch located at the bottom of the stairs. One luminaire (fixture) is controlled by the three-way switches that also control the stairwell luminaire (light fixture) hung from the ceiling above the stair landing midway up the stairs.

IMPORTANT: TO PREVENT THE LUMINAIRE (FIXTURE) FROM INADVERTENTLY FALLING, *410.16(C)* OF THE CODE REQUIRES THAT (1) SUSPENDED CEILING FRAMING MEMBERS THAT SUPPORT RECESSED LUMINAIRES (FIXTURES) MUST BE SECURELY FASTENED TO EACH OTHER, AND MUST BE SECURELY ATTACHED TO THE BUILDING STRUCTURE AT APPPROPRIATE INTERVALS, AND (2) RECESSED LUMINAIRES (FIXTURES) MUST BE SECURELY FASTENED TO THE SUSPENDED CEILING FRAMING MEMBERS BY BOLTS, SCREWS, RIVETS, OR SPECIAL LISTED CLIPS PROVIDED BY THE MANUFACTURER OF THE LUMINAIRE (FIXTURE) FOR THE PURPOSE OF ATTACHING THE LUMINAIRE (FIXTURE) TO THE FRAMING MEMBER.

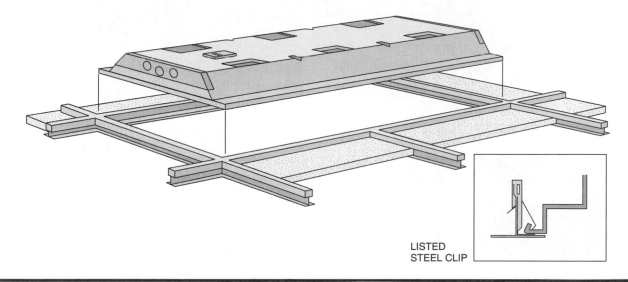

LISTED
STEEL CLIP

Figure 17-1 Typical lay-in fluorescent luminaire (fixture) commonly used in conjunction with dropped suspended ceilings. See *410.16(C)*.

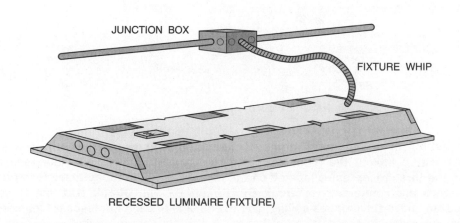

JUNCTION BOX

FIXTURE WHIP

RECESSED LUMINAIRE (FIXTURE)

Figure 17-2 Typical suspended ceiling "lay-in" fluorescent luminaire (fixture) showing flexible fixture whip connection. Conductors in whip must have temperature rating as required on label in luminaire (fixture); for example, "For Supply Use 194°F (90°C) conductors."

This was done so that the entire recreation room lighting would not be on continually whenever someone came down the stairs to go into the workshop. This is a practical energy-saving feature.

Circuit B12 feeds the fluorescent luminaires (fixtures) in the recreation room. See the cable layout in Figure 17-3. Table 17-1 summarizes the outlets and estimated load for the recreation room.

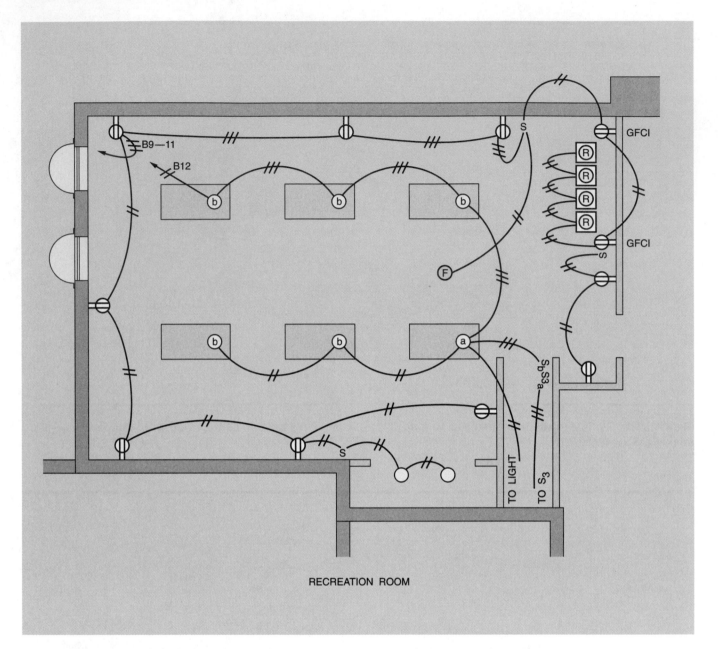

RECREATION ROOM

Figure 17-3 Cable layout for recreation room. The six fluorescent luminaires (fixtures) are connected to Circuit B12. The receptacles located around the room are connected to Circuit B11. The four recessed luminaires (fixtures) over the bar, the three receptacles located on the bar wall, and the receptacles located to the right as you come down the steps are connected to Circuit B9. Circuit B9 and Circuit B11 are three-wire circuits. Junction boxes are located above the dropped ceiling, and to one side of each fluorescent luminaire (fixture). The electrical connections are made in these junction boxes. A flexible fixture "whip" connects between each junction box and each luminaire (fixture). See Figure 17-2.

How to Connect Recessed and Lay-In Luminaires (Fixtures)

Figure 17-1 and Figure 17-2 show typical lay-in fluorescent luminaires (lighting fixtures). The installation and wiring connections for recessed luminaires (lighting fixtures) and lay-in suspended-ceiling luminaires (fixtures) are covered in Unit 7.

RECEPTACLES AND WET BAR (B9–11)

The circuitry for the recreation room wall receptacles and the wet-bar lighting area introduces a new type of circuit. This is termed a *multiwire branch-circuit* or a *three-wire branch-circuit*.

In many cases the use of a multiwire branch-circuit can save money in that one three-wire

DESCRIPTION	QUANTITY	WATTS	VOLT-AMPERES
Circuit Number B11			
Receptacles @ 120 W each	7	840	840
Closet luminaires (fixtures) @ 75 W each	2	150	150
Exhaust fan (2.5 A × 120 V)	1	160	300
TOTALS	10	1150	1290
Circuit Number B12			
Luminaire (fixture) on stairway	1	100	100
Recessed fluorescent luminaires (fixtures) with four 40-W lamps each	6	960	1008
TOTALS	7	1060	1108
Circuit Number B9			
Receptacles @ 120 W each	4	480	480
Wet-bar recessed luminaires (fixtures) @ 75 W each	4	300	300
TOTALS	8	780	780

Table 17-1   Recreation room outlets and estimated load (three circuits).

branch-circuit will do the job of two two-wire branch-circuits.

Also, if the loads are nearly balanced in a three-wire branch-circuit, the neutral conductor carries only the unbalanced current. This results in less voltage drop and watts loss for a three-wire branch-circuit as compared to similar loads connected to separate two-wire branch-circuits.

Figure 17-4 and Figure 17-5 illustrate the benefits of a multiwire branch-circuit relative to watts loss and voltage drop. The example, for simplicity, is a purely resistive circuit, and loads are exactly equal.

The distance from the load to the source is 50 ft. The conductor size is 14 AWG solid, uncoated copper. From *Table 8* in *Chapter 9* in the *NEC,* we find that the resistance of 1000 ft (300 m) of a 14 AWG uncoated copper conductor is 3.07 ohms (10.1 ohms/km). The resistance of 50 ft (15 m) is:

$$\frac{3.07}{1000} \times 50 = 0.1535 \text{ ohms (round off to 0.154)}$$

EXAMPLE 1:
TWO TWO-WIRE BRANCH-CIRCUITS

Watts loss in each current-carrying conductor is

$$\text{Watts} = I^2R = 10 \times 10 \times 0.154 = 15.4 \text{ watts}$$

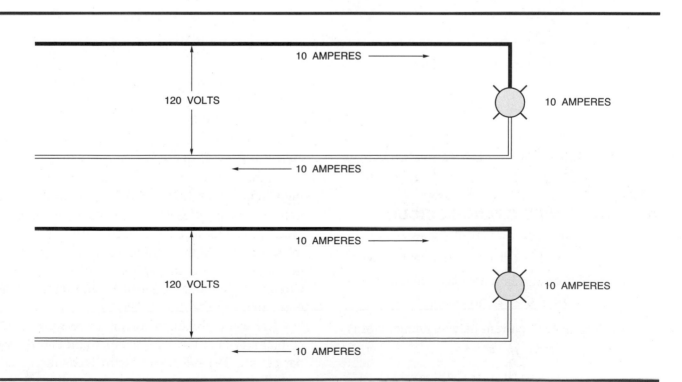

Figure 17-4   Two two-wire branch-circuits.

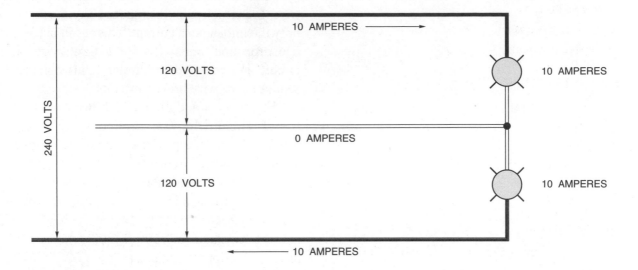

Figure 17-5 A three-wire branch-circuit.

Watts loss in all four current-carrying conductors is

$$15.4 \times 4 = 61.6 \text{ watts}$$

Voltage drop in each current-carrying conductor is

$$E_d = IR$$
$$= 10 \times 0.154$$
$$= 1.54 \text{ volts}$$

E_d for both current-carrying conductors in each circuit:

$$2 \times 1.54 = 3.08 \text{ volts}$$

Voltage available at load:

$$120 - 3.08 = 116.92 \text{ volts}$$

EXAMPLE 2:
ONE THREE-WIRE BRANCH-CIRCUIT

Watts loss in each current-carrying conductor is

$$\text{Watts} = I^2R = 10 \times 10 \times 0.154 = 15.4 \text{ watts}$$

Watts loss in both current-carrying conductors is

$$15.4 \times 2 = 30.8 \text{ watts}$$

Voltage drop in each current-carrying conductor is

$$E_d = IR$$
$$= 10 \times 0.154$$
$$= 1.54 \text{ volts}$$

E_d for both current-carrying conductors is:

$$2 \times 1.54 = 3.08 \text{ volts}$$

Voltage available at loads:

$$240 - 3.08 = 236.92 \text{ volts}$$

Voltage available at each load:

$$\frac{236.92}{2} = 118.46 \text{ volts}$$

It is readily apparent that a multiwire circuit can greatly reduce watts loss and voltage drop. This is the same reasoning that is used for the three-wire service-entrance conductors and the three-wire feeder to Panel B in this residence.

In the recreation room, a three-wire cable carrying Circuit B9 and Circuit B11 feeds from Panel B to the receptacle closest to the panel. This three-wire cable continues to each receptacle wall box along the same wall. At the third receptacle along the wall, the multiwire circuit splits, where the wet-bar Circuit B9 continues on to feed into the receptacle outlet to the left of the wet bar.

The white conductor of the three-wire cable is common to both the receptacle circuit and the wet-bar circuit. The black conductor feeds the receptacles. The red conductor is spliced straight through the boxes and feeds the wet bar area.

Care must be used when connecting a three-wire circuit to the panel. The black and red conductors *must* be connected to the opposite phases in the panel to prevent heavy overloading of the neutral grounded (white) conductor.

The neutral (grounded) conductor of the three-wire cable carries the unbalanced current. This current is the difference between the current in the black wire and the current in the red wire. For example, if one load is 12 amperes and the other load is 10 amperes, the neutral current is the difference between these loads (2 amperes), Figure 17-6.

If the black and red conductors of the three-wire cable are connected to the same phase in Panel A, Figure 17-7, the neutral conductor must carry the total current of both the red and black conductors

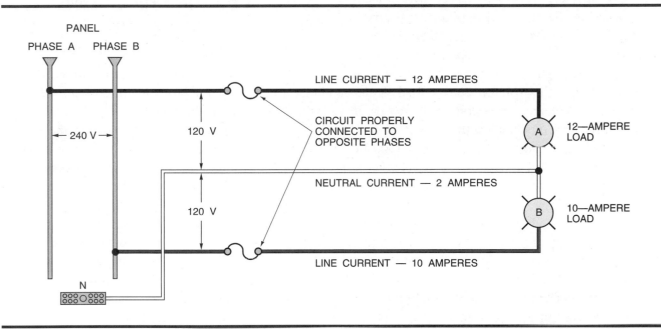

Figure 17-6   Correct wiring connections for a three-wire (multiwire) branch-circuit.

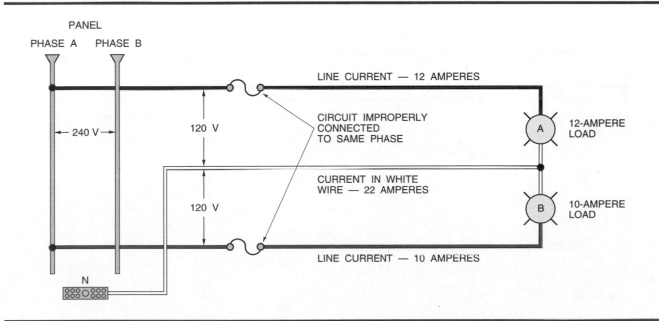

Figure 17-7   Improperly connected three-wire (multiwire) branch-circuit.

rather than the unbalanced current. As a result, the neutral conductor will be overloaded. All single-phase, 120/240-volt panels are clearly marked to help prevent an error in phase wiring. The electrician must check all panels for the proper wiring diagrams.

Figure 17-7 shows how an improperly connected three-wire branch-circuit results in an overloaded neutral conductor.

If an open neutral occurs on a three-wire branch-circuit, some of the electrical appliances in operation may experience voltages higher than the rated voltage at the instant the neutral opens.

For example, Figure 17-8 shows that for an open neutral condition, the voltage across load A decreases and the voltage across load B increases. If the load on each circuit changes, the voltage on each circuit also changes. According to Ohm's law, the voltage drop across any device in a series circuit is directly proportional to the resistance of that device. In other words, if load B has twice the resistance of load A, then load B will be subjected to twice the voltage of load A for an open neutral condition. To ensure the proper connection, care must be used when splicing the conductors.

An example of what can occur should the neutral of a three-wire multiwire branch-circuit open is shown in Figure 17-9. Trace the flow of current from phase A through the television set, then through the

toaster, then back to phase B, thus completing the circuit. The following simple calculations show why the television set (or stereo or home computer) can be expected to burn up.

$$R_t = 8.45 + 80 = 88.45 \text{ ohms}$$

$$I = \frac{E}{R} = \frac{240}{88.45} = 2.71 \text{ amperes}$$

Voltage appearing across the toaster:

$$IR = 2.71 \times 8.45 = 22.9 \text{ volts}$$

Voltage appearing across the television:

$$IR = 2.71 \times 80 = 216.8 \text{ volts}$$

This example illustrates the problems that can arise with an open neutral on a three-wire, 120/240-volt multiwire branch-circuit.

The same problem can arise when the neutral of the utility company's incoming service-entrance conductors (underground or overhead) opens. The problem is minimized because the neutral of the service is solidly grounded to the metal water piping system within the building. However, there are cases on record where poor service grounding has resulted in serious and expensive damage to appliances within the home because of an open neutral in the incoming service-entrance conductors.

For example, poor service grounding results when relying on a driven iron pipe (that rusts in a short period of time) as the only means of obtaining

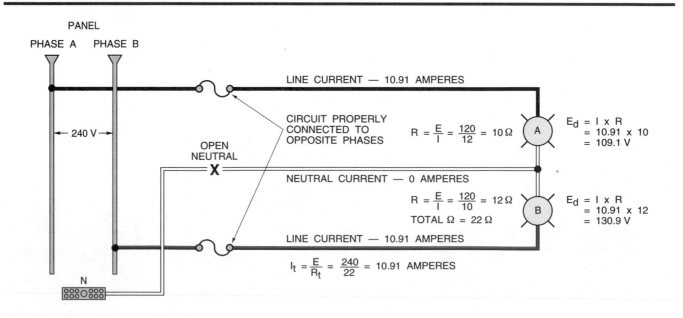

Figure 17-8   Example of an open neutral conductor.

the service equipment ground. This was quite often done years ago when the electrical code was not as stringent as today's *NEC.* When working on older homes, check the service ground. It could very well be inadequate, unsafe, and in need of updating to the present *NEC.*

The integrity of using only ground rods is highly questionable, whether the rod is an iron pipe, a galvanized pipe, or a copper-clad rod. The Code discourages the use of ground rods as the only means of obtaining a good service-entrance equipment ground.

See Unit 27 for details on the grounding and bonding of service-entrance equipment.

In summary, three-wire branch-circuits are used occasionally in residential installations. The two "hot" conductors of three-wire cable are connected to opposite phases. One white grounded (neutral) conductor is common to both "hot" conductors.

There are two receptacle outlets above the wet bar. Receptacles installed to serve countertop surfaces that are within 6 ft (1.8 m) of the outer edge of a wet bar sink must be GFCI protected, *210.8(A)(7)*, and must not be installed face-up on the countertop surfaces or work surfaces, *406.4(E)*.

A single-pole switch to the right of the wet bar controls the four recessed luminaires (fixtures)

above it. See Unit 7 for details pertaining to recessed luminaires (fixtures).

An exhaust fan of a type similar to the one installed in the laundry is installed in the ceiling of the recreation room to exhaust stale, stagnant, and smoky air from the room.

The basic switch, receptacle, and luminaire (fixture) connections are repetitive of most wiring situations discussed in previous units.

The wall boxes for the receptacles are selected based upon the measurements of the furring strips on the walls. Two-by-two furring strips will require the use of 4-in. square boxes with suitable raised covers. If the walls are furred with 2 × 4s, then possibly sectional device boxes could be used. Select a box that can contain the number of conductors, devices, and clamps to "meet Code." After all, the installation must be safe.

Probably the biggest factor in deciding which wiring method (cable or EMT) will be used to wire the recreation room is "what size and type of wall furring will be used?" See Unit 4 for the in-depth discussion on the mechanical protection required for nonmetallic sheathed cables where the cables will be less than 1¼ in. (32 mm) from the edge of the framing members.

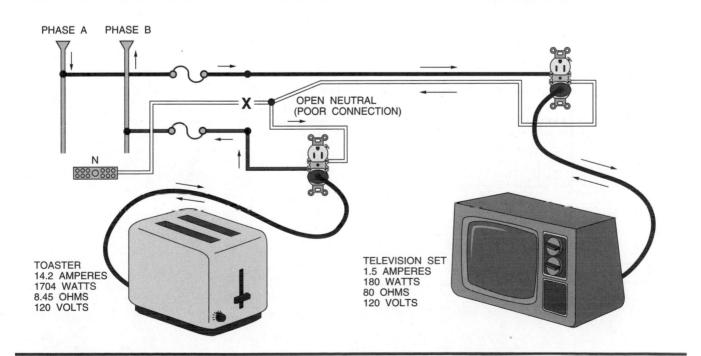

Figure 17-9 Problems that can occur with an open neutral on a three-wire (multiwire) branch-circuit.

REVIEW

Note: Refer to the Code or plans where necessary.

1. What is the total current draw when all six fluorescent luminaires (fixtures) are turned on?

   _____

   _____

2. The junction box that will be installed above the dropped ceiling near the fluorescent luminaires (fixtures) closest to the stairway will have _____ (number of) 14 AWG conductors.

3. Why is it important that the "hot" conductor in a three-wire branch-circuit be properly connected to opposite phases in a panel? _____

   _____

4. In the diagram, Load A is rated at 10 amperes, 120 volts. Load B is rated at 5 amperes, 120 volts.

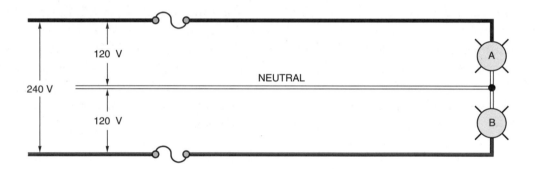

   a. When connected to the three-wire branch-circuit as indicated, how much current will flow in the neutral? _____

   b. If the neutral should open, to what voltage would each load be subjected, assuming both loads were operating at the time the neutral opened? Show all calculations.

5. Calculate the watts loss and voltage drop in each conductor in the following circuit. Show calculations.

```
┌──────────────────────────────────────────────────────────────┐
│         50 FT 12 AWG SOLID COPPER CONDUCTOR                    │
│                                                               ╳
│                                              15 AMPERE      ─(○)─
│         50 FT 12 AWG SOLID COPPER CONDUCTOR                   ╳
└──────────────────────────────────────────────────────────────┘
```

6. Unless specifically designed, all recessed incandescent luminaires (fixtures) must be provided with factory-installed protection (see Unit 7). _____

7. Where in the *NEC*® would you look for the installation requirements regarding flush and recessed lighting luminaires (fixtures)?

_____

8. What is the current draw of the recessed luminaires (fixtures) above the bar? _____

_____

_____

_____

9. a. VA load for Circuit B9 is _____ volt-amperes.

   b. VA load for Circuit B11 is _____ volt-amperes.

   c. VA load for Circuit B12 is _____ volt-amperes.

10. Calculate the total current draw for Circuit B9, Circuit B11, and Circuit B12.

11. Complete the wiring diagram for the recreation room. Follow the suggested cable layout. Use colored pencils or pens to identify the various colors of the conductor insulation.

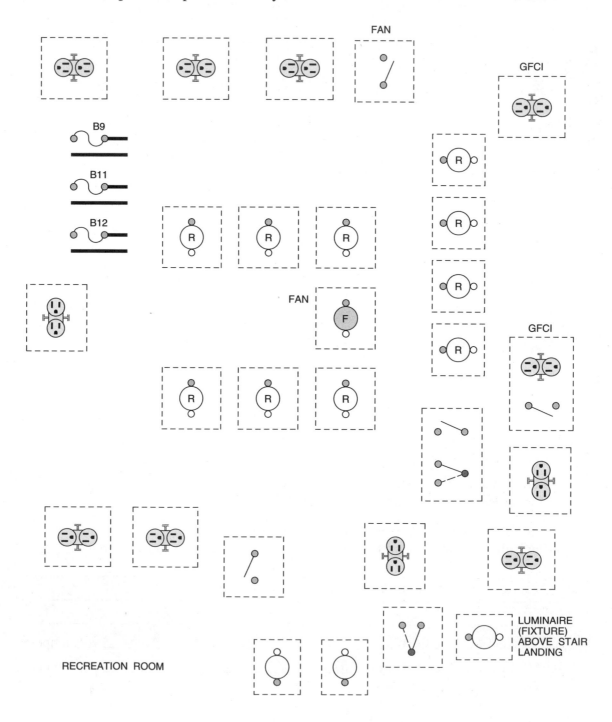

12. a. May a fluorescent luminaire (fixture) that is marked "Recessed Fluorescent Luminaire (Fixture)" be installed in a suspended ceiling? _____

    b. May a fluorescent luminaire (fixture) that is marked "Suspended Ceiling Fluorescent Luminaire (Fixture)" be installed in a recessed cavity of a ceiling? _____

13. The *NEC*® requires GFCI protection for receptacles that serve countertop areas within 6 ft (1.8 m) of the edge of a wet-bar sink, as in the case of the two recreation room receptacles. This requirement is found in *NEC*® _____ .

14. Lay-in fluorescent luminaires (fixtures) that rest on the metal framing members of a suspended ceiling cannot just lay on the framing members. *NEC*® *410.16(C)* requires that (1) framing members used to support recessed luminaires (lighting fixtures) must be securely fastened to each other and to the building structure at appropriate intervals, and (2) the luminaires (fixtures) must _____

    _____ .

Lighting Branch-Circuit, Receptacle Circuits for Workshop

OBJECTIVES

After studying this unit, the student should be able to

- understand the meaning, use, and installation of multioutlet assemblies.
- understand where GFCI protection is required in basements.
- understand the Code requirements for conduit installations.
- select the proper outlet boxes to use for surface mounting.
- make conduit fill calculations based upon the number of conductors in the conduit.
- make use of derating and correction factors for determining conductor current-carrying capacity.

The workshop area is supplied by more than one circuit:

A13 Separate circuit for freezer

A17 Lighting

A18 Plug-in strip (multioutlet assembly)

A20 Two receptacles on window wall

The wiring method in the workshop is EMT. Figure 18-1 shows the conduit layout for the workshop.

A single-pole switch at the entry controls all five ceiling porcelain lampholders. However, note on the plans that three of these lampholders have pull-chains, which would allow the homeowner to turn these lampholders on and off as needed (Figure 18-2). They would not have to be on all of the time, which would save energy.

In addition to supplying the lighting, Circuit A17 also feeds the smoke detectors, chime transformer, and ceiling exhaust fan. The current draw of the

smoke detectors and chime transformer is extremely small. Refer to Table 18-1. Refer to Table 18-2 for estimated loads on Circuit A13, Circuit A18, and Circuit A20.

Smoke detectors are discussed in detail in Unit 26. Smoke detectors are *not* to be connected to

DESCRIPTION	QUANTITY	WATTS	VOLT-AMPERES
Ceiling lights @ 100 W each	5	500	500
Fluorescent luminaires (fixtures) Two 40-W lamps each	2	160	200
Chime transformer	1	8	10
Exhaust fan	1	80	90
Smoke detectors @ ½ W each	4	2	2
TOTALS	13	750	802

Table 18-1 Workshop: outlet count and estimated load. Lighting load (Circuit A17).

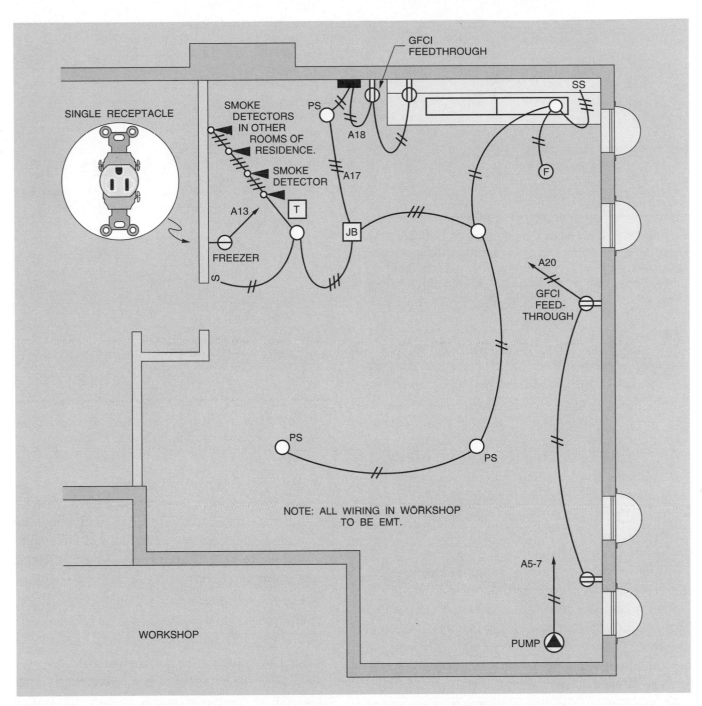

Figure 18-1 Conduit layout for workshop. Lighting is Circuit A17. Freezer single receptacle is Circuit A13. The two receptacles on the window wall are connected to Circuit A20. Plug-in strip receptacles and receptacle to the right of Main Panel are connected to Circuit A18. All receptacles are GFCI protected except the single receptacle outlet serving the freezer. A junction box mounted on the ceiling near the fluorescent luminaires (fixtures) above the workbench will provide a convenient place to make up the necessary electrical connections.

GFCI-protected circuits. Smoke detectors should be connected to a branch-circuit that supplies other lighting outlets so it will be obvious to the homeowner that if the lighting is not working, then the smoke detectors are also not working.

WORKBENCH LIGHTING

Two two-lamp, 40-watt fluorescent luminaires (fixtures) are mounted above the workbench to reduce shadows over the work area. The electrical plans show that these luminaires (fixtures) are controlled

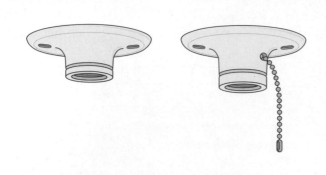

Figure 18-2 Porcelain lampholders of both the key-less and pull-chain types. These lampholders are also available with a receptacle outlet, which must be GFCI protected when installed in locations where GFCI protection is required by the Code.

by a single-pole wall switch. A junction box is mounted immediately above or adjacent to the fluorescent luminaire (fixture) so that the connections can be made readily.

When armored cable or flexible metal conduit is used as the wiring method between the luminaires (fixtures) and the junction box or boxes on the ceiling, either wiring method provides the necessary equipment grounding for the fluorescent luminaires (fixtures), provided the metal outlet box is properly grounded. See *250.118* for the many acceptable means to accomplish equipment grounding.

Many workshop fluorescent luminaires (lighting fixtures) are furnished with a three-wire (black, white, green) flexible cord suitable for the purpose that has the attachment plug cap factory-attached to the cord. These luminaires (fixtures) can be plugged into a grounding-type receptacle mounted on the ceiling immediately above the fluorescent luminaires (fixtures). This provides proper grounding of the luminaire (lighting fixture).

RECEPTACLE OUTLETS

NEC® 210.8(A)(4) and *210.8(A)(5)* require that all 125-volt, single-phase, 15- or 20-ampere receptacles installed in unfinished basements—such as workshop and storage areas that are not considered habitable, and crawl spaces that are at or below ground level—must be GFCI protected.

Exceptions to this requirement are:

- Receptacles that are not readily accessible.

DESCRIPTION	QUANTITY	WATTS	VOLT-AMPERES
20-A Circuit, Number A18			
Receptacles next to main panel	1	180	180
A six-receptacle plug-in multioutlet assembly at 1½ A per outlet (180 VA)	1	1080	1080
TOTALS	2	1260	1260
20-A Circuit, Number A20			
Receptacles on window wall @ 180 W each	2	360	360
TOTALS	2	360	360
20-A Circuit, Number A13			
Single receptacle for freezer (*not* GFCI)	1	—	696
TOTALS	1	1620	2316

Table 18-2 Workshop: outlet count and estimated load for three 20-A circuits A13, A18, A20.

- A single receptacle or a duplex receptacle for two appliances that occupy a dedicated space for each appliance, are cord- and plug-connected, and are not easily moved. Examples include freezers, refrigerators, water softeners, and central vacuum equipment.

- For the receptacle(s) installed for the laundry equipment as required by *210.52(F)* and *210.11(C)(2)*.

See Unit 6 for a full discussion of GFCIs.

NEC® 210.52(G) makes it mandatory to install at least one receptacle outlet in each unfinished basement area. If the laundry area is in the basement, then a receptacle outlet must *also* be installed for the laundry equipment, *210.52(F)*. This additional receptacle must be within 6 ft (1.8 m) of the intended location of the washer, *210.50(C)*, and must be a 20-ampere circuit.

The circuit feeding the required laundry receptacle shall have no other outlets connected to it, *210.11(C)(2)*.

When a single receptacle is installed on an individual branch-circuit, the receptacle rating must not be less than the rating of the branch-circuit, *210.21(B)(1)*.

CABLE INSTALLATION IN BASEMENTS

Unit 4 covers the installation of nonmetallic-sheathed cable (Romex) and armored cable (BX).

Some additional rules apply when nonmetallic-sheathed cable and armored cable are run exposed, because of possible physical damage to the cables.

In unfinished basements, nonmetallic-sheathed cables made up of conductors smaller than two 6 AWG or three 8 AWG must be run through holes bored in joists, or on running boards, *334.15(C)*. In unfinished basements, exposed runs of armored cable must closely follow the building surface or running boards to which it is fastened, *320.15*. *NEC® 320.15* also permits armored cable (BX) to be run on the undersides of joists in basements, attics, or crawl spaces, where supported on each joist and so located as not to be subject to physical damage. This exception is not permitted when using non-metallic-sheathed cable. The local inspection authority usually interprets the meaning of "subject to physical damage."

Review Unit 4 and Unit 15 for additional discussion regarding protection of cables installed in exposed areas of basements and attics.

CONDUIT INSTALLATION IN BASEMENTS

Where metal raceways are used in house wiring, in most cases the raceways will be EMT, *Article 358*, as in Figure 18-3. In some instances, rigid metal conduit (*Article 344*) will be used for a mast service; refer to Unit 27. The following discussion only briefly touches upon metal raceways, because most house wiring across the country is done with non-metallic-sheathed cable, which is covered in detail in Unit 4.

Many local electrical codes do not permit cable wiring in unfinished basements other than where absolutely necessary. For example, a cable may be dropped into a basement from a switch at the head of the basement stairs. Transitions between cable and conduit also must be made. Usually this will be a junction box. The electrician must check the local codes before selecting conduit or cable for the installation, to prevent costly wiring errors.

The electrician should never assume that the *NEC®* is the recognized standard everywhere. Any city or state may pass electrical installation and licensing laws. In many cases these laws are more stringent than the *NEC®.*

Outlet boxes may be fastened to masonry walls with lead or plastic anchors, shields, concrete nails, or power-actuated studs.

Conduit may be fastened to masonry or wood surfaces using straps, such as the one-hole strap illustrated in Figure 18-4.

The conduits on the workshop walls and on the ceiling are exposed to view. The electrician must install the exposed wiring in a neat and skillful manner while complying with the following practices:

- The conduit runs must be straight.

- Bends and offsets must be true.

- Vertical runs down the surfaces of the walls must be plumb.

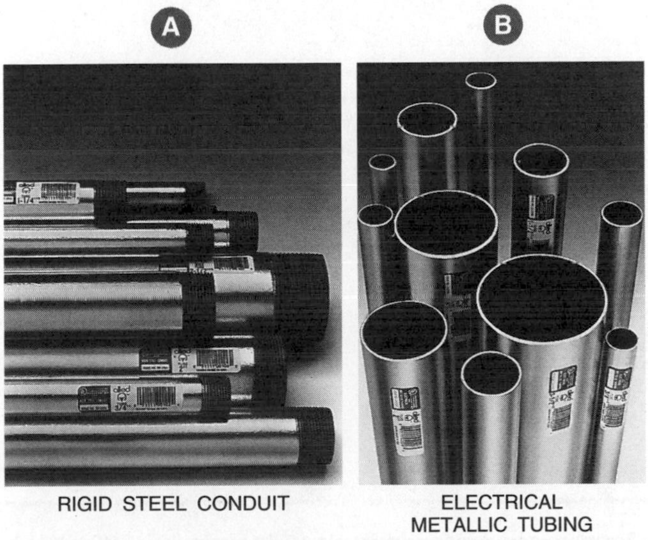

RIGID STEEL CONDUIT · ELECTRICAL METALLIC TUBING

Figure 18-3 Types of electrical raceways. *Courtesy of* **Allied Tube & Conduit.**

Figure 18-4 A one-hole conduit strap.

• To ensure that the conductor insulation will not be injured, be sure to ream the cut ends of rigid conduit and electrical metallic tubing to remove rough edges and burrs.

All conduits and boxes on the workshop ceiling are fastened to the underside of the wood joists, as in Figure 18-5. Raceways shall be installed as a complete system between pull points before pulling in the wires, as shown in Figure 18-6.

If the electrician can get to the residence before the basement concrete floor is poured, he might find it advantageous to install conduit runs in or under the concrete floor from the main panel to the freezer outlet, the window wall receptacles, and the water pump disconnect switch location. This can be a cost-saving (labor and material) benefit.

Metal raceways and metal cables (BX) are acceptable equipment grounding conductors, *250.118.* The runs would have to be continuous between other effectively grounded boxes, panels, and/or equipment. When a nonmetallic wiring method (NM, NMC, UF, etc.) is installed where short sections of metal raceway are provided for mechanical protection—or for appearance—these short sections of raceway need not be grounded, *250.86, Exception No. 2.*

Outlet Boxes for Use in Exposed Installations

Figure 18-7 illustrates outlet boxes and plaster rings that are used for concealed wiring, such as

for the recreation room and first floor wiring in the residence. For surface wiring such as the workshop, various types and sizes of outlet boxes may be used. For example, a handy box, Figure 18-8, could be used for the freezer receptacle. The remaining workshop boxes could be 4-in. square outlet boxes with raised covers, Figure 18-9. The box size and type is determined by the maximum number of conductors contained in the box, *Article 314.*

NEC® 406.4(C) prohibits mounting and supporting a receptacle to a cover (flat or raised) by only one screw unless the device assembly or box cover is listed and identified for securing by a single screw.

The listing and identifying for the purpose is a decision that is made by an NRTL, such as Underwriters Laboratories, Inc. In the past, securing a receptacle with a single screw often resulted in a loose receptacle when used over a long period of time. Many short-circuits and ground-faults have been reported because of this.

For the typical raised cover, as illustrated in Figure 18-9, the duplex receptacle would be fastened to the raised cover by two small No. 6-32 screws and nuts, usually furnished with the raised cover. The raised cover has two small "knock-out" holes for these screws. The center No. 6-32 screw could also be used.

▶If a receptacle is installed on a surface-mounted box where there is direct metal-to-metal contact between the grounded metal box and the metal yoke

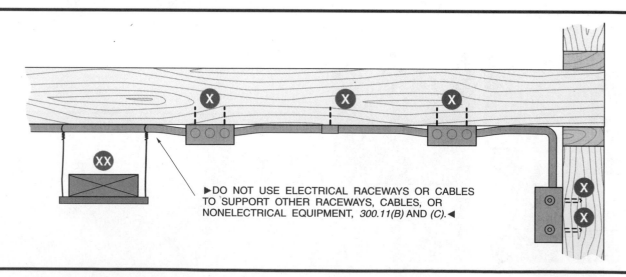

▶DO NOT USE ELECTRICAL RACEWAYS OR CABLES TO SUPPORT OTHER RACEWAYS, CABLES, OR NONELECTRICAL EQUIPMENT, *300.11(B)* AND *(C).*◀

Figure 18-5 Raceways, boxes, fittings, and cabinets must be securely fastened in place (points *X*), *300.11(A).* ▶It is not permissible to hang other raceways, cables, or nonelectrical equipment from an electrical raceway or cable (point *XX*), *300.11(B)* and *(C).*◀ Refer to Figure 24-12 for an exception that allows low-voltage thermostat cables to be fastened to the conduit that feeds a furnace or other similar equipment. See Unit 17 for discussion of the use of suspended-ceiling support wire for supporting equipment.

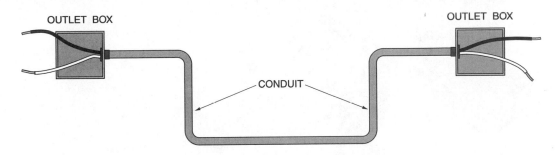

INSTALL CONDUIT COMPLETELY BETWEEN BOXES BEFORE PULLING IN CONDUCTORS, *300.18(A)*. NO MORE THAN THE EQUIVALENT OF FOUR QUARTER BENDS (360° TOTAL), IS PERMITTED BETWEEN PULL POINTS. THIS RULE IS FOUND IN *SECTION 26* IN *ARTICLES 342, 344, 348, 352, 358, 350, 356*, AND *362*.

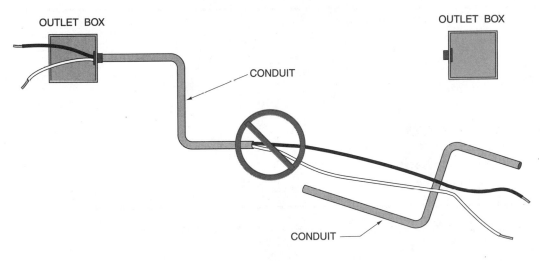

DO NOT PULL CONDUCTORS INTO A PARTIALLY INSTALLED CONDUIT SYSTEM, AND THEN ATTEMPT TO SLIDE THE REMAINING SECTION OF RACEWAY OVER THE CONDUCTORS. THIS IS A VIOLATION OF *300.18(A)*.

Figure 18-6 Do not pull wires into the raceway until the conduit system is completely installed between pull points.

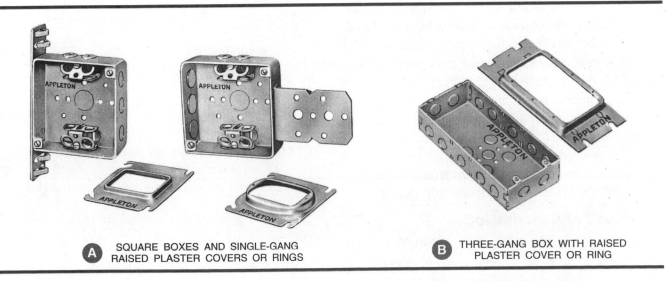

A SQUARE BOXES AND SINGLE-GANG RAISED PLASTER COVERS OR RINGS

B THREE-GANG BOX WITH RAISED PLASTER COVER OR RING

Figure 18-7 Typical outlet boxes and raised covers. *Courtesy of* Appleton Electric Company.

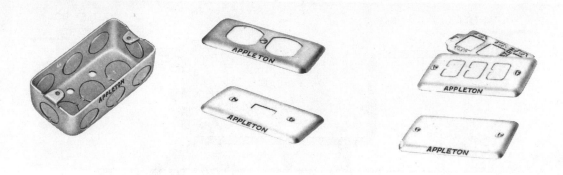

Figure 18-8　Handy box and covers. *Courtesy of* Appleton Electric Company.

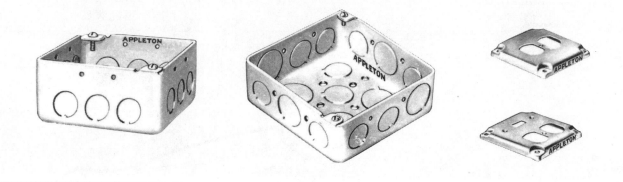

Figure 18-9　Four-in. square boxes and raised covers. *Courtesy of* Appleton Electric Company.

of the receptacle, the receptacle is considered to be properly grounded without connecting an additional equipment grounding conductor to the equipment grounding terminal on the receptacle, *250.146.*◀ Be sure to remove the small cardboards from the No. 6-32 screws to attain metal-to-metal contact.

Most commonly used outlet, device, and junction boxes have knockouts for trade size ½, ¾, and 1 connectors. The knockout size selected depends upon the size of the conduits entering the box. The conduit size is determined by the maximum percentage of fill of the conduit's total cross-sectional area by the conductors pulled into the conduit.

Supporting equipment and/or raceways to the joists, walls, and so on, is accomplished with proper nails, screws, anchors, and straps. Be careful when trying to support other equipment from raceways as this is not always permitted. Refer to Figure 18-5.

Conduit Fill Calculations

For simplicity, the following conduit fill calculations show in. and in.² To show calculations in in., in.², mm, and mm² would be very confusing.

An electrician must be able to select the proper size raceways based upon the number, size, and type of conductors to be installed in a particular type of raceway. To do this, the electrician must become familiar with the use of certain tables found in *Chapter 9* and *Annex C* of the *NEC*.®

Percent Fill.　The *NEC*® has established the following "percent fill" valves for conduit and tubing.

- one conductor . . . . 53% fill
- two conductors . . . . 31% fill
- three or more conductors . . . . . 40% fill
- conduit or nipples not over
  24 in.(600 mm) . . . . 60% fill
  Derating of conductor ampacity not required.

These percentages are taken from *Chapter 9, Table 1* of the *NEC*.®

All Conductors the Same Size and Type. When all conductors are the same size and type:

- Determine how many conductors of one size and type are to be installed in the raceway.

- Refer to *Tables C1* through *C12A* in the Annex of the *NEC®* for the type of conduit, tubing, or raceway, and for the specific size and type of conductors.

- Select the proper size raceway directly from the table.

These tables are not difficult to use. Here are a few examples of conductor fill for EMT. Look these up in the tables to confirm the accuracy of the examples.

- Not over five 10 AWG THHN conductors may be pulled into a trade size ½ EMT.

- Not over eight 12 AWG THHN conductors may be pulled into a trade size ¾ EMT.

- Not over three 8 AWG THHN conductors may be pulled into a trade size ½ EMT.

- Not over six 8 AWG THHN conductors may be pulled into a trade size 1 EMT.

Conductors of Different Sizes and/or Types. When conductors are different sizes or types:

- Refer to *Chapter 9, Table 5* of the *NEC®*.

- Find the size and type of conductors used to determine the approximate cross-sectional area in in.^{2} of the conductors.

TYPE	SIZE	APPROX. DIAM. IN.	APPROX. AREA IN.^{2}
THHN	14	0.111	0.0097
THHN	12	0.130	0.0133
THHN	10	0.164	0.0211
THHN	8	0.216	0.0366
THHN	6	0.254	0.0507
THHN	3	0.352	0.0973
THW	6	0.304	0.0726
THW	8	0.266	0.0556

Table 18-3 This sampling of typical conductors and their dimensions is taken from *Table 5, Chapter 9, NEC*.® See *Table 5* for metric values.

- Refer to *Chapter 9, Table 8* for the in.^{2} area of bare conductors.

- Total the in.^{2} areas of all conductors.

- Refer to *Chapter 9, Table 4* for the type and size of raceway used.

- Determine the minimum size raceway based on the total in.^{2} area of the conductors and the allowable in.^{2} area fill for the particular type of raceway used.

- The sum must not exceed the allowable percentage fill of the cross-sectional in.^{2} area of the raceway used.

For illustrative purposes, only the data for a few types of wires from *Table 5* are shown in Table 18-3. The dimensions and areas for some EMT from *Table 4* are shown in Table 18-4. You will need to

		ELECTRICAL METALLIC TUBING			
Trade Size (in.)	Internal Diameter (in.)	Total Area 100% (in.^{2})	2 Wires 31% (in.^{2})	Over 2 Wires 40% (in.^{2})	1 Wire 53% (in.^{2})
½	0.622	0.304	0.094	0.122	0.161
¾	0.824	0.533	0.165	0.213	0.283
1	1.049	0.864	0.268	0.346	0.458
1¼	1.380	1.496	0.464	0.598	0.793
1½	1.610	2.036	0.631	0.814	1.079
2	2.067	3.356	1.040	1.342	1.778
2½	2.731	5.858	1.816	2.343	3.105
3	3.356	8.846	2.742	3.538	4.688
3½	3.834	11.545	3.579	4.618	6.119
4	4.334	14.753	4.573	5.901	7.819

Table 18-4 Dimensional and percent fill data for EMT. This is one of the twelve charts found in *Chapter 9, Table 4, NEC*,® for various types of raceways. See *Table 4* for metric values.

refer to *Chapter 9* and *Annex C* of the *NEC®* for the many other sizes and types of conductors and raceways.

EXAMPLE: If three 6 AWG THW conductors and two 8 AWG THW conductors are to be installed in one EMT, determine the proper size EMT.

SOLUTION:

1. Find the in.^{2} area of the conductors in *Chapter 9, Table 5*.

Three 6 AWG THW conductors	0.0726 in.^{2}
	0.0726 in.^{2}
	0.0726 in.^{2}
Two 8 AWG THW conductors	0.0556 in.^{2}
	0.0556 in.^{2}
Total Area	0.3290 in.^{2}

2. In *Chapter 9, Table 4, NEC®* look at the fifth column (over 2 wires, 40% fill) for an EMT that has an area of 0.3290 in.^{2} or more. Note that a trade size ¾ EMT holds up to 0.213 in.^{2} of conductor fill, and a trade size 1 EMT holds up to 0.346 in.^{2} of conductor fill. Therefore, a trade size 1 EMT is the minimum size for the combination of conductors in the example.

Many electrical ordinances require that a separate equipment grounding conductor be installed in all raceways, whether rigid or flexible, whether metallic or nonmetallic. This requirement resulted from inspectors finding loose or missing set screws on set-screw type connectors and couplings. Corrosion of fittings also contributes to poor grounding. A separate equipment grounding conductor does take up space, and therefore must be counted when considering conduit fill.

EXAMPLE: A feeder to an electric furnace is fed with two 3 AWG THHN conductors plus one green insulated equipment grounding conductor. This feeder is installed in EMT. The feeder overcurrent protection is 100 amperes. What is the minimum size EMT to be installed?

SOLUTION: According to *Table 250.122*, the minimum size equipment grounding conductor for a 100-ampere feeder is 8 AWG copper. Find the in.^{2} area of the conductors in *Table 5, Chapter 9, NEC®*:

Two 3 AWG THHN	0.0973 in.^{2}
	0.0973 in.^{2}
One 8 AWG THHN	0.0366 in.^{2}
Total Area	0.2312 in.^{2}

From *Table 4, Chapter 9, NEC®*:

40% fill for trade size ¾ EMT = 0.213 in.^{2}
40% fill for trade size 1 EMT = 0.346 in.^{2}

Therefore, install trade size 1 EMT.

Remember, *300.18(A)* states that a conduit system must be completely installed before pulling in the conductors. Refer to Figure 18-6.

Conduit Bodies *[314.16(C)]*

Conduit bodies are used with conduit installations to provide an easy means to turn corners, terminate conduits, and mount switches and receptacles. They are also used to provide access to conductors, to provide space for splicing (when permitted), and to provide a means for pulling conductors (Figure 18-10).

A conduit body:

- must have a cross-sectional area not less than twice the cross-sectional area of the largest con-

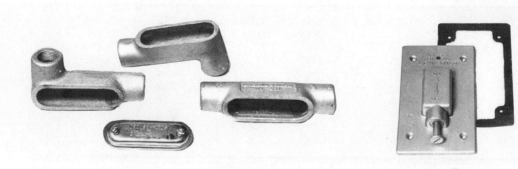

Figure 18-10   A variety of common conduit bodies and covers.

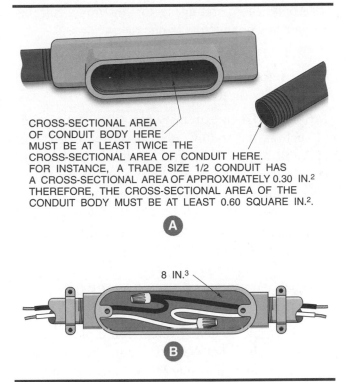

CROSS-SECTIONAL AREA
OF CONDUIT BODY HERE
MUST BE AT LEAST TWICE THE
CROSS-SECTIONAL AREA OF CONDUIT HERE.
FOR INSTANCE, A TRADE SIZE 1/2 CONDUIT HAS
A CROSS-SECTIONAL AREA OF APPROXIMATELY 0.30 IN.²
THEREFORE, THE CROSS-SECTIONAL AREA OF THE
CONDUIT BODY MUST BE AT LEAST 0.60 SQUARE IN.².

A

8 IN.³

B

Figure 18-11   Requirements for conduit bodies.

NUMBER OF CURRENT-CARRYING CONDUCTORS	PERCENT OF VALUES IN TABLES. IF HIGH AMBIENT TEMPERATURES ARE PRESENT, ALSO APPLY CORRECTION FACTORS FOUND AT THE BOTTOM OF *Table 310.16.*
4 through 6	80
7 through 9	70
10 through 20	50
21 through 30	45
31 through 40	40
41 and above	35

Table 18-5   Derating factors necessary to determine conductor ampacity when there are more than three current-carrying conductors in a raceway or cable. See *310.15(B)(2)(a)*.

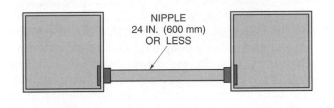

NIPPLE
24 IN. (600 mm)
OR LESS

Figure 18-12   ▶The derating factors in Table 18-5 are not required for short lengths of conduit that do not exceed 24 in. (600 mm), *310.15(B)(2)(a), Exception No. 3*. Conductor fill not to exceed 60 percent according to *Chapter 9, Table 1, Note 4*.◀

duit to which it is attached, as shown in Figure 18-11(A). This is a requirement only when the conduit body contains 6 AWG conductors or smaller.

- may contain the same maximum number of conductors permitted for the size raceway attached to the conduit body.

- must not contain splices, taps, or devices unless the conduit body is marked with its in.³ capacity, as in Figure 18-11(B), and the conduit body must be supported "in a rigid and secure manner."

- the conductor fill volume must be properly calculated according to *Table 314.16(B)*, which lists the free space that must be provided for each conductor within the conduit body. If the conduit body is to contain splices, taps, or devices, it must be supported rigidly and securely.

DERATING FACTORS

For more than three current-carrying wires in conduit or cable, refer to Table 18-5.

NEC® 310.15(B)(2)(a) states that for the conductors listed in the tables, the maximum allowable

ampacity must be reduced when more than three current-carrying conductors are installed in a raceway. The reduction factors based on the number of conductors are noted in Table 18-5. When the conduit nipple does not exceed 24 in. (600 mm), the reduction factor as shown in Figure 18-12 does not apply.

According to *310.15(B)(4)(a)*, a neutral conductor of a multiwire branch-circuit carrying only unbalanced currents shall not be counted in determining current-carrying capacities, as shown in Figure 18-13.

EXAMPLE: What is the correct ampacity for size 6 AWG THHN when there are six current-carrying conductors in the conduit?

$$75 \times 0.80 = 60 \text{ amperes}$$

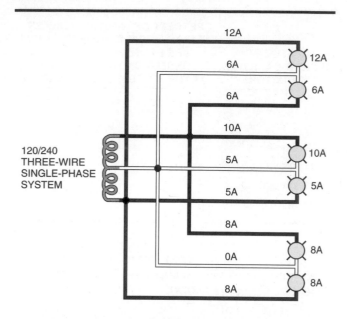

Figure 18-13 A neutral conductor carrying the unbalanced currents from the other conductors need *not* be counted when determining the ampacity of the conductor, according to *310.15(B)(4)(a)*. The Code would recognize this example as *two current-carrying conductors*. Thus, in the three multiwire circuits shown, the actual total is nine conductors, but only six current-carrying conductors are considered when the derating factor of Table 18-5 is applied.

CORRECTION FACTORS (DUE TO HIGH TEMPERATURES)

When conductors are installed in locations where the temperature is higher than 86°F (30°C), the correction factors noted below *Table 310.16* must apply.

For example, consider that four current-carrying 3 AWG THWN copper conductors are to be installed in one raceway in an ambient temperature of approximately 90°F (32°C). "Ambient" temperature means "surrounding" temperature. This temperature is frequently exceeded in attics where the sun shines directly onto the roof.

The maximum ampacity for these conductors is determined as follows:

- The allowable ampacity of 3 AWG THWN copper conductors from *Table 310.16* = 100 amperes.

- Apply the correction factor for 90°F (32°C) found below *Table 310.16*. The first column shows Celsius; the last column shows Fahrenheit.

$$100 \times 0.94 = 94 \text{ amperes}$$

- Apply the derating factor for four current-carrying conductors in one raceway.

$$94 \times 0.80 = 75.2 \text{ amperes}$$

- Therefore, 75.2 amperes is the new corrected ampacity for the example given.

Some AHJs in extremely hot areas of the country, such as the southwestern desert climates, will require that service-entrance conductors running up the side of a building or above a roof exposed to direct sunlight be "corrected" per the correction factors below *Table 310.16*. Check this out before proceeding with an installation where this situation might be encountered.

OVERCURRENT PROTECTION FOR BRANCH-CIRCUIT CONDUCTORS

The basic rule is that overcurrent protection for conductors shall not be more than the allowable ampacity of the conductor, *240.4*. Allowable ampacities for conductors are found in *Table 310.16*.

If the allowable ampacity of a conductor does not match a standard ampere rating of a fuse or circuit breaker, *240.4(B)* permits the use of the next higher rated fuse or circuit breaker, 800 amperes or less. For example, a conductor having an ampacity of 55 amperes could be protected by a 60-ampere overcurrent device. This permission does not apply if the conductors of the branch-circuit supply a number of receptacles for cord- and plug connection of portable loads because it would be impossible to know just how much load might be plugged in.

Small size branch-circuit conductors must be protected according to the values shown in Table 18-6. This table is based on *240.4(D)*.

An exception to this basic maximum size overcurrent protection rule is the protection of motor branch-circuit conductors. For motor branch-circuits, the maximum size overcurrent protection values are found in *Table 430.52*.

CONDUCTOR SIZE (Copper)	MAXIMUM AMPERE RATING OF OVERCURRENT DEVICE
14 AWG	15 amperes
12 AWG	20 amperes
10 AWG	30 amperes

Table 18-6 Maximum size overcurrent protection for small size conductors, *240.4(D)*.

If the allowable ampacity of a given conductor is *derated* and/or *corrected*, the overcurrent protection must be based on the new *derated* or *corrected* ampacity.

BASIC CODE CONSIDERATIONS FOR CONDUCTOR SIZING AND OVERCURRENT PROTECTION

The following is a quick review of Code requirements for conductor sizing and conductor overcurrent protection encountered in house wiring.

- *210.3:* The rating of a branch-circuit is based on the rating of the overcurrent device.

- *210.19(A):* Branch-circuit conductors shall have an ampacity not less than the maximum load to be served.

- *210.20:* An overcurrent device shall not be less than 125% for continuous loads on branch-circuits.

- *210.20(B):* This section refers us to *240.4*, where we find that overcurrent protection for conductors shall not exceed the ampacity of a conductor. See *240.4* for exceptions such as "next standard size permitted" if the OCD and the conductor ampacity do not match.

- *210.23:* In no case shall the load exceed the branch-circuit ampere rating.

- *215.2(A):* Feeder conductors shall have an ampacity not less than required to supply the load.

- *215.3:* An overcurrent device shall not be less than 125% for continuous loads on feeders. This section refers us to *Part 1* of *240.4*, where we find that overcurrent protection for feeders shall not exceed the ampacity of a conductor. Refer to *240.4* for specific requirements and exceptions to this basic rule.

- *220.4:* The total load shall not exceed the rating of the branch-circuit.

- *240.4:* This section contains all of the overcurrent protection requirements for conductors.

- *230.42:* An overcurrent device shall not be less than 125% for continuous loads on services.

- *424.3(B):* For fixed electric space heating equipment such as an electric furnace, the branch-circuit conductors and the over-current protective devices shall not be less than 125 percent of the total load of the motor(s) and heater(s).

- *426.4:* For fixed outdoor electric deicing and snow melting equipment such as heating cables buried in the concrete of a driveway, the branch-circuit conductors and the overcurrent protective devices shall not be less than 125 percent of the total load of the heaters.

- *430.22:* Branch-circuit conductors for motors shall not be less than 125 percent of the motor's full-load current. Motor branch-circuit short-circuit and ground-fault protection is sized according to the percentages listed in *Table 430.52*.

- *Article 440, Part III:* Branch-circuit short-circuit and ground-fault protection for air-conditioning equipment is covered in Unit 23.

- *Article 440, Part IV:* Conductor sizing for air-conditioning equipment is covered in Unit 23.

A few of the above Code sections make reference to *continuous loads*. The Code defines a continuous load as: *"where the maximum current is expected to continue for three hours or more."*

Loads in homes are generally not considered to be continuous. However, an electric furnace or snow melting heating cables buried in the concrete of a driveway can be classified as a continuous load, because they could be on for periods of three hours or more. The 125 percent factor for these has already been included in *424.3(B)* and *426.4*.

Many loads in commercial and industrial installations can be considered to be continuous loads. For example, an electric range in a restaurant might be used continuously 24 hours a day, seven days a week. An electric range in a home would not be used in this manner.

For continuous loads, the overcurrent protective device and the conductors must not be less than 125 percent of the continuous load. For example, a continuous load is 40 amperes. The minimum overcurrent device rating and the minimum conductor ampacity is $40 \times 1.25 = 50$ amperes.

For combination continuous and noncontinuous loads, the overcurrent protective device and the conductors must not be less than 125 percent of the continuous load plus the noncontinuous load. For example, a feeder supplies a continuous load of 40 amperes and a noncontinuous load of 30 amperes. The minimum overcurrent device rating and the minimum conductor ampacity is $(40 \times 1.25) + 30 = 80$ amperes.

EXAMPLES OF DERATING, OVERCURRENT PROTECTION, AND CONDUCTOR SIZING

To illustrate what has been learned so far about derating and maximum overcurrent protection, suppose that eight THW copper conductors are installed in one raceway. All are considered to be current-carrying. The load on each conductor is 18 amperes. Let us consider the possibilities.

- 14 AWG THW copper conductors
  a. The maximum overcurrent protection from Table 18-6 is 15 amperes.
  b. The ampacity of 14 AWG THW copper conductors from *Table 310.16* is 20 amperes.
  c. Apply the derating factor (8 current-carrying wires in the raceway) from Table 18-5.

    $$20 \times 0.70 = 14 \text{ amperes}$$
    $$\text{(derated ampacity)}$$

    Thus, 14 AWG THW conductors are *too small* to supply the 18-ampere load.

- 12 AWG THW copper conductors
  a. The maximum overcurrent protection from Table 18-6 is 20 amperes.
  b. The ampacity of 12 AWG THW copper conductors from *Table 310.16* is 25 amperes.
  c. Apply the derating factor (8 current-carrying conductors in the raceway) from Table 18-5.

    $$25 \times 0.70 = 17.5 \text{ amperes}$$
    $$\text{(derated ampacity)}$$

    Thus, 12 AWG THW conductors are *too small* to supply the 18-ampere load.

- 10 AWG THW copper conductors
  a. The maximum overcurrent protection from Table 18-6 is 30 amperes.
  b. The ampacity of 10 AWG THW copper wire from *Table 310.16* is 35 amperes.

  c. Apply the derating factor (8 current-carrying conductors in the raceway) from Table 18-5.

    $$35 \times 0.70 = 24.5 \text{ amperes}$$
    $$\text{(derated ampacity)}$$

    Thus, 10 AWG THW copper conductors are adequate to supply the 18-ampere load.

MULTIOUTLET ASSEMBLY

A multioutlet assembly has been installed above the workbench. For safety reasons, as well as for adequate wiring, it is recommended that any workshop receptacles be connected to a separate circuit. In the event of a malfunction in any power tool commonly used in the home workshop (saw, planer, lathe, or drill), only receptacle Circuit A18 is affected. The lighting in the workshop is not affected by a power outage on the receptacle circuit.

The GFCI feedthrough receptacle to the left of the workbench provides GFCI protection for the entire plug-in strip, *210.8(A)(5)*, as shown in Figure 18-14.

The installation of multioutlet assemblies must conform to the requirements of *Article 380* of the Code.

Load Considerations for Multioutlet Assemblies

NEC® 220.3(A) gives the general lighting loads in volt-amperes per ft² (0.093 m²) for types of occupancies. *NEC® 220.3(B)(10)* indicates that *all* general-use receptacle outlets in one, two, and multifamily dwellings are to be considered outlets for general lighting, and as such have been included in the volt-amperes per ft² calculations. Therefore, no additional load need be added.

Because the multioutlet assembly in Figure 18-15 has been provided in the workshop for the purpose of plugging in portable tools, a separate circuit, A18, is provided. This circuit has been shown in the service-entrance calculations at 1500 volt-amperes, similar to the load requirements for small appliance circuits.

Many types of receptacles are available for various sizes of multioutlet assemblies. For example, duplex, single-circuit grounding, split-circuit grounding, and duplex split-circuit receptacles can be used with multioutlet assemblies.

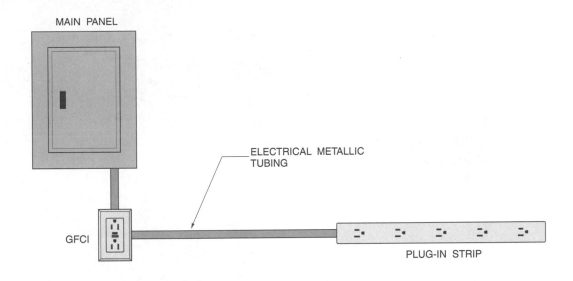

Figure 18-14 Detail of how the workbench receptacle plug-in strip is connected by feeding through a GFCI feedthrough receptacle located below the Main Panel. This is Circuit A18.

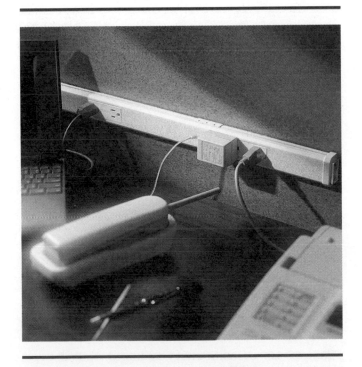

Figure 18-15 A typical metal multioutlet assembly. Multioutlet assemblies are permitted to be run through (not within) a dry partition when no outlets are located within the partition, and all exposed portions of the assembly can have the cover removed. Nonmetallic multioutlet assemblies are also available. *Courtesy of* The Wiremold Company.

Wiring a Multioutlet Assembly

A multioutlet assembly can be connected by bringing the supply into one end as in Figure 18-14, or into the back of the assembly as in Figure 18-16.

Receptacles for multioutlet assemblies can be spaced and wired on the job, in which case the covers are cut to the desired length. Receptacles can also be factory prewired, with six receptacles in a 3-ft (900-mm) long strip, and ten receptacles in a 5-ft (1.5-m) long strip.

For other strip lengths, standard spacing of receptacles for precut assemblies are 6 ft (1.8 m), 9 in. (225 mm), 12 in. (300 mm), or 18 in. (450 mm). These measurements are center to center. Prewired multioutlet assemblies are generally wired with 12 AWG THHN conductors, and have either 15- or 20-ampere receptacles, depending upon how they are specified and ordered.

Multioutlet assemblies are not difficult to install. There is a wide variety of fittings, such as connectors, couplings, ground clamps, blank and end fittings, elbows for turning corners (flat, inside, outside, twist), flush plate adapters, starter boxes, raceway adapters, mounting clips, and so on, to fit just about any application.

Figure 18-16 A multioutlet assembly back plate with the armored cable supply feeding into the back of the assembly.

Figure 18-17 A cord-connected, prewired multioutlet assembly installed above a workbench. *Courtesy of The Wiremold Company.*

Figure 18-17 shows a cord-connected multioutlet assembly of the type used in workshops or other locations where there is a need for the convenience of having a number of receptacles in close proximity. These prewired assemblies are available with six receptacles and a circuit breaker, 40 in. long, or ten receptacles and a circuit breaker, 52 in. long. These can also be used under kitchen cabinets. Figure 18-16 shows a multioutlet assembly back plate with the electrical supply feeding into the back of the assembly. Figure 18-18 shows a multioutlet assembly above the backsplash of a kitchen countertop.

EMPTY CONDUITS

Although not truly part of the workshop wiring, *Note 8* to the first-floor electrical plans indicates that two empty trade size 1 EMT raceways are to be installed, running from the workshop to the attic. This installation is unique, but certainly welcomed if at some later date additional wiring for telephone, security, computer, structured wiring, home automation cables, or other similar future needs might be required. The empty raceways will provide the

Figure 18-18 A multioutlet assembly installed above the backsplash of a kitchen countertop. *Courtesy of The Wiremold Company.*

necessary route from the basement to the attic, thus eliminating the need to fish through the wall partitions.

REVIEW

1. a. What circuit supplies the workshop lighting? _____

   b. What circuit supplies the plug-in strip over the workbench? _____

   c. What circuit supplies the freezer receptacle? _____

2. Approximately how much trade size ½ EMT is used for connecting Circuit A13, Circuit A17, Circuit A18, and Circuit A20? _____

3. a. How is EMT fastened to masonry? _____

   _____

   b. The cut ends of rigid and electrical metallic tubing must be _____ to prevent the insulation on conductors from being damaged.

4. What type of luminaires (lighting fixtures) are installed on the workshop ceiling?

   _____

   _____

5. A check of the plans indicates that two empty trade size 1 EMT conduits are installed between the basement and the attic. In your opinion, is this is a good idea? Briefly explain your thoughts. _____

   _____

   _____

   _____

   _____

   _____

   _____

6. To what circuit are the smoke detectors connected? _____

7. Is it a Code requirement to connect smoke detectors to a GFCI-protected circuit?

   _____

8. When a freezer is plugged into a receptacle, be sure that the receptacle is protected by a GFCI. Is this statement true or false? Explain.

   _____

   _____

   _____

   _____

   _____

9. A sump pump is plugged into a single receptacle outlet. This receptacle shall

   a. be GFCI protected

   b. not be GFCI protected

   Circle the correct answer.

10. When a 120-volt receptacle outlet is provided for the laundry equipment in an unfinished basement, what is the *minimum* number of additional receptacle outlets required in that basement? What Code reference? _____

    _____

11. Derating factors for conductor ampacities must always be taken into consideration when (a) _____ are contained in one raceway. Correction factors are applied when (b) _____ .

12. What current is flowing in the neutral conductor at:

    a. ____ amperes?   b. ____ amperes?   c. ____ amperes?   d. ____ amperes?

13. When a single receptacle is installed on an individual branch-circuit, the receptacle must have a rating _____ than the rating of the branch-circuit.

14. Calculate the total current draw of Circuit A17 if all luminaires (lighting fixtures) and the exhaust fan were turned on.

15. A conduit body must have a cross-sectional area not less than (two) (three) (four) times the cross-sectional area of the (largest) (smallest) conduit to which it is attached. Circle correct answers.

16. A conduit body may contain splices if marked with:

    a. the UL logo.

    b. its in.^{3} area.

    c. the size of conduit entries.

    Circle the correct answer.

17. What size box would you use where Circuit A17 enters the junction box on the ceiling? _____

    _____

18. List the proper trade size of electrical metallic tubing for the following. Assume that it is new work and that THHN insulated conductors are used. See *Chapter 9, Table C1, NEC.®*

    a. Three 14 AWG _____   f. Four 12 AWG _____

    b. Four 14 AWG _____   g. Five 12 AWG _____

    c. Five 14 AWG _____   h. Six 12 AWG _____

    d. Six 14 AWG _____   i. Six 10 AWG _____

    e. Three 12 AWG _____   j. Four 8 AWG _____

19. According to the Code, what trade size EMT is required for each of the following combinations of conductors? Assume that these are new installations. Show all calculations. Refer to *Chapter 9*, *NEC,® Table 4*, *Table 5*, and *Table 8*.

    a. Three 14 AWG THHN, four 12 AWG THHN

    b. Two 12 AWG TW, three 8 AWG TW

    c. Three 3/0 AWG THHN, two 8 AWG THHN

20. a. When more than three current-carrying conductors are installed in one raceway, their allowable ampacities must be reduced according to *NEC®* _____

    b. Four 10 AWG THW current-carrying conductors are installed in one raceway. The room temperature will not exceed 86°F (30°C). The connected load is nonmotor, non-continuous. (1) What is the ampacity of these conductors before derating? (2) What is the maximum overcurrent protection permitted to protect these conductors? (3) What is the ampacity of these conductors after derating? _____

    _____

    _____

    _____

    _____

    _____

    c. Why is it not necessary to count equipment grounding conductors and/or all of the "travelers" when installing three-way and four-way switches for the purposes of derating? _____

    _____

    _____

    _____

21. What is the minimum number of receptacles required by the Code for basements?

    _____

22. For laundry equipment in basements, what sort of electrical circuitry is required by the Code? _____

    _____

    _____

    _____

    _____

23. How many receptacles are required to complete the multioutlet assembly above the workbench? _____

24. A Type NM cable carries a branch-circuit to the outside of a building, where it terminates in a metal weatherproof junction box. From this box, a short length of EMT is installed upward to another metal weatherproof junction box on which is attached a cluster of outdoor flood lights. The first box is located approximately 24 in. (600 mm) above the ground. To "meet Code," do the metal boxes and metal raceway have to be grounded?   Yes _____   No _____   *NEC®* _____ .

25. When raised covers are used for receptacles, does the Code permit fastening the receptacle to the cover with the one center No. 6-32 screw? Yes _____ No _____
    NEC® _____ .

26. The following is a layout of the lighting circuits for the workshop. Using the conduit layout in Figure 18-1, make a complete wiring diagram. Use colored pencils or pens to indicate conductors.

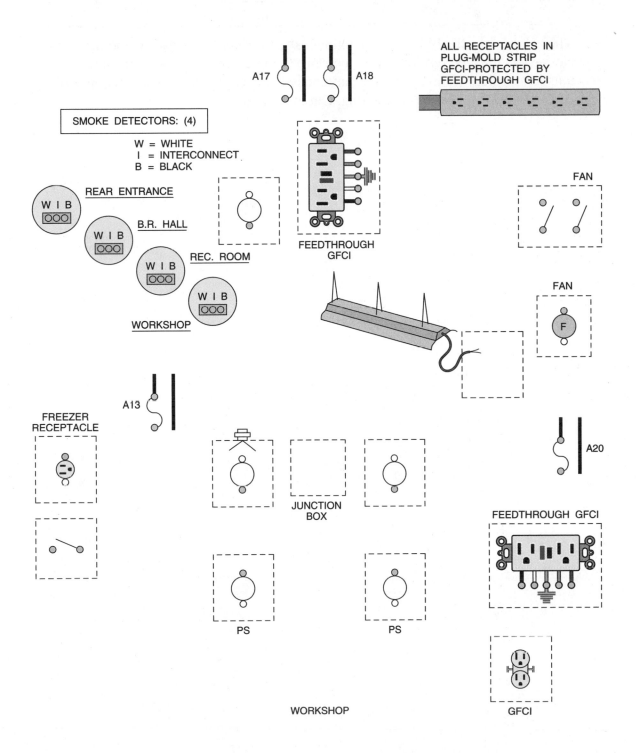

27. *NEC® 210.8(A)(5)* requires that GFCI protection be provided for all receptacles in unfinished basements. There are some exceptions. What are these exceptions?

_____

_____

_____

28. Where loads are likely to be on continuously, the *computed* load for branch-circuits and feeders must be figured at (100 percent) (125 percent) (150 percent) of the continuous load, (plus) (minus) the noncontinuous load. This requirement generally (does) (does not) apply to residential wiring. Circle the correct answers.

Special-Purpose Outlets–
Water Pump, Water Heater

OBJECTIVES

After studying this unit, the student should be able to

- list the requirements for deep-well jet pump and submersible pump installations.
- calculate (given the rating of the motor) the conductor size, conduit size, and overcurrent protection required for pump circuits.
- list some of the electrical circuits used for connecting and metering electric water heaters.
- describe the basic operation of water heaters.
- understand the meaning of the term "Time-of-Use."
- list the functions of the tank, the heating elements, and the thermostat for water heaters.
- understand "recovery" time for electric water heaters.
- make the proper electrical and grounding connections for water heaters, water pumps, and metal well casings.
- understand the hazards of possible scalding.
- discuss heat pump water heaters.

WATER PUMP CIRCUIT Ⓐ_B

All dwellings need a good water supply. City dwellings generally are connected to the city water system. In rural areas, where there is no public water supply, each dwelling has its own water supply system in the form of a well. The residence plans show that a deep-well jet pump is used to pump water from the well to the various plumbing outlets. The outlet for this jet pump is shown by the symbol Ⓐ_B. An electric motor drives the jet pump. A circuit must be installed for this system.

JET PUMP OPERATION

Figure 19-1 shows the major parts of a typical jet pump. The pump impeller wheel ① forces water

down a drive pipe ② at a high velocity and pressure to a point just above the water level in the well casing. Some well casings may be driven to a depth of more than 100 ft (30 m) before striking water. Just above the water level, the drive pipe curves sharply upward and enters a larger vertical suction pipe ③. The drive pipe terminates in a small nozzle or jet ④ in the suction pipe. The water emerges from the jet with great force and flows upward through the suction pipe.

Water rises in the suction pipe, drawn up through the tailpipe ⑤ by the action of the jet. The water rises to the pump inlet and passes through the impeller wheel of the pump. Part of the water is forced down through the drive pipe again. The

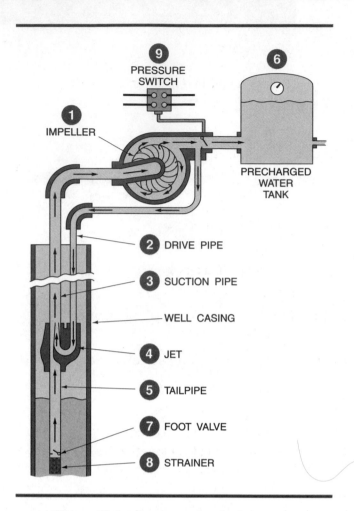

Figure 19-1 Components of a jet pump.

remaining water passes through a check valve and enters the storage tank ⑥.

The tailpipe is submerged in the well water to a depth of about 10 ft (3.0 m). A foot valve ⑦ and strainer ⑧ are provided at the end of the tailpipe. The foot valve prevents water in the pumping equipment from draining back into the well when the pump is not operating.

When the pump is operating, the lower part of the storage tank fills with water and air is trapped in the upper part. As the water rises in the tank, the air is compressed. When the air pressure is 40 pounds per in.², a pressure switch ⑨ disconnects the pump motor. The pressure switch is adjusted so that as the water is used and the air pressure falls to 20 pounds per in.², the pump restarts and fills the tank again.

The Pump Motor

The jet pump motor in this residence is a 3600-rpm, 1-horsepower (hp) single-phase, capacitor-start,

dual-voltage (115/230) volt motor. A capacitor-start motor is designed to produce a starting torque adequate for pumps with comparatively low-starting current. In pumping operations, a high-starting torque helps to overcome the back pressure within the tank and the weight of the water being lifted from the well.

A dual-voltage motor can be connected to either voltage by changing the connections in the junction box on the motor. When connected for the higher voltage, the current draw will be half that when connected for the lower voltage. For example, in *Table 430.148*, the full-load current rating of a 1-hp, 230-volt, single-phase motor is 8 amperes. The same motor connected for 115 volts has a current draw of 16 amperes. Current is a big factor in determining conductor size, watts loss in the conductors, voltage drop, disconnect switch size, and motor controller size.

Note that the current values shown in *Table 430.148* for 115-volt motors cover the voltage range of 110 to 120 volts. For 230-volt motors, the voltage range is 220 to 240 volts. Many single-phase motors are marked 115/208-230 volt. The 120/208-volt wye systems are popular in commercial buildings, where the line-to-line voltage is 208 volts, and the line-to-neutral voltage is 120 volts.

The NEMA *Information Guide for General Purpose Industrial AC Small and Medium Squirrel-Cage Induction Motor Standards*, which is a condensation of the much larger NEMA Standard MG 1-1988 *Motors and Generators*, shows that most standard NEMA-rated motors are designed to operate at ± 10 percent of the motor's rated voltage.

The Pump Motor Circuit

Figure 19-2 shows the electrical circuit that operates the jet pump motor and pressure switch. This circuit originates at Main Service Panel A, Circuit A5-7. *Table 430.148* indicates that a 1-hp single-phase motor has a rating of 8 amperes at 230 volts. Using this information, the conductor size, conduit size, branch-circuit overcurrent protection, and running overcurrent protection can be determined for the motor.

Conductor Size (430.22)

The motor has a rating of 8 amperes. To calculate conductor size, use 125% of 8 amperes.

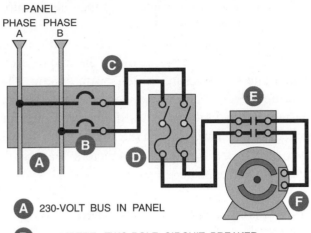

PANEL
PHASE   PHASE
A   B

A 230-VOLT BUS IN PANEL

B 20-AMPERE, TWO-POLE CIRCUIT BREAKER

C 12 AWG THHN CONDUCTORS

D TWO-POLE DISCONNECT SWITCH AND MOTOR OVERLOAD PROTECTION

E TWO-POLE PRESSURE SWITCH

F 1-HP, 230-VOLT, SINGLE-PHASE MOTOR

Figure 19-2   The pump circuit.

$$1.25 \times 8 = 10 \text{ amperes}$$

Table 310.16 of the Code indicates that the minimum conductor size permissible is 14 AWG. However, the Schedule of Special Purpose Outlets found in this text, which is part of the Specifications for the residence, indicates that 12 AWG conductors are to be used for the pump circuit. Installing larger size conductors can minimize voltage drop, and might be suitable at a later date should the need arise to install a larger capacity pump.

Conduit Size

Two 12 AWG Type THHN conductors require trade size ½ EMT, *Annex C, Table C1, NEC.*®

Motor Branch-Circuit Short-Circuit and Ground-Fault Protection
(430.52, Table 430.52)

Fuses and circuit breakers are the most commonly used forms of motor branch-circuit short-circuit and ground-fault protection. Should the motor windings or branch-circuit conductors short circuit or go to ground, the motor's branch-circuit short-circuit and ground-fault overcurrent device, either fuses or circuit breakers, will open.

For the jet pump motor in this residence connected for 230 volts, the various choices for the motor branch-circuit short-circuit and ground-fault protection are found in *430.52* and in *Table 430.52*. They are:

- Non-Time-Delay Fuses
  $8 \times 3 = 24$ amperes (use 25-ampere fuses)
  Maximum size: 400% (4.0)

- Dual-Element, Time-Delay Fuses
  $8 \times 1.75 = 14$ amperes
  (use 15-ampere fuses)
  Maximum size: 225% (2.25)

- Instant Trip Breakers
  $8 \times 8 = 64$ amperes setting
  Maximum setting: 1300% (13)

- Inverse Time Breakers
  $8 \times 2.5 = 20$-ampere setting
  Maximum setting:
    400% for F.L.A. of 100 or less.
    300% for F.L.A. of over 100.

The motor branch-circuit overcurrent device must be capable of allowing the motor to start, *430.52(B)*.

Where the values for branch-circuit protective devices (fuses or breakers) as determined by *Table 430.52* do not correspond to the standard sizes, ratings, or settings as listed in *240.6*, the next higher size, rating, or setting is permitted, *430.52(C)(1), Exception No. 1*.

If, after applying the values found in *Table 430.52* and after selecting the next higher standard size, rating, or setting of the branch-circuit overcurrent device, the motor still will not start, *430.52(C)(1), Exception No. 2* permits using an even larger size, rating, or setting. The maximum values (percentages) are shown in this *Exception*.

A nuisance opening of a fuse, or tripping of a breaker, can easily be predicted by referring to time/current curves for the intended fuse or breaker. This topic is covered briefly in Unit 28.

Motor Overload Protection
(430.31 through 430.44)

Overload protection is needed to keep the motor from dangerous overheating due to overloads or failure of the motor to start. *NEC*® *430.31* through *430.44* cover virtually every type and size of motor, starting characteristics, and duty (continuous or noncontinuous) in use today. The overload protection for a motor might be thermal overloads, electronic sensing overload devices, built-in (inherent) thermal protection, or time-delay fuses.

Although there are exceptions, most of the common installations require motor overload protection not to exceed 125 percent of the motor's full load current draw, as indicated on the nameplate. For the jet pump motor, this would be:

$$1.25 \times 8 = 10 \text{ amperes.}$$

Instructions furnished with a particular motor or motor operated appliance will probably indicate the maximum size overload protection and maximum size branch-circuit protection. These values must be followed.

Wiring for the Pump Motor Circuit

The water pump circuit is run in trade size ½ EMT from Main Panel A to a disconnect switch and controller that are mounted on the wall next to the pump (see Figure 19-3 and Figure 19-4). The EMT may be run across the ceiling of the workshop or be embedded in the concrete floor. Thermal overload devices (sometimes called *heaters*) are installed in the controller to provide running overcurrent protection. Two 12 AWG THHN conductors are connected to the two-pole, 20-ampere branch-circuit breaker in Panel A, Circuit A(5-7). Dual-element, time-delay fuses are installed in the disconnect switch located near the pump. We have the following choices for sizing these fuses:

430.32(A)(1):	$8 \times 1.25 = 10$ amperes
Table 430.52:	$8 \times 1.75 = 14$ amperes
430.52(C)(1), Exception No. 1:	Next standard size is 15 amperes.
430.52(C)(1), Exception No. 2:	If 15-ampere fuses do not allow the motor to start, it is permitted to size up, not to exceed 225%. $8 \times 2.25 = 18$ amperes maximum This puts us back to the 15-ampere rating. However, 17½-ampere, dual-element, time-delay fuses are available.

Any of these ampere ratings of dual-element, time-delay fuses are suitable, and they all fit into a 30-ampere disconnect switch.

A short length of flexible metal conduit is connected between the load side of the thermal overload switch and the pressure switch on the pump. This pressure switch is usually connected to the motor at the factory. In Figure 19-2, a double-pole pressure switch is shown at Ⓔ. When the contacts of the pressure switch are open, all conductors feeding the motor are disconnected.

The Importance of a Disconnecting Means

Figure 19-3 and Figure 19-4 show a typical submersible pump installation. Depending on the installation, another disconnect should be mounted on the outside of the house so as to be clearly visible from the pump location. Although the disconnect inside the house might be capable of being locked in the "OFF" position, this "LOCK-OFF" feature is not always used. This sets the stage for a possible electrocution. Anyone working on the pump needs to be assured that the power is "OFF," and stays "OFF" until he is ready to turn the power back on. See *430.102(A)* and *(B)*.

SUBMERSIBLE PUMP

A submersible pump, shown in Figure 19-3, consists of a centrifugal pump driven by an electric motor. The pump and the motor are contained in one housing, submersed below the permanent water level within the well casing. When running, the pump raises the water upward through the piping to the water tank. Proper pressure is maintained in the system by a pressure switch. The disconnect switch, pressure and limit switches, and controller are in-stalled in a logical and convenient location near the water tank.

Most water pressure tanks have a precharged air chamber in an elastic bag (bladder) that separates the air from the water. This ensures that air is not absorbed by the water. Water pressure tanks compress air to maintain delivery pressure in a usable range without cycling the pump at too fast a rate. When in direct contact with the water, air is gradually absorbed into the water, and the cycling rate of the pump will become too rapid for good pump life, unless the air is periodically replaced. With the elas-

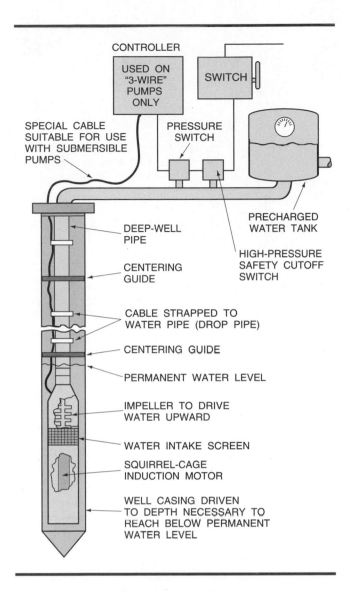

CONTROLLER
USED ON "3-WIRE" PUMPS ONLY

SWITCH

SPECIAL CABLE SUITABLE FOR USE WITH SUBMERSIBLE PUMPS

PRESSURE SWITCH

PRECHARGED WATER TANK

DEEP-WELL PIPE

HIGH-PRESSURE SAFETY CUTOFF SWITCH

CENTERING GUIDE

CABLE STRAPPED TO WATER PIPE (DROP PIPE)

CENTERING GUIDE

PERMANENT WATER LEVEL

IMPELLER TO DRIVE WATER UPWARD

WATER INTAKE SCREEN

SQUIRREL-CAGE INDUCTION MOTOR

WELL CASING DRIVEN TO DEPTH NECESSARY TO REACH BELOW PERMANENT WATER LEVEL

Figure 19-3 Submersible pump.

tic bag, the initial air charge is always maintained, so recharging is unnecessary.

Most pumps commonly called "two-wire pumps" contain all required starting and protection components and connect directly to the pressure switch—with no aboveground controller. Generally, there are no moving electrical parts within the pump, such as the centrifugal starting switch found in a typical single-phase, split-phase induction motor.

Other pumps, referred to as "three-wire pumps," require an aboveground controller that contains any required components not in the motor, such as a starting relay, overload protection, starting and running capacitors, lightning arrester, and terminals for making the necessary electrical connections.

Rating data are on the pumps themselves, on the "three-wire" controllers, and on a separate nameplate furnished with each "two-wire" pump for aboveground identification.

The calculations for sizing the conductors, and the requirements for the disconnect switch, the motors' branch-circuit fuses, and the grounding connections are the same as those for the jet pump motor. The nameplate data and instructions furnished with the pump must be followed. The manufacturer's recommendations for conductor size, type, and length are extremely important if the pump is to operate properly.

See Unit 16 for information about underground wiring.

Submersible pumps are covered by UL Standard 778.

Submersible Pump Cable

Power to the motor is supplied by a cable especially designed for use with submersible pumps. This cable is marked "submersible pump cable," and it is generally supplied with the pump and the pump's various other components. The cable is cut to the proper length to reach between the pump and its controller, as shown in Figure 19-3, or between the pump and the well-cap, Figure 19-4. When needed, the cable may be spliced according to the manufacturer's specifications.

Submersible water pump cable is "tag-marked" for use within the well casing for wiring deep-well water pumps where the cable is not subject to repetitive handling caused by frequent servicing of the pump units.

Submersible pump cable is *not* designed for direct burial in the ground unless it is marked Type USE or Type UF. Should it be required to run the pump's circuit underground for any distance, it is necessary to install underground feeder cable (UF), Type USE, or a raceway with suitable conductors, and then make up the necessary splices in a "listed" weatherproof junction box, as in Figure 19-4.

GROUNDING

Figure 19-4 illustrates how to provide proper grounding of a submersible water pump and the well casing. Proper grounding of the well casing and submersible pump motor will minimize or eliminate

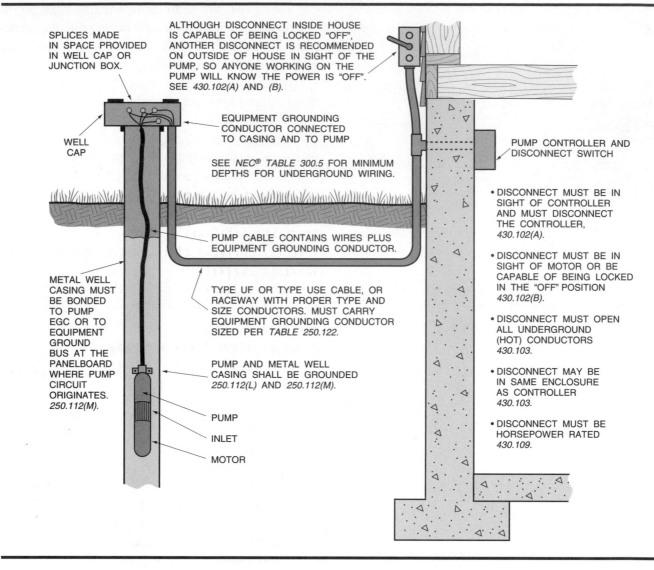

SPLICES MADE
IN SPACE PROVIDED
IN WELL CAP OR
JUNCTION BOX.

ALTHOUGH DISCONNECT INSIDE HOUSE
IS CAPABLE OF BEING LOCKED "OFF",
ANOTHER DISCONNECT IS RECOMMENDED
ON OUTSIDE OF HOUSE IN SIGHT OF THE
PUMP, SO ANYONE WORKING ON THE
PUMP WILL KNOW THE POWER IS "OFF".
SEE *430.102(A)* AND *(B)*.

EQUIPMENT GROUNDING
CONDUCTOR CONNECTED
TO CASING AND TO PUMP

SEE *NEC®* TABLE 300.5 FOR MINIMUM
DEPTHS FOR UNDERGROUND WIRING.

WELL
CAP

PUMP CONTROLLER AND
DISCONNECT SWITCH

• DISCONNECT MUST BE IN
SIGHT OF CONTROLLER
AND MUST DISCONNECT
THE CONTROLLER,
430.102(A).

PUMP CABLE CONTAINS WIRES PLUS
EQUIPMENT GROUNDING CONDUCTOR.

• DISCONNECT MUST BE IN
SIGHT OF MOTOR OR BE
CAPABLE OF BEING LOCKED
IN THE "OFF" POSITION
430.102(B).

METAL WELL
CASING MUST
BE BONDED
TO PUMP
EGC OR TO
EQUIPMENT
GROUND
BUS AT THE
PANELBOARD
WHERE PUMP
CIRCUIT
ORIGINATES.
250.112(M).

TYPE UF OR TYPE USE CABLE, OR
RACEWAY WITH PROPER TYPE AND
SIZE CONDUCTORS. MUST CARRY
EQUIPMENT GROUNDING CONDUCTOR
SIZED PER *TABLE 250.122*.

• DISCONNECT MUST OPEN
ALL UNDERGROUND
(HOT) CONDUCTORS
430.103.

PUMP AND METAL WELL
CASING SHALL BE GROUNDED
250.112(L) AND *250.112(M)*.

• DISCONNECT MAY BE
IN SAME ENCLOSURE
AS CONTROLLER
430.103.

PUMP

INLET

MOTOR

• DISCONNECT MUST BE
HORSEPOWER RATED
430.109.

Figure 19-4   Grounding and disconnecting means requirements for submersible water pumps.

stray voltage problems that could occur if the pump motor is not grounded.

Key code sections that relate to grounding water pumping equipment are:

- *NEC® 250.86:* Metal enclosures for circuit conductors shall be grounded, other than short sections of metal enclosures that provide support or protection.

- *NEC® 250.112(L):* Motor-operated water pumps, including submersible pumps, must be grounded.

- *NEC® 250.112(M):* Bond metal well casing to the pump circuit's equipment grounding conductor.

- *NEC® 250.134(B):* Equipment fastened in place or connected by permanent wiring must have the equipment grounding conductor run with the circuit conductors.

- *NEC® 250.4(A)(5):* The last sentence states ►The earth shall not be used as the sole equipment grounding conductor or effective ground fault current path. ◄

- *NEC® 430.141:* Grounding of motors—general.

- *NEC® 430.142:* Grounding of motors—stationary motors.

- *NEC® 430.144:* Grounding of controllers.

- *NEC® 430.145:* Acceptable methods of grounding.

Because so much nonmetallic (PVC) water piping is used today, some electrical codes require that a grounding electrode conductor be installed between the neutral bar of the main service disconnecting means and the metal well casing. This would satisfy the requirements of *250.104* (bonding),

250.50 (grounding electrode system), *250.52* (made and other electrodes), and *250.56* (resistance of electrodes). This then is the service ground and is sized per *Table 250.66* of the Code. At the metal casing, the grounding electrode conductor is attached to a lug termination, or by means of exothermic (Cadweld) welding.

WATER HEATER CIRCUIT ⓐc

Residential electric water heaters are listed under UL Standard 174, *Household Electric Storage Tank Water Heaters*.

Electrical contractors many times are also in the plumbing, heating, and appliance business. They need to know more than how to do the electrical hookup. The following text discusses electrical as well as other data about electric water heaters that will prove useful.

All homes require a supply of hot water. To meet this need, one or more automatic water heaters are generally installed as close as practical to the areas having the greatest need for hot water. Water piping carries the heated water from the water heater to the various plumbing fixtures and to appliances such as dishwashers and clothes washers.

For safety reasons, in addition to the regular temperature control thermostat that can be adjusted by the homeowner, the *NEC®* and UL Standards require that electric water heaters be equipped with a high temperature cut-off. This high temperature control limits the maximum water temperature to 190°F (88°C). This is 20°F (11°C) below the 210°F (99°C) temperature that causes the "temperature/pressure" relief valve to open. The high temperature cutoff is factory preset, and should never be tampered with or modified in the field. In conformance to *422.47* of the *NEC®*, the high temperature limiting control must disconnect all ungrounded conductors.

In most residential electric water heaters, the high temperature cutoff and upper thermostat are combined into one device, as in Figure 19-5.

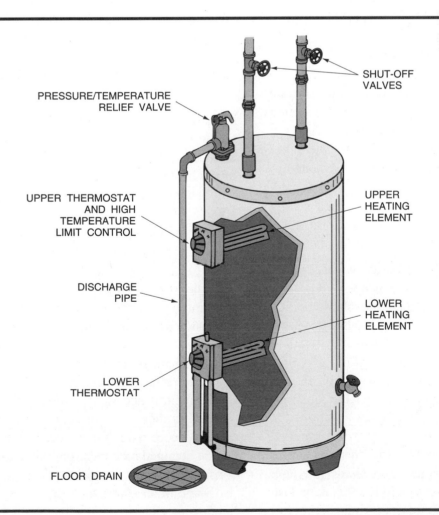

Figure 19-5 Typical electric water heater showing location of heating elements and thermostats.

Combination Pressure/Temperature Safety Relief Valves

Look at Figure 19-5. A pressure/temperature relief valve is an important safety device installed into an opening in the water heater tank within 6 in. (150 mm) of the top. The opening is provided for and clearly marked by the manufacturer. Many water heater pressure/temperature relief valves are factory installed, although some models still require installation of the valve by the installing contractor. Proper discharge piping must be installed from the pressure/temperature relief valve downward to within 6 in. (150 mm) of the floor, preferably near a floor drain. Terminating the discharge pipe to within 6 in. (150 mm) of the floor is important to protect people from injury from scalding water or steam discharge should the relief valve activate for any reason. This also protects electrical equipment in and around the water heater from becoming soaked should the relief valve operate.

Pressure/temperature relief valves are installed for two reasons. Their pressure/temperature sensing features are interrelated.

- *Pressure:* Water expands when heated. Failure of the normal temperature thermostat and the high-temperature limiting device to operate could result in pressures that exceed the rated working pressure of the water heater tank and the plumbing system components. The relief valve, properly installed and properly rated to local codes, will open to relieve the excess pressure in the tank to the predetermined setting marked on the relief valve. The rating is stamped on the valve's identification plate.

- *Temperature:* Should runaway water heating conditions occur within the water heater, temperatures far exceeding the atmospheric boiling point of water (212°F [100°C]) can take place. Under high pressures, water will stay liquid at extremely high temperatures. For a typical municipal water supply pressure of 45 psi, water will not boil until it reaches approximately 290°F (143°C). At 290°F (143°C), opening a faucet, a broken pipe, or a ruptured water heater tank would allow the pressure to instantly drop to normal atmospheric pressure. The superheated water would immediately flash into steam, having about 1600 times more vol-

ume than the liquid water occupied. This blast of steam can be catastrophic, and could result in severe burns or death.

Thus, the function of the temperature portion of the pressure/temperature relief valve is to open when a temperature of 210°F (99°C) occurs within the water heater tank. Properly rated and maintained, the pressure/temperature relief valve will allow the water to flow from the water heater through the discharge pipe at a rate faster than the heating process, thereby keeping the water temperature below the boiling point. Should the cold water supply be shut off, the relief valve will allow the steam to escape without undue buildup of temperature and pressure in the tank.

▶Requirements for relief valves for hot water supply systems are found in the ANSI Standard Z21.22.◀

Sizes

Residential electric water heaters are available in many sizes, such as 6, 12, 15, 20, 30, 32, 40, 42, 50, 52, 66, 75, 80, 82, 100, and 120 gallons.

Corrosion of Tank

To reduce corrosion of the steel tank, most water heater tanks are glass lined. Glass lining is a combination of silica and other minerals (rasorite, rutile, zircon, cobalt, nickel oxide, and fluxes). This special mix is heated, melted, cooled, crushed, then sprayed onto the interior steel surfaces and baked on at about 1600°F (871°C). To further reduce corrosion, aluminum or magnesium anode rods are installed in special openings, or as part of the hot water outlet fitting, to allow for the flow of a protective current *from* the rod(s) *to* any exposed steel of the tank. This is commonly referred to as *cathodic protection.* Corrosion problems within the water heater tank are minimized as long as the aluminum or magnesium rod(s) remain in an active state. Anodes should be inspected at regular intervals to determine when replacement is necessary, usually when the rod(s) is reduced to about ⅓ of its original diameter, or when the rod's core wire is exposed. Further information can be obtained by contacting the manufacturer, and consulting the installation manual.

Some water heater tanks are fiberglass or plastic lined.

Heating Elements

The wattage ratings of electric water heaters can vary greatly, depending upon the size of the heater in gallons, the speed of recovery desired, local electric utility regulations, and codes. Typical wattage ratings are 1500, 2000, 2500, 3000, 3800, 4500, and 5500 watts.

A resistance heating element might be dual rated. For example, the element might be marked 5500 watts at 250 volts, and 3800 watts at 208 volts.

Most residential-type water heaters are connected to 240 volts except for the smaller 2-, 4-, and 6-gallon *point of use* sizes generally rated 1500 watts at 120 volts. Commercial electric water heaters can be rated single-phase or three-phase, 208, 240, 277, or 480 volts.

UL requires that the power (wattage) input must not exceed 105% of the water heater's nameplate rating. All testing is done with a supply voltage equal to the heating element's *rated* voltage. Most heating elements may burn out prematurely if operated at voltages 5% higher than for which they are rated.

To help reduce the premature burnout of heating elements, some manufacturers will supply 250-volt heating elements, yet will mark the nameplate 240 volts with its corresponding wattage at 240 volts. This allows a safety factor if slightly higher than normal voltages are experienced.

Heating Element Construction

Heating elements generally contain a nickel-chrome (nichrome) resistance wire embedded in compacted powdered magnesium oxide to ensure that no grounds occur between the wire and the sheath. The magnesium oxide is an excellent insulator of electricity, yet it effectively conducts the heat from the nichrome wire to the sheath.

The sheath can be made of tin-coated copper, stainless steel, or an iron-nickel-chromium alloy. The later alloy comes under the trade name of Incoloy.® The stainless steel and Incoloy® types can withstand higher operating temperatures than the tin-coated copper elements. They are used in premium water heaters because of their resistance to deterioration, lime scaling, and dry firing burnout.

High watt density elements generally have a U-shaped tube, and give off a great amount of heat per square inch of surface. Figure 19-6(A) illustrates a flange-mounted high-density heating element. Figure 19-6(B) illustrates a high-density heating element that screws into a threaded opening in the tank. High-density elements are very susceptible to lime-scale burnout.

Low watt density elements, as in Figure 19-6(C), generally have a double loop, like a U tube bent in half. They give off less heat per in^2, have a longer life, and are less noisy than high-density elements.

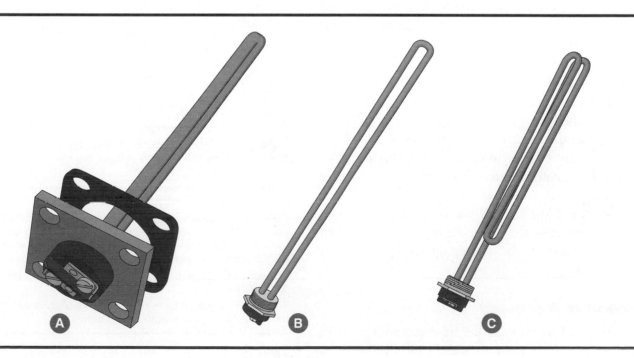

Figure 19-6 Heating elements.

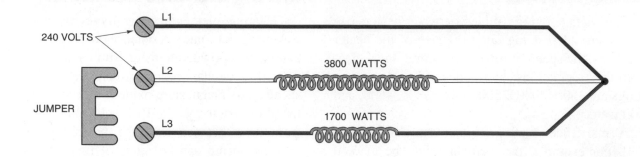

Figure 19-7 A dual-wattage heating element. The two wires of the incoming 240-volt circuit are connected to L1 and L2. These leads connect to the 3800-watt element. To connect the 1700-watt element, the jumper is connected between L2 and L3. With both heating elements connected, the total wattage is 5500 watts.

An option offered by at least one manufacturer of electric water heaters is a dual-wattage heating element. A dual-wattage heating element actually contains two heating elements in the one jacket. One element is rated 3800 watts and one element is rated 1700 watts. Three leads are brought out to the terminal block. The 3800-watt heating element is connected initially, with the option of connecting the second 1700-watt heating element if the wiring to the water heater is capable of handling the higher total combined wattage of 5500 watts. To attain the higher wattage, a jumper (provided by the manufacturer) is connected between two terminals on the terminal block, Figure 19-7.

Any electric heating element coated with lime scale may be noisy during the heating cycle. Limed up heating elements should be cleaned or replaced, following the manufacturer's instructions.

Do not energize the electrical supply to a water heater unless you are sure it is full of water. Most water heater heating elements are designed to operate only when submersed in water.

Residential electric water heaters are available with one or two heating elements. Figure 19-5 shows two heating elements. Single-element water heaters will have the heating element located near the bottom of the tank. If the water heater has two heating elements, the upper element will be located about two-thirds of the way up the tank.

Speed of Recovery

Speed of recovery is the time required to bring the water temperature to satisfactory levels in a given length of time. The current accepted industry standard is based on a 90°F (32.22°C) rise. For example, referring to Table 19-1, we find that a 3500-watt element can raise the water temperature 90°F (32.22°C) of a little more than 16 gallons an

HEATING ELEMENT WATTAGE	GALLONS PER HOUR RECOVERY FOR INDICATED TEMPERATURE RISE					
	60°F	70°F	80°F	90°F	100°F	Equivalent NET Btu output
750	5.2	4.4	3.9	3.5	3.1	2559
1000	6.9	5.9	5.2	4.6	4.1	3412
1250	8.6	7.4	6.5	5.7	5.2	4266
1500	10.3	8.9	7.8	6.9	6.2	5119
2000	13.8	11.8	10.3	9.2	8.3	6825
2250	15.5	13.3	11.6	10.3	9.3	7678
2500	17.2	14.8	12.9	11.5	10.3	8531
3000	20.7	17.7	15.5	13.8	12.4	10,238
3500	24.1	20.7	18.1	16.1	14.5	11,944
4000	27.6	23.6	20.7	18.4	16.6	13,650
4500	31.0	26.6	23.3	20.7	18.6	15,356
5000	34.5	29.6	25.9	23.0	20.7	17,063
5500	37.9	32.5	28.4	25.3	22.5	18,769
6000	41.4	35.5	31.0	27.6	24.6	20,475

NOTES: If the incoming water temperature is 40°F (Northern states), use the 80° column to raise the water temperature to 120°F.

If the incoming water temperature is 60°F (Southern states), use the 60° column to raise the water temperature to 120°F.

For example, approximately how long would it initially take to raise the water temperature in a 42-gallon electric water heater to 120°F where the incoming water temperature is 40°F? The water heater has a 2500-watt heating element.

Answer: 42 ÷ 12.9 = 3.26 hours (approximately 3 hours and 16 minutes)

To keep this table simple, non-cluttered, and easy to read, only Fahrenheit temperatures are shown.

Table 19-1 Electric water heater recovery rate per gallon of water for various wattages.

hour. A 5500-watt element can raise water temperature 90°F (50°C) of about 25 gallons an hour. Speed of recovery is affected by the type and amount of insulation surrounding the tank, the supply voltage, and the temperature of the incoming cold water.

Table 19-2 shows the approximate time it takes to run out of hot water using typical shower heads with various capacity electric water heaters. Regular residential-type shower heads deliver 5 to 6 gallons per minute. Water saver shower heads deliver 1 to 2 gallons per minute. This varies with water pressure. An easy way to determine a shower head's flow rate is to turn on the water, hold a large bucket of known capacity under it, and time how long it takes to fill the bucket.

Scalding from Hot Water

To avoid scalding, many states have laws regulating the maximum temperature settings for water heaters. This maximum setting is usually 120°F (49°C). Where such laws do not exist, the Consumer Product Safety Commission suggests a maximum temperature setting of 130°F (54°C). Manufacturers post caution labels on their water heaters that read something like this:

SCALD HAZARD: WATER TEMPERATURE OVER 125°F (51°C) CAN CAUSE SEVERE BURNS INSTANTLY OR DEATH FROM SCALDS. SEE INSTRUCTION MANUAL BEFORE CHANGING TEMPERATURE SETTING.

| Shower head(s) rate in gallons per minute | Hot water flow rate in gallons per minute based on 120°F. Stored hot water to provide 105°F. mixed water out of shower head. | Hot water flow rate in gallons per minute based on 120°F. Stored hot water to provide 105°F. mixed water out of shower head. | LENGTH OF CONTINUOUS SHOWER TIME—IN MINUTES—FOR TYPICAL SIZES OF RESIDENTIAL ELECTRIC WATER HEATERS. CHART BASED ON 70% DRAW EFFICIENCY. RECOVERY RATE IS NOT FIGURED IN BECAUSE IT IS NORMALLY NOT A FACTOR IN CONTINUOUS DRAW OF WATER SITUATIONS. | | | | | | | | | | | | |
| | | | 100 GALLON | | 80 GALLON | | 66 GALLON | | 50 GALLON | | 40 GALLON | | 30 GALLON | |
	40°F. INCOMING WATER	60°F. INCOMING WATER	40°F. INCOMING WATER	60°F. INCOMING WATER	40°F. INCOMING WATER	60°F. INCOMING WATER	40°F. INCOMING WATER	60°F. INCOMING WATER	40°F. INCOMING WATER	60°F. INCOMING WATER	40°F. INCOMING WATER	60°F. INCOMING WATER	40°F. INCOMING WATER	60°F. INCOMING WATER
2.0	1.6	1.5	43	47	34	37	28	31	22	23	17	19	13	14
2.5	2.0	1.9	34	38	28	30	23	25	17	19	14	15	10	11
3.0	2.4	2.3	29	31	23	25	19	21	14	16	11	12	9	9
4.0	3.3	3.0	22	23	17	19	14	15	11	12	9	9	6	7
5.0	4.0	3.8	17	19	14	15	11	12	9	9	7	7	5	6
6.0	4.9	4.5	14	16	11	12	9	10	7	8	6	6	4	5
8.0	6.5	6.0	11	12	9	9	7	8	5	6	4	5	3	4
12.0	9.8	9.0	7	8	6	6	5	5	4	4	3	3	2	2

Example #1: A 40-gallon electric water heater has a 40°F incoming water supply. How many minutes can a 3 GPM shower head be used to obtain 105°F hot water?

Answer: 11 minutes

Example #2: A residence has an 80-gallon electric water heater. The incoming water supply is approximately 60°F. How many minutes can two 2 GPM shower heads be used at the same time to obtain 105°F hot water? Hint: Two 2 GPM shower heads are the same as one 4 GPM shower head.

Answer: 19 minutes

Table 19-2 This chart shows the approximate time in minutes it takes to run out of hot water when using various shower heads, different incoming water temperatures, and different capacity water heaters.

Certain states do not accept the water heater thermostat as an acceptable water temperature control device, and require that showers and shower/bath combinations be provided with an automatic safety mixing device to prevent sudden unanticipated changes in water temperatures. Some manufacturers use nonadjustable upper thermostats, factory set at 120°F (49°C). Because 120°F (49°C) is not hot enough to dissolve grease and to activate power detergents, most newer dishwashers boost the water temperature in the dishwasher from 120°F (49°C) to 140°F–145°F (60°C–63°C) using the same heating element that provides the heat for the drying cycle.

Table 19-3 shows time/temperature relationships relating to scalding.

Thermostats/High Temperature Limit Controls

All electric water heaters have a thermostat(s) and a high temperature limit control(s).

Thermostats and high temperature limit controls are available in many configurations. They are available as separate controls and as combination controls.

Figure 19-8 illustrates a combination thermostat/high temperature limit control. This particular control is an interlocking type sometimes referred to as a snap-over type. It is used on two-element water heaters and controls the upper heating element as well as providing the safety feature of limiting the

water temperature to a factory preset value. Another thermostat is used to control the lower heating element. When all of the water in the water heater is cold, the upper element heats the upper third of the tank. The lower thermostats are closed, calling for heat, but the interlocking characteristic of the upper thermostat/high limit will not allow the lower heating element to become energized. With this connection, both heating elements cannot be on at the same time. This is referred to as *nonsimultaneous* or *limited demand* operation.

When the water in the upper third of the tank reaches the temperature setting of the upper thermostat, the upper thermostat "snaps over" completing the circuit to the lower element. The lower element then begins to heat the lower two-thirds of the tank. Most of the time, it is the lower heating element that keeps the water at the desired temperature. Should a large amount of water be used in a short time, the upper element comes on. This provides fast recovery of the water in the upper portion of the tank, which is where the hot water is drawn from the tank.

Safety Note: Most residential water heater thermostats are of the single-pole type. They do not open both ungrounded conductors of the 240-volt supply. They open one conductor only, which is all

TEMPERATURE	TIME TO PRODUCE SERIOUS BURN
120°F (48.89°C)	Approx. 9½ minutes
125°F (51.67°C)	Approx. 2 minutes
130°F (54.44°C)	Approx. 30 seconds
135°F (57.22°C)	Approx. 15 seconds
140°F (60°C)	Approx. 5 seconds
145°F (62.78°C)	Approx. 2½ seconds
150°F (65.76°C)	Approx. 1⁸⁄10 seconds
155°F (68.33°C)	Approx. 1 second
160°F (71.11°C)	Approx. ½ second

Table 19-3 Time/temperature relationships related to scalding of an adult subjected to moving water. For children, the time to produce a serious burn is less than for adults. *Courtesy of* Shriners Burn Institute, Cincinnati Unit.

Figure 19-8 Typical electric water heater controls. The seven terminal control combines water temperature control plus high temperature limit control. This is the type commonly used to control the upper heating element as well as providing a high temperature limiting feature. The two terminal control is the type used to control water temperature for the lower heating element. *Courtesy of* THERM-O-DISC, Inc., subsidiary of Emerson Electric.

that is needed to open the 240-volt circuit to the heating element. What this means is that a reading of 120 volts to ground will always be present at the terminals of the heating element on a 240-volt, single-phase system such as found in residential wiring. If the heating element becomes grounded to the metal sheath, the element can continue to heat even when the thermostat is in the "OFF" position because of the presence of the 120 volts.

The high temperature limit control opens both ungrounded conductors.

Various Types of Electrical Connections

There are many variations for electric water heater hookups. Every electric utility has its own unique time-of-use programs and specific electrical connection requirements for electric water heaters. It is absolutely essential that you contact your local electric utility for this information.

Because of the cumulative large power consumption of electric water heater loads, electric utilities across the country have been innovative in creating special lower rate structures (programs) for electric water heaters, air conditioners, and heat pumps. Electric utilities want to be able to shed some of this load during their peak periods. They will offer the homeowner choices of how the water heater, air conditioner, or heat pump is to be connected, and in some cases, will provide financial incentives (lower rates) to those who are willing to use these restricted "time-of-day" connections. "Time-of-day," "time-of use," and "off-peak" are terms used by different utilities. If homeowners have a large storage capacity electric water heater, they will have an adequate supply of hot water during those periods of time the power to the water heater is off. If they cannot live with this, they have other options. Check with the local power company for details on their particular programs.

The following text discusses some of the methods in use around the country for connecting electric water heaters, and in some cases, air-conditioning and heat pump equipment.

Utility Controlled

The world of electronics has opened up many new ways for an electric utility to meter and control residential electric water heater, air-conditioning,

and heat pump loads. Electronic watt-hour meters can be read remotely, programmed remotely, and can detect errors remotely. The options are endless.

The utility can install watt-hour meters that combine a watt-hour meter and a set of switching contacts. The contacts are switched on and off by the utility during peak hours using a power line carrier signal. Older style versions of this type of watt-hour meter incorporated a watt-hour meter, time clock, and switching contacts. The time clock would be set by the utility for the desired times of the on-off cycle. Today, the metering and the programming for timing and control of the on-off cycling is accomplished through electronic solid-state devices. Electronic meters contain a microprocessor that records, and an optical assembly that scans the rotating disc to record the number of turns and speed of the disc. These electronic digital meters can record the energy consumption during normal hours and during peak hours, allowing the electric utility to bill the homeowner at different rates for different periods of time. One type of residential single-phase watt-hour meter can register four different "time-of-use" rates. This type of meter is shown in Figure 29-2.

One major utility programs their residential meters for "peak periods" from 9 AM to 10 PM, Monday through Friday—except for holidays. Their "off-peak" periods are weekend days and all other hours. Programming of electronic meters can extend out for more than five years. The possibilities for programming electronic meters are endless. Older style mechanical meters used gears to record this data, and did not have the programming capabilities that electronic meters do as in Figure 19-9, Figure 19-10, and Figure 19-11.

The switching contacts in the meter are connected in *series* with the water heater, air-conditioner, or heat pump load that is to be controlled for the special "time-of-day" rates, Figure 19-9, Figure 19-10, and Figure 19-11. The utility determines when they want to have the electric water heater on or off.

Figure 19-9 shows a separate combination meter/time clock for the electric water heater load. The utility sets the clock to be "OFF" during their peak hours of high power consumption. Thus the term *off-peak metering*. Off-peak metering is oftentimes called "time-of-use" or "time-of-day" metering. In exchange for lower electric rates, the

A 240/120-VOLT SERVICE

B MAIN WATT-HOUR METER

C WATER HEATER WATT-HOUR METER

D SERVICE CONDUCTORS TO MAIN PANEL

E TWO-POLE DISCONNECT SWITCH OR CIRCUIT BREAKER

F COMBINATION HIGH TEMPERATURE LIMIT AND UPPER INTERLOCKING THERMOSTAT

G UPPER HEATING ELEMENT

H LOWER THERMOSTAT

I LOWER HEATING ELEMENT

Notations used in Figure 19-9, Figure 19-10, Figure 19-12, and Figure 19-13.

homeowner runs the risk of being without hot water. Should homeowners run out of hot water during these off periods, they have no recourse but to wait until the power to the water heater comes on again. For this type of installation, large storage capacity water heaters are needed so as to be able to have hot water during these periods.

Figure 19-10 illustrates a type of meter that contains an integral set of contacts, controlled by the utility via signals sent over the power lines. These meters are available with either 30-ampere or 50-ampere contacts, the contact rating selected to switch the load to be controlled. The electric utility determines when to shed the load. The homeowner must be willing to live with the fact that the power to the water heater, air-conditioner, or heat pump might be turned off during the peak periods.

In Figure 19-11, the utility mounts a radio-controlled relay to the side of the meter socket. One major utility using this method makes use of ten different radio frequencies. When the utility needs to shed load, it can signal one or all of the relays to turn the water heater, air conditioner, or heat pump load off and on. This might only be a few times each year, as opposed to time-clock programs that operate daily. This method is a simple one. One watt-hour meter registers the total power consumption. The homeowner signs up for this program. The program offers a financial incentive, and

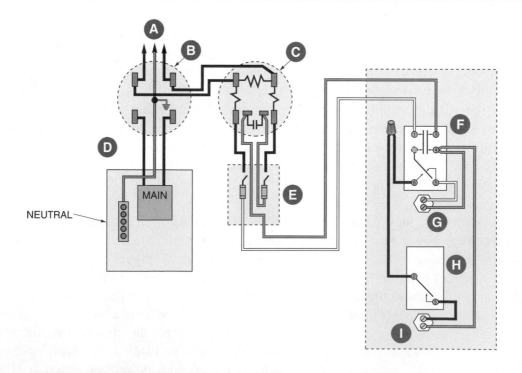

Figure 19-9 Wiring for a typical "off-peak" electric water heater circuit. A separate watt-hour meter/time clock controls the circuit to the water heater. The utility sets the time clock to turn the power off during certain peak periods of the day. This type of circuitry is called "time-of-day" programming.

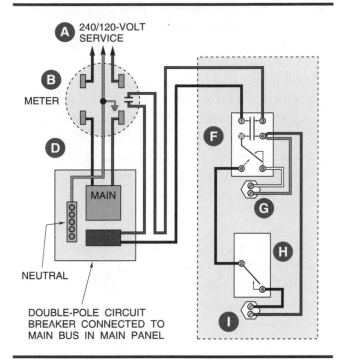

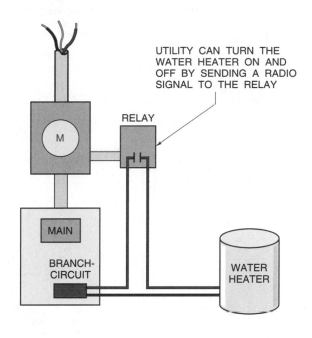

UTILITY CAN TURN THE WATER HEATER ON AND OFF BY SENDING A RADIO SIGNAL TO THE RELAY

Figure 19-10 The meter contains a set of contacts that can be controlled by the utility. The water heater circuit is fed from the main panel, through the contacts, then to the water heater.

Figure 19-11 In this diagram, the utility sends a radio signal to the relay, which in turn can "shed" the load of a residential water heater, air conditioner, or heat pump. If the relay has two sets of contacts, four conductors would be required. Low-voltage relays are also used, in which case low-voltage conductors are run between the radio control unit and a relay in the appliance. For the control of an air conditioner or heat pump, the radio control unit is connected "in series" with the room thermostat. This arrangement can be used with one watt-hour meter using a sliding rate schedule, with a "time-of-day" control. Contact the electric utility for technical data regarding how they want the electrical connections to be made for their particular programs.

the homeowner knows up front that the utility might occasionally turn off the load during crucial peak periods.

Customer Controlled

There are some rate schedules where customers can somewhat control their energy costs by doing their utmost to use electrical energy during off-peak hours. For example, they would not wash clothes, use the electric clothes dryer, take showers, or use the dishwasher during peak (premium) hours. The choice is up to homeowners. To accomplish this, watt-hour meters can be installed that have two sets of dials: one set showing the *total kilowatt hours*, and the second set showing *premium time kilowatt hours*. See Figure 29-2. With this type of meter, the utility will have the total kilowatt hours used as well as the premium time kilowatt hours used. With this information, the utility bills customers at one rate for the energy used during peak hours, and at another rate for the difference between total kilowatt hours and premium time kilowatt hours. The electric utility will determine and define *premium time*.

Some homeowners have installed time-clocks on their water-heater circuit to shut off during peak premium time hours to be sure they are not using energy during the high rate per kilowatt-hour periods.

Uncontrolled

The simplest uncontrolled method is shown in Figure 19-12. There are no extra meters, switching contacts in the meter, extra disconnect switches, or controllable relays. The water heater circuit is connected to a two-pole branch-circuit breaker in the main distribution panel. The electrical power to the water heater is on 24 hours a day. This is the connection for the water heater in the residence in this text.

KILOWATT-HOUR POWER CONSUMPTION OF WATER HEATER IS METERED BY SAME METER USED FOR NORMAL HOUSEHOLD LOADS. ONLY ONE METER IS REQUIRED. POWER RATES ARE USUALLY ON A SLIDING SCALE; THAT IS, THE MORE POWER IN KILOWATT-HOURS USED, THE LOWER RATE PER KILOWATT-HOUR.

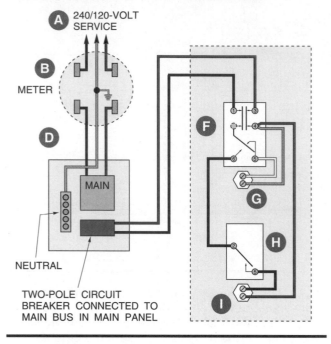

Figure 19-12 Water heater connected to two-pole circuit in Main Panel.

Instead of separate "low-rate" metering that requires an additional watt-hour meter, the utility offers a sliding scale.

EXAMPLE:

METER READING 06-01-2002	00532
METER READING 05-01-2002	00007
TOTAL KILOWATT-HOURS USED	00525
BASIC MONTHLY ENERGY CHARGE	$11.24
1ST. 400 kWh × 0.10206	40.82
NEXT 125 kWh × 0.06690	8.36
AMOUNT DUE	$60.42

State, local, and applicable regulatory taxes would be added to the amount due.

Certain utilities will use the higher energy charge during the June, July, August, and September summer months.

Another uncontrolled method is to use two watt-hour meters, Figure 19-13. The water heater has power 24 hours per day. One meter will register the normal lighting load at the regular residential energy charge. The second meter will register the energy consumed by the water heater, air conditioner, or heat pump at some lower energy rate.

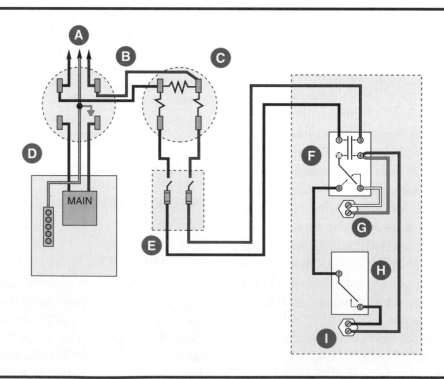

Figure 19-13 Twenty-four-hour water heater installation. The total energy demand for the water heater is fed through a separate watt-hour meter at a lower kWh rate than the regular meter.

Are Water Heater Control Conductors Permitted in the Same Raceway as Service-Entrance Conductors?

Conductors other than service-entrance conductors are not permitted in the same raceway as the service-entrance conductors, *230.7*. However, *Exception No. 2* to this section does permit load management control conductors that have overcurrent protection to be installed in the service raceway.

ELECTRIC WATER HEATER BRANCH-CIRCUITS

Branch-Circuit Rating

NEC® 422.13 applies to all fixed storage electric water heaters having a capacity of 120 gallons (454.2 L) or less. This would include most typical residential electric water heaters. The branch-circuit rating must not be less than 125 percent of the nameplate rating of the water heater. The rating of a branch-circuit is based on the rating of the overcurrent device, *210.3*.

Disconnecting Means

NEC® 422.30 requires that the appliance must have a means of disconnect. This would normally be located in the panel (load center), or it could be provided by a separate disconnect switch. Although not normally done, *422.34* does allow the unit switch that is an integral part of an appliance having an "OFF" position that disconnects all ungrounded conductors from the source to be considered the required **disconnecting means**. In many local codes, this is not acceptable. *NEC® 422.34* further requires that there be another means of disconnect for the appliance, which in the case of a one-family dwelling is the main service disconnect.

NEC® 422.35 requires that switches and circuit breakers used as the disconnecting means must be of the indicating type, meaning they must clearly show that they are in the "ON" or "OFF" position.

NEC® 422.31(B) requires that for permanently connected appliances rated greater than 300 volt-amperes or ⅛ horsepower, the disconnecting means must be within sight of the appliance, or be capable of being locked in the "OPEN" position. The electric water heater in the residence provides a good example of this Code requirement.

Overcurrent Protection and Conductor Size

There are two separate issues that must be considered.

- The overcurrent protective device ampere rating must be calculated.
- The conductor size must be determined.

Do both calculations. Then compare the results of both calculations to make sure that both overcurrent device and conductors are Code compliant.

Branch-Circuit Overcurrent Protection

NEC® 422.11(E) states that for single non-motor-operated electrical appliances, the overcurrent protective device shall not exceed the protective device rating marked on the appliance. If the appliance has no such marking, then the overcurrent device is to be sized as follows:

- If appliance does not exceed 13.3 amperes— 20 amperes
- If appliance draws more than 13.3 amperes— 150 percent of the appliance rating

For a single non-motor-operated appliance, if the 150 percent sizing does not result in a standard size overcurrent device rating as listed in *240.6*, then it is permitted to go to the next standard size.

EXAMPLE: A water heater nameplate indicates 4500 watts, 240 volts. What is the maximum size fuse permitted by the Code?

SOLUTION:

$$I = \frac{W}{E} = \frac{4500}{240} = 18.75 \text{ amperes}$$

NEC® 422.13 requires that the branch-circuit rating must not be less than 125 percent of the water heater's nameplate rating. The minimum branch-circuit rating is:

$$18.75 \times 1.25 = 23.44 \text{ amperes}$$

This would require a minimum 25-ampere fuse or circuit breaker, which is the next standard size larger than the calculated 23.44 amperes. See *240.6(A)* for standard ampere ratings for fuses and circuit breakers.

The maximum overcurrent device for the water heater is:

$$18.75 \times 1.5 = 28.125 \text{ amperes}$$

This would be a 30-ampere fuse or circuit breaker, which is the next standard size larger than the calculated 28.125 amperes. But there is no reason to use a 30-ampere branch circuit fuse or circuit breaker because we have already determined that a 25-ampere fuse or circuit breaker is suitable for the 4500-watt load.

Conductor Size

The conductors supplying the 4500-watt water heater will be protected by the 25-ampere or breaker as determined above.

In *Table 310.16*, we find the allowable ampacity of conductors in the 60°C column as required by *110.14(C)*. In *240.4(D)*, we find the maximum overcurrent protection for small conductors. Putting this all together, we have the following shown in Table 19-4.

For the example, we selected a 25-ampere overcurrent device. A 30-ampere OCD would have been acceptable. Next, we need to find a conductor that is properly protected by a 25- or 30-ampere OCD. The conductor would have to be a minimum 10 AWG Type THHN, which has an allowable ampacity of 30 amperes, more than adequate for the minimum branch-circuit rating of 23.44 amperes. A 10 AWG conductor is properly protected by a maximum 30-ampere overcurrent device.

If a 12 AWG Type THHN had been selected from the above table, it would have been suitable for the load but would not have been properly protected by the 25-ampere OCD. The maximum OCD for a 12 AWG is 20 amperes.

You might want to review Unit 4 for a refresher as to why we use the 60°C column of *Table 310.16* for selecting the conductor size.

CONDUCTOR SIZE (THHN Copper)	ALLOWABLE AMPACITY FROM 60°C COLUMN OF *TABLE 310.16*	MAXIMUM AMPERE RATING OF OVERCURRENT DEVICE
14 AWG	20 amperes	15 amperes
12 AWG	25 amperes	20 amperes
10 AWG	30 amperes	30 amperes

Table 19-4 Table showing typical small conductors, their allowable ampacities, and the maximum rating overcurrent devices permitted for these conductors.

The Water Heater for This Residence

The water heater for the residence is connected to Circuit A6-8. This is a 20-ampere, two-pole circuit breaker located in the main panel in the workshop. This circuit is straight 240 volts and does not require a neutral conductor.

Checking the Schedule of Special-purpose Outlets, we find that the electric water heater has two heating elements: a 2000-watt element and a 3000-watt element. Because of the thermostats on the water heater, both elements cannot be energized at the same time. Therefore, the maximum load demand is

$$\text{amperes} = \frac{\text{watts}}{\text{volts}} = \frac{3000}{240} = 12.5 \text{ amperes}$$

For water heaters having two heating elements connected for nonsimultaneous operation, the nameplate on the water heater will be marked with the largest element's wattage at rated voltage. For water heaters having two heating elements connected for simultaneous operation, the nameplate on the water heater will be marked with the total wattage of both elements at rated voltage.

Conductor Size and Raceway Size for the Water Heater in This Residence

The specifications for this residence call for two 12 AWG Type THHN conductors. The branch-circuit overcurrent device is rated 20 amperes.

Table C1 in *Annex C* of the *NEC®* indicates that trade size ½ EMT will be okay for two 12 THHN conductors. For ease of installation, a short length of trade size ½ flexible metal conduit 18 to 24 in. (450 to 600 mm) long is attached to the EMT, and is connected into the knockout provided for the purpose on the water heater.

Cord/Plug Connections Not Permitted for Water Heaters

NEC® *400.7(A)(8)* and *422.16(C)* state the uses where flexible cords are permitted. A key code requirement, often violated, is that flexible cords shall be used only "where the fastening means and mechanical connections are specifically designed to permit ready removal for maintenance and repair, and the appliance is intended or identified for flexible cord connection." Certainly the plumbing does not allow for the "ready removal" of the water heater. The conductors in flexible cords cannot handle the high temperatures encountered on the water heater terminals. Check the instruction manual for proper installation methods.

Equipment Grounding

The electric water heater is grounded through the trade size ½ EMT and the flexible metal conduit (FMC). Code references are *250.110*, *250.118*, *250.134(B)*, and *348.60*. The subject of equipment grounding is covered in Unit 4.

EFFECT OF VOLTAGE VARIATION

As previously stated, the heating elements in the heater in this residence are rated 240 volts. A resistive heating element will only produce rated wattage at rated voltage. They will operate at a lower voltage with a reduction in wattage. If connected to voltages above their rating, they will have a very short life.

Ohm's law and the wattage formula show how the wattage and current depend upon the applied voltage. The following calculations are theoretical. The "cold" resistance of a metal can be considerably lower than its "hot" resistance, as in the case of tungsten used to make lamp filaments, which is why a tungsten filament lamp has such a high initial inrush current and the need for "T" Rated switches. Nichrome wire commonly used to make heating elements on the other hand has a "hot" resistance as much as 10% higher than its "cold" resistance. Always use a *rated* voltage to calculate the resistance, current draw, and wattage of a given resistive heating element. Check manufacturers technical data for specifications.

CAUTION: When actually testing a heating element, never take an ohms reading on an energized circuit. Remove any other connected wires from the heating element terminal block on a water heater to prevent erratic ohms readings. Always check from the heating element leads to ground to confirm that there is no ground fault in the heating element.

$$R = \frac{E^2}{W} = \frac{240 \times 240}{3000} = 19.2 \text{ ohms}$$

$$I = \frac{E}{R} = \frac{240}{19.2} = 12.5 \text{ amperes}$$

If a different voltage is substituted, the wattage and current values change accordingly.

At 220 volts:

$$W = \frac{E^2}{R} = \frac{220 \times 220}{19.2} = 2521 \text{ watts}$$

$$I = \frac{W}{E} = \frac{2521}{220} = 11.46 \text{ amperes}$$

Another way to calculate the effect of voltage variance is:

$$\text{Correction Factor} = \frac{\text{Applied Voltage Squared}}{\text{Rated Voltage Squared}}$$

Using the previous example:

$$\text{Correction Factor} = \frac{220 \times 220}{240 \times 240} = \frac{48,400}{57,600} = 0.84$$

then $3000 \times 0.84 = 2521$ watts

In resistive circuits, current is directly proportional to voltage and can be simply calculated using a ratio and proportion formula. For example, if a 240-volt heating element draws 12.7 amperes at rated voltage, the current draw can be determined at any other applied voltage, say 208 volts.

$$\frac{208}{240} = \frac{X}{12.7}$$

$$\frac{208 \times 12.7}{240} = 11.006 \text{ amperes}$$

To calculate this example if the applied voltage is 120 volts:

$$\frac{120}{240} = \frac{X}{12.7}$$

$$\frac{120 \times 12.7}{240} = 6.35 \text{ amperes}$$

Also, in a resistive circuit, wattage varies as the square of the current. Therefore, when the voltage on a heating element is doubled, the current also doubles and the wattage increases four times. When the voltage is reduced to one-half, the current is halved and the wattage is reduced to one-fourth.

Using the above formulae, it will be found that a 240-volt heating element when connected to 208 volts will have approximately three-quarters of the wattage rating than at rated voltage.

The above formulas work only for resistive circuits. Formulas for inductive circuits such as an electric motor are much more complex than this book is intended to cover. For typical electric motors, the following information will suffice (see Table 19-5).

VOLTAGE VARIATION	FULL-LOAD CURRENT	STARTING CURRENT
110%	7% decrease	10–12% Increase
90%	11% decrease	10–12% Decrease

Table 19-5 Table showing the change in full-load current and starting-current for typical electric motors when operated at under voltage (90%) and over voltage (110%) conditions.

As previously mentioned, NEMA rated electric motors are designed to operate at ±10% of their nameplate voltage.

HEAT PUMP WATER HEATERS

With so much attention given to energy savings, heat pump water heaters have entered the scene.

Residential heat pump water heaters can save 50 to 60 percent over the energy consumption of resistance-type electric water heaters, depending upon energy rates. The heat energy transferred to the water is three to four times greater than the electrical energy used to operate the compressor, fan, and pump.

Heat pump water heaters work in conjunction with the normal resistance-type electric water heater. Located indoors near the electric water heater and properly interconnected, the heat pump water heater becomes the primary source for hot water. The resistance electric heating elements in the electric water heater are the back-up source should it be needed.

Heat pump water heaters remove low-grade heat from the surrounding air, but instead of transferring the heat outdoors as does a regular air-conditioning unit, it transfers the heat to the water. The unit shuts off when the water reaches 130°F (54.44°C). Moisture is removed from the air at approximately one pint per hour, reducing the need to run a dehumidifier. The exhausted cool air from the heat pump water heater is about 10°–15°F (−22.22°C–9.44°C) below room temperature. The cool air can be used for limited cooling purposes, or it can be ducted outdoors.

A typical residential heat pump water heater rating is 240 volts, 60 Hz, single-phase, 500 watts.

Heat pump water heaters operate with a hermetic refrigerant motor compressor. The electric hookup is similar to a typical 240-volt, single-phase, residential air-conditioning unit. *Article 440* of the *NEC®* applies. This is covered in Unit 23.

Electric utilities are a good source to learn more about the economics of installing heat pump water heaters to save energy.

REVIEW

Note: Refer to the Code or the plans where necessary.

WATER PUMP CIRCUIT Ⓐ ᴮ

1. Does a jet pump have any electrical moving parts below the ground level? _____

2. Which is larger, the drive pipe or the suction pipe? _____

3. Where is the jet of the pump located? _____

4. What does the impeller wheel move? _____

5. Where does the water flow after leaving the impeller wheel?

    a. _____

    b. _____

6. What prevents water from draining back into the pump from the tank? _____

7. What prevents water from draining back into the well from the equipment? _____

_____

8. What is compressed in the water storage tank? _____

9. Explain the difference between a "two-wire" submersible pump and a "three-wire" submersible pump. _____

_____

_____

_____

10. What is a common speed for jet pump motors? _____

11. Why is a 240-volt motor preferable to a 120-volt motor for use in this residence?

_____

_____

12. How many amperes does a 1-hp, 240-volt, single-phase motor draw? (See *Table 430.148.*) _____

13. What size are the conductors used for this circuit? _____

14. What is the branch-circuit protective device? _____

15. What provides the running overload protection for the pump motor? _____

_____

_____

16. What is the maximum ampere setting permitted for running overload protection of the 1-hp, 240-volt pump motor? _____

17. Submersible water pumps operate with the electrical motor and actual pump located (Circle correct answer.)
    a. above permanent water level
    b. below permanent water level
    c. half above and half below permanent water level

18. Because the controller contains the motor starting relay and the running and starting capacitors, the motor itself contains _____

19. What type of pump moves the water upward inside of the deep-well pipe? _____

_____

20. Proper pressure of the submersible pump system is maintained by a _____

21. Fill in the data for a 16-ampere electric motor, single-phase, no Code letters.
    a. Branch-circuit protection, non-time-delay fuses.
       Normal size _____ A   Switch size _____ A
       Maximum size _____ A   Switch size _____ A
    b. Branch-circuit protection dual-element, time-delay fuses.
       Normal size _____ A   Switch size _____ A
       Maximum size _____ A   Switch size _____ A
    c. Branch-circuit protection instant-trip breaker.
       Normal setting _____ A
       Maximum setting _____ A

d. Branch-circuit protection—inverse time breaker.

Normal rating _____ A

Maximum rating _____ A

e. Branch-circuit conductor size, Type THHN _____

Ampacity _____ A

f. Motor overload protection using dual-element time-delay fuses.

Maximum size _____ A

22. The *NEC®* is very specific in its requirement that submersible electric water pump motors be grounded. Where is this specific requirement found in the Code?

_____

_____

23. Does the *NEC®* allow submersible pump cable to be buried directly in the ground?

_____

_____

24. Must the disconnect switch for a submersible pump be located next to the well?

_____

_____

25. A metal well casing (shall) (shall not) be bonded to the pump's equipment grounding conductor or by grounding it with a separate equipment grounding conductor run all the way back to the same ground bus in the panel that supplies the pump circuit, *NEC®* _____ .

WATER HEATER CIRCUIT Ⓐc

1. According to *422.47* the high temperature limiting control must disconnect _____ of the ungrounded conductors. The high temperature control limits the maximum water temperature to _____°F (_____°C).

2. A major hazard involved with water heaters is that they operate under whatever pressure the serving water utility supplies. Water stays liquid at temperatures higher than the normal boiling point of 212°F (100°C). If the high temperature limit control failed to operate for whatever reason, should a pipe burst or a faucet be opened, the pressure would instantly drop to normal atmospheric pressure, causing the superheated water to turn into _____ that could result in _____ . To prevent this from happening, water heaters are equipped with pressure/temperature relief valves.

3. Magnesium rods are installed inside the water tank to reduce _____ .

4. The heating elements in electric water heaters are generally classified into two categories. These are _____ density and _____ density elements.

5. An 80-gallon electric water heater is energized for the first time. Approximately how many hours would it take to raise the water temperature 80°F (26.67°C) (from 40°F to 120°F [4.44°C to 48.89°C])? The water heater has a 3000-watt heating element.

_____

_____

6. What term is used by utilities when the water heater power consumption is measured using different rates during different periods of the day? _____

7. Explain how the electric utility in your area meters residential electric water heater loads.

_____

_____

_____

8. For residential water heaters, the Consumer Product Safety Commission suggests a maximum temperature setting of _____ °F (_____°C). Some states have laws stating a maximum temperature setting of _____ °F (_____°C).

9. An 80-gallon electric water heater has 60°F (15.56°C) incoming water. How many minutes can a 3 gallon/minute showerhead be used to draw 105°F (40.56°C) water? The water heater thermostat had been set at 120°F (48.89°C). _____

10. Approximately how long would it take to produce serious burns to an adult with 140°F (60°C) water? _____

11. Two thermostats are generally used in an electric water heater.

    a. What is the location of each thermostat? _____

    _____

    b. What type of thermostat is used at each location? _____

    _____

12. a. How many heating elements are provided in the heater in the residence discussed in this text? _____

    b. Are these heating elements allowed to operate at the same time? _____

13. When does the lower heating element operate? _____

    _____

14. The Code states that water heaters having a capacity of 120 gallons (450 L) or less shall be considered _____ duty and, as such, the circuit must have a rating of not less than _____ percent of the rating of the water heater.

15. Why does the storage tank hold the heat so long? _____

16. The electric water heater in this residence is connected for "limited demand" so that only one heating element can be on at one time.

    a. What size wire is used to connect the water heater? _____

    b. What size overcurrent device is used? _____

17. a. If both elements of the water heater in this residence are energized at the same time, how much current will they draw? (Assume the elements are rated at 240 volts.)

b. What size and Type THHN wire is required for the load of both elements? Show calculations.

18. a. How much power in watts would the two elements in question 17 use if connected to 220 volts? Show calculations.

b. What is the current draw at 220 volts? Show calculations.

19. A 240-volt heater is rated at 1500 watts.
   a. What is its current rating at rated voltage? _____
   b. What current draw would occur if connected to a 120-volt supply?

   c. What is the wattage output at 120 volts?

20. For a single, non-motor-operated electrical appliance rated greater than 13.3 amperes that has no marking that would indicate the size of branch-circuit overcurrent device, what percentage of the nameplate rating is used to determine the maximum size branch-circuit overcurrent device? _____

21. A 7000-watt resistance-type heating appliance is rated 240 volts. What is the maximum size fuse permitted to protect the branch-circuit supplying this appliance?

_____

22. The Code requirements for standard resistance electric water heaters are found in *Article 422* of the Code. Because heat pump water heaters contain a hermetic refrigerant motor compressor, the Code requirements are found in *Article* _____ of the Code.

Special-Purpose Outlets for Ranges, Counter-Mounted Cooking Unit ▲G, and Wall-Mounted Oven ▲F

OBJECTIVES

After studying this unit, the student should be able to

- interpret electrical plans to determine special installation requirements for counter-mounted cooking units, wall-mounted ovens, and freestanding ranges.

- understand the latest *NEC*® connection requirements for electric ranges, ovens, and counter-mounted cooking units.

- identify and understand the purpose of the National Electrical Manufacturers Association (NEMA) blade and slot configurations on 30-ampere and 50-ampere receptacles and cords.

- identify the terminals of 30-ampere and 50-ampere receptacles.

- understand when to use 3-wire and 4-wire receptacles and cords for electric range hookups.

- compute loads and select proper conductor sizes for electric ranges, wall-mounted ovens, and counter-mounted cooking units.

- properly ground electric ranges, wall-mounted ovens, and counter-mounted cooking units in conformance to the *NEC*® for both new installations and existing installations.

- describe a seven-heat control and a three-heat control for a dual-element heating unit for a cooktop.

- understand how infinite heat temperature controls operate.

- supply counter-mounted cooking units and wall-mounted ovens by connecting them to one feeder using the "tap" rule.

- understand how to install a feeder to a load center, then divide the feeder to individual circuits to supply the appliances.

BASIC CIRCUIT REQUIREMENTS FOR ELECTRIC RANGES, COUNTER-MOUNTED COOKING UNITS, AND WALL-MOUNTED OVENS

Electric ranges are covered in *Article 422* of the *NEC*® and by UL Standard 858, *Household Electric Ranges*.

UL Standard 1082 covers *Electric Household Cooking and Food Serving Appliances*.

Connection Methods

Any recognized wiring method can be used to hook up appliances, such as nonmetallic-sheathed cable, armored cable, EMT, flexible metal or non-metallic conduit, or service-entrance cable. Local electrical codes can contain information as to the restriction on the use of any of the wiring methods listed above.

Here are some of the more common methods for connecting electric ranges, ovens, and cooktops.

Direct Connection. One method is to run the branch-circuit wiring directly to a junction box on the appliance provided by the manufacturer of the appliance. When "roughing-in" the branch-circuit wiring, sufficient length must be provided so as to allow flexibility to make up the connections, and to place the appliance into position, and for servicing.

Junction Box. Another method is to run the branch-circuit wiring to a junction box (an outlet box) behind, under, or adjacent to the range, oven, or cooktop location. By definition, this junction box is accessible because should the need arise, the junction box can be reached by removing the appliance. The "whip" on the appliance is then connected to the junction box where the splicing of the tap conductors in the whip and the branch-circuit conductors are made. According to UL Standard No. 858, the flexible metal whip must not be less than 3 ft (900 mm) and no longer than 6 ft (1.8 mm). The conductors in the whip are rated for the high temperatures encountered in the appliance.

Depending on the location of the appliances, it might be possible to connect more than one cooking appliance to a single branch-circuit from a single junction box or from two junction boxes. This is illustrated in Figure 20-1.

Cord- and Plug-Connection. Still another very common method is to use a cord- and plug-connection, *422.33(A)*. This is particularly true for free-standing ranges. A cord- and plug connection also serves as the disconnecting means for household cooking equipment. The range receptacle is permitted to be at the rear base of an electric range. Removing the bottom drawer makes this connection accessible, *422.33(B)*. The receptacle must have an

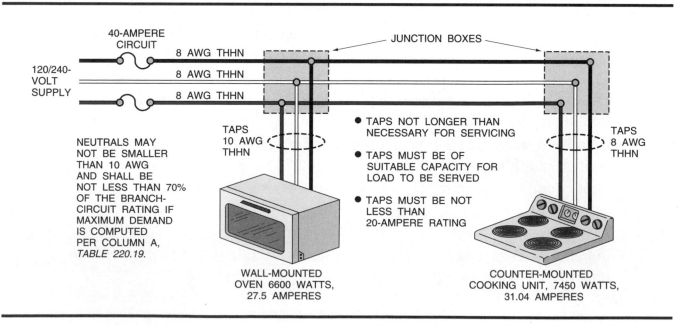

Figure 20-1 A counter-mounted cooktop unit and a wall-mounted oven connected to one branch-circuit. Depending on the location of the appliances, it might be possible to make the installation with one junction box.

ampere rating not less than the rating of the appliance, *422.33(C)*. For household cooking equipment, the demand factors of *Table 220.19* are also applicable to the receptacle, *422.33(C), Exception*. Range and dryer receptacles are identical, and are discussed later in this unit.

The load of an appliance that is cord- and plug-connected shall not exceed 80 percent of the ampere rating of the branch-circuit, *210.23(A)* and *(B)*.

Cord sets rated 30 amperes are referred to as *dryer cords*. Cord sets rated 40, 45, or 50 amperes are referred to as *range cords*.

Receptacles rated 30 amperes are referred to as *dryer receptacles*. Receptacles rated 50 amperes are referred to as *range receptacles*.

Nameplate on Appliance

NEC® 422.60(A) requires that the nameplate must show the appliance's rating in volts and amperes, or in volts and watts.

Load Calculations

In *210.19(C)* and *Exceptions*, we find that:

- the branch-circuit conductors must have an ampacity not less than the branch-circuit rating.

- the branch-circuit conductors' ampacity must not be less than the maximum load being served.

- the branch-circuit rating must not be less than 40 amperes for ranges of 8¾ kW or more.

- for ranges 8¾ kW or more, the neutral of a three-wire branch-circuit may be 70 percent of the branch-circuit rating, but not smaller than 10 AWG.

- tap conductors connecting ranges, ovens, and cooking unit to a 50-ampere branch-circuit must be rated at least 20 amperes, be adequate for the load, and must not be any longer than necessary for servicing the appliance.

- *NEC® 220.19* permits the use of the values found in *Table 220.19* and its footnotes for the load computation.

- *Table 220.19* shows the kW-demand for different kW rated household cooking appliances that are over 1¾ kW. Column A is used in most situations. Note that the maximum demand is considerably less than the actual appliance's nameplate rating. This is because rarely, if ever,

will all of the heating elements be used at the same time. *Note 3* to the table tells us when we are permitted to use Columns B and C. This is discussed later in this unit.

- *NEC® 220.22* states that for calculating a feeder or service, the neutral load can be figured at 70 percent of the load on the ungrounded conductors for ranges, ovens, and counter-mounted cooktops using *Table 220.19* as the basis for the load calculation.

In *422.10(A)* we find that the branch-circuit rating shall not be less than the marked rating of the appliance. However, there is an exception for household cooking appliances. The last paragraph of this section states that *"Branch circuits for household cooking appliances shall be permitted to be in accordance with Table 220.19."*

The demand load for individual wall-mounted ovens or individual counter-mounted cooking units must be calculated at the appliance's full nameplate rating, *Table 220.19, Note 4*.

Wire Size

After making the load calculations to determine the appliance's computed load in amperes, we then refer to *Table 310.16* to select a conductor that has an allowable ampacity equal to or greater than the computed load. As mentioned many times in this text, *110.14(C)* as well as the UL Standards require that we use the 60°C column, regardless of the conductor's insulation rating.

An exception to this basic rule is when the equipment is marked for 75°C. This was discussed in detail in Unit 4 in the section entitled *"The Weakest Link of The Chain."*

Overcurrent Protection

NEC® 422.11(A) states that the branch-circuit overcurrent protection shall not exceed the rating marked on the appliance.

For appliances that have no marked maximum overcurrent protection, then the branch-circuit overcurrent protection is sized at the ampacity of the branch-circuit conductors supplying the appliance. See *240.4(B), (D)*, and *(E)*.

Some large commercial electric ranges might have integral overcurrent protection that subdivides the range into two or more circuits, each overcurrent device rated not over 50 amperes, *422.11(B)*.

Disconnecting Means

All appliances require a means of disconnecting it from the power source, *422.30*.

NEC® 422.33(A) accepts a cord-and-plug arrangement as the required disconnecting means.

A branch-circuit switch or circuit breaker is acceptable as the disconnecting means if it is within sight of the appliance or is capable of being locked in the "OFF" position, *422.31(B)*. Breaker manufacturers supply "lock-off" devices that fit over the circuit breaker. Individual disconnect switches also have a "lock-off" provision.

NEC® 422.34 accepts the "unit switches" that have a marked "OFF" position to be an acceptable disconnect for the appliance because they disconnect all of the ungrounded conductors. This enables a person to work on the appliance knowing that the power is off to the various electrical components of the range. However, the line side terminals of the appliance would still be "hot." The branch-circuit switch or circuit breaker serves as the disconnect that will turn off the total power to the appliance. This branch-circuit breaker disconnecting means is very important when appliances are "hard wired" as opposed to being cord- and plug-connected.

GROUNDING FRAMES OF ELECTRIC RANGES, WALL-MOUNTED OVENS, AND COUNTER-MOUNTED COOKING UNITS

As with most electrical equipment, the frames of electric ranges, wall-mounted ovens, and counter-mounted cooking units must be grounded. Grounding is accomplished through the equipment grounding conductor in nonmetallic-sheathed cable, the metal armor of BX, or with metal raceways such as EMT and flexible metal conduit. See *250.118* for a list of acceptable equipment grounding means.

Electric ranges, wall-mounted ovens, and counter-mounted cooking units are not permitted to be grounded to the neutral branch-circuit conductor, *250.140*. The small copper grounding strap (link)(jumper) furnished with these appliances *must not* be connected between the neutral terminal and the metal frame of the appliance. It is best to remove and discard this grounding jumper.

However, unusual as it may seem, grounding the frames of these appliances to the neutral was permit-

ted prior to the 1996 *NEC®.* Beginning with the 1996 edition of the *NEC®,* the Code recognized this previously permitted practice with restrictions. Read the next sentence carefully! When installing a new electric range, wall-mounted oven, or counter-mounted cooking unit on an *existing* installation branch-circuit where the wiring method did not contain a separate equipment grounding conductor, it is permitted to ground their frames to the neutral circuit conductor. In this situation, the small copper grounding strap (link)(jumper) furnished with these appliances *must* be used. Note that this practice is permitted only for existing installations.

This subject was covered in detail in Unit 15, and will not be repeated here.

Receptacles for Electric Ranges and Dryers

Because of their similarities, the following text applies to both range and dryer receptacles.

Residential electric ranges, electric dryers, wall-mounted ovens, and surface-mounted cooking units may be connected with cords designed specifically for this use. Depending upon the appliance's wattage rating, the receptacles are usually rated 30 amperes or 50 amperes, 125/250 volts. Cord sets rated 40 amperes or 45 amperes contain 40- or 45-ampere conductors, yet the plug cap is rated 50 amperes. It becomes a cost issue to use a full 50-ampere rated cord set in all cases, or to match the cord ampere rating as close as possible to that of the appliance being connected.

Range and dryer receptacles are available in both surface mount and flush mount.

Blade and Slot Configuration for Plug Caps and Receptacles

Figure 20-2 shows the different configurations for receptacles commonly used for electric range and dryer hookups. NEMA uses the letter "R" to indicate a receptacle, and the letter "P" for plug cap.

- **30-ampere, Three-wire:** The L-shaped slot on the receptacle (NEMA 10-30R) and the L-shaped blade on the plug cap (NEMA 10-30P) are for the white neutral wire. Frames of electric ranges, ovens, and clothes dryers installed prior to the 1996 *NEC®* were permitted to be grounded to the neutral conductor, as stated in

50-AMPERE
3-POLE, 3-WIRE
125/250-VOLT
NEMA 10-50R
PERMITTED
PRIOR TO 1996 *NEC®*

50-AMPERE
4-POLE, 4-WIRE
125/250-VOLT
NEMA 14-50R
REQUIRED BY 1996 *NEC®*

30-AMPERE
3-POLE, 3-WIRE
125/250-VOLT
NEMA 10-30R
PERMITTED
PRIOR TO 1996 *NEC®*

30-AMPERE
4-POLE, 4-WIRE
125/250-VOLT
NEMA 14-30R
REQUIRED BY 1996 *NEC®*

Figure 20-2 Illustrations of 30-ampere and 50-ampere receptacles used for cord- and plug-connection of electric ranges, ovens, counter-mounted cooking units, and electric clothes dryers. Four-wire receptacles and four-wire cord sets are required for installations according to the *NEC.®* Three-wire receptacles and three-wire cord sets were permitted prior to the 1996 *NEC.®* See *250.140* and *250.142*.

250.140 and *250.142* of the 1996 edition and previous editions of the *NEC®* The neutral conductor served a dual purpose, that of the neutral and that of the appliance's required equipment ground. Today, this permission is for existing branch-circuits only.

- **30-ampere, Four-wire:** The L-shaped slot on the receptacle (NEMA 14-30R) and the L-shaped blade on the plug cap (NEMA 14-30P) are for the white neutral wire. The horseshoe-shaped slot on the receptacle and the round or horseshoe-shaped blade on the plug cap are for the equipment ground.

- **50-ampere, Three-wire:** The wide flat slot on the receptacle (NEMA 10-50R) and the matching wide blade on the plug cap (NEMA 10-50P) are for the white neutral wire. Frames of electric ranges, ovens, and clothes dryers installed prior to the 1996 *NEC®* were permitted to be grounded to the neutral conductor as stated in *250.140* and *250.142*. This permission is for existing branch-circuits only. The neutral connection served a dual-purpose: that of the neutral and that of the appliance's required equipment ground.

- **50-ampere, Four-wire:** The wide flat slot on the receptacle (NEMA 14-50R) and the matching wide blade on the plug cap (NEMA 14-50P) are for the white neutral wire. The horseshoe-shaped slot on the receptacle and the round or horseshoe-shaped blade on the plug cap are for the equipment ground.

Cord Sets for Electric Ranges and Dryers

A 30-ampere cord set contains:

- **Three-wire:** three 10 AWG conductors. The attachment plug cap (NEMA 10-30P) is rated 30 amperes.

- **Four-wire:** four 10 AWG conductors. The attachment plug cap (NEMA 14-30P) is rated 30 amperes.

A 40-ampere cord set contains:

- **Three-wire:** three 8 AWG conductors. The attachment plug cap (NEMA 10-50P) is rated 50 amperes.

- **Four-wire:** three 8 AWG conductors and one 10 AWG conductor. The attachment plug cap (NEMA 14-50P) is rated 50 amperes.

A 45-ampere cord set contains:

- **Three-wire:** two 6 AWG conductors and one 8 AWG conductor. The attachment plug cap (NEMA 10-50P) is rated 50 amperes.

- **Four-wire:** three 6 AWG conductors and one 8 AWG conductor. The attachment plug cap (NEMA 14-50P) is rated 50 amperes.

A 50-ampere cord set contains:

- **Three-wire:** two 6 AWG conductors and one 8 AWG conductor. The attachment plug cap (NEMA 10-50P) is rated 50 amperes.

- **Four-wire:** three 6 AWG conductors and one 8 AWG conductor. The attachment plug cap (NEMA 14-50P) is rated 50 amperes.

Terminal Identification for Receptacles and Cords

Receptacle and cord terminals are marked:

- "X" and "Y" for the ungrounded conductors.

- "W" for the white grounded conductor. The "W" terminals will generally be whitish or silver (tinned) in color, in accordance with *200.9* and *200.10* of the *NEC*.®

- "G" for the equipment grounding conductor. This terminal is green colored, hexagon shaped, and is marked "G," "GR," "GRN," or "GRND" in accordance with *250.126* and *406.9(B)* of the *NEC*.® The grounding blade on the plug cap must be longer than the other blades so that its connection is made before the ungrounded conductors make connection in accordance with *406.9(D)* of the *NEC*.®

WALL-MOUNTED OVEN CIRCUIT Ⓐꜰ

The wall-mounted oven in this residence is located to the right of the kitchen sink.

The branch-circuit wiring, using nonmetallic-sheathed cable, is run from Panel B to the oven location, coming into the cabinet space where the oven will be installed from behind or from the bottom. Earlier in this unit we discussed the various methods for hooking up appliances. The choices are direct connection, junction box, or cord- and plug connection. In most cases, a built-in wall-mounted oven will have a flexible metal conduit attached to it, which is run to a junction box where the electrical connections are completed.

Circuit Number. The circuit number is B (6-8). This information is found in the Schedule of Special Purpose Outlets contained in the Specifications at the rear of this text, and on the Directory for Panel B found in Unit 27.

Nameplate Data. This appliance is rated 6.6 kW (6000 watts) @ 120/240 volts.

Load Calculations. (Use nameplate rating per *220.19, Note 4*.)

$$I = \frac{W}{E} = \frac{6600}{240} = 27.5 \text{ amperes}$$

Wire Size. The conductors in nonmetallic-sheathed cable are rated for 194°F (90°C). However, in conformance to *334.80*, the allowable ampacity is determined using the 140°F (60°C) column of *Table 310.16*. Here we find that a 10 AWG conductor has an allowable ampacity of 30 amperes—more than adequate for the appliance's calculated load of 27.5 amperes. This will be a 10 AWG three-conductor plus equipment grounding conductor nonmetallic-sheathed cable.

Overcurrent Protection. A two-pole, 30-ampere circuit breaker located in Panel B.

Disconnecting Means. The disconnecting means is the 30-ampere, two-pole circuit breaker in Panel B. If a cord- and plug connection were to be made, that also would serve as a disconnecting means.

Grounding. The equipment grounding conductor contained in the nonmetallic-sheathed cable provides the means for grounding the frame of the oven. According to *Table 250.122*, the equipment grounding conductor for a circuit protected by a 30-ampere overcurrent device is a 10 AWG copper conductor. That is the size EGC contained in 10 AWG nonmetallic-sheathed cable.

Figure 20-3 shows a typical wiring diagram for a standard oven unit.

COUNTER-MOUNTED COOKING UNIT CIRCUIT Ⓐɢ

The cooktop in this residence is located in the island.

Counter-mounted cooking units are available in many styles. They may have two, three, or four surface heating elements. Instead of individual exposed surface heating elements, some have a smooth ceramic surface where the electric heating elements are in direct contact with the underside of the ceramic top. These units have the same controls as conventional surface heating elements and are wired according to the general *NEC*® guidelines previously discussed.

Most of these ceramic-top cooking units have a faint design to indicate the location of the heating elements. The ceramic is rugged, attractive, and easy to clean.

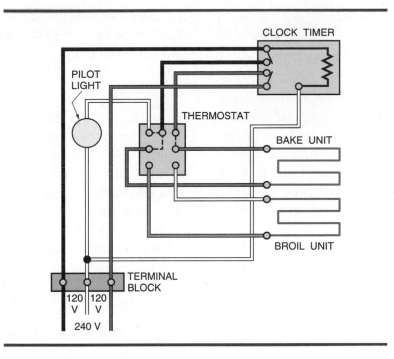

Figure 20-3 Wiring diagram for an oven unit.

The branch-circuit wiring is run from Panel B to the island. The nonmetallic-sheathed cable is concealed above the recreation room ceiling, upward into the island cabinetry, ending in a junction box below the cooking unit. Earlier in this unit we discussed the various methods for hooking up appliances. The choices are direct connection, junction box, or cord-and-plug connection. In most cases, the cooktop will have a flexible metal conduit attached to it, which also runs to the junction box where the electrical connections are completed.

Note on the plans that a receptacle outlet is to be installed on the side of the island. This receptacle outlet is supplied by small appliance branch-circuit B16.

Circuit Number. The circuit number is B (2-4). This information is found in the Schedule of Special Purpose Outlets contained in the Specifications at the rear of this text, and on the Directory for Panel B found in Unit 27.

Nameplate Data. This appliance is rated 7450 watts @ 120/240 volts.

Load Calculations. (Use nameplate rating per *Table 220.19, Note 4*.)

$$I = \frac{W}{E} = \frac{7,450}{240} = 31.04 \text{ amperes}$$

Wire Size. The conductors in nonmetallic-sheathed cable are rated for 90°C. However, according to *334.80*, the allowable ampacity is determined using the 60°C column of *Table 310.16*. Here we find that an 8 AWG conductor has an allowable ampacity of 40 amperes—more than adequate for the appliance's calculated load of 31.04 amperes. This will be an 8 AWG three-conductor plus equipment grounding conductor nonmetallic-sheathed cable.

Overcurrent Protection. A two-pole, 40-ampere circuit breaker located in Panel B.

Disconnecting Means. The disconnecting means is the 40-ampere, two-pole circuit breaker in Panel B. A cord- and plug connection would also be a disconnecting means.

Grounding. The equipment grounding conductor contained in the nonmetallic-sheathed cable provides the means for grounding the frame of the appliance. According to *Table 250.122*, the equipment grounding conductor for a circuit protected by a 40-ampere overcurrent device is a 10 AWG copper conductor. That is the size EGC contained in an 8 AWG nonmetallic-sheathed cable.

FREESTANDING RANGE

Load Calculations

The load calculations for a freestanding electric range are simple.

According to *Table 220.19*, the demand is permitted to be figured at 8 kW for an electric range that is rated 12 kW or less. For example, if an electric range is marked 11.4 kW, the maximum demand can be figured at 8 kW because under normal situations, the surface heating elements, broiler element, baking element, and lights would not all be on at the same time.

Larger double-oven freestanding ranges are probably rated more than 12 kW. Let us consider a range that has a rating of 14,050 watts. This is the same wattage as the previously discussed wall-mounted oven and counter-mounted cooking unit.

The total connected load of this freestanding electric range is:

$$I = \frac{W}{E} = \frac{14,050}{240} = 58.5 \text{ amperes}$$

However, *Table 220.19* permits the computed load to be considerably less than the connected load.

Table 220.19 and the footnotes provide the information we need to make the load calculation.

1. 14,050 − 12,000 watts = 2050 watts (2 kW)

2. According to *Note 1*, *Table 220.19*, a 5% increase for each kW and major fraction in excess of 12 kW must be added to the load demand in Column A. 2 kW (the kW over 12 kW) × 5% per kW = 10%.

3. 8 kW (from Column A) × 0.10 = 0.8 kW

4. The calculated load is: 8 kW + 0.8 kW = 8.8 kW

5. In amperes, the calculated load is:

$$I = \frac{W}{E} = \frac{8,800}{240} = 36.7 \text{ amperes}$$

Wire Size.   The conductors in nonmetallic-sheathed cable are rated for 90°C. However, in conformance to *334.80*, the allowable ampacity is determined using the 60°C column of *Table 310.16*. Here we find that an 8 AWG conductor has an allowable ampacity of 40 amperes—more than adequate for the appliance's calculated load of 36.7 amperes. This will be an 8 AWG three-conductor plus equipment grounding conductor nonmetallic-sheathed cable.

NEC® 210.19(B) states that for ranges of 8¾ kW or more, where the maximum demand has been computed using Column A of *Table 220.19*, the neutral may be reduced to not less than 70 percent of the branch-circuit rating, but never smaller than 10 AWG.

$$40 \times 0.70 = 28 \text{ amperes}$$

Checking *Table 310.16*, we find that a 10 AWG conductor has an allowable ampacity of 30 amperes—more than adequate to carry the computed neutral load. If the wiring method were a raceway method, this electric range could be served by two 8 AWG phase conductors and one 10 AWG neutral conductor.

Using the 70 percent multiplier is a moot point when installing nonmetallic-sheathed cable or armored cable, because the insulated phase conductors in these cables are all the same size.

Why The Reduced Neutral?

The heating elements are connected across the 240-volt source. Therefore, no neutral current is flowing in the 240/120-volt supply to the range.

Some loads in a range are 120-volt loads. These might be controls, timers, convention fans, rotisseries, and similar items. These loads do result in a current flow in the neutral conductor of the 240/120-volt supply to the range.

The permission to reduce the neutral conductor supplying an electric range to 70 percent is more than adequate to carry the above mentioned "line-to-neutral" connected loads within the range.

Some manufacturers supply a small 240/120-volt transformer inside the range that is connected across the two "hot" 240-volt conductors. The 120-volt secondary supplies the 120-volt loads within the range. With this arrangement, there is no neutral current flowing in the 240/120-volt supply to the range. In fact, the branch-circuit to this type of electric range really only needs the two "hot" ungrounded conductors. Because you have no way of knowing what type of electric range will be installed, you must install a 4-wire branch-circuit consisting of two "hot" ungrounded conductors, one ungrounded neutral conductor, and one equipment grounding conductor.

Overcurrent Protection.   A two-pole 40-ampere circuit breaker.

Disconnecting Means.   The disconnecting means would be the 40-ampere, two-pole circuit breaker. A cord-and-plug connection behind the base of a freestanding electric range would also serve as a disconnecting means, *422.33(B)*.

Grounding.   The equipment grounding conductor contained in the nonmetallic-sheathed cable provides the means for grounding the frame of the appliance. According to *Table 250.122*, the equipment grounding conductor for a circuit protected by a 40-ampere overcurrent device is a 10 AWG copper conductor. That is the size EGC contained in an 8 AWG nonmetallic-sheathed cable.

CALCULATIONS WHEN MORE THAN ONE WALL-MOUNTED OVEN AND COUNTER-MOUNTED COOKING UNIT ARE SUPPLIED BY ONE BRANCH-CIRCUIT

The following discusses how to make the load calculations when more than one wall-mounted oven and counter-mounted cooking unit that are supplied by one branch-circuit.

Load Calculations

Note 4 to *Table 220.19* states that when a single branch-circuit supplies a counter-mounted cooking unit, and not more than two wall-mounted ovens, all located in the same room, the calculation is made by adding up the nameplate ratings of the individual appliances, then treating this total as if it were one electric range.

Figure 20-1 shows a wall-mounted oven and a counter-mounted cooking unit supplied by one 40-ampere branch-circuit.

The total connected load of these two appliances is:

$$I = \frac{W}{E} = \frac{14{,}050}{240} = 58.5 \text{ amperes}$$

However, since we are permitted to treat the two individual cooking appliances as one, we use the same calculations as we did for the freestanding range previously discussed.

The resulting conductor size, overcurrent protection, disconnecting means, and grounding are the same as for the equivalent freestanding range.

The cost of installing one 40-ampere branch-circuit might be higher than the cost of installing separate branch-circuits to each appliance. There will be additional junction boxes, different size cables, cable connectors, conduit fittings, and splices. In addition, there are restrictions in the *NEC®* on how to size the smaller "tap" conductors.

Factory-installed flexible metal conduit furnished and connected to these built-in cooking appliances will have the proper size and temperature-rated conductors for the particular appliance.

If you provide the flexible connections (whips) in the field, the minimum conductor sizing would be calculated as discussed above for each appliance. After the minimum conductor size is determined, we then must make sure that the conductors meet the "tap" rule. This is important because the "taps" will have overcurrent protection rated greater than the ampacity of the "tap" conductors.

NEC® 210.23(C) states that fixed cooking appliances that are fastened in place may be connected to 40- or 50-ampere branch-circuits. If "taps" are to be made to a 40- or 50-ampere branch-circuit, the taps must be able to carry the load, and in no case may the taps be less than 20 amperes, *210.19(C)*.

For example, the 40-ampere circuit shown in Figure 20-1 has 10 AWG THHN tap conductors for the oven, and 8 AWG THHN tap conductors for the counter-mounted cooking unit. Each of these taps has an ampacity of greater than 20 amperes.

According to *Table 310.16*, an 8 AWG THHN conductor has an allowable ampacity of 40 amperes, using the 60°C column of the table.

USING A LOAD CENTER

In this residence, the built-in oven and range are very close to Panel B, in which case it is economical to install a separate branch-circuit to each appliance. However, there are situations where the panel and the built-in cooking appliances are far apart.

The decision becomes one of installation cost (labor and material) as to whether to run separate branch-circuits, or to run one larger branch-circuit and use the "tap" rules.

We just finished discussing the issues involved when installing one branch-circuit for more than one built-in cooking appliance, then applying the rather restrictive "tap" rules.

The tap rules can be ignored if a smaller load center is installed as illustrated in Figure 20-4. There will be no "taps"—just branch-circuits to each appliance sized for that particular appliance.

The calculations for the feeder from the main panel to another smaller load center are based on the two cooking appliances being treated as one. The conductors from the smaller load center are calculated for each appliance as previously discussed.

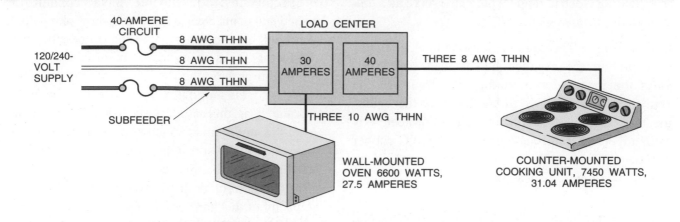

Figure 20-4 Appliances connected to a load center.

In homes with a basement, the load center could be installed below the kitchen where the appliances are installed.

So now you have it! Run separate branch-circuits to each appliance, run one large branch-circuit and make "taps" to serve each appliance, or run one large branch-circuit and install a small load center. The choice is yours. Whichever method you choose, make sure the installation "Meets Code."

CALCULATIONS WHEN MORE THAN ONE ELECTRIC RANGE, WALL-MOUNTED OVEN, OR COUNTER-MOUNTED COOKING UNIT ARE SUPPLIED BY A FEEDER OR SERVICE

Although generally not an issue for wiring one- and two-family dwellings, the calculations for services and feeders of larger multifamily dwellings is more complex where a number of household cooking units are installed.

We need to refer to *Table 220.19*. This table has three columns. Column A is always used, with one exception. *Exception No. 3* shows that for household cooking units having a nameplate rating of more than 1¾ kW through 8¾ kW, Columns B and C are permitted to be used.

Column B is for household cooking units rated more than 1¾ kW but less than 3½ kW. Column C is for household cooking units rated 3½ kW to 8¾ kW.

When using Columns B and C, apply the demand factors for the number of household cooking units whose kW ratings fall within that column. Then add the results of each column together.

EXAMPLE: Calculate the demand for a feeder that has four 3-kW wall-mounted ovens and four 6-kW counter-mounted cooktops connected to it.

Step 1: 4 ovens × 3 kW = 12 kW
12 kW × 66% = 7.92 kW

Step 2: 4 cooktops × 6 kW = 24 kW
24 kW × 50% = 12 kW

Step 3: 7.92 kW + 12 kW = 19.92 kW

Therefore, 19.92 kW is the demand load for the four ovens and four cooktops, even though the actual connected load is 36 kW. Diversity plays a large factor in arriving at the computed demand load.

Review of Choices for Hooking Up Electric Cooking Appliances

- run a separate branch-circuit from the main panel to each appliance,

- run one large branch-circuit to a junction box, then make "taps" to serve each appliance. See Figure 20-1.

- run one large branch-circuit from the main panel to a load center located near the appliances, then run a separate branch-circuit to each appliance. See Figure 20-4.

When one considers the complicated computations, additional splices, taps, working with different size conductors, extra time and material, the best and simplest choice is probably to run a separate branch-circuit from the main panel to each appliance.

HEATING ELEMENTS

Electric heating elements generate a large amount of radiant heat. Heat is measured in British thermal units (Btu). A Btu is defined as the amount of heat required to raise the temperature of one pound of water by one degree Fahrenheit. A 1000-watt heating element generates approximately 3412 Btu of heat per hour.

Surface Heating Elements for Cooking Units

The heating elements used in surface-type cooking units are manufactured in several steps. First, spiral-wound nichrome resistance wire is impacted in magnesium oxide (a white, chalklike powder). The wire is then encased in a nickel-steel alloy sheath, flattened under very high pressure, and formed into coils. The flattened surface of the coil makes good contact with the bottoms of cooking utensils. Thus, efficient heat transfer takes place between the elements and the utensils.

Surface unit heating elements with one-coil (one heating element) construction are used in conjunction with infinite heat switches that "pulse" the circuit to the heating element. Two-coil (two heating elements) surface unit heating elements are used with fixed-heat 3-, 5-, and 7-position switches. Three-coil (three heating elements) sur-

face unit heating elements are used with 3-position switches.

Figure 20-5 and Figure 20-6 illustrate typical electric range surface units.

TEMPERATURE CONTROLS

Infinite-Position Temperature Controls

Modern electric ranges are equipped with infinite-position temperature knobs (controls). Newer switches have a rotating cam that provides an infinite number of heat positions. Older style heat controls connect the heating elements of a surface unit to 120 volts or 240 volts in series or in parallel to attain a certain number of specific heat levels.

The contacts open and close as a result of a heater-bimetal device that is an integral part of the rotary control. When the control is turned on, a separate set of pilot light contacts closes and remains closed, while the contacts for the heating element open and close to attain and maintain the desired temperature of the surface heating unit, Figure 20-7.

Inside these infinite-position temperature controls arc contacts that open and close repetitively. The circuit to the surface heating element is regulated by the ratio of the time the contacts are in

Figure 20-5 Typical 240-volt electric range surface heating element, used with infinite heat controls.

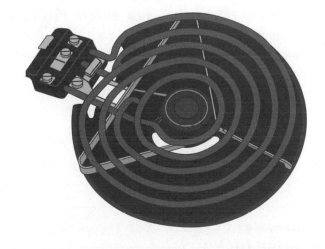

Figure 20-6 Typical electric range surface unit of the type used with seven-position controls.

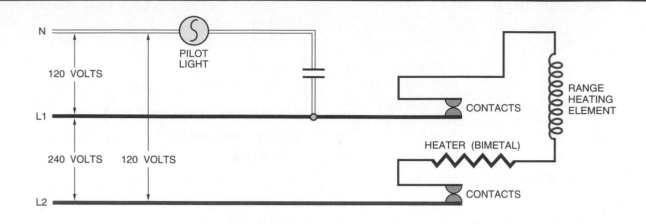

Figure 20-7 **Typical internal wiring of an infinite heat surface element control.**

a closed position versus the open position. This is termed *input percentage*.

The contacts in a temperature control of this type have an expected life of over 250,000 automatic cycles. The knob rotation expected life exceeds 30,000 operations.

One-coil (single-element) surface units are used with infinite heat switches.

Fixed-Heat Temperature Controls

Still available are the older style rotary controls that have specific "indents" that lock the switch contacts into place as the knob is rotated to the various heat positions. Similar switching arrangements are found in push-button-type controls, where one button is pushed at a time. Pushing another button will raise and cancel the first button. Fixed-heat types of rotary and push-button switches, with or without pilot lights, are used in conjunction with two-coil surface unit heating elements, as in Figure 20-8 and Figure 20-9. The different heat levels are obtained by switching between different voltages and coil circuitry. Their unique switching operation offers an excellent opportunity to make electrical calculations for series, parallel, and series/parallel circuits.

Seven-Heat Control for Two-Coil Heating Unit

One method of obtaining several levels of heat is to use a two-element unit, as in Figure 20-8. Figure

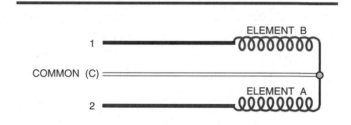

Figure 20-8 **Typical surface unit wiring.**

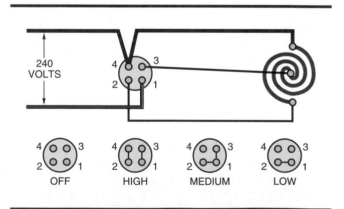

Figure 20-9 **Surface unit with three-heat series-parallel switch.**

20-10 shows that seven levels of heat can be obtained from such a unit if a seven-point "fixed-heat" switch is used. Note that the two elements of the surface heating unit are both contained in the same steel jacket, and they are connected together at the inner end of the steel jacket. A return wire (not a

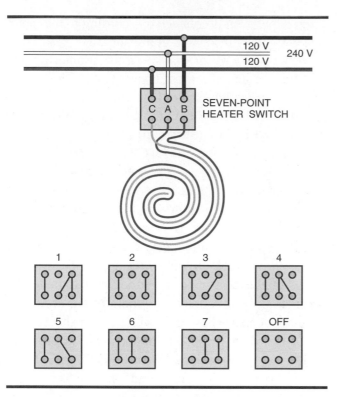

Figure 20-10 Surface heating element with seven levels of heat.

resistance wire) is carried back to the terminal block. This arrangement means that the elements can be operated singly, together, or in series.

Figure 20-8 shows the heating element connected with element A as lead 2, element B as lead 1, and the return wire as the common (C). Resistance readings can be taken between these leads as follows:

Test between C and 2   90 ohms (element A)

Test between C and 1
(single operation)   74 ohms (element B)

Test between 2 and 1   164 ohms (elements A
(elements in series)   and B in series)

Test between C and 1–2
connected together
(elements connected   40.6 ohms (elements A
in parallel)   and B in parallel)

Element B allows more current to flow than does element A because element B has 74 ohms of resistance as compared to 90 ohms of resistance for element A. As the current increases, the wattage also increases. The eight switch positions shown in Figure 20-10 result in the wattages shown in the

following list. For each switch position, the formula $W = E^2/R$ is used to find the wattage value.

1. A and B parallel,
   240 volts   40.6 ohms   1418 watts

2. B only, 240 volts   74 ohms   778 watts

3. A only, 240 volts   90 ohms   640 watts

4. A and B parallel,
   120 volts   40.6 ohms   ____ watts*

5. B only, 120 volts   74 ohms   ____ watts*

6. A only, 120 volts   90 ohms   ____ watts*

7. A and B series,
   120 volts   164 ohms   ____ watts*

8. "OFF" position   Infinity   0 watts

*These values are omitted because the student is required to calculate the values in the Review section. A sample calculation is completed as follows:

$$W = \frac{E^2}{R} = \frac{240 \times 240}{40.6} = \frac{57,600}{40.6} = 1418 \text{ watts}$$

Three-Heat and Five-Heat Control for Two-Coil Heating Units

Three-heat and five-heat switches are for use with two-coil heating elements. These switches function in much the same manner as the seven-heat switch, except for the fewer heat levels available.

A simple, three-heat series-parallel switch is shown in Figure 20-9. A five-heat switch is not shown.

Coil Size Selector Switch

Coil size selector switches are used with three-coil heating elements. They control the heat by surface area rather than by varying the heat over the entire heating element. In the first switch position, the center 4-in. heating element comes on. In the second switch position, the center 4-in. element stays on, and the middle 6-in. element also comes on. In the third switch position, the center 4-in., the middle 6-in., and the outer 8-in. element are all on. These units supposedly save energy because the heating can be focused on the contact area directly beneath a cooking utensil.

Automatic Sensor Control

Infinite heat is attained using a one-coil surface heating element that has a sensor in the center of the surface unit. A special 12-volt control circuit is provided through a small transformer. The control knob is set to the desired temperature. A "responder" pulses the heating element off and on to maintain the set temperature. To operate properly, the cooking utensil must maintain good contact with the sensor. Use flat bottom utensils only.

Caution When Making Connections

After determining the current rating ($I = W/E$), refer to *Table 310.16*. We find that for 27.5 amperes, a 10 AWG copper TW, THW, or THHN may be used to supply this wall-mounted oven. As previously discussed, for nonmetallic-sheathed cable, even though the conductor insulation is rated 90°C, the allowable ampacity must be determined according to the 60°C column of *Table 310.16*. This is stated in *336.80(B)*.

When using armored cable, if the conductors are rated for 75°C as in Type ACTH, the ampacity is permitted to be based upon the 60°C or 75°C column of *Table 310.16*, depending upon the equipment and/or device terminal marking. If the conductors are rated for 90°C as in Type ACTHH, the ampacity would still have to be based upon the 60°C or 75°C column of *Table 310.16*, again depending upon the terminal marking. When armored cable is buried in insulation, it must have 90°C rated conductors, and its ampacity is based upon that of 60°C conductors, *320.80*.

Recall from Unit 4 that, according to UL, and *110.14(C)* of the *NEC*,® when conductor sizes 14 AWG through 1 AWG are installed, their allowable current-carrying capacity (ampacity) is to be that of the 60°C column of *Table 310.16* unless the terminals are marked with some other temperature rating, such as 75°C or 90°C.

That is why it is so important to carefully read the appliance manufacturer's installation data for recommended conductor size and type. Check particularly for the temperature rating.

Microwave Ovens

Microwave ovens are available in countertop models, under-the-cabinet models, and as part of a freestanding electric range.

The electrician's primary concern is to install the proper type and size circuit.

A freestanding electric range that includes a microwave oven is connected according to the methods previously discussed in this unit.

Microwave ovens that sit on the countertop will be plugged into the most convenient receptacle. If the receptacles that serve the countertops have been wired according to the *NEC*® minimum, they would be supplied by at least two 20-ampere, small-appliance branch-circuits, *210.11(C)(1)* and *210.52(B)(1)*. Operating a microwave oven and another appliance at the same time might be too much load for a 20-ampere circuit, and could trip the circuit breaker or open the fuse. That is why microwave cooking appliance manufacturers suggest a dedicated 15- or 20-ampere branch-circuit for the appliance.

Microwave ovens fastened to the underside of a cabinet above a countertop generally are cord- and plug-connected. They come with a rather short power supply cord. A receptacle outlet must be provided inside the upper cabinet. Instructions furnished with these "under-cabinet mounted" microwave ovens describe where to position this receptacle. A receptacle inside the cabinet is *in addition* to the receptacles required to serve countertop surfaces, *210.52*, and is not to be connected to any of the 20-ampere, small-appliance branch-circuits that serve the countertop areas, *210.52(B)(2)*. ▶Furthermore, *210.52(C)(5)* states that receptacles that serve countertops shall not be located more than 20 in. (500 mm) above the countertop.◀

A cord- and plug-connected appliance must not exceed 80 percent of the branch-circuit rating, *210.23(A)(1)*. A 20-ampere branch circuit may supply a maximum cord connected load of $20 \times 0.80 = 16$ amperes.

Single Receptacle on a 20-Ampere Branch-Circuit

As mentioned above, it might be necessary to run a separate 120-volt, 20-ampere branch-circuit for a microwave oven. When a single receptacle is installed on an individual 20-ampere branch-circuit, the receptacle must be rated 20 amperes, *210.21(B)(1)*.

Lightwave Energy Ovens

A recent innovation for ovens is one that combines and cycles "ON" and "OFF" lightwave energy from three 1500-watt halogen lamps and microwave energy. Cooking time is about four times faster than a conventional oven. These ovens are mounted to the underside of an upper cabinet. They require a dedicated three-wire, 30-ampere, 240/120-volt circuit. A four-wire, 30-ampere cord-and-plug is furnished with these ovens. A 30-ampere, four-wire, NEMA 14-30R receptacle must be installed in the cabinet above the oven.

Grounding Microwave Ovens

Grounding of cord-connected microwave and lightwave energy ovens is accomplished by the equipment grounding conductor in the cord furnished with the appliance.

Grounding of hard-wired ovens is established by the equipment grounding conductor in nonmetallic-sheathed cable, or through the use of a metallic wiring method such as Type AC or MC cable, or flexible metal conduit.

UL Standard 923 covers microwave and lightwave energy cooking appliances.

Connecting Foreign-Made Appliances

Be careful when connecting foreign-made appliances. It is quite possible that the color coding of the terminals and conductors will be different than what you are accustomed to. Throughout this text, we have discussed color coding as white for the grounded circuit conductor, black and red for the ungrounded "hot" conductor, and green or bare for the equipment grounding conductor. In this country, we use the term "ground." In Europe, "ground" is referred to as "earth." Instructions might indicate something like this:

Green, green/yellow, or bare = earth

Blue = neutral

Brown = live

The equipment grounding terminal might be marked with the letter "E."

Just be aware that there could be differences in the identification of the terminals and conductors.

REVIEW

Note: Refer to the Code or the plans where necessary.

COUNTER-MOUNTED COOKING UNIT CIRCUIT Ⓐ_G

1. a. What circuit supplies the counter-mounted cooking unit in this residence? _____

   b. What is the rating of this circuit? _____

2. What three methods may be used to connect counter-mounted cooking units? _____

   _____

   _____

   _____

   _____

   _____

3. Is it permissible to use standard 60°C insulated conductors to connect all counter-mounted cooking units? Why? _____

   _____

   _____

   _____

   _____

4. What is the maximum operating temperature (in degrees Celsius) for a
    a. Type TW conductor? _____
    b. Type THW conductor? _____
    c. Type THHN conductor? _____

5. The *NEC®* permits the following wiring methods to supply household built-in ovens and ranges. True or False (Circle the correct answer.)
    a. Nonmetallic-sheathed cable   (True) (False)
    b. Armored cable   (True) (False)
    c. Flexible metal conduit   (True) (False)

6. For electric ranges that have a computed demand of 8¾ kW or more, the neutral conductor can be reduced to _____ percent of the branch-circuit rating, but shall not be smaller than _____ AWG.

7. In the illustration of a typical seven-heat, eight-position switch, Figure 20-10, the wattage values for positions 4, 5, 6, and 7 are omitted. Calculate these values. Show all calculations.

    Position 4

    Position 5

    Position 6

    Position 7

8. One kilowatt equals _____ Btu per hour.

9. Older style electric ranges had heat control knobs that had "indents" that could be felt when rotating the knob to adjust temperature to three, five, or seven different heat positions. Most new electric ranges have control knobs that do not have "indents." This type of heat control is known as an _____ heat control.

WALL-MOUNTED OVEN CIRCUIT ⒶF

1. To what circuit is the wall-mounted oven connected? _____

2. An oven is rated at 7.5 kW. This is equal to

   a. _____ watts.

   b. _____ amperes at 240 volts.

3. a. What section of the Code governs the grounding of a wall-mounted oven? _____

   _____

   b. By what methods may wall-mounted ovens be grounded? _____

   _____

   _____

   _____

   _____

   _____

   _____

4. What is the type and ampere rating of the overcurrent device protecting the wall-mounted oven in this residence? Hint: Refer to the Schedule of Special Purpose Outlets in the Specifications. _____

   _____

   _____

   _____

5. Approximately how many ft (m) of cable are required to connect the oven in the residence? _____

   _____

6. When connecting a wall-mounted oven and a counter-mounted cooking unit to one feeder, how long are the taps to the individual appliances? _____

   _____

   _____

7. The branch-circuit load for a single wall-mounted self-cleaning oven or counter-mounted cooking unit shall be the _____ rating of the appliance.

8. A 6-kW counter-mounted cooking unit and a 4-kW wall-mounted oven are to be installed in a residence. Calculate the maximum demand according to *Column A, Table 220.19*. Show all calculations. Both appliances will be connected to the same branch-circuit.

9. The size of the neutral conductor supplying an electric range may be based on _____ percent of the ampacity of the ungrounded conductors.

10. A freestanding electric range is rated 11.8 kW, 120/240 volts. The wiring method is nonmetallic-sheathed cable. Use the 60°C column of *Table 310.16* in conformance to *110.14(C)*. Answer the following questions.

   a. According to Column A, *Table 220.19*, what is the maximum demand? _____ kW.

   b. What is the minimum size for the ungrounded copper conductors?   _____ AWG

   c. What is the minimum size neutral copper conductor?   _____ AWG

   d. What is the correct rating for the branch-circuit overcurrent protection?
      _____ amperes

11. A double-oven electric range is rated at 18 kW, 120/240 volts. Calculate the maximum demand according to *Table 220.19*. Show all calculations.

12. For the range discussed in question 11:
    a. What size 140°F (60°C) ungrounded conductors are required? _____
    b. What size 140°F (60°C) neutral conductor is required? _____
    c. What size 194°F (75°C) ungrounded conductors are required? The terminal block and lugs on the range, the branch-circuit breaker, and the panelboard are marked as suitable for 194°F (75°C) wire. _____
    d. What size 194°F (75°C) neutral conductor is required? Terminations are the same as in c. _____

13. For ranges of 8¾ kW or higher rating, the minimum branch-circuit rating is _____ amperes.

14. When a separate circuit supplies a counter-mounted cooking unit, what Code reference tells us that it is "against Code" to apply the demand factors of *Table 220.19*, and requires us to compute the load based upon the appliance's actual nameplate rating? _____

15. A nonmetallic-sheathed cable is used to connect a wall-mounted oven. The insulated conductors are 10 AWG. What is the size of the equipment grounding conductor in this cable? _____

16. Match the following statements with the correct letter. Some letters may be used more than once.

_____ A 30-ampere, three-wire receptacle	a. for the equipment grounding
_____ A 50-ampere, four-wire receptacle	conductor
_____ The L-shaped blade on a	b. for the white neutral conductor
30-ampere plug cap	c. for the "hot" ungrounded
_____ The horseshoe-shaped slot	conductors
on a 50-ampere receptacle	d. NEMA 10-30R
_____ A 50-ampere, four-wire plug cap	e. NEMA 14-50R
_____ Terminals "X" and "Y"	f. NEMA 14-50P
_____ Terminal "W"	
_____ Terminal "G"	

17. A receptacle is mounted inside of a kitchen cabinet for the purpose of plugging in a microwave oven. Circle the correct answer.

   a. The receptacle may be connected to the same circuit that supplies the other receptacles that serve the countertop areas.

   b. The receptacle shall not be connected to the same circuit that supplies the other receptacles that serve the countertop areas.

Special-Purpose Outlets – Food Waste Disposer ⬲ᴴ, Dishwasher ⬲ᵢ

OBJECTIVES

After studying this unit, the student should be able to

- install circuits for kitchen appliances such as a food waste disposer and a dishwasher.
- provide adequate running overcurrent protection for an appliance when such protection is not furnished by the manufacturer of the appliance.
- describe the difference in wiring for a food waste disposer with and without an integral "ON-OFF" switch.
- determine the maximum power demand of a dishwasher.
- make the proper grounding connections to the appliances.
- describe disconnecting means for kitchen appliances.

This unit discusses in detail the circuit requirements for food waste disposers and dishwashers as installed in residences. However, because there are many other types of appliances found in homes, the electrician needs to read carefully *Article 422* of the Code. *Article 422* covers all types of appliances in general, branch-circuit requirements, and installation requirements. It also discusses specific types of appliances.

CAUTION: Keep in mind that these 120-volt appliances are *not* permitted to be connected to the small appliance circuits provided for in the kitchen. The small appliance circuits are intended to be used for the numerous types of portable appliances used in the kitchen area.

FOOD WASTE DISPOSER ⬲ᴴ

The food waste disposer outlet is shown on the plans by the symbol ⬲ᴴ. The food waste disposer, rated at 7.2 amperes, is connected to Circuit B19, a sepa-

rate 20-ampere, 120-volt branch-circuit in Panel B. To meet the specifications, 12 AWG THHN conductors are used. The conductors terminate in a junction box provided on the disposer.

Overload Protection

The food waste disposer is a motor-operated appliance. Food waste disposers normally are driven by a 1/4- or 1/3-hp, split-phase, 120-volt motor. Therefore, running overcurrent protection must be provided.

The manufacturers of food waste disposers provide the required motor overload protection with an integral thermal overload protector. Usually this is a small manual reset button on the unit. In the event of an overload caused by prolonged use of the appliance, or if something falls into the drain and jams the appliance, the thermal overload protector trips, turning off power to the appliance. After the problem has been taken care of, merely push the reset button.

For cord- and plug-connected food waste disposers, a conventional grounding-type receptacle can be installed under the sink. A fuse/receptacle box cover unit could also be installed under the sink as illustrated in Figure 21-1. A dual-element, time-delay plug fuse sized at approximately 125 percent of the motor's full load current is inserted into the fuse holder of the box-cover unit. This fuse serves as additional back-up overload protection and short-circuit protection for the appliance.

Disconnecting Means

All electrical appliances must be provided with some means of disconnecting the appliance, *Article 422, Part III*. The disconnecting means may be a separate "ON-OFF" switch, Figure 21-2, a cord connection, Figure 21-1, or a branch-circuit switch or circuit breaker, within sight of the appliance or capable of being locked in the "OFF" position.

Some local codes require that food waste disposers, dishwashers, and trash compactors be cord connected, as illustrated in Figure 21-1, to make it easier to disconnect the unit, to replace it, to service it, and to reduce noise and vibration. The appliance

shall be intended or identified for flexible cord connection, *NEC® 400.7* and *422.16(A)* and *(B)*.

UL lists food waste disposers with and without a cord-and-plug assembly.

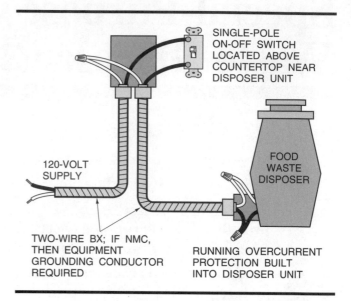

SINGLE-POLE ON-OFF SWITCH LOCATED ABOVE COUNTERTOP NEAR DISPOSER UNIT

FOOD WASTE DISPOSER

120-VOLT SUPPLY

TWO-WIRE BX; IF NMC, THEN EQUIPMENT GROUNDING CONDUCTOR REQUIRED

RUNNING OVERCURRENT PROTECTION BUILT INTO DISPOSER UNIT

Figure 21-2   Wiring for a food waste disposer operated by a separate switch located above countertop near the sink. Many local codes require this wall switch for all installations, to be sure that there is a safe means of easily disconnecting the appliance.

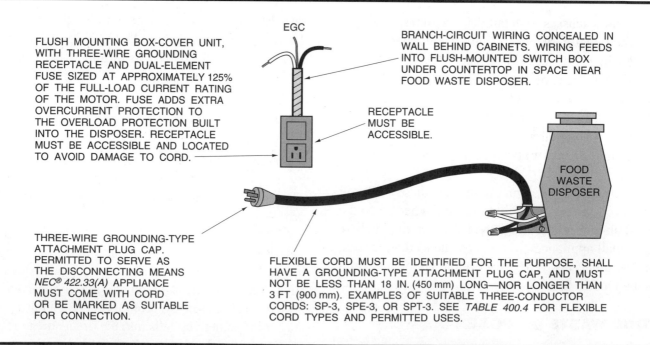

FLUSH MOUNTING BOX-COVER UNIT, WITH THREE-WIRE GROUNDING RECEPTACLE AND DUAL-ELEMENT FUSE SIZED AT APPROXIMATELY 125% OF THE FULL-LOAD CURRENT RATING OF THE MOTOR. FUSE ADDS EXTRA OVERCURRENT PROTECTION TO THE OVERLOAD PROTECTION BUILT INTO THE DISPOSER. RECEPTACLE MUST BE ACCESSIBLE AND LOCATED TO AVOID DAMAGE TO CORD.

EGC

BRANCH-CIRCUIT WIRING CONCEALED IN WALL BEHIND CABINETS. WIRING FEEDS INTO FLUSH-MOUNTED SWITCH BOX UNDER COUNTERTOP IN SPACE NEAR FOOD WASTE DISPOSER.

RECEPTACLE MUST BE ACCESSIBLE.

FOOD WASTE DISPOSER

THREE-WIRE GROUNDING-TYPE ATTACHMENT PLUG CAP. PERMITTED TO SERVE AS THE DISCONNECTING MEANS *NEC® 422.33(A)* APPLIANCE MUST COME WITH CORD OR BE MARKED AS SUITABLE FOR CONNECTION.

FLEXIBLE CORD MUST BE IDENTIFIED FOR THE PURPOSE, SHALL HAVE A GROUNDING-TYPE ATTACHMENT PLUG CAP, AND MUST NOT BE LESS THAN 18 IN. (450 mm) LONG—NOR LONGER THAN 3 FT (900 mm). EXAMPLES OF SUITABLE THREE-CONDUCTOR CORDS: SP-3, SPE-3, OR SPT-3. SEE *TABLE 400.4* FOR FLEXIBLE CORD TYPES AND PERMITTED USES.

Figure 21-1   Typical cord connection for a food waste disposer, *422.16(A)* and *(B)(1)*. The same rules apply to built-in dishwashers and trash compactors, except the cord length must be 3 to 4 ft (900 mm to 1.2 m).

Turning the Food Waste Disposer On and Off

Separate "ON-OFF" Switch. When a separate ON-OFF switch is used, a simple circuit arrangement can be made by running a two-wire supply cable to a switch box located next to the sink at a convenient location above the countertop. Not only is this convenient, but it positions the switch out of the reach of children, as in Figure 21-2. A second cable is run from the switch box to the junction box of the food waste disposer. A single-pole switch in the switch box provides "ON-OFF" control of the disposer.

In the kitchen of this residence, the "continuous feed" food waste disposer is controlled by a single-pole switch located to the right of the kitchen sink above the countertop, in combination with the light switch and receptacle.

Another possibility, but not as convenient, is to install the switch for the disposer inside the cabinet space directly under the sink near the food waste disposer.

Integral "ON-OFF" Switch. A "batch-feed" food waste disposer is equipped with an integral prewired control switch. This integral control starts and stops the disposer when the user twists the drain cover into place. An extra "ON-OFF" switch is not required with this type of control. To connect the disposer, the electrician runs the supply conductors directly to the junction box on the disposer and makes the proper connections, as in Figure 21-3.

Disposer with Flow Switch in Cold Water Line. Some plumbers like to install a flow switch in the cold water line under the sink, as in Figure 21-4. This switch is connected in series with the disposer

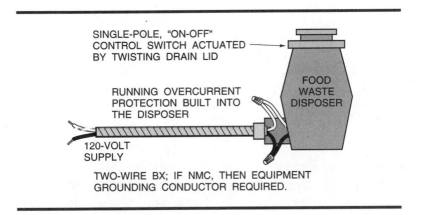

Figure 21-3   Wiring for a "batch-feed" food waste disposer with an integral "ON-OFF" switch.

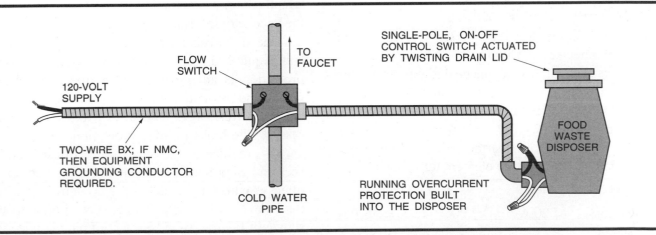

Figure 21-4   "Batch-feed" disposer with a flow switch in the cold water line.

motor. The flow switch prevents the disposer from operating until a predetermined quantity of water is flowing through the disposer. The cold water helps prevent clogged drains by solidifying any grease in the disposer. Thus, the addition of a flow switch means that the disposer cannot be operated without water.

For commercial installations, controllers are installed that let the cold water flow for 45 seconds after the food waste disposer switch is shut off, assuring that the food waste will be flushed through the sewer line.

Grounding

- *Article 250, Part VI* covers equipment grounding and equipment grounding conductors.

- *Article 250, Part VII* covers methods of equipment grounding.

- *NEC® 250.118:* Lists all of the acceptable methods of grounding equipment, such as the equipment grounding conductor in nonmetallic-sheathed cable, armored cable, or flexible metal conduit.

- *NEC® 250.134*: Discusses grounding of equipment fastened in place, or "hard-wired." Refers back to *250.118*.

- *NEC® 250.138:* Accepts the equipment grounding conductor in flexible cords for cord- and plug-connected appliances.

- *NEC® 250.142(B):* Grounding appliances to the grounded circuit conductor is not permitted.

DISHWASHER ▲₁

The dishwasher is supplied by a separate 20-ampere circuit connected to Circuit B5. 12 AWG THHN conductors are used to connect the appliance. The dishwasher has a ⅓-hp motor rated at 7.2 amperes and 120 volts. The outlet for the dishwasher is shown by the symbol ▲₁. During the drying cycle of the dishwasher, a thermostatically controlled, 1000-watt electric heating element turns on.

Actual connected load		Circuit calculation	
Motor	7.20 A	$7.20 \times 1.25 = 9.00$ A	
Heater	8.33 A		8.33 A
Total	15.53 A		17.33 A

For most dishwashers, the motor does not run during the drying cycle. Thus, the actual maximum demand on the branch-circuit would only be the larger of the two loads, which in the previous example is the 1000-watt heating element.

Water heater manufacturers ship their water heaters with the thermostat set as low as possible. When the water heater is installed, the thermostat must be set at recommended settings, such as 120°F (49°C). In fact, because of the problem of people being scalded, many municipalities and states have implemented code requirements that water heater thermostats not be set over 120°F (49°C). This temperature is acceptable for baths and showers, but 120°F (49°C) is not hot enough to properly clean dishes and glasses. The manufacturers of the dishwashers provide a feature that turns on the heating element for the final rinse cycle. This raises the water temperature from 120°F (49°C) to the desired 140°F (60°C) to 145°F (63°C).

See Unit 19 for a discussion on water heaters and the dangers of scalding with hot water of varying temperatures.

Wiring the Dishwasher

The dishwasher manufacturer supplies a terminal or junction box for the electrical hookup. The electrician must make sure the supply conductors (cable or flex conduit) are brought in at the proper location. Usually, the electrician brings the supply conductors into the space under the sink, but the supply conductors could also be brought up from underneath or from behind the dishwasher. The manufacturer's installation manual provides all of the necessary information.

The dishwasher is a motor-operated appliance and requires motor overload protection. Overload protection is an integral part of the dishwasher's motor, and prevents the motor from burning out if it becomes overloaded, stalled, or overheated for any reason. The branch-circuit protection for the appliance is provided by the branch-circuit fuse or circuit breaker. Supplemental overcurrent protection is often provided by installing a box-cover "switch/fuse unit" under the sink of a type similar to that shown in Figure 16-25. For a cord/plug connection, the box-cover unit would be a "receptacle/fuse" type, as in Figure 21-1. Time-delay, dual-element

Type S fuses sized at approximately 125 percent of the nameplate rating of the dishwasher would be installed in the box-cover unit. This also serves as the disconnecting means required by *NEC® 422.30.*

Common Feed for Dishwasher and Food Waste Disposer

In this residence, a separate branch-circuit is run to each appliance: B5 for the dishwasher, and B19 for the food waste disposer. These branch-circuits could be two-wire branch-circuits to each appliance, or a three-wire multi-wire branch-circuit to a junction box under the sink where two-wire connections are made to hook up each appliance.

Another option might make sense where the distance from the kitchen sink area to the electrical panel is very long. Instead of running two separate branch-circuits, it could be more economical to run one branch-circuit to a junction box in the cabinet space under the sink then split the single circuit into two circuits, using a box-cover unit that has two fuseholders and two switches similar to the single switch type illustrated in Figure 16-25. This provides a disconnecting means and supplemental overcurrent protection. Size the time-delay, dual-element fuses to approximately 125 percent of the appliance's full-load ampere rating.

Because of their relatively low current rating, the food waste disposer and dishwasher in this residence could have been supplied by one 20-ampere branch-circuit. The calculations are:

Food waste disposer
$$= 7.2 \times 1.25 \quad = \quad 9.00 \text{ amperes}$$

Heater in dishwasher
$$= 1000 \div 120 \quad = \quad \underline{8.33 \text{ amperes}}$$

Maximum calculated demand $= 17.33$ amperes

But since we usually do not know what the ratings of the disposer or dishwasher will be, the surest method is to run a separate branch-circuit to each appliance.

Do not connect the food waste disposer or dishwasher to any of the required 20-ampere small appliance branch-circuits in the kitchen. This is a violation of *210.52(B)(2).* The small appliance circuits are intended to serve countertops for cord-connected *portable* appliances only. See Unit 12.

Grounding

The *NEC®* requirements for grounding food waste disposers and dishwashers are the same. The subject is discussed in the food waste disposer text.

PORTABLE DISHWASHERS

In addition to built-in dishwashers, portable models are available. These portable units have one hose that is attached to the water faucet and a water drainage hose that hangs in the sink. The obvious location for a portable dishwasher is near the sink. Thus, the dishwasher probably will be plugged into the receptacle nearest the sink.

Portable dishwashers are supplied with a three-wire cord containing two circuit conductors, one equipment grounding conductor, and a three-wire grounding-type attachment plug cap. If the three-wire plug cap is plugged into the three-wire grounding-type receptacle, the dishwasher is adequately grounded.

In older homes that do not have grounding-type receptacles, the use of a portable cord- and plug-connected dishwasher will require the installation of a grounding-type receptacle(s). Assuming the appliance will be used in the kitchen, the receptacle will also have to be GFCI protected, *210.8(A)(6).* This topic is covered in Unit 6.

Cord- and plug-connected portable dishwashers are generally rated 10 amperes or less at 120 volts.

CORD CONNECTION OF FIXED APPLIANCES

As previously discussed, food waste disposers, built-in dishwashers, and trash compactors in residential occupancies are permitted to be cord-connected, as in Figure 21-1.

The Code electrical installation requirements for dishwashers and trash compactors are very similar to the requirements for food waste disposers, except the cord is required to be 3 to 4 ft (900 mm to 1.2 m) long. The receptacle outlet should be located in the under-cabinet space adjacent to the appliance, which generally would be under the sink, allowing for easy access if the appliance needs to be disconnected. It is not recommended that the receptacle outlet be installed behind the appliance. See *422.16(A)* and *(B)(2).*

Receptacle outlets installed beneath the sink for the purpose of cord connecting a dishwasher, trash compactor, or food waste disposer are not required to be of the GFCI type, because these receptacles are not intended to serve the countertop area. Repeating what was stated at the beginning of this unit, these receptacle outlets must not be connected to the required small appliance 20-ampere branch-circuits that serve the kitchen and/or dining room areas.

Some electrical inspectors feel that using the cord-connected method of connecting these appliances is better than direct connection (sometimes called "hard-wired") because servicing these appliances is generally done by appliance repair people.

NEC® 400.7 covers the uses permitted for flexible cords. Specifically, *400.7(A)(8)* states that the appliance must be fastened in such a manner as to allow ready removal of the appliance for maintenance and repairs. The appliance must be intended or identified for flexible cord connection. Similar wording on the use of flexible cord connections for appliances is covered in *422.16(A)* and *(B)*.

Simply stated, these *NEC®* requirements as well as the UL standards mean that unless the appliance is clearly marked as "identified for flexible cord connection," a connection in the field would be a violation and would invalidate the appliance's listing. Some appliances are identified for both flexible and direct connection. As always, pay attention to what the labeling and instructions have to say.

REVIEW

Note: Refer to the Code or the plans where necessary.

FOOD WASTE DISPOSER CIRCUIT ⓐₕ

1. How many amperes does the food waste disposer draw? _____

2. a. To what circuit is the food waste disposer connected? _____

   b. What size wire is used to connect the food waste disposer? _____

3. Means must be provided to disconnect the food waste disposer. The homeowner need not be involved in electrical connections when servicing the disposer if the disconnecting means is _____

4. How is running overcurrent protection provided in most food waste disposer units?

5. When running overcurrent protection is not provided by the manufacturer or if additional backup overcurrent protection is wanted, dual-element, time-delay fuses may be installed in a separate box-cover unit. These fuses are sized at not over _____ percent of the full-load rating of the motor.

6. Why are flow switches sometimes installed on food waste disposers? _____

7. What Code sections relate to the grounding of appliances? _____

8. Do the plans show a wall switch for controlling the food waste disposer? _____

9. A separate circuit supplies the food waste disposer in this residence. Approximately how many ft (m) of cable will be required to connect the disposer?

_____

10. If a receptacle is installed underneath the sink for the purpose of plugging in a cord-connected food waste disposer, must the receptacle be GFCI protected? _____

DISHWASHER CIRCUIT Ⓐ₁

1. a. To what circuit is the dishwasher in this residence connected? _____

   b. What size wire is used to connect the dishwasher? _____

2. The motor on the dishwasher is (circle one)

   a. ¼ hp   b. ⅓ hp   c. ½ hp

3. The heating element is rated at (circle one)

   a. 750 watts   b. 1000 watts   c. 1250 watts

4. How many amperes at 120 volts do the following heating elements draw?

   a. 750 watts _____

   b. 1000 watts _____

   c. 1250 watts _____

5. How is the dishwasher in this residence grounded? _____

_____

_____

_____

_____

6. What type of cord is used on most portable dishwashers? _____

7. How is a portable dishwasher grounded? _____

_____

_____

8. Who is to furnish the dishwasher? _____

_____

9. What article of the Code specifically addresses electrical appliances? _____

Special-Purpose Outlets for the Bathroom Ceiling Heat/Vent/ Lights ▲ₖ ▲ⱼ, the Attic Fan ▲ʟ, and the Hydromassage Tub ▲ₐ

OBJECTIVES

After studying this unit, the student should be able to

- explain the operation and switching sequence of the heat/vent/light.
- describe the operation of a humidistat.
- install attic exhaust fans with humidistats, both with and without a relay.
- list the various methods of controlling exhaust fans.
- understand advantages of installing exhaust fans in residences.
- understand the electrical circuit and Code requirements for hydromassage bathtubs.
- understand grounding requirements for electrical equipment.

BATHROOM CEILING HEATER CIRCUITS ▲ₖ ▲ⱼ

Both bathrooms contain a combination heater, light, and exhaust fan installed in the ceiling. The heat/vent/light is shown in Figure 22-1. The symbols ▲ₖ and ▲ⱼ represent the outlets for these units.

Each heat/vent/light contains a heating element similar to the surface burners of electric ranges. The appliances also have a single-shaft motor with a blower wheel, a lamp with a diffusing lens, and a means of discharging air to the outside of the dwelling.

The heat/vent/lights specified for this residence are rated 1500 watts at 120 volts. The unit in the master bedroom bathroom ▲ⱼ is connected to Cir-

cuit A12, a 20-ampere, 120-volt circuit. In the front bedroom bathroom, the unit is connected to Circuit A11, also a 20-ampere, 120-volt circuit.

The current rating of the 1500-watt unit is

$$I = \frac{W}{E} = \frac{1500}{120} = 12.5 \text{ amperes}$$

Article 424 of the *NEC*® covers fixed electric space heating. *NEC*® *424.3(B)* requires that the branch-circuit conductors and the overcurrent protective device (fuses or circuit breakers) shall not be less than 125 percent of the appliances' ampere rating. Thus, 20-ampere, 120-volt branch-circuits rather than 15-ampere branch-circuits have been chosen to supply the electric ceiling heaters in the bathroom.

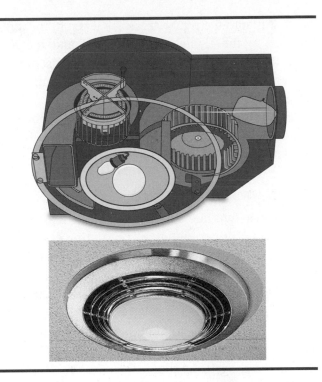

Figure 22-1   Heat/vent/light. *Photo courtesy of NuTone.*

Wiring

In Figure 22-2, we see the wiring diagram for the heat/vent/light. The 120-volt line is run from Panel A to the wall box. A raceway (flexible or electrical metallic tubing) is run between the wall box where the switches are located to the heat/vent/light

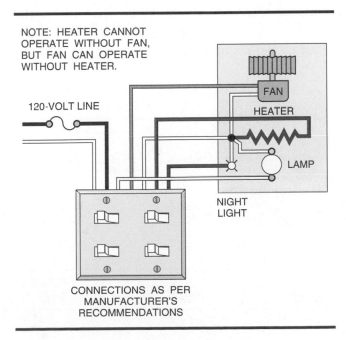

NOTE: HEATER CANNOT OPERATE WITHOUT FAN, BUT FAN CAN OPERATE WITHOUT HEATER.

120-VOLT LINE

FAN
HEATER
LAMP
NIGHT LIGHT

CONNECTIONS AS PER MANUFACTURER'S RECOMMENDATIONS

Figure 22-2   Wiring for heat/vent/light.

unit. In this raceway, five wires are needed—one common white grounded circuit conductor, and one conductor each for the heating element, fan motor, 100-watt lamp, and the 7-watt night light. The four ungrounded switch leg wires may be any color except white, gray, or green. Color coding of conductors was discussed in detail in Unit 5. Because the circuit rating is 20 amperes, 12 AWG conductors are installed. See the Schedule of Special Purpose Outlets contained in the Specifications for this text.

Nontrained electricians often violate the code by running one three-wire and one two-wire cable between the wall box and heat/vent/light. *NEC® 300.3(B)* states that "all conductors of the same circuit and, where used, the neutral and all equipment grounding conductors shall be contained within the same raceway, cable tray, trench, cable, or cord."

The switch location requires a two-gang opening, with the wall box being large enough to accommodate seven 12 AWG conductors, four switches, and cable clamps. For this installation, a 4-in. square, 2⅛ in. deep box with a two-gang raised plaster ring is used. This provides ample room for the conductors and switches. Box sizing was covered in Unit 2 and Unit 8.

Operation of the Heat/Vent/Light

The heat/vent/light is controlled by four switches, as shown in Figure 22-2:

- one switch for the heater
- one switch for the light
- one switch for the exhaust fan
- one switch for the night light

When the heater switch is turned on, the heating element begins to give off heat. The heat activates a bimetallic coil attached to a damper section in the housing of the unit. The heat-sensitive coil expands until the damper closes the discharge opening of the exhaust fan. Air is taken in through the outer grille of the unit. The air is blown downward over the heating element. The blower wheel circulates the heated air back into the room area.

If the heater is turned off so that only the exhaust fan is on, the air is pulled into the unit and is exhausted to the outside of the house by the blower wheel.

Exhaust fans and luminaires (fixtures) are available with integral automatic moisture sensors and motion detectors. Step into a dark bathroom, and the light goes on. Step into and run the shower, and the exhaust fan turns on at a predetermined moisture level.

Grounding the Heat/Vent/Light

The heat/vent/light must be grounded!

The grounding of this appliance is accomplished by using any of the accepted equipment grounding methods found in *250.118*. For residential wiring, the equipment grounding conductor will be in nonmetallic-sheathed cable, armored cable, or flexible metal conduit. Never attempt to ground an appliance to the white grounded circuit conductor, *250.142(B)*.

ATTIC EXHAUST FAN CIRCUIT ⒶL

An exhaust fan is mounted in the hall ceiling between the master bedroom and the front bedroom, as in Figure 22-3(A). When running, this fan removes hot, stagnant, humid, or smoky air from the dwelling. It will draw in fresh air through open windows. Such an exhaust fan can be used to lower the indoor temperature of the house as much as 10°F to 20°F (6°C to 11°C).

The exhaust fan in this residence is connected to Circuit A10, a 120-volt, 15-ampere circuit in the main panel utilizing 14 AWG THHN wire.

Exhaust fans can also be installed in a gable of a house, as in Figure 22-3(B), or in the roof.

On hot days, the air in the attic may reach a temperature of 150°F (66°C) or more. Thus, it is desirable to provide a means of taking this attic air to the outside. The heat from the attic can radiate through the ceiling into the living areas. As a result, the temperature in the living area is raised and the air-conditioning load increases. Personal discomfort increases as well. These problems are minimized by properly vented exhaust fans.

Figure 22-4 shows a typical ceiling-mounted exhaust fan, sometimes referred to as whole-house ventilator fans.

Sizing an Exhaust Fan

Exhaust fans are rated in CFM, which means cubic feet of air per minute of ventilating capability.

To determine the minimum amount of CFM needed, multiply the square footage of the house by 3. For example, an 1800-ft^2 house would need an exhaust fan rated: $1800 \times 3 = 5400$ CFM.

To determine the minimum amount of attic exhaust louver square footage area, divide the CFM by 750. In the above example, $5400 \div 750 = 7.2$ ft^2.

The exhaust fan manufacturers' literature contains sizing recommendations for their particular product.

Fan Operation

Many sizes and types of attic exhaust fans are available. The attic exhaust fan in this residence has a two-speed, ¼-horsepower direct-drive, permanent split capacitor motor. Direct-drive means that the fan blade is attached directly to the shaft of the motor. It is rated 120 volts, 60 Hz (60 cycles per second), 5.8 amperes, 696 VA, and operates at a maximum speed of 1050 rpm. The motor is protected against

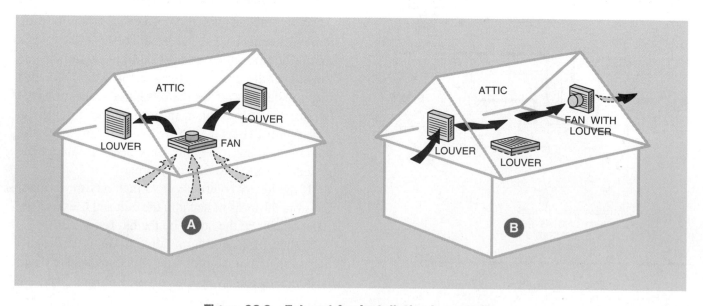

Figure 22-3   Exhaust fan installation in an attic.

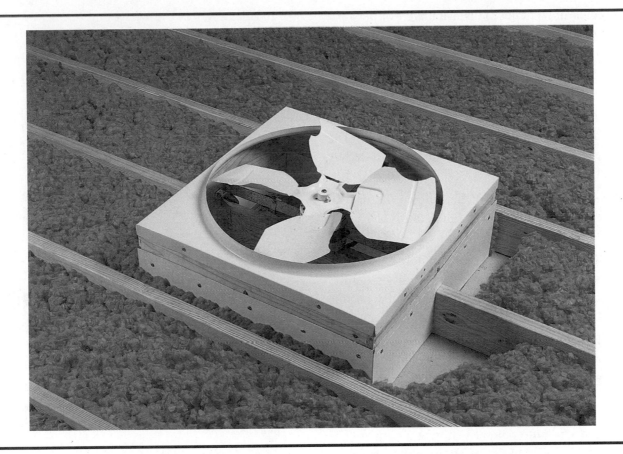

Figure 22-4 Ceiling exhaust fan as viewed from the attic. *Courtesy of* NuTone.

overload by internal thermal protection. The motor is mounted on rubber cushions for quiet operation.

When the fan is not running, the louver remains in the "CLOSED" position. When the fan is turned on, the louver opens. When the fan is turned off, the louver automatically closes to prevent the escape of air from the living quarters.

Fan Control

Typical residential exhaust fans can be controlled by a variety of switches.

Some electricians, architects, and home designers prefer to have fan switches (controls), as in Figure 22-5 and Figure 22-6, mounted 6 ft (1.8 m) above the floor so that they will not be confused

Figure 22-5 Speed control switches. *Courtesy of* NuTone.

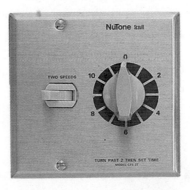

Figure 22-6 Timer switch. *Courtesy of* NuTone.

with other wall switches. This is a matter of personal preference and should be verified with the owner.

Figure 22-7 shows some of the ways that an exhaust fan can be connected.

A. A simple "ON-OFF" switch.

B. A speed control switch that allows multiple and/or an infinite number of speeds. (The exhaust fan in this residence is controlled by this type of control.) See also Figure 22-5.

C. A timer switch (S_T) that allows the user to select how long the exhaust fan is to run, up to 12 hours. See also Figure 22-6.

D. A humidity control (H) switch that senses moisture buildups. See "Humidity Control" later in this unit for further details on this type of switch.

E. Exhaust fans mounted into end gables or roofs of residences are available with an adjustable temperature control on the frame of the fan. This thermostat generally has a start range of 70°F to 130°F (21°C to 54°C), and will automatically stop at a temperature 10°F (6°C) below the start setting, thus providing totally automatic "ON-OFF" control of the exhaust fan.

F. A high temperature automatic heat sensor that will shut off the fan motor when the temperature reaches 200°F (93°C). This is a safety feature so that the fan will not spread a fire. Connect in series with other control switches.

G. A combination of controls. This circuit shows an exhaust fan controlled by an infinite-speed switch, a humidity control (H), and a high temperature heat sensor. Note that the speed control and the humidity control are connected in parallel so that either can start the fan. The high temperature heat sensor is in series so that it will shut off the power to the fan when it senses 200°F (93°C), even if the speed control or humidity control is in the "ON" position.

Overload Protection for the Fan Motor

Several methods can be used to provide running overload protection for the ¼-hp attic fan motor. For

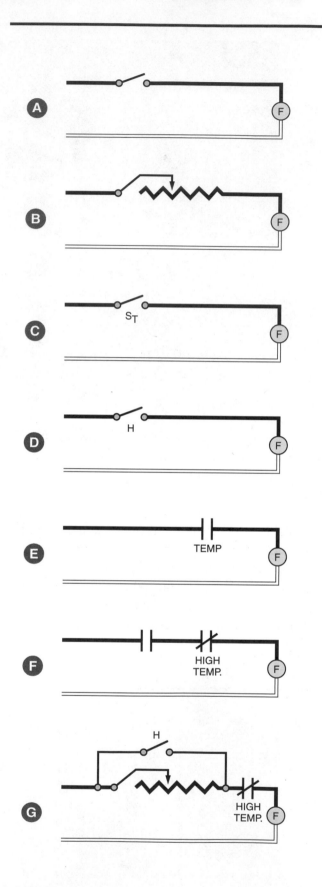

Figure 22-7 Different options for the control of exhaust fans.

example, an overload device can be built into the motor or a combination overload tripping device and thermal element can be added to the switch assembly. Several manufacturers of motor controls also make switches with overcurrent devices. Such a switch may be installed in any standard flush switch box opening (2-in. × 3-in.). A pilot light may be used to indicate that the fan is running. Thus a two-gang switch box or 1½-in. × 4-in. square outlet box with a two-gang raised plaster cover may be used for both the switch and the pilot light.

NEC® 430.32(C) states that for automatically started motors, the running overload device must not exceed 125 percent of the full-load running current of the motor. *Table 430.148* shows that the ¼-hp exhaust fan motor has a full-load current rating of 5.8 amperes. Thus, the overload device rating is $5.8 \times 1.25 = 7.25$ amperes.

Branch-Circuit Short-Circuit and Ground-Fault Protection for the Fan Motor

The attic exhaust fan is connected to Circuit A10, which is a 15-ampere, 120-volt circuit. This branch-circuit supplies the attic exhaust fan only. It supplies no other loads.

The fan motor is equipped with built-in thermal overload protection, and has a full-load current rating of 5.8 amperes at 115 volts. We have already discussed the subject of motor branch-circuit short-circuit and ground-fault protection in Unit 19. Here are the basics.

NEC® 430.32(A)(1): $5.8 \times 1.25 = 7.25$ amperes for overload protection.
(7½-ampere, dual-element, time-delay fuses are available.)

Table 430.52: $5.8 \times 1.75 = 11+$ ampere dual-element, time-delay fuses. (12-ampere dual-element, time-delay fuses are available.)

NEC® 430.52(C)(1), Exception No. 1: Next standard size is 15 amperes.
(12-ampere time-delay, dual-element fuses are available.)

NEC® 430.52(C)(1), Exception No. 2: The maximum size must not exceed 225 percent. $5.8 \times 2.25 = 13+$ amperes maximum. This puts us back to the 12-ampere, dual-element, time-delay fuses.

A switch/fuse box-cover unit can be mounted near the attic exhaust fan.

Since the motor has built-in thermal overload protection, installing either a 7½-, 8-, or 10-ampere, time-delay, dual-element fuse would be a good choice. Should the motor short-circuit or go to ground, the fuse would open without tripping the branch-circuit breaker back in the Panel A.

HUMIDITY CONTROL

An electrically heated dwelling can experience a problem with excess humidity due to the "tightness" of the house, because of the care taken in the installation of the insulation and vapor barriers. High humidity is uncomfortable. It promotes the growth of mold and the deterioration of fabrics and floor coverings. Framing members, wall panels, and plaster or drywall of a dwelling may deteriorate because of the humidity. Insulation must be kept dry or its efficiency decreases. A low humidity level can be maintained by automatically controlling the exhaust fan. One type of automatic control is the *humidistat*. This device starts the exhaust fan when the relative humidity reaches a high level. The fan exhausts air until the relative humidity drops to a comfortable level. The comfort level is about 50 percent relative humidity. Adjustable settings are from 0 percent to 90 percent relative humidity.

The electrician must check the maximum current and voltage ratings of the humidistat before the device is installed. Some humidistats are low-voltage devices and require a relay. Other humidistats are rated at line voltage and can be used to switch the motor directly (a relay is not required). However, a relay must be installed on the line-voltage humidistat if the connected load exceeds the maximum allowable current rating of the humidistat.

The switching mechanism of a humidistat is controlled by a nylon element, as in Figure 22-8. This element is very sensitive to changes in humidity. A bimetallic element cannot be used because it reacts to temperature changes only.

Wiring

The humidistat and relay in the dwelling are installed using two-wire 14 AWG cable. The cable

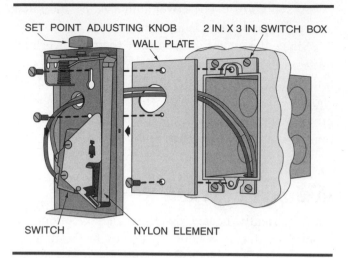

SET POINT ADJUSTING KNOB

WALL PLATE

2 IN. X 3 IN. SWITCH BOX

SWITCH

NYLON ELEMENT

Figure 22-8   Details of a humidity control used with an exhaust fan.

runs from Circuit A10 to a 4-in. square, 2⅛-in.-deep outlet box located in the attic near the fan, as in Figure 22-9. A box-cover unit is mounted on this outlet box. The box-cover unit serves as a disconnecting means within sight of the motor as required by *422.30* and *430.102*.

Mounted next to the box-cover unit is a relay that can carry the full-load current of the motor. Low-voltage thermostat wire runs from this relay to the humidistat. The humidistat is located in the hall between the bedrooms. The thermostat cable is a two-conductor cable because the humidistat has a single-pole, single-throw switching action. The line-voltage side of the relay provides single-pole switching to the motor. Normally the white grounded conductor is not broken.

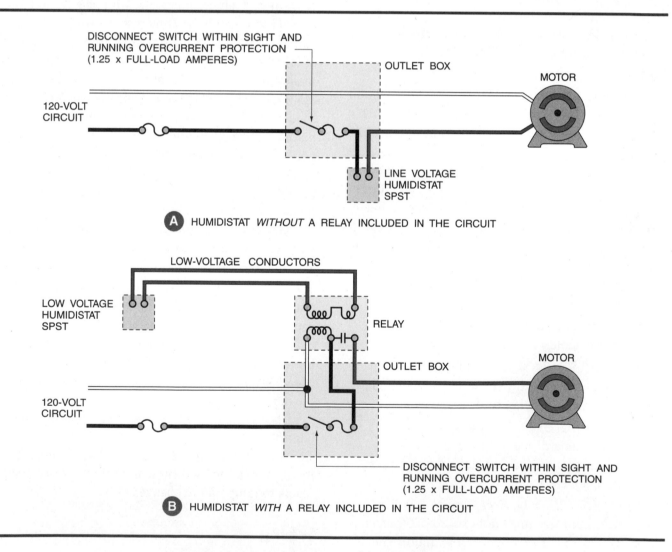

DISCONNECT SWITCH WITHIN SIGHT AND
RUNNING OVERCURRENT PROTECTION
(1.25 x FULL-LOAD AMPERES)

OUTLET BOX

MOTOR

120-VOLT
CIRCUIT

LINE VOLTAGE
HUMIDISTAT
SPST

A HUMIDISTAT *WITHOUT* A RELAY INCLUDED IN THE CIRCUIT

LOW-VOLTAGE CONDUCTORS

LOW VOLTAGE
HUMIDISTAT
SPST

RELAY

OUTLET BOX

MOTOR

120-VOLT
CIRCUIT

DISCONNECT SWITCH WITHIN SIGHT AND
RUNNING OVERCURRENT PROTECTION
(1.25 x FULL-LOAD AMPERES)

B HUMIDISTAT *WITH* A RELAY INCLUDED IN THE CIRCUIT

Figure 22-9   Wiring for the humidistat control.

Although the example in this residence utilizes a relay in order to use low-voltage wiring, it is certainly proper to install line-voltage wiring and use line-voltage thermostats and humidity controls. These are becoming more and more popular, with the switches actually being solid-state speed controllers permitting operation of the fan at an infinite number of speeds between high and low settings.

Grounding the Fan

The metal frame and motor must be grounded. It is possible that this appliance could be within reach of metal ducts in the attic.

Equipment grounding is accomplished by using any of the accepted equipment grounding methods found in *250.118*. For residential wiring, the equipment grounding conductor will be in nonmetallic-sheathed cable, armored cable, or flexible metal conduit. Never ground an appliance to the white grounded circuit conductor, *250.142(B)*.

APPLIANCE DISCONNECTING MEANS

To clean, adjust, maintain, or repair an appliance, it must be disconnected to prevent personal injury. *NEC® 422.30* through *422.35* outline the basic methods for disconnecting fixed, portable, and stationary electrical appliances. Note that each appliance in the dwelling conforms to one or more of the disconnecting methods listed. The more important Code rules for appliance disconnects are as follows:

- each appliance must have a disconnecting means.

- the disconnecting means may be a separate disconnect switch if the appliance is permanently connected.

- the disconnecting means may be the branch-circuit switch or circuit breaker if the appliance is permanently connected and not over ⅛ hp or 300 volt-amperes.

- the disconnecting means may be an attachment plug cap if the appliance is cord connected.

- the disconnecting means may be the unit switch on the appliance only if other means for dis-

connection are also provided. In a single-family residence, the service disconnect serves as the other means.

- the disconnect must have a positive "ON-OFF" position.

- the disconnect must be capable of being locked in the "OFF" position or be within sight of a motor-driven appliance where the motor is more than ⅛ hp.

Read *422.30* through *422.35* for the details of appliance disconnect requirements.

HYDROMASSAGE TUB CIRCUIT ⒜A

The master bedroom is equipped with a hydromassage tub. A hydromassage bathtub is sometimes referred to as a whirlpool bath.

NEC® 680.2 defines a hydromassage bathtub as "a permanently installed bathtub equipped with a recirculating piping system, pump and associated equipment. It is designed so that it can accept, circulate, and discharge water upon each use."

In other words, fill—use—drain.

The significant difference between a hydromassage bathtub and a regular bathtub is the recirculating piping system and the electric pump that circulates the water. Both types of tubs are drained completely after each use.

Spas and hot tubs are intended to be filled, then used. They are *not* drained after each use because they have a filtering and heating system.

Electrical Connections

The *NEC®* requirements for hydromassage tubs are found in Unit 30.

The hydromassage tub is fed with a separate Circuit A9, a 15-ampere, 120-volt circuit using 14 AWG THHN conductors.

The Schedule of Special-Purpose Outlets indicates that the hydromassage bathtub in this residence has a ½-hp motor that draws 10 amperes.

The hydromassage bathtub electrical control is prewired by the manufacturer. All the electrician needs to do is run the separate 15-ampere, 120-volt, GFCI-protected circuit to the end of the tub where

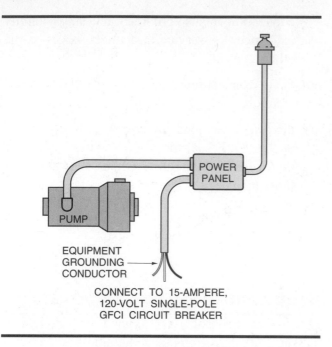

Figure 22-10 Typical wiring diagram of a hydromassage tub showing the motor, power panel, and electrical supply leads.

the pump and control are located. Generally the manufacturer will supply a 3-ft (900 mm) length of watertight flexible conduit that contains one black, one white, and one green equipment grounding conductor, as in Figure 22-10. Make sure that the equipment grounding conductor of the circuit is properly connected to the green grounding conductor of the hydromassage tub.

Do not connect the green grounding conductor to the grounded (white) circuit conductor!

Proper electrical connections are made in the junction box where the branch-circuit wiring is to be connected to the hydromassage wires. This junction box, because it contains splices, must be accessible.

The pump and power panel may also need servicing. Access may be from underneath or the end, whichever is convenient for the installation, as in Figure 22-11.

Figure 22-12 is a photograph of a typical hydromassage tub.

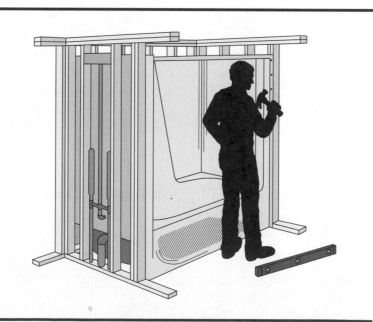

Figure 22-11 The basic roughing-in of a hydromassage bathtub is similar to that of a regular bathtub. The electrician runs a separate 15-ampere, 120-volt GFCI-protected circuit to the area where the pump and control are located. Check the manufacturer's specifications for this data. An access panel from the end or from below is necessary to service the wiring, the pump, and the power panel.

Figure 22-12 A typical hydromassage bathtub, sometimes referred to as a whirlpool. *Courtesy of* The Kohler Company.

REVIEW

Note: Refer to the Code or the plans where necessary.

BATHROOM CEILING HEATER CIRCUITS Ⓐ_K Ⓐ_J

1. What is the wattage rating of the heat/vent/light? _____

2. To what circuits are the heat/vent/lights connected? _____

3. Why is a 4-in. square, 2⅛-in. deep box with a two-gang raised plaster ring or similar deep box used for the switch assembly for the heat/vent/light? _____

_____

_____

_____

_____

4. a. How many wires are required to connect the wall switches and the heat/vent/light?

_____

   b. What size wires are used? _____

5. Can the heating element be energized when the fan is not operating? _____

6. Can the fan be turned on without the heating element? _____

7. What device can be used to provide automatic control of the heating element and the fan of the heat/vent/light? _____

8. Where does the air enter the heat/vent/light? _____

9. Where does the air leave this unit? _____

10. Who is to furnish the heat/vent/light? _____

_____

11. For a ceiling heater rated 1200 watts at 120 volts, what is the current draw? _____

_____

12. An electrician wired up a heat/vent/light as described in this unit. The wiring diagram showed that the electrician needed to run five conductors between the switches and the heat/vent/light. The electrician used one 12/3 nonmetallic-sheathed cable and one 12/2 nonmetallic-sheathed cable. Both cables had equipment grounding conductors. During the rough-in inspection, the electrical inspector cited "Noncompliance with *NEC*® 300.3(B)." What was the problem? _____

_____

_____

ATTIC EXHAUST FAN CIRCUIT Ⓐ ʟ

1. What is the purpose of the attic exhaust fan? _____

_____

_____

2. At what voltage does the fan operate? _____

3. What is the horsepower rating of the fan motor? _____

4. Is the fan direct- or belt-driven? _____

5. How is the fan controlled? _____

6. What is the setting of the running overcurrent device? _____

7. What is the rating of the running overcurrent protection if the motor is rated at 10 amperes? _____

8. What is the basic difference between a thermostat and a humidistat? _____

_____

_____

9. What size conductors are to be used for this circuit? _____

10. How many ft (m) of cable are required to complete the wiring for the attic exhaust fan circuit? _____

11. May the metal frame of the fan be grounded to the grounded circuit conductor?

_____

12. Which section of the Code prohibits grounding equipment to a grounding circuit conductor? _____

HYDROMASSAGE BATHTUB CIRCUIT Ⓐ_A

1. Which circuit supplies the hydromassage bathtub? _____

2. Which of the following statements "meets Code"?

   a. Hydromassage tubs and all of their electrical components shall be GFCI protected.

   b. Hydromassage tubs and all of their electrical components shall not be GFCI protected.

   Circle the correct statement.

3. Which conductor size feeds the hydromassage tub in this residence? _____

4. What is the fundamental difference between a hydromassage bathtub and a spa?

   _____

   _____

5. Which sections of the Code reference hydromassage bathtubs? _____

6. Must the metal parts of the pump and power panel of the hydromassage tub be grounded? _____

   _____

   _____

   _____

7. Is it permissible to connect the hydromassage tub's green equipment grounding conductor to the branch-circuit's grounded (white) conductor? _____

8. All 120-volt, single-phase receptacles within (5 ft [1.5 m]) (10 ft [3.0 m]) (15 ft [4.5 m]) of a hydromassage tub must be GFCI protected. Circle the correct answer.

Special-Purpose Outlets– Electric Heating ◭M, Air Conditioning ◉N

There are many types of electric heat available for heating homes (e.g., heating cable, unit heaters, boilers, electric furnaces, duct heaters, baseboard heaters, heat pumps, and radiant heating panels). This unit discusses many of these types. The residence discussed in this text is heated by an electric furnace located in the workshop.

Detailed Code requirements are found in *Article 424*.

This text cannot cover in detail the methods used to calculate heat loss and the wattage required to provide a comfortable level of heat in the building. For this residence, the total estimated wattage is 13,000 watts. Depending upon the location of the residence (in the Northeast, Midwest, or South, for example), the heating load will vary.

Electric heating has gained wide acceptance when compared with other types of heating systems. It has a number of advantages. Electric heating is flexible when baseboard heating is used because each room can have its own thermostat. Thus, one room can be kept cool while an adjoining room is warm. This type of zone control for an electric, gas- or oil-fired central heating system is more complex and more expensive.

Electric heating is safer than heating with fuels. The system does not require storage space, tanks, or chimneys. Electric heating is quiet. Electric heat does not add or remove anything from the air. As a result, electric heat is cleaner. This type of heating is considered to be healthier than fuel heating systems that remove oxygen from the air. The only moving part of an electric baseboard heating system is the thermostat. This means that there is a minimum of maintenance.

As with all heating systems, adequate insulation must be provided. Proper insulation can keep electric bills to a minimum. Insulation also helps to keep the residence cool during the hot summer months. The cost of extra insulation is offset through the years by the decreased burden on the air-conditioning equipment. Energy conservation measures require the installation of proper and adequate insulation.

TYPES OF ELECTRIC HEATING SYSTEMS

Electric heating units are available in baseboard, wall-mounted, and floor-mounted styles. These units may or may not have built-in thermal overload protection. The type of unit to be installed depends on structural conditions and the purpose for which the room is to be used.

Another method of providing electric heating is to embed resistance-type cables in the plaster, or between two layers of drywall on the ceiling, *424.34* through *424.45.* Because premise wiring (non-metallic-sheathed cable, etc.) will be located above these ceilings and will be subjected to the heat created by the heating cables, the Code requires the following:

- keep the wiring not less than 2 in. (50 mm) above the ceiling, and *also* reduce the ampacity of the wiring to the 50°C correction factors found at the bottom of *Table 310.16,* or

- keep the wiring above insulation that is at least 2 in. (50 mm) thick—in which case the correction factors do not have to be applied.

For ease in identification, *424.35* requires that the leads on heating cables be color coded as follows:

- 120 volts   yellow
- 208 volts   blue
- 240 volts   red
- 277 volts   brown
- 480 volts   orange

Electric heat can also be supplied by an electric furnace and a duct system similar to the type used on conventional hot-air central heating systems. The heat is supplied by electric heating elements rather than the burning of fuel. Air conditioning, humidity control, air circulation, electronic air cleaning, and zone control can be provided on an electric furnace system.

CIRCUIT REQUIREMENTS FOR ELECTRIC FURNACES

In an electric furnace, the source of heat is the resistance heating element(s). The blower motor assembly, filter section, condensate coil, refrigerant line connections, fan motor speed control, high temperature limit controls, relays, and similar components are much the same as those found in gas furnaces.

You probably have noticed that there is a repetition of *NEC®* requirements for various appliances. As with all appliances, many of the *NEC®* requirements have already been met by virtue of the product being UL listed. We might look at this in this manner: do the premise wiring according to the *NEC®* requirements and let the UL requirements take care of the safety requirements of the appliance itself. Your big concern when hooking up an electric furnace is to size the conductors, the branch-circuit overcurrent protection, and the disconnecting means properly.

Rather than repeating the Code requirements in detail, here is a summary of the basic code rules:

- *NEC® 220.15:* When making the calculation for sizing a service or a feeder, include 100 percent of the rating of the total connected fixed space heating load. This is shown in Unit 29.

- *Article 250, Part VII:* Contains most of the requirements for which equipment and appliances are required to be grounded.

- *NEC® 250.118:* Lists the acceptable methods for equipment and appliance grounding. This is generally the equipment grounding conductor in nonmetallic-sheathed cable, the metal armor and bonding strip in armored cable, or a metal raceway such as EMT or FMC.

- *NEC® 422.10(A):* The branch-circuit rating must not be less than the marked rating. This governs the overcurrent protection and conductor sizing issues.

- *NEC® 422.11:* The branch-circuit overcurrent protection must be sized according to the value indicated on the nameplate. The electrician in the field need do nothing more than size the conductors and overcurrent protection according to the nameplate data and instructions furnished with the furnace. The manufacturer of the furnace has done all of the calculations according to UL standards to determine the actual current draw of the appliance.

  The motor has built-in thermal overload protection. The overcurrent protection for other internal components is provided by the manufacturer. This is why it is important to purchase a UL-listed electric furnace.

- *NEC® 422.12:* Central heating equipment must be supplied by a separate branch-circuit. This includes electric, gas, and oil furnaces. It also includes heat pumps. Associated equipment such as humidifiers and electrostatic air cleaners are permitted to be connected to the same branch-circuit. The logic behind this "separate circuit" requirement is that if the heating system were to be connected to some other branch-circuit, such as a lighting branch-circuit, a fault on that circuit would shut off the power to the heating system. This could cause freezing of water pipes. We find this same requirement in *5.6.4* of *NFPA 54*, the *National Fuel Gas Code.®*

- *NEC® 422.30:* A disconnecting means must be provided.

- *NEC® 422.31(B):* The disconnecting means must be within sight of the furnace. For electric furnaces, the disconnect might be mounted on the side of the furnace, or on an adjacent wall. For gas furnaces, the disconnect is simply a snap switch or switch/fuse mounted in a handy box on the side of the furnace. In some instances, the manufacturer of the furnace incorporates the disconnecting means right into the equipment.

- *NEC® 422.32:* Another reminder is that the disconnect for motor-driven appliances having a motor of greater than ⅛ hp must be within sight of the appliance.

- *NEC® 430.60(B):* The nameplate must show volts and watts, or volts and amperes.

- *NEC® 422.62(B)(1):* The nameplate shall specify the minimum size supply circuit conductor's ampacity and the maximum branch-circuit overcurrent protection.

- *NEC® 430.102(A):* The disconnect shall be within sight of the controller and shall disconnect the controller.

- ▶*NEC® 430.102(B):* The disconnect shall be located within sight of the motor and driven machinery. If it is impractical to locate the disconnect within sight of the motor and driven machinery, then it is permissible to use a disconnect that has a permanent method of locking the disconnect in the "OFF" position. Examples of what might be considered "impractical" are given in the *Fine Print Note No. 1.*◀

Figure 23-1 illustrates the key *NEC®* requirements for hooking up an electric furnace.

Supply Voltage

Because proper supply voltage is critical to ensure full output of the furnace's heating elements, voltage must be maintained at not less than 98 percent of the furnace's rated voltage. The effect of voltage drop has already been discussed in Units 4 and 19. Mathematically, for every 1 percent drop in voltage, there will be a 2 percent drop in wattage output of a resistance heating element. Manufacturers' installation instructions usually have tables that show the wattage or kilowatt output at different voltages. Table 23-1 is a typical table.

Therefore, the actual voltage at the supply terminals of an electric furnace or electric baseboard heating unit will determine the true wattage output of the heating elements.

KILOWATT RATING	
@ 240 volts	**@ 208 volts**
12.0	9.0
15.0	11.3
20.0	15.0
25.0	18.8
30.0	22.5

Table 23-1

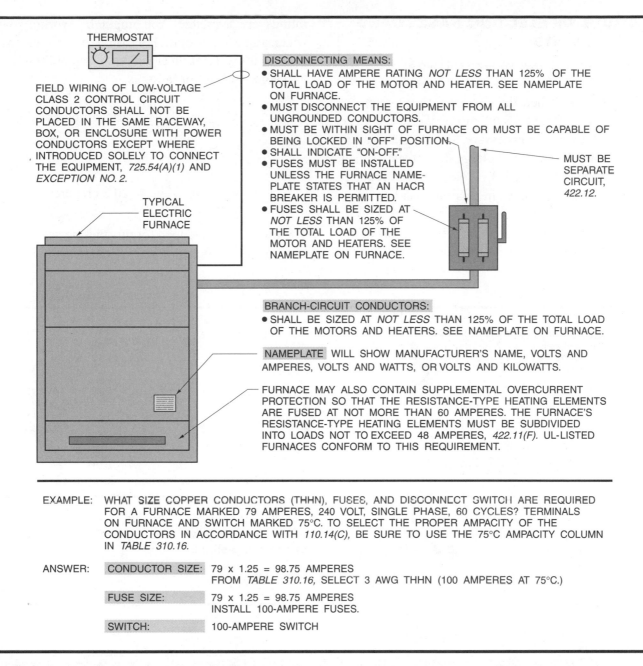

THERMOSTAT

FIELD WIRING OF LOW-VOLTAGE CLASS 2 CONTROL CIRCUIT CONDUCTORS SHALL NOT BE PLACED IN THE SAME RACEWAY, BOX, OR ENCLOSURE WITH POWER CONDUCTORS EXCEPT WHERE INTRODUCED SOLELY TO CONNECT THE EQUIPMENT, *725.54(A)(1)* AND *EXCEPTION NO. 2.*

TYPICAL ELECTRIC FURNACE

DISCONNECTING MEANS:
• SHALL HAVE AMPERE RATING *NOT LESS* THAN 125% OF THE TOTAL LOAD OF THE MOTOR AND HEATER. SEE NAMEPLATE ON FURNACE.
• MUST DISCONNECT THE EQUIPMENT FROM ALL UNGROUNDED CONDUCTORS.
• MUST BE WITHIN SIGHT OF FURNACE OR MUST BE CAPABLE OF BEING LOCKED IN "OFF" POSITION.
• SHALL INDICATE "ON-OFF."
• FUSES MUST BE INSTALLED UNLESS THE FURNACE NAME-PLATE STATES THAT AN HACR BREAKER IS PERMITTED.
• FUSES SHALL BE SIZED AT *NOT LESS* THAN 125% OF THE TOTAL LOAD OF THE MOTOR AND HEATERS. SEE NAMEPLATE ON FURNACE.

MUST BE SEPARATE CIRCUIT, *422.12.*

BRANCH-CIRCUIT CONDUCTORS:
• SHALL BE SIZED AT *NOT LESS* THAN 125% OF THE TOTAL LOAD OF THE MOTORS AND HEATERS. SEE NAMEPLATE ON FURNACE.

NAMEPLATE WILL SHOW MANUFACTURER'S NAME, VOLTS AND AMPERES, VOLTS AND WATTS, OR VOLTS AND KILOWATTS.

FURNACE MAY ALSO CONTAIN SUPPLEMENTAL OVERCURRENT PROTECTION SO THAT THE RESISTANCE-TYPE HEATING ELEMENTS ARE FUSED AT NOT MORE THAN 60 AMPERES. THE FURNACE'S RESISTANCE-TYPE HEATING ELEMENTS MUST BE SUBDIVIDED INTO LOADS NOT TO EXCEED 48 AMPERES, *422.11(F)*. UL-LISTED FURNACES CONFORM TO THIS REQUIREMENT.

EXAMPLE: WHAT SIZE COPPER CONDUCTORS (THHN), FUSES, AND DISCONNECT SWITCH ARE REQUIRED FOR A FURNACE MARKED 79 AMPERES, 240 VOLT, SINGLE PHASE, 60 CYCLES? TERMINALS ON FURNACE AND SWITCH MARKED 75°C. TO SELECT THE PROPER AMPACITY OF THE CONDUCTORS IN ACCORDANCE WITH *110.14(C)*, BE SURE TO USE THE 75°C AMPACITY COLUMN IN *TABLE 310.16.*

ANSWER:
CONDUCTOR SIZE: 79 x 1.25 = 98.75 AMPERES
FROM *TABLE 310.16*, SELECT 3 AWG THHN (100 AMPERES AT 75°C.)

FUSE SIZE: 79 x 1.25 = 98.75 AMPERES
INSTALL 100-AMPERE FUSES.

SWITCH: 100-AMPERE SWITCH

Figure 23-1 Basic Code requirements for an electric furnace. These "package" units have all of the internal components, such as limit switches, relays, motors, and contactors prewired. The electrician generally need only provide the branch-circuit supply and disconnecting means. See *Article 424* for additional Code information.

Cord- and Plug-Connection Not Permitted

NEC® *400.7(A)(8)* and *422.16(A)* describe the permitted uses for flexible cords. A key requirement, oftentimes violated, is that flexible cords shall be used only *"where the fastening means and mechanical connections are specifically designed to permit ready removal for maintenance and repair, and the appliance is intended or identified for flexible cord connection."* Certainly the gas piping and the size of a gas or electric furnace do not allow for "ready removal" of the furnace. Furthermore, the conductors in flexible cords cannot withstand the high temperature encountered on the terminals of a furnace.

ANSI Standard Z21.47 does not allow cord- and plug-connection for gas furnaces.

CONTROL OF ELECTRIC BASEBOARD HEATING UNITS

Line-voltage and low-voltage thermostats can be used to control electric baseboard heating units, as in Figure 23-2. When the current rating or ampere rating of the load exceeds the rating of a line-voltage thermostat, then it is necessary to utilize a relay in conjunction with the low-voltage thermostat. This is illustrated in Figure 23-3. Color-coding requirements are explained later in this unit, and are covered in detail in Unit 5.

Common ratings for line-voltage thermostats are 2500 watts, 3000 watts, and 5000 watts. Other ratings are available. You might be working with watts, amperes, or both. For example, if a thermostat has a 2500-watt rating at 120 volts and 5000 watts at 240 volts, its current rating is:

$$I = \frac{W}{E} = \frac{2500}{120} = 20.8 \text{ amperes}$$

—or—

$$I = \frac{W}{E} = \frac{5000}{240} = 20.8 \text{ amperes}$$

The total connected load for this thermostat must not exceed 20.8 amperes.

Figure 23-2 Thermostats for electric heating systems. A thermostat or other switching device that is marked with an "OFF" position must open all ungrounded conductors of the circuit when turned to the "OFF" position. Typical mounting height is 52 in. (1.3 m) to center. *Courtesy of* Honeywell, Inc.

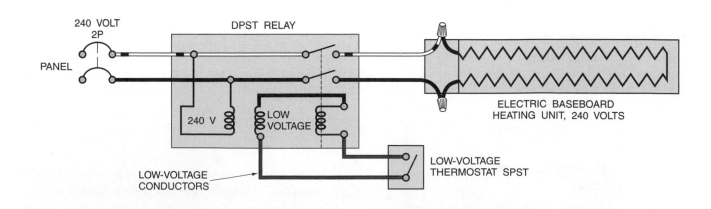

Figure 23-3 Wiring for an electric baseboard heating unit having a low-voltage thermostat and relay. *NEC® 200.7* requires that a white wire used as an ungrounded conductor must be permanently reidentified at points where connections are made or where it is visible and accessible. This becomes necessary when nonmetallic-sheathed cable or armored cable is the wiring method. Reidentification can be accomplished with black paint or black tape. Color coding and reidentification of conductors is discussed in Unit 5 and later in this unit.

CIRCUIT REQUIREMENTS FOR ELECTRIC BASEBOARD HEATING UNITS

Figure 23-3 shows an electric baseboard heater controlled by a low-voltage thermostat. The low-voltage contacts of the relay are connected to the thermostat. The line-voltage contacts of the relay are used to switch the actual heater load. Relays are used whenever the load exceeds the ampere or wattage limitations of a line-voltage thermostat. Color coding for conductors is explained in Unit 5 and is also discussed later in this unit.

Figure 23-4 shows typical electric baseboard heating units. The heater to the right has an integral thermostat on it. Electric baseboard heating units are available in both 240-volt and 120-volt ratings and in a variety of wattage ratings.

The wiring for an individual baseboard heating unit or group of units is shown in Figure 23-5 and Figure 23-6. A two-wire cable (armored cable or nonmetallic-sheathed cable with ground) would be run from a 240-volt, two-pole circuit in the main panel to the outlet box or switch box installed at

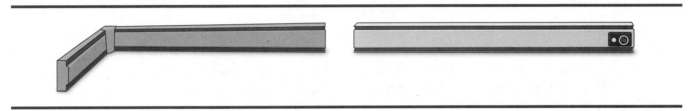

Figure 23-4   Electric baseboard heating systems.

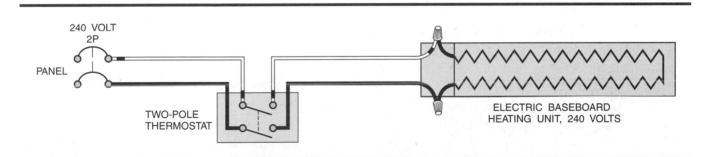

Figure 23-5   Wiring for a single electric baseboard heating unit. The thermostat is a line-voltage thermostat. See Figure 23-3 caption for color-coding requirements.

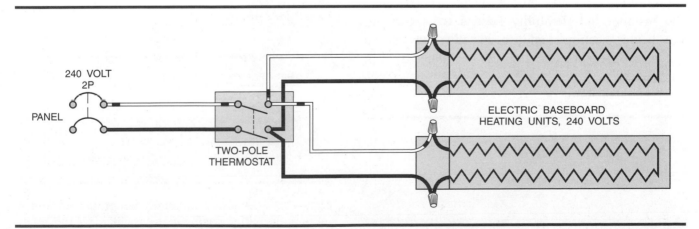

Figure 23-6   Wiring for two electric baseboard heating units. Note the circuit connections at the thermostat location. The thermostat is a line-voltage thermostat. See Figure 23-3 caption for color-coding requirements.

the thermostat location. A second two-wire cable runs from the thermostat to the junction box on the heater unit. The proper connections are made in this junction box. Most heating unit manufacturers provide knockouts at the rear and on the bottom of the junction box. The supply conductors can be run through these knockouts. Most baseboard units also have a channel or wiring space running the full length of the unit, usually at the bottom. When two or more heating units are joined together, the conductors are run in this wiring channel. Most manufacturers indicate the type of wire required for these units because of conductor temperature limitations.

Supply conductors generally must be rated 194°F (90°C). This would be Type THHN conductors, or nonmetallic-sheathed cable or armored cable that has 194°F (90°C) conductors. The nameplate and/or instructions furnished with the electric baseboard heating unit provides this information.

A variety of fittings such as internal and external elbows (for turning corners) and blank sections are available from the manufacturer.

Most wall and baseboard heating units are available with built-in thermostats, as in Figure 23-4. The supply cable for such a unit runs from the main panel to the junction box on the unit.

Some heaters have receptacle outlets. UL states that when receptacle sections are included with the other components of baseboard heating systems, they must be supplied separately using conventional wiring methods.

Manufacturers of electric baseboard heaters list "blank" sections ranging in length from 2 to 10 ft (600 mm to 3.0 m). These blank sections give the installer the flexibility needed to spread out the heater sections, and to install these blanks where wall receptacle outlets are encountered, as in Figure 23-7.

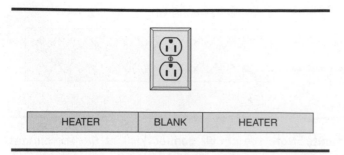

Figure 23-7 Use of blank baseboard heating sections.

LOCATION OF ELECTRIC BASEBOARD HEATERS IN RELATION TO RECEPTACLE OUTLETS

Listed electric baseboard heaters shall not be installed below wall receptacle outlets unless the instructions furnished with the baseboard heaters indicate that they may be installed below receptacle outlets. See the *Fine Print Note* to *210.52*. The reasoning for generally not allowing electric baseboard heaters to be installed below receptacle outlets is the possible fire and shock hazard resulting when a cord hangs over and touches the heated electric baseboard unit. The insulation on the cord can melt from the heat.

Electricians must pay close attention to the location of receptacle outlets and the electric baseboard heaters. They must study the plans and specifications carefully.

Figure 23-8 *Code violation.* Unless the instructions furnished with the electric baseboard heater specifically state that the unit is listed for installation below a receptacle outlet, then this installation is a violation of *110.3(B)*. In this type of installation, cords attached to the receptacle outlet would hang over the heater, creating a fire hazard and possible shock hazard should the insulation on the cord melt. See *Fine Print Note* to *210.52*.

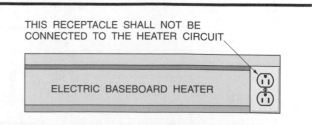

Figure 23-9 Factory-installed receptacle outlets or receptacle outlet assemblies provided by the manufacturer for use with its electric baseboard heaters may be counted as the required receptacle outlet for the space occupied by a permanently installed heater. See the second paragraph of *210.52*.

They may have to fine-tune the location of the receptacle outlets and the electric baseboard heaters, and/or install blank spacer sections, keeping in mind the Code requirements for spacing receptacle outlets and just how much (wattage and length) baseboard heating is required (Figure 23-8 and Figure 23-9).

The Code does permit factory-installed or factory-furnished receptacle outlet assemblies as part of permanently installed electric baseboard heaters. These receptacle outlets shall not be connected to the heater circuit (see second paragraph of *210.52* and Figure 23-10).

HEAT PUMPS

A heat pump is simply an air conditioner operating in reverse. A "reversing valve" inside the unit changes the direction of the flow of the system's refrigerant.

A "split" heat pump system consists of an outdoor unit with a hermetic motor compressor, a fan, and a coil. The indoor unit consists of a coil inside of the furnace. The movement of air inside the home is accomplished by the blower fan on the furnace. The "split system" is the most common type for residential use.

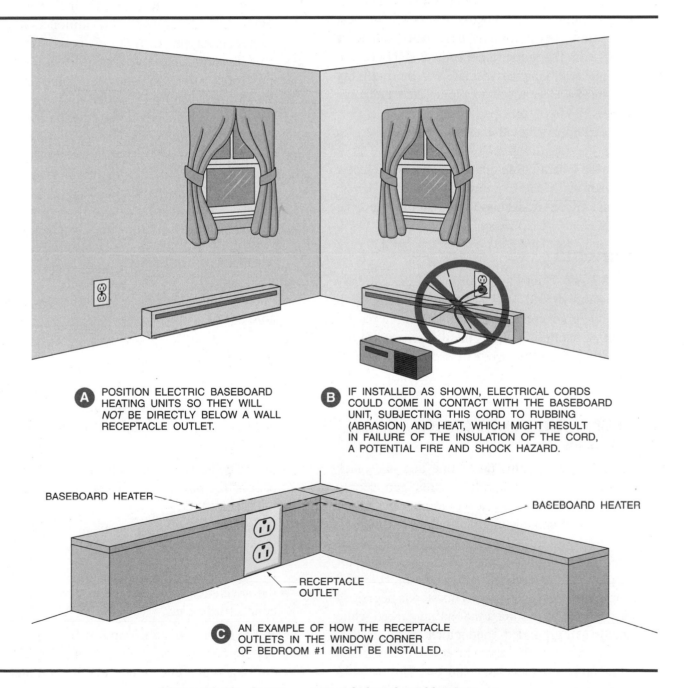

A POSITION ELECTRIC BASEBOARD HEATING UNITS SO THEY WILL *NOT* BE DIRECTLY BELOW A WALL RECEPTACLE OUTLET.

B IF INSTALLED AS SHOWN, ELECTRICAL CORDS COULD COME IN CONTACT WITH THE BASEBOARD UNIT, SUBJECTING THIS CORD TO RUBBING (ABRASION) AND HEAT, WHICH MIGHT RESULT IN FAILURE OF THE INSULATION OF THE CORD, A POTENTIAL FIRE AND SHOCK HAZARD.

BASEBOARD HEATER

BASEBOARD HEATER

RECEPTACLE OUTLET

C AN EXAMPLE OF HOW THE RECEPTACLE OUTLETS IN THE WINDOW CORNER OF BEDROOM #1 MIGHT BE INSTALLED.

Figure 23-10 Position of electric baseboard heaters.

When all of the components are housed in one outdoor unit, the system is referred to as a "packaged system."

The hermetic motor compressor is the heart of the system. The refrigerant circulates between the indoor and outdoor units, through the coils, tubing, and compressor, absorbing and releasing heat as it travels through the system.

In the winter, the heat pump (the outdoor coil) extracts heat from the outdoor air and distributes it through the warm air ducts in the home. In extremely cold climates, a heat pump is supplemented by electric heating elements in the air handler.

In the summer, the process reverses. The heat pump (indoor coil) absorbs heat and condenses humidity from the indoor air. The heat is transferred to the outside by the refrigerant, and the humidity (condensation) is removed by piping it to a suitable drain.

Heat pump water heaters are discussed in Unit 19.

Because a heat pump contains a hermetic motor compressor, the Code requirements found in *Article 440* apply. These are discussed later in this unit.

GROUNDING

Grounding of electrical equipment has already been discussed and will not be repeated here. Grounding is covered in *Article 250* of the *NEC*®. To repeat a caution, *never* ground electrical equipment to the grounded (white) circuit conductor of the circuit.

MARKING THE CONDUCTORS OF CABLES

Two-wire cable contains one white and one black conductor. Because 240-volt circuits can supply electric baseboard heaters, it would appear that the use of two-wire cable is in violation of the Code.

According to *200.7(C)*, two-wire cable may be used for the 240-volt heaters if the white conductor is permanently re-identified by paint, colored tape, or other effective means. This step is necessary because it may be assumed that an unmarked white conductor is a grounded conductor having no voltage to ground. Actually, the white wire is connected to a hot phase in the panel and has 120 volts to

ground. *Thus, a person can be subject to a harmful shock by touching this wire and the grounded baseboard heater (or any other grounded object) at the same time*. Most electrical inspectors accept black paint or tape as a means of changing the color of the white conductor. The white conductor must be permanently re-identified at the electric heater terminals and at the panels where these cables originate. Re-read the section on "Color Coding" in Unit 5.

ROOM AIR CONDITIONERS

For homes that do not have central air conditioning, window or through-the-wall air conditioners may be installed. These types of room air conditioners are available in both 120-volt and 240-volt ratings. Because room air conditioners are plug-and-cord connected, the receptacle outlet and the circuit capacity must be selected and installed according to applicable Code regulations. The Code rules for air conditioning are found in *Article 440*. The Code requirements for room air-conditioning units are found in *440.60* through *440.64* of the *NEC*.®

The basic Code rules for installing these cord-and plug-connected units and their receptacle outlets are as follows:

- the air conditioners must be grounded.

- the air conditioners must be connected using a cord and attachment plug.

- the air-conditioner rating may not exceed 40 amperes at 250 volts, single phase.

- the rating of the branch-circuit overcurrent device must not exceed the branch-circuit conductor rating or the receptacle rating, whichever is less.

- the air-conditioner load shall not exceed 80 percent of the branch-circuit ampacity if no other loads are served.

- the air-conditioner load shall not exceed 50 percent of the branch-circuit ampacity if other loads are served.

- the attachment plug cap may serve as the disconnecting means.

- ▶The maximum cord length is 10 ft (3.0 m) for 120-volt units, and 6 ft (1.8 m) for 208- or 240-volt units. This is also a UL requirement.◀

RECEPTACLES FOR AIR CONDITIONERS

Figure 23-11 shows two types of receptacles that can be used for 240-volt installations. A combination receptacle is shown in Figure 23-11(B). The lower portion of the receptacle is for 120-volt use and the upper portion is for 240-volt use. Note that the tandem and grounding slot arrangement for the 240-volt receptacle meets the requirements of *406.3(F)*. This section states that receptacle outlets for different voltage levels must be noninterchangeable with each other.

CENTRAL HEATING AND AIR CONDITIONING

The residence discussed throughout this text has central electric heating and air conditioning consisting of an electric furnace and a central air-conditioning unit.

The wiring for central heating and cooling systems is shown in Figure 23-12. Basic circuit requirements are shown in Figure 23-13. Note that one branch-circuit runs to the electric furnace and another branch-circuit runs to the air conditioner or heat pump outside the dwelling. Low-voltage wiring

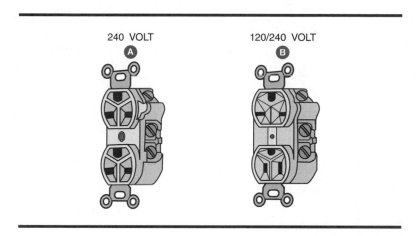

240 VOLT Ⓐ   **120/240 VOLT** Ⓑ

Figure 23-11 Types of 240-volt receptacles.

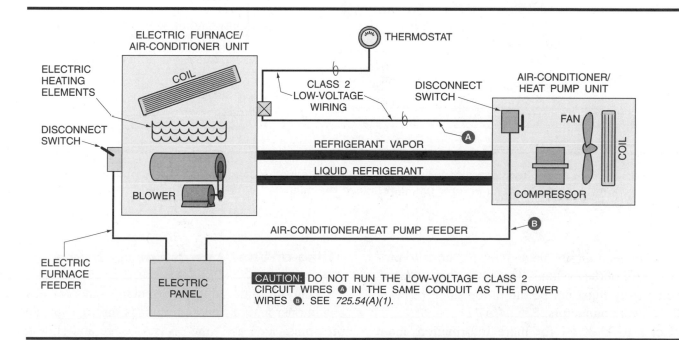

Figure 23-12 Connection diagram showing typical electric furnace and air-conditioner/heat pump installation.

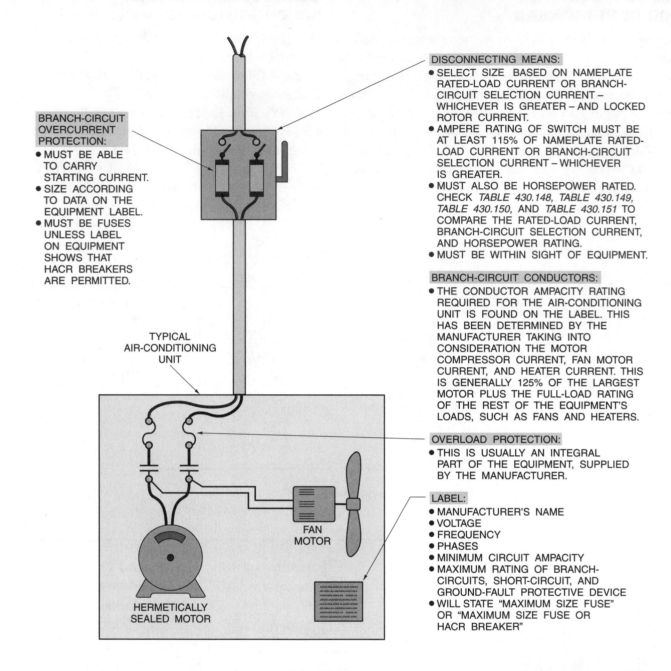

BRANCH-CIRCUIT OVERCURRENT PROTECTION:
- MUST BE ABLE TO CARRY STARTING CURRENT.
- SIZE ACCORDING TO DATA ON THE EQUIPMENT LABEL.
- MUST BE FUSES UNLESS LABEL ON EQUIPMENT SHOWS THAT HACR BREAKERS ARE PERMITTED.

TYPICAL AIR-CONDITIONING UNIT

HERMETICALLY SEALED MOTOR

FAN MOTOR

DISCONNECTING MEANS:
- SELECT SIZE BASED ON NAMEPLATE RATED-LOAD CURRENT OR BRANCH-CIRCUIT SELECTION CURRENT – WHICHEVER IS GREATER – AND LOCKED ROTOR CURRENT.
- AMPERE RATING OF SWITCH MUST BE AT LEAST 115% OF NAMEPLATE RATED-LOAD CURRENT OR BRANCH-CIRCUIT SELECTION CURRENT – WHICHEVER IS GREATER.
- MUST ALSO BE HORSEPOWER RATED. CHECK *TABLE 430.148*, *TABLE 430.149*, *TABLE 430.150*, AND *TABLE 430.151* TO COMPARE THE RATED-LOAD CURRENT, BRANCH-CIRCUIT SELECTION CURRENT, AND HORSEPOWER RATING.
- MUST BE WITHIN SIGHT OF EQUIPMENT.

BRANCH-CIRCUIT CONDUCTORS:
- THE CONDUCTOR AMPACITY RATING REQUIRED FOR THE AIR-CONDITIONING UNIT IS FOUND ON THE LABEL. THIS HAS BEEN DETERMINED BY THE MANUFACTURER TAKING INTO CONSIDERATION THE MOTOR COMPRESSOR CURRENT, FAN MOTOR CURRENT, AND HEATER CURRENT. THIS IS GENERALLY 125% OF THE LARGEST MOTOR PLUS THE FULL-LOAD RATING OF THE REST OF THE EQUIPMENT'S LOADS, SUCH AS FANS AND HEATERS.

OVERLOAD PROTECTION:
- THIS IS USUALLY AN INTEGRAL PART OF THE EQUIPMENT, SUPPLIED BY THE MANUFACTURER.

LABEL:
- MANUFACTURER'S NAME
- VOLTAGE
- FREQUENCY
- PHASES
- MINIMUM CIRCUIT AMPACITY
- MAXIMUM RATING OF BRANCH-CIRCUITS, SHORT-CIRCUIT, AND GROUND-FAULT PROTECTIVE DEVICE
- WILL STATE "MAXIMUM SIZE FUSE" OR "MAXIMUM SIZE FUSE OR HACR BREAKER"

Figure 23-13   Basic circuit requirements for a typical residential-type air conditioner or heat pump. Reading the label is important in that the manufacturer has determined the minimum conductor size and branch-circuit fuse size, *Article 440*.

is used between the inside and outside units to provide control of the systems. The low-voltage Class 2 circuit wiring must not be run in the same raceway as the power conductors, *725.54(A)(1)*.

Refer to Unit 24 for more information about low-voltage circuitry.

"Time-of-Use" Lower Energy Rates

Many utilities offer lower energy rates for air-conditioner and/or heat pump loads through special programs, such as "time-of-use" usage and sliding scale energy rates. This is covered in Unit 19.

UNDERSTANDING THE DATA FOUND ON AN HVAC NAMEPLATE

The letters HVAC stand for heating, ventilating, and air conditioning.

Article 440 applies when HVAC equipment employs a hermetic refrigerant motor-compressor(s). *Article 440* is supplemental to the other articles of the *NEC*, and is needed because of the unique characteristics of hermetic refrigerant motor-compressors. The most common examples are air conditioners, heat pumps, and refrigeration equipment.

Hermetic refrigerant motor-compressors:

- combine the motor and compressor into one unit
- have no external shaft
- operate in the refrigerant
- do not have horsepower and full-load current (often referred to as full-load amperes or FLA) ratings like standard electric motors. This is because as the compressor builds up pressure, the current increases, causing the windings to get hot. At the same time, the refrigerant gets colder, passes over the windings, and cools them. Because of this, a hermetically sealed motor can be "worked" much harder than a conventional electric motor of the same size.

Special terminology is needed to understand how to properly make the electrical installation. These special terms are found in *440.2* and in UL Standard 1995. An air-conditioning nameplate from Shiver Manufacturing Company help explain these special terms.

- **Rated-Load Current (RLC):** Rated-load current is established by the hermetic motor-compressor manufacturer under actual operation at rated refrigerant pressure and temperature, rated voltage, and rated frequency. RLC is marked on the nameplate. In most instances, the marked RLC is at least equal to 64.1 percent of the hermetic refrigerant motor-compressor's maximum continuous current (MCC). In our example, the RLC is 17.8 amperes. See *440.2*.

- **Maximum Continuous Current (MCC):** This is the maximum current that the hermetic refrigerant motor-compressor can draw before damage occurs. Manufacturers know the damage capabilities of their equipment, and generally provide the necessary internal overcurrent pro-tection that keeps the hermetic motor-compressor from burning out. This value is 156 percent of the marked RLC or branch-circuit selection current (BCSC). The overload protective system for the hermetic refrigerant motor-compressor must operate for currents in excess of the 156 percent value. Do not look for the MCC value on the nameplate of the end-use equipment. The MCC value might be marked on the motor-compressor. It is nice, but not necessary for the electrician to know the MCC value of the motor-compressor. See *440.52*.

- **Branch-Circuit Selection Current (BCSC):** The manufacturer of the HVAC end-use equipment might design better cooling and better heat dissipation into the equipment, in which case the hermetic refrigerant motor-compressor might be capable of being continuously "worked" harder than other equipment not so designed. "Working" the hermetic refrigerant motor-compressor harder will result in a higher current draw. This is safe insofar as the motor-compressor is concerned, but the conductors, disconnect switch, and branch-circuit overcurrent devices must also be capable of safely carrying this higher current draw. This higher value of current is marked on the nameplate BCSC.

Because the BCSC ampere value is always equal to or greater than the RLC ampere value, this BCSC value must be used instead of the RLC value when selecting conductors, disconnects, overcurrent devices, and other associated electrical equipment. Table 23-2 shows that the BCSC is 19.9 amperes. See *440.2* and *440.4(C)*.

SHIVER MANUFACTURING COMPANY						
MODEL NO. XYZ – ELECTRICAL RATINGS –						
	VAC	HZ	PH	RLC	LRC	FLA
Compressor	230	60	1	17.8	107	—
Outdoor fan motor ¼ hp	230	60	1	—	—	1.5
Branch circuit selection current			19.9 amperes			
Minimum circuit ampacity			26.4 amperes			
Maximum fuse or HACR type brkr.			45.0 amperes			
Operating voltage range:			197 min. 253 max.			

Table 23-2

• **Minimum Circuit Ampacity (MCA):** This is one of two things (the other is MOP) the electrician needs to look for on the nameplate. MCA is the minimum circuit ampacity requirement to determine the conductor size and switch rating. MCA data is always marked on the nameplate of the end-use equipment, and is determined by the manufacturer of the end-use equipment as follows:

MCA = (RLC or BCSC × 1.25) + other loads

Other loads would include condensing fans, electric heaters, coils, and so forth, that operate concurrently (at the same time).

In our example, the marked MCA is 26.4 amperes, calculated by the manufacturer of the end-use equipment as follows:

$$(19.9 \times 1.25) + 1.5 = 26.375 \text{ amperes}$$
$$(\text{round off to } 26.4)$$

Referring to *Table 310.16*, using the 60°C column, 10 AWG copper conductors would be adequate for this air-conditioning unit.

See *Article 440, Part IV.*

A mistake that electricians often make is to multiply the MCA by 1.25 again. Certainly there is no hazard in doing this, but it might result in an unnecessary higher installation cost. For long runs, larger conductors would keep voltage drop to a minimum.

• **Maximum Overcurrent Protection (MOP):** The MOP value is the maximum size overcurrent protection, fuse, or HACR circuit breaker. The MOP value is marked on the nameplate, and is determined by the manufacturer of the end-use equipment as follows:

(RLC or BCSC) × 2.25 + other loads

Other loads include condensing fans, electric heaters, coils, and so on, that operate at the same time. In our example, the marked MOP is 45 amperes, determined by the manufacturer as follows:

$$(19.9 \times 2.25) + 1.5 = 46.275 \text{ amperes}$$

Rounding down to the next lower standard size, the MOP is 45 amperes. See *Article 440, Part III.*

• **Type of Overcurrent Protection:** A typical nameplate will show the "maximum size fuse," or "maximum size fuse or **HACR**-type circuit breaker." If maximum size fuse is specified, then circuit breakers are not permitted. *110.3(B)*

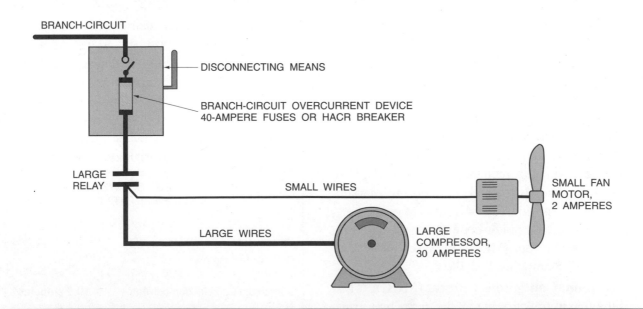

Figure 23-14 In a typical air-conditioner unit, the branch-circuit overcurrent device must protect the large components (large wires, large relay, hermetic motor compressor) as well as the small components (small wires, fan motor, crankcase heater) under short-circuit and/or ground fault situations. It is extremely important to install the proper *size* and *type* of overcurrent device. Read the nameplate on the equipment and the instructions furnished with the equipment!

of the Code states that "listed or labeled equipment shall be used or installed in accordance with any instructions included in the listing or labeling." Figure 23-14 illustrates the importance of using the proper type of overcurrent protective devices. In our example, the nameplate indicates "maximum fuse or HACR-type breaker 40 amperes."

In recent years, most 240-volt circuit breakers rated 100 amperes or less of the types used in residential wiring bear the letters HACR. But be very careful when connecting an air conditioner or heat pump to older style breakers that were not tested to the HACR requirements and are not marked with the letters HACR. To make such a hookup would be in violation of *110.3(B)*.

See Figure 23-14, Figure 23-15, Figure 23-16, and Figure 23-17.

- **Disconnecting Means to Be in Sight**: *NEC®* *440.14* of the Code requires that the disconnecting means must be within sight of the unit, and it must be readily accessible. See Figure 23-15, Figure 23-16, and Figure 23-17. Mounting the disconnect on the side of the house instead of

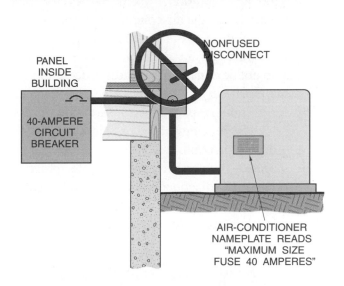

Figure 23-16 This installation *violates* the Code, *110.3(B)*. Although the disconnect is within sight of the air conditioner, it does not contain fuses. Note that the branch-circuit protection is provided by the 40-ampere circuit breaker inside the building. Note also that the nameplate requires that the branch-circuit protection be 40-ampere fuses maximum. If fused branch-circuit protection were provided at the panel inside the building, the installation would meet Code requirements.

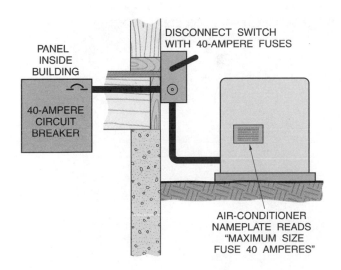

Figure 23-15 This installation conforms to the Code, *440.14*. The disconnect switch is within sight of the unit and contains the 40-ampere fuses called for on the air-conditioner nameplate as the branch-circuit protection.

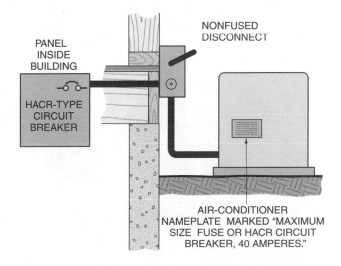

Figure 23-17 This installation "meets Code" because the circuit breaker in the panel is marked "HACR." This is permitted when the label on the breaker and the label on the air-conditioner bear the letters "HACR."

on the air conditioner itself allows for easy replacement of the air-conditioning unit, should that become necessary. Mounting the disconnect behind the air conditioner or heat pump would not be considered "readily accessible" by most electrical inspectors. Generally, there is just not enough space to work safely on the disconnect. Squeezing behind or leaning across the top of the air conditioner to gain access to the disconnect is certainly not a safe practice. Mounting the disconnect on the outside wall of the house, to one side of the AC unit, is the accepted practice. See *Article 440, Part II.*

- **Disconnecting Means Rating:** Figure 23-18 shows a fusible and a non-fusible "pull-out" disconnect commonly used for residential air-conditioning and/or heat pump installations. The horsepower rating of the disconnecting means must be at least equal to the sum of all of the individual loads within the end-use equipment, at rated load conditions, and at locked rotor conditions, *440.12(B)(1).*

  The ampere rating of the disconnecting means must be at least 115 percent of the sum of all of the individual loads within the end-use equipment, at rated load conditions,

440.12(B)(2). Installing a disconnect switch that equals or exceeds the MOP value will in almost all cases be the correct choice. The MOP in our example is 40 amperes. Thus, a 60-ampere disconnect switch is the correct size. In cases where the MOP value is close to the ampere rating of the disconnect switch (i.e., 30, 60, 100, 200, etc.), *locked-rotor current* values must be considered.

- **Locked-Rotor Current (LRC):** This is the maximum current draw when the motor is in a "LOCKED" position. When the rotor is locked, it is not turning. The disconnect switch and controller (if used) must be capable of safely interrupting locked-rotor current. See *440.12* and *440.41.* In our example, the nameplate indicates that the hermetic refrigerant motor-compressor has an LRC of 107 amperes. The fan motor has an FLA of 1.5 amperes, and would have an LRC approximately six times higher. According to *440.12(B)(1)(b),* all locked-rotor currents and other loads are added together when combined loads are involved. Therefore, the total locked-rotor current that the disconnect switch in our example would be called upon to interrupt is:

$$(1.5 \times 6) + 107 = 116 \text{ locked-rotor amperes}$$

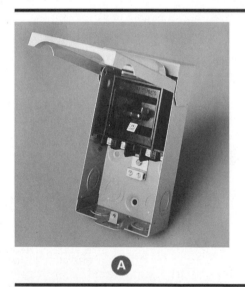
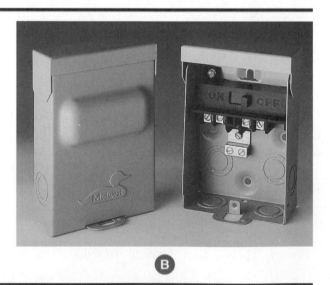

Figure 23-18 (A) is a pull-out fusible disconnect. Fuses are inserted into fuse clips on the pull-out device. Insert the pull-out device for "ON," and remove the pull-out device for "OFF." (B) is a non-fusible disconnect, available in 30- and 60-ampere ratings at 240 volts. Both have padlock provisions to prevent unauthorized tampering. *Courtesy of* Midwest Electric Products.

Checking *Table 430.151(A)* for the conversion of locked-rotor current to horsepower, we find that the locked-rotor current for a 230-volt, single-phase, 3-hp motor is 102 amperes, and the locked-rotor current for a 230-volt, single-phase, 5-hp motor is 168 amperes. Selecting a disconnect switch on the basis of the locked-rotor current in our example, the air-conditioning unit is considered to be a 5-hp unit, because 116 falls between 102 and 168. Manufacturers' technical literature indicates that a 30-ampere, 240-volt heavy-duty, single-phase disconnect switch has a 3-hp rating, and a 60-ampere, 240-volt heavy-duty, single-phase disconnect switch has a 10-hp rating. Thus, a 60-ampere disconnect is selected. Note that this is the same size disconnect selected previously on the basis of MOP.

ENERGY RATINGS

Energy ratings indicate the efficiency of heating and cooling equipment. Basically, these ratings are a comparison of output (heating or cooling) to input (electricity, gas, or oil).

Energy Efficiency Rating (EER): Cooling efficiency rating for room air conditioners. The ratio of the rated cooling capacity in Btu per hour divided by the amount of electrical power used in kilowatt hours. The higher the EER number, the greater the efficiency.

Seasonal Energy Efficiency Rating (SEER): Cooling efficiency rating for central air conditioners and heat pumps. SEER is determined by the total cooling of an air conditioner or heat pump in Btu during its normal usage period for cooling divided by the total electrical energy input in kilowatt-hours during the same period. The higher the SEER number, the greater the efficiency.

Annual Fuel Utilization Efficiency (AFUE): Tells you how efficiently a furnace converts fuel (gas or oil) to heat. For example, an AFUE of 85 percent means that 85 percent of the fuel is used to heat your home, and the other 15 percent goes up the chimney. The higher the efficiency, the lower the operating cost. Old furnaces might have an AFUE rating as low as 60 percent. Mid-efficiency ratings are approximately 80 percent. High-efficiency ratings are 90 percent or higher. Maximum furnace efficiency available is approximately 96.6 percent. The higher the AFUE number, the greater the efficiency.

Heating Seasonal Performance Factor (HSPF): Heating efficiency of a heat pump. HSPF is determined by the total heating of a heat pump in Btu during its normal usage period for heating divided by the total electrical energy input in kilowatt-hours during the same period. The higher the HSPF rating, the greater the efficiency.

NONCOINCIDENT LOADS

Loads such as heating and air-conditioning are not likely to operate at the same time. The *NEC*® recognizes this diversity in *220.21*. Therefore, when calculating a feeder that supplies both types of loads, or when sizing a service-entrance, only the larger of the two loads need be considered. This is discussed in Unit 29 where service-entrance calculations are presented. Of course, the branch-circuit supplying the heating load is sized for that particular load, and the branch-circuit supplying the air-conditioning load is sized for that particular load.

Gas Explosion Hazard

Although it might seem a bit out of place in an electrical book to talk about gas explosions, we must talk about it.

Often overlooked is a requirement in *2.7.2(c)* of *NFPA 54*, *National Fuel Gas Code*, that gas meters be located at least 3 ft (900 mm) from sources of ignition. An electric meter or a disconnect switch are possible sources of ignition. Some utility regulations require a minimum of 3 ft (900 mm) clearance between electric metering equipment and gas meters and gas-regulating equipment. It is better to be safe than sorry! Check this issue out with the local electrical inspector and/or the local electric utility before installing the air conditioner or heat pump disconnect on the outside of the house. Also consider the location of the service watt-hour meter.

REVIEW

Note: Refer to the Code or the plans where necessary.

ELECTRIC HEAT

1. a. What is the allowance in watts made for electric heat in this residence? _____

    b. What is the value in amperes of this load? _____

    _____

2. What are some of the advantages of electric heating? _____

    _____

    _____

    _____

3. List the different types of electric heating system installations. _____

    _____

    _____

    _____

4. There are two basic voltage classifications for thermostats. What are they? _____

    _____

5. What device is required when the total connected load exceeds the maximum rating of
    a thermostat? _____

6. The electric heat in this residence is provided by what type of equipment? _____

    _____

7. At what voltage does the electric furnace operate? _____

8. A certain type of control connects electric heating units to a 120-volt supply or a 240-volt
    supply, depending upon the amount of the temperature drop in a room. These controls
    are supplied from a 120/240-volt, three-wire, single-phase source. Assuming that this
    type of device controls a 240-volt, 2000-watt heating unit, what is the wattage produced
    when the control supplies 120 volts to the heating unit? Show all calculations.

9. What advantages does a 240-volt heating unit have over a 120-volt heating unit?

    _____

    _____

    _____

10. The white wire of a cable may be used to connect to a "hot" ungrounded circuit conductor only if _____

_____

11. Receptacle outlets furnished as part of a permanently installed electric baseboard heater, when not connected to the heater's branch-circuit, (may) (may not) be counted as the required receptacle outlet for the space occupied by the baseboard heater. Electric baseboard heaters (shall) (shall not) be installed beneath wall receptacle outlets unless the instructions furnished with the heaters indicate that it is acceptable to install the heater below a receptacle outlet. Circle the correct answers.

12. The branch-circuit supplying a fixed electric space heater must be sized to at least _____ percent of the heater's rating according to *NEC*® _____ .

13. Compute the current draw of the following electric furnaces. The furnaces are all rated 240 volts.
    a. 7.5 kW
    b. 15 kW
    c. 20 kW

14. For "ballpark" calculations, the wattage output of a 240-volt electric furnace connected to a 208-volt supply will be approximately 75 percent of the wattage output had the furnace been connected to a 240-volt supply. In question 13, calculate the wattage output of (a), (b), and (c) if connected to 208 volts. _____

_____

_____

_____

15. A central electric furnace heating system is installed in a home. The circuit supplying this furnace:
    a. may be connected to a circuit that supplies other loads.
    b. shall be connected to a circuit that supplies other loads.
    c. shall be connected to a separate circuit.

16. What section of the Code provides the correct answer to question 15? _____

17. Electric heating cable embedded in plaster, or "sandwiched" between layers of dry wall, creates heat. Therefore, any nonmetallic-sheathed cable run above these ceilings and buried in insulation must have its current-carrying capacity derated according to the (40°C) (50°C) (60°C) correction factors found at the bottom of *Table 310.16* in the *NEC*® Circle the correct answer.

AIR CONDITIONING

1. a. When calculating air-conditioner load requirements and electric heating load requirements, is it necessary to add the two loads together to determine the combined load on the system? _____

   b. Explain the answer to part (a). _____

_____

_____

2. The total load of an air conditioner shall not exceed what percentage of a separate branch-circuit? (circle one)

   a. 75 percent   b. 80 percent   c. 125 percent

3. The total load of an air conditioner shall not exceed what percentage of a branch-circuit that also supplies lighting? (circle one)

   a. 50 percent   b. 75 percent   c. 80 percent

4. a. Must an air conditioner installed in a window opening be grounded if a person on the ground outside the building can touch the air conditioner? _____

   b. What Code section governs the answer to part (a)? _____

5. A 120-volt air conditioner draws 13 amperes. What size is the circuit to which the air conditioner will be connected? _____

6. What is the Code requirement for receptacles connected to circuits of different voltages and installed in one building? _____

7. When a central air-conditioning unit is installed and the label states "Maximum size fuse 50-amperes," is it permissible to connect the unit to a 50-ampere circuit breaker?

   _____

8. When the nameplate on an air-conditioning unit states "Maximum Size Fuse or HACR Circuit Breaker," what type of circuit breaker must be used? _____

   _____

9. Briefly, what is the reasoning behind the UL requirement that air conditioners and other multimotor appliances be protected with fuses or special HACR circuit breakers?

   _____

   _____

   _____

   _____

   _____

   _____

   _____

10. What section of the *NEC*® prohibits the installing of Class 2 control circuit conductors in the same raceway as the power conductors? _____

    _____

11. Match the following terms with the statement that most closely defines the term. Enter the letters in the blank space.

    RLC _____   a) The maximum ampere rating of the overcurrent device.

    MCA _____   b) The current value to be used instead of the rated-load ampere.

    BCSC _____   c) The value used to determine the minimum ampere rating for the branch-circuit.

    MCC _____   d) A current value for the hermetic motor compressor that was determined by operating it at rated temperature, voltage, and frequency.

    MOP _____   e) The maximum current that the compressor can draw continuously without damage.

12. The disconnect for an air-conditioner or heat pump must be installed _____
    _____ of the unit.

Gas and Oil Central Heating Systems

OBJECTIVES

After studying this unit, the student should be able to

- understand the basics of typical home heating systems.
- understand some of the more important components of typical home heating systems.
- interpret basic schematic wiring diagrams.
- understand how a thermocouple and a thermopile operate.
- understand and apply the Code requirements for control-circuit wiring, *Article 725, NEC.*®

In Unit 23, we discussed electrical branch-circuit requirements for electric heating (furnace, baseboard, ceiling cable). In this unit, typical residential gas and oil heating systems are discussed.

Every manufacturer of residential heating systems provides installation manuals that must be followed.

WARM AIR FURNACES

By far, the most popular method for heating a home is with a gas-fired, forced-warm air furnace. Oil-fired, forced-warm air furnaces are also available. Warm air furnaces are available with upflow, downflow, and horizontal air flow. A fan forces hot air through the hot air ducts to various rooms in the house. Cold air ducts, usually referred to as "cold air returns" bring the air back to the furnace where the air is again filtered, heated, and forced out through the hot air duct.

Humidity control is another feature commonly installed in home heating systems.

Some systems bring fresh air into the home. This ventilation equipment brings fresh air into the home, then filters and circulates this fresh air through the ductwork in the home.

For larger homes, by installing special dampers in the hot air ducts and the necessary associated controls, heating systems can provide zone control so that different areas of the home can be controlled by their own thermostats.

HOT WATER SYSTEM

Hot water is circulated through the piping system by a circulating pump. This is often referred to as a *hydronic* system. The heating unit is referred to as a boiler, and can be gas or oil fired. A hot water system is very adaptable to zone control by using a combination of one or more circulating pumps and/or valves for the different zones. Hot water baseboard heaters look very much like electric baseboard heaters.

Hot water piping systems are not easily adaptable to central air conditioning. Additional ductwork is required.

Upscale homes might have special plastic tubing secured to the underside of the floors to heat the floors from below. Hot water is circulated through this tubing when the thermostat calls for heat. This type of system is used under tile floors in bathrooms, kitchens, and similar areas.

PRINCIPLE OF OPERATION

Most residential heating systems operate as follows. A room thermostat is connected to the proper electrical terminals on the furnace or boiler. Inside the equipment, the various controls and valves are interconnected in a manner that will provide safe and adequate operation of the furnace.

Most residential heating systems are "packaged units" in which the burner, hot surface igniter or hot spark igniter, safety controls, valves, fan controls, sensors, high temperature limit switches, fan blower motor, blower fan speed control, draft inducer motor, primary and secondary heat exchanger, and so on, are preassembled, prewired parts of the furnace or boiler. Many of these items are included in a solid-state integrated printed circuit board module, which is the "brain" of the system. This module provides proper sequencing and safety features.

Older models have standing gas pilot lights and thermocouples.

Similar electronic circuitry is used in oil burners where the ignition, the oil burner motor, blower fan motor, high temperature limit controls, hot water circulating pump in the case of hot water systems, and the other devices previously listed are all interconnected to provide a packaged oil burner unit.

Evaporator coils for air-conditioning purposes might be an integral part of the unit, or they might be mounted in the ductwork above the furnace.

Wiring a Residential Central Heating System

For the electrician, the "field wiring" for a typical residential furnace or boiler consists of:

1. Installing and connecting the low-voltage wiring between the furnace and the thermostat. This might be a cable consisting of two, three, four, or five small 18 AWG or 20 AWG conductors. Small knockouts are provided on the furnace through which the low-voltage Class 2 wires are brought into the furnace's wiring compartment (see Figure 24-1). Class 2 wiring must not be run through the same raceway or cable as power wiring, *725.54(A)(1)*.

2. Installing and connecting the branch-circuit power supply to the furnace. Knockouts are provided on the furnace through which the line voltage power supply is brought into the

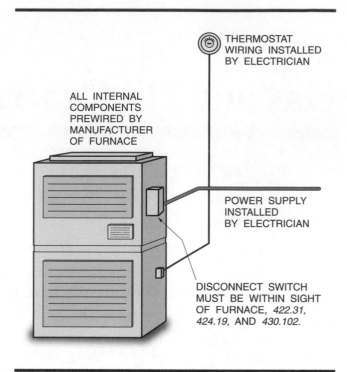

Figure 24-1 Diagram showing thermostat wiring, power supply wiring, and disconnect switch. ANSI Standard Z21.47 does not allow cord-and-plug connections for a gas furnace.

wiring compartment of the furnace. Depending on local codes, the line voltage power supply wiring might be in EMT, armored cable, or other accepted wiring methods (Figure 24-1).

The disconnecting means must be within sight of the furnace, *422.31, 424.19*, and *430.102*. Some furnaces come with a disconnect switch. If not, a toggle switch can be mounted on a box on the side of the furnace. Some electric furnaces come complete with circuit breakers that provide the overcurrent protection for the heating elements as required by UL, and to serve as the disconnecting means required by the *NEC*.

Separate Branch-Circuit Required

NEC® 422.12 requires that central heating equipment must be supplied by a separate branch-circuit.

The size of the branch-circuit depends on the requirements of the particular furnace. A typical gas-forced warm air furnace might require only a 120-volt, 15-ampere branch-circuit, whereas an

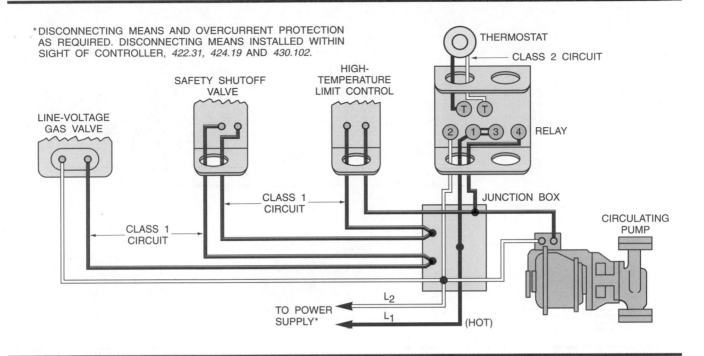

Figure 24-2 Typical wiring diagram for a gas burner, forced hot water system. Always consult the manufacturer's wiring diagrams relating to a particular model heating unit.

electric furnace requires a 240-volt branch-circuit and much larger conductors. Instructions furnished with a given furnace will indicate the branch-circuit requirements.

Some typical wiring diagrams are shown in Figure 24-2, Figure 24-3, Figure 24-4A, and Figure 24-4B. These wiring diagrams show the individual electrical components of a typical residential heating system. In older systems, many of these components were wired by the electrician. In newer systems, most components other than the thermostat are prewired as part of a packaged furnace or boiler. The manufacturer's literature includes detailed wiring diagrams for a given model.

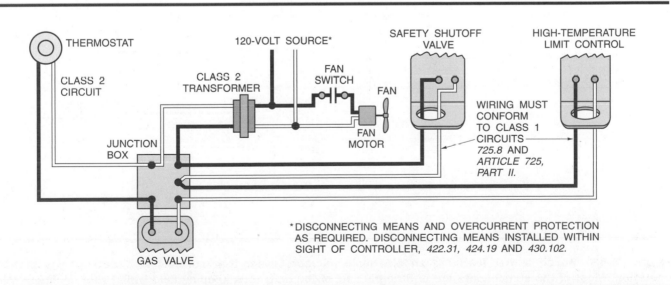

Figure 24-3 Typical wiring diagram for a gas burner, forced warm air system. Always consult the manufacturer's wiring diagrams relating to a particular model heating unit.

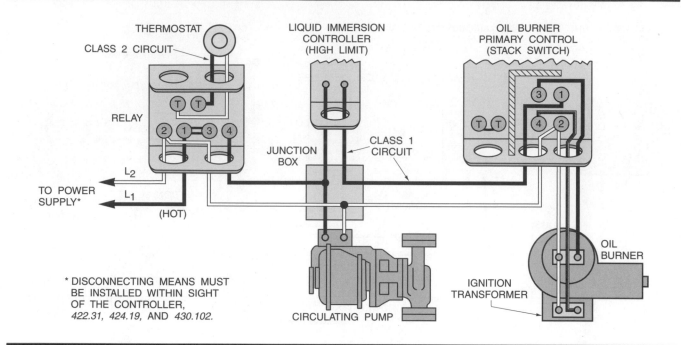

Figure 24-4A Typical wiring diagram of an oil burner heating installation. Always consult the manufacturer's wiring diagrams relating to a particular model heating unit.

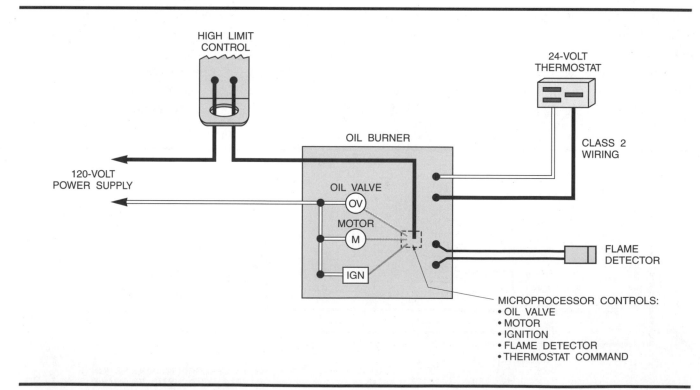

Figure 24-4B An oil burner featuring an electronic microprocessor that controls all facets of the burner operation. Most of the components are an integral part of the unit, thus keeping field wiring to a minimum. All manufacturers provide detailed wiring diagrams for their products. Always consult the manufacturer's wiring diagrams relating to a particular model heating unit.

MAJOR COMPONENTS

Gas- and oil-fired warm air systems and hot water boilers contain many individual components. Here is a brief description of these components as well as definitions for terms unique to the heating industry. Always read and thoroughly understand the manufacturer's installation manuals.

Aquastat

An aquastat regulates the temperature of boiler water to ensure that the circulating water maintains a proper, satisfactory temperature.

Cad Cell

A cad cell is a cadmium sulfide sensor for sensing flame. Cad cells have pretty much replaced the older style stack-mounted bimetal switches.

Combustion Chamber

The combustion chamber surrounds the flame, and radiates heat back into the flame to aid in combustion.

Combustion Head

A combustion head creates a specific pattern of air at the end of the air tube. The air is directed in such a way as to force oxygen into the oil spray so the oil can burn. A combustion head might also be referred to as the turbulator, fire ring, retention ring, or end cone.

Draft Regulator

A counterweighted swinging door that opens and closes to help maintain a constant level of draft over the fire.

Fan Control

A fan control is used to control the blower fan on forced warm air systems. On newer furnaces, the fan control is actually an electronic timer causing the fan to start running so many seconds after the main burner has come on. This is sufficient time to allow the burner to warm the air so that the blower does not blow cold air. Where adjustable, fan controls are usually set to turn on between 120°F and 140°F (49°C and 60°C), and to shut off at 90°F to 100°F (32°C to 38°C).

Fan Motor

The fan motor is the electric motor that forces the warm air through the ducts of a heating system. It may be single, multiple, or variable speed. The blower fan itself might be directly connected to the shaft or it may be belt driven. In many cases, a squirrel-cage fan is connected to the shaft of the motor. The motor in most cases is provided with integral overload protection, in conformance to the *NEC.*®

Heat Exchanger

A heat exchanger transfers the heat energy from the combustion gases to the air in the furnace or to the water in a boiler.

High Temperature Limit Control

Should the temperature in the plenum of the furnace reach dangerously high temperatures (approximately 200°F, or 93°C), the high-temperature limit control will shut off the power to the burner. The fan will continue to run. When the temperature in the plenum returns to a safe level, the high-temperature limit control will allow the burner to resume normal operation. In newer furnaces, the high-temperature limit control is part of the integrated circuit board, and is communicated to by a sensor strategically located in the plenum. Older furnaces might have individual high-temperature limit controls and a separate fan control, or they might have combination high-temperature limit/fan controls.

Hot Surface Igniter

In furnaces that do not have a standing constant pilot light, the hot surface igniter heats to a cherry red/orange color. When it reaches the proper temperature, the main gas valve is allowed to open. If the hot surface igniter does not come on, the main gas valve will not open. Energy-efficient furnaces may use this concept for ignition.

Ignition

Constant: The igniter is designed to be able stay on continuously.

Intermittent Duty: Defined by UL 296 as *"ignition by an energy source that is continuously maintained throughout the time the burner is firing."* In other words, the igniter is on the entire time the burner is firing.

Interrupted Duty: Defined by UL 296 as *"an ignition system that is energized each time the main burner is to be fired and de-energized at the end of a timed trial for ignition period or after the main flame is proven to be established."* In other words, the igniter comes on to light the flame; then, after the flame is established, the igniter is turned off and the flame keeps burning.

Induced Draft Blower

When the thermostat calls for heat, the induced draft blower starts first to expel any gases remaining in the combustion chamber from a previous burning cycle. It continues to run, pulling hot combustion gases through the heat exchangers, then vents the gases to the outdoors.

Integrated Control

An integrated control is a solid-state, printed circuit board containing many electronic components. When the thermostat calls for heat, the integrated circuit board takes over to manage the sequence of events that allow the burner to operate safely.

Liquid Immersion Control

A liquid immersion control controls high water and circulating water temperatures in boilers. When acting as a high-temperature limit control, it will shut off the burner if the water temperature in the boiler reaches dangerous levels. When acting as a circulating water temperature control, it makes sure that the water being circulated through the piping system is at the desired temperature, not cold. A liquid immersion control is mounted so that its sensing tube is either in direct contact with the water, or inserted into an immersion well that is in direct contact with the boiler water. Temperature settings must be in conformance to the manufacturer's instructions. In packaged boilers, the settings are usually set for the particular boiler. Field inspection to make sure the setting is correct is recommended.

Low Water Control

Sometimes referred to as a low-water cutoff, this device senses low water levels in a boiler. When a low-water situation occurs, the low-water control shuts off the electrical power to the burner.

Main Gas Valve

The main gas valve is the valve that allows the gas to flow to the main burner of a gas-fired furnace. It may also function as a pressure regulator, a safety shutoff, and as the pilot and main gas valve. Should there be a "flame-out," if the pilot light fails to operate, or in the event of a power failure, this valve will shut off the flow of gas to the main burner.

Nozzle

A nozzle produces the desired spray pattern for the particular appliance in which the burner is used.

Oil Burner

The function of an oil burner is to break fuel oil into small droplets, mix the droplets with air, and ignite the resulting spray to form a flame.

Primary Control

A primary control controls the oil burner motor, ignition, and oil valve in response to commands from a thermostat. If the oil fails to ignite, the controller shuts down the oil burner. A primary control is sometimes referred to as a "stack switch." Primary controllers are available for both constant and intermittent ignition burners. Both types may recycle once if the burner fails to ignite the oil, in which case the controller trips off to a safety mode.

Pump and Zone Controls

Pump and zone controls regulate the flow of water or steam in boiler systems.

Safety Controls

Safety controls such as pressure relief valves, high-temperature limit controls, low-water cut-offs, and burner primary controls protect against appliance malfunction.

Solenoid Valve

An electrically operated device in which, when the valve coil is energized, a magnetic field is developed, causing a spring-loaded steel valve piston to overcome the resistance of the spring and immediately pull the piston into the stem. The valve is now in the open position. When de-energized, the solenoid coil magnetic field instantly dissipates, and the spring-loaded valve piston snaps closed, stopping oil flow or gas flow to the nozzle. The flame is extinguished, allowing fuel to flow. Usually a spring returns the valve back to the closed position.

Spark Ignition

In spark ignition models, the spark comes on while pilot gas flows to the pilot orifice. Once the pilot flame is "proven" through an electronic sensing circuit, the main gas valve opens. Once ignition takes place, the sensor will monitor and prove existence of a main flame. If for some reason the main flame goes off, the furnace shuts down. Energy-efficient furnaces may use this concept for ignition.

Switching Relay

A switching relay is used between the low-voltage wiring and line-voltage wiring. When a low-voltage thermostat calls for heat, the relay "pulls in," closing the line voltage contacts on the relay, which turns on the main power to the furnace or boiler. Usually they have a built-in transformer that provides the low voltage for the low-voltage Class 2 control circuit.

Thermocouple

A thermocouple is used in systems having a standing pilot to hold open a gas valve pilot solenoid magnet. A thermocouple consists of two dissimilar metals connected together to form a circuit. The metals might be iron and copper, copper and iron constantan, copper-nickel alloy and chrome-iron alloy, platinum and platinum-rhodium alloy, or chromel and alumel. The types of metals used depend on the temperatures involved. When one of the junctions is heated, an electrical current flows (Figure 24-5). To operate, there must be a temperature difference between the metal junctions. In a gas

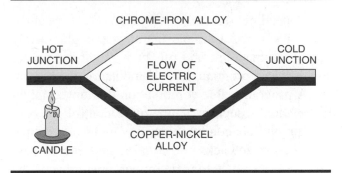

Figure 24-5 Principle of a thermocouple.

burner, the source of heat for the thermocouple is the pilot light. The cold junction of the thermocouple remains open and is connected to the pilot safety shutoff gas valve circuit.

A single thermocouple develops a voltage of about 25 to 30 dc millivolts. Because of this extremely low voltage, circuit resistance is kept to a very low value.

Thermopile

When thermocouples are connected in series, a thermopile is formed (Figure 24-6). The power output of a thermopile is greater than that of a single thermocouple—typically either 250 or 750 dc millivolts.

For example, 10 thermocouples (25 dc millivolts each) connected in series result in a 250 dc millivolt thermopile. Twenty-six thermocouples connected in series results in a 750 dc millivolt thermopile. For comparison, 750 millivolts is one-half the voltage of a 1½-volt "D" size flashlight battery.

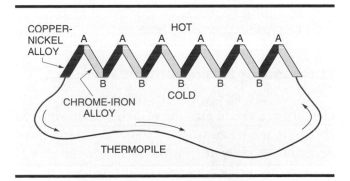

Figure 24-6 Principle of a thermopile.

Thermostat

Depending on the complexity of the heating/cooling system, a simple thermostat might merely turn the furnace on and off, or it might be a more sophisticated programmable combination thermostat.

A programmable combination thermostat might have features such as temperature control for both heating and air conditioning, day/night or multiple time period setbacks for comfort and energy savings, a fan switch for continuous operation of the air circulating fan motor, a clock, a thermometer, outdoor temperature sensing, humidity control, ventilation and filtration control, and indoor air quality control. A programmable thermostat is shown in Figure 24-7. Quality thermostats generally are accurate to ±1 degree.

Most residential thermostats have some kind of adjustment feature to control proper cycling.

Some thermostats feature an integral heat anticipator and a small adjusting lever. The adjusting lever points to the numbers .10, .20, .25, .30, etc., which correspond to the current draw of the circuit controlled by the thermostat.

Electronic thermostats use a logic circuit to control the cycling rate instead of a manual adjusting lever.

Thermostats in residences should be mounted approximately 52 in. (1.3 m) above the finished floor and not mounted where influenced by drafts, concealed hot or cold water pipes or ducts, on outside walls, luminaires (fixtures), direct rays from the sun, heat from fireplaces and heating appliances, or air currents from hot or cold air registers.

Some thermostats contain a small vial of mercury. When the vial "tips," the mercury rolls downward, opening and/or closing the circuit by bridging the gap between the electrical contacts inside the vial. To ensure accuracy, these types of thermostats must be mounted perfectly level.

Thermostat Replacement Warning:

If you are replacing a thermostat that contains mercury in a sealed glass tube, do *not* place your old control in the trash. Contact your local waste management authority for instructions regarding recycling and the proper disposal of this thermostat. You will want to be sure that disposal is in compliance with local, state, and federal regulations.

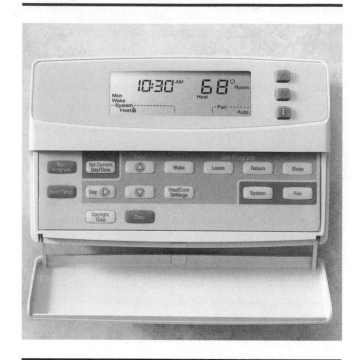

Figure 24-7   A digital electronic programmable thermostat that provides proper cycling of the heating system, resulting in comfort as well as energy savings. This type of thermostat, depending on the model, can be programmed to control temperature, time of operation, humidity, ventilation, filtration, circulation of air, and zone control. Installation and operational instructions are furnished with the thermostat. *Courtesy of Honeywell, Inc.*

Transformer

The transformer converts (transforms) the branch-circuit voltage to low voltage. For example, the furnace's branch-circuit might be 120 volts. The primary of the transformer is connected to the 120-volt supply. The transformer steps down the 120-volt primary voltage to 24 volts on the secondary side of the transformer. Class 2 circuits for the low-voltage wiring between the thermostat and a furnace or boiler are generally 24 volts.

Water Circulating Pump

In hydronic (wet) systems, this pump(s) circulates hot water through the piping system. Multizone systems may have a number of circulating pumps, each controlled by a thermostat for a particular zone. In multizone systems, a high-temperature limit control might be set to maintain the boiler water temperature to a specific temperature.

Some multizone systems have one circulating pump and use electrically operated valves that open and close on command from the thermostat(s) located in a particular zone(s).

CONTROL CIRCUIT WIRING

Article 725 covers most of the control-circuit wiring requirements. *Article 430, Part VI* covers motor control circuits.

Class 1, 2, and 3 circuits offer alternative wiring methods regarding wire size, derating factors, overcurrent protection, insulation, and materials that differentiate these types of circuits from those of regular light and power wiring as covered in *Chapter 1* through *Chapter 4* of the *NEC.*®

Figure 24-8 defines remote control, signaling, and power-limited terminology. More discussion regarding this is found in Unit 25 and Unit 26 of this text.

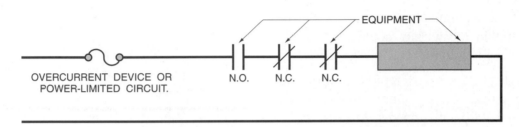

OVERCURRENT DEVICE OR POWER-LIMITED CIRCUIT. N.O. N.C. N.C. EQUIPMENT

A A REMOTE-CONTROL, SIGNALING, OR POWER-LIMITED CIRCUIT IS THAT PORTION OF THE WIRING SYSTEM BETWEEN THE LOAD SIDE OF THE OVERCURRENT DEVICE OR THE POWER-LIMITED SUPPLY AND ALL CONNECTED EQUIPMENT, *725.2.*

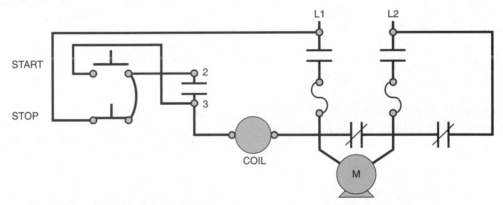

START STOP COIL L1 L2 2 3 M

B A REMOTE-CONTROL CIRCUIT IS ANY ELECTRICAL CIRCUIT THAT CONTROLS ANY OTHER CIRCUIT THROUGH A RELAY OR EQUIVALENT DEVICE, *ARTICLE 100, DEFINITIONS.*

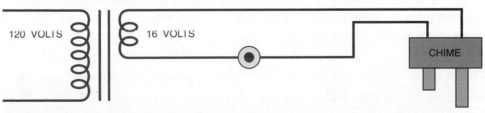

120 VOLTS 16 VOLTS CHIME

C A SIGNAL CIRCUIT IS ANY ELECTRICAL CIRCUIT THAT ENERGIZES SIGNALING EQUIPMENT, *ARTICLE 100, DEFINITIONS.*

Figure 24-8 The above diagrams explain a remote-control circuit and a signaling circuit.

Article 725 discusses three classes of control circuits. Because electricians will come across other types of control-circuit wiring in their career, the following text provides a brief description of the three classes.

Class 1 Circuits

- Class 1 wiring for safety control equipment, such as high-temperature limit and low-water controls, must be installed in rigid or intermediate metal conduit, rigid nonmetallic conduit, EMT, Type MI or MC cable, or be otherwise suitably protected from physical damage. A failure of the equipment, or a short circuit or open circuit in the wiring, could introduce a direct fire or life hazard. See *725.8(A)* and *(B)*.

- A Class 1 power-limited remote-control and signaling circuit output source is limited to 30 volts and 1000 volt-amperes, *725.21(A)*.

- Class 1 remote-control and signaling circuits must not exceed 600 volts, *725.21(B)*.

- Class 1 conductor maximum overcurrent protection, *725.23*:

  14 AWG and larger: size according to the conductor's ampacity.

  16 AWG: 10 amperes
  18 AWG: 7 amperes

- Class 1 overcurrent protection must be located where the conductors receive their supply, *725.24(A)*.

- Class 1 circuits may be run in the same raceway only if the interconnected equipment is *functionally associated, 725.26(B)*. See Figure 24-9.

- Class 1 conductors must be rated 600 volts, *725.27(B)*.

Class 1 circuits are rarely found in residential equipment, but are mentioned here as part of the overall understanding for control-circuit wiring.

Class 2 Circuits

In residential wiring, most low-voltage wiring comes under the Class 2 circuit category. Some examples of Class 2 circuits are the thermostat wiring for heating, air conditioning, and heat pumps. The wiring for security systems, intercom systems, and chime wiring is Class 2.

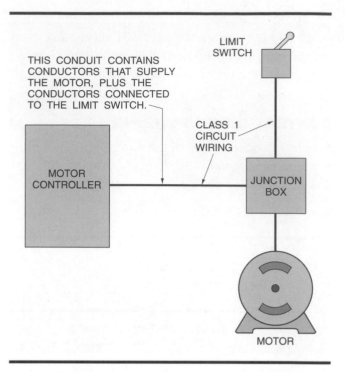

THIS CONDUIT CONTAINS CONDUCTORS THAT SUPPLY THE MOTOR, PLUS THE CONDUCTORS CONNECTED TO THE LIMIT SWITCH.

LIMIT SWITCH

CLASS 1 CIRCUIT WIRING

MOTOR CONTROLLER

JUNCTION BOX

MOTOR

Figure 24-9 The power conductors supplying the motor and the Class 1 circuit remote-control conductors to the limit switch and controller may be run in the same conduit when the interconnected equipment is "functionally associated," 725.26(B). The insulation on the Class 1 circuit conductors must be insulated for the maximum voltage of any other conductors in the raceway. The minimum voltage rating for Class 1 conductors is 600 volts, 725.27(B).

- Class 2 circuits are "power-limited," and are considered to be safe from fire hazard and from electric shock hazard.

- Class 2 wiring does not have to be in a raceway or conduit because should a failure in the wiring occur, such as a short circuit or an open circuit in a thermostat cable, there would not be a direct fire or shock hazard. The operation of the safety controls in the heating equipment would prevent the occurrence of a fire or life hazard (see Figure 24-10).

- Class 2 circuits shall not be run in the same raceway, cable, compartment, outlet box, or fitting with electric light and power conductors, *725.54(A)(1)*. Even if the Class 2 conductors have the same 600-volt insulation as the power conductors, this is not permitted (Figure 24-11). A barrier inside an enclosure that keeps the

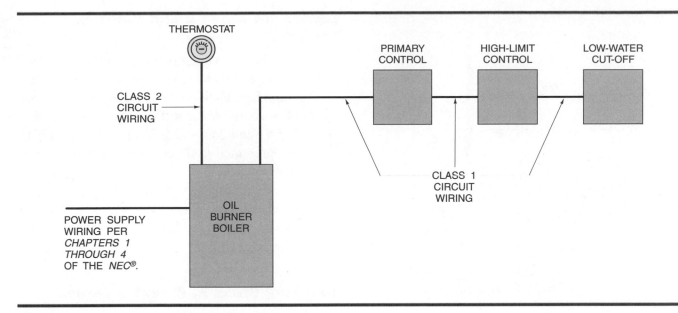

Figure 24-10 Any failure in Class 1 circuit wiring would introduce a direct fire or life hazard. A failure in the Class 2 circuit wiring would *not* introduce a direct fire or life hazard. See *725.8(A)* and *(B)*.

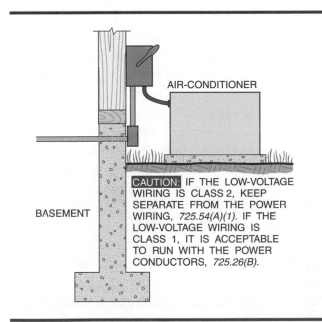

Figure 24-11 Most residential air-conditioning units, furnaces, and heat-pumps have their low-voltage circuitry classified as Class 2. *NEC® 725.54(A)(1)* prohibits installing low-voltage Class 2 conductors in the same raceway as the power conductors, even if the Class 2 conductors have 600-volt insulation. Therefore, the conduit running out of the basement wall in the above diagram would not be permitted to contain both the 240-volt power conductors and the 24-volt low-voltage Class 2 conductors. Always read the instructions furnished with these types of appliances to be sure your wiring is in compliance with these instructions, per *110.3(B)* and the *NEC.* Review Unit 4 for grounding methods.

Class 2 wiring and power conductors apart is permitted.

- Class 2 circuits found in homes are generally supplied by a small transformer that is marked "Class 2 Transformer."

- Class 2 circuits that are inherently current-limited, such as through a Class 2 transformer, do not require separate overcurrent protection.

  Manufacturers of heating and cooling appliances provide separate openings for bringing in the Class 2 wiring, and separate knockouts for bringing in the power supply.

- Class 2 wiring may be fastened to the conduit that contains the power supply conductors to specific equipment, *725.54(D)* and *300.11(B)(2)* (Figure 24-12).

- Transformers that are intended to supply Class 2 and Class 3 circuits are listed by UL and are marked "Class 2 Transformer" or "Class 3 Transformer." These transformers have built-in protection against the possibility of overheating should a short circuit occur somewhere in the secondary wiring. These transformers may be connected to circuits having overcurrent protection not more than 20 amperes, *725.51.*

- Class 2 cables shall have a voltage rating of not less than 150 volts, *725.71(G).*

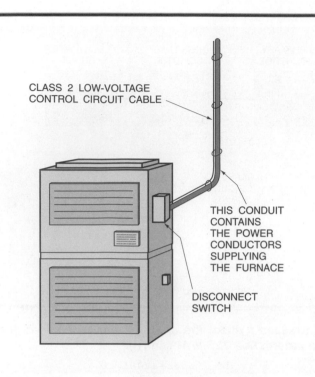

CLASS 2 LOW-VOLTAGE
CONTROL CIRCUIT CABLE

THIS CONDUIT
CONTAINS
THE POWER
CONDUCTORS
SUPPLYING
THE FURNACE

DISCONNECT
SWITCH

Figure 24-12 The Code in *300.11(B)(2)* and *725.54(D)*, allows Class 2 control circuit conductors (cables) to be supported by the raceway that contains the power conductors supplying electrical equipment such as the furnace shown.

Class 3 Circuits

- Class 3 circuits are power limited and are considered to be safe from fire hazard, but could present an electrical shock hazard because of the higher voltage and current levels permitted, *725.2*.

- Class 3 circuits are similar to Class 2 circuits, except that they operate at higher voltage and current levels.

- Class 3 cables shall have a voltage rating of not less than 300 volts, *725.71(G)*.

Class 2 and Class 3 Power Source Requirements

Class 2 and Class 3 power source voltage and current limitations, and overcurrent protection requirements, are found in *Chapter 9*, *Table 11(A)* and *Table 11(B)* of the *NEC*.® You should refer to the nameplate on Class 2 and Class 3 transformers, power supplies, and other equipment to confirm that they are UL listed for a particular class.

Cables for Class 2 and Class 3 Circuits

Low-voltage conductors for Class 2 and Class 3 wiring are rated in a hierarchy based on their ability to withstand fire. Their permitted uses and permitted substitutions between the different types are listed in *Table 725.61* and *Table 725.71* of the *NEC*.® These tables are presented in descending order relative to their fire resistance.

The low cost of Class 2 and Class 3 conductors generally installed in homes are Types CL2X and CL3X. When you ask for "bell wire," you are probably getting Type CL2X.

Securing Class 1, 2, and 3 Cables

Class 1, 2, and 3 cables must be installed in a neat and workmanlike manner, and shall be adequately supported by the building structure in such a manner that they will not be damaged by normal building use, *725.7*.

▶*NEC*® *725.7* states that "*Such cables shall be attached to structural components by straps, staples, hangers or similar fittings designed and installed so as not to damage the cable.*"◄

NEC® *300.11(B)* states that "*raceways shall not be used as a means of support for other raceways, cables, or nonelectric equipment.*" Taping, strapping, tie wrapping, or hanging or securing Class 1, 2, or 3 cables to electrical raceways, piping, or ducts is generally prohibited.

However, the exception to *725.54(D)* refers back to *300.11(B)*, which allows Class 2 conductors to be supported by the raceway that contains the power supply conductors for the same equipment that the Class 2 conductors are connected to (see Figure 24-12).

Figure 24-13(A) shows the control circuit totally contained within the equipment. Factory internal wiring is not governed by *Article 725*. This wiring is covered in the UL Standards.

Figure 24-13(B) shows the control wiring leaving the equipment. The external wiring in this illustration comes under the requirements of *Article 430, Part VI*. If the control circuit is supplied other than being connected to the load side of the motor's branch-circuit overcurrent device, the control-circuit wiring would be governed by the requirements of *Article 725*.

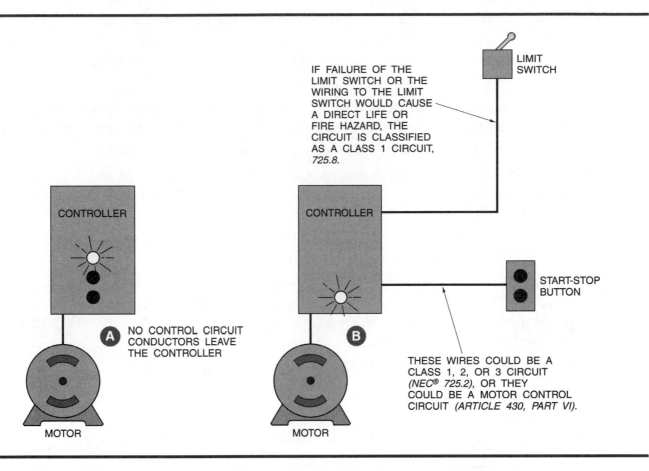

IF FAILURE OF THE LIMIT SWITCH OR THE WIRING TO THE LIMIT SWITCH WOULD CAUSE A DIRECT LIFE OR FIRE HAZARD, THE CIRCUIT IS CLASSIFIED AS A CLASS 1 CIRCUIT, *725.8.*

LIMIT SWITCH

CONTROLLER

CONTROLLER

START-STOP BUTTON

A NO CONTROL CIRCUIT CONDUCTORS LEAVE THE CONTROLLER

B

THESE WIRES COULD BE A CLASS 1, 2, OR 3 CIRCUIT *(NEC® 725.2)*, OR THEY COULD BE A MOTOR CONTROL CIRCUIT *(ARTICLE 430, PART VI)*.

MOTOR

MOTOR

Figure 24-13   Diagram showing how Class 1, 2, and 3 circuits are categorized.

REVIEW

1. The residence in this text is heated with:

    a. Gas

    b. Electricity

    c. Oil

    Circle the correct answer.

2. The *NEC®* in _____, _____, and _____ requires that a disconnecting means be _____ _____ of the furnace.

3. The *NEC®* in _____ requires that a central heating system be supplied by a _____ branch-circuit.

4. In a home, the low-voltage wiring between the thermostat and furnace is:

    a. Class 1

    b. Class 2

    c. Class 3

    Circle the correct answer.

5. Class 2 circuits are _____-limited and are considered to be _____ from _____ hazard and from _____ _____ .

6. The nameplate on a transformer will indicate whether or not it is a Class 2 transformer. (True) (False) Circle the correct answer.

7. If Class 2 conductors are insulated for 32 volts, and the power conductors are insulated for 600 volts, does the Code permit pulling these conductors through the same raceway? (Yes) (No) Circle the correct answer. Give the *NEC*® section number. _____ .

8. If you use 600-volt rated conductors for Class 2 wiring, does the Code permit pulling these conductors through the same raceway as the power conductors? (Yes) (No) Circle the correct answer. Give the *NEC*® section number. _____ .

9. The *NEC*® is very strict about securing anything to electrical raceways. Name the Code sections that prohibit this practice. *NEC*® _____

_____

10. Does the Code permit attaching the low-voltage thermostat cable to the EMT that brings power to a furnace? (Yes) (No) Circle the correct answer. Give the *NEC*® section number. _____

11. Explain what a thermocouple is and how it operates. _____

_____

_____

_____

_____

Television, Telephone, and Low-Voltage Signal Systems

OBJECTIVES

After studying this unit, the student should be able to

- **install television outlets, antennas, cables, and lead-in wires.**
- **list CATV installation requirements.**
- **describe the basic operation of satellite antennas.**
- **install telephone conductors, outlet boxes, and outlets.**
- **define what is meant by a signal circuit.**
- **describe the operation of a two-tone chime and a four-note chime.**
- **install a chime circuit with one main chime and one or more extension chimes.**

INSTALLING THE WIRING FOR HOME TELEVISION

This unit discusses the basics of installing outlets, cables, receivers, antennas, amplifiers, multiset couplers, and some of the *NEC®* rules for home television. Television is a highly technical and complex field. To ensure a trouble-free installation of a home system, it should be done by a competent television technician. These individuals receive training and certification through the Electronic Technicians Association in much the same way that electricians receive their training through apprenticeship and journeyman training programs, learning about the *NEC®* via membership in the International Association of Electrical Inspectors organization. Some cities and states require that, when more than three television outlets are to be installed in a residence, the television technician making the installation be licensed and certified. Just as electricians can experience problems because of poor connections, terminations, and splices, poor reception problems can arise as a result of poor terminations. In most cases, poor crimping is the culprit. Electrical shock hazards are present when components are improperly grounded and bonded.

According to the plans and for this residence, television outlets are installed in the following rooms:

Front Bedroom	1
Kitchen	1
Laundry	1
Living Room	3
Master Bedroom	1
Recreation Room	3
Study/Bedroom	1
Workshop	1

Because TV is a low-voltage system, metallic or nonmetallic standard single-gang device boxes, 4-in. square outlet boxes with a plaster ring, plaster rings only, or special mounting brackets can be installed during the rough-in stage at each likely location of a television set. (More on this later.) See Figure 25-1. For new construction, shielded coaxial cables are

Figure 25-1   Nonmetallic boxes and a nonmetallic raised cover.

installed concealed in the walls. For remodel work, cables can be concealed in the walls by fishing the cables through the walls and installing mounting brackets that are inserted and snapped into place through a hole cut into the wall.

Shielded 75-ohm RG-6 coaxial cable is most often used to hook up television sets to minimize interference and keep the color signal strong. There are different kinds of shielded coaxial cable. Double-shielded cable that has a 100 percent foil shield covered with a 40 percent or greater woven braid is recommended. This coaxial cable has a PVC outer jacket. The older style, flat 300-ohm twin-lead cable, can still be found on existing installations, but the reception might be poor. To improve reception, 300-ohm cables can be replaced with shielded 75-ohm coaxial cables.

For this residence, 12 television outlets are to be installed. For this many outlets, it is recommended that a shielded 75-ohm RG-6 coaxial cable be run from each outlet location back to one central point, as in Figure 25-2.

This central point could be where the incoming Community Antenna Television (CATV) cable comes in, where the cable from an antenna comes in, or where the TV output of a satellite receiver is carried. There are so many possibilities that each installation must be analyzed to determine what is best for that particular installation. For rooftop or similar antenna mounting, the cable is run down the outside of the house, then into the basement, garage, or other convenient location where the proper connections are made. CATV companies generally run their incoming coaxial cables underground, again to

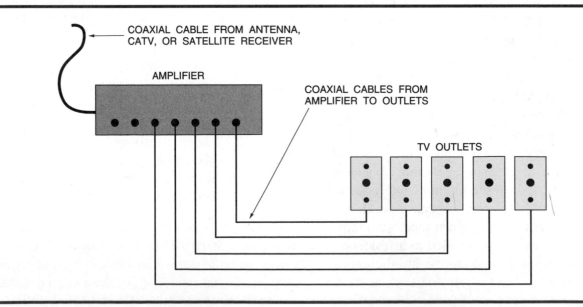

Figure 25-2   A television master amplifier distribution system may be needed where many TV outlets are to be installed. This will minimize the signal loss. In simple installations, a multiset coupler can be used, as shown in Figure 25-3.

some convenient point just inside of the house. Here, the technician can hook up all of the coax cables using a multiset splitter, as shown in Figure 25-3, or an amplifier, as shown in Figure 25-2, to boost possible lost signals. An amplifier may not be necessary if *only* those coaxial cables that will be used are hooked up.

Cable television does not require an antenna on the house because the cable company has the proper antenna to receive signals from satellites. The cable company then distributes these signals throughout

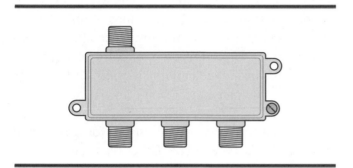

Figure 25-3 A 3-way CATV cable splitter: One in, three out. Two-way and four-way splitters are also commonly available.

the community they have contractually agreed to serve. These contracts usually require the cable company to run their coaxial cable to a point just inside of the house. Inside the house, they will complete the installation by furnishing cable boxes, controls, and the necessary wiring. In geographical areas where cable television is not available, an antenna mounted, as shown in Figure 25-4, or a satellite "dish" must be installed.

Faceplates for the TV outlets are available in styles to match most types of electrical faceplates in the home.

For safety reasons, video/voice/data wiring must be separated from the 120-volt wiring, *810.18(C)*. If both 120-volt and low-voltage circuits are run into the same box, then a permanent barrier must be provided in the box to separate the two systems. A better choice is to keep the two systems totally separate using two wall boxes.

Another popular choice is where video/voice/data cables are run to a location right next to a 120-volt receptacle. Mounting brackets that are commercially available can be installed around the electrical device box during the rough-in stage. This

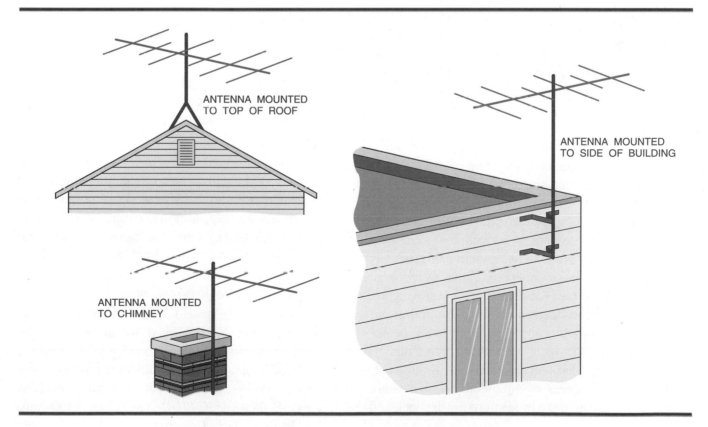

ANTENNA MOUNTED TO TOP OF ROOF

ANTENNA MOUNTED TO SIDE OF BUILDING

ANTENNA MOUNTED TO CHIMNEY

Figure 25-4 Typical mounting of television antennas.

Figure 25-5 A special mounting bracket fastened to the wood stud. The bracket fits nicely over the electrical wall box to accommodate a 120-volt receptacle, two coax outlets, and one telephone outlet. A single two-gang faceplate is used for this installation. Many other combinations are possible.

lets you trim out a 120-volt receptacle and video/ voice/data jacks with one faceplate, as shown in Figure 25-5. These mounting brackets keep the center to center measurements of the device mounting holes for the electrical box and the video/voice/data wiring precisely in alignment. With this method, a wall box is not provided for the video/voice/data wiring. The video/voice/data cables merely come out of the drywall next to the electrical box, and the faceplate takes care of trimming out the receptacle and the video/voice/data wiring jacks.

When wiring a new house, it certainly makes sense to run at least one coaxial cable and one Category 5 cable to a single wall box wherever TV, telephone, and/or computers might be used. This could be a single box located close to a 120-volt receptacle outlet, or it could be a mounting bracket as shown in Figure 25-5. Run a separate coaxial cable and a separate Category 5 cable from each location to a common point in the house. You will use a lot of cable, but this will enable you to interconnect the cables as required, such as for a local area network (LAN) to serve more than one computer located in different rooms, or other home automation systems. This is the beginning of "structured wiring" as discussed in Unit 31.

For adding video/voice/data wiring to an existing home, many types of nonmetallic brackets are available that snap into a hole cut in the drywall, and lock in place.

More *NEC*® rules are discussed later in this unit.

Unit 31 discusses structured wiring using Category 5 unshielded twisted pair (UTP) cable and coaxial shielded cable.

Although shielded coaxial cable minimizes interference, it is good practice to keep the coaxial cables on one side of the stud space, and keep the light and power cables on the other side of the stud space.

Code Rules for Cable Television (CATV) *(Article 820)*

CATV systems are installed both overhead and underground in a community. Then, to supply an individual customer, the CATV company runs coaxial cables through the wall of a residence at some convenient point. Up to this point of entry *(Article 820, Part II)* and inside the building *(Article 820, Part V)*, the cable company must conform to the requirements of *Article 820,* plus local codes if applicable.

Coaxial cable generally used for residential installations is Type CATV. Other types are listed in *Table 820.53* of the *NEC.*®

Here are some of the key rules to follow when making a coaxial cable installation.

1. The outer conductive shield of the coaxial cables must be grounded as close to the point of entry as possible, *820.33.*

2. Coaxial cables shall not be run in the same conduits or box with electric light and power conductors, *820.52(A)(1)(b).*

3. Do not support coaxial cables from raceways that contain electrical light and power conductors, *820.52(E).*

4. Keep the coaxial cable at least 2 in. (50 mm) from light and power conductors unless the conductors are in a raceway, nonmetallic-sheathed cable, armored cable, or UF cable, *820.52(A)(2).* This clearance requirement really pertains to old knob-and-tube wiring. The 2-in. (50-mm) clearance is not required if the coaxial cable is in a raceway.

5. Where underground coaxial cables are run, they must be separated by at least 12 in. (300 mm) from underground light and power conductors, unless the underground conductors are in a raceway, Type UF cable, or Type USE cable, *820.11(B).* The 12-in. (300-mm) clearance is not required if the coaxial cable has a metal cable armor.

6. The grounding conductor (*Article 820, Part III* and *Part IV*):

   a. ►outer conductive shield shall be grounded as close as possible to the coaxial cable entrance or attachment to the building. ◄

   b. shall not be smaller than a 14 AWG copper or other corrosion-resistant conductive material. ►It need not be larger than 6 AWG copper. The grounding conductor must be insulated. It shall have a current-carrying capacity not less than that of the coaxial cable's outer metal shield. ◄ See the section in this unit on *NEC® 810.21*.

   c. ►length shall be as short as possible, but not longer than 20 ft (6.0 m). If impossible to keep the grounding conductor to this maximum permitted length, then the *Exception* to *820.40(D)* permits installing a separate grounding electrode. These two electrodes shall be bonded together with a bonding jumper not smaller than 6 AWG copper. ◄

   d. may be solid or stranded.

   e. shall be guarded from physical damage.

   f. shall be run in a line as straight as practical.

   g. shall be connected to the nearest accessible location on one of the following:

- the building grounding electrode.
- the grounded interior metal water piping pipe, within 5 ft (1.5 m) of where the pipe enters the building. Refer to *250.52* for more details.
- the power service accessible means external to the enclosure.
- the metallic power service raceway.
- the service equipment enclosure.
- the grounding electrode conductor or its metal enclosure.
- the grounding conductor or grounding electrode of a disconnecting means that is grounded to an electrode according to *250.32*. This pertains to a second building on the same property served by a feeder or branch circuit from the main electrical service in the first building.

   h. If *none* of the options in *g* is available, then ground to any of the electrodes per *250.52*, such as metal underground water pipes, the metal frame of the building, a concrete encased electrode, or a ground ring. Watch out for the maximum length permitted for the grounding conductor, as mentioned above.

CAUTION: DO NOT simply drive a separate ground rod to ground the metal shielding tape on the coax cable. Difference of potential between the coax cable shield and the electrical system ground during lightning strikes could result in a shock hazard as well as damage to the electronic equipment. If for any reason one grounding electrode is installed for the CATV shield grounding and another grounding electrode for the electrical system, bonding these two electrodes together with no smaller than a 6 AWG copper conductor will minimize the possibility that a difference of potential might exist between the two electrodes.

SATELLITE ANTENNAS

A satellite antenna is often referred to as a "dish" (Figure 25-6). A satellite dish has a parabolic shape that concentrates and reflects the signal beamed down from one of the many satellites located 22,247 miles above the equator. The orbit in which satellites travel is called the "Clarke belt," as shown in Figure 25-7.

Satellite reception has made it possible for the homeowner to:

- use satellite reception only, in which case the homeowner would not receive local programs.
- use satellite reception *and* use CATV for local channels. By law, CATV companies must supply this basic service and cannot require the homeowner to sign up for premium programs.
- use satellite reception for all of the programs offered by the satellite company, *and* a rooftop or similar antenna for local channels.

Satellite antennas are available in solid or "see-through" types. See-through types are constructed of perforated metal, and have less wind resistance and

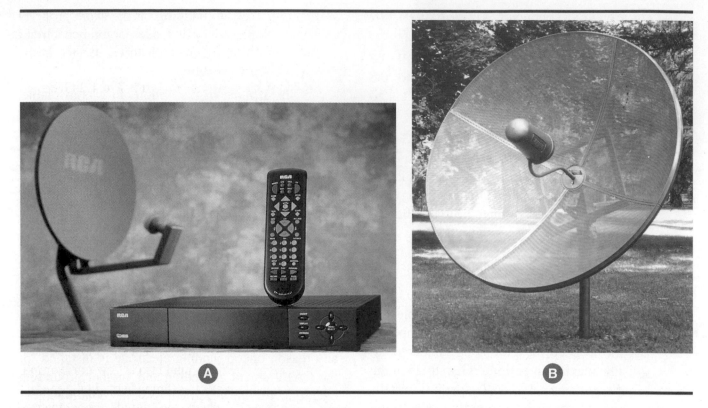

Figure 25-6 (A) A digital satellite system 18-in. (450-mm) antenna, a receiver, and a remote control. *Courtesy of* Thomson Consumer Electronics. (B) A large 6-ft (1.8-m) satellite antenna securely mounted to a post that is anchored in the ground.

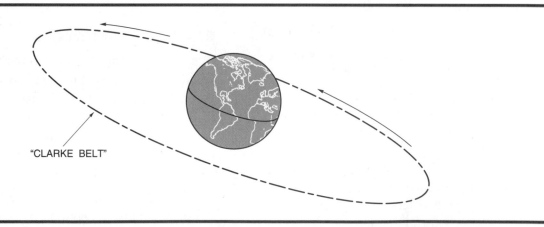

"CLARKE BELT"

Figure 25-7 All satellites travel around the earth in the same orbit called the "Clarke belt." They appear to be stationary in space because they are rotating at the same speed at which the earth rotates. This is called "geosynchronous orbit (geostationary orbit)." The satellite receives the uplink signal from earth, amplifies the signal, and transmits it back to earth. The downlink signal is picked up by the satellite antenna.

less weight than solid dishes. The size of the antenna shown in Figure 25-6(B) can be from 6 to 10 ft (1.8 to 3.0 m) in diameter. Depending upon local codes, satellite antennas can be mounted on the roof, a chimney, a post, a pedestal, or on the side of the house. Some localities do not permit these large size antennas.

The line-of-sight from the antenna to the satellite must not be obstructed by trees, buildings, telephone poles, or other structures.

The latest satellite technology is the *digital satellite system* (DSS). A digital system allows the use of small antennas, approximately 18 in. (450 mm) in diameter, that can easily be mounted on the roof, the chimney, the side of the house, on a pipe, or on a pedestal, using the proper mounting hardware, Figure 25-6(A). A digital system offers clearer pictures and clearer sound, greater choices of channel selection, and neater appearance insofar as the size of the antenna is concerned. The receiver of a digital system requires a connection to the telephone system for program billing, updating the program access, and ensuring continuous program service. Once each month the satellite company will call the receiver and download all of the information that has been stored. Any and all pay-per-view and other premium services used during that month will be recorded during this very brief downloading. The digital satellite receiver has a modular telephone phone jack on the back of the receiver for this connection.

The basic operation of a satellite system is for a transmitter on earth to beam an "uplink" signal to a satellite in space. Electronic devices on the satellite reamplify and convert this signal to a "downlink" signal, then retransmit the downlink signal back to earth, as shown in Figure 25-8.

C-band satellites transmit at 3.7 to 4.2 GHz.

KU-band (low and medium power) transmit at 11.7 to 12.2 GHz.

KU-band (high power) transmit at 12.2 to 12.7 GHz. This high frequency makes it possible to use a small 18-in. (450 mm) antenna.

The term *gigahertz* (GHz) means billions of cycles per second. The term *megahertz* (MHz) means millions of cycles per second.

On earth, the downlink signal is picked up by the "feedhorn" located on the front of a satellite antenna, as shown in Figure 25-6.

A feedhorn is located at the exact focal point in relationship to the parabolic reflector to gather the maximum amount of signal. The feedhorn funnels the downlink satellite signal into a "low noise amplifier" where the signal is amplified, then fed to a "downconverter" where the signals are converted from high frequency to a low frequency (70 MHz)

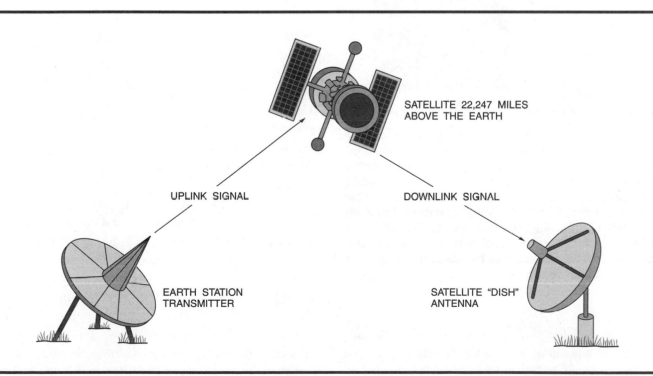

SATELLITE 22,247 MILES ABOVE THE EARTH

UPLINK SIGNAL

DOWNLINK SIGNAL

EARTH STATION TRANSMITTER

SATELLITE "DISH" ANTENNA

Figure 25-8   The earth station transmitter beams the signal up to the satellite where the signal is amplified, converted, and beamed back to earth. The power requirements of the satellite are in the range of 5 watts and are provided by the use of solar cells.

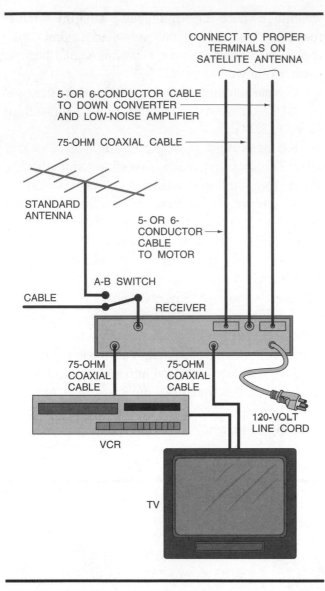

Figure 25-9 Typical satellite antenna/cable/ standard antenna wiring. Connection to a standard antenna cable, CATV cable, or to a satellite cable enables the homeowner to watch one channel while recording another channel. The installation manual for the receiver usually has a number of different hookup diagrams. In this diagram, the A-B switch allows a choice of connecting to a standard antenna or to the CATV cable. This hookup can be desirable should the CATV or satellite reception fail.

that is proper for the receiver. All of these are mounted in a housing in front of the antenna.

The receiver is located inside the home and is connected to the downconverter on the dish with a 75-ohm coaxial cable. The receiver in turn is connected to the television set, as shown in Figure 25-9 and Figure 25-10.

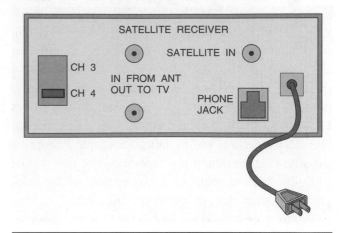

Figure 25-10 The terminals on the back of a digital satellite receiver. The connections are similar to those in Figure 25-9. Note the difference in that the digital satellite receiver has a modular telephone plug, a feature that uses a toll-free number to update the access card inside the receiver to ensure continuous program service. The telephone can also handle program billing.

More and more satellite programmers are scrambling their uplink signals. The home satellite dish owner needs a special descrambling device to be able to decode and view these scrambled channels.

Small 18-in. (450-mm) antennas are easily focused following the manufacturer's instructions. Large antennas are more difficult to focus, and in some instances are motor driven. One manufacturer utilizes a reversible 36-volt dc motor, controllable from inside the home for fine-tune focusing of the antenna to the satellite.

For the more complicated large dish installations, the cables that are installed between the receiver inside the home and the satellite antenna consist of color-coded conductors, making them easy to identify when doing the hook-up. The number of conductors necessary depends upon circuit designs of the receiver and the satellite. For the newer digital satellite systems, the hook-up is greatly simplified, using 75-ohm coaxial cable between the receiver and the antenna. Always follow the manufacturer's installation instructions.

Figure 25-11 shows one method of installing a large satellite antenna to a metal post securely positioned in concrete.

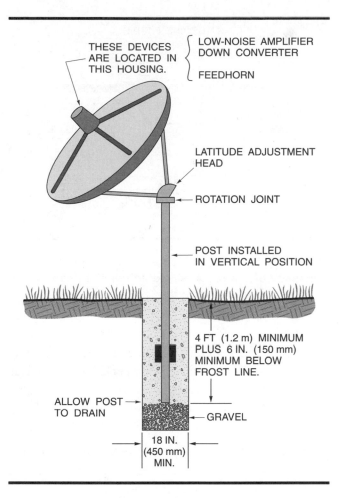

THESE DEVICES
ARE LOCATED IN
THIS HOUSING.

LOW-NOISE AMPLIFIER
DOWN CONVERTER

FEEDHORN

LATITUDE ADJUSTMENT
HEAD

ROTATION JOINT

POST INSTALLED
IN VERTICAL POSITION

4 FT (1.2 m) MINIMUM
PLUS 6 IN. (150 mm)
MINIMUM BELOW
FROST LINE.

ALLOW POST
TO DRAIN

GRAVEL

18 IN.
(450 mm)
MIN.

Figure 25-11 Typical satellite antenna of the type installed in remote areas where reception is poor, or in areas served by CATV, but where the homeowner does not wish to subscribe and pay for the service of the CATV company. Always check with the local inspectors to be sure that all installation code rules are met.

CODE RULES FOR THE INSTALLATION OF ANTENNAS AND LEAD-IN WIRES
(ARTICLE 810)

Home television, video cassette recorders (VCRs), compact disc equipment, and AM and FM radios generally come complete with built-in antennas. For those locations in outlying areas on the fringe or out of reach of strong signals, it is quite common to install a separate antenna system.

Either indoor or outdoor antennas may be used with black-and-white and color televisions. The front of an outdoor antenna is aimed at the television transmitting station. When there is more than one transmitting station and they are located in different directions, a rotor is installed. A rotor turns the antenna on its mast so that it can face in the direction of each transmitter. The rotor is controlled from inside the building. The rotor controller's cord is plugged into a regular 120-volt receptacle to obtain power. A four-wire cable is usually installed between the rotor motor and the control unit. The wiring for a rotor may be installed during the roughing-in stage of construction, running the rotor's four-wire cable into a device wall box, allowing 5 to 6 ft (1.5 to 1.8 m) of extra cable, then installing a regular single-gang switchplate when finishing.

Article 810 covers radio and television equipment. Although instructions are supplied with antennas, the following key points of the Code regarding the installation of antennas and lead-in wires should be followed.

1. Antennas and lead-in wires shall be securely supported, *810.12*.

2. Antennas and lead-in wires shall not be attached to the electric service mast, *810.12*.

3. Antennas and lead-in wires shall be kept away from all light and power conductors to avoid accidental contact with the light and power conductors, *810.13*.

4. Antennas and lead-in wires shall not be attached to any poles that carry light and power wires over 250 volts between conductors, *810.12*.

5. Lead-in wires shall be securely attached to the antenna, *810.12*.

6. Outdoor antennas and lead-in conductors shall not cross over light and power wires, *810.13*.

7. Outdoor antennas and lead-in conductors shall be kept at least 24 in. (600 mm) away from open light and power conductors, *810.13*.

8. Where practicable, antenna conductors shall not be run under open light and power conductors, *810.13*.

9. On the outside of a building:

   a. Position and fasten lead-in wires so they cannot swing closer than 24 in. (600 mm) to light and power wires having *not* over 250 volts between conductors; 10 ft (3.0 m) if *over* 250 volts between conductors, *810.18(A)*.

b. Keep lead-in conductors at least 6 ft (1.8 m) away from a lightning rod system, or bonded together according to *250.60*.

Note: The clearances in (a) and (b) are not required if the light and power conductors or the lead-in conductors are in a metal raceway or metal cable armor.

10. On the inside of a building:

a. Keep the antenna and lead-in wires at least 2 in. (50 mm) from other open wiring (as in old houses) unless the other wiring is in a metal raceway or cable, *810.18(B)*.

b. Keep lead-in wires out of electric boxes unless there is an effective, permanently installed barrier to separate the light and power wires from the lead-in wire, *810.18(C)*.

11. Grounding:

a. All metal masts and metal structures that support antennas shall be grounded, *810.21*.

b. The grounding wire must be copper, aluminum, copper-clad steel, bronze, or a similar corrosion-resistant material, *810.21(A)*.

c. The grounding wire need not be insulated. It must be securely fastened in place, may be attached directly to a surface without the need for insulating supports, shall be protected from physical damage or be large enough to compensate for lack of protection, and shall be run in as straight a line as is practicable, *810.21(B), (C), (D), and (E)*.

d. The grounding conductor shall be connected to the nearest accessible location on one of the following:

- the building or structure grounding electrode. Refer to *250.50* for more details.

- ►the grounded interior metal water pipe (as shown in Figure 25-12), within 5 ft (1.5 m) of where the water pipe enters the building.◄ Refer to *250.52* for more details.

- the metallic power service raceway.

- the service equipment enclosure.

- the grounding electrode conductor or its metal enclosure.

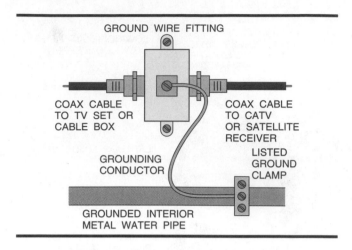

Figure 25-12 Typical connection for the required grounding of the metal shield on coaxial cable to the interior metal water piping.

e. If none of *d* is available, then ground to any one of the electrodes per *250.52*, such as metal underground water pipe, metal frame of building, concrete-encased electrode, or ground ring, *Section 810.21(F)(2)*.

f. If neither *d* nor *e* are available, then ground to an effectively grounded metal structure or to any one of the electrodes per *250.52*, a driven rod or pipe, or a metal plate, *810.21(F)(3)*.

g. The grounding conductor may be run inside or outside of the building, *810.21(G)*.

h. The grounding conductor shall not be smaller than 10 AWG copper or 8 AWG aluminum, *810.21(H)*.

CAUTION: DO NOT simply drive a separate ground rod to ground the metal mast, structure, or antenna. Difference of potential between the metal mast, structure, or antenna and the electrical system ground during a lightning strike could result in a shock hazard as well as damage to the electronic equipment. If any reason permitted by the Code results in one grounding electrode for the antenna and another grounding electrode for the electrical system, bond the two electrodes together with a bonding jumper not smaller than 6 AWG copper or equivalent, *810.21(J)*.

The objective of grounding and bonding to the same grounding electrode as the main service ground is to reduce the possibility of having a difference of voltage potential between the two systems.

TELEPHONE WIRING (SYMBOL ▶|)
(ARTICLE 800)

Since the deregulation of telephone companies, residential "do-it-yourself" telephone wiring has become quite common. Following is an overview of residential telephone wiring.

The telephone company will install the service line to a residence and terminate at a protector device, as in Figure 25-13. The protector protects the system from hazardous voltages. The protector may be mounted either outside or inside the home. Different telephone companies have different rules. Always check with the phone company before starting your installation.

The point where the telephone company ends and the homeowner's interior wiring begins is called the *demarcation point*, Figure 25-13. The preferred demarcation point is outside of the home, generally near the electric meter where proper grounding and

bonding can be done. A *network interface* might be of the types depicted in Figure 25-13, at (A) and (B). In the past, interior telephone wiring in homes was installed by the telephone company. Today, the responsibility for roughing-in the interior wiring is usually left up to the electrician. Telephone wiring is *not* to be run in the same raceway with electrical wiring. In addition, it cannot be installed in the same outlet and junction boxes as electrical wiring. Whatever the case, the rules found in *Article 800* apply. Refer to Figure 25-13, Figure 25-14, and Figure 25-15.

For a typical residence, the electrician will rough-in wall boxes wherever a telephone outlet is wanted. The boxes can be of the "single-gang" types illustrated in Figure 25-1, Figure 2-11, Figure 2-17, Figure 2-19, and Figure 2-32. Concealed telephone wiring in new homes is generally accomplished using multipair cables that are run from each tele-

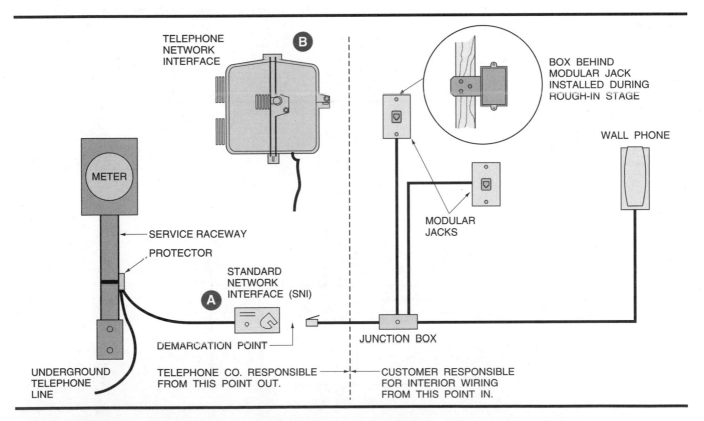

Figure 25-13 Typical residential telephone installation. The telephone company installs and connects the underground cable to the protector. In this illustration, the protector is mounted to the metal service-entrance raceway. This establishes the ground that offers protection against hazardous voltages such as a lightning surge on the incoming line. A cable is then run to a Standard Network Interface, referred to as SNI or just NI. These are shown in (A) and (B) in the diagram. The interior telephone wiring is then connected to the SNI, either with a modular plug as shown in (A), or "hard-wired" through a neoprene grommet on the bottom of (B). The SNI of the type depicted by (B) are preferred because they are weatherproof and can be mounted on the outside of the house.

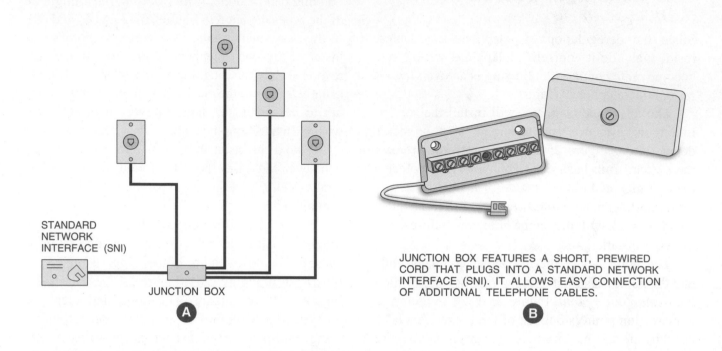

STANDARD
NETWORK
INTERFACE (SNI)

JUNCTION BOX

A

JUNCTION BOX FEATURES A SHORT, PREWIRED
CORD THAT PLUGS INTO A STANDARD NETWORK
INTERFACE (SNI). IT ALLOWS EASY CONNECTION
OF ADDITIONAL TELEPHONE CABLES.

B

Figure 25-14   (A) shows individual telephone cables run to each telephone outlet from the common connection point. This is sometimes referred to as a star-wired system. (B) shows a junction box for easy connection of multiple telephone cables.

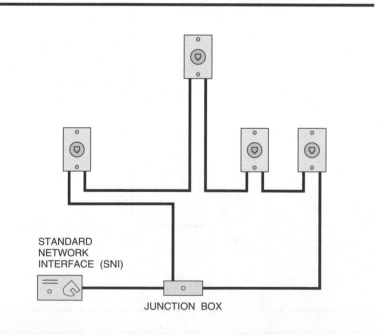

STANDARD
NETWORK
INTERFACE (SNI)

JUNCTION BOX

Figure 25-15   A complete "loop system" (daisy-chain) of the telephone cable. If something happens to one section of the cable, the circuit can be fed from the other direction.

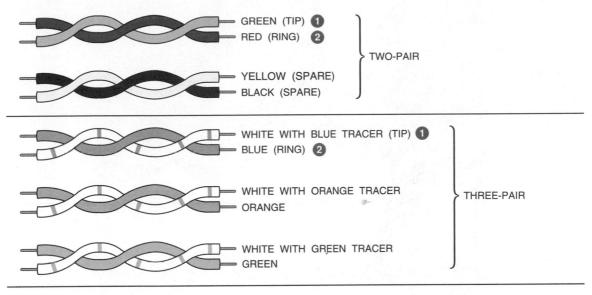

GREEN (TIP) **1**
RED (RING) **2**
} TWO-PAIR

YELLOW (SPARE)
BLACK (SPARE)

WHITE WITH BLUE TRACER (TIP) **1**
BLUE (RING) **2**

WHITE WITH ORANGE TRACER
ORANGE
} THREE-PAIR

WHITE WITH GREEN TRACER
GREEN

1 *TIP* IS THE CONDUCTOR THAT IS CONNECTED TO THE TELEPHONE COMPANY'S "POSITIVE" TERMINAL. IT IS SIMILAR TO THE NEUTRAL CONDUCTOR OF A RESIDENTIAL WIRING CIRCUIT.

2 *RING* IS THE CONDUCTOR THAT IS CONNECTED TO THE TELEPHONE COMPANY'S "NEGATIVE" TERMINAL. IT IS SIMILAR TO THE "HOT" UNGROUNDED CONDUCTOR OF A RESIDENTIAL WIRING CIRCUIT.

Figure 25-16   Color coding for a two-pair and a three-pair, six-conductor twisted telephone cable. The color coding for two-pair telephone cable stands alone. For three or more pair cables, the color coding becomes WHITE/BLUE, WHITE/ORANGE, WHITE/GREEN, WHITE/BROWN, WHITE/SLATE, RED/BLUE, RED/ORANGE, RED/GREEN, RED/BROWN, RED/SLATE, BLACK/BLUE, BLACK/ORANGE, BLACK/GREEN, BLACK/ BROWN, BLACK/SLATE, YELLOW/BLUE, YELLOW/ORANGE, YELLOW/GREEN, YELLOW/BROWN, YELLOW/SLATE, VIOLET/BLUE, VIOLET/ORANGE, VIOLET/GREEN, VIOLET/BROWN, VIOLET/SLATE. In multipair cables, the additional pairs can be used for more telephones, fax machines, security reporting, speakerphones, dialers, background music, etc.

phone outlet to a common point, as in Figure 25-14(A). The cable can also take the form of a "loop system," as in Figure 25-15. A three-pair, six-conductor twisted cable is shown in Figure 25-16. There are some localities that require that these cables be run in a metal raceway, such as EMT.

The outer jacket is usually of a thermoplastic material. This outer jacket most often is a neutral color such as white, light gray, or beige that blends in with decorator colors in the home. This is particularly important in existing homes where the telephone cable may have to be exposed. Cables, mounting boxes, junction boxes, terminal blocks, jacks, adaptors, faceplates, cords, hardware, plugs, and so on, are all available through electrical distributors, builders' supply outlets, telephone stores, electronic stores, hardware stores, and similar wholesale and retail outlets. The selection of types of the preceding components is endless.

Telephone Conductors

The color coding of telephone cables is shown in Figure 25-16. Figure 25-17 shows some of the many types of telephone cords available.

Cables are available with various numbers of conductors. Here are a few.

2 pair	4 conductors
3 pair	6 conductors
4 pair	8 conductors
6 pair	12 conductors
12 pair	24 conductors

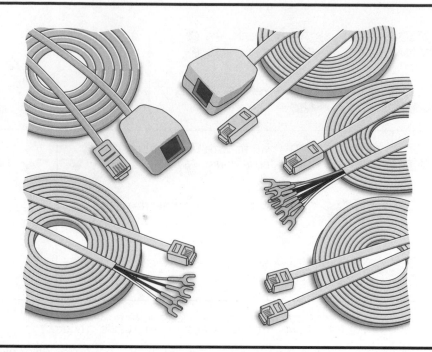

Figure 25-17 Some of the many types of telephone cable available are illustrated. Note that for ease of installation, the modular plugs and terminals have been attached by the manufacturer of the cables. These cords are available in round and flat configurations, depending upon the number of conductors in the cable.

Telephone circuits require a separate pair of conductors from the telephone all the way back to the phone company's central switching center. To keep to a minimum interference that could come from other electrical equipment, such as electric motors and fluorescent fixtures, each pair of telephone wires (two-wires) is twisted.

Cross-Talk Problems

A common telephone interference problem in homes is "cross-talk." This is apparent if you hear someone faintly talking in the background when you are using the telephone. Another problem is losing an on-line connection to your computer. The signals traveling in one pair of conductors are picked up by the adjacent pair of conductors in the cord or cable. Quite often, these problems can be traced to older style flat (untwisted) two-line telephone lines. Older style cables could carry audio signals quite well, but they are unsatisfactory for data transmission. These problems are virtually eliminated by using cables and cords that have conductors twisted at proper intervals, such as Category 5 cables. In Figure 25-16, each pair of conductors is twisted,

then all of the pairs are twisted again. Generally, you should avoid flat two-line cords in lengths over 7 ft (2.13 m).

The recommended circuit lengths are:

24 AWG gauge – – – – not over 200 ft (60 m)

22 AWG gauge – – – – not over 250 ft (75 m)

Ringer Equivalence Number (REN)

Telephones, fax machines, answering machines, and so on, that have a sounder (ringer) are assigned an REN number. Generally, the maximum number of RENs on any single telephone line is 5. This maximum number of RENs assures that all telephones will ring. When too many telephones are connected, the ring signal can become unreliable, resulting in devices failing to respond.

Older style electromechanical ringers were considered to be 1-REN. Electronic ringers have a minimal REN value. Mixing electromechanical and electronic ringers could present a problem as the electromechanical ringers tend to hog the current. It all depends upon your telephone line. Try it and see what happens.

Pay close attention to the total RENs connected in your home. REN values are found on the label of the device.

EXAMPLE:

Four electronic telephones:	$4 \times 0.3 = 1.2$ RENs
One Fax machine:	1.5 RENs
One desk top telephone/ Answering machine:	0.7 REN
One desk top telephone:	0.4 RENs
Total	3.8 RENs

Call your telephone company if you think you have a problem because of too many telephones connected to one telephone line.

Plug-in or Permanent Connection?

Most residential telephones plug into a jack of the type illustrated in Figure 25-18 and Figure 25-19. It might be a good idea to have at least one permanently connected telephone, such as a wall phone in the kitchen. If all of the phones were of the plug-in type, it is possible (but highly unlikely) that all of them could be unplugged. As a result, there would be no audible signal.

The electrical plans show that nine telephone outlets are provided as follows:

Front Bedroom	1
Kitchen	1
Laundry	1
Living Room	1
Master Bedroom	1
Rear Outdoor Patio	1
Recreation Room	1
Study/Bedroom	1
Workshop	1

Figure 25-18 Telephone modular jack.

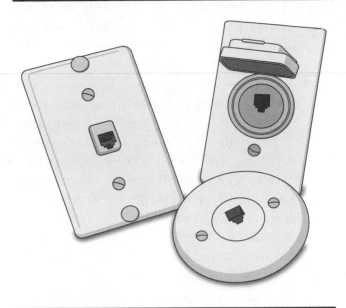

Figure 25-19 Three styles of wall plates for modular telephone jacks: rectangular stainless steel, weatherproof for outdoor use, and circular.

Installation of Telephone Conductors
(800.51 and *800.52)*

Telephone conductors:

1. are generally Type CM or Type CMX (for dwellings only), listed for telephone installation as being resistant to the spread of fire. Refer to *Table 800.50* for other types.

2. shall be separated by at least 2 in. (50 mm) from light and power conductors unless the light and power conductors are in a raceway, or in nonmetallic-sheathed cable, Type AC cable, or Type UF cable, *800.52(A)(2)*.

3. shall not be placed in any conduit or boxes with electric light and power conductors unless the conductors are separated by a partition, *800.52(A)(1)(c)*.

4. do not support telephone cables from raceways that contain light and power conductors, *800.52(E)*.

5. should not share the same bored holes as electrical wiring, plumbing, gas pipes, and so on.

6. may be terminated in either metallic or nonmetallic boxes or plaster rings.

7. should be secured using rounded or depthstop plastic staples. Do not use metal staples.

8. do not use nail guns. Do not crush the cable.

9. should be kept away from hot water pipes, hot air ducts, and other heat sources that might harm the insulation.

10. do not run telephone cables in the same stud space as electrical branch-circuit wiring, and keep cables at least 12 in. (300 mm) from power wiring where the cables are run parallel to the power wiring. These cautions were written before twisted pairs and category-rated cables entered the scene. With the advent of *properly* installed twisted and shielded cables, these "old wives' tale" recommendations from the past are probably not an issue.

Follow the specifications and instructions for the cable you are installing.

Conduit

In some communities, electricians prefer to install regular device boxes at a telephone outlet location, then "stub" a trade size ½ EMT to the basement or attic from this box. Then, at a later date, the telephone cable can be fished through the conduit.

Grounding *(Article 800, Part IV)*

The telephone company will provide the proper grounding of their incoming cable sheath and primary protector, generally using an insulated conductor not smaller than 14 AWG copper or other corrosion resistant material, and not longer than 20 ft (6.0 m).

►For one- and two-family dwellings, where it might be impractical to keep the grounding conductor 20 ft (6.0 m) or less, an additional communications grounding electrode must be installed. This additional ground rod must be bonded to the power grounding electrode system with a bonding conductor not smaller than a 6 AWG copper. In just about all instances, all grounding and bonding of telephone equipment is done by the telephone company personnel.◄ Many times, they clamp the primary protector to the grounded metal service raceway conduit to establish "ground."

Safety

Open-circuit voltage between conductors of an idle pair of telephone conductors is approximately 48 volts dc. The superimposed ringing voltage can reach 90 volts ac. Therefore, always work carefully with insulated tools and stay clear of bare terminals and grounded surfaces. Disconnect the interior telephone wiring if work must be done on the circuit, or take the phone off the hook, in which case the dc voltage level will drop to approximately 7 to 9 volts dc, and there should be no ac ringing voltage delivered.

Wiring for Computers and Internet Access

Wiring for telephones, high-speed Internet access, computers, television, printers, modems, security systems, intercoms, and similar home automation equipment could be thought of as one big project. Planning ahead is critical. Deciding on where this equipment is likely to be located will help design an entire video/voice/data system.

You will want to install Category 5 or Category 5e cables between each telephone outlet and a central distribution point so as to have the flexibility to connect the video/voice/data equipment in any configuration.

In addition to the previously mentioned equipment, there might very well be a scanner, answering machine, desk lamp, floor lamp, adding machine/calculator, radio, CD player, CD "burner," cassette tape players, and VCR. All need to be plugged into a 120-volt receptacle! That many power cords is a major problem. Give consideration to installing multiple 120-volt receptacles at these locations. Multi-outlet plug-in strips will probably be needed. Unit 31 contains much more information on this.

Bringing Technology Into Your Home

In some areas, the telephone company provides the entire "package" for homes. This might include digital local and long distance telephone service, internet access, cable modems for personal computers, cable television, digital subscriber lines (DSL) (through existing copper telephone lines), and whatever other great new things may come next.

SIGNAL SYSTEM (CHIMES)

A signal circuit is described in the *NEC*® as any electrical circuit that energizes signaling equipment. Signaling equipment includes such devices as chimes, doorbells, buzzers, code-calling systems, and signal lights.

Door Chimes (Symbol [CH])

Present-day dwellings use chimes rather than bells or buzzers to announce that someone is at a door. A musical tone is sounded rather than a harsh ringing or buzzing sound. Chimes are available in single-note, two-note, eight-note (four-tube), and repeater tone styles. In a repeater tone chime, both notes sound as long as the push button is depressed. In an eight-note chime, contacts on a motor-driven cam are arranged in sequence to sound the notes of a simple melody when the chime button is pushed. This type of chime is usually installed in dwellings having three entrances. The chime can be connected so that the eight-note melody sounds for the front door, two notes sound for the side door, and a single note sounds for the rear door. Chimes are also available with clocks and lights.

Electronic chimes may relay their chime tones through the various speakers of an intercom system. When any chime is installed, the manufacturer's instructions must be followed.

The plans show that two chimes are installed in the residence. Two-note chimes are used. Each chime has two solenoids and two iron plungers. When one solenoid is energized, the iron plunger is drawn into the opening of the solenoid. A plastic peg in the end of the plunger strikes one chime tone bar. When the solenoid is de-energized, spring action returns the plunger where it comes to rest against a soft felt pad so that it does not strike the other chime tone bar. Thus, a single chime tone sounds. As the second solenoid is energized, one chime tone bar is struck. When the second solenoid is de-energized, the plunger returns and strikes the second tone bar. A two-tone signal is produced. The plunger then comes to rest between the two tone bars. Generally, two notes indicate front door signaling, and one note indicates rear or side door signaling.

Figure 25-20 shows various push-button styles used for chimes. Many other styles are available. Figure 25-21 shows several types of typical residential-type wall-mounted chimes. The symbols used to indicate push buttons and audible signals on the plans are shown in Figure 25-22.

Figure 25-23 shows how to provide proper backing for chimes. This is done during the rough-in stages of the electrical installation.

Chimes for the Hearing Impaired

Chimes for the hearing impaired are available with accessory devices that can turn on a dedicated lamp at the same time the chime is sounded. Figure 25-24 shows the devices needed to accomplish this. Figure 25-25 shows the wiring details.

Chime Transformers

Chime transformers are covered under UL Standard 1585. Because of their low voltage rating and power limitation, they are listed as Class 2 transformers. The low voltage wiring on the secondary side of the transformer is Class 2 wiring. Class 2 circuits are discussed in Unit 24 and in *Article 725*.

Figure 25-26 shows two views of a chime transformer. The top view shows the 120-volt leads. Note the small set-screw for ease of mounting into a conduit knockout in an outlet box. The bottom view shows the low voltage terminals.

Figure 25-20 Push buttons for door chimes. *Courtesy of* Nutone, Inc.

Figure 25-21   Typical residential door chimes. *Courtesy of* NuTone, Inc.

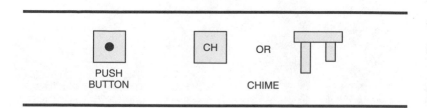

Figure 25-22   Symbols for chimes and push buttons.

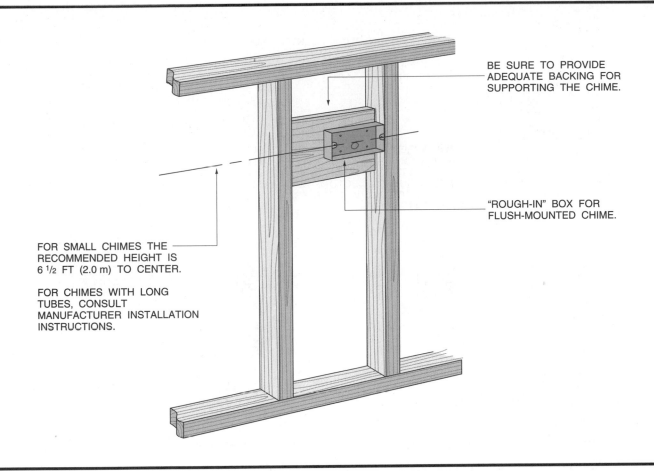

BE SURE TO PROVIDE
ADEQUATE BACKING FOR
SUPPORTING THE CHIME.

"ROUGH-IN" BOX FOR
FLUSH-MOUNTED CHIME.

FOR SMALL CHIMES THE
RECOMMENDED HEIGHT IS
6 ½ FT (2.0 m) TO CENTER.

FOR CHIMES WITH LONG
TUBES, CONSULT
MANUFACTURER INSTALLATION
INSTRUCTIONS.

Figure 25-23   Roughing-in for a flush-mounted chime.

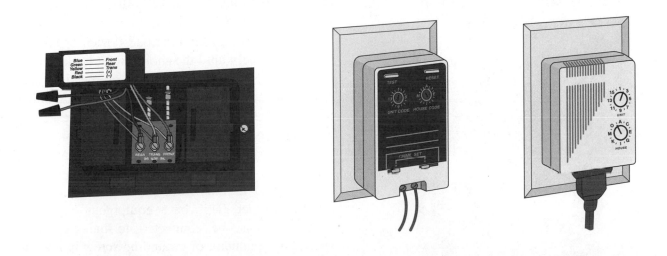

Figure 25-24   Accessory devices can be attached to new or existing chimes to provide visual signals when the chime is sounded. The first photo shows the module that is connected and mounted inside the chime. The second photo shows a transmitter that is plugged into a nearby receptacle. The third photo shows a receiver into which a lamp is plugged.

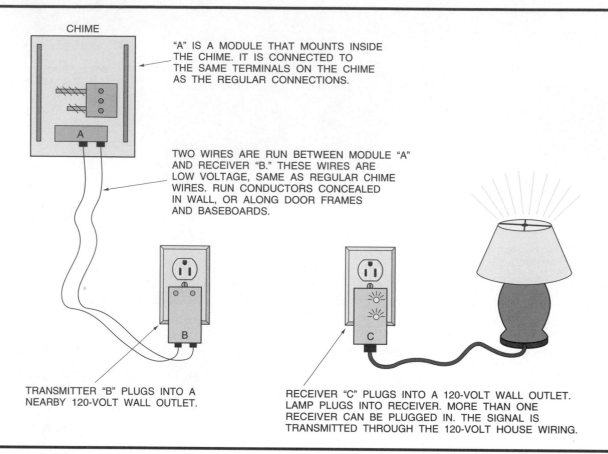

CHIME

"A" IS A MODULE THAT MOUNTS INSIDE THE CHIME. IT IS CONNECTED TO THE SAME TERMINALS ON THE CHIME AS THE REGULAR CONNECTIONS.

TWO WIRES ARE RUN BETWEEN MODULE "A" AND RECEIVER "B." THESE WIRES ARE LOW VOLTAGE, SAME AS REGULAR CHIME WIRES. RUN CONDUCTORS CONCEALED IN WALL, OR ALONG DOOR FRAMES AND BASEBOARDS.

TRANSMITTER "B" PLUGS INTO A NEARBY 120-VOLT WALL OUTLET.

RECEIVER "C" PLUGS INTO A 120-VOLT WALL OUTLET. LAMP PLUGS INTO RECEIVER. MORE THAN ONE RECEIVER CAN BE PLUGGED IN. THE SIGNAL IS TRANSMITTED THROUGH THE 120-VOLT HOUSE WIRING.

Figure 25-25 Installation diagram for auxiliary devices for use with chimes that can help the hearing impaired. When the chime sounds, the signal is transmitted to the receiver, turning on the lamp that is plugged into the receiver. The "ON" time is adjustable at "C". The lamp should be a "dedicated" lamp that lights up only when the chime sounds.

Figure 25-26 Chime transformers. *Courtesy of NuTone.*

Chime transformers used in dwellings are generally rated 16 volts. They have built-in thermal overload protection. Should a short circuit occur in the low voltage wiring, the overload device opens and closes repeatedly until the short circuit is cleared.

The UL standard requires that all metal parts of a transformer be properly grounded. To accomplish this, some chime transformers have a bare copper equipment grounding conductor in addition to the black and white supply conductors. Internal to the transformer, this equipment grounding conductor is connected to the metal parts (laminations) of the transformer. This bare equipment grounding conductor must be connected to the equipment grounding conductor or grounding screw in the outlet box. Do not connect this equipment grounding conductor to the branch-circuit grounded (white) conductor.

Some chime transformers are marked "Install in Metal Box Only."

Additional Chimes

To extend a chime system to cover a larger area, a second or third chime may be added. In the residence, two chimes are used: one chime is mounted in the front hall and a second (extension) chime is mounted in the recreation room. The extension chime is wired in parallel to the first chime. The wires are run from one chime terminal board to the terminal board of the other chime. The terminals are connected as follows: transformer to transformer, front to front, and rear to rear, Figure 25-27.

When more than one chime is installed, it will probably be necessary to install a transformer with a higher volt-ampere (wattage) rating. Do not install a transformer with a higher voltage rating. Check the manufacturers' instructions furnished with the chime transformer.

Do not add another transformer to an existing transformer to solve the problem of multiple chimes not responding properly. The *NEC®* in *725.41(B)* prohibits connecting transformers in parallel unless specifically listed for interconnection. Residential chime transformers are not listed for interconnection.

TYPE OF CHIME	POWER CONSUMPTION
Standard two-note	10 watts
Repeating chime	10 watts
Internally lighted, two lamps	10 watts
Internally lighted, four lamps	15 watts
Combination chime and clock	15 watts
Motor-driven chime	15 watts
Electronic chime	15 watts
Programmable musical	15 watts

Table 25-1   Typical chimes with their power consumption

If a buzzer (or bell) and a chime are connected to a single transformer and are used at the same time, the transformer will put out a fluctuating voltage. This condition does not allow either the buzzer or the chime to operate properly. The use of a transformer with a larger rating may solve this problem.

The wattage consumption of chimes varies with the manufacturer. Typical ratings are shown in Table 25-1.

Transformers with ratings of 5, 10, 15, 20, and 30 watts (VA) are available. For a multiple chime installation, wattage ratings for the individual chimes are added. The total value is the minimum transformer rating needed to do the job properly.

Wiring for Chime Installation

Bell Wire and Cable.   The wire and cable used for low-voltage bell and chime wiring, and for connecting low-voltage thermostats, is called *bell wire, annunciator wire,* or *thermostat wire.* Low-voltage wiring for chimes is classified as Class 2 wiring.

These wires are generally made of copper and have a thermoplastic insulation suitable for use on 30 volts or less.

Because the current required for bell and chime circuits is rather small, 18 AWG conductors are quite often used. If the length of the circuit is very long, a larger conductor should be used.

Multiconductor cables consist of two, three, or more single wires covered with a single protective insulation. This type of cable is often used for electrical installations because there is less danger of dam-

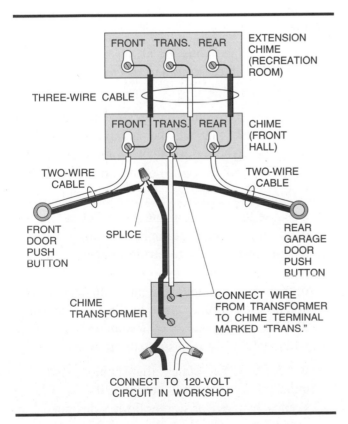

Figure 25-27   Circuit for chime installation.

age to individual wires and it gives a neat appearance to the wiring. The conductors within the cables are color coded to make circuit identification easy.

Bell wire and cable may be fastened directly to surfaces with insulated staples or cleats, or may be installed in the raceways. Be careful not to pierce or crush the wires. Do not pull bell wires or other low-voltage cables through the same holes in studs and joists that contain nonmetallic-sheathed cable, BX, conduit, or other piping because they cannot stand the physical abuse that they would be subjected to.

Wiring Circuit. The circuit shown in Figure 25-27 is recommended for this chime installation because it provides a hot low-voltage circuit at the front hall location. However, it is not the only way in which these chimes may be connected.

Figure 25-27 shows that a two-wire cable runs from the transformer in the utility room to the front hall chime. A two-wire cable then runs from the chime to both the front and rear door push buttons. A three-wire cable also runs between the front hall chime and the recreation room chime. Because of the hot, low-voltage circuit at the front hall location, a chime with a built-in clock can be used. A four-conductor cable may be run to the extension chime so that the owner may install a clock-chime at this location also.

The plans show that the chime transformer is mounted into the knockout on one of the ceiling boxes in the utility room. The line-voltage connections are easy to make here. Some electricians prefer to mount the chime transformer on the top or side of the distribution panel or load center. The electrician decides where to mount the transformer after considering the factors of convenience, economy, and good wiring practice. Don't conceal boxes!

It may seem obvious, but the issue of accessibility bears repeating. Electrical equipment junction boxes and conduit bodies must be accessible, *370.29.* It is permitted to have certain electrical equipment such as a chime transformer, junction boxes and conduit bodies above a dropped lay-in ceiling because these are accessible by dropping out a ceiling panel. Never install electrical equipment, junction boxes, or conduit boxes above a permanently closed-in ceiling or within a wall. There may come a time when the electrical equipment and junction boxes need to be accessed. This includes chime transformers.

NEC® Rules for Signal Systems

Article 725 covers the requirements for low-voltage, Class 2 circuits, discussed in Unit 24.

Repeating some of the most basic Code installation requirements:

- Do *not* install Class 2 wiring in the same raceway or cable as light and power wiring, *725.54(A)(1).*

- Keep Class 2 bell wires at least 2 in. (50 mm) away from light or power wiring. This really pertains to old, open-knob and tube wiring. Today, where light and power wiring is installed in Type NMC, Type AC, or in conduit, there is no problem, *725.54(A)(3).*

- Class 2 wiring may not enter the same enclosure, unless there is a barrier in the enclosure that separates the Class 2 wiring from the light and power conductors, *725.54(A)(1).*

- Chime transformers are listed as Class 2 transformers. Thus, all of the wiring connected to the secondary of the transformer is Class 2 wiring.

- Do *not* run Class 2 cables through the same holes as light and power cables and conduits. There is too much chance of physical damage to the Class 2 cables.

- Do not strap, tape, or attach by any means Class 2 cables to other electrical raceways, *725.54(D).*

- Class 2 wiring must be installed in a neat and workmanlike manner, and shall be supported by the building structure in such a manner that the cable will not be damaged by normal building use, *725.7.*

- Because the insulation and the jacket of Class 2 cables and conductors are rather thin, be careful during installation to avoid damage. Secure these cables and conductors with care, using the proper type of staples. Staple "guns" are available that use rounded staples that nicely straddle the cable instead of flattening the cable, which might result in shorted out wires.

- ▶*NEC® 725.7* states that cables shall be attached to structural components by straps, staples, hangers, or similar fittings designed and installed so as not to damage the cable. ◀

REVIEW

Note: Refer to the Code or the plans where necessary.

TELEVISION CIRCUIT

1. How many television outlets are installed in this residence? _____

2. Which type of television cable is commonly used and recommended? _____
   _____

3. What determines the design of the faceplates used? _____
   _____

4. What must be provided when installing a television outlet and receptacle outlet in one
   wall box? _____
   _____
   _____
   _____
   _____
   _____
   _____

5. From a cost standpoint, which system is more economical to install: a master amplifier
   distribution system or a multiset coupler? Explain the basic differences between these
   two systems. _____
   _____
   _____
   _____
   _____
   _____

6. How many wires are in the cable used between a rotor and its controller? _____

7. Digital satellite systems use an antenna that is approximately (18 in. [450 mm]) (36 in.
   [900 mm]) (72 in. [1.8 mm]) in diameter. Circle the correct answer.

8. List the requirements for cable television inside the house. _____
   _____
   _____
   _____
   _____
   _____

9. Which article of the Code references the requirements for cable community television installation? _____

10. It is generally understood that grounding and bonding together all metal parts of an electrical system and the metal shield of the cable television cable to the same grounding reference point in a residence will keep both systems at the same voltage level should a surge, such as lightning, occur. Therefore, if the incoming cable that has been installed by the CATV cable company installer has the metal shield grounded to a driven ground rod, does this installation conform to the *NEC®*? _____

_____

11. The basics of cable television are that a transmitter on earth sends an _____ signal to a _____ , where the signal is reamplified and sent back to earth via the _____ signal. This signal is picked up by the _____ on the satellite antenna, which is sometimes referred to as _____ . The feedhorn funnels the signal into a _____ _____ _____ , then on to a _____ , where the high-frequency signal is converted to a _____-frequency signal suitable for the television receiver.

12. All television satellites rotate above the earth in (the same orbit) (different orbits). Circle correct answer.

13. Television satellites are set in orbit (10,000) (18,000) (22,247) miles above the earth, which results in their rotating around the earth at (precisely the same) (different) rotational speed as the earth rotates. This is done so that the satellite "dish" can be focused on a specific satellite (once) (one time each month) (whenever the television set is used). Circle correct answer.

14. Which section of the Code prohibits supporting coaxial cables from raceways that contain light or power conductors? *NEC®* _____

15. When hooking up one CATV cable and another cable from an outdoor antenna to a receiver that has only one antenna input terminal, an _____ switch is usually installed.

TELEPHONE SYSTEM

1. How many locations are provided for telephones in the residence? _____

2. At what height are the telephone outlets in this residence mounted? Give measurement to center.

_____

_____

_____

_____

3. Sketch the symbol for a telephone outlet.

4. Is the telephone system regulated by the *NEC®*? _____

5. a. Who is to furnish the outlet boxes required at each telephone outlet? _____

   b. Who is to furnish the faceplates? _____

6. Who is to furnish the telephones? _____

7. Who does the actual installation of the telephone equipment? _____

8. How are the telephone cables concealed in this residence? _____

_____

_____

_____

9. The point where the telephone company's cable ends and the interior telephone wiring meets is called the _____ point. The device installed at this point is called a _____ .

10. What are the colors contained in a four-conductor telephone cable assembly and what are they used for? _____

_____

_____

_____

_____

_____

11. Itemize the Code rules for the installation of telephone cables in a residence. _____

_____

_____

_____

12. If finger contact were made between the red conductor and green conductor at the instant a "ring" occurs, what shock voltage would be felt? _____

13. What section of the Code prohibits supporting telephone wires from raceways that contain light or power conductors? *NEC®* _____

14. The term *cross-talk* is used to define the hearing of the faint sound of voices in the background when you are using the telephone. Cross-talk can be reduced significantly by running (flat cables) (twisted pair cables) to the telephones. Circle the correct answer.

15. Which section of the Code prohibits telephone cables from being installed in the same box or enclosure, or from being pulled into the same raceway as light and power circuits? _____

SIGNAL SYSTEM

1. What is a signal circuit? _____

_____

2. What style of chime is used in this residence? _____

3. a. How many solenoids are contained in a two-tone chime? _____

   b. What closes the circuit to the solenoid of a chime? _____

4. Explain briefly how two notes are sounded by depressing one push button (when two solenoids are provided). _____

_____

_____

_____

_____

_____

_____

_____

5. a. Sketch the symbol for a push button. _____

   b. Sketch the symbol for a chime. _____

6. a. At what voltage do residence chimes generally operate? _____

   b. How is this voltage obtained? _____

7. What is the maximum volt-ampere rating of transformers supplying Class 2 systems?

_____

8. What two types of chime transformers for Class 2 systems are listed by UL?

_____

_____

9. Is the extension chime connected with the front hall chime in series or in parallel?

_____

10. How many bell wires terminate at

    a. the transformer? _____

    b. the front hall chime? _____

    c. the extension chime? _____

    d. each push button? _____

11. a. What change in equipment may be necessary when more than one chime is connected to sound at the same time on one circuit? _____

    _____

    _____

    _____

    b. Why? _____

    _____

    _____

    _____

    _____

12. What type of insulation is usually found on low-voltage wires? _____

13. What size wire is installed for signal systems of the type in this residence? _____

14. a. How many wires are run between the front hall chime and the extension chime in the recreation room? _____

 b. How many wires are required to provide a "hot" low-voltage circuit at the extension chime? _____

15. Why is it recommended that the low-voltage secondary of the transformer be run to the front hall chime location and separate two-wire cables be installed to each push button? _____

_____

16. a. Is it permissible to install low-voltage Class 2 systems in the same raceway or enclosure with light and proper wiring? _____

 b. Which Code section covers this? _____

17. a. Where is the transformer in the residence mounted? _____

 b. To which circuit is the transformer connected? _____

18. a. How many feet of two-conductor bell wire cable are required? _____

 b. How many feet of three-conductor bell wire cable are required? _____

19. How many insulated staples are needed for the bell wire if it is stapled every 24 in. (600 mm)? _____

20. Should low-voltage bell wire be pulled through the same holes in studs and joists that also contain nonmetallic-sheathed cable? _____

21. What section of the Code prohibits supporting fire-protective signaling circuits from raceways that contain light or power conductors? *NEC®* _____ .

Heat, Smoke, and Carbon Monoxide Detectors, Fire Alarms, and Security Systems

OBJECTIVES

After studying this unit, the student should be able to

- understand the *NFPA Standard 72*, which is the *National Fire Alarm Code,* and *NFPA Standard 720*, which is the *Recommended Practice for the Installation of Household Carbon Monoxide Warning Equipment.*

- name the two basic types of smoke detectors.

- discuss the location requirements for the installation of heat and smoke detectors for minimum acceptable levels of protection.

- discuss the location requirements for the installation of heat and smoke detectors for increased protection that exceed the minimum acceptable levels of protection.

- list the major components of typical residential smoke, heat, and security systems.

- discuss general *NEC®* requirements for the installation of residential smoke, heat, and security systems.

- understand the basics of carbon monoxide detectors.

HEAT, SMOKE, AND CARBON MONOXIDE DETECTORS

This unit will touch upon the basic requirements for protection against the hazards of fire, heat, smoke, and carbon monoxide.

Fire is the third leading cause of accidental death. Home fires account for the biggest share of these fatalities, most of which occur at night during sleeping hours.

Heat, smoke, and carbon monoxide detectors are installed in a residence to give the occupants early warning of the presence of fire or toxic fumes. Fires produce smoke and toxic gases that can overcome the occupants while they sleep. Most fatalities result from the inhalation of smoke and toxic gases, rather than from burns. Heavy smoke reduces visibility.

In nearly all home fires, detectable smoke precedes detectable levels of heat. People sleeping are less likely to smell smoke than people who are awake. The smell of smoke probably will not awaken a sleeping person. Therefore, smoke detectors are considered to be the primary devices for protecting lives. Heat detectors should be installed in addition to smoke detectors, but should not replace them. A rather recent entry into the arena are carbon monoxide detectors, which sense dangerous carbon monoxide emitted from a malfunctioning furnace or other source. *NFPA 720* is the *Recommended Practice for the Installation of Household Carbon Monoxide Warning Equipment*. It covers carbon monoxide detection and protection. UL Standard 2034 covers single- and multiple-station carbon monoxide detectors.

FIRE ALARM CIRCUITS

The *NEC®* in *760.2* defines a fire alarm circuit as follows: "*The portion of the wiring system between the load side of the overcurrent device or the power-limited supply and the connected equipment of all circuits powered and controlled by the fire alarm system. Fire alarm circuits are classified as either nonpower-limited or power-limited.*"

Fire protective signaling systems installed in homes have power-limited fire alarm (PLFA) circuits where the power output is limited by the listed power supply for the system. Here is a brief overview of some of the *NEC®* requirements in *Article 760*.

- must have power output capabilities limited to the values specified in *Chapter 9*, *Table 12(A)* and *Table 12(B)*. Refer to *760.41*.

- do not connect the equipment to a GFCI-protected branch-circuit, *760.41*.

- wiring on the supply side of equipment is installed according to the conventional wiring methods found in *Chapter 1* through *Chapter 4* of the *NEC®*.

- the branch-circuit overcurrent protection supplying the fire alarm system shall not exceed 20 amperes, *760.51*.

- cables may be run on the surface or be concealed, and shall be adequately supported and protected against physical damage, *760.51(B)(1)*.

- when run exposed within 7 ft (2.1 m) from the floor, the cable shall be securely fastened at intervals not over 18 in. (450 mm), *760.52(B)(1)*.

- Cables shall not be run in the same raceway, or be in the same enclosure as light and power wiring unless separated by a barrier or in the enclosure to connect that particular piece of equipment. Refer to *760.54(A)(1)* for exceptions to this basic rule.

- cables shall not be supported by taping, strapping, or attaching by any means to any electrical conduits, or other raceways, *760.54(C)*.

- wiring on the load side shall be insulated solid or stranded copper conductors of the types listed in *Table 760.61*.

- *Table 760.61* is a table showing different types of cable that can be substituted for one another when necessary.

- if installed in a duct, the cable must be plenum rated, *760.61(A)*.

It makes sense to contact your local inspection authority before installing a fire alarm system to be sure the system you intend to install meets the requirements of the local Codes!

NATIONAL FIRE PROTECTION ASSOCIATION (NFPA) CODE

The *NFPA 72*, the *National Fire Alarm Code,®* presents the minimum requirements for the proper selection, installation, operation, and maintenance of fire warning equipment. The standard also includes many recommendations that exceed the minimum requirements. Chapter 2 specifically covers household fire warning equipment.

NFPA 72 defines a household fire alarm system as "*A system of devices that produces an alarm signal in the household for the purpose of notifying the occupants of the presence of a fire so that they will evacuate the premises.*"

Fire warning devices commonly used in a residence are heat detectors (Figure 26-1), and smoke detectors (Figure 26-2).

The more elaborate systems connect to a central monitoring customer service center through a telephone line. These systems offer instant contact with police or fire departments, a panic button for emergency medical problems, low-temperature detection, flood (high-water) detection, perimeter protection, interior motion detection, and other features. All detections are transmitted first to the company's customer service center, where personnel on duty

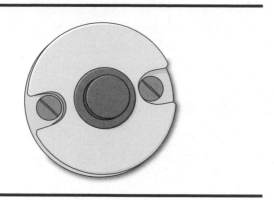

Figure 26-1   Heat detector.

Figure 26-2   Smoke detector.

monitor the system 24 hours a day, 7 days a week. They will verify that the signal is valid. After verification, they will contact the police department, the fire department, and individuals whose names appear on a previously agreed-upon list that the homeowner prepared and submitted to the company's customer service center.

NFPA standards are *not* law until adopted by a city, state, or other governmental body. Some NFPA standards are adopted in total, and others are adopted in part. Installers must be aware of all requirements in the locality in which they are doing work. Most communities require a special permit for the installation of fire warning equipment and security systems in homes, and require registration of the system with the police and/or fire departments. In most instances, fire protection requirements are found in the building codes of a community, and not necessarily spelled out in that

community's electrical code. Remember, the *NEC*® sets forth the requirements of how to physically install a fire protective signaling system. It does *not* say where to install the detectors. The "where to install" information is found in *NFPA Standard No. 72*, which is the standard adopted by most building officials.

A sufficient number of detectors, properly located, must be installed to provide adequate protection. These requirements and recommendations are discussed later in this unit.

TYPES OF SMOKE DETECTORS

Two common types of smoke detectors are the photoelectronic type and the ionization type. They usually contain an indication light to show that the unit is functioning properly. They also may have a test button that simulates smoke, so that when the button is pushed, the detector's smoke detecting ability as well as its circuitry and alarm are tested. Some systems are tested with a magnet.

Smoke detectors generally will not sense heat, flame, or gas. However, some smoke detectors can be set off by acetylene and propane gas.

Photoelectronic Type

The photoelectronic type of smoke detector has a light sensor that measures the amount of light in a chamber. When smoke is present, an alarm is sounded that indicates a reduction in light due to the obstruction of the smoke. This type of sensor detects smoke from burning materials that produce great quantities of smoke, such as furniture, mattresses, and rags. This type of detector is less effective for gasoline and alcohol fires, which do not produce heavy smoke.

Ionization Type

The ionization type of detector contains a low-level radioactive source, which supplies particles that ionize the air in the detector's smoke chamber. Plates in this chamber are oppositely charged. Because the air is ionized, an extremely small amount of current (millionths of an ampere) flows between the plates. Smoke entering the chamber impedes the movement of the ions, reducing the current flow, which causes an alarm to sound.

The ionization type of detector is effective for detecting small amounts of smoke, as in gasoline and alcohol fires, which are fast-flaming with little or no smoke.

TYPES OF HEAT DETECTORS

Some of the types of heat detectors are described in the following.

- *Fixed* temperature heat detectors are available that sense a specific *fixed* temperature, such as 135°F (57°C) or 200°F (93°C). They shall have a temperature rating at least 25°F (14°C) above the normal temperature expected, but not to exceed 50°F (28°C) above the expected temperature.

- *Rate-of-Rise* heat detectors sense rapid changes in temperature (12°F to 15°F per minute), such as those caused by flash fires.

- *Fixed/Rate-of-Rise* temperature detectors are available as a combination unit.

- *Fixed/Rate-of-Rise/Smoke* combination detectors are also available in one unit.

INSTALLATION REQUIREMENTS

The following information includes specific recommendations for installing smoke and heat detectors in homes. Some of these requirements are illustrated in Figure 26-3, Figure 26-4, and Figure 26-5. Complete data is found in *NFPA Standard No. 72* and in the instructions furnished by the manufacturer of the equipment.

Individual detectors as well as complete systems can be purchased at electrical supply houses, home centers, hardware stores, and similar retail outlets. These are packaged with the necessary mounting hardware and detailed installation instructions.

Once installed, detectors should be checked periodically to make sure they are operating

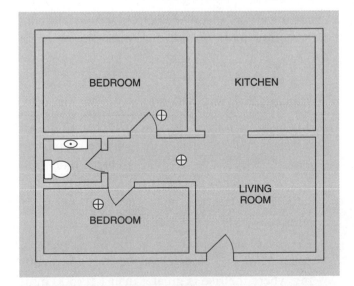

Figure 26-3 Recommended location of heat or smoke detector *between* sleeping areas and rest of house. Locate outside of the bedroom, but near the bedrooms. In new construction, a smoke detector must be installed in each bedroom.

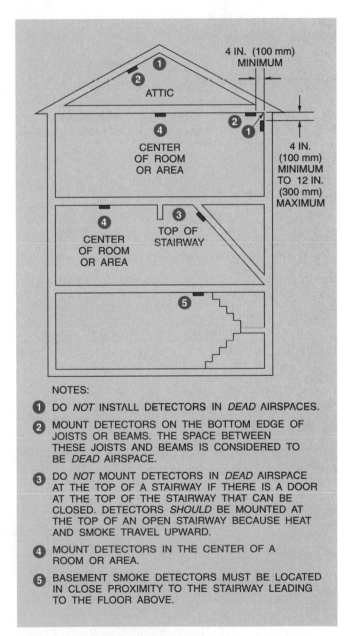

NOTES:

1. DO *NOT* INSTALL DETECTORS IN *DEAD* AIRSPACES.

2. MOUNT DETECTORS ON THE BOTTOM EDGE OF JOISTS OR BEAMS. THE SPACE BETWEEN THESE JOISTS AND BEAMS IS CONSIDERED TO BE *DEAD* AIRSPACE.

3. DO *NOT* MOUNT DETECTORS IN *DEAD* AIRSPACE AT THE TOP OF A STAIRWAY IF THERE IS A DOOR AT THE TOP OF THE STAIRWAY THAT CAN BE CLOSED. DETECTORS *SHOULD* BE MOUNTED AT THE TOP OF AN OPEN STAIRWAY BECAUSE HEAT AND SMOKE TRAVEL UPWARD.

4. MOUNT DETECTORS IN THE CENTER OF A ROOM OR AREA.

5. BASEMENT SMOKE DETECTORS MUST BE LOCATED IN CLOSE PROXIMITY TO THE STAIRWAY LEADING TO THE FLOOR ABOVE.

Figure 26-4 Recommendations for the installation of heat and smoke detectors.

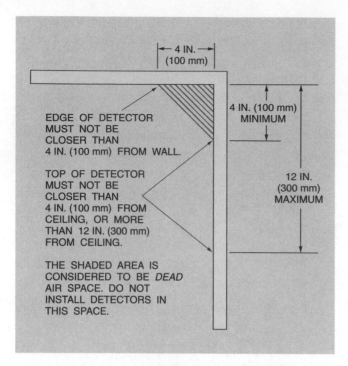

EDGE OF DETECTOR MUST NOT BE CLOSER THAN 4 IN. (100 mm) FROM WALL.

TOP OF DETECTOR MUST NOT BE CLOSER THAN 4 IN. (100 mm) FROM CEILING, OR MORE THAN 12 IN. (300 mm) FROM CEILING.

THE SHADED AREA IS CONSIDERED TO BE *DEAD* AIR SPACE. DO NOT INSTALL DETECTORS IN THIS SPACE.

4 IN. (100 mm)

4 IN. (100 mm) MINIMUM

12 IN. (300 mm) MAXIMUM

Figure 26-5 Do not mount smoke or heat detectors in the *dead* airspace where the ceiling meets the wall.

properly. Testing every six months is quite common. Refer to the manufacturer's recommendations.

Many companies specialize in installing complete fire and security systems. They also offer system monitoring at a central office and can notify the police or fire department when the system gives the alarm. How elaborate the system should be is up to the homeowner.

Many local building codes do have requirements for the location, number, and type of smoke detectors to be installed in residential occupancies.

The Absolute Minimum Level of Protection

To meet the absolute minimum level of protection, do the following things:

- Install smoke detectors on each level of a residence.

- Install a smoke detector in each bedroom or sleeping area of a residence.

- In new construction, interconnect smoke detectors so that when one operates, all others will also operate. See "Wiring Requirements."

- Make sure that the detectors are listed by a recognized testing agency.

Exceeding Minimum Levels of Protection

The following are recommendations for attaining levels of protection that exceed the minimum level stated above and include guidelines for installing smoke and heat detectors in a home.

- Install smoke detectors for protection that *exceeds* the minimum acceptable protection levels, in *all* rooms, basements, hallways, heated attached garages, and storage areas. Installing detectors in these locations will increase escape time, particularly if the room or area is separated by a door(s). In some instances, smoke detectors are installed in attics and crawl spaces. Check the manufacturer's recommendations for installing smoke detectors in attics, crawl spaces, and garages.

- Consider special detectors that light up or vibrate when the occupants are hard of hearing.

- Consider low temperature detectors (i.e., 45°F or 7°C) that can detect low temperatures should the heating system fail. The damage caused by frozen water pipes bursting can be extremely costly.

Installation

Following are general recommendations for installing smoke and heat detectors.

- *Do* install only listed detectors, and install them according to the manufacturer's instructions.

- *Do* install smoke detectors on the ceiling, as close as possible to the center of the room or hallway.

- *Do* install smoke detectors at the high end of a room that has a sloped, gabled, or peaked ceiling where the rise is greater than 1 ft (300 mm) per 8 ft (2.6 m) measured horizontally. Do not install in the dead air space, as shown in Figure 26-4 and Figure 26-5.

- *Do* install *heat* detectors at the high end of a room that has a sloped, gabled, or peaked ceiling where the rise is greater than 1 ft (300 mm) per 8 ft (2.6 m) measured horizontally. Mount them within 3 in. (75 mm) of the peak, but do not install in the dead air space, as shown in Figure 26-4 and Figure 26-5.

- *Do* make sure that the path of rising smoke will reach the smoke detector when the detector is installed in a stairwell. This will usually be at the top of the stairway. The detector must not be located in a dead air space created by a closed door at the head of the stairway.

- *Do* install smoke detectors on a basement ceiling close to the stairway to the first floor.

- *Do* install smoke detectors in split-level homes. A smoke detector installed on the ceiling of an upper level can suffice for the protection of an adjacent lower level, provided the two levels are not separated by a door. Better protection is to install detectors for each level.

- *Do* install smoke detectors that are hard-wired directly to a 120 VAC source when installing detectors in *new* construction. Dual-powered detectors having both 120 VAC and battery power provide the best protection. In existing homes, battery-powered detectors are the most common, but 120 VAC detectors or dual-powered detectors can be installed. Remember, battery-powered units will not operate with dead batteries. AC-powered units will not operate when the power supply is off.

- *Do* install interconnected detectors. In *2-2.2.1* of *NFPA 72* we find that "*In new construction, where more than one smoke detector is required by 2-2.1, detectors shall be arranged so that operation of any smoke detector causes the alarm in all smoke detectors within the dwelling to sound.*"

- *Do* install smoke detectors so they are not the only load on the branch circuit. Connect smoke detectors to a branch-circuit that supplies other lighting outlets in habitable spaces.

- *Do* always consider the fact that doors, beams, joists, walls, partitions, and similar obstructions will interfere with the flow of heat and smoke and, in most cases, create new areas needing additional smoke and heat detectors.

- *Do* consider that the maximum distance between heat detectors mounted on flat ceilings is 50 ft (15 m), and 25 ft (7.5 m) from detector to wall. This information is explained in detail by *NFPA 72*. Where beams, joists, etc., will obstruct the flow of heat, the 50-foot distance is reduced to 25 ft (7.5 m), and the 25-foot distance is reduced to 12½ ft (3.75 m).

- *Do not* install smoke or heat detectors in the dead air space at the top of a stairway that can be closed off by a door.

- *Do not* place the edge of a ceiling-mounted smoke or heat detector closer than 4 in. (100 mm) to the wall.

- *Do not* place the top edge of a wall-mounted smoke or heat detector closer than 4 in. (100 mm) to the ceiling and farther than 12 in. (300 mm) down from the ceiling. Some manufacturers recommend placement not farther down than 6 in. (150 mm).

- *Do not* install smoke or heat detectors where the ceiling meets the wall because this is considered dead air space where smoke and heat may not reach the detector.

- *Do not* connect smoke or heat detectors to wiring that is controlled by a wall switch.

- *Do not* connect smoke or heat detectors to a circuit that is protected by a GFCI.

- *Do not* install smoke or heat detectors where the relative humidity exceeds 85 percent, such as in bathrooms with showers, laundry areas, or other areas where large amounts of visible water vapor collect. Check the manufacturer's instructions.

- *Do not* install smoke or heat detectors in front of air ducts, air conditioners, or any high-draft areas where the moving air will keep the smoke or heat from entering the detector.

- *Do not* install smoke detectors in kitchens where the accumulation of household smoke can result in setting off an alarm, even though there is no real hazard. The person in the kitchen will know why the alarm sounded, but other people in the house may panic. This problem exists in multi-family dwelling units, where unwanted triggering of the alarm in one dwelling unit might cause panic for people in the other units. The photoelectronic type *may* be installed in kitchens, but must *not* be installed directly over the range or cooking appliance. A better choice in a kitchen would be to install a heat detector. Consult the manufacturer's recommendations.

- *Do not* install smoke detectors where the temperature can fall below 32°F (0°C) or rise above 120°F (49°C) unless the detector is specifically identified for this application.

- *Do not* install smoke detectors in garages where vehicle exhaust might set off the detector. Instead of a smoke detector, install a heat detector.

- *Do not* install smoke detectors in airstreams that will pass air originating at the kitchen cooking appliances across the detector. False alarms will result.

- *Do not* install smoke or heat detectors on a ceiling where the ceiling will be excessively cold or hot. Smoke and heat will have difficulty reaching the detectors. This could be the case in older homes that are not insulated—or are poorly insulated. Instead, mount the detectors on the side wall, as in Figure 26-5.

- *Do not* install smoke or heat detectors on an outside wall that is not insulated, or poorly insulated. Instead, mount the detectors on an inside wall.

- *Do not* install smoke or heat detectors on a ceiling that employs radiant heating.

Many of the installation general recommendations are also found in the *International Residential Code*, published by the International Code Council, Inc.

Building Codes

In most instances the various building codes published by BOCA, ICBO, SBCCI, and ICC adopt *NFPA 72* by reference, but it would be wise to check with your local AHJ to find out if there are any differences between these codes that might affect your installation. See Unit 1 for information about these building code organizations. Also refer to the Website list following the Appendix of this text.

Manufacturers' Requirements

Here are a few of the responsibilities of the manufacturers of smoke and heat detectors. Complete data is found in *NFPA Standard No. 72*.

- The power supply must be capable of operating the signal for at least 4 minutes continuously.

- Battery-powered units must be capable of a low battery warning "beep" of at least one beep per minute for seven consecutive days.

- Direct-connected 120V ac detectors must have a visible indicator that shows "power-on."

- Detectors must not signal when a power loss occurs or when power is restored.

Wiring Requirements

Direct-Connected Units. These units are connected to a 120-volt circuit. The black and white wires on the unit are spliced to the black and white wires of the 120-volt circuit, as in Figure 26-6.

Feedthrough (Tandem) Units. These units may be connected in tandem up to 10 units. Any one of these detectors can sense smoke, then send a signal to the remaining detectors, setting them off so that all detectors sound an alarm. These units contain three wires: one black, one white, and one yellow, as shown in Figure 26-7. The yellow wire is the interconnecting wire.

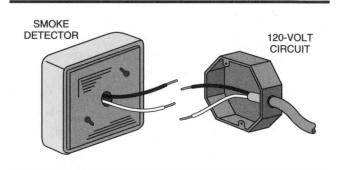

Figure 26-6   Wiring of direct-connected detector units.

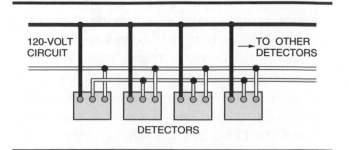

Figure 26-7   Wiring of feedthrough (tandem) detector units. These detectors are "interconnected."

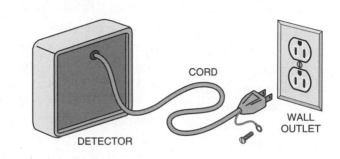

Figure 26-8   Cord-connected detector units are plugged into wall outlet.

Cord-Connected Units.   These units operate in the same way as direct-connected units, except that they are plugged into a wall outlet that is not controlled by a switch. These units are mounted to the wall on a bracket or screws. A clip on the cord is attached to the wall outlet with the same screw that fastens the faceplate to the outlet, as in Figure 26-8. This prevents the cord from being pulled out, rendering the unit inoperative.

Battery-Operated Units.   These units require no wiring; they are simply mounted in the desired locations. Batteries last for about a year. Battery-operated units should be tested periodically, as recommended by the manufacturer.

COMBINATION DIRECT/BATTERY/ FEED-THROUGH DETECTORS

Combination direct/battery/feed-through units are connected directly to a 120-volt ac circuit (not GFCI protected) and are equipped with a 9-volt dc battery, as in Figure 26-9. Thus, in the event of a power outage, these detectors will continue to operate on the 9-volt dc battery. The ability to interconnect a number of detectors offers "whole house" protection, because when one unit senses smoke, all the other interconnected units will sound an alarm.

Read carefully the application and installation manuals that are prepared by the manufacturers of the smoke and heat detectors.

In this residence we find four smoke detectors connected to Circuit A17. These units are hardwired directly to the 120-volt ac branch-circuit.

Figure 26-9   An AC/DC smoke alarm that operates on both 120 volts AC power and a 9-volt battery. In the event of a power outage, the detector continues to operate on the 9-volt battery. A constant green LED indicates that the unit is operating on 120-volt AC power. An intermittent red LED indicates that the unit has switched to the 9-volt battery back-up mode. *Courtesy of* BRK Electronics.

They are interconnected feed-through units that will set off signaling in all units should one unit trigger. These interconnected smoke detectors are also powered by a 9-volt dc battery that allows the unit to operate in the event of a power failure in the 120-volt ac supply.

POWER SUPPLIES

For new homes, *NFPA 72, Section 2-3.1.2* requires that smoke detectors be powered by 120-volt ac supply as the main power source, and by a battery as the secondary (standby) power source. In existing homes, *Exception No. 5* to *Section 2-3.1.2* permits the detectors to be either ac only or battery powered only. Check with your local building officials to find out what they require in your community.

CARBON MONOXIDE DETECTORS

When a heat or smoke detector goes off, it is generally not difficult to determine the source of the problem. But when a carbon monoxide detector goes off, it is hard to determine the source.

Carbon monoxide (CO) is odorless, invisible, and tasteless. It is undetectable to any of a person's five senses—taste, smell, sight, touch, hear. Carbon monoxide replaces oxygen in the bloodstream, resulting in brain damage or total suffocation.

Carbon monoxide results when fuel burns incompletely. Common sources of carbon monoxide in a home can come from a malfunctioning furnace, gas appliances, kerosene heaters, automobile or other gas engines, charcoal grills, fireplaces, or a clogged chimney.

Symptoms of carbon monoxide poisoning are similar to flulike illnesses such as dizziness, fatigue, headaches, nausea, and irregular breathing.

Carbon monoxide detectors detect carbon monoxide from any source of combustion. They are not designed to detect smoke, heat, or gas. The Consumer Product Safety Commission (CPSC) recommends that at least one carbon monoxide detector be installed in each home. If only one carbon monoxide detector is installed, it should be in the area just outside of individual bedrooms. The CPSC recommends that a carbon monoxide detector be installed on each floor of a multilevel house.

Carbon monoxide detectors monitor the air in the house, and sound a loud alarm when carbon monoxide above a predetermined level is detected. They provide early warning before deadly gases build up to dangerous levels.

For residential applications, carbon monoxide detectors are available for hard wiring (direct connection to the branch-circuit wiring), plug-in (the male attachment plug is built into the back of the detector for plugging directly into a 120-volt wall receptacle outlet), cord connected (has a cord and plug), battery operated, and combination 120-volt and battery-operated units.

NFPA 720 is the *Recommended Practice for the Installation of Household Carbon Monoxide (CO) Warning Equipment.* This standard contains the installation recommendations for carbon monoxide detectors. Here is a summary of these recommendations.

Installation

- *Do* buy carbon monoxide detectors that are tested and listed in conformance to UL Standard 2034.

- *Do* install carbon monoxide detectors according to the manufacturer's instructions.

- *Do* install carbon monoxide detectors in or near bedrooms and living areas.

- *Do* install carbon monoxide detectors in locations where smoke detectors are installed.

- *Do* test carbon monoxide detectors once each week, or as recommended by the manufacturer.

- *Do* remove carbon monoxide detectors before painting, stripping, wall-papering, or using aerosol sprays. Store in a plastic bag until project is completed.

- *Do not* connect carbon monoxide detectors to a switched circuit, or a circuit controlled by a dimmer.

- *Do not* install carbon monoxide detectors closer than 6 in. (150 mm) from a ceiling. This is considered dead air space.

- *Do not* install carbon monoxide detectors in garages, kitchens, or furnace rooms. This could lead to nuisance alarms, may subject the detector to substances that can damage or contaminate the detector, or the alarm might not be heard.

- *Do not* install carbon monoxide detectors within 15 ft (4.5 m) of a cooking or heating gas appliance.

- *Do not* install carbon monoxide detectors in dusty, dirty, or greasy areas. The sensor inside of the detector can become coated or contaminated.

- *Do not* install carbon monoxide detectors where the air will be blocked from reaching the detector.

- *Do not* install carbon monoxide detectors in dead air spaces such as in the park of a vaulted ceiling, or where a ceiling and wall meet.

- *Do not* install carbon monoxide detectors where the detector will be in the direct air stream of a fan.

- *Do not* install carbon monoxide detectors where temperatures are expected to drop below 40°F (4°C) or get hotter than 100°F (38°C). Its sensitivity will be affected.

- *Do not* install carbon monoxide detectors in damp or wet locations such as showers and steamy bathrooms. Its sensitivity will be affected.

- *Do not* mount carbon monoxide detectors directly above or near a diaper pail. The high methane gas will cause the detector to register, and this is not carbon monoxide.

- *Do not* clean carbon monoxide detectors with detergents or solvents.

- *Do not* spray carbon monoxide detectors with hair spray, paint, air fresheners, or other aerosol sprays.

- *Do* carefully vacuum carbon monoxide detectors once each month or as recommended by the manufacturer.

UL Standard 2034 *Single and Multiple Station Carbon Monoxide Detectors* contains the requirements that manufacturers must conform to for household carbon monoxide alarms.

SECURITY SYSTEMS

It is beyond the scope of this text to cover all types of residential security systems. Instead, we will focus on some of the features available today from various manufacturers of these systems.

Professionally installed and homeowner-installed security systems can range from a simple system to a very complex system. Features and options include master control panels, remote controls, perimeter sensors for doors and windows, motion sensors, passive infrared sensors, wireless devices, interior and exterior sirens, bells, electronic buzzers, strobe lights to provide audio (sound) as well as visual detection, heat sensors, carbon monoxide sensors, smoke detectors, glass break sensors, flood sensors, low temperature sensors, and lighting modules. Figure 26-10 shows some of these devices. Figure 26-11 shows a schematic one-line diagram for a typical security system.

Monitored security systems are connected by telephone lines to a central station, a 24-hour monitoring facility. If the alarm sounds, security professionals will contact the end user to verify the alarm before sending the appropriate authorities (fire, police, ambulance) to the home.

The decision about how detailed an individual residential security system should be generally begins with a meeting between the homeowner and the security consultant or licensed electrical contractor before the actual installation. Most security system installers, consultants, and licensed electrical contractors are familiar with a particular manufacturer's system.

The systems are usually explained and even demonstrated during an in-home security presentation. They are then professionally installed by the security system company or licensed electrical contractor.

Do-it-yourself kits that contain many of the features provided by the professionally installed systems can be found at home centers, hardware stores, electrical distributors, and electronics stores. Depending on the system, it may or may not be possible to connect to a central monitoring facility.

The wiring of a security system consists of small, easy to install, low-voltage, multiconductor cables made up of 18 AWG conductors. These conductors should be installed after the regular house wiring is completed to prevent damage to these smaller cables. Usually the wiring can be done at the same time as the chime wiring is being installed. Security system wiring comes under the scope of *Article 725*.

The wireless key allows the end user to disarm and arm the system with the press of a button. There is no need to remember a security code and it can also be used to operate lights and appliances.

Monitored smoke detectors are on 24 hours a day, even if the burglary protection is turned off.

Magnetic contacts are often located at vulnerable entry points such as doors.

Sirens, such as these placed inside the home, alert families to emergencies.

Motion detectors are interior security devices that sense motion inside the home.

A wireless panic button allows the user to press a button to signal for help.

The wireless keypad is portable so you can arm or disarm the system from the yard or driveway.

Glassbreak detectors, available in a variety of sizes, will sound the alarm when they detect the sound of breaking glass.

Figure 26-10   *Photos courtesy of* First Alert Professional Security Systems.

When wiring detectors such as door entry, glass break, floor mat, and window foil detectors, circuits are electrically connected in series so that if any part of the circuit is opened, the security system will detect the open circuit. These circuits are generally referred to as "closed" or "closed loop." Alarms, horns, and other signaling devices are connected in parallel, because they will all signal at the same time

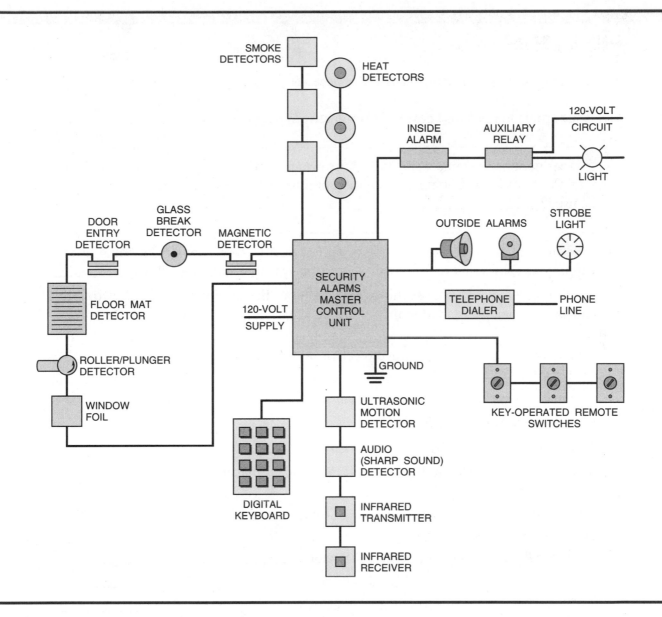

Figure 26-11 Diagram of typical residential security system showing some of the devices available. Complete wiring and installation instructions are included with these systems. Check local code requirements in addition to following the detailed instructions furnished with the system. Most of the interconnecting conductors are 18 AWG.

when the security system is set off. Heat detectors and smoke detectors are generally connected in parallel because any of these devices will "close" the circuit to the security master control unit, setting off the alarms.

The instructions furnished with all security systems cover the installation requirements in detail, alerting the installer to Code regulations, clearances, suggested locations, and mounting heights of the systems components.

Always check with the local electrical inspector to determine if there are any special requirements in your locality relative to the installation of security systems.

Sprinkling Systems

Home sprinkling systems are covered by *NFPA 13D* which is the *Standard for the Installation of Sprinkler Systems in One- and Two-Family Dwellings and Manufactured Homes.*

REVIEW

Note: Refer to the Code or the plans where necessary.

1. What article in the *NEC*® covers fire protective signaling systems? _____
_____

2. Is it permissible to connect smoke and fire detectors with the same kind of low-voltage wire that is used to hook up door chimes, or must cables that are listed for the purpose be used? _____
_____

3. Find in *Article 760* the section that prohibits electrical raceways from being used to support the cables of a fire protective signaling system installation. *NEC*® _____

4. Find in *Article 760* the sections that prohibit fire protective system wires and cables from being installed in the same raceway as light and power conductors. *NEC*®
_____

5. If it is necessary to install fire protective system cables through a duct, the cables (circle the correct answer):
   a. must be listed for the purpose.
   b. must be at least 14 AWG.
   c. are not permitted to be run through ducts.

6. What is the name and number of the NFPA standard that covers smoke and heat detectors for homes? _____

7. Has the standard that covers smoke and heat detectors for homes been adopted by the community in which you live? _____

8. a. Name the two basic types of heat detectors. _____
_____

   b. Name the two basic types of smoke detectors. _____
_____

9. Why is it important to mount a smoke or heat detector on the ceiling not closer than 4 in. (100 mm) from a wall? _____
_____
_____

10. In new residential construction, a smoke detector must be installed (circle the correct answer):
    a. in each bedroom or sleeping area.
    b. at the top of a basement stairway that has a closed door at the top.
    c. in a garage that is subject to subzero temperatures.

11. Always follow the installation instructions of (circle the correct answer):
    a. the manufacturer of the system.
    b. your neighbor.
    c. the man at the hardware store.

12. Smoke detectors, heat detectors, and security system wires and cables are much smaller than regular house wiring. Would it be better to install these wires (before) or (after) installing the regular house wiring to avoid possible damage to the smaller cables? Circle the correct answer.

13. Fire protection system conductors are generally _____ AWG.

14. When more than one smoke detector is installed in a new home, these units must be
    a. connected to separate 120-volt circuits.
    b. connected to one 120-volt circuit.
    c. interconnected so that if one detector is set off, the others would also sound a warning.
    Circle the correct answer(s).

15. It is advisable to connect smoke and heat detectors to circuits that are protected by GFCI devices. (True) (False)   Circle the correct answer.

16. Because a smoke detector might need servicing or cleaning, be sure to connect it in such a manner that it is controlled by a wall switch. (True) (False)   Circle the correct answer.

17. Smoke detectors of the type installed in homes must be able to sound a continuous alarm when set off for at least
    a. 4 minutes
    b. 30 minutes
    c. 60 minutes

18. In addition to detecting carbon monoxide, these detectors also detect heat and smoke. (True) (False)   Circle the correct answer.

19. Carbon monoxide is the result of fuel burning _____ .

20. In a small home, the best place to install a carbon monoxide detector is:
    a. in the basement.
    b. near the bedrooms.
    c. in the kitchen.

21. Where would you look to learn all of the installation requirements for carbon monoxide detectors? Circle the correct answer(s).
    a. NFPA 70
    b. NFPA 720
    c. UL Standard 2034

Service-Entrance Equipment

OBJECTIVES

After studying this unit, the student should able to

- define electrical service, overhead service, service drop, service lateral, and underground service.
- understand the various Code sections covering the installation of a mast-type overhead service and an underground service.
- discuss the Code requirements for the disconnecting means for the electrical service.
- discuss the grounding and bonding requirements for the main service equipment.
- understand the meaning of a "grounding electrode system."
- understand the different terms used when discussing grounding and bonding.
- discuss the "UFER" ground.
- understand conductor "withstand rating."
- calculate the cost of using electrical energy.
- understand the requirements for proper grounding of electrical equipment in a separate building.

An electric service is required for all buildings containing an electrical system and receiving electrical energy from a utility company. The *NEC®* defines the term *service* as "*the conductors and equipment for delivering electric energy from the serving utility to the wiring system of the premises served.*" The utility company generally must be contacted to determine where they want the meter to be located.

OVERHEAD SERVICE

The Code defines a **service drop** as the overhead service conductors, including splices, if any, which are connected from an outdoor support to the service-entrance conductors at the structure.

An overhead service includes the service raceways, fittings, meter, meter socket, the main service disconnecting means, and the service conductors between the main service disconnecting means and the point of attachment to the utility's service-drop conductors. The point of attachment is called the **service point**. Usually, residential-type watt-hour meters are located on the exterior of a building. In some cases, the entire service-entrance equipment may be mounted outside the building. This includes the watt-hour meter and the disconnecting means. Figure 27-1 illustrates the Code terms for the various components of a service entrance. Figure 27-2 illustrates the service entrance, main panel, subpanel, and grounding for this service.

Overhead service drop conductors might be attached to a through-the-roof mast-type service, or to the side of the house. In either case, proper clearance is required above the ground, roof, fences, windows, sidewalks, decks, and so on.

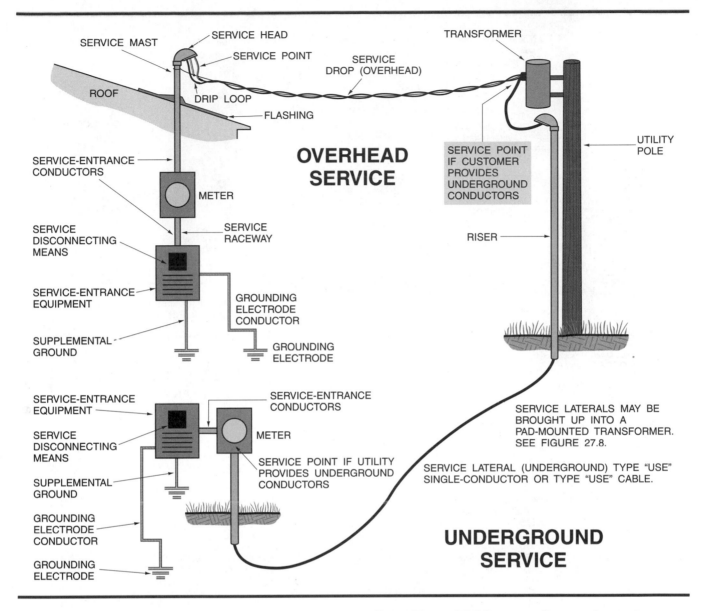

SERVICE MAST — SERVICE HEAD — SERVICE POINT — SERVICE DROP (OVERHEAD) — TRANSFORMER — ROOF — DRIP LOOP — FLASHING — UTILITY POLE — SERVICE-ENTRANCE CONDUCTORS — METER — **OVERHEAD SERVICE** — SERVICE POINT IF CUSTOMER PROVIDES UNDERGROUND CONDUCTORS — SERVICE DISCONNECTING MEANS — SERVICE RACEWAY — RISER — SERVICE-ENTRANCE EQUIPMENT — GROUNDING ELECTRODE CONDUCTOR — SUPPLEMENTAL GROUND — GROUNDING ELECTRODE

SERVICE-ENTRANCE EQUIPMENT — SERVICE-ENTRANCE CONDUCTORS — SERVICE DISCONNECTING MEANS — METER — SUPPLEMENTAL GROUND — SERVICE POINT IF UTILITY PROVIDES UNDERGROUND CONDUCTORS — SERVICE LATERALS MAY BE BROUGHT UP INTO A PAD-MOUNTED TRANSFORMER. SEE FIGURE 27.8. — GROUNDING ELECTRODE CONDUCTOR — SERVICE LATERAL (UNDERGROUND) TYPE "USE" SINGLE-CONDUCTOR OR TYPE "USE" CABLE. — GROUNDING ELECTRODE — **UNDERGROUND SERVICE**

Figure 27-1   Code terms for services. See Figure 27-2 and Figure 27-7 for grounding requirements.

MAST-TYPE SERVICE

The mast service, Figure 27-1, is a commonly used method of installing an overhead service entrance. The mast service is often used on buildings with low roofs, such as ranch-style dwellings, to ensure adequate clearance between the ground and the lowest service conductor.

The service raceway is run through the roof, as shown in Figure 27-1, using a roof flashing and neoprene seal fitting, as illustrated in Figure 27-3(A). This fitting keeps water from seeping in where the raceway penetrates the roof. The conduit is securely fastened to comply with code requirements. Many types of fittings are readily available at electrical supply houses for securely fastening raceways and other electrical equipment to wood, brick, masonry, and other surfaces. Figure 27-3(B) shows a typical through-the-wall support fitting.

Clearance Requirements for Mast Installations

Several factors determine the maximum length of conduit that can be installed between the roof support and the point where the service-drop conductors are attached. These factors are the service-drop length, the system voltage, the roof pitch, and

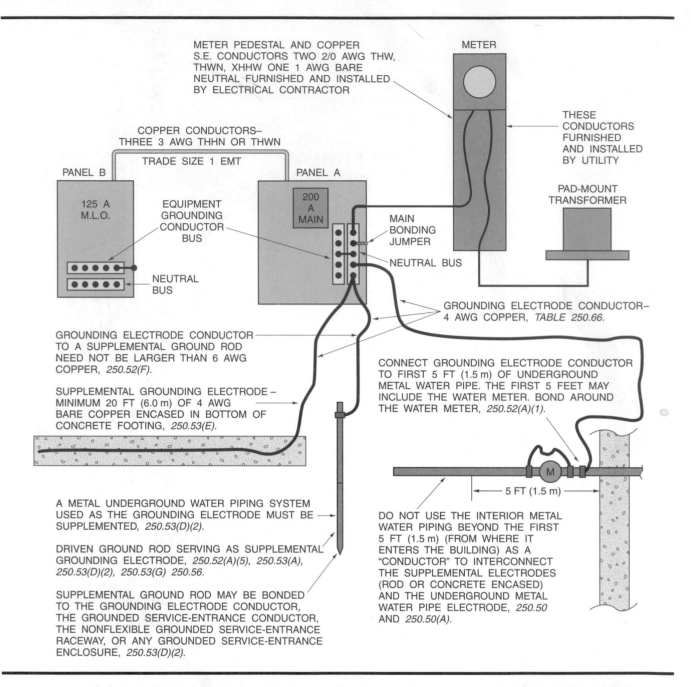

METER PEDESTAL AND COPPER
S.E. CONDUCTORS TWO 2/0 AWG THW,
THWN, XHHW ONE 1 AWG BARE
NEUTRAL FURNISHED AND INSTALLED
BY ELECTRICAL CONTRACTOR

METER

THESE
CONDUCTORS
FURNISHED
AND INSTALLED
BY UTILITY

COPPER CONDUCTORS–
THREE 3 AWG THHN OR THWN

TRADE SIZE 1 EMT

PANEL B

PANEL A

PAD-MOUNT
TRANSFORMER

125 A
M.L.O.

EQUIPMENT
GROUNDING
CONDUCTOR
BUS

200
A
MAIN

MAIN
BONDING
JUMPER

NEUTRAL BUS

NEUTRAL
BUS

GROUNDING ELECTRODE CONDUCTOR–
4 AWG COPPER, *TABLE 250.66*.

GROUNDING ELECTRODE CONDUCTOR
TO A SUPPLEMENTAL GROUND ROD
NEED NOT BE LARGER THAN 6 AWG
COPPER, *250.52(F)*.

SUPPLEMENTAL GROUNDING ELECTRODE –
MINIMUM 20 FT (6.0 m) OF 4 AWG
BARE COPPER ENCASED IN BOTTOM OF
CONCRETE FOOTING, *250.53(E)*.

CONNECT GROUNDING ELECTRODE CONDUCTOR
TO FIRST 5 FT (1.5 m) OF UNDERGROUND
METAL WATER PIPE. THE FIRST 5 FEET MAY
INCLUDE THE WATER METER. BOND AROUND
THE WATER METER, *250.52(A)(1)*.

5 FT (1.5 m)

A METAL UNDERGROUND WATER PIPING SYSTEM
USED AS THE GROUNDING ELECTRODE MUST BE
SUPPLEMENTED, *250.53(D)(2)*.

DRIVEN GROUND ROD SERVING AS SUPPLEMENTAL
GROUNDING ELECTRODE, *250.52(A)(5)*, *250.53(A)*,
250.53(D)(2), *250.53(G)* *250.56*.

SUPPLEMENTAL GROUND ROD MAY BE BONDED
TO THE GROUNDING ELECTRODE CONDUCTOR,
THE GROUNDED SERVICE-ENTRANCE CONDUCTOR,
THE NONFLEXIBLE GROUNDED SERVICE-ENTRANCE
RACEWAY, OR ANY GROUNDED SERVICE-ENTRANCE
ENCLOSURE, *250.53(D)(2)*.

DO NOT USE THE INTERIOR METAL
WATER PIPING BEYOND THE FIRST
5 FT (1.5 m) (FROM WHERE IT
ENTERS THE BUILDING) AS A
"CONDUCTOR" TO INTERCONNECT
THE SUPPLEMENTAL ELECTRODES
(ROD OR CONCRETE ENCASED)
AND THE UNDERGROUND METAL
WATER PIPE ELECTRODE, *250.50*
AND *250.50(A)*.

Figure 27-2 The service-entrance, main panel, subpanel, and grounding for the service in this residence. Note that because the integrity of underground water pipes is questionable, the Code in *250.53(D)(2)* requires a supplemental grounding electrode. Shown are two common types of supplemental grounding electrodes: a ground rod and a concrete-encased grounding electrode. In many instances, the concrete-encased grounding electrode is better than the water pipe ground, particularly since the advent of nonmetallic (PVC) water piping systems.

the size and type of conduit (aluminum or steel), as shown in Figure 27-4.

Service Mast as Support *(230.28)*

The bending force on the conduit increases with an increase in the distance between the roof support and the point where the service-drop conductors are attached. The pulling force of a service drop on a mast service conduit increases as the length of the service drop increases. As the length of the service drop decreases, the pulling force on the mast service conduit decreases.

If extra support is not to be provided, the mast service conduit must be at least 2 in. (50 mm) in diameter. This size prevents the conduit from bending due to the strain of the service-drop conductors.

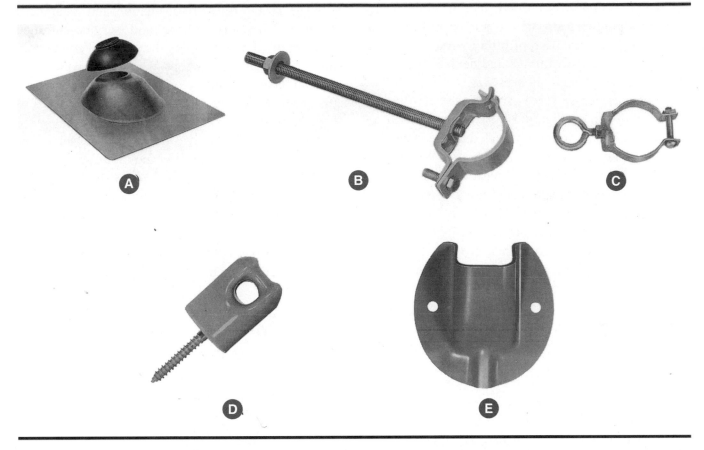

Figure 27-3 (A) is a neoprene seal with metal flashing to use where the service raceway comes through the roof. (B) is a through-the-wall support for securing the service raceway. (C) is a clamp that fastens to the service raceway to which a guy wire can be attached for support of the service riser. (D) is an insulator that can be screwed through the roof into a roof truss or rafter. A guy wire is attached to it and to the clamp (C). Commonly referred to as a "sill plate," (E) is used to protect service-entrance cable where it enters the building.

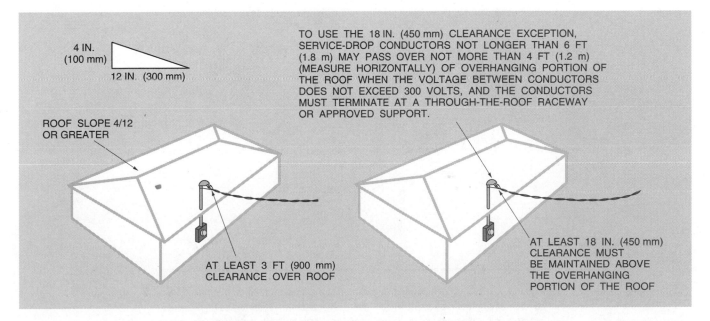

Figure 27-4 Clearance requirements for service-drop conductors passing over residential roofs, *230.24(A)*, where voltage between conductors does not exceed 300 volts.

If extra support is needed, it is usually in the form of a guy wire attached to a mast fitting, as in Figure 27-3(C), and to a roof fitting, Figure 27-3(D). Again, many support fittings and devices are available at electrical supply houses.

When service-entrance cable is used, a fitting of the type shown in Figure 27-3(E) can be used to protect the cable from physical damage where it penetrates the outside wall of the house. When used as a through-the-roof service mast, rigid metal conduit or intermediate metal conduit is not required to be supported within 3 ft (900 mm) of the service head, *344.30(A)* and *342.30(A)*. The roof mast kit provides adequate support when properly installed. A guy wire might be necessary if the service-drop conductors are long or heavy.

Intermediate metal conduit is also permitted to be used as a service mast. It *does not* require secure fastening within 3 ft (900 mm) of the service head, *342.30(A)*.

Only power service-drop conductors are permitted to be attached to and supported by the service mast, *230.28*. Refer to Figure 27-5.

All fittings used with raceway-type service masts must be identified for use with service masts, *230.28*.

Consult the utility company and electrical inspection authority for information relating to their specific requirements for clearances, support, and so on, for service masts.

The *NEC®* rules for insulation and clearances apply to the service-drop and service-entrance conductors. For example, the service conductors must be insulated, except where the voltage to ground does not exceed 300 volts. In this case, the grounded neutral conductors are not required to have insulation.

NEC® 230.24(A) gives clearance allowances for the service drop passing over the roof of a dwelling, as in Figure 27-4.

The installation requirements for a typical service entrance are shown in Figure 27-6 and Figure 27-7. Figure 27-6 shows the required clearances above the ground. The wiring connections, grounding requirements, and Code references are given in Figure 27-7.

UNDERGROUND SERVICE

Underground service means the cable installed underground from the point of connection to the system provided by the utility company.

New residential developments often include underground installations of the high-voltage electrical systems. The conductors in these distribution systems end in the bases of pad-mounted transformers, Figure 27-8. These transformers are placed at the rear lot line or in other inconspicuous locations in the development. The transformer and primary high-voltage conductors are generally

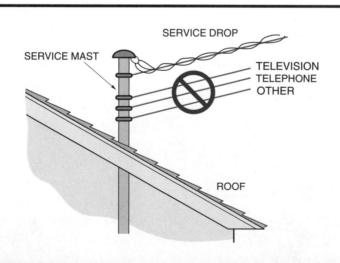

Figure 27-5 Only power service-drop conductors are permitted to be attached to and supported by the service mast. Do *not* attach or support television cables, telephone cables, or anything else to the mast, *230.28*.

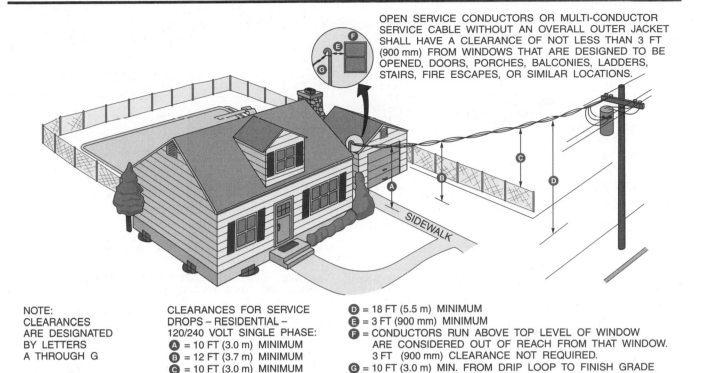

OPEN SERVICE CONDUCTORS OR MULTI-CONDUCTOR SERVICE CABLE WITHOUT AN OVERALL OUTER JACKET SHALL HAVE A CLEARANCE OF NOT LESS THAN 3 FT (900 mm) FROM WINDOWS THAT ARE DESIGNED TO BE OPENED, DOORS, PORCHES, BALCONIES, LADDERS, STAIRS, FIRE ESCAPES, OR SIMILAR LOCATIONS.

NOTE:
CLEARANCES
ARE DESIGNATED
BY LETTERS
A THROUGH G

CLEARANCES FOR SERVICE DROPS – RESIDENTIAL – 120/240 VOLT SINGLE PHASE:
Ⓐ = 10 FT (3.0 m) MINIMUM
Ⓑ = 12 FT (3.7 m) MINIMUM
Ⓒ = 10 FT (3.0 m) MINIMUM

Ⓓ = 18 FT (5.5 m) MINIMUM
Ⓔ = 3 FT (900 mm) MINIMUM
Ⓕ = CONDUCTORS RUN ABOVE TOP LEVEL OF WINDOW ARE CONSIDERED OUT OF REACH FROM THAT WINDOW. 3 FT (900 mm) CLEARANCE NOT REQUIRED.
Ⓖ = 10 FT (3.0 m) MIN. FROM DRIP LOOP TO FINISH GRADE

NOTE: ELECTRIC UTILITIES FOLLOW THE NATIONAL ELECTRICAL SAFETY CODE (NESC). THE CLEARANCE REQUIREMENTS IN THE NESC ARE DIFFERENT THAN THOSE IN THE *NATIONAL ELECTRICAL CODE®* (*NEC®*). THE DECIDING FACTOR MIGHT BE: IS THE INSTALLATION CUSTOMER INSTALLED, OWNED, AND MAINTAINED, OR IS THE INSTALLATION UTILITY INSTALLED, OWNED, AND MAINTAINED?

Figure 27-6   Clearances for a typical residential service in accordance with *230.24* and *230.26*. Clearances (A) and (C) of 10 feet (3.05 m) permitted only if the service-entrance drop conductors are supported on and cabled together with a grounded bare messenger wire where the voltage to ground *does not* exceed 150 volts. This 10 ft (3.0 m) clearance must also be maintained from the lowest point of the drip-loop. Clearance (B) of 12 ft (3.7 m) is permitted over residential property such as lawns and driveways where the voltage to ground *does not* exceed 300 volts. The circled insert shows clearances from building openings, *230.9*.

installed by, and are the responsibility of, the utility company.

The conductors installed between the pad-mounted transformer and the meter are called **service lateral conductors**. Normally, the electric utility furnishes and installs the service laterals. In many areas, the utility will install both the service lateral conductors and the communication cables in the same trench. In some areas, but not too commonly, service laterals, communication cables, and gas lines are run in the same trench. Figure 27-8 shows a typical underground installation.

The wiring from the external meter to the main service equipment is the same as the wiring for a service connected from overhead lines, as in Figure 27-7. Some local codes may require conduit to be installed underground from the pole to the service-entrance equipment. *NEC®* requirements for underground services are given in *230.30, 230.31,* and

230.32. The underground conductors must be suitable for direct burial in the earth, Type USE single-conductor or Type USE cable containing more than one conductor, *Article 338*.

If the electric utility installs the underground service conductors, the work must comply with the rules established by the utility. Utilities follow the National Electrical Safety Code, NESC. These rules may not be the same as those given in *NEC® 90.2*.

▶The *NEC®* in *90.2(B)(5)* excludes "*Installations under the exclusive control of an electric utility where such installations consist of wiring for service drops or laterals or are located in legally established easements, or right-of-ways either designated by or recognized by Public Service Commissions, Utility Commissions, or other regulatory agency having jurisdiction for such installations, or on property owned or leased by the electric utility for the purpose of communications, metering,*

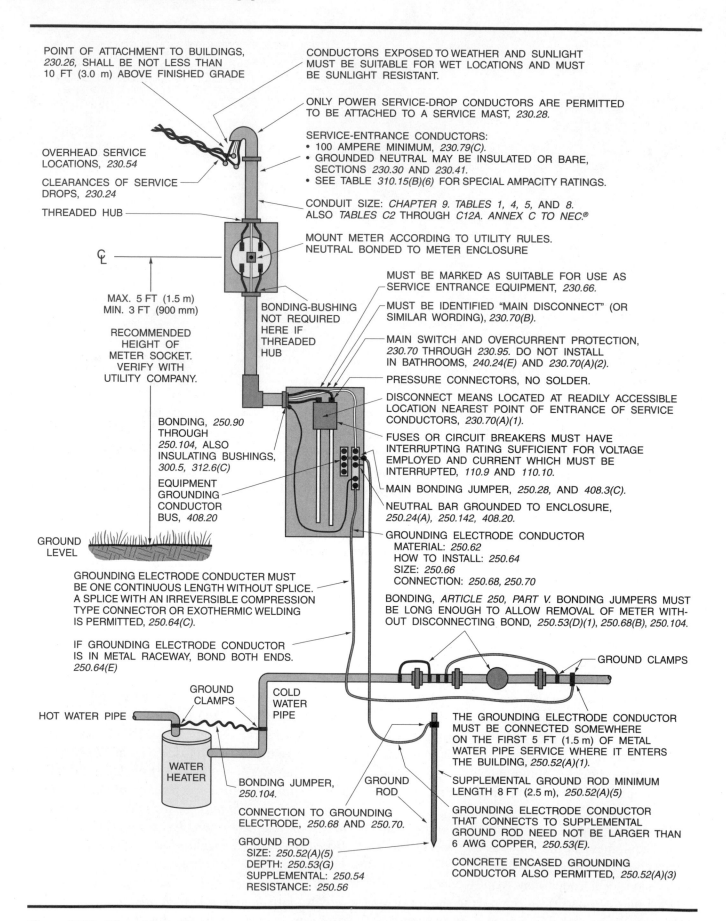

POINT OF ATTACHMENT TO BUILDINGS, *230.26*, SHALL BE NOT LESS THAN 10 FT (3.0 m) ABOVE FINISHED GRADE

CONDUCTORS EXPOSED TO WEATHER AND SUNLIGHT MUST BE SUITABLE FOR WET LOCATIONS AND MUST BE SUNLIGHT RESISTANT.

ONLY POWER SERVICE-DROP CONDUCTORS ARE PERMITTED TO BE ATTACHED TO A SERVICE MAST, *230.28*.

OVERHEAD SERVICE LOCATIONS, *230.54*

SERVICE-ENTRANCE CONDUCTORS:
• 100 AMPERE MINIMUM, *230.79(C)*.
• GROUNDED NEUTRAL MAY BE INSULATED OR BARE, SECTIONS *230.30* AND *230.41*.
• SEE TABLE *310.15(B)(6)* FOR SPECIAL AMPACITY RATINGS.

CLEARANCES OF SERVICE DROPS, *230.24*

CONDUIT SIZE: *CHAPTER 9. TABLES 1, 4, 5, AND 8.* ALSO *TABLES C2 THROUGH C12A. ANNEX C TO NEC.®*

THREADED HUB

MOUNT METER ACCORDING TO UTILITY RULES. NEUTRAL BONDED TO METER ENCLOSURE

C̱L

MAX. 5 FT (1.5 m) MIN. 3 FT (900 mm)

MUST BE MARKED AS SUITABLE FOR USE AS SERVICE ENTRANCE EQUIPMENT, *230.66*.

BONDING-BUSHING NOT REQUIRED HERE IF THREADED HUB

MUST BE IDENTIFIED "MAIN DISCONNECT" (OR SIMILAR WORDING), *230.70(B)*.

RECOMMENDED HEIGHT OF METER SOCKET. VERIFY WITH UTILITY COMPANY.

MAIN SWITCH AND OVERCURRENT PROTECTION, *230.70* THROUGH *230.95*. DO NOT INSTALL IN BATHROOMS, *240.24(E)* AND *230.70(A)(2)*.

PRESSURE CONNECTORS, NO SOLDER.

DISCONNECT MEANS LOCATED AT READILY ACCESSIBLE LOCATION NEAREST POINT OF ENTRANCE OF SERVICE CONDUCTORS, *230.70(A)(1)*.

BONDING, *250.90* THROUGH *250.104*, ALSO INSULATING BUSHINGS, *300.5, 312.6(C)*

FUSES OR CIRCUIT BREAKERS MUST HAVE INTERRUPTING RATING SUFFICIENT FOR VOLTAGE EMPLOYED AND CURRENT WHICH MUST BE INTERRUPTED, *110.9* AND *110.10*.

EQUIPMENT GROUNDING CONDUCTOR BUS, *408.20*

MAIN BONDING JUMPER, *250.28*, AND *408.3(C)*.

NEUTRAL BAR GROUNDED TO ENCLOSURE, *250.24(A), 250.142, 408.20*.

GROUND LEVEL

GROUNDING ELECTRODE CONDUCTOR MATERIAL: *250.62* HOW TO INSTALL: *250.64* SIZE: *250.66* CONNECTION: *250.68, 250.70*

GROUNDING ELECTRODE CONDUCTER MUST BE ONE CONTINUOUS LENGTH WITHOUT SPLICE. A SPLICE WITH AN IRREVERSIBLE COMPRESSION TYPE CONNECTOR OR EXOTHERMIC WELDING IS PERMITTED, *250.64(C)*.

BONDING, *ARTICLE 250, PART V.* BONDING JUMPERS MUST BE LONG ENOUGH TO ALLOW REMOVAL OF METER WITHOUT DISCONNECTING BOND, *250.53(D)(1), 250.68(B), 250.104*.

IF GROUNDING ELECTRODE CONDUCTOR IS IN METAL RACEWAY, BOND BOTH ENDS. *250.64(E)*

GROUND CLAMPS

GROUND CLAMPS

COLD WATER PIPE

HOT WATER PIPE

THE GROUNDING ELECTRODE CONDUCTOR MUST BE CONNECTED SOMEWHERE ON THE FIRST 5 FT (1.5 m) OF METAL WATER PIPE SERVICE WHERE IT ENTERS THE BUILDING, *250.52(A)(1)*.

WATER HEATER

BONDING JUMPER, *250.104*.

GROUND ROD

SUPPLEMENTAL GROUND ROD MINIMUM LENGTH 8 FT (2.5 m), *250.52(A)(5)*

CONNECTION TO GROUNDING ELECTRODE, *250.68* AND *250.70*.

GROUNDING ELECTRODE CONDUCTOR THAT CONNECTS TO SUPPLEMENTAL GROUND ROD NEED NOT BE LARGER THAN 6 AWG COPPER, *250.53(E)*.

GROUND ROD SIZE: *250.52(A)(5)* DEPTH: *250.53(G)* SUPPLEMENTAL: *250.54* RESISTANCE: *250.56*

CONCRETE ENCASED GROUNDING CONDUCTOR ALSO PERMITTED, *250.52(A)(3)*

Figure 27-7   The wiring of a typical service-entrance installation and Code rules for the system grounding. Refer also to Figure 27-2 and Figure 27-21. Today, most services are installed underground, as shown in Figure 27-8.

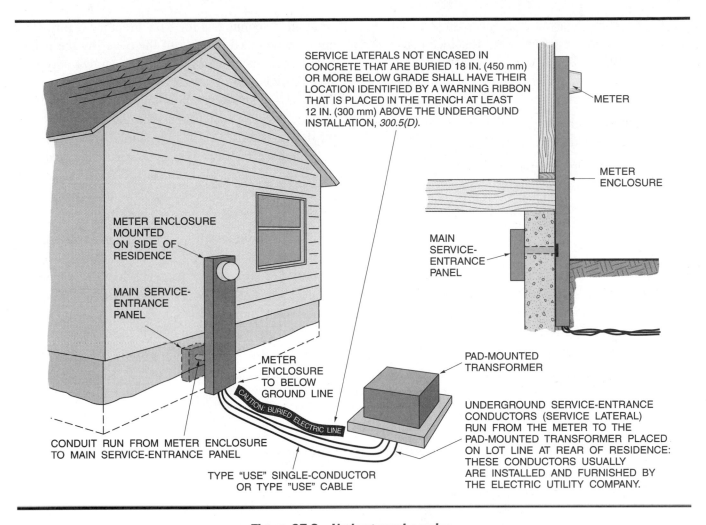

SERVICE LATERALS NOT ENCASED IN CONCRETE THAT ARE BURIED 18 IN. (450 mm) OR MORE BELOW GRADE SHALL HAVE THEIR LOCATION IDENTIFIED BY A WARNING RIBBON THAT IS PLACED IN THE TRENCH AT LEAST 12 IN. (300 mm) ABOVE THE UNDERGROUND INSTALLATION, *300.5(D)*.

METER

METER ENCLOSURE

MAIN SERVICE-ENTRANCE PANEL

METER ENCLOSURE MOUNTED ON SIDE OF RESIDENCE

MAIN SERVICE-ENTRANCE PANEL

PAD-MOUNTED TRANSFORMER

METER ENCLOSURE TO BELOW GROUND LINE

CAUTION: BURIED ELECTRIC LINE

CONDUIT RUN FROM METER ENCLOSURE TO MAIN SERVICE-ENTRANCE PANEL

TYPE "USE" SINGLE-CONDUCTOR OR TYPE "USE" CABLE

UNDERGROUND SERVICE-ENTRANCE CONDUCTORS (SERVICE LATERAL) RUN FROM THE METER TO THE PAD-MOUNTED TRANSFORMER PLACED ON LOT LINE AT REAR OF RESIDENCE: THESE CONDUCTORS USUALLY ARE INSTALLED AND FURNISHED BY THE ELECTRIC UTILITY COMPANY.

Figure 27-8 Underground service.

generation, control, transformation, or distribution of electric energy." ◀

When the underground conductors are installed by the electrician, *230.49* applies. This section deals with the protection of conductors against physical damage. *NEC® 230.49* also refers to *300.5*, which covers all situations involving underground wiring, such as the sealing of raceways where they enter a building; see the "Meter/Meter Base" section in this unit.

MAIN SERVICE DISCONNECT LOCATION

In conformance to *230.70(A)(1)* and *230.70(A)(2)*, the main service disconnecting means shall be located:

* inside the building at the nearest point of entrance of the service conductors, Figure 27-9, or

* outside the building on or within site of the building.

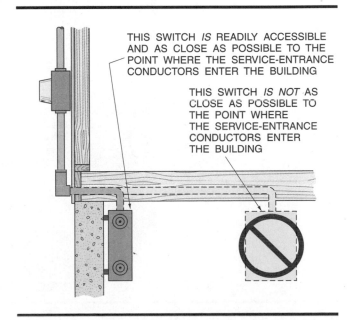

THIS SWITCH *IS* READILY ACCESSIBLE AND AS CLOSE AS POSSIBLE TO THE POINT WHERE THE SERVICE-ENTRANCE CONDUCTORS ENTER THE BUILDING

THIS SWITCH *IS NOT* AS CLOSE AS POSSIBLE TO THE POINT WHERE THE SERVICE-ENTRANCE CONDUCTORS ENTER THE BUILDING

Figure 27-9 Main service disconnect location. Do *not* install in a bathroom. Refer to *230.70(A)(2)* and *240.24(E)*.

If for some reason it is necessary to locate service-entrance equipment some distance inside the building, *NEC® 230.6* considers service-entrance conductors to be outside of a building if the service-entrance conductors are one of the following:

- installed under not less than 2 in. (50 mm) of concrete beneath the building, or

- installed within the building in a raceway encased in at least 2 in. (50 mm) of concrete or brick, or

- installed in conduit, and covered by at least 18 in. (450 mm) of earth beneath the building.

Panelboard Directory

NEC® 408.4 states that all panels must have a "legibly identified" circuit directory indicating what the circuits are for. *NEC® 110.22* also requires that disconnect switches be "legibly marked" as to

what the disconnect is for. The directory is furnished and attached to the inside of the cover by the panel-board manufacturer. Figure 27-10 shows a typical 240/120-volt, single-phase, three-wire load center. Figure 27-11 shows the circuit identification for Main Service Panel A. Many times, the electrician will find the directory furnished by the manufacturer too small, will type a directory on a larger size card, and will attach this to the inside of the panel cover.

Disconnect Means (Panel A)

The requirements for disconnecting electrical services are covered in *230.70* through *230.82*. *NEC® 230.71(A)* requires that the service disconnecting means consist of not more than six switches or six circuit breakers mounted in a single enclosure, in a group of separate enclosures in the same location, or in or on a switchboard. This permits the disconnection of all

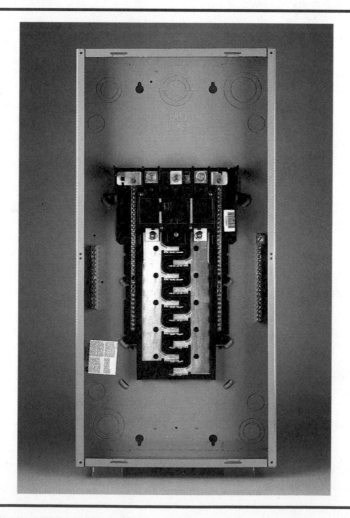

Figure 27-10   A typical 120/240-volt, single-phase, three-wire load center. This load center has a main circuit breaker. Note the neutral terminal strips on which to terminate the white branch-circuit grounded conductors, and the equipment grounding terminal strips on which to terminate the equipment grounding conductors of nonmetallic-sheathed cable. See *230.66. Courtesy of* Square D Company.

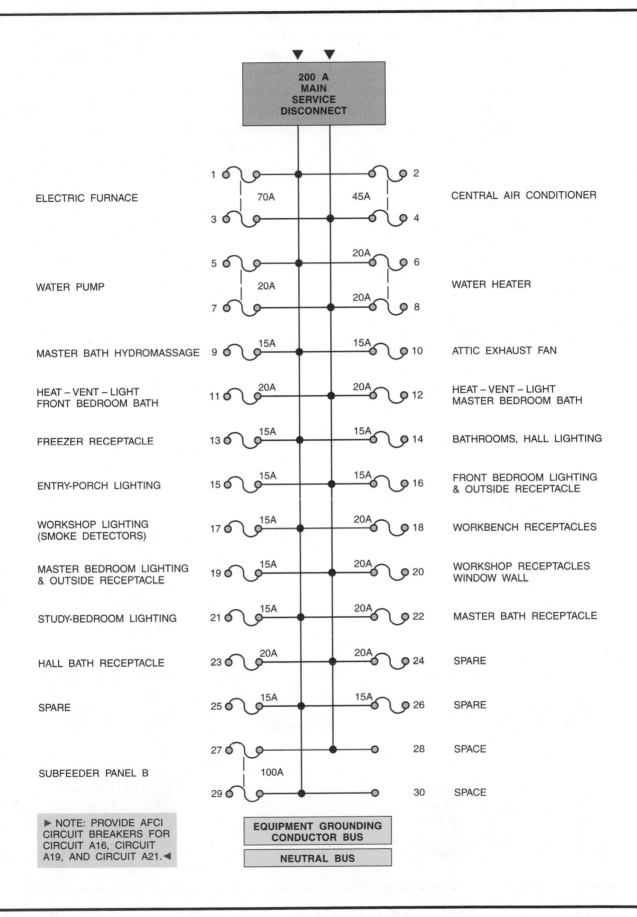

Figure 27-11   Circuit schedule of Main Service Panel "A."

electrical equipment in the house with not more than six hand operations. Some local codes take exception to the six-disconnect rule, and require a single main disconnect. Some localities require the main disconnecting means to be located outside. ▶If located outside, *230.70(A)* requires that the service disconnect must be located on the building, and must be readily accessible.◀

You need to check this out with the local authority that has jurisdiction in your area.

In damp and wet locations, to prevent the accumulation of moisture that could lead to rusting of the enclosure and damage to the equipment inside the enclosure, *312.2(A)* specifies that there be at least a ¼-in. (6-mm) air space between the wall and a surface-mounted enclosure. Most disconnect switches, panelboards, meter sockets, and similar equipment have raised mounting holes that provide the necessary clearance. This is clearly visible in the open-meter socket in Figure 27-15 of the "Meter/Meter Base" section in this unit. The *NEC®* (in Definitions) considers masonry in direct contact with the earth to be a wet location. The *Exception* to *312.2(A)* allows nonmetallic enclosures to be mounted without an air space on concrete, masonry, tile, and similar surfaces.

NEC® 408.14(A) defines a lighting and appliance branch-circuit panelboard as one that has more than

10 percent of the overcurrent devices rated less than 30 amperes with provisions for the connection of neutral conductors. *NEC® 408.16(A)* states that a lighting and appliance branch-circuit panelboard *shall not* have more than two main circuit breakers or two sets of fuses for the panelboards overcurrent protection and disconnecting means.

A lighting and appliance branch-circuit panelboard *shall not* have more than 42 overcurrent devices in it, *408.15*. A two-pole breaker is counted as two overcurrent devices.

NEC® 230.79 tells us that the service disconnect shall have a rating not less than the load to be carried, determined in accordance with *Article 220*. We show how these calculations are made in Unit 29.

For a one-family residence, the minimum size main service disconnecting means is 100 amperes, three-wire, *NEC® 230.79(C)*.

Service equipment must be marked that it is suitable for use as service equipment, *230.66*.

NEC® 230.70(B) requires that the main service disconnect must be permanently marked to identify it as service disconnect.

When nonmetallic-sheathed cable is used, a panelboard must have a terminal bar for attaching the equipment grounding conductors, *408.20*. See Figure 27-12.

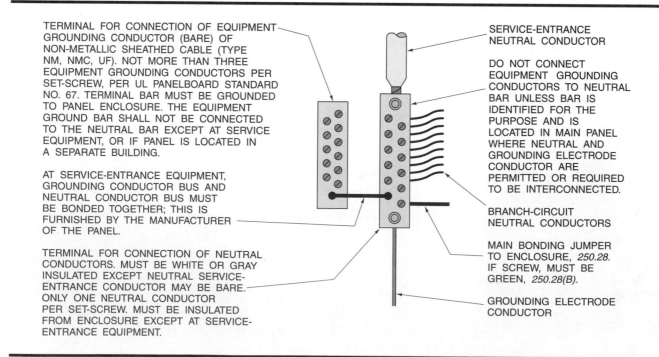

TERMINAL FOR CONNECTION OF EQUIPMENT GROUNDING CONDUCTOR (BARE) OF NON-METALLIC SHEATHED CABLE (TYPE NM, NMC, UF). NOT MORE THAN THREE EQUIPMENT GROUNDING CONDUCTORS PER SET-SCREW, PER UL PANELBOARD STANDARD NO. 67. TERMINAL BAR MUST BE GROUNDED TO PANEL ENCLOSURE. THE EQUIPMENT GROUND BAR SHALL NOT BE CONNECTED TO THE NEUTRAL BAR EXCEPT AT SERVICE EQUIPMENT, OR IF PANEL IS LOCATED IN A SEPARATE BUILDING.

AT SERVICE-ENTRANCE EQUIPMENT, GROUNDING CONDUCTOR BUS AND NEUTRAL CONDUCTOR BUS MUST BE BONDED TOGETHER; THIS IS FURNISHED BY THE MANUFACTURER OF THE PANEL.

TERMINAL FOR CONNECTION OF NEUTRAL CONDUCTORS. MUST BE WHITE OR GRAY INSULATED EXCEPT NEUTRAL SERVICE-ENTRANCE CONDUCTOR MAY BE BARE. ONLY ONE NEUTRAL CONDUCTOR PER SET-SCREW. MUST BE INSULATED FROM ENCLOSURE EXCEPT AT SERVICE-ENTRANCE EQUIPMENT.

SERVICE-ENTRANCE NEUTRAL CONDUCTOR

DO NOT CONNECT EQUIPMENT GROUNDING CONDUCTORS TO NEUTRAL BAR UNLESS BAR IS IDENTIFIED FOR THE PURPOSE AND IS LOCATED IN MAIN PANEL WHERE NEUTRAL AND GROUNDING ELECTRODE CONDUCTOR ARE PERMITTED OR REQUIRED TO BE INTERCONNECTED.

BRANCH-CIRCUIT NEUTRAL CONDUCTORS

MAIN BONDING JUMPER TO ENCLOSURE, 250.28. IF SCREW, MUST BE GREEN, 250.28(B).

GROUNDING ELECTRODE CONDUCTOR

Figure 27-12 Connections of service neutral, branch-circuit neutral, and equipment grounding conductors at panelboards, *384.20*.

For this residence, Panel A serves as the main disconnecting means for the service. This load center consists of a main circuit breaker and many branch-circuit breakers. Figure 27-10 is a photo of this type of load center. Figure 27-11 is the schedule of the circuits connected to Panel A. Panel A is located in the Workshop.

Working Space. To provide for safe working conditions around electrical equipment, *110.26* contains a number of rules that must be followed. For residential installations, there are some guidelines for working space.

- A working space not less than 30 in. (750 mm) wide and 3 ft (900 mm) deep must be provided in front of electrical equipment, such as Panel A and Panel B in this residence.
- This working space must be kept clear and not used for storage.
- Do not install electrical panels inside of cabinets, nor above shelving, washers, dryers, freezers, work benches, and so forth.
- Do not install electrical panels in bathrooms.
- Do not install electrical panels above or close to sump pump holes.
- The hinged cover (door) on the panel shall be able to be opened to a full 90 degrees.
- A "zone" equal to the width and depth of the electrical equipment, from the floor to a height of 6 ft (1.8 m) above the equipment or to the structural ceiling, whichever is lower, shall be dedicated to the electrical installation. No piping, ducts, or equipment foreign to the electrical installation shall be located in this zone. This zone is dedicated space intended for electrical equipment only! See *110.26(F)*.

Mounting Height. Disconnects and panelboards must be mounted so the center of the main disconnect handle in its highest position is not higher than 6 ft 7 in. (2.0 m) above the floor, *404.8*.

Headroom. To provide for safe adequate working space and easy access to service equipment and electrical panels, the *NEC®* in *110.26(E)* requires that there be at least 6½ ft (2.0 m) of headroom. ▶If you are installing service equipment in an existing dwelling, services of 200 amperes or less are permitted to be located in areas where the headroom is less than 6½ ft (2.0 m). ◀

Lighting Required for Service Equipment and Panels. Working on electrical equipment with inadequate lighting can result in injury or death. In *110.26(D)*, there is a requirement that illumination be provided for service equipment and panelboards. But, the Code does not spell out how much illumination is required. It becomes a judgment call on the part of the electrician and/or electrical inspector. Note that the electrical plans for the workshop show a porcelain pull-chain lampholder to the front of and off to one side of the main service Panel A. In many cases, adjacent lighting, such as the fluorescent luminaires (fixtures) in the recreation room, is considered to be the required lighting for equipment such as Panel B.

Load Center (Panel B)

When a second load center is installed, it is oftentimes called a subpanel.

For installation where the main service panel is a long distance from areas that have many circuits and/or heavy load concentration, as in the case of the kitchen and laundry of this residence, it is recommended that at least one additional panel be installed near these loads. This results in the branch-circuit wiring being short. Line losses (voltage and wattage) are considerably less than if the branch-circuits had been run all the way back to the main panel.

The cost of material and labor to install an extra panel should be compared to the cost of running all of the branch-circuit "home runs" back to the main panel.

Panel B in this residence is located in the recreation room. It is fed by three 3 AWG THHN or THWN conductors run in a trade size 1 EMT originating at Panel A. This feeder is protected by a 100-ampere, 2-pole circuit breaker located in Panel A.

Do not install panelboards or load centers in closets, *240.24(D)*. There is a fire hazard because of the presence of ignitable materials.

Do not install panelboards or load centers in bathrooms, *240.24(E)*. Damaging moisture problems may result from the presence of water, and shock hazard may result from both the presence of water and the close proximity of metal faucets, and other plumbing fixtures.

Panel B is shown in Figure 27-13. The circuit schedule for Panel B is found in Figure 27-14.

Figure 27-13 Typical 240/120-volt, single-phase, three-wire panelboard, sometimes called a load center. This panel does not have a main disconnect, and is referred to as a *main lug only (MLO)* panel. The branch circuit-breakers are not shown. This is the type of panel that could be used as Panel B in the residence discussed in this text. Most panels have a separate terminal bus on which to terminate the equipment grounding conductors of nonmetallic-sheathed cable, as in Figure 27-12. *Courtesy of* Square D Company, Group Schneider.

SERVICE-ENTRANCE CONDUCTOR SIZING

NEC® 230.42(A) states that the minimum size service-entrance conductors must be of sufficient size to carry the load as computed according to *Article 220.*

NEC® 230.42(B) states that the ungrounded service-entrance conductors must have an ampacity of not less than the minimum rating of the service disconnect.

NEC® 230.79(C) calls for a minimum service of 100 amperes for a one-family residence.

The calculations for sizing service-entrance conductors are discussed in detail in Unit 29.

RUNNING CABLES INTO TOP OF SERVICE PANEL

See Figure 4-20 for running nonmetallic-sheathed cables into the top of a panel (load center).

SERVICE-ENTRANCE OVERCURRENT PROTECTION

Overcurrent protection (fuses and circuit breakers) is discussed in Unit 28.

SERVICE-ENTRANCE RACEWAY SIZING

This residence is served with an underground service. The conduit running from the meter to Main Panel A must be sized correctly for the conductors it will contain. In the case of a through-the-roof mast service, the conduit might be sized for mechanical strength reasons rather than conductor fill. All utilities have requirements for size and guying of mast services. Table 27-1 shows the calculation for sizing the service-entrance conduit. This calculation is similar to other examples in this text. The conductor dimensions are found in *Chapter 9* of the *NEC.®*

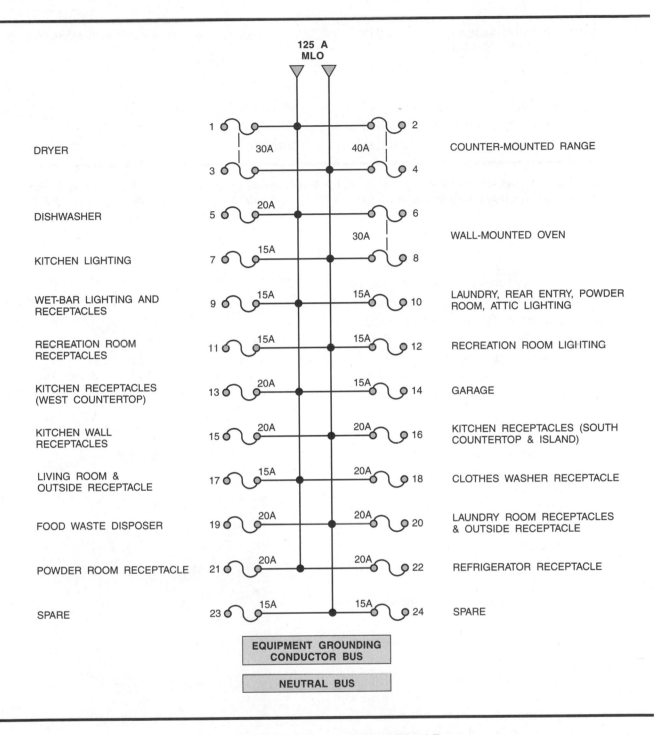

125 A MLO

DRYER — 1, 2 — COUNTER-MOUNTED RANGE (30A / 40A)
— 3, 4

DISHWASHER — 5, 6 (20A)
WALL-MOUNTED OVEN — 8 (30A)

KITCHEN LIGHTING — 7 (15A)

WET-BAR LIGHTING AND RECEPTACLES — 9 (15A) / 10 (15A) — LAUNDRY, REAR ENTRY, POWDER ROOM, ATTIC LIGHTING

RECREATION ROOM RECEPTACLES — 11 (15A) / 12 (15A) — RECREATION ROOM LIGHTING

KITCHEN RECEPTACLES (WEST COUNTERTOP) — 13 (20A) / 14 (15A) — GARAGE

KITCHEN WALL RECEPTACLES — 15 (20A) / 16 (20A) — KITCHEN RECEPTACLES (SOUTH COUNTERTOP & ISLAND)

LIVING ROOM & OUTSIDE RECEPTACLE — 17 (15A) / 18 (20A) — CLOTHES WASHER RECEPTACLE

FOOD WASTE DISPOSER — 19 (20A) / 20 (20A) — LAUNDRY ROOM RECEPTACLES & OUTSIDE RECEPTACLE

POWDER ROOM RECEPTACLE — 21 (20A) / 22 (20A) — REFRIGERATOR RECEPTACLE

SPARE — 23 (15A) / 24 (15A) — SPARE

EQUIPMENT GROUNDING CONDUCTOR BUS

NEUTRAL BUS

Figure 27-14 Circuit schedule of Panel B.

METER/METER BASE

The electric utility must be contacted before the installation of the service equipment begins. They will determine when, where, and how they intend to connect to the homeowner's service. Because of the ever-increasing number of wood decks and concrete patios being added to homes, many utilities prefer that the meter be mounted on the side of the house rather than in the rear. If there is a raised deck, the meter could be only a short distance above the deck, making it both subject to physical damage and hard to read. If concrete patios are poured over underground utility cables, it is very difficult to repair or replace these cables should there be problems. Utilities are happy to furnish manuals and brochures detailing their requirements for services that are not found in the *NEC*®.

CONDUCTOR SIZE BASED UPON AMPACITY PER *310.15(B)(6)*		CONDUCTOR SIZE BASED UPON AMPACITY PER *TABLE 310.16*	
TWO 2/0 AWG THWN COPPER	0.2223 in²	TWO 3/0 AWG THWN COPPER	0.2679 in²
	0.2223 in²		0.2679 in²
ONE 1 AWG BARE COPPER	0.0870 in²	ONE 1 AWG BARE COPPER	0.0870 in²
TOTAL	0.5316 in²	TOTAL	0.6228 in²
CONDUIT	TRADE SIZE 1¼	CONDUIT	TRADE SIZE 1½

Table 27-1 Calculations for determining the size of the EMT service-entrance raceway between Main Panel A and the meter base. Dimensional data found in *Chapter 9, Table 1, Table 4, Table 5,* and *Table 8* of the *NEC.*

The electric utility furnishes the watt-hour meter. The meter base, also referred to as a meter socket, is usually furnished and installed by the electrical contractor, although some utilities furnish the meter base to the electrical contractor for installation.

Figure 27-8 shows a service supplied from an underground service. Figure 27-9 discusses the requirement that the main disconnect must be as close as possible to the point of entrance of the service-entrance conductors. Figure 27-15 shows a typical meter socket for an overhead service, usually mounted at eye level on the outside wall of the house. The service raceway or service-entrance cable is connected to the threaded hub (boss) on top of the meter socket. Figure 27-16 shows how to seal the raceway where it enters the building to keep out moisture.

Probably more common for residential services is the "pedestal," as shown in Figure 27-17. Usually mounted on the outside wall of the house, it could also be installed on the lot line between residential properties. Contact the electric utility on this issue.

For underground services, the utility generally runs the underground lateral service-entrance conductors in a trench from a pad-mount transformer to the line-side terminals of the meter base. The electrician then takes over to complete the service installation by running service-entrance conductors from the load side of the meter to the line side of the main service disconnect.

Quite often, the electric utility and the telephone company have agreements whereby the electric utility will lay both power cables and telephone cables in the same trench. The electric utility carries both types of cables on their underground installation

Figure 27-15 Typical meter socket. *Courtesy of Milbank Mfg. Co.*

vehicles. This is a much more cost effective way to do this, rather than having each utility dig and then backfill their own trenches.

Often overlooked is a requirement in *NFPA Standard No. 54 (Fuel Gas Code®), Section 2.7.2(c),*

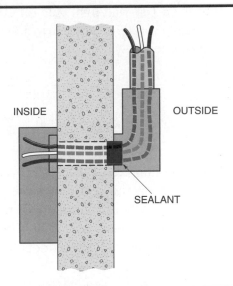

NEC® 300.5(G) AND 300.7 STATE THAT WHEN RACEWAYS PASS THROUGH AREAS HAVING GREAT TEMPERATURE DIFFERENCES, SOME MEANS MUST BE PROVIDED TO PREVENT PASSAGE OF AIR BACK AND FORTH THROUGH THE RACEWAY. NOTE THAT OUTSIDE AIR IS DRAWN IN THROUGH THE CONDUIT WHENEVER A DOOR OPENS. COLD OUTSIDE AIR MEETING WARM INSIDE AIR CAUSES THE CONDENSATION OF MOISTURE. THIS CAN RESULT IN RUSTING AND CORROSION OF VITAL ELECTRICAL COMPONENTS. EQUIPMENT HAVING MOVING PARTS, SUCH AS CIRCUIT BREAKERS, SWITCHES, AND CONTROLLERS, IS ESPECIALLY AFFECTED BY MOISTURE. THE SLUGGISH ACTION OF THE MOVING PARTS IN THIS EQUIPMENT IS UNDESIRABLE.

▶ SEALANT SHALL BE APPROVED FOR SUCH USE, AND SHALL NOT HAVE AN ADVERSE EFFECT ON THE CONDUCTOR INSULATION.◀

NEC® 230.8 REQUIRES SEALS WHERE SERVICE RACEWAYS ENTER FROM AN UNDERGROUND DISTRIBUTION SYSTEM.

Figure 27-16   Installation of conduit through a basement wall.

THE ELECTRIC UTILITY PROVIDES THE UNDERGROUND SERVICE LATERAL

Figure 27-17 Installing a metal "pedestal" allows for ease of installation of the underground service-entrance conductors by the utility. *Photo Courtesy of* Milbank Manufacturing Co.

requiring that gas meters be located at least 3 ft (900 mm) from sources of ignition. Examples of sources of ignition might be an air-conditioner disconnect where an arc can be produced when the disconnect is opened under load. Furthermore, some electric utilities require a minimum of 3 ft (900 mm) clearance between electric metering equipment and gas meters and gas regulating equipment. It is better to be safe than sorry. Check this issue with the local electrical inspector if there might be a problem with the installation you are working on.

COST OF USING ELECTRICAL ENERGY

A watt-hour meter is always connected into some part of the service-entrance equipment to record the amount of energy used. Billing by the utility is generally done on a monthly basis. The meter might be mounted on the side of the house, or on a pedestal somewhere on the premise on the lot line. The utility makes this decision. The residence discussed in this text has one meter, mounted on the back of the house near the sliding doors of the master bedroom.

The kilowatt (kW) is a convenient unit of electrical power. One thousand watts (w) is equal to one kilowatt. The watt-hour meter measures and records both wattage and time.

For residential metering, most utilities have rate schedules based on "cents per kilowatt-hour." Stated another way: how much wattage is being used and for how long?

Burning a 100-watt light bulb for 10 hours is the same as using a 1000-watt electric heater for 1 hour. Both equal one kilowatt-hour.

$$kWh = \frac{watts \times hours}{1000} = \frac{100 \times 10}{1000} = 1 \ kWh$$

$$kWh = \frac{watts \times hours}{1000} = \frac{1000 \times 1}{1000} = 1 \ kWh$$

In these examples, if the electric rate is 8 cents per kilowatt-hour, the cost to operate the 100-watt light bulb for ten hours and the cost to operate the electric heater for one hour are the same—8 cents. Both loads use one kilowatt-hour of electricity.

Simple Formula for Calculating the Cost of Using Electricity

The cost of electrical energy used can be calculated as follows:

$$Cost = \frac{watts \times hours \ used \times cost \ per \ kWh}{1000}$$

EXAMPLE: Find the cost of operating a color television set for 8 hours. The label on the back of the television set indicates 175 watts. The electric rate is $0.10494 per kilowatt-hour.

$$Cost = \frac{175 \times 8 \times \$0.10494}{1000} = \frac{\$0.1469}{(approx. \ 15¢)}$$

EXAMPLE: Find the approximate cost of operating a central air conditioner per day that on average runs 50 percent of the time during a 24-hour period on a typical hot summer day. The unit's nameplate is marked 240 volts, single-phase, 23 amperes. The electric rate is $.09 per kilowatt-hour. The steps are:

1. The time the air conditioner operates each day is: $24 \times 0.50 = 12$ hours.

2. Convert the nameplate data to use in the calculations:

$$Watts = volts \times amperes = 240 \times 23$$
$$= 5520 \ watts$$

3. $$Cost = \frac{watts \times hours \ used \times cost \ per \ kWh}{1000}$$

$$= \frac{5520 \times 12 \times \$0.09}{1000} = \$5.96$$

Note: The answers to these examples are approximate because they were based on watts. Power factor and efficiency factors were not included as they generally are unknown when coming up with rough estimates of an electric bill. For all practical purposes, the answers are acceptable.

Table 27-2 is an example of what a typical monthly electric bill might look like.

Some utilities increase their rates during the hot summer months when the air-conditioning load is high. Other utilities provide a second meter for specific loads such as electric water heaters, air conditioners, heat pumps, or total electric heat. Other utilities use electronic watt-hour meters that have the capability of registering kilowatt-hour consumption during a specific "time-of-use." These electronic watt-hour meters might have up to four different "time-of-use" periods, each period having a different "cents per kilowatt-hour" rate.

Other charges that might appear on a "light bill" might be a fuel adjustment charge based on a "per kWh" basis. Such charges enable a utility to recover from the consumer extra expenses it might incur for fuel costs used in generating electricity. Fuel charges can vary with each monthly bill without the utility having to apply to the regulatory agency for a rate change.

GENERIC ELECTRIC COMPANY
Anyplace, USA

Days of Service	from 06-01-2002	to 06-28-2002	Due Date 07-25-2002
Present reading			84,980
Previous reading			83,655
Kilowatt-hours used			1,325
Rate/kWh	1st 400 kWh @ $0.10494		$41.98
	remaining 925 kWh @ $0.06168 . .		57.05
Energy charge			99.03
Basic service charge (single-dwelling)			8.91
State tax			4.24
Total current charges			$112.18
Total amount due by 07-25-2002			$112.18
Total amount due after 07-25-2002			$117.79

Table 27-2 A typical monthly electric bill.

GROUNDING

Article 250 covers grounding and bonding. To better understand grounding and bonding, it is necessary to understand the definitions found in *250.2*.

►**Ground-Fault:** An unintentional, electrically conducting connection between an ungrounded conductor of an electrical circuit and the normally non-current carrying conductors, metallic enclosures, metallic raceways, metallic equipment, or earth.

Ground-Fault Current Path: An electrically conductive path from the point of a ground-fault on a wiring system through normally non-current-carrying conductors, equipment, or the earth to the electrical supply source.

Effective Ground-Fault Current Path: An intentionally constructed, permanent, low-impedance electrically conductive path designed and intended to carry current under ground-fault conditions from the point of a ground-fault on a wiring system to the electrical supply source. ◄

We find in *250.4* the fundamental reasons for grounding electrical systems and electrical equipment. All residential wiring involves a grounded electrical system. In *250.4(A)*:

►1. *Electrical System Grounding.* Electrical systems that are grounded shall be connected to earth in a manner that will limit the voltage imposed by lightning, line surges, or unintentional contact with higher voltage lines and that will stabilize the voltage to earth during normal operation.

2. *Grounding of Electrical Equipment.* Non-current-carrying conductive materials enclosing electrical conductors or equipment, or forming part of such equipment, shall be connected to earth so as to limit the voltage to ground on these materials.

3. *Bonding of Electrical Equipment.* Non-current-carrying conductive materials enclosing electrical conductors or equipment, or forming part of such equipment, shall be connected together and to the electrical supply source in a manner that establishes an effective ground-fault current path.

4. *Bonding of Electrically Conductive Materials and Other Equipment.* Electrically conductive materials that are likely to become energized shall be connected together and to the electrical supply source in a manner that establishes an effective ground-fault current path.

5. *Effective Ground-Fault Current Path.* Electrical equipment and wiring and other electrically conductive material likely to become energized shall be installed in a manner that creates a permanent, low-impedance circuit capable of safely carrying the maximum ground-fault current likely to be imposed on it from any point on the wiring system where a ground-fault may occur to the electrical supply source. The earth shall not be used as the sole equipment grounding conductor or effective ground-fault current path. ◄

A few other key Code sections relating to system grounding are:

250.20(B)(1): The electrical system must be grounded so the maximum voltage to ground on the ungrounded conductors does not exceed 150 volts. Since home electrical systems are 120/240 volts, the voltage to ground is 120 volts.

250.26(2): The neutral conductor must be grounded for three-wire, single-phase systems. Residential electrical systems are 120/240, three-wire, single-phase. The neutral is grounded at the main service.

Proper grounding means that overcurrent protective devices will operate fast when responding to ground faults. Effective grounding occurs when a low-impedance (ac resistance) ground path is provided. Ohm's law verifies that, in a given circuit, "the lower the impedance, the higher the value of current." As the value of ground-fault current increases, there is an increase in the speed with which a fuse will open or a circuit breaker will trip. This is called an *inverse time* relationship.

Clearing a ground fault or short-circuit is important because arcing damage to electrical equipment as well as conductor insulation damage is closely related to a value called *ampere-squared-seconds* (I^2t), where

I = the current in amperes flowing phase to ground, or phase to phase.

t = the time in seconds that the current is flowing.

Thus, there will be less equipment and/or conductor damage when the fault current is kept to a low value, and when the time that the fault current is allowed to flow is kept to a minimum. The impedance of the circuit determines the *amount* of fault current that will flow. The speed of operation of a fuse or circuit breaker determines the amount of *time* the fault current will flow.

Grounding electrode conductors and equipment grounding conductors carry an insignificant amount of current under normal conditions. However, when a ground-fault occurs, these grounding conductors as well as the actual circuit conductors must be capable of carrying whatever value of fault current might flow (how much?) for the time (how long?) it takes for the overcurrent protective device to clear the fault.

This potential hazard is recognized in the *Note* below *Table 250.122*, which states that *"Where necessary to comply with 250.4(A)(5), the equipment grounding conductor shall be sized larger than this table."* This note calls attention to the fact that equipment grounding conductors may need to be sized larger than indicated in the table when high-level ground-fault currents are possible.

What is a high-level ground-fault? To determine the available short-circuit and ground-fault current, a short-circuit calculation is necessary. It is also necessary to know the time/current characteristic of the overcurrent protective device. This is discussed in *Electrical Wiring—Commercial* (©Delmar, a division of Thomson Learning).

This is referred to as the conductor's "withstand rating."

Table 250.66 and *Table 250.122* are based upon the fact that copper conductors and their bolted connections can withstand:

- one ampere
- for 5 seconds
- for every 30 circular mils

Thermoplastic insulation on a copper conductor rated 75°C can safely withstand:

- one ampere
- for 5 seconds
- for every 42.25 circular mils

To exceed these values will result in damage to the conductors, with possible burn-off and loss of the grounding or bonding path.

When bare equipment grounding conductors are in the same raceway or cable as insulated conductors, always apply the insulated conductor withstand rating limitation.

Properly selected fuses and circuit breakers for normal residential installations generally will protect conductors and other electrical equipment against the types of ground-faults and short-circuits to be expected in homes. Fault currents can be quite high in single-family homes where the pad-mounted transformer is located close to the service equipment, and in multifamily dwellings (apartments, condos, town houses) making it necessary to take equipment short-circuit ratings and conductor withstand rating into consideration.

The subject of conductor and equipment withstand rating is covered in greater depth in *Electrical Wiring—Commercial* (©Delmar, a division of Thomson Learning).

Grounding Electrode System

Article 250, Part III covers the requirements for grounding electrode systems, grounding electrode conductors, and ground rods. It is very lengthy. A grounding electrode system "ties" everything together. The service-entrance neutral conductor is grounded, this neutral is bonded to the main panel, a grounding electrode conductor is connected to a grounding electrode and ▶all metal water piping systems installed in or attached to a building◀ and all metal gas pipes are bonded together. Should any one of these become disconnected, the integrity of the grounding system is maintained through the other paths. Here are a few key points.

- *250.90* states that "bonding shall be provided where necessary to assure electrical continuity and the capacity to conduct safely any fault current likely to be imposed."

- *250.92(A)* explains what parts of a service must be bonded together.

- *250.94* explains what is acceptable as a bonding means.

- *250.95(A)* states in part that bonding of metal raceways, cable armor, enclosures, frames, fittings, and so on, that serve as the grounding path "shall be effectively bonded where necessary to assure electrical continuity and the capacity to conduct safely any fault current likely to be imposed on them."

By having all metal parts bonded together for a grounding electrode system, potential differences between non-current-carrying metal parts is virtually eliminated. In addition, the grounding electrode system serves as a means to bleed off lightning, stabilize the system voltage, and ensure that the overcurrent protective devices will operate.

Figure 27-18 and the following steps illustrate what can happen if the electrical system is not properly grounded and bonded:

1. A "live" wire contacts the gas pipe. The bonding jumper "A" has not been installed.

2. The gas pipe now has 120 volts on it. The pipe is energized. It is "hot."

3. The insulating joint in the gas pipe results in a poor path to ground. Assume a resistance (impedance) of 8 ohms.

4. The 20-ampere overcurrent device does not open:

$$I = \frac{E}{R} = \frac{120}{8} = 15 \text{ amperes}$$

5. If a person touches the "live" gas pipe and the water pipe at the same time, current flows through the person's body. If the body resistance is 12,000 ohms, the current is:

$$I = \frac{E}{R} = \frac{120}{12,000} = 0.01 \text{ ampere}$$

This amount of current passing through a human body can cause death. See Unit 6 for a discussion regarding electric shock.

6. The overcurrent device is now "seeing" $15 + 0.01 = 15.01$ amperes and does not open.

7. If the grounding electrode system concept had been used, bonding jumper "A" would have kept the voltage difference between the water pipe and the gas pipe at zero. The overcurrent device would have opened. Checking *Table 8, Chapter 9, NEC,* we find the dc resistance of 1000 feet (305 m) of an uncoated 4 AWG copper wire is 0.308 ohms. The resistance of 100 ft (30 m) is 0.0308 ohms. The resistance of 10 ft (3.0 m) is 0.00308 ohms. The current would be:

$$I = \frac{E}{R} = \frac{120}{0.00308} = 38,961 \text{ amperes}$$

In an actual installation, the total impedance of *all* parts of the circuit would be much higher than these simple calculations. A much lower current would result. The value of current would be enough to cause the overcurrent device to open.

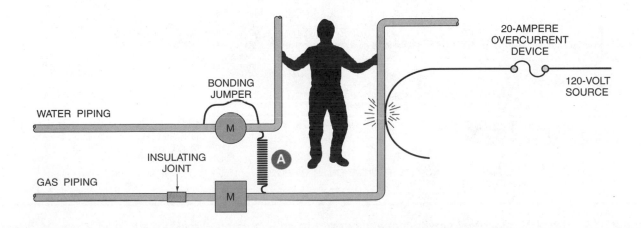

Figure 27-18 System grounding.

Grounding and Bonding The Electrical System in a Typical Residence

Figure 27-2, Figure 27-7, Figure 27-19, and Figure 27-20 show the *NEC®* code rules applicable to residential electrical service grounding and bonding as required by *Article 250, Part III*.

A metal underground water piping system 10 ft (3.0 m) or longer, in direct contact with the earth, is acceptable as the primary grounding electrode,

250.52(A)(1). The connection of the grounding electrode conductor must be made on the first 5 ft (1.52 m) of where the metal underground water piping enters the building, *250.52(A)(1)*.

When metal underground water piping is used as the primary grounding electrode, it must be supplemented by at least one additional grounding electrode, *250.53(D)(2)*. In this residence, a driven ground rod is the supplemental grounding electrode, as permitted by *250.52(A)(5)*.

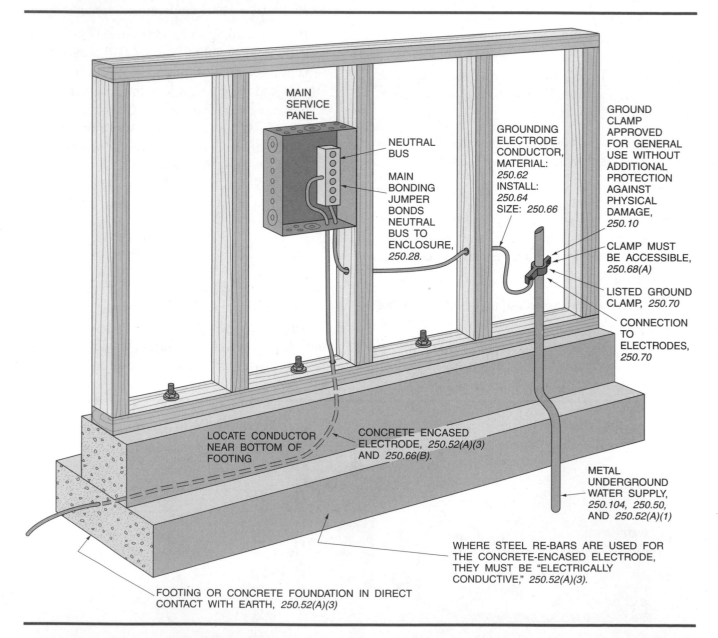

Figure 27-19   A typical electrical service panel grounded to an underground metal water piping supply. A water pipe ground must be supplemented by at least one additional electrode, *250.53(D)(2)*. Here we show a concrete-encased 4 AWG bare copper conductor laid near the bottom of the concrete footing. The minimum length of the bare conductor is 20 ft (6.0 m). This is quite often referred to as a UFER ground, named after the individual who years ago came up with this concept.

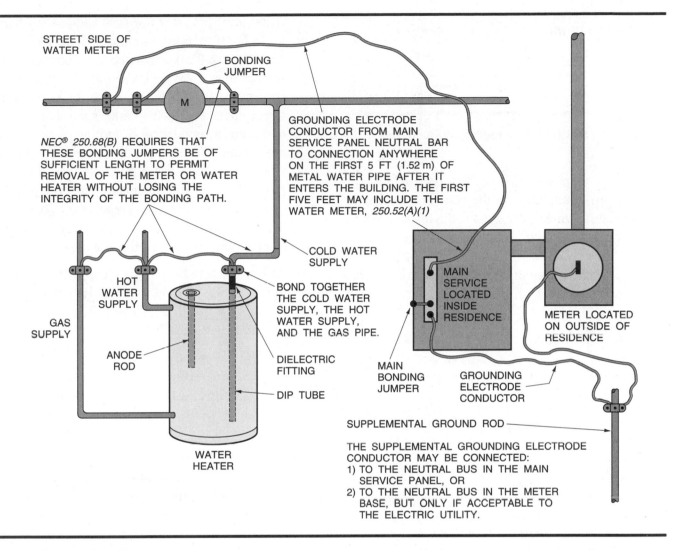

STREET SIDE OF
WATER METER

BONDING
JUMPER

NEC® 250.68(B) REQUIRES THAT
THESE BONDING JUMPERS BE OF
SUFFICIENT LENGTH TO PERMIT
REMOVAL OF THE METER OR WATER
HEATER WITHOUT LOSING THE
INTEGRITY OF THE BONDING PATH.

GROUNDING ELECTRODE
CONDUCTOR FROM MAIN
SERVICE PANEL NEUTRAL BAR
TO CONNECTION ANYWHERE
ON THE FIRST 5 FT (1.52 m) OF
METAL WATER PIPE AFTER IT
ENTERS THE BUILDING. THE FIRST
FIVE FEET MAY INCLUDE THE
WATER METER, *250.52(A)(1)*

COLD WATER
SUPPLY

HOT
WATER
SUPPLY

GAS
SUPPLY

ANODE
ROD

BOND TOGETHER
THE COLD WATER
SUPPLY, THE HOT
WATER SUPPLY,
AND THE GAS PIPE.

DIELECTRIC
FITTING

DIP TUBE

WATER
HEATER

MAIN
SERVICE
LOCATED
INSIDE
RESIDENCE

METER LOCATED
ON OUTSIDE OF
RESIDENCE

MAIN
BONDING
JUMPER

GROUNDING
ELECTRODE
CONDUCTOR

SUPPLEMENTAL GROUND ROD

THE SUPPLEMENTAL GROUNDING ELECTRODE
CONDUCTOR MAY BE CONNECTED:
1) TO THE NEUTRAL BUS IN THE MAIN
   SERVICE PANEL, OR
2) TO THE NEUTRAL BUS IN THE METER
   BASE, BUT ONLY IF ACCEPTABLE TO
   THE ELECTRIC UTILITY.

Figure 27-20   One method that may be used to provide proper grounding and bonding of service-entrance equipment for a typical one-family residence.

Watch out when using ground rods! If the resistance to ground of the first ground rod is more than 25 ohms, at least one additional ground rod must be added, *250.56*. There are a number of manufacturers of testers that measure ground/earth resistance.

More Ground Rod Rules For Residential Application

• The most common ground rods are copper-clad steel. Copper makes for an excellent connection between the rod and the ground clamp. The steel gives it strength to withstand being driven into the ground. Galvanized and stainless steel rods are also available. Aluminum rods are not permitted.

• Must be at least ½ in. (13 mm) in diameter.

• Must be installed below the permanent moisture level if possible.

• Must not be less than 8 ft (2.5 m) in length.

• Must be driven to a depth so that at least 8 ft (2.5 m) is in contact with the soil.

• If solid rock is encountered, drive rod at an angle not greater than 45 degrees from vertical, or lay it in a trench that is at least 2½ ft (750 mm) deep.

• Drive rod so upper end is flush with or just below ground level. If upper end is exposed, the ground rod, the ground clamp, and the grounding electrode conductor must be protected from physical damage.

NUMBER OF RODS	MULTIPLIER
2	1.16
3	1.29
4	1.36

Table 27-3 Multipliers for multiple ground rods.

- If more than one rod is needed, keep at least 6 ft (1.8 m) apart. Driving the rods close together does not significantly change the resistance values of a single rod. Actually, it is better to space multiple rods the length of the rods. For example, when driving two 8-ft (2.5-m) ground rods, space them 8 ft (2.5 m) apart.

Table 27-3 shows accurate multipliers for multiple ground rods spaced rod length apart. To use this chart, divide the resistance value of one rod by the number of rods used, then apply the multiplier.

EXAMPLE: The ground/earth resistance reading of one 8 ft (2.5 m) ground rod is 30 ohms. This exceeds the *NEC®* maximum of 25 ohms. A second ground rod is driven and connected in parallel to the first ground rod. What is the approximate ground/earth resistance of these two ground rods when spaced 8 ft (2.5 m) apart?

$$\frac{30}{2} \times 1.16 = 17.4 \text{ ohms}$$

- A grounding electrode conductor that is the sole (only) connection to a driven ground rod need not be larger than 6 AWG copper or 4 AWG aluminum, *250.53(E)*.

In some parts of the country, a water pipe ground is considered unreliable. The concern is the prolific use of insulating fittings and nonmetallic water piping services. These parts of the country have found that a concrete encased electrode provides excellent grounding capabilities. Refer to *250.52(A)(3)*.

Other Acceptable Supplemental Grounding Electrodes

If the driven ground rod is not selected as the required supplemental grounding electrode, then any one of the following is acceptable:

- The metal frame of a building (this residence is constructed of wood), *250.50(A)(2)*.

- At least 20 ft (6.0 m) of steel reinforcing bars, ½ inch (12.7 mm) minimum diameter, or at least 20 ft (6.0 m) of bare copper conductor not smaller than 4 AWG. These must be encased in concrete at least 2 in. (50 mm) thick that is in direct contact with the earth, such as near the bottom of a foundation or footing. A bare copper conductor encased in the concrete footing of a building is commonly referred to as a UFER ground, named after the man who developed the concept, *250.52(A)(3)*. Reinforcing bars are permitted to be bonded together by the usual steel tie wires or other effective means.

- At least 20 ft (6.0 m) of bare copper wire, minimum size 2 AWG, buried directly in the earth at least 2½ ft (750 mm) deep. This is called a *ground ring*. See *250.52(A)(4)* and *250.52(G)*.

- Ground plates must be at least 2 ft² (0.186 m²), *250.52(A)(6)*. At least one additional plate must be laid if the resistance of one plate is found to exceed 25 ohms, *250.56*. The plate must be buried at least 2½ ft (750 mm) deep.

- *250.53(D)(2)* tells us that a metal underground water piping system must have a supplemental electrode installed. An underground metal water piping system is the only grounding electrode that needs to be supplemented. Other types of grounding electrodes, such as a concrete-encased electrode, do not need a supplemental electrode.

Some Issues Regarding Bonding of Metal Piping Inside the Home or Attached to the Outside of the Home

Many homes are served with copper underground gas pipes. Over the years, many of these underground metal pipes have corroded and leaked gas, setting up a hazardous situation. Gas utilities now install insulated fittings where the piping enters the building. This breaks the metallic connection and prevents the galvanic corrosion of the pipe. Many homes are served by PVC underground gas piping from the street to the home.

Here are a few key issues.

- *250.52(B)(1)*: Metal underground gas piping is not permitted to be used as a grounding electrode.

- *250.104(A)*: requires that interior metal water piping be bonded and that the bonding conductor be sized according to *Table 250.66*.

Figure 27-20 illustrates the proper bonding together of the cold and hot water pipes and the metal gas piping. The bonding jumper is sized per *Table 250.66*. This installation eliminates the questionable bonding/grounding of the many interconnections of the metal water and gas piping through various appliances.

Some electrical inspectors will accept the interconnection of metallic mixer faucets, water heaters, clothes washers, and so on, as an acceptable bonding jumper between the metal hot and cold water pipes. This relies on plumbing to do a job that is electrical in nature. Do it right! Follow Figure 27-20 for proper bonding.

- *250.104(B), FPN*: Bonding all piping and metal air ducts within the premises will provide additional safety.

- *250.104(B)*: the bonding conductor for the metal piping shall be sized according to *Table 250.122* for the ampere rating of the circuit that might energize the metal piping.

- ▶*250.104(B)* makes it clear that metal gas piping must also be bonded.◀

Along the same lines, the *National Fuel Gas Code*, *NFPA 54*, in *Section 3-14(a)* requires that "*Each aboveground portion of a gas piping system upstream from the equipment shutoff valve shall be electrically continuous and bonded to the grounding electrode system.*" The Code Making Panel for *NFPA 54* has stated that a properly electrically grounded gas appliance such as a gas furnace meets the intent of this requirement because the appliance is connected to the metal gas piping system.

Grounding Conductors

A *grounding conductor* is defined as "*a conductor used to connect equipment or the grounded circuit of a wiring system to a grounding electrode or electrodes.*" There are two types of grounding conductors.

1. An *equipment grounding conductor* is the conductor that is used to connect noncurrent-carrying metal parts of equipment, raceways,

and other enclosures to the system grounded conductor, to the grounding electrode conductor, or to both—at the service equipment. *NEC®* 250.118 recognizes many types of equipment grounding conductors. Copper or aluminum conductors, rigid metal conduit, EMT, the metal armor of armored cable, and flexible metal conduit all qualify as equipment grounding conductors. Figure 5-34, Figure 5-35, Figure 5-36, Figure 27-2, Figure 27-12, and Figure 30-3 illustrate equipment grounding conductors.

Equipment grounding is discussed throughout this text in the appropriate unit where a specific piece of electrical equipment or appliance requires grounding. Some examples of this are the range, cooktop, dryer, food waste disposer, dishwasher, and water pump. Equipment grounding conductors are sized according to *Table 250.122*.

Table 250.66 Grounding Electrode Conductor for Alternating-Current Systems

Size of Largest Ungrounded Service-Entrance Conductor or Equivalent Area for Parallel Conductors[a] (AWG/kcmil)		Size of Grounding Electrode Conductor (AWG/kcmil)	
Copper	Aluminum or Copper-Clad Aluminum	Copper	Aluminum or Copper-Clad Aluminum[b]
2 or smaller	1/0 or smaller	8	6
1 or 1/0	2/0 or 3/0	6	4
2/0 or 3/0	4/0 or 250	4	2
Over 3/0 through 350	Over 250 through 500	2	1/0
Over 350 through 600	Over 500 through 900	1/0	3/0
Over 600 through 1100	Over 900 through 1750	2/0	4/0
Over 1100	Over 1750	3/0	250

Notes:

1. Where multiple sets of service-entrance conductors are used as permitted in 230.40, Exception No. 2, the equivalent size of the largest service-entrance conductor shall be determined by the largest sum of the areas of the corresponding conductors of each set.

2. Where there are no service-entrance conductors, the grounding electrode conductor size shall be determined by the equivalent size of the largest service-entrance conductor required for the load to be served.

[a]This table also applies to the derived conductors of separately derived ac systems.

[b]See installation restrictions in 250.64(A).

(Reprinted with permission from NFPA 70-2002)

2. ►A *grounding electrode conductor* is the conductor that connects the grounding electrode to the equipment grounding conductor, to the grounded conductor of the circuit, or to both—at the service or at a second building supplied from the electrical service in the first building.◄ Acceptable grounding electrodes are listed in *250.52(A)(1)* through *(7)*. Reference to grounding electrode conductors is made in Figure 27-1, Figure 27-2, Figure 27-7, Figure 27-12, Figure 27-19, Figure 27-20, Figure 27-35, and Figure 30-3. See also the section on "Grounding Electrical Equipment at a Second Building" in this unit.

Grounding electrode conductors:

• are sized according to *Table 250.66*.

• shall not be spliced unless done so with an irreversible compression-type connector listed for the purpose, or by exothermic welding, *250.28, 250.52(D)(3)*, and *250.64(C)*.

In this residence, a 4 AWG armored copper grounding electrode conductor runs from Main Panel A, across the workshop ceiling to the water pipe, where it is terminated with a listed ground clamp.

Ground Clamps

Ground clamps used for bonding and grounding must be listed for the purpose. The use of solder to make up bonding and grounding connections is not acceptable because under high levels of fault current, the solder would probably melt, resulting in loss of the integrity of the bonding and/or grounding path.

Various types of ground clamps are shown in Figure 27-21, Figure 27-22, Figure 27-23, and Figure 27-24. These clamps and their attachment to the grounding electrode must conform to *250.70*.

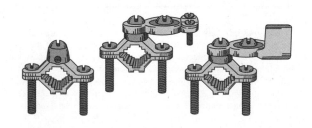

Figure 27-21 Typical ground clamps used in residential systems.

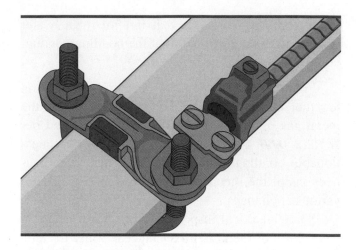

Figure 27-22 Armored grounding conductor connected with ground clamp to water pipe. *Courtesy of* Thomas & Betts.

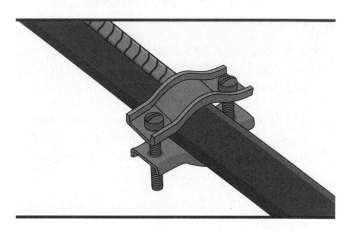

Figure 27-23 Ground clamp of the type used to bond (jumper) around water meter. *Courtesy of* Thomas & Betts.

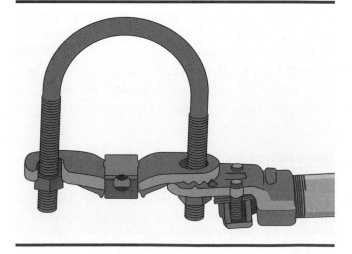

Figure 27-24 Ground clamp of the type used to attach ground wire to well casings. *Courtesy of* Thomas & Betts.

Connection of Grounding and Grounded Conductors in Main Panel and Subpanel

According to the second paragraph of *408.20*, equipment *grounding* conductors (the bare copper equipment grounding conductors found in non-metallic-sheathed cable) shall not be connected to the *grounded* conductor (the neutral conductor) terminal bar (neutral bar) unless the bar is identified for the purpose, and is located where the *grounded* conductor is connected to the *grounding* electrode conductor. In this residence, this occurs in Main Panel A. The green bonding screw furnished with Panel A is installed in Panel A. This bonds the neutral bar, the ground bus, the grounded neutral conductor, the grounding electrode conductor, and the panel enclosure together, as in Figure 27-12. *Grounding* conductors and *grounded* conductors are not to be connected together anywhere on the load side of the main service disconnect, *250.142(B)*. The green bonding screw furnished with Panel B will *not* be installed in Panel B. More on this a little later in this unit.

Sheet Metal Screws Not Permitted for Connecting Grounding Conductors

Sheet metal screws are *not* permitted to be used as a means of connecting grounding conductors to enclosures, *250.8*. Sheet metal screws do not have the same fine thread that No. 10-32 machine screws have, which match the tapped No. 10-32 threaded holes in outlet boxes, device boxes, and other enclosures. Sheet metal screws "force" themselves into a hole instead of nicely threading themselves into a pretapped matching hole. Sheet metal screws have *not* been tested for their ability to safely carry ground-fault currents as required by *250.4(A)(5)*.

BONDING

The *NEC®* defines **bonding** as "*the permanent joining of metallic parts to form an electrically conductive path that will ensure electrical continuity and the capacity to conduct safely any current likely to be imposed.*"

Bonding jumpers accomplish this in a number of ways:

1. Equipment bonding "ties" equipment together to keep voltage buildup on the equipment the same, so there will be no difference in potential between the various pieces of equipment. For example, to bond two metal raceways together, an **equipment bonding jumper** is installed where there is a break in the metal conduit run, as in Figure 27-25. Bonding equipment together does *not* necessarily mean that the equipment is properly grounded.

2. A **circuit bonding jumper** connects conductors of a circuit together. This jumper must have the same or greater current-carrying ability than the circuit conductors, Figure 27-26.

3. A **main bonding jumper** is defined as "the connection between the grounded circuit conductor and the equipment grounding conductor at the service." Main bonding jumpers are clearly illustrated in Figure 27-2, Figure 27-7, Figure 27-12, Figure 27-19, Figure 27-20, and Figure 30-3.

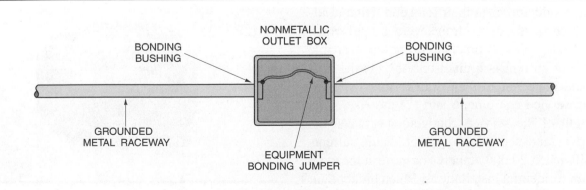

Figure 27-25   Bonding jumper (equipment).

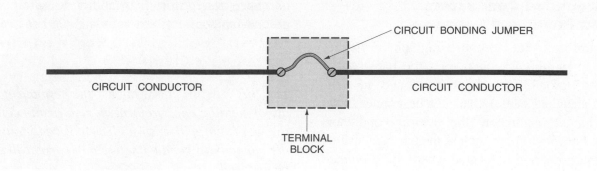

CIRCUIT BONDING JUMPER

CIRCUIT CONDUCTOR

CIRCUIT CONDUCTOR

TERMINAL
BLOCK

Figure 27-26   Bonding jumper (circuit).

4. Bonding for swimming pools is shown in Figure 30-4.

At the main service-entrance equipment, the grounded neutral conductor must be bonded to the metal enclosure. For most residential panels, this main bonding jumper is a bonding screw that is furnished with the panel. This bonding screw is inserted through the neutral bar into a threaded hole in the back of the panel. This bonding screw must be green and must be clearly visible after it is in place, *250.28(B)*.

The required main bonding jumpers for services must not be smaller than the grounding electrode conductor, *250.28(D)*. The lugs on bonding bushings are based on the ampacity of the conductors that would normally be installed in that particular size of raceway. The lugs become larger as the trade size of the bushing increases.

Bonding at service-entrance equipment is very important because service-entrance conductors do not have overcurrent protection at their line side, other than the electric utility company's primary transformer fuses. Overload protection for service-entrance conductors is at their load end. High available fault currents can result in severe arcing, which is a fire hazard. For all practical purposes, available short-circuit current is limited only by (1) the kVA rating and impedance of the transformer that supplies the service equipment, and (2) the size, type, and length of the service conductors between the transformer and service equipment. Fault currents can easily reach 20,000 amperes or more at the main service in residential installations. Much higher fault current is available in multifamily dwellings such as apartments and condominiums.

Figure 27-7 and Figure 27-20 show the hot and cold metal water pipes bonded together. Some electrical inspectors will consider the hot/cold mixing faucets and the water heater as an acceptable means of bonding the hot and cold water metal pipes together. Most, however, will require a bonding jumper as illustrated in the figures, consistent with the thinking that electrical grounding and bonding shall not be dependent upon the plumbing trade.

Fault current calculations are presented later in this unit. The text *Electrical Wiring—Commercial* (©Delmar Publishers, a division of Thomson Learning) covers fault current calculations in greater depth.

NEC® 250.92(A) and *(B)* lists the parts of the service-entrance equipment that must be bonded. These include the meter base, service raceways, service cable armor, and main disconnect. *NEC® 250.94* lists the methods acceptable for bonding all of the previous equipment. Bonding bushings, shown in Figure 27-27, and bonding jumpers are installed to ensure a low-impedance path to ground

Figure 27-27   Insulating bonding bushing with grounding lug.

if a fault occurs. Bonding jumpers are required where there are concentric or eccentric knockouts, or where reducing washers are installed. *NEC® 300.4(F)* and *312.6(C)* state that if 4 AWG or larger conductors are installed in a raceway, an insulating bushing or equivalent must be used, shown in Figure 27-28. Insulating bushings protect the conductors from abrasion where they pass through the bushing. Combination metal/insulating bushings can be used. Some EMT connectors have an insulated throat.

If the bushing is made of insulating material only, as in Figure 27-28, then two locknuts must be used, as shown in Figure 27-29.

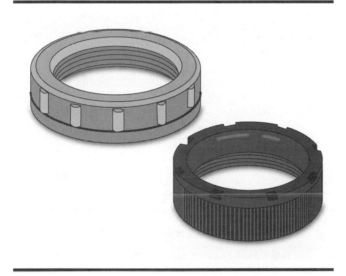

Figure 27-28   Insulating bushings.

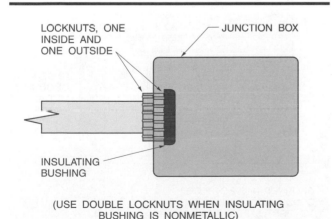

(USE DOUBLE LOCKNUTS WHEN INSULATING BUSHING IS NONMETALLIC)

Figure 27-29   The use of double locknuts plus an insulating bushing. See *300.4(F)* and *312.6(C)*.

Installing a Grounding Electrode Conductor from Main Service Disconnect to Metal Underground Water Pipe that Will Serve as the Grounding Electrode

The Code specifies in *NEC® 250.64(A)* how a grounding electrode conductor is to be installed, as shown in Table 27-4.

In order to carry high values of ground-fault current, the ground path must have as low an impedance as possible. *NEC® 250.4(A)(5)* requires that the ground path be *effective*. This means that the path must be permanent and continuous, must have the capability of carrying any value of fault current that it might be called upon to carry, and shall have an impedance low enough to limit the voltage to ground and to ensure that the protective overcurrent device will operate. *NEC® 250.90* and *250.96* have similar requirements. For example, the metal raceway enclosing a grounding electrode conductor must be continuous from the main disconnect

4 AWG OR LARGER	6 AWG	SMALLER THAN 6 AWG
May be attached to the surface on which it is carried. Does not need additional mechanical protection unless exposed to severe physical damage, which is highly unlikely in the case of residential installations.	If not subject to physical damage, may be run along the surface of the building structure without additional protection. If subject to physical damage, install in rigid metal conduit, electrical metallic tubing, intermediate metal conduit, rigid nonmetallic conduit, or cable armor.	Must be installed in rigid metal conduit, electrical metallic tubing, intermediate metal conduit, rigid nonmetallic conduit, or cable armor.

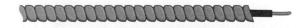

A TYPICAL ARMORED GROUNDING ELECTRODE CONDUCTOR.

Table 27-4   Installing a grounding electrode conductor.

to the ground clamp, or be made continuous by proper bonding.

NEC® 250.92(A)(3) requires that the metal raceway be bonded at both ends. At first thought, it might appear that simply installing a grounding electrode conductor in a metal raceway makes for a neat and workmanlike installation. However, unless the enclosing metal raceway is properly bonded on both ends, the ground path will be a high impedance path. Bonding the enclosing metal raceway at one end only will result in a ground path impedance approximately twice that of bonding the metal raceway on both ends. Remember—the higher the impedance, the lower the current flow and the longer it takes the overcurrent device to clear the fault.

Past testing (Soares, IAEI, and IEEE data) has shown that for a 300-ampere ground fault, approximately 295 amperes flowed in the metal raceway, and 5 amperes flowed in the enclosed grounding electrode conductor. Another test revealed that for a 590-ampere ground fault, 585 amperes flowed in the metal raceway, and 5 amperes flowed in the enclosed grounding electrode conductor. The tendency for current to flow along the surface of a conductor is called "skin effect." Proper bonding at both ends of a metal raceway is necessary to ensure a low impedance path that will safely transfer the high magnitudes of ground-fault current flowing in the

metal conduit to the ground bus in the main panel, and/or to the ground clamp.

Figure 27-30, Figure 27-31, Figure 27-32, Figure 27-33, and Figure 27-34 illustrate accepted methods of installing a grounding electrode conductor between the main panel and the ground clamp. Additional information concerning grounding electrode conductors can be found in Unit 4.

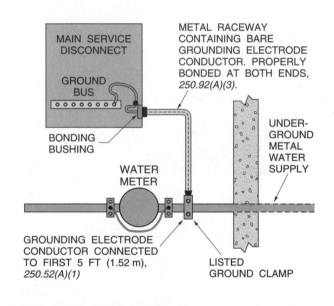

Figure 27-31   This installation "meets Code."

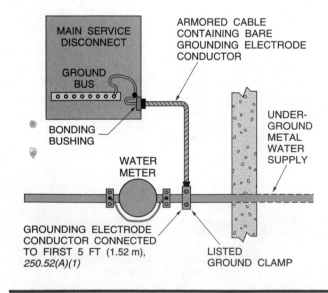

Figure 27-30   This installation "meets Code."

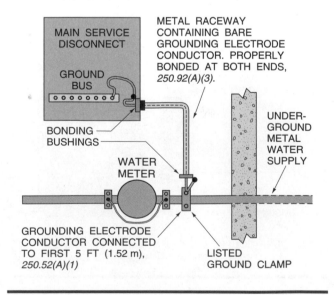

Figure 27-32   This installation "meets Code."

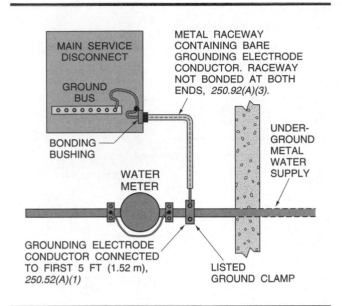

Figure 27-33   This installation *does not* "meet Code."

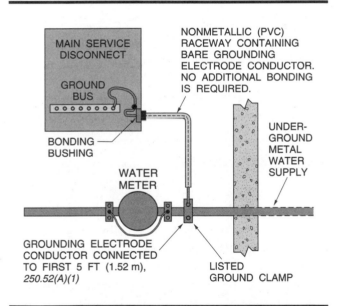

Figure 27-34   This installation "meets Code."

GROUNDING ELECTRICAL EQUIPMENT AT A SECOND BUILDING

Detached Garage—Grounding

A detached garage or other second building on the same residential property is almost always served by the same electrical service.

For dwellings, a detached garage is not required to have any electrical supply. However, if you do run electric power to a detached garage, there are specific Code requirements.

We discussed the receptacle and lighting requirements for garages in Unit 3. We will now discuss the grounding requirements for a second building, *250.32*. See Figure 27-35.

Branch Circuit to Garage

If only one branch-circuit is run to a detached garage or other second building, the simplest way to provide proper and adequate grounding for the equipment in that detached garage is to install an underground cable Type UF that contains an equipment grounding conductor. A grounded metal raceway between the house and the out-building may also serve as the equipment grounding conductor.

With either of these two methods, a separate grounding electrode at the second building is not required, *250.32(A), Exception*.

Feeder to Garage

When a feeder supplies a panel in a garage or other second building that has more than one branch-circuit, proper grounding at the second building is accomplished by:

- installing a grounded metal raceway, which is an acceptable equipment grounding conductor, between the main building panel and the panel in the second building, or

- installing a separate equipment grounding conductor (in a cable or in a nonmetallic raceway) between the main building panel and the panel in the second building.

In either case, a grounding electrode is required at the second building. A grounding electrode conductor is run between the grounding electrode and the equipment grounding bus in the panel in the second building.

▶The grounding electrode conductor is sized according to *Table 250.66*, but does not have to be larger than the largest ungrounded supply conductor, *250.32(E)*.◀

The neutral bus and the equipment grounding bus in the second building are not connected together. If these two buses were tied together in the second building, there would be more than one

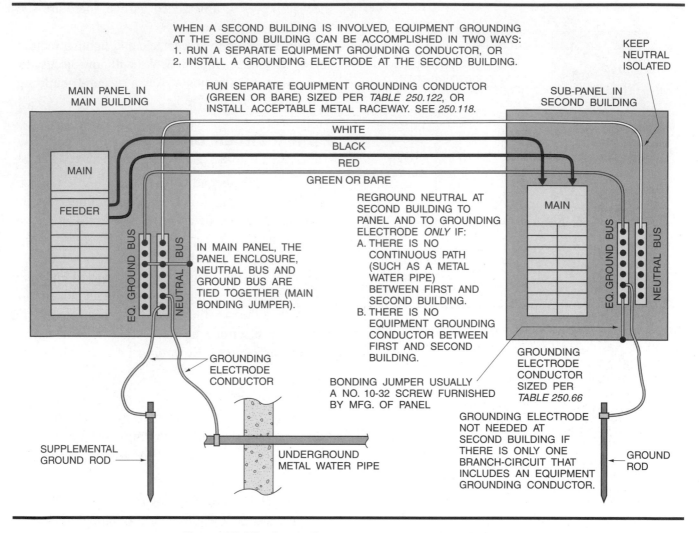

WHEN A SECOND BUILDING IS INVOLVED, EQUIPMENT GROUNDING AT THE SECOND BUILDING CAN BE ACCOMPLISHED IN TWO WAYS:
1. RUN A SEPARATE EQUIPMENT GROUNDING CONDUCTOR, OR
2. INSTALL A GROUNDING ELECTRODE AT THE SECOND BUILDING.

KEEP NEUTRAL ISOLATED

MAIN PANEL IN MAIN BUILDING

RUN SEPARATE EQUIPMENT GROUNDING CONDUCTOR (GREEN OR BARE) SIZED PER *TABLE 250.122*, OR INSTALL ACCEPTABLE METAL RACEWAY. SEE *250.118*.

SUB-PANEL IN SECOND BUILDING

WHITE
BLACK
RED
GREEN OR BARE

MAIN

FEEDER

MAIN

EQ. GROUND BUS

NEUTRAL BUS

EQ. GROUND BUS

NEUTRAL BUS

IN MAIN PANEL, THE PANEL ENCLOSURE, NEUTRAL BUS AND GROUND BUS ARE TIED TOGETHER (MAIN BONDING JUMPER).

REGROUND NEUTRAL AT SECOND BUILDING TO PANEL AND TO GROUNDING ELECTRODE *ONLY* IF:
A. THERE IS NO CONTINUOUS PATH (SUCH AS A METAL WATER PIPE) BETWEEN FIRST AND SECOND BUILDING.
B. THERE IS NO EQUIPMENT GROUNDING CONDUCTOR BETWEEN FIRST AND SECOND BUILDING.

GROUNDING ELECTRODE CONDUCTOR

GROUNDING ELECTRODE CONDUCTOR SIZED PER *TABLE 250.66*

BONDING JUMPER USUALLY A NO. 10-32 SCREW FURNISHED BY MFG. OF PANEL

SUPPLEMENTAL GROUND ROD

UNDERGROUND METAL WATER PIPE

GROUNDING ELECTRODE NOT NEEDED AT SECOND BUILDING IF THERE IS ONLY ONE BRANCH-CIRCUIT THAT INCLUDES AN EQUIPMENT GROUNDING CONDUCTOR.

GROUND ROD

Figure 27-35 Grounding at a second building, *250.32*.

path (a parallel path) for neutral current to flow back on. Take a look at Figure 27-35. Imagine a tie between the neutral bus and the equipment grounding bus. Now imagine that there is some value of neutral current. You can readily see that this neutral current would have many ways to return to the source at the main building panel. Some current would return on the neutral conductor, some on the equipment grounding conductor, some on the grounding electrode conductor—to the grounding electrode—and back through the earth, and some through a metal water pipe if one is present. These multiple paths of neutral return current are not permitted.

If there is no metal path such as a metal water pipe between the two buildings, and no equipment grounding conductor is run between the two buildings, a grounding electrode at the second building is required, and the neutral bus and the equipment grounding bus are connected together. A grounding electrode conductor is run between the grounding bus and the grounding electrode.

▶If you do not run a separate equipment grounding conductor to a residential garage (or other second building), the grounded conductor of the feeder must not be smaller than the sizes given in *Table 250.122*. This is to ensure a good ground-fault return path. For a feeder to a detached residential garage or other second building, you probably would size the grounded neutral conductor the same size as the two ungrounded conductors. The grounded feeder conductor need not be larger than the largest ungrounded supply conductor.◀

All of the above requirements are found in *230.32*, and are illustrated in Figure 27-35.

REVIEW

Note: Refer to the *NEC,*® the text, and/or the plans where necessary.

1. Define the "service point." _____

   _____

   _____

   _____

2. Define "service-drop conductors." _____

   _____

   _____

   _____

   _____

3. Who is responsible for determining the service location? _____

4. a. The service head must be located (above) (below) the point where the service-drop conductors are spliced to the service-entrance conductors. Circle one.

   b. What Code section provides the answer to Part (a)? _____

5. What is a mast-type service entrance? _____

   _____

   _____

6. When a conduit is extended through a roof, must it be guyed? _____

7. a. What size and type of conductors are installed for this service? _____

      _____

   b. What size conduit is installed? _____

   c. What size grounding electrode conductor is installed? (not neutral) _____

   d. Is the grounding electrode conductor insulated, armored, or bare? _____

8. How and where is the grounding electrode conductor attached to the water pipe?

   _____

   _____

9. What are the minimum distances or clearances for the following?

   a. Service-drop clearance over private driveway _____

   b. Service-drop clearance over private sidewalks _____

   c. Service-drop clearance over alleys _____

   d. Service-drop clearance over a roof having a roof pitch of not less than 4/12. (Voltage between conductors does not exceed 300 volts.) _____

   e. Service-drop horizontal clearance from a porch _____

   f. Service-drop clearance from a fence that can be climbed upon _____

10. What are the minimum size ungrounded conductors using Type THW copper for the following residential electrical services? The panels are marked 75°C. For the 400-ampere service, install two conductors in parallel per phase. Use two raceways. Install three conductors in each raceway.

AMPERE RATING OF SERVICE	SERVICE-ENTRANCE CONDUCTORS SIZED PER *TABLE 310.16*	SERVICE-ENTRANCE CONDUCTORS SIZED PER *TABLE 310.15(B)(6)*
100		
200		
400		

11. What is the minimum size copper grounding electrode conductor for each of the following residential electrical services? Refer to *Table 250.66*. The ungrounded conductors are Type THW copper.

AMPERE RATING OF SERVICE	GROUNDING ELECTRODE CONDUCTOR WHEN THE SERVICE-ENTRANCE CONDUCTORS ARE SIZED PER *TABLE 310.16*	GROUNDING ELECTRODE CONDUCTOR WHEN THE SERVICE-ENTRANCE CONDUCTORS ARE SIZED PER *TABLE 310.15(B)(6)*
100		
200		
400		

*Table 310.15(B)(6) is only for 120/240-volt, single-phase residential services and feeders. This table does not apply to services and feeders other than residential.

12. What is the recommended height of a meter socket from the ground? _____

13. a. May the bare grounded neutral conductor of a service be buried directly in the ground? _____

  b. What section of the Code covers this? _____

14. How far must mechanical protection be provided when underground service conductors are carried up a pole? _____

15. a. A service disconnect may consist of not more than how many switches or circuit breakers? _____

  b. Must these devices be in one enclosure? _____

  c. What type of main disconnect is provided in this residence? _____

  d. Does your city permit more than one service disconnect? _____

16. Complete the following table by filling in the columns with the appropriate information.

	CIRCUIT NUMBER	AMPERE RATING	POLES	VOLTS	CONDUCTOR SIZE
A. Living room receptacle outlets					
B. Workbench receptacle outlets					
C. Water pump					
D. Attic exhaust fan					
E. Kitchen lighting					
F. Hydromassage tub					
G. Attic lighting					
H. Counter-mounted cooking unit					
I. Electric furnace					

17. a. What size conductors supply Panel B? _____

    b. What size raceway? _____

    c. Is this raceway run in the form of electrical metallic tubing or rigid conduit?

    _____

    d. What size overcurrent device protects the feeders to Panel B? _____

18. How many electric meters are provided for this residence? _____

19. a. According to the *NEC*® is it permissible to ground a residential rural electrical service to driven ground rods only when a metallic water system is available?

    _____

    b. What sections of the Code apply? _____

20. What table in the *NEC*® lists the sizes of grounding electrode conductors required for electrical services of various sizes? Table _____

21. *NEC*® 250.53(D)(2) requires that a supplemental ground be provided if the available grounding electrode is: (Circle correct answer.)
    a. metal underground water pipe
    b. building steel
    c. concrete encases ground

22. Do the following conductors require mechanical protection?

    a. 8 AWG grounding electrode conductor _____

    b. 6 AWG grounding electrode conductor _____

    c. 4 AWG grounding electrode conductor _____

23. Why is bonding of service-entrance equipment necessary? _____
_____
_____

24. What special types of bushings are required on service entrances? _____
_____

25. When conductors 4 AWG or larger are installed in conduit, what additional provision is required on the conduit ends? _____
_____

26. What minimum size copper bonding jumpers must be installed to bond properly the electrical service for the residence discussed in this text? _____
_____

27. a. Where is Panel A located? _____
    b. On what type of wall is Panel A fastened? _____
    c. Where is Panel B located? _____
    d. On what type of wall is Panel B fastened? _____

28. When conduits pass through the wall from outside to inside, the conduit must be _____ to prevent air circulation through the conduit.

29. Briefly explain why electrical systems and equipment are grounded. _____
_____
_____
_____
_____

30. In general, systems are grounded so that the maximum voltage to ground does not exceed (Circle the correct response.)
    a. 120 volts.   b. 150 volts.   c. 300 volts.

31. To ensure a complete grounding electrode system (Circle the correct response.)
    a. everything must be bonded together.
    b. all metal pipes and conduits must be isolated from one another.
    c. the service neutral is grounded to the water pipe only.

32. An electric clothes dryer is rated at 5700 watts. The electric rate is 10.091 cents per kilowatt-hour. The dryer is used continuously for 3 hours. Find the cost of operation, assuming the heating element is on continuously.

33. A heating cable rated at 750 watts is used continuously for 72 hours to prevent snow from freezing in the gutters of the house. The electric rate is 8.907 cents per kilowatt-hour. Find the cost of operation.

34. When used as service equipment, a panelboard (load center) must be _____ that is suitable for use as service equipment, *230.66*. If the panelboard (load center) contains the main service disconnect, it must be clearly marked _____ _____ , *230.70(B)*.

35. Here are five commonly used terms in the electrical industry. Enter the letter of the term that corresponds to its definition.
    a. Grounding electrode conductor
    b. Main bonding jumper
    c. Grounded circuit conductor
    d. Equipment grounding conductor
    e. Lateral service conductors

    _____ The neutral conductor.

    _____ The term used to define underground service-entrance conductors that run between the meter and the utilities connection.

    _____ The conductor (sometimes a large threaded screw) that connects the neutral bar in the service equipment to the service-entrance enclosure.

    _____ The conductor that runs between the neutral bar in the main service equipment to the grounding electrode (water pipe, ground rod, etc.).

    _____ The bare copper conductor found in nonmetallic-sheathed cable.

36. The electrician mounted the disconnect switch for a central air-conditioning unit 8 ft (2.5 m) above the ground. This was easy to do because all he had to do was run the conduit across the ceiling joists in the basement, through the outside wall, and directly into the back of the disconnect switch. The electrical inspector turned the job down, citing *NEC®* _____ that requires that the disconnecting operating handle must not be higher than _____ feet above the ground.

37. a. According to the *NEC®* definition of a wet location, a basement cement wall that is in direct contact with the earth is considered to be a wet location. A panelboard mounted on this wall must have at least [¼ in. (6 mm)], [½ in. (13 mm)] [1 in. (25 mm)] space between the wall and the panel. Circle the correct answer.

    b. How is this required spacing accomplished? _____
    _____

38. When using re-bars as the concrete encased electrode, the re-bars must be (a) insulated with plastic material so they will not rust, or (b) bonded together by the usual steel tie wires or other effective means. Underline the correct answer.

39. The circuit directory in a panelboard (may) (shall) be filled out according to *NEC®* _____ . Circle the correct answer, and enter the correct *NEC®* section number.

40. A main service panel is located in a dark corner of a basement, far from the basement light. In your opinion, does this installation meet the requirements of *110.26(D)*? Explain your answer. _____

_____

41. Does the electric utility in your area allow the location of the meter to be on the back side of a residence? _____

42. The term used when the utility charges different rates during different periods during the day is _____.

43. A ground rod is driven below the meter outside of the house. An armored grounding electrode conductor connects between the meter base and the ground rod. The ground clamp is buried under the surface of the soil. This is permitted if the ground clamp is _____ for direct burial according to *NEC®* _____.

44. Copper-coated steel ground rods are the most commonly used grounding electrodes. These rods must:

    a. be at least _____ in. (_____ mm) in diameter, *250.52(A)(5)*.

    b. be driven to a depth of at least _____ ft (_____ m) unless solid rock is encountered, in which case the rod may be driven at a _____-degree angle, or it may be laid in a trench that is at least _____ ft (_____ mm) deep, *250.53(G)*.

    c. be separated by at least _____ ft (_____ m) when more than one rod is driven, *250.56*.

    d. have a ground resistance of not over _____ ohms for one rod, *250.56*.

45. What section of the Code prohibits the use of sheet metal screws as a means of attaching grounding conductors to enclosures? *NEC®* _____

46. Which section of the *NEC®* prohibits using the space below and in front of an electrical panel as storage space? *NEC®* _____.

Overcurrent Protection— Fuses and Circuit Breakers

OBJECTIVES

After studying this unit, the student should able to

- understand the basics of fuses and circuit breakers.
- study normal, overload, short circuit, and ground-fault conditions.
- understand various types of fuses.
- understand circuit breakers.
- know when to use single-pole breakers and two-pole breakers.
- understand the term *interrupting rating*.
- determine available short-circuit current using a simple formula.
- understand *series-rated* panelboards.
- understand the meaning of *selective coordination* and *nonselective coordination*.

THE BASICS

The *NEC®* covers overcurrent protection in *Article 240*. Overcurrent protection for residential main services, branch-circuits, and feeders is provided by circuit breakers or fuses. These are the "safety valves" of an electrical circuit.

Fuses and circuit breakers are sized by matching their ampere ratings to conductor ampacities and connected load currents. They sense overloads, short-circuits, and ground faults and protect the wiring and equipment from reaching dangerous temperatures.

A fuse will function only one time. It does its job of protecting the circuit. When a fuse opens, find out what caused it to open, fix the problem, and replace it with the size and type that will provide proper overcurrent protection. Do not keep replacing a fuse without finding out why the fuse blew in the first place.

When a circuit breaker trips, find out what caused it to trip, fix the problem, then reset it. Do not repeatedly reset the breaker again and again into a fault. This is looking for trouble.

KEY *NEC®* REQUIREMENTS FOR OVERCURRENT PROTECTION

- *NEC® Table 210.24:* This table shows the maximum overcurrent protection for branch-circuit conductors.
- *NEC® 240.4:* Overcurrent protection is sized according to the ampacity of a conductor.
- *NEC® 240.4(B):* Where the standard ampere rating of fuses or circuit breaker does not match the ampacity of the conductor, the next higher ampere rating is permitted.

- *NEC® 240.4(D):* The maximum overcurrent protection of small conductors is:

CONDUCTOR SIZE	MAXIMUM OVERCURRENT PROTECTION
14 AWG copper	15 amperes
12 AWG copper	20 amperes
10 AWG copper	30 amperes

- *NEC® 240.6:* Lists the standard ampere ratings for fuses and circuit breakers.

- *NEC® 240.20(A):* Overcurrent protection must be provided for each ungrounded ("hot") conductor.

- *NEC® 240.21:* Overcurrent protection shall be at the point where a conductor receives its supply. Refer to this section for exceptions. Note that the overcurrent protection in the main service panelboard is located at the load end of the service-entrance conductors. This is permitted because the *NEC®* accepts the fact that short-circuit protection of service-entrance conductors is provided by the utility's primary transformer fuse.

- *NEC® 240.22:* Overcurrent protection devices not permitted in the grounded conductor.

- *NEC® 240.24(A):* Overcurrent protection devices shall be accessible.

- *NEC® 240.24(D):* Overcurrent protective devices must not be near easily ignitable material, such as in clothes closets.

- *NEC® 240.24(E):* Overcurrent devices must not be located in bathrooms.

- *NEC® 230.70(A)(2):* Service disconnect shall not be located in bathrooms.

- *NEC® 230.79(C):* 100-ampere minimum size service for a one-family dwelling.

- *NEC® 230.90(A):* Overcurrent protection must be provided for each ungrounded ("hot") service-entrance conductor.

- *NEC® 230.91:* The main service overcurrent device is usually an integral part of the service disconnecting means. For large services, the service disconnecting means could be a separate disconnect switch.

- *NEC® Table 310.15(B)(6):* This table shows special ampacities permitted for service-entrance conductors for three-wire, single-phase, 120/240-volt services for dwellings. Overcurrent protection for these conductors is not to exceed the special ampacities.

FOUR CIRCUIT CONDITIONS

Normal loading of a circuit is when the current flowing is within the capability of the circuit and/or the connected equipment. In Figure 28-1 we have a 15-ampere circuit carrying 10 amperes.

An **overload** is a condition where the current flowing is more than the circuit and/or connected equipment is designed to safely carry. Figure 28-2 shows a 15-ampere circuit carrying 20 amperes. A momentary overload is harmless, but a continuous overload condition will cause the conductors and/or equipment to overheat—a potential cause for fire. The current flows through the "intended path"—the conductors and/or equipment.

A **short circuit** is a condition when two or more conductors come in contact with one another, resulting in a current flow that bypasses the connected load. A short circuit might be two "hot" conductors coming together, or it might be a "hot" conductor and a "grounded" conductor coming together. In either case, the current flows outside of the "intended path." The only resistance (impedance) is that of the conductors, the source, and the arc. This low resistance results in high levels of short circuit current. The heat generated at the point of the arc can result in a fire, as in Figure 28-3.

A **ground fault** is a condition when a "hot" conductor comes in contact with a grounded surface, such as a grounded metal raceway, metal water pipe, sheet metal, etc., as in Figure 28-4. The current flows outside of the "intended path." A ground-fault can result in a flow of current greater than normal, in which case the overcurrent device will open. A ground-fault can also result in a current flow less than normal, in which case the overcurrent device may not open.

A GFCI is an example of a device that protects against very low levels of ground-fault passing through the human body when a person touches a "live" wire and a grounded surface. See Unit 6.

Ⓐ NORMAL CIRCUIT

THE CONDUCTORS CAN SAFELY CARRY THE CURRENT. THEY DO NOT GET HOT. THE 15-AMPERE FUSES DO NOT OPEN.

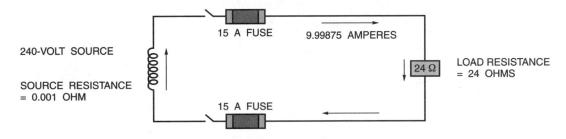

CONDUCTORS RATED 15 AMPERES
CONDUCTOR RESISTANCE = 0.001 OHM

15 A FUSE   9.99875 AMPERES

240-VOLT SOURCE

SOURCE RESISTANCE = 0.001 OHM

15 A FUSE

24 Ω   LOAD RESISTANCE = 24 OHMS

CONDUCTORS RATED 15 AMPERES
CONDUCTOR RESISTANCE = 0.001 OHM

$$\text{I (THROUGH CIRCUIT)} = \frac{E}{R} = \frac{240}{24 + 0.001 + 0.001 + 0.001} = \frac{240}{24.003} = 9.99875 \text{ AMPERES}$$

Figure 28-1   A normally loaded circuit.

Ⓑ OVERLOADED CIRCUIT

THE CONDUCTORS BEGIN TO GET HOT, BUT THE 15-AMPERE FUSES WILL OPEN BEFORE THE CONDUCTORS ARE DAMAGED.

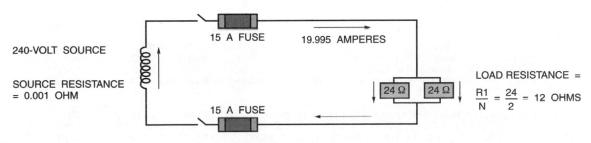

CONDUCTORS RATED 15 AMPERES
CONDUCTOR RESISTANCE = 0.001 OHM

15 A FUSE   19.995 AMPERES

240-VOLT SOURCE

SOURCE RESISTANCE = 0.001 OHM

15 A FUSE

24 Ω   24 Ω   LOAD RESISTANCE = $\frac{R1}{N} = \frac{24}{2} = 12$ OHMS

CONDUCTORS RATED 15 AMPERES
CONDUCTOR RESISTANCE = 0.001 OHM

$$\text{I (THROUGH CIRCUIT)} = \frac{E}{R} = \frac{240}{12 + 0.001 + 0.001 + 0.001} = \frac{240}{12.003} = 19.995 \text{ AMPERES}$$

Figure 28-2   An overloaded circuit.

Ⓒ SHORT CIRCUIT

THE CONDUCTORS GET EXTREMELY HOT. THE INSULATION WILL MELT OFF AND THE CONDUCTORS WILL MELT UNLESS THE FUSES OPEN IN A VERY SHORT (FAST) PERIOD OF TIME. CURRENT-LIMITING OVERCURRENT DEVICES WILL LIMIT THE AMOUNT OF "LET-THROUGH" CURRENT, AND WILL LIMIT THE TIME THE FAULT CURRENT IS PERMITTED TO FLOW.

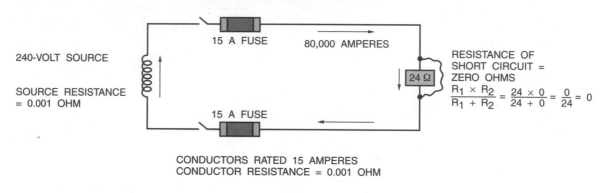

CONDUCTORS RATED 15 AMPERES
CONDUCTOR RESISTANCE = 0.001 OHM

15 A FUSE   80,000 AMPERES

240-VOLT SOURCE

SOURCE RESISTANCE = 0.001 OHM

24 Ω

RESISTANCE OF SHORT CIRCUIT = ZERO OHMS

$$\frac{R_1 \times R_2}{R_1 + R_2} = \frac{24 \times 0}{24 + 0} = \frac{0}{24} = 0$$

15 A FUSE

CONDUCTORS RATED 15 AMPERES
CONDUCTOR RESISTANCE = 0.001 OHM

$$I \text{ (THROUGH CIRCUIT)} = \frac{E}{R} = \frac{240}{0.001 + 0.001 + 0.001} = \frac{240}{0.003} = 80,000 \text{ AMPERES}$$

Figure 28-3 Note that the connected load is short-circuited.

Ⓓ GROUND-FAULT

"HOT" CONDUCTOR

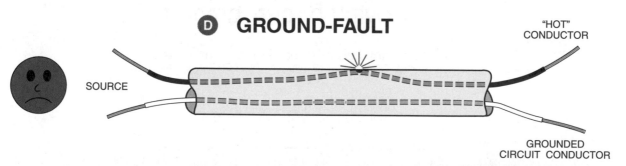

SOURCE

GROUNDED CIRCUIT CONDUCTOR

"HOT" CONDUCTOR COMES IN CONTACT WITH METAL RACEWAY OR OTHER METAL OBJECT. IF THE RETURN GROUND PATH HAS LOW RESISTANCE (IMPEDANCE), THE OVERCURRENT DEVICE PROTECTING THE CIRCUIT WILL CLEAR THE FAULT. IF THE RETURN GROUND PATH HAS HIGH RESISTANCE (IMPEDANCE), THE OVERCURRENT DEVICE WILL NOT CLEAR THE FAULT. THE METAL OBJECT WILL THEN HAVE A VOLTAGE TO GROUND THE SAME AS THE "HOT" CONDUCTOR HAS TO GROUND. IN HOUSE WIRING, THIS VOLTAGE TO GROUND IS 120 VOLTS. PROPER GROUNDING AND GROUND-FAULT CIRCUIT INTERRUPTER PROTECTION IS DISCUSSED ELSEWHERE IN THIS TEXT. THE CALCULATION PROCEDURE FOR A GROUND FAULT IS THE SAME AS FOR A SHORT CIRCUIT, HOWEVER, THE VALUES OF "R" CAN VARY GREATLY BECAUSE OF THE UNKNOWN IMPEDANCE OF THE GROUND RETURN PATH. LOOSE LOCKNUTS, BUSHINGS, SET SCREWS ON CONNECTORS AND COUPLINGS, POOR TERMINATIONS, RUST, ETC., ALL CONTRIBUTE TO THE RESISTANCE OF THE RETURN GROUND PATH, MAKING IT EXTREMELY DIFFICULT TO DETERMINE THE ACTUAL GROUND-FAULT CURRENT VALUES.

Figure 28-4 The insulation on the "hot" conductor has come in contact with the metal conduit. This is termed a "ground-fault."

FUSES

Fuses are available in two basic types: plug and cartridge. The following is a discussion on those types of fuses used in residential application.

Plug Fuses, Fuseholders, and Adapters *(Article 240, Part V)*

Fuses are a reliable and economical form of overcurrent protection. *NEC® 240.50* through *240.54* give the requirements for plug fuses, fuseholders, and adapters. These requirements include the following:

- Plug fuses may be used (1) in circuits that do not exceed 125 volts between conductors, and (2) in circuits having a grounded neutral where no conductor operates at over 150 volts to ground. In a residential 120/240-volt electrical system, the voltage between the two "hot" conductors is 240 volts, and the voltage from either "hot" conductor to ground is 120 volts.

- The fuses shall have ampere ratings of 0 to 30 amperes.

- Plug fuses shall have a hexagonal configuration somewhere on the fuse when rated at 15 amperes or less.

- The screw shell of the fuseholder must be connected to the load side of the circuit.

- Edison-base plug fuses may be used only to replace fuses in existing installations where there is no sign of overfusing or tampering.

- All new installations shall be in Type S fuses.

- Type S fuses are classified at 0 to 15 amperes, 16 to 20 amperes, and 21 to 30 amperes. The reason for this classification is given in the following paragraph.

When the electrician installs fusible equipment, the ampere rating of the various circuits must be determined. Based on this rating, an adapter of the proper size is inserted into the Edison-base fuseholder. The proper Type S fuse is then screwed into the adapter. Because of the adapter, the fuseholder is nontamperable and noninterchangeable. For example, assume that a 15-ampere adapter is inserted for the 15-ampere branch-circuits. It is impossible to substitute a Type S fuse with a larger rating without removing the 15-ampere adapter.

Type S fuses and adapters, shown in Figure 28-5, are available with ratings from $3/10$ of an ampere to 30 amperes. A Type S fuse may have a time-delay feature. If it does, it is known as a dual-element fuse. Momentary overloads do not cause a dual-element fuse to blow. An example of such an overload is the current surge as an electric motor is started. However, a dual-element fuse opens rapidly if there is a heavy overload or a short-circuit. Figure 28-6 shows a 15-ampere Edison-base plug fuse and a 20-ampere time-delay Edison-base plug fuse.

Table 28-1 is a chart showing the various Type S fuse and adapter ampere ratings, and the range of Type S fuse ampere ratings that fit into a given ampere-rated adapter.

Figure 28-5   Type S fuse and adapter. *Courtesy of Cooper Bussmann, Inc.*

Figure 28-6   A 15-ampere Type W Edison-base plug fuse and a 20-ampere Type T time-delay Edison-base plug fuse. *Courtesy of* **Cooper Bussmann, Inc.**

TYPE S FUSE INFORMATION

Type S Fuse Ampere Ratings	Type S Adapter Rating	Type S Fuse Ampere Ratings That Fit Into This Adapter
³⁄₁₀, ½, ⁸⁄₁₀, ⁴⁄₁₀, ⁶⁄₁₀, 1	1	1 ampere & smaller
1⅛, 1¼	1¼	All smaller
1⁴⁄₁₀, 1⁶⁄₁₀	1⁶⁄₁₀	All smaller
1⁸⁄₁₀, 2	2	1⁸⁄₁₀, 2
2¼, 3½	2½	1⁸⁄₁₀, 2, 2¼, 2½
2⁸⁄₁₀, 3²⁄₁₀	3²⁄₁₀	1⁸⁄₁₀, 2, 2¼, 2½, 2⁸⁄₁₀, 3²⁄₁₀
3½, 4	4	3½, 4
4½, 5	5	3½, 4, 4½, 5
5⁶⁄₁₀, 6¼	6¼	3½, 4, 4½, 5, 5⁶⁄₁₀, 6¼
7, 8	8	7, 8
9, 10	10	7, 8, 9, 10
12, 14	14	7, 8, 9, 10, 12, 14
15	15	15
20	20	20
25	30	20, 25, 30
30	30	20, 25, 30

Table 28-1 This table shows the many ampere rating combinations for Type S fuses and adapters.

Dual-Element Fuse (Cartridge and Plug Type)

The most common type of dual-element cartridge fuse is shown in Figure 28-7. These fuses are available in 250-volt and 600-volt sizes with ratings from 0 to 600 amperes.

A dual-element fuse has two fusible elements connected in series. These elements are known as the *overload element* and the *short-circuit element*. When an excessive current flows, one of the elements opens. The amount of excess current determines which element opens.

The electrical characteristics of these two elements are very different. Thus, a dual-element fuse has a greater range of protection than the single-element fuse.

Overload Element. Under overload conditions, dual-element fuses open in the spring-loaded trigger section. Each trigger section has a fusing alloy which melts at 286°F (141°C), that is connected in a series with a short-circuit element. The mass of the trigger assembly absorbs a great deal of heat and provides a time delay of about 10 seconds at a current of 500 percent of the fuse rating before it opens. This provides accurate protection for prolonged overloads. Harmless, short-term overloads do not melt the fusing alloy. A dual-element fuse may have more than one fusible element, depending on its ampere rating. Dual-element fuses are also referred to as time-delay fuses.

Short-Circuit Element. The capacity of the short-circuit element is high enough to prevent it from opening on low overloads. This element is designed to clear the circuit quickly of short circuits or heavy overloads above 500 percent of the fuse rating.

Dual-element fuses are used on motor and appliance circuits where the time-lag characteristic of the fuse is required. Single-element fuses do not have this time lag.

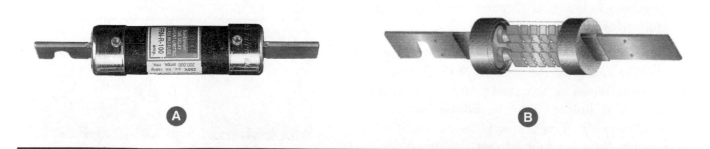

Figure 28-7 Cartridge-type dual-element fuse (A) is a 250-volt, 100-ampere fuse. The cutaway view in (B) shows the internal parts of the fuse. When this type of fuse has a rejection slot in the blades of 70–600 ampere sizes, or a rejection ring in the end ferrules of the 0–60 ampere sizes, it is a UL Class RK1 or RK5 fuse. *Courtesy of Cooper Bussmann, Inc.*

Sometimes poor connections at the terminals or loose fuse clips can be found because the fuse tubing will appear to be charred. This problem can be dangerous and should be corrected as soon as possible. Quite often, the heat resulting from poor connections in the fuseclips or fuseholder will cause the brass or copper fuseclips or fuseholders to turn from a bright, shiny appearance to that of antique brass or copper.

Class G Fuses

Class G fuses rated ½ through 20 amperes are rated 600 volts AC; those with ratings of 25 through 60 amperes are rated 480 volts AC. They are available in many ampere ratings from ½ through 60 amperes. The physical size of these fuses is smaller than that of standard cartridge fuses, as shown in Figure 28-8.

At 200 percent load, they have a minimum opening time of 25 seconds. This time-delay characteristic means that they can handle harmless current surges of momentary overloads without nuisance blowing. However, these fuses open very quickly under short-circuit or ground-fault conditions.

Class G fuses prevent overfusing because they are size limiting in four different case sizes:

½ through 15 amperes

20 amperes

25 and 30 amperes

35 through 60 amperes

For example, a fuseholder designed to accept a 15-ampere, Class G fuse will reject a 20-ampere,

Figure 28-8 Class G fuses. *Courtesy of* **Cooper Bussmann, Inc.**

Class G fuse. In the same manner, a fuseholder designed to accept a 20-ampere, Class G fuse will reject a 30-ampere, Class G fuse.

Class T Fuses

Another type of physically small fuse is the Class T fuse, shown in Figure 28-9. This type of fuse has a high-interrupting rating and is current limiting.

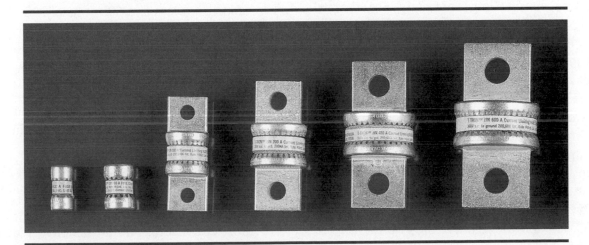

Figure 28-9 Class T fuses. Illustrated are the 30-, 60-, 100-, 200-, 400-, and 600-ampere, 300-volt ratings. *Courtesy of* **Cooper Bussmann, Inc.**

It cannot be interchanged with standard cartridge fuses because of its unique physical size. Available in ampere ratings through 1200 amperes in 300-volt ratings, Class T fuses open extremely fast under short-circuit or ground-fault conditions. In "series-rated" load centers, Class T fuses are used as main fuses in panels that have circuit breakers for the branch-circuits. These fuses protect the low-interrupting rating branch-circuit breakers against high-level short circuits or ground-faults, as described in the section on "Interrupting Ratings for Fuses and Circuit Breakers."

Most manufacturers list service equipment, metering equipment, and disconnect switches that make use of Class T fuses. This allows the safe installation of this equipment where fault-currents exceed 10,000 amperes, such as on large services, and any service equipment located close to pad-mount transformers, such as are found in apartments, condominiums, shopping centers, and similar locations.

Fuse Ampere Ratings

The standard ampere ratings for fuses are given in *240.6(A)*. In addition to the standard ratings listed, there are many other "in-between" ampere ratings available to match specific load requirements. Check a fuse manufacturer's catalog for other available ampere ratings.

CIRCUIT BREAKERS

Installations in dwellings normally use thermal-magnetic circuit breakers. On a continuous overload, a bimetallic element in such a breaker moves until it unlatches the inner tripping mechanism of the breaker. Momentary small overloads do not cause the element to trip the breaker. If the overload is heavy or if there is a short circuit, a magnetic coil in the breaker causes it to interrupt the branch-circuit almost instantly. Unit 13 covers the effect of tungsten lamp loads on circuit breakers. Figure 28-10 illustrates a typical single-pole and a two-pole circuit breaker.

NEC® 240.80 through *240.86* give the requirements for circuit breakers. The following points are taken from these sections.

Figure 28-10 Typical molded-case circuit breakers.

- Circuit breakers shall be trip-free so that even if the handle is held in the "ON" position, the internal mechanism will trip to the "OFF" position.

- Breakers shall indicate clearly whether they are on or off.

- A breaker shall be nontamperable so that it cannot be readjusted (trip point changed) without dismantling the breaker or breaking the seal.

- The rating shall be durably marked on the breaker. For small breakers rated at 100 amperes or less and 600 volts or less, the rating must be molded, stamped, or etched on the handle or on another part of the breaker that will be visible after the cover of the panel is installed.

- Every breaker with an interrupting rating other than 5000 amperes shall have this rating marked on the breaker.

- Circuit breakers rated at 120 volts and 277 volts and used for fluorescent loads, shall not be used as switches unless marked "SWD."

Most circuit breakers are ambient temperature compensated. This means that the tripping point of the breaker is not affected by an increase in the surrounding temperature. An ambient-compensated breaker has two elements. One element heats up due to the current passing through it and the heat in the surrounding area. The other element heats up because of the surrounding air only. The actions of these elements oppose each other. Thus, as the tripping element tends to lower its tripping point because of external heat, the second element opposes the tripping element and stabilizes the tripping point. As a result, the current through the tripping element is the only factor that causes the element to open the circuit. It is a good practice to turn the breaker on and off periodically to "exercise" its moving parts.

NEC® 240.6(A) gives the standard ampere ratings of circuit breakers.

One Two-Pole or Two Single-Pole Circuit Breakers

NEC® 240.20(B) states that a circuit breaker must open all ungrounded ("hot") conductors, but there are a few exceptions.

Generally, use two-pole circuit breakers on three-wire 120/240-volt branch-circuits (two "hots" and one "neutral") for loads such as electric ranges, electric ovens, and electric clothes dryers.

Generally, use two-pole circuit breakers for straight 240-volt branch-circuits (two "hots") for such loads as electric water heaters, electric furnaces, electric baseboard heaters, air conditioners, and heat pumps.

The use of "handle ties," which connect the handles of two single-pole circuit breakers together so both poles trip together, is permitted for a few very specific applications. Handle ties are not recommended, and are generally used as a "last resort." If you have to use handle ties, be sure to use listed ones!

Handle ties *do* allow two single-pole circuit breakers to be switched simultaneously to the "OFF" or "ON" positions.

Handle ties generally *do not* simultaneously trip both handle tied single-pole circuit breakers on overloads, short circuits, or ground-faults on one of the circuit breakers. This could result in a false sense of security for someone working on an appliance where two single-pole circuit breakers with a handle tie has tripped. He thinks that the branch-circuit is totally in the "OFF" position, when in fact it is

quite possible that one of the circuit breakers may still be "ON."

One permission to use handle ties is found in *240.20(B)(1)*, which states that single-pole circuit breakers with handle ties are permitted on multiwire branch-circuits that serve only single-phase, line-to-neutral loads. However, *210.4(B)*, states that *In dwelling units, a multiwire branch-circuit supplying more than one device or equipment on the same yoke shall be provided with a means to disconnect simultaneously all ungrounded conductors at the panelboard where the branch-circuit originated.*

Another permission to use handle ties is given in *240.20(B)(2)* where we find permission to use handle ties between individual single-pole circuit breakers for line-to-line single-phase loads.

To repeat, handle ties should only be used as "the last resort."

INTERRUPTING RATINGS FOR FUSES AND CIRCUIT BREAKERS

Here are some very important *NEC®* code sections regarding interrupting ratings.

- *NEC® 110.9:* **Interrupting Rating.** *"Equipment intended to interrupt current at fault levels shall have an interrupting rating sufficient for the nominal circuit voltage and the current that is available at the line terminals of the equipment. Equipment intended to interrupt current at other than fault levels shall have an interrupting rating at nominal circuit voltage sufficient for the current that must be interrupted."*

- *NEC® 110.10:* **Circuit Impedance and Other Characteristics.** *"The overcurrent protective devices, the total impedance, the component short-circuit current ratings, and other characteristics of the circuit to be protected shall be selected and coordinated to permit the circuit protective devices used to clear a fault to do so without extensive damage to the electrical components of the circuit. This fault shall be assumed to be either between two or more of the circuit conductors, or between any circuit conductor and the grounding conductor or enclosing metal raceway. Listed products applied in accordance with their listing shall be considered to meet the requirements of this section."*

• UL Standard No. 67 requires that panelboards be marked with their short-circuit current rating.

Panelboards are available in two types: "fully-rated" and "series-rated." Figure 28-11, Figure 28-12, Figure 28-13, and Figure 28-14 explain the meaning of these terms.

The overcurrent protective device must be able to interrupt the current that may flow under any condition (overload or short circuit). Such interruption must be made with complete safety to personnel and without damage to the panel or switch in which the overcurrent device is installed.

CAUTION: Overcurrent devices with inadequate interrupting ratings are, in effect, bombs waiting for a short-circuit to trigger them into an explosion. Personal injury may result and serious damage will be done to the electrical equipment.

Interrupting Ratings

Interrupting ratings of overcurrent devices, as shown in Table 28-2, are given in root-mean-square (RMS) values.

The interrupting rating of a fuse or circuit breaker does not indicate the short-circuit rating of a panelboard. These are stand-alone ratings. The short-circuit rating of a panelboard is derived by testing the panel as an assembly with breakers and/or fuses installed in the panel. The short-circuit rating of a panelboard is marked on its label.

Short-Circuit Currents

This text does not cover in detail the methods of calculating short-circuit currents. For a detailed discussion, see *Electrical Wiring—Commercial* (©Delmar, a division of Thomson Learning). The ratings required to determine the maximum available short-circuit current delivered by a transformer are the kVA and impedance values of the transformer. The size and length of wire installed between the transformer and the overcurrent device must be considered as well.

The transformers used in modern electrical installations are efficient and have very low impedance values. A low-impedance transformer having a given kVA rating delivers more short-circuit current than a transformer with the same kVA rating and a higher impedance. When an electrical service is connected to a low-impedance transformer, the problem of available short-circuit current is very serious.

The examples given in Figure 28-11, Figure 28-12, Figure 28-13, and Figure 28-14 show that the available short-circuit current at the transformer secondary terminals is 20,800 amperes.

HOW TO CALCULATE SHORT-CIRCUIT CURRENT

Short-circuit current is also referred to as fault current.

The local electric utility and the electrical inspector are good sources of information when short-circuit current is to be determined.

A simplified method is given as follows to determine the approximate available short-circuit current at the terminals of a transformer. This method assumes that there is an infinite amount of primary

TYPE OF OVERCURRENT DEVICE	INTERRUPTING RATING
Circuit breakers:	are marked with their interrupting rating if other than 5000 amperes.
Plug fuses (Edison base & Type S)	10,000 amperes
Class G fuses	100,000 amperes
Class R fuses	200,000 amperes
Class T fuses	200,000 amperes
Class L fuses	200,000 amperes. Some have a 300,000-ampere interrupting rating.

Table 28–2 The interrupting ratings of overcurrent devices.

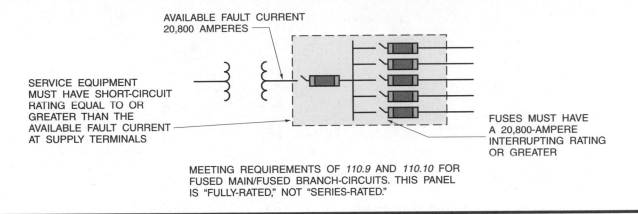

AVAILABLE FAULT CURRENT
20,800 AMPERES

SERVICE EQUIPMENT
MUST HAVE SHORT-CIRCUIT
RATING EQUAL TO OR
GREATER THAN THE
AVAILABLE FAULT CURRENT
AT SUPPLY TERMINALS

FUSES MUST HAVE
A 20,800-AMPERE
INTERRUPTING RATING
OR GREATER

MEETING REQUIREMENTS OF *110.9* AND *110.10* FOR
FUSED MAIN/FUSED BRANCH-CIRCUITS. THIS PANEL
IS "FULLY-RATED," NOT "SERIES-RATED."

Figure 28-11   Fused main/fused branch-circuits, fully-rated.

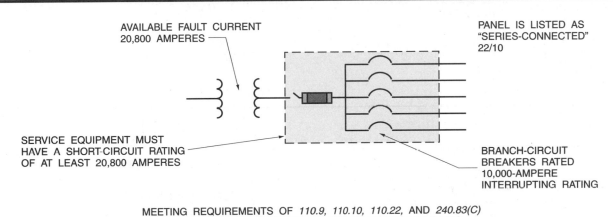

AVAILABLE FAULT CURRENT
20,800 AMPERES

PANEL IS LISTED AS
"SERIES-CONNECTED"
22/10

SERVICE EQUIPMENT MUST
HAVE A SHORT-CIRCUIT RATING
OF AT LEAST 20,800 AMPERES

BRANCH-CIRCUIT
BREAKERS RATED
10,000-AMPERE
INTERRUPTING RATING

MEETING REQUIREMENTS OF *110.9, 110.10, 110.22,* AND *240.83(C)*
FOR SERIES RATED FUSED MAIN/BREAKER BRANCH-CIRCUITS.
THIS PANEL IS "SERIES-RATED," NOT "FULLY-RATED."

Figure 28-12   Fused main/breaker branch-circuits, series-rated.

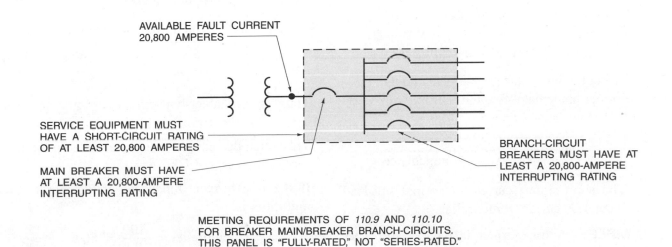

AVAILABLE FAULT CURRENT
20,800 AMPERES

SERVICE EQUIPMENT MUST
HAVE A SHORT-CIRCUIT RATING
OF AT LEAST 20,800 AMPERES

MAIN BREAKER MUST HAVE
AT LEAST A 20,800-AMPERE
INTERRUPTING RATING

BRANCH-CIRCUIT
BREAKERS MUST HAVE AT
LEAST A 20,800-AMPERE
INTERRUPTING RATING

MEETING REQUIREMENTS OF *110.9* AND *110.10*
FOR BREAKER MAIN/BREAKER BRANCH-CIRCUITS.
THIS PANEL IS "FULLY-RATED," NOT "SERIES-RATED."

Figure 28-13   Breaker main/breaker branch-circuits, fully-rated.

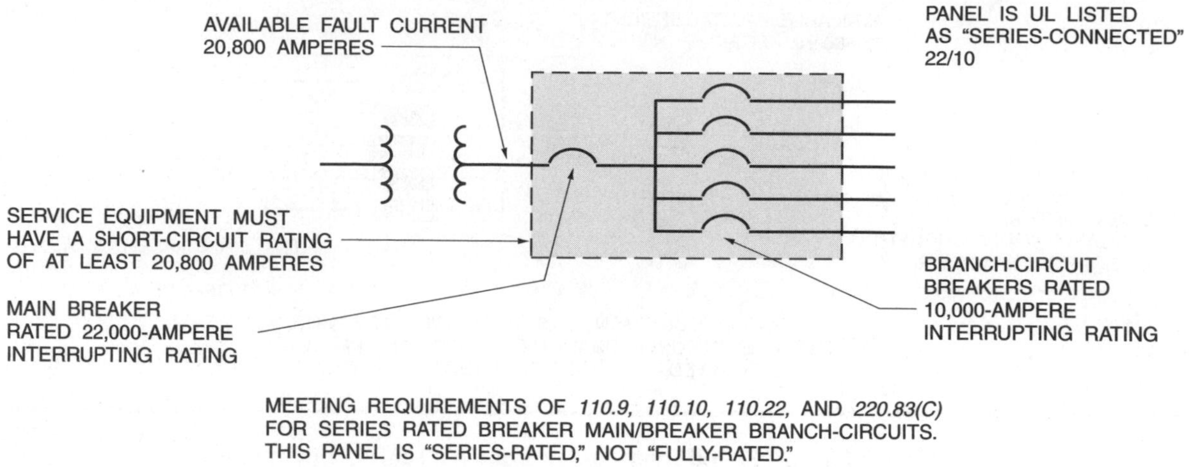

AVAILABLE FAULT CURRENT
20,800 AMPERES

PANEL IS UL LISTED
AS "SERIES-CONNECTED"
22/10

SERVICE EQUIPMENT MUST
HAVE A SHORT-CIRCUIT RATING
OF AT LEAST 20,800 AMPERES

MAIN BREAKER
RATED 22,000-AMPERE
INTERRUPTING RATING

BRANCH-CIRCUIT
BREAKERS RATED
10,000-AMPERE
INTERRUPTING RATING

MEETING REQUIREMENTS OF *110.9, 110.10, 110.22,* AND *220.83(C)*
FOR SERIES RATED BREAKER MAIN/BREAKER BRANCH-CIRCUITS.
THIS PANEL IS "SERIES-RATED," NOT "FULLY-RATED."

Figure 28-14   Breaker main/breaker branch-circuits, series-rated.

short-circuit current available (infinite bus), and a bolted fault at the secondary transformer terminals.

1. Determine the normal full-load secondary current delivered by the transformer.

   For single-phase transformers:

   $$I = \frac{kVA \times 1000}{E}$$

   For three-phase transformers:

   $$I = \frac{kVA \times 1000}{E \times 1.73}$$

   Where I = current, in amperes
       kVA = kilovolt-amperes (transformer nameplate rating)
       E = secondary line-to-line voltage (transformer nameplate rating)

2. Using the impedance value given on the transformer nameplate, find the multiplier to determine the short-circuit current.

   $$\text{multiplier} = \frac{100}{\text{percent impedance}}$$

3. The short-circuit current = normal full-load secondary current × multiplier.

EXAMPLE: A transformer is rated at 100 kVA and 120/240 volts. It is a single-phase transformer with an impedance of 1 percent (from the transformer nameplate). Find the short-circuit current.

For a single-phase transformer:

$$I = \frac{kVA \times 1000}{E}$$

$$I = \frac{100 \times 1000}{240}$$

= 416 amperes, full-load current

The multiplier for a transformer impedance of one percent is:

$$\text{multiplier} = \frac{100}{\text{percent impedance}}$$

$$= \frac{100}{1}$$

multiplier = 100

The short-circuit current = I × multiplier

= 416 amperes × 100

= 41,600 amperes

Thus, the available short-circuit current at the terminals of the transformer is 41,600 amperes.

This value decreases as the distance from the transformer increases.

If the transformer impedance is 1.5 percent, the multiplier is:

$$\text{multiplier} = \frac{100}{1.5} = 66.6$$

The short-circuit current = 416 × 66.6

= 27,706 amperes

If the transformer impedance is 2 percent:

$$\text{multiplier} = \frac{100}{2} = 50$$

$$\begin{aligned}\text{The short-circuit current} &= 416 \times 50 \\ &= 20{,}800 \text{ amperes}\end{aligned}$$

(*Note:* A short-circuit current of 20,800 amperes is used in Figure 28-11, Figure 28-12, Figure 28-13, and Figure 28-14.)

CAUTION: The line-to-neutral short-circuit current at the secondary of a single-phase transformer is approximately 1½ times greater than the line-to-line short-circuit current. For the previous example,

Line-to-line short-circuit current = 20,800 amperes

$$\begin{aligned}&\text{Line-to-neutral short-circuit current} \\ &= 20{,}800 \times 1.5 = 31{,}200 \text{ amperes}\end{aligned}$$

It is significant to note that UL Standard 1561 allows the marked impedance on the nameplate of a transformer to vary plus or minus (±) 10% from the actual impedance value of the transformer. Thus, for a transformer marked "2%Z," the actual impedance could be as low as 1.8%Z to as much as 2.2%Z. Therefore, the available short-circuit current at the secondary of a transformer as calculated previously should be increased by 10 percent if one wishes to be on the safe side when considering the interrupting rating of the fuses and breakers to be installed.

An excellent point-to-point method for calculating fault currents at various distances from the secondary of a transformer that uses different size conductors is covered in detail in *Electrical Wiring—Commercial* (©Delmar, a division of Thomson Learning). It is simple and accurate. The point-to-point method can be used to compute both single-phase and three-phase faults.

Calculating fault-currents is just as important as calculating load-currents. Overloads will cause conductors and equipment to run hot, will shorten their life, and will eventually destroy them. Fault-currents that exceed the interrupting rating of the overcurrent protective devices can cause violent electrical explosions of equipment with the potential of serious injury to people standing near it. Fire hazard is also present.

PANELBOARDS AND LOAD CENTERS

The terms panel, panelboard, and load center are all used, and for all practical purposes mean the same thing. The *NEC®* and the UL Standards use the term **panelboard**. Neither the *NEC®* nor the UL Standards contain the term **load center**. A load center is really a panelboard with less gutter (wiring) space and less depth and width, and generally does not have features such as integral relays, remote-controlled circuit breakers, load-shedding devices, and other similar optional features. However, some load centers are available for home automation applications that do have certain remote control features. Circuit breakers used in load centers are usually of the "plug-on" type as opposed to the "bolt-on" type found in commercial and industrial panelboards. Load centers generally are for use in a shallow wall (2 × 4 studs) such as found in residential and light commercial construction. They are available with a main circuit breaker or main fuses, or with main lugs only (MLO).

Panelboards are tested and listed under UL Standard 67.

Table 28-3 compares the requirements for different types of panelboards.

Selective Coordination/Nonselective Coordination

These are terms used by the real pros!

Selective Coordination: Under overload, short-circuit, or ground-fault conditions, only the overcurrent device nearest the fault opens. The main fuses or circuit breaker remain closed. Refer to *240.2*.

Nonselective Coordination: Under overload, short-circuit, or ground-fault conditions, the branch breaker or fuses might open, the main breaker or fuses might open, or both might open. For example, a short-circuit in a branch-circuit could also cause the main breaker to trip, resulting in a total power outage. Series-rated systems are susceptible to nonselectivity when available fault-currents are high.

The theory behind selective and nonselective systems is covered in detail in *Electrical Wiring—Commercial* (©Delmar, a division of Thomson Learning). Time-current curves for fuses and circuit breakers, circuit-breaker "unlatching time" data, and fuse melting/total clearing data are discussed.

	FUSED MAIN/FUSED BRANCHES (Fully-Rated) FIGURE 28-11	FUSED MAIN/BREAKER BRANCHES (Series-Rated) FIGURE 28-12	BREAKER MAIN/BREAKER BRANCHES (Fully-Rated) FIGURE 28-13	BREAKER MAIN/BREAKER BRANCHES (Series-Rated) FIGURE 28-14
Must be listed and marked as suitable for use as service equipment when used as service equipment. Does not need this marking when used as a subpanel.	YES	YES	YES	YES
Must have short-circuit current rating adequate for the available fault-current at the line side of the panelboard.	YES	YES	YES	YES
Must be marked with its short-circuit current rating.*	YES	YES	YES	YES
Must be marked "Series-Rated."	NO	YES Marking: Caution—Series Combination System Rated _____ Amperes.	NO	YES Marking: Caution—Series Combination System Rated _____ Amperes.
Main fuses or circuit breaker must have interrupting rating adequate for the available fault-current on the line side of the panelboard.	YES	YES	YES	YES
Branch fuses or breakers must have interrupting rating adequate for the available fault-current on the line side of the panelboard.	YES	NO. Main fuses might have 200,000-ampere interrupting rating. Branch breakers might have 10,000-ampere interrupting rating. Other combinations available. Check manufacturers data.	YES	NO. Main breaker might have 22,000-ampere interrupting rating. Branch breakers might have 10,000-ampere interrupting rating. Other combinations available. Check manufacturers data.
Main fuses are current-limiting such as Class T. See 240.2.	YES	YES	NOT APPLICABLE	NOT APPLICABLE
Must be sized as determined by service-entrance calculations and/or local codes.	YES	YES	YES	YES
Check manufacturer's data for time-current characteristic curves for fuses and breakers, and for "unlatching times" of breakers.	ALWAYS A GOOD IDEA.	ALWAYS A GOOD IDEA.	ALWAYS A GOOD IDEA.	ALWAYS A GOOD IDEA.

*If a panelboard has no marked short-circuit rating, then its short-circuit rating is based upon the lowest interrupting-rated device installed.

Table 28-3   Residential-type panelboards (load centers).

REVIEW

1. What is the minimum size service for a one-family dwelling? _____

2. a. What is a Type S fuse? _____
   _____

   b. Where must Type S fuses be installed? _____
   _____

3. a. What is the maximum voltage permitted between conductors when using plug fuses?
   _____

   b. May plug fuses (Type S) be installed in a switch that disconnects a 120/240-volt clothes dryer? _____

   c. Give a reason for the answer to (b). _____
   _____
   _____

4. Will a 20-ampere, Type S fuse fit properly into a 15-ampere adapter? _____

5. A time-delay, dual-element fuse will hold 500 percent of its rating for approximately (5 seconds) (10 seconds) (20 seconds). Circle the correct answer.

6. What part of a circuit breaker causes the breaker to trip

   a. on an overload? _____   b. on a short circuit? _____

7. What is meant by an ambient-compensated circuit breaker? _____
   _____
   _____
   _____

8. List the standard sizes of fuses and circuit breakers up to and including 100 amperes.
   _____
   _____

9. When hooking up a 240-volt electric baseboard heater, you should use a:
   a. two-pole circuit breaker.
   b. use two single-pole breakers and a handle tie.

10. Using the method shown in this unit, what is the approximate short-circuit current available at the terminals of a 50-kVA single-phase transformer rated 120/240 volts? The transformer impedance is 1 percent.
    a. line-to-line?
    b. line-to-neutral?

11. State four possible combinations of service equipment that meet the requirements of *110.9* and *110.10* of the Code.

    a. _____   c. _____

    b. _____   d. _____

12. Which Code section states that all overcurrent devices must have adequate interrupting ratings for the available fault-current to be interrupted? _____

    _____

13. All electrical components have some sort of "withstand rating." The *NEC*® and UL standards refer to this as the components short-circuit current rating. This rating indicates the ability of the component to withstand a specified amount of fault-current for the time it takes the overcurrent device upstream of the component to open the circuit. *NEC*® _____ refers to short-circuit current rating.

14. Arcing fault damage is closely related to the value of _____ .

15. The utility company has provided a letter to the contractor stating that the available fault-current at the line side of the main service-entrance equipment in a residence is 17,000 amperes RMS symmetrical, line-to-line. In the space following each statement, write in "Meets Code" or "Violation" of *110.9* and *110.10* of the Code.

    a. Main breaker has a 10,000-ampere interrupting rating; branch breakers have a 10,000-ampere interrupting rating. _____

    b. Main current-limiting fuse has a 200,000-ampere interrupting rating; branch breakers have a 10,000-ampere interrupting rating. The panel is "Series-Rated."

    _____

    c. Main breaker has a 22,000-ampere interrupting rating; branch breakers have a 10,000-ampere interrupting rating. The panel is marked "Series-Connected."

    _____

16. While working on the main panel with the panel energized, the electrician inadvertently causes a direct short circuit (line-to-ground) on one of the branch-circuit breakers. The available fault-current at the main service equipment is rather high. The panel is labeled "Series-Connected." The main breaker is rated 100 amperes. Both branch-circuit breaker and main breaker trip "OFF." This condition is referred to as being (selective) (nonselective). Circle the correct answer.

17. Repeat question 13 on a main service panel that consists of 100-ampere main current-limiting fuses and breakers for the branch-circuits. The branch circuit trips; the 100-ampere fuses do not open. This condition is referred to as being (selective) (non-selective). Circle the correct answer.

18. *NEC*® _____ prohibits the connecting of fuses and/or circuit breakers into the grounded circuit conductor.

19. In general, overcurrent protective devices are inserted where the conductor receives its

    _____ .

20. a. Does the *NEC*® permit panelboards to be installed in clothes closets? _____

    b. Does the *NEC*® permit panelboards to be installed in bathrooms? _____

21. Overcurrent devices must be accessible. (True) (False). Circle the correct response.

Service-Entrance Calculations

OBJECTIVES

After studying this unit, the student should be able to

- determine the total calculated load of the residence.

- calculate the size of the service entrance, including the size of the neutral conductors.

- understand the Code requirements for services, *Article 230*.

- understand how to read a watt-hour meter.

- fully understand special Code rules for single-family dwelling service-entrance conductor sizing.

- derate the service-entrance conductor if installation is located in an extremely hot climate.

- do an optional calculation for computing the required size of service-entrance conductors for residence.

The various load values determined in earlier units of this text are now used to illustrate the proper method of determining the size of the service-entrance conductors for the residence. The calculations are based on *NEC®* requirements. The student must check local and state electrical codes for any variations in requirements that may take precedence over the *NEC®*

SIZE OF SERVICE-ENTRANCE CONDUCTORS AND SERVICE DISCONNECTING MEANS

NEC® *230.42* tells us that service-entrance conductors must be sufficient to carry the load as computed according to *Article 220*, and shall have an ampacity not less than the rating of the service disconnect.

For a one-family residence, the minimum size main service disconnecting means is 100 amperes, three-wire, *230.79(C)*.

Two methods for calculating services on homes are permitted by the *NEC®* Method 1 is outlined in *Article 220*, *Parts I* and *II*. Method 2 is given in *Article 220*, *Part III*.

Method 1 *(Article 220, Parts I and II)*

All of the volt-ampere values in the following calculations are taken from previous units of this text.

Single-Family Dwelling Service-Entrance Calculations

1. General Lighting Load, *220.3(A)*

   3232 ft² @ 3 VA per ft² = 9696 VA

2. Minimum Number of Lighting Branch-Circuits

   9696 VA ÷ 120 volts = 80.8 amperes

   then, $\dfrac{amperes}{15} = \dfrac{80.8}{15}$ = 5.387 (round up to 6) 15-ampere branch-circuits

3. Small Appliance Load, *210.11(C)(1)* and *220.16(A)*
   (Minimum of two 20-ampere branch-circuits)

Kitchen	3
Clothes Washer	1*
Workshop	2**

   6 branch-circuits @ 1500 VA each = 9000 VA

4. Laundry Branch-Circuit, *210.11(C)(2)* and *220.16(B)*
   (Minimum of one 20-ampere branch-circuit)

   1 branch-circuit(s) @ 1500 VA = 1500 VA

5. Total General Lighting, Small Appliance, and Laundry Load

   Lines 1 + 3 + 4 = 20,196 VA

6. Net computed General Lighting, Small Appliance, and Laundry Loads (less ranges, ovens, and "fastened-in-place" appliances). Apply demand factors from *Table 220.11*.

   a. First 3000 VA @ 100% = 3000 VA

   b. Line 5 20,196 − 3000 = 17,196 @ 35% = 6019 VA

   Total a + b = 9019 VA

7. Electric Range, Wall-Mounted Ovens, Counter-Mounted Cooking Units *(Table 220.19)*

Wall-mounted oven	7450 VA
Counter-mounted cooking unit	6600 VA
Total	14,050 VA (14 kW)

   14 kW exceeds 12 kW by 2 kW

   2 kW × 5% = 10% increase, therefore: 8 kW + 0.8 kW = 8.8 kW = 8,800 VA

---

* This is in addition to the required minimum of one laundry branch-circuit. Thus, the laundry is served by a 20-ampere branch-circuit B18 for the automatic washer and another 20-ampere branch-circuit B20 supplying two receptacles for plugging in an electric iron, steamer, or clothes press.

** Although not true small appliance circuits as defined by the *NEC*,® it is better to add these circuits in as 1500-VA small-appliance circuits rather than including them in the general lighting load calculations. This results in a more adequate load calculation.

8. Electric Dryer, *220.18* = 5700 VA

9. Electric Furnace, *220.15*, Air Conditioner, Heat Pump, *Article 440*

   Air conditioner: 26.4 × 240 = 6336 VA
   Electric Furnace: 13,000 VA
   (Enter largest value, *220.21*) = 13,000 VA

10. Net Computed General Lighting, Small Appliance, Laundry, Ranges, Ovens, Cooktop Units, HVAC.

    Lines 6 + 7 + 8 + 9 = 36,519 VA

11. List "Fastened-in-Place" Appliances *in addition* to
    Electric Ranges, Air Conditioners, Clothes Dryers, Space Heaters.

Appliance	VA Load
Water Heater	3000 VA
Dishwasher*	1000 VA

    * Motor: 7.2 × 120 = 864 VA
      Heater: 1000 VA (watts)
      Enter largest value, *220.21*.

Garage Door Opener 5.8 × 120 =	696 VA
Food Waste Disposer 7.2 × 120 =	864 VA
Water Pump 8 × 240 =	1920 VA
Hydromassage Tub 10 × 120 =	1200 VA
Attic Exhaust Fan 5.8 × 120 =	696 VA
Heat/Vent/Lights 1500 × 2 =	3000 VA
Freezer 5.8 × 120 =	696 VA
Total	13,072 VA

12. Apply 75% Demand Factor, *220.17,* if Four or More "Fastened-in-Place" Appliances. If Less than Four, Figure at 100%. Do not include electric ranges, clothes dryers, space heating, or air conditioning equipment.

    Line 11 13,072 × 0.75 = 9804 VA

13. Total Computed Load (Lighting, Small Appliance, Ranges, Dryers, HVAC, "Fastened-in-Place" Appliances)

    Line 10 36,519 + Line 12 9804 = 46,323 VA

14. Add 25% of Largest Motor *(220.14)*
    This is the water pump motor:

    8 × 240 × 0.25 = 480 VA

15. Grand Total Line 13 + Line 14 = 46,803 VA

16. Minimum Ampacity for Ungrounded Service-Entrance Conductors

$$\text{Amperes} = \frac{\text{Line } 15}{240} = \frac{46{,}803}{240} = 195 \text{ amperes}$$

17. Minimum Ampacity for Neutral Service-Entrance Conductor, *220.22* and *310.15(B)(6)*. Do not include straight 240-volt loads.

a. Line 6			9019 VA
b. Line 7 8800 × 0.70		=	6160 VA
c. Line 8 5700 × 0.70		=	3990 VA

    d. Line 11 (include only 120-volt loads)

Freezer	696 VA
Food Waste Disposer	864 VA
Garage Door Opener	696 VA
Heat/Vent/Light	3000 VA
Attic Fan	696 VA
Hydromassage tub	1200 VA
Dishwasher heater	1000 VA
Total	8152 VA

    e. Line d total @ 75% Demand Factor: 8152 × 0.75 = 6114 VA

    f. Add 25% of largest 120-volt motor
       This is food waste disposer:
       7.2 × 120 × 0.25   = 216 VA

    g. Total a + b + c + e + f   = 25,499 VA

$$\text{Amperes} = \frac{\text{volt} - \text{amperes}}{\text{volts}} = \frac{25{,}499}{240} = 106.245 \text{ amperes}$$

18. Ungrounded Conductor Size (copper)   2/0 AWG THWN

    Note: This could be a 3/0 AWG THW, THHW, THWN, XHHW, or THHN per *Table 310.16*, or a AWG 2/0 (same types) using *Table 310.15(B)(6)*. *Table 310.15(B)(6)* may only be used for 120/240-volt, three-wire, residential single-phase service-entrance conductors, service lateral conductors, and feeder conductors that serve as the main power feeder to a dwelling unit. This table *cannot* be applied to the feeder to Panel B because that feeder carries only part of the load in the residence. We have assumed that the equipment is marked 75°C. We read the ampacity from the 75°C column in *Table 310.16* as stated in *110.14(C)*.

19. Neutral Conductor Size (copper), *220.22*   1 AWG

    Note: The calculations indicate that a 2 AWG copper neutral conductor would be adequate based upon the unbalanced load calculations. *NEC® 310.15(B)(6)* states that the neutral conductor is permitted to be smaller than the ungrounded "hot" conductors if the requirements of *215.2, 220.22,* and *230.42* are met. The specifications for this residence specify that a 1 AWG neutral conductor be installed. The neutral conductor must not be smaller than the grounding electrode conductor, *250.24(B)(1)*. Where bare conductors are used with insulated conductors, their ampacity is based on the ampacity of the other insulated conductors in the raceway, *310.15(B)(3)*.

20. Grounding Electrode Conductor Size (copper), *Table 250.66*   4 AWG

21. Conduit (See Table 27-1 for calculations) trade size 1¼.

Table 310.15(B)(6) provides special consideration for 120/240-volt, three-wire, single-phase service entrance, service lateral, and feeder conductors that serve as the main power supply to a dwelling unit. Table 29-1 in this text replicates *Table 310.15(B)(6)* from the *NEC*. The current ratings for a given size wire are higher than the allowable ampacities found in *Table 310.16* because of the tremendous diversity of electricity use in homes.

Table 310.15(B)(6) cannot be used to determine the feeder size to Panel B because this feeder does not carry the main power to the home.

Be careful. Locations exposed to the weather are considered to be wet locations, per the definition in the *NEC*. Therefore, most electrical inspectors will require that service-entrance conductors, from the line side terminals of the service disconnect to the service head, be suitable for wet locations. This would mean that the conductors must have a "W" designation in their type lettering. See *310.8(C)*.

SPECIAL AMPACITY RATINGS FOR RESIDENTIAL 120/240-VOLT, THREE-WIRE, SINGLE-PHASE SERVICE-ENTRANCE CONDUCTORS, SERVICE LATERAL CONDUCTORS, FEEDER CONDUCTORS, AND CABLES THAT SERVE AS THE MAIN POWER FEEDER TO A DWELLING UNIT.

Copper Conductor Size (AWG) for RHH, RHW-2, THHN, THHW, THW, THW-2, THWN, THWN-2, XHHW, XHHW-2, SE, USE, USE-2	Aluminum or Copper-Clad Aluminum Conductors (AWG)	Allowable Special Ampacity
4	2	100
3	1	110
2	1/0	125
1	2/0	150
1/0	3/0	175
2/0	4/0	200
3/0	250 kcmil	225
4/0	300 kcmil	250
250 kcmil	350 kcmil	300
350 kcmil	500 kcmil	350
400 kcmil	600 kcmil	400

Table 29-1 This table is similar to the table in *Table 310.15(B)(6)*, and is *only* for residential 120/240-volt, three-wire, single-phase service-entrance conductors, service lateral conductors, and residential feeder conductors that carry the main power to the dwelling.

▶Make sure that any insulated conductors that are exposed to the direct rays of the sun are of a type listed for sunlight resistance, or are listed and marked for sunlight resistance. See *310.8(D)*.◀

High Temperatures

In certain parts of the country, such as the southwestern desert climates where extremely hot temperatures are common, the AHJ may require that service-entrance conductors in raceways or cables exposed to direct sunlight be corrected (derated) according to the "Correction Factors" found below *Table 310.16*. See Figure 29-1.

EXAMPLE: If the U.S. weather bureau lists the average summer temperature as 113°F (45°C), then a correction factor of 0.82 must be applied. For instance, a 3/0 XHHW copper conductor per *Table 310.16* is 200 amperes at 86°F (30°C). At 113°F (45°C) the conductor's ampacity is

$$200 \times 0.82 = 164 \text{ amperes}$$

A properly sized conductor capable of carrying 200 amperes safely requires

$$\frac{200}{0.82} = 243.9 \text{ amperes}$$

Therefore, according to *Table 310.16* and the applied correction factor, a 250-kcmil Type XHHW copper conductor is required.

Check this with the electrical inspection department where this situation might be encountered.

Neutral Conductor

Why is the neutral conductor permitted to be smaller than the "hot" ungrounded conductors?

The neutral conductor must be able to carry the computed maximum neutral current, *NEC* 220.22. Straight 240-volt loads are not included in the calculation of neutral conductors. The calculations resulted in a computed load of 195 amperes for the ungrounded conductors, and 106 amperes for the neutral conductor. Referring to *Table 310.16*, this could be two 3/0 AWG THW, THHW, THWN, XHHW, or THHN copper conductors and one 2 AWG neutral conductor. The neutral conductor may be insulated or bare per *230.30* and *230.41*.

NEC 310.15(B)(6) also permits the neutral conductor to be smaller than the "hot" ungrounded

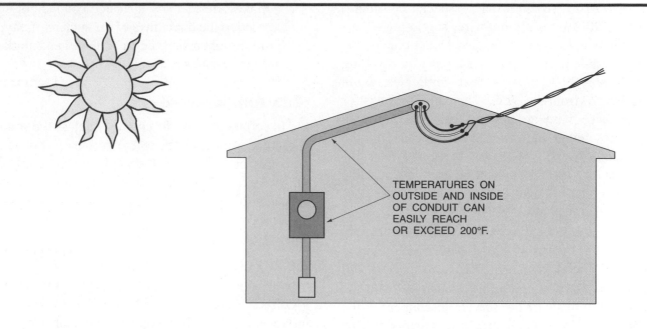

TEMPERATURES ON OUTSIDE AND INSIDE OF CONDUIT CAN EASILY REACH OR EXCEED 200°F.

Figure 29-1 Example of high-temperature location.

conductors, but only if it can be verified by calculations that the requirements of *215.2, 220.22,* and *230.42* are met. We verified this in our calculations for the residence discussed in this text.

In general, service-entrance conductors must be insulated, *230.30, 230.41,* and *310.2.* However, the exceptions to these rules permit the use of bare neutral conductors.

Where bare conductors are used with insulated conductors in raceways or cables, their allowable ampacity is limited to the allowable ampacity for the adjacent insulated conductors, *310.15(B)(3).*

For example, for a service consisting of two 3/0 AWG Type THW conductors and one 1/0 AWG bare neutral conductor, the allowable ampacity of the bare neutral conductor is found in *Table 310.16* in the THW column.

3/0 AWG THW phase conductors	200 amperes
1/0 AWG bare neutral conductor	150 amperes

The specifications for this residence call for two 2/0 AWG ungrounded "hot" conductors and one 1 AWG neutral conductor.

The service neutral conductor serves another purpose. In addition to carrying the unbalanced current, the neutral conductor must also be capable of safely carrying any fault current that it might

be called upon to carry, such as a line-to-ground fault in the service equipment. Sizing the neutral conductor only for the normal neutral unbalance current could result in a neutral conductor too small to safely carry fault currents. The neutral conductor might burn off, causing serious voltage problems, damage to equipment, and the creation of fire and shock hazards.

In sizing neutral conductors for dwellings, this generally is not a problem. But installing reduced size neutrals on commercial and industrial services can present a major problem. The grounded conductor of *any* service must never be smaller than the required grounding electrode conductor for that particular service. Refer to *230.23, 230.31, 230.42(C), 250.2, 250.4,* and *250.24(B).*

Subpanel B Calculations

General lighting load *(220.3(A), 220.11)*	
1420 sq. ft. @ 3 volt-amperes per ft² =	4260 volt-amperes
(kitchen, living room, laundry, rear entry hall, powder room, recreation room)	

Small appliance load
(220.16)

Kitchen	3	
Laundry	1	
Automatic washer	1	
Total	5 @ 1500 VA per circuit =	7500 volt-amperes

Total general lighting and
small appliance load = 11,760 volt-amperes

Application of demand
factors (Table 220.11)

3000 volt-amperes
@100% =   3000 volt-amperes

11,760 − 3000 = 8760
@ 35% =   3066 volt-amperes

A Net computed load (less
range and "fastened-in-
place" appliances)   6066 volt-amperes

Wall-mounted oven and
counter-mounted range:
Table 220.19, Note 4:

Wall-mounted oven	7450 volt-amperes
Counter-mounted range	6600 volt-amperes
Total	14,050 volt-amperes

14 kW exceeds 12 kW by 2 kW:
2 kW × 5% = 10%
increase
8 kW × 0.10 = 0.8 kW
(Column A, Table 220.19)

B 8 + 0.8 = 8.8 kW =   8800 volt-amperes

C Net computed load (with
range) = A + B = C   14,866 volt-amperes

Dryer (220.18)   5700 volt-amperes
Dishwasher
Motor: 7.2 × 120 =
864 volt-amperes
Heater: 1000 watts
(volt-amperes)
(maximum demand
is 1000 VA because

motor and heater do
not operate at the
same time, 220.21)   1,000 volt-amperes

Food waste disposer
7.2 × 120 =   864 volt-amperes

Garage door opener
5.8 × 120 =   696 volt-amperes

Total "fastened-in-place"
appliance load   8260 volt-amperes

Lighting, small appliance
circuits, ranges, dryer = 14,866 volt-amperes

D Add 25% of largest motor.
7.2 × 120 × 0.25 =   216 volt-amperes

E Net computed load
(lighting, small appli-
ance circuits, ranges,
dryer, dishwasher, food
waste disposer, garage
door opener)   23,342 volt-amperes

$$\text{Amperes} = \frac{\text{volt-amperes}}{\text{volts}} = \frac{23,342}{240} = 97.3 \text{ amperes}$$

The feeder conductors supplying Panel B could
be 3 AWG THHW, THW, THHN, or XHHW per
Table 310.16. We have not shown the calculations
for the neutral to Panel B. The specifications and
also Figure 27-2 require that three 3 AWG THHN
feeder conductors supply Panel B.

Method 2 (Optional Calculations — Dwellings) (Article 220, Part III)

A second method for determining the load for a
one-family dwelling is given in 220.30. This method
simplifies the calculations. But remember, the minimum size service for a one-family dwelling is 100
amperes, 230.42(B) and 230.79(C).

Let's take a look at Part III of Article 220. This
is an alternative method of computing service loads
and feeder loads. It is referred to as the optional
method.

NEC® 220.30(A) tells us that we are permitted to
calculate service-entrance conductors and feeder
conductors (both phase conductors and the neutral
conductor) using an optional method. NEC® 220.30
addresses single-family dwellings. NEC® 220.32 addresses multifamily dwellings.

NEC® 220.30(B) tells us to

1. include 1500 volt-amperes for each 20-ampere small appliance circuit.

2. include 3 volt-amperes per ft^2 for lighting and general-use receptacles.

3. include the nameplate rating of appliances:
   - that are fastened in place, or
   - that are permanently connected, or
   - that are located to be connected to a specific circuit, such as: ranges, wall-mounted ovens, counter-mounted cooking units, clothes dryers, and water heaters.

4. include nameplate ampere rating or kVA for motors and all low-power-factor loads. The intent of the reference to low power factor is to address such loads as low-cost, low-power-factor fluorescent ballasts. It is always recommended that high-power-factor ballasts be installed.

NEC® 220.30(C) tells us that for heating and air-conditioning loads, select the largest load from a list of six types of loads:

1. 100 percent of air conditioning and cooling.

2. 100 percent of heat pump compressors and supplemental heating unless both cannot operate at the same time.

3. 100 percent electric thermal storage heating if expected to be continuous at full nameplate value.

4. 65 percent of central electric space heating that also has supplemental heating in heat pumps where the compressor and supplemental heating will not operate at the same time.

5. 65 percent of electric space heating of less than four separately controlled units.

6. 40 percent of electric space heating if four or more separately controlled units.

So let's begin our optional calculation for the residence discussed in this text. The residence has an air conditioner and an electric furnace.

Air conditioner

$26.4 \times 240 = 6336$ volt-amperes

Electric furnace

$13,000 \times 0.65 = 8450$ volt-amperes

Therefore we will select the electric furnace load for our calculations because it is the largest load. It is also a noncoincidental load, as defined in *220.21*. We can omit the air-conditioner load from our calculations from here on.

We now add up all of the other loads:

General lighting load 3232 ft^2 @ 3 volt-amperes per ft^2 =	9696 volt-amperes
Small appliance circuits (7) @ 1500 volt-amperes each	10,500 volt-amperes
Wall-mounted oven (nameplate rating)	6600 volt-amperes
Counter-mounted cooking unit (nameplate rating)	7450 volt-amperes
Water heater (nameplate rating)	3000 volt-amperes
Clothes dryer (nameplate rating)	5700 volt-amperes
Dishwasher (maximum demand, heater only)	1000 volt-amperes
Food waste disposer 7.2 × 120	864 volt-amperes
Water pump 8 × 240	1920 volt-amperes
Garage door opener 5.8 × 120	696 volt-amperes
Heat-vent-lights (2) 1500 × 2	3000 volt-amperes
Attic exhaust fan 5.8 × 120	696 volt-amperes
Hydromassage tub 10 × 120	1200 volt-amperes
Total other loads	52,322 volt-amperes

We can now complete our optional calculation:

Enter electric furnace load $13,000 \times 0.65 =$	8450 volt-amperes
Plus first 10 kVA of all other loads at 100%	10,000 volt-amperes
Plus remainder all other loads at 40%:	
$52,322 - 10,000 = 42,322$	
$42,322 \times 0.40$	16,929 volt-amperes
Total	35,379 volt-amperes

$$\text{Amperes} = \frac{\text{volt-amperes}}{\text{volts}} = \frac{35,379}{240} = 147.4 \text{ amperes}$$

Using the optional calculation method, and referring to *Table 310.15(B)(6)*, we find that for the 147+ amperes computed load, a 1 AWG copper conductor with any of the insulation types listed has a special ampacity of 150 amperes and could be installed for this service-entrance. This seems a bit small, and that is why some local electrical codes do not permit the use of this special ampacity table.

The specifications for the residence discussed in this text call for a full 200-ampere service consisting of two AWG 2/0 THW, THHN, THWN, or XHHW copper phase conductors and one 1 AWG bare copper neutral conductor.

Existing Homes

When adding new loads in an existing home, *220.31* provides an optional calculation method that may be used to confirm that the existing service is large enough to handle the additional load. The procedure is a mirror image of the optional calculation method we discussed above.

Service-Entrance Conductor Size Table

Table 29-2 has been taken from one major city that prefers to show minimum service-entrance conductor size requirements rather than having electrical contractors make calculations each and every time they install a service.

It is within the realm of local authorities to publish code requirements specifically for their communities.

Main Disconnect

The main service disconnecting means in this residence is rated 200 amperes, discussed in Unit 27.

Grounding Electrode Conductor

Grounding electrode conductors are sized according to *Table 250.66* of the Code.

As an example, this residence is supplied by 2/0 AWG copper service-entrance conductors. Checking *Table 250.66*, we find that the minimum grounding electrode conductor must be a 4 AWG copper conductor.

The grounding electrode conductor for the service in this residence is a 4 AWG copper armored ground cable, discussed in Unit 27.

TYPES OF WATT-HOUR METERS

Figure 29-2 shows typical single-phase watt-hour meters. The first, (A), is a conventional five-dial meter. The second, (B), is a five-dial meter where one set of dials registers total kilowatt-hours and the second set of dials registers premium time kilowatt-hours. The third, (C), is a programmable electronic digital display watt-hour meter that can register kilowatt-hour consumption for up to four different rate schedules. Utilities that offer special "time-of-use" rates install meters of the second and third types.

Wiring diagrams for the metering and control of electric water heaters, air conditioners, and heat pumps are discussed in Unit 19. Electric utilities

	COPPER			ALUMINUM		
Size (amperes)	Phase (Hot) Conductors	Neutral Conductor	Conduit Trade Size	Phase (Hot) Conductors	Neutral Conductor	Conduit Trade Size
100	2 AWG	4 AWG	1½	1 AWG	2 AWG	1½
125	1 AWG	4 AWG	1½	00 AWG	1 AWG	1½
150	0 AWG	4 AWG	1½	000 AWG	0 AWG	2
175	00 AWG	2 AWG	2	0000 AWG	00 AWG	2
200	000 AWG	2 AWG	2	250 kcmil	000 AWG	2

SERVICE-ENTRANCE CONDUCTOR SIZE

Table 29-2 Service-entrance conductor sizing used in some cities so that service-entrance calculations do not have to be made for each service-entrance installation.

Table 250.66 Grounding Electrode Conductor for Alternating-Current Systems

Size of Largest Ungrounded Service-Entrance Conductor or Equivalent Area for Parallel Conductors[a] (AWG/kcmil)		Size of Grounding Electrode Conductor (AWG/kcmil)	
Copper	Aluminum or Copper-Clad Aluminum	Copper	Aluminum or Copper-Clad Aluminum[b]
2 or smaller	1/0 or smaller	8	6
1 or 1/0	2/0 or 3/0	6	4
2/0 or 3/0	4/0 or 250	4	2
Over 3/0 through 350	Over 250 through 500	2	1/0
Over 350 through 600	Over 500 through 900	1/0	3/0
Over 600 through 1100	Over 900 through 1750	2/0	4/0
Over 1100	Over 1750	3/0	250

Notes:

1. Where multiple sets of service-entrance conductors are used as permitted in 230.40, Exception No. 2, the equivalent size of the largest service-entrance conductor shall be determined by the largest sum of the areas of the corresponding conductors of each set.

2. Where there are no service-entrance conductors, the grounding electrode conductor size shall be determined by the equivalent size of the largest service-entrance conductor required for the load to be served.

[a]This table also applies to the derived conductors of separately derived ac systems.

[b]See installation restrictions in 250.64(A).

(Reprinted with permission from NFPA 70-2002)

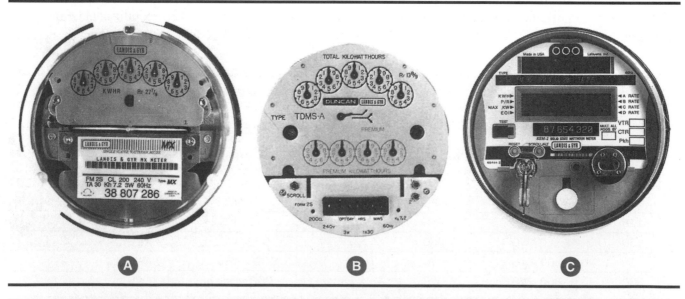

Figure 29-2　Photo A shows a typical single-phase watt-hour meter. Photo B shows a single-phase watt-hour meter that has two sets of dials: one registers the total kilowatt-hours, the second registers premium time kilowatt-hours. Photo C shows a programmable electronic digital display watt-hour meter that is capable of registering four different time-of-use rates. *Courtesy of* Landis & Gyr Energy Management, Inc.

across the country have different requirements for hooking up these loads. Always consult your local utility for their latest information regarding electric service.

READING WATT-HOUR METERS

Reading a digital watt-hour meter is easy. Just read the numbers.

For dial-type meters, starting with the first dial, record the last number the pointer has just passed. Continue doing this with each dial until all dial readings have been recorded. The reading on the five-dial watt-hour meter in Figure 29-3 is 18,672 kilowatt-hours.

If the meter reads 18,975 one month later, as indicated in Figure 29-4, by subtracting the previous reading of 18,672, it is found that 303 kilowatt-hours of electricity were used during the month. The 303 kilowatt-hours of electricity used is multiplied by the rate per kilowatt-hour, and the power company bills the consumer for the energy used. The utility may also add a fuel adjustment charge or other charges.

Figure 29-3   The reading of this five-dial meter is 18,672 kilowatt-hours.

Figure 29-4   One month later, the meter reads 18,975 kilowatt-hours, indicating that 303 kilowatt-hours were used during the month.

REVIEW

Note: Refer to the Code or the plans where necessary.

1. According to *NEC*® _____ , the minimum size service required for a one-family dwelling is _____ amperes.

2. a. What is the unit load per ft^2 for the general lighting load of a residence?

   _____

   b. What are the demand factors for the general lighting load in dwellings? _____

   _____

   _____

   _____

3. a. What is the ampere rating of the circuits that are provided for the small appliance loads? _____

   b. What is the minimum number of small appliance circuits permitted by the Code?

   _____

   c. How many small appliance circuits are included in this residence? _____

4. Why is the air-conditioning load for this residence omitted in the service calculations?

   _____

   _____

   _____

   _____

   _____

   _____

5. What demand factor may be applied when four or more fixed appliances are connected to a service, in addition to an electric range, air conditioner, clothes dryer, or space heating equipment? *(220.17)* _____

   _____

6. What load may be used for an electric range rated at not over 12 kW? *(Table 220.19)*

   _____

7. What is the computed load for an electric range rated at 16 kW? *(Table 220.19)* Show calculations.

8. What is the computed load when fixed electric heating is used in a residence? *(220.15)*

   _____

9. On what basis is the neutral conductor of a service entrance determined? _____

   _____

   _____

   _____

   _____

10. Why is it permissible to omit an electric space heater, water heater, and certain other 240-volt equipment when calculating the neutral service-entrance conductor for a residence? _____

    _____

    _____

11. a. What section of the Code contains an optional method for determining residential service-entrance loads? _____

    b. Is this section applicable to a two-family residence? _____

12. Calculate the minimum size of copper service-entrance conductors and grounding electrode conductor required for a residence containing the following: floor area is 24 ft × 38 ft (7.3 m × 11.6 m). The dwelling will have:

    • a 12-kW, 120/240-volt electric range

    • a 5-kW, 120/240-volt electric clothes dryer consisting of a 4-kW, 240-volt heating element, a 120-volt motor, and 120-volt lamp that have a combined load of 1 kW

    • a 2200-watt, 120-volt sauna heater

    • six individually controlled 2-kW, 240-volt baseboard electric heaters

    • a 12-ampere, 240-volt air conditioner

    • a 3-kW, 240-volt electric water heater

    • a 1000-watt, 120-volt dishwasher

    • a 7.2-ampere, 120-volt food waste disposer

    • a 5.8-ampere, 120-volt attic exhaust fan

    Determine the sizes of the ungrounded "hot" conductors and the neutral conductor. Use Type THWN. Use *Table 310.15(B)(6)*. Also determine the correct size grounding electrode conductor.

    Do not forget to include the required two 20-ampere small appliance circuits and the laundry circuit.

    The panelboard is marked 167°F (75°C).

    You may use the blank *Single-Family Dwelling Service-Entrance Calculation* form in the Appendix of this text, or use the form as a guide to make your calculations in the proper steps.

    Two _____ AWG THWN ungrounded "hot" conductors

    One _____ AWG THWN or bare neutral conductor

    One _____ AWG grounding electrode conductor

STUDENT CALCULATIONS

13. Read the meter shown. Last month's reading was 22,796. How many kilowatt-hours of electricity were used for the current month?

14. After studying this unit, we realize that electric utilities across the country have many ways to meter residential customers. They may have lower rates during certain hours of the day. A common term used to describe these programs is _____

_____

Swimming Pools, Spas, Hot Tubs, and Hydromassage Baths

OBJECTIVES

After studying this unit, the student should be able to

- recognize the importance of proper swimming pool wiring with regard to human safety.
- discuss the hazards of electrical shock associated with faulty wiring in, on, or near pools.
- understand and apply the basic Code requirements for the wiring of swimming pools, spas, hot tubs, and hydromassage bathtubs.

Water and electricity do not mix!

To protect people against the hazards of electric shock associated with swimming pools, spas, hot tubs, and hydromassage bathtubs, special Code requirements are necessary.

A picture is worth a million words. A detailed drawing of the *NEC®* requirements for swimming pools appears on Sheet 10 of 10 in the Plans found in the back of this text. Refer to this drawing often as you study this unit.

ELECTRICAL HAZARDS

A person can suffer an immobilizing or lethal shock in a residential-type pool in either of two ways.

1. **Direct contact:** An electrical shock can be deadly to someone in a pool who touches a "live" wire or touches an object such as a metal ladder, metal fence, metal parts of a luminaire (fixture), metal enclosures of electrical equipment, or appliances that have become "live" by an ungrounded conductor coming in contact with the metal enclosure or exposed metal parts of an appliance. Figure 30-1 shows a person in the water touching a faulty appliance.

2. **Indirect contact:** An electrical shock can be lethal to someone in a pool if the person is merely in the water when voltage gradients in the water are present. This is illustrated in Figure 30-2.

As shown in Figure 30-2, "rings" of voltage radiate outward from the radio to the pool walls and bottom of the pool. These rings can be likened to the rings that form when a rock is thrown into the water. The voltage rings or gradients range from 120 volts at the radio to zero volts at the pool walls and bottom. The pool walls and bottom are assumed to be at ground or zero potential. The gradients, in varying degrees, are found throughout the body of water. Figure 30-2 shows voltage gradients in the pool of 90 volts and 60 volts. This figure is a simplification of the actual situation in which there are many voltage gradients. In this case, the voltage differential, 30 volts, is an extremely hazardous value. The person in the pool immersed in these voltage

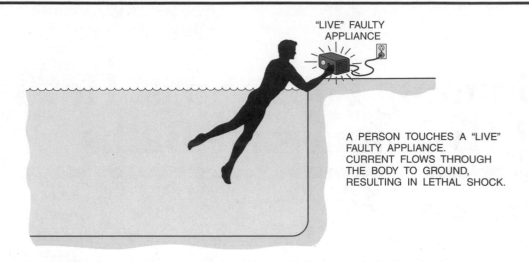

Figure 30-1 Touching a "live" faulty appliance can cause lethal shock.

Figure 30-2 Voltage gradients surrounding a person in the pool can cause severe shock, drowning, or electrocution.

gradients, is subject to severe shock, immobilization (which can result in drowning), or actual electrocution. Tests conducted over the years have shown that a voltage gradient of 1½ volts per foot can cause paralysis.

The shock hazard to a person in a pool is quite different from that of the normal "touch" shock hazard to a person not submersed in water. The water makes contact with the entire skin surface of the body rather than just at one "touch" point. Skin wounds, such as cuts and scratches, reduce the body's resistance to shock to a much lower value

than that of the skin alone. Body openings such as ears, nose, and mouth further reduce the body resistance. As Ohm's law states, for a given voltage, the lower the resistance, the higher the current.

KEY *NEC*® REQUIREMENTS— WIRING FOR SWIMMING POOLS

Article 680 covers swimming pools, wading pools, hydromassage bathtubs, fountains, therapeutic and decorative pools, hot tubs, and spas.

Here are some of the general Code rules. Comprehensive details for swimming pool requirements are found on Sheet 10 of 10 of the Plans at the back of this text.

General, Part I: Contains general requirements that apply to all types of swimming pools, spas, hot tubs, hydromassage tubs, and similar water related equipment. *Parts II, III, IV, V, VI,* and *VII* discuss the wiring requirements for specific equipment.

Scope, *680.1:* The scope tells us exactly what is covered by *Article 680.*

Definitions, *680.2:* There are many one-of-a-kind terms listed in *680.2.* These terms might seem like a foreign language. It makes sense to read these definitions carefully to make it easier to understand the rest of *Article 680.*

Other Articles, *680.3:* Many of the basic requirements of Chapter 1 through Chapter 4 in the *NEC®* are modified by special rules in *Article 680* that are unique to swimming pools, wading pools, hydromassage bathtubs, therapeutic pools, decorative pools, hot tubs, and spas. This section of the *NEC®* contains a convenient table that lists six specific articles that apply.

Ground-Fault Circuit Interrupters, *680.5:* GFCIs shall be listed circuit breakers or receptacles, or other listed types. GFCIs were discussed in Unit 6. Throughout *Article 680,* you will find the need for GFCIs.

Be careful when connecting underwater luminaires (fixtures). Once you run conductors beyond (on the load side of) a GFCI device protecting that wiring, you are not permitted to have other conductors in the same raceway, box, or enclosure. The exceptions to this rule are: (1) the other conductors are protected by GFCIs; (2) the other conductors are grounding conductors; (3) the other conductors are supply conductors to a feedthrough type GFCI; or (4) the GFCI is installed in a panel-board, where obviously there will be a "mix" of other circuit conductors that are not GFCI protected. This is covered in *680.23(F)(3).*

Grounding, *680.6:* In general, grounding shall be done as required in *Article 250.* But there are modifications that specifically address swimming pools.

Cord- and Plug-Connected Equipment, *680.7:* Here we find permission to use flexible cords to connect certain equipment. Underwater luminaires (fixtures) are not included in this permission.

Overhead Conductors, *680.8:* Overhead conductors must be kept out of reach. Clearances are shown in Plan 10 of 10 in the back of this text.

Electric Pool Water Heaters, *680.9:* The branch-circuit conductor ampacity and the rating of the branch-circuit overcurrent devices must be not less than 125 percent of the total name-plate rated-load ampere rating of the electric heater. Figure 30-3 shows the requirements for unit heaters, radiant heaters, and embedded heating cables.

Underground Wiring Location, *680.10:* Underground wiring is not permitted under pools or withing 5 ft (1.5 m) horizontally from the inside wall of the pool. There are a few exceptions.

▶**Maintenance Disconnecting Means, *680.12:*** One or more disconnecting means must be provided for pool-associated equipment. They must be accessible and within sight of the equipment they disconnect. Pool lighting is not covered by *680.12.*◀

Permanently Installed Pools, Part II. Here we find the rules governing permanently installed pools such as for motors; clearances for overhead lighting; underwater lighting; existing installations; receptacles; requirements for switching; equipment associated with the pool; bonding, grounding, and audio equipment; and electric heaters. Electric deck heater Code requirements are shown in Figure 30-3.

Grounding and bonding together all metal parts in and around pools ensures that these metal parts are at the same voltage potential to ground, greatly reducing the shock hazard. Stray voltage and voltage gradient (Figure 30-2) problems are kept to a minimum. Proper grounding and bonding also facilitates the opening of the overcurrent protection devices should a fault occur in the circuit(s).

You will see on Sheet 10 of 10 of the Plans that a metal conduit by itself is not considered an adequate

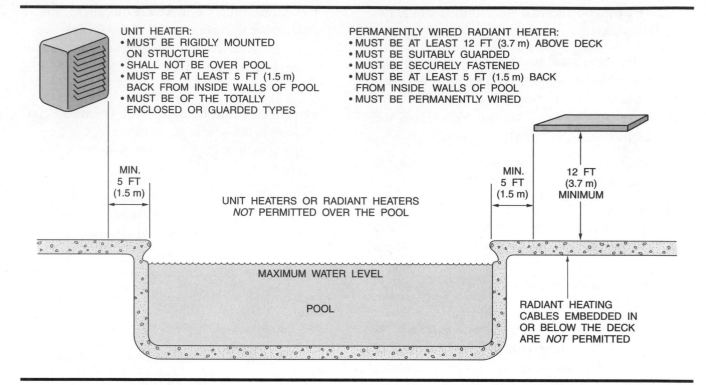

UNIT HEATER:
• MUST BE RIGIDLY MOUNTED
  ON STRUCTURE
• SHALL NOT BE OVER POOL
• MUST BE AT LEAST 5 FT (1.5 m)
  BACK FROM INSIDE WALLS OF POOL
• MUST BE OF THE TOTALLY
  ENCLOSED OR GUARDED TYPES

PERMANENTLY WIRED RADIANT HEATER:
• MUST BE AT LEAST 12 FT (3.7 m) ABOVE DECK
• MUST BE SUITABLY GUARDED
• MUST BE SECURELY FASTENED
• MUST BE AT LEAST 5 FT (1.5 m) BACK
  FROM INSIDE WALLS OF POOL
• MUST BE PERMANENTLY WIRED

MIN.
5 FT
(1.5 m)

UNIT HEATERS OR RADIANT HEATERS
NOT PERMITTED OVER THE POOL

MIN.
5 FT
(1.5 m)

12 FT
(3.7 m)
MINIMUM

MAXIMUM WATER LEVEL

POOL

RADIANT HEATING
CABLES EMBEDDED IN
OR BELOW THE DECK
ARE *NOT* PERMITTED

Figure 30-3   Deck-area electric heating within 20 ft (6.0 m) from inside edge of swimming pools, *680.27(C)*.

grounding means for equipment in and around pools.

Figure 30-4 is a detailed illustration showing the grounding of important metal parts of a permanently installed pool.

Figure 30-5 shows a common bonding grid of a permanently installed swimming pool. Note that the bonding conductor does not need to be connected to any service equipment or any grounding electrode. It merely ties all metal parts together.

Storable Pools, Part III.   Here we find the Code rules for storable pools. Storable pools are just that: they are used on or above the ground, and can be dismantled and stored.

Spas and Hot Tubs, Part IV.   What's the difference? A hot tub is constructed of wood such as redwood, teak, cypress or oak. A spa is made of plastic, fiberglass, concrete, tile, or other man-made products. Code rules for spas and hot tubs are discussed here because they are not shown on Sheet 10 of 10 of the Plans.

Outdoor Installations, *680.42*: Shall be wired according to *Parts I* and *II*, except for the following:

(A) Flexible connections using flexible conduit and cord-and-plug connections are permitted.

(B) Different metal components where there is metal-to-metal mounting of the same metal frame are acceptable as the required bond.

(C) ▶For one-family dwellings, a self-contained spa or hot tub or a packaged spa or hot tub assembly may be connected using conventional wiring methods that contain a copper equipment grounding conductor not smaller than 12 AWG. Conventional wiring methods are found in Chapter 3 of the *NEC*.®◀

Nonmetallic-sheathed cable with 14 AWG conductors has a 14 AWG equipment grounding conductor.

Nonmetallic-sheathed cable with 12 AWG conductors has a 12 AWG equipment grounding conductor.

Indoor Installations, *680.43*: Shall be wired according to *Parts I* and *II*, except for the following:

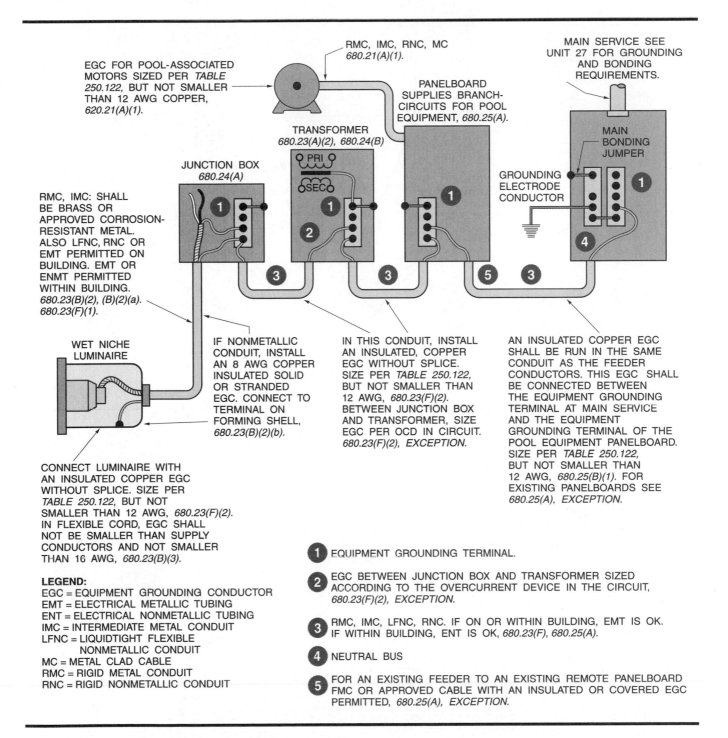

Figure 30-4 A "trail" of the grounding requirements from a wet-niche luminaire back to the main service panelboard.

(A) Receptacles:

- At least one 125-volt, 15- or 20-ampere receptacle connected to a general-purpose branch-circuit must be installed at least 5 ft (1.5 m) but not more than 10 ft (3.0 m) from the inside wall of the spa or hot tub.
- All other 125-volt, 15- or 20-ampere receptacles must be located at least 5 ft (1.5 m), measured horizontally, from the inside wall of the spa or hot tub.
- All 125-volt receptacles rated 30 amperes or less located within 10 ft (3.0 m) of the inside walls of a spa or hot tub must be GFCI protected.

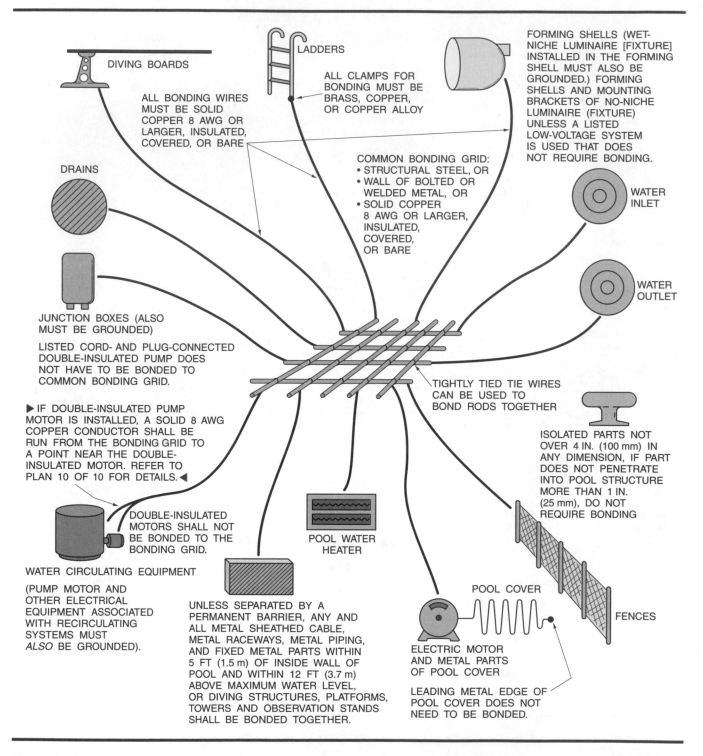

Figure 30-5   Bonding the metal parts of a permanently installed swimming pool, 680.26. Also see Plan 10 of 10.

- A receptacle that supplies power to a spa or hot tub must be GFCI protected.

(B) Luminaires (fixtures), lighting outlets, and ceiling-suspended paddle fans above spa or hot tub within 5 ft (1.5 m) from the inside walls of the spa or hot tub:

- not less than 12 ft (3.7 m) above if not GFCI protected.

- not less than 7 ft 6 in. (2.3 m) if GFCI protected.

- if less than 7 ft 6 in. (2.3 m), must be suitable for damp locations and be:

– recessed luminaires (fixtures) with a glass or plastic lens and nonmetallic or isolated metal trim.

– surface-mounted luminaires (fixtures) with a glass or plastic globe, a nonmetallic body, or a metallic body isolated from contact.

- if underwater lighting is installed, the rules for regular swimming pools apply, in which case refer to Sheet 10 of 10 of the Plans.

(C) Switches:

- wall switches must be kept at least 5 ft (1.5 m) measured horizontally from the inside edge of the spa or hot tub.

(D) Bonding: Bonding requirements are similar to that of conventional swimming pools. Bonding keeps everything at the same potential. Excluded are small conductive surfaces not likely to become energized such as drains and air and water jets. Bond together:

- all metal fittings within or attached to the spa or hot tub structure.

- metal parts of associated electrical equipment including pump motors.

- metal conduits, metal piping, and other metal surfaces within 5 ft (1.5 m) of the inside walls of the spa or hot tub, unless separated by a permanent barrier, such as a wall or building.

- electrical devices and controls *not* associated with the pool or hot tub and located not less than 5 ft (1.5 m) from the spa or hot tub. Bonding to the spa or hot tub system also is acceptable.

(E) Bonding all of the above things together can be accomplished by means of threaded piping and fittings, by metal-to-metal mounting on a common base or frame, or by a copper 8 AWG or larger solid, insulated, covered or bare bonding jumper.

(F) Ground all electrical equipment within 5 ft (1.5 m) of the inside wall of a spa or hot tub. Also ground all electrical equipment associated with the circulating system of the spa or hot tub.

Protection, 680.44: The word "protection" refers to GFCI protection. Outlets for spas and hot tubs shall be GFCI protected, unless the unit has integral GFCI protection. A few other exceptions are mentioned in *680.44.*

Fountains, Part V: The residence discussed in this text does not have a fountain. Permanently installed fountains come under the requirements of *Part V.* For fountains that share water with a regular swimming pool, the wiring must conform to the requirements found in *Part II* of *Article 680,* which covers permanently installed swimming pools.

Small self-contained portable fountains that have no dimension over 5 ft (1.5 m) do not come under the requirements of *Part V.*

Pools and Tubs for Therapeutic Use, Part VI: The residence discussed in this text does not have any therapeutic equipment.

Hydromassage Bathtubs, Part VII: A hydromassage tub is intended to be filled, used, then drained after each use. These are also known as whirlpool bathtubs. Figure 22-12 illustrates one manufacturer's hydromassage bathtub.

- Protection, *680.71:* The word "protection" refers to GFCI protection. Hydromassage bathtubs, together with their associated electrical components, must be GFCI protected.

- Protection, *680.71:* All 125-volt, single-phase receptacles that are located within 5 ft (1.5 m) measured horizontally from the inside walls of the hydromassage bathtub must be GFCI protected.

- Other Electrical Equipment, *680.72:* A hydromassage bathtub does not constitute any more of a shock hazard than a regular bathtub. All wiring (fixtures, switches, receptacles, and other equipment) in the same room but not directly associated with the hydromassage bathtub is installed according to all the normal Code requirements covering installation of that equipment in bathrooms.

- Accessibility, *680.73:* Electrical equipment associated with a hydromassage bathtub must be accessible without damaging the building structure or finish.

- Bonding, *680.74*: All metal piping, metal parts of electrical equipment, and pump motors associated with the hydromassage bathtub shall be bonded together by a copper 8 AWG or larger solid, insulated, covered or bare bonding jumper.

This residence has a hydromassage tub in the master bathroom. The wiring of it is covered in Unit 22.

GETTING TRAPPED UNDER WATER

There is another real hazard, not truly electrical in nature, but solvable electrically.

Drownings have occurred when a person is trapped under water by the tremendous suction at an unprotected drain. The grate over the drain may have been removed, or is broken. One report indicated that three adult males could not pull a basketball from a drain because of the suction!

To address this hazard, *680.38* requires that a clearly labeled emergency shutoff switch for the control of the recirculation system and jet system for spas and hot tubs shall be installed not less than 5 ft (1.5 m) away. adjacent to, and within sight of the spa or hot tub. Although this requirement does not apply to single-family dwellings, some states require this disconnect for single-family dwelling pools.

Another hazard associated with spas and hot tubs is prolonged immersion in hot water. If the water is too hot and/or the immersion too long, lethargy (drowsiness, hyperthermia) can set in. This can increase the risk of drowning. UL Standard 1563 establishes the maximum water temperature at 104°F (40°C). The suggested maximum time of immersion is generally 15 minutes. Instructions furnished with spas and hot tubs specify the maximum temperature and time permitted for using the spa or hot tub. Hot water, in combination with drugs and alcohol, presents a real hazard to life. Caution must be observed at all times.

UNDERWRITERS LABORATORIES STANDARDS

UL standards of interest are:

UL 676   Underwater Lighting Fixtures

UL 943   Ground-Fault Circuit Interrupters

UL 1081   Swimming Pool Pumps, Filters, and Chlorinators

UL 1241   Junction Boxes for Underwater Pool Luminaires (Fixtures)

UL 1563   Electric Spas, Equipment Assemblies, and Associated Equipment

UL 1795   Hydromassage Bathtubs

REVIEW

Note: Refer to the Code or the plans where necessary.

1. Most of the requirements for the wiring of swimming pools are covered in *NEC*® *Article* _____ .

2. Name the two ways in which a person may sustain an electrical shock when in a pool.

   a. _____

   b. _____

3. Using the text, the Plans in the back of this text, and the *NEC*,® determine if the following items must be grounded. Place an "X" in the correct space.

	True	False
a. Wet-and-dry-niche luminaires (fixtures).	_____	_____
b. Electrical equipment located within 5 ft (1.52 m) of inside walls of pool.	_____	_____
c. Electrical equipment located within 10 ft (3.0 m) of inside walls of pool.	_____	_____
d. Recirculating equipment and pumps.	_____	_____
e. Luminaires (fixtures) and ceiling fans installed more than 15 ft (4.5 m) from inside walls of pool.	_____	_____
f. Junction boxes, transformers, and GFCI enclosures that contain conductors that supply wet-niche luminaires (fixtures).	_____	_____
g. Panelboards that supply the electrical equipment for the pool.	_____	_____
h. Panelboards 20 ft (6.0 m) from the pool that do not supply the electrical equipment for the pool.	_____	_____

4. Is it permitted to run other conductors in the same conduit as the GFCI protected conductors that run to underwater luminaires (fixtures)? _____

5. *NEC*® _____ prohibits running electrical conduits under a swimming pool.

6. What is the purpose of grounding and bonding? _____

_____

_____

7. Does the *NEC*® require that, after all of the metal parts of a swimming pool are bonded together to form a common grounding grid, all of this grid must be connected to a grounding electrode? _____

8. May conductors be run above the pool? Explain. _____

_____

_____

9. What is the closest distance from the inside wall of a pool that a receptacle may be installed ? _____

10. Receptacles located within 20 ft (6.0 m) from the inside wall of a pool must be _____ protected.

11. Luminaires (fixtures) installed above a new outdoor swimming pool must be mounted not less than :

  a. 10 ft. (3.0 m)   b. 12 ft. (3.7 m)

above the maximum water level. Circle correct answer.

12. Junction boxes and enclosures for tranformers and GFCIs have one thing in common. They all (are made of brass) (have threaded hubs) (must be mounted at least 8 in. [200 mm] above the deck). Circle the correct answer.

13. For indoor spas and hot tubs, the following statements are either true or false. Check one.

		True	False
a.	Receptacles may be installed within 5 ft (1.5 m) from the edge of the spa or hot tub.	_____	_____
b.	All receptacles within 10 ft (3.0 m) of the spa or hot tub must be GFCI protected.	_____	_____
c.	Any receptacles that supply power to pool equipment must be GFCI protected.	_____	_____
d.	Wall switches must be located at least 5 ft (1.5 m) from the pool.	_____	_____
e.	In general, luminaires (fixtures) above a spa shall not be less than 7 ft 6 in. (2.3 m) above the maximum water level if the circuit is GFCI protected.	_____	_____

14. Bonding and grounding of electrical equipment in and around spas and hot tubs (is required by the *NEC*®) (is decided by the electrician). Circle correct answer.

15. Where a spa or hot tub is installed in an existing bathroom or other suitable location, an existing receptacle outlet within 5 ft (1.5 m) of the tub is permitted to remain, but only if the circuit feeding the receptacle has _____ protection.

16. Underwater pool luminaires (fixtures) shall be installed so that the top of the lens is not less than _____ in. below the normal water line, unless they are listed for a lesser depth.

17. a. Hydromassage bathtubs and their associated electrical components (shall) (shall not) be protected by a GFCI. Circle the correct answer.

   b. All receptacles located within 5 ft (1.5 m) of a hydromassage tub (shall) (shall not) be protected by a GFCI. Circle the correct answer.

Wiring for the Future: Home Automation Systems

OBJECTIVES

After studying this unit, the student should be able to

- understand the basics of X10 systems.
- understand the basics of a "structured wiring" system.
- understand the types of category-rated cables.
- understand some of the terminology used in the home automation industry.
- understand the basics of "wireless."

An increasing trend in upper-end, upscale residential wiring is the installation of advanced home automation systems. These systems are generally installed in larger, more expensive homes, but are also suitable for the typical home on a limited basis.

Unit 31 is an introduction to advanced home automation wiring. It is beyond the scope of this book to cover the subject in detail because of the many manufacturer's products available today. Some of these products can be interchanged; others cannot. There are texts, videos, and Web sites devoted specifically to advanced home wiring. Web sites (listed after the Appendix of this text) identified with an * are related to home automation. You will want to browse these Web sites and download specific information that fits your needs.

We can hear about, read articles about, and even attend seminars about home automation wiring. Names such as ActiveHome, Avaya's HomeStar, CEBus, Decora Home Controls (Leviton), Echelon, Elan, Enerlogic, Gemini, Greyfox Systems, Home Base, Honeywell Total-Home, IBM Home Director, IES Technologies, LightTouch, LonWorks, Lutrons RadioRA, OnQ, Radio Shack, "Plug 'N Power," Stargate, Smart-House, SmartLinc, Stanley Light-

minder, and X-10 Powerhouse are associated with home automation systems. The list is endless!

Everyone is familiar with individual control devices. Examples are wall switches, thermostats, motion detectors, occupancy sensors, timers, TV remote controllers, and similar devices. They control one "thing."

Now we have additional systems for controlling lighting (on, off, and dimming), heating, cooling, ventilation, appliances, entertainment, and security in a home. Control of single or multiple loads from any number of locations is possible. Modules are available that can be controlled through a computer.

Today, there are different types of advanced home systems for residential wiring.

THE X10 SYSTEM

X10 technology was invented in Scotland in 1978. The original X10 patent expired in 1997, so it is now an open standard for any manufacturer to apply to their products.

X10 technology is simple and is the least costly home automation system, especially for retrofit wiring in existing homes. For the most part, special wiring is not required for basic functions. For many applications, all that is necessary is to plug in the various components as needed.

This system transmits carrier wave signals that are superimposed on the regular 120/240-volt, 60-cycle branch-circuit wiring in the home. Components communicate with each other over the existing electrical wiring without a central controller.

Transmitting devices take a small amount of power from the 120-volt power line, modulate or change it to a higher frequency, and then superimpose this high-frequency signal back onto the ac circuit. At the other end, a receiving device responds to the high-frequency burst. Xl0 is basically a one-way communications system. Some devices "talk" and others "listen."

The control of appliances, audio/video equipment, outdoor and indoor lighting, and receptacles, as well as the arming of a security system, is possible from just about any place in the home. You might call it a "plug-n-play" system. Components are easily added as needed.

An X10 system might include programmers; receiving modules (Figure 31-1); transmitters; keypad

wall or tabletop controller (Figure 31-2); dimmers; single-pole, three-way, four-way, and double-pole switches; on/off/dim lamp modules; on/off appliance modules; receptacle modules; thermostats; timers; surge protection devices; burglar alarm devices; motion sensors (turn on lights, sound an alarm); wireless controllers; handheld remotes; photocells for light/darkness sensing; drapery controllers; telephone responders; tie-in with a personal computer; voice activation; and so on. A typical system of this type will have the capability of 256 easily adjusted different "addresses," more than adequate for just about any size home. Receivers and switches generally mount in standard wall boxes. X10 wall switches and receptacles can replace conventional switches and receptacles. Make sure that the wall box has ample space for the larger X10 devices so the wires in the wall box will not be jammed.

To ensure trouble-free operation, a phase-coupler (Figure 31-3) connected at the main panel is used to provide proper strength command signals to both sides of the ac wiring.

Limitations of X10 Technology

X10 modules are susceptible to interference from an ac power line. Many things can generate "noise" on 120-volt power lines, including fluorescent luminaire (fixture) ballasts, appliances, and wireless intercoms. And because standard home wiring (most often Type NM cable) is unshielded, it can pick up unintended induced signals in the same

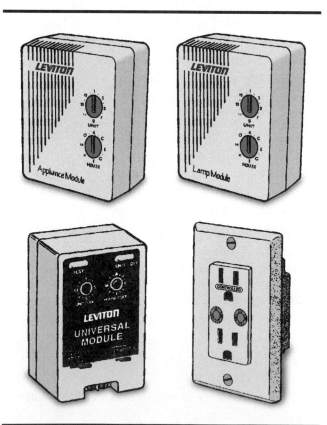

Figure 31-1   Various types of typical receiver modules used with X10 systems.

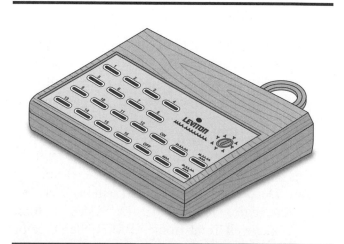

Figure 31-2   A typical X10 tabletop keypad controller makes it easy to transmit preprogrammed signals to receiver modules.

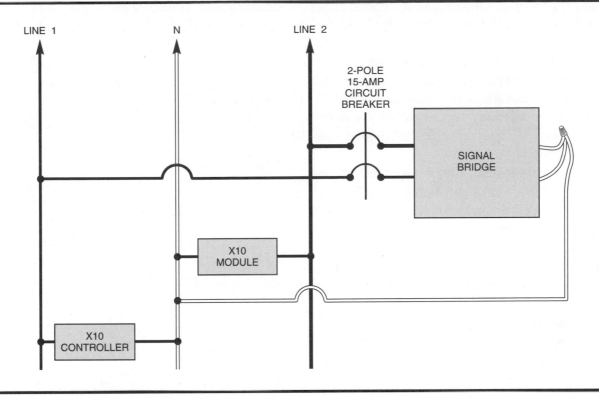

LINE 1   N   LINE 2

2-POLE
15-AMP
CIRCUIT
BREAKER

SIGNAL
BRIDGE

X10
MODULE

X10
CONTROLLER

Figure 31-3   A signal bridge phase coupler connected to the main panel to ensure proper signals on both sides of an ac line.

way that an antenna does, even from devices that are not attached to the same circuit.

In fact, where two or more houses are connected to the secondary of the same utility transformer, signals could be transmitted from one house to another. X10 filters are available that install at the service panel to block interference from outside the house.

X10 signals are relatively slow. The maximum data rate is about 60 bits per second as compared to systems that handle 10,000 bps. This means there can be a noticeable delay between the time you activate a control and the time a controlled action takes place.

STRUCTURED RESIDENTIAL WIRING SYSTEMS

Another type of system is referred to as "structured wiring" or "infrastructure wiring," and is completely separate from the conventional branch-circuit wiring in the home. It is nicely suited for new construction. Some electricians refer to this concept as "future proofing," which means wiring the new home during the original construction (rough-in) stage for all the applications that the homeowner may want now or at some later date.

These systems are wired with special cables such as unshielded twisted pair (UTP), shielded twisted pair (STP), speaker wire, coaxial, fiber optic (FO), and cables that contain multiple types of conductors, as shown in Figure 31-4. The UL Standard refers to FO cables as *optical fiber cables*. A single pair of FO cables can handle the same amount of voice traffic as 1400 pairs of copper conductors.

Structured premises wiring systems, sometimes called "home LANs" (for local area networks), are intended to allow distribution of high-speed data,

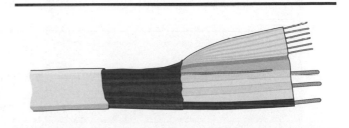

Figure 31-4   A typical cable consisting of ac power, telephone, coaxial cable, and low-voltage control conductors.

audio, and video signals throughout the home. They offer such advantages as faster Internet access for home computers and convenient networking of data devices (computers, printers, and fax machines) in home offices. Structured premises wiring systems get their name from ANSI/TIA/EIA standards 568 and 570, which describe "structured" or standardized cabling systems for businesses and homes (see box on home wiring standards).

While different systems may not be interchangeable, they all provide the same wiring infrastructure and similar product interfaces such as cables, FO cables, modules, receptacles, switches, outlets, connectors, jacks, keypads, and so forth.

Here are a few of the common devices used in a "structured" wiring installation.

Combination Outlets

Telecommunications, coaxial, and sometimes FO connectors can be combined in the same wall plates, with multiple cables to each outlet location, as shown in Figure 31-5.

Figure 31-5 A single-gang combination wall outlet with two telecommunications outlets and two coaxial cable outlets. Many other combinations are available.

Central Distribution

Cables from room outlets are run to a central cabinet, sometimes referred to as a "service-center." Except for outlets, most system components are located at this cabinet, which also functions as a patch panel for making most wiring changes at a central location, as shown in Figure 31-6.

Standards

The EIA/TIA 570-A *"Telecommunications Cabling Standard"* is available from Global Engineering, 15 Inverness Way East, Englewood, CO 80112-5776 (Phone numbers are 1-800-854-7179 and 1-303-397-7956). This standard was jointly developed by the Electronic Industries Association (EIA) and the Telecommunications Industry Association (TIA). It describes a wiring system that uses twisted

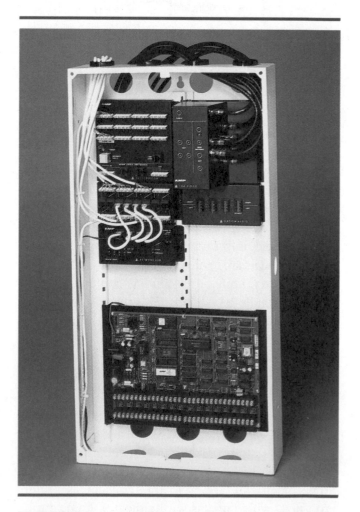

Figure 31-6 A "service-center" is the hub of a structured wiring system with telecommunications, audio, video, and home automation controllers installed.

four-pair telecommunications cabling, modular jack outlets, and star wiring (home runs) to each room from the distribution panel.

The NEMA Standard WC 63.1-2000 entitled *Performance Standard for Twisted Pair Premise Voice and Data Communications Cable* contains the minimum electrical performance as well as material and mechanical specifications, definitions, and test methods for these types of cables.

CABLE TYPES AND INSTALLATION RECOMMENDATIONS

Communication cable is tested and listed in accordance with UL Standard 444. Optical fiber cable is tested and listed in accordance to UL Standard 1651. These standards are mirror images of the EIA and TIA standards. Cables that have passed the UL testing will bear a "listing" marking.

If the cable has also passed the data transmission performance category program, the cable will be surface-marked and tag-marked, "Verified to UL Performance Category Program."

Some communities are installing underground hybrid FO/coaxial cable systems that combine telephone, CATV, and community intranet capabilities.

The categories are shown in Table 31-1.

Proper selection and installation of any type of audio/video/data cable is extremely important. The final performance will be that of the weakest link in the system. Improper installation, connection hardware, outlets, and other components can reduce the performance to that of an old-fashioned phone system.

Although you may not experience problems with audio transmission, using the proper cable for data transmission is critical. Losing your on-line Internet connection for no apparent reason could very well be pinpointed to older style untwisted-pair connecting cables that are unable to handle high-speed data transmission. Using high-quality cables is extremely important.

For residential applications, a minimum of one Category 5 (or 5e) unshielded twisted-pair (UTP) cable with four twisted pairs of 24 AWG copper conductors or two Category 5 (or 5e) UTP cables with two twisted pairs of 24 AWG copper conductors should be installed for the "home runs" between

Category 1:	Four-pair non-twisted cable. Referred to as *quad wire*. OK for audio and low-speed data. Not suitable for modern audio, data, and imaging applications. This is the old fashioned Plain Old Telephone Service (POTS) cable. If you have any, get rid of it!
Category 2:	Usually four-pair with slight twist to each pair. OK for audio and low-speed data.
Category 3:	UTP*. Data networks up to 16 MHz.
Category 4:	UTP*. Data networks up to 20 MHz.
Category 5:	UTP*. Data networks up to 100 MHz. By far the most popular, having four pairs with 24 AWG copper conductors, solid (for structured wiring) and stranded (for patch cords). Cable has an outer PVC jacket.
Category 5e:	UTP*. Basically an enhanced Cat 5 manufactured to tighter tolerances for higher performance. Data networks up to 350 MHz. This is detailed in the new TIA/EIA 568A-5 Standard. Will eventually replace Cat 5 for new installations.
Category 6:	UTP*. Four twisted pairs. Data networks up to 250 MHz. Has a spline between the four pairs to minimize interference between the pairs.
Category 7:	Four twisted pairs surrounded by a metal shield. Data networks up to 750 MHz. Specifications not yet standardized at the time of publishing this text.

* UTP means unshielded twisted-pair.

Table 31–1 Types of category-rated cables.

each outlet location and a central distribution panel where all of the audio/video/data cables are run.

There are still widespread problems with existing telephone lines outside of the home causing havoc with on-line Internet connections. There is not much you can do about this, other than complain to the telephone company.

Installation Recommendations

Because it is so easy to damage the cable, thus affecting its transmission properties, here are some cautions for installing any audio/video/data cables:

- Keep the runs as short as possible.
- Run the cables in one continuous length from Point A to Point B without splices.
- Do not to make sharp bends in the cable. Make slow sweeping bends: radius 4D for Category 5 cable, radius 10D for coaxial cable.

- Do not pull so hard as to damage the conductor insulation and the cable jacket (25 lb. is the maximum it can stand).

- Do not kink or knot the cable.

- Use rounded or depth-stop plastic staples.

- Where a number of cables are bundled, use tie-wraps loosely around the bundle, then nail the tie-wrap to the framing member.

- Do not use nail guns or metal staples. This can only lead to trouble.

- Always untwist the least amount, about ½ in. (13 mm) of the conductors at point of termination. The tighter the twist in a cable, the less distortion and interference.

- Do not run these cables through the same holes as power wiring, conduit, pipes, and ducts. There is too much possibility of damage to the cables.

- Keep cables away from hot water pipes, hot air ducts, and other sources of high heat.

- Keep cables at least 12 in. (300 mm) from power wiring where the cables are run parallel to the power wiring. This caution was written before twisted pairs and category-rated cables entered the scene. With the advent of properly installed twisted and shielded Category 5 and 5e cables, this recommendation from the past is probably not an issue. Category 5 and 5e cables are highly immune to interference.

- Run a dedicated 15-ampere 120-volt branch-circuit to the central distribution panel that will be hub of the audio/video/data system.

- Always check the specifications and recommendations of the cable you are installing.

- Consider installing a trade size 2 PVC conduit from the basement to the attic. Additional cables can be installed at a later date without time-consuming fishing of cables or damaging walls and ceilings.

Many types of bundled cables are available that contain a multitude of cables. Examples are one RG-6 (75-ohm) coaxial cable and one Cat 5 cable; two RG-6 (75-ohm) coaxial cables, two Cat 5 cables, and one ground; two FO cables, two RG-6 (75-ohm) coaxial cables, two Cat 5 cables, and one ground; or two RG-6 (75-ohm) coaxial cables, and one Cat 5 cable. The combinations are endless, and you need to check out manufacturers' catalogs to find the cable that meets your needs.

TERMINOLOGY

The letters RG stand for *Radio Grade*.

The jacks (connectors) for Cat 5 cables are generally the RJ-11 six-pin jack and the RJ-45 eight-pin jack. The letters RJ stand for *Registered Jack*.

Connectors for coaxial cable are referred to as *F* connectors. The threaded type is preferred over the push-on type.

The term M means *mega*, which is one million (for short: *meg*).

The term G means *giga*, which is one billion (for short: *gig*).

The letters *bps* mean bits per second.

The term Hz means *hertz*, which are cycles per second.

The term MHz means *megahertz*, which are millions of cycles per second.

A bit is the smallest unit of measurement in the binary system. The expression is derived from the term *binary digit*.

It takes eight bits to make one byte. A *byte* is the amount of data required to describe a single character of text.

The term Mbps means *megabits per second*. For example, 20 Mbps means that twenty million (20,000,000) *bits* of data are being transmitted per second over copper or FO cable.

There is no direct correlation between *megahertz* and *megabits*. Different encoding schemes represent information in different ways.

WIRELESS

X10 technology uses the existing wiring in your home. Structured wiring involves separate conductors and cables. Both systems use wireless remote control devices using infrared technology. Infrared technology is the heart of wireless remote control devices. These devices can turn TVs, radios, and VCRs "OFF" and "ON"; change channels and stations; control ceiling paddle fans, garage door openers, small relays that plus into a wall receptacle, or car locking systems; read bar codes at the checkout counter; "read" CDs; and so on. Sound familiar? You may not have thought about it, but you are already in the world of home automation.

SUMMARY

Since home automation systems vary, visit the many Web sites marked with an * listed after the Appendix of this text.

An excellent text on the subject is *Premises Cabling*, available from Delmar, a division of Thomson Learning. See Data and Voice Communications Cabling and Fiber Optics in the front material of this text for other related books.

REVIEW

1. One type of home automation system is referred to as:
   a. X10   b. X20   c. HAS

2. When using an X10 system, which of the following statements is most correct?
   a. A totally independent wiring system must be installed.
   b. The devices generally are "plug-n-play," requiring no additional wiring.
   c. An X10 system is ideally suited for installations in new construction.

3. A home automation system that requires a totally separate wiring system is referred to as:
   a. an X10 system.
   b. a structured system.
   c. a DC-powered system.

4. The most recommended "category-rated" cable for wiring from various outlets back to the main "service-center" for the interconnection of computers is:
   a. Category 1.
   b. Category 2.
   c. Category 3.
   d. Category 5 or 5e.

5. When installing Category 5 rated cable, the minimum bending radius generally recommended is:
   a. 4 times the diameter of the cable.
   b. 6 times the diameter of the cable.
   c. 10 times the diameter of the cable.

6. When installing coaxial cable, the minimum bending radius generally recommended is:
   a. 4 times the diameter of the cable.
   b. 6 times the diameter of the cable.
   c. 10 times the diameter of the cable.

7. When selecting cables for a structured wiring system or computer hookups, the best choice is to select a cable that has:
   a. untwisted pairs of conductors.
   b. twisted pairs of conductors.

8. Match each of the following terms with its description by drawing a line between them.

   a. RG   * cyles per second

   b. RJ   * the smallest unit of measurement in the binary system.

   c. bps   * registered jack

   d. mbps   * radio grade

   e. Hz   * bits per second

   f. bit   * megabits per second

Standby Power Systems

OBJECTIVES

After studying this unit, the student should be able to

- understand some of the safety issues concerning optional standby power systems.
- understand the basics of standby power.
- understand the types of standby power systems.
- understand wiring diagrams for portable and standby power systems.
- understand transfer switches, disconnecting means, and sizing recommendations.
- understand the *National Electric Code®* requirements for standby power systems.

Caution: The text that follows is a general discussion and overview of home-type temporary generator systems. It is not possible in this unit to cover all the variables and details involved with the many types of systems available in the marketplace today. Always follow the manufacturer's installation instructions. Always turn off the power when hooking up standby systems. Do not work on "live" equipment. Do not stand in water or otherwise work on electrical equipment with wet hands. Do not touch bare wires. Operate the generator only outside in open air. Exhaust gases contain deadly carbon monoxide, the same as an automobile engine. Carbon monoxide can cause severe nausea, fainting, or death. It is particularly dangerous because it is an odorless, colorless, tasteless, nonirritating gas. Typical symptoms of carbon monoxide poisoning are dizziness, headache, light-headedness, vomiting, stomachache, blurred vision, inability to speak clearly. If you suspect carbon monoxide poisoning, remain active. Do not sit down, lie down, or fall asleep. Breathe fresh air—fast!

Store gasoline only in approved red containers clearly marked "GASOLINE."

Use only UL-listed equipment.

WHY STANDBY (TEMPORARY) POWER?

Everyone has experienced a power outage, sometimes more often than one would like. You are left in the dark. Your furnace does not operate. You are concerned about water pipes freezing. Your refrigerator and freezer are not running. Your sump pump is not operating, and your basement is flooding. Your overhead garage door does not operate.

What can you do? The Y2K crisis made everyone aware of the consequences of a major power outage. Generators were carried out of home centers and other dealers at a furious pace. Now that the Y2K crisis is history, it still makes sense to consider a standby power system.

A few questions need to be answered. In your home, which loads are critical and must continue to operate in the event of a utility power outage? You need to know this so you can select a generator panel that will have enough branch-circuits. Only you can make this decision. Another question is, How large must the generator be to serve the loads that are considered critical in your home? Still another question is, How simple or complicated a standby system do you want?

WHAT TYPES OF STANDBY POWER SYSTEMS ARE AVAILABLE?

The Simplest

Home centers usually carry the simplest and most economical portable generators, as shown in Figure 32-1. These consist of a gasoline-driven motor/generator set that must be started manually. Some models have the recoil "pull-to-start" feature; others have battery electric start. These are the types construction workers use for temporary power on job sites. Depending upon the size, these smaller generators might be capable of running for four to eight hours. This type of portable generator is commonly referred to as "back-up" power.

These small portable generators consist of a gasoline engine driving a small electrical generator, just like a gasoline engine drives the blades of a lawn mower or snow thrower. Some generator sets have one or more standard 15- or 20-ampere single or duplex grounding-type receptacles, some have GFCI receptacles, and others have 20- and 30-ampere twistlock-type receptacles. Some are rated 120 volts, while others are rated 120/240 volts. These generators are available in sizes up to about 5000 watts. Also available are models with more "bells and whistles" such as oil alerts, adjustable output voltage, larger fuel tanks, and so on.

The procedure for this type is to start up the generator, then plug an extension cord into the receptacle and run it to whatever critical cord- and plug-connected load needs to operate.

If you are not home when a power outage occurs, you are out of luck!

The Next Step Up

The next step up involves both permanent wiring and cord- and plug-connected wiring.

This is by far the most popular when it comes to standby power for homes. In the event of a power outage, you must manually start the generator, flip the transfer switching device over from normal power to standby power, and plug in the power "patch" cord (see Figure 32-2).

The permanent wiring part of the installation involves connecting specific branch-circuits to a sep-

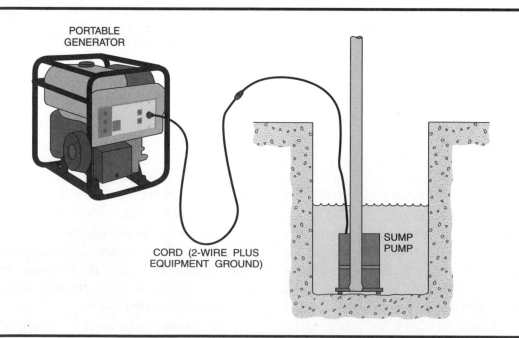

Figure 32-1 A portable generator serving standby power to a 120-volt sump pump.

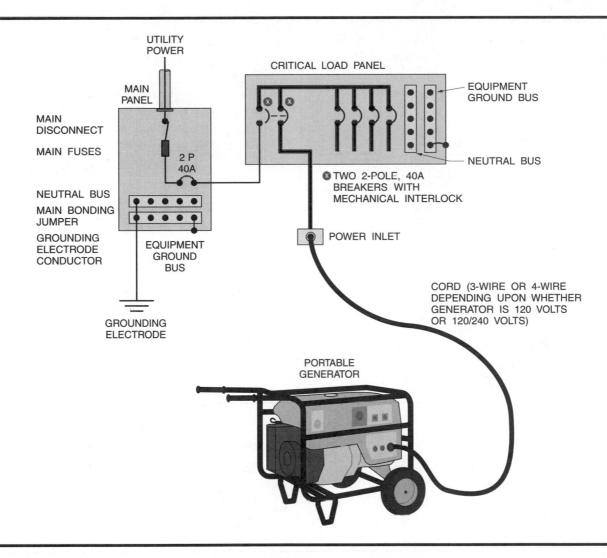

UTILITY POWER

CRITICAL LOAD PANEL

MAIN PANEL

MAIN DISCONNECT

MAIN FUSES

2 P 40A

NEUTRAL BUS

MAIN BONDING JUMPER

GROUNDING ELECTRODE CONDUCTOR

EQUIPMENT GROUND BUS

GROUNDING ELECTRODE

EQUIPMENT GROUND BUS

NEUTRAL BUS

⊗ TWO 2-POLE, 40A BREAKERS WITH MECHANICAL INTERLOCK

POWER INLET

CORD (3-WIRE OR 4-WIRE DEPENDING UPON WHETHER GENERATOR IS 120 VOLTS OR 120/240 VOLTS)

PORTABLE GENERATOR

Figure 32-2   A portable generator supplying standby power to a critical load panel. The circuit breakers in the critical load panel must be manually turned to the temporary power position. These breakers are equipped with a mechanical interlock so there will be no feedback of electricity between the generator power and the utility normal power.

arate generator panel (usually located right next to the main panel), installing and properly connecting a transfer switch (discussed later), and installing a power inlet receptacle (discussed later). The generator panel serves those circuits that you have selected as critical to operate in the event of a power outage. This panel is sometimes referred to as a generator panel, a selected load panel, an emergency panel, and sometimes a critical load panel (see Figure 32-3).

The cord- and plug-connected part of the installation consists of merely a flexible power patch cord that runs between the generator and the power inlet receptacle. This cord and the generator are set up as needed.

These are gasoline-powered generators that use a flexible four-wire rubber cord (12 AWG for 5000-watt generators, 10 AWG for 7500-watt generators) that plugs into a male polarized twistlock receptacle on the generator. The other end of the cord plugs into a polarized twistlock receptacle mounted on the outside of the house in a weatherproof power inlet box. This polarized twistlock receptacle is permanently connected to a special electrical generator panel in which the critical branch-circuits are connected. The generator panel will have a transfer mechanism plus from four to twenty branch-circuit breakers. It all depends upon the wattage rating of the generator. Some generator panels have the polarized twistlock

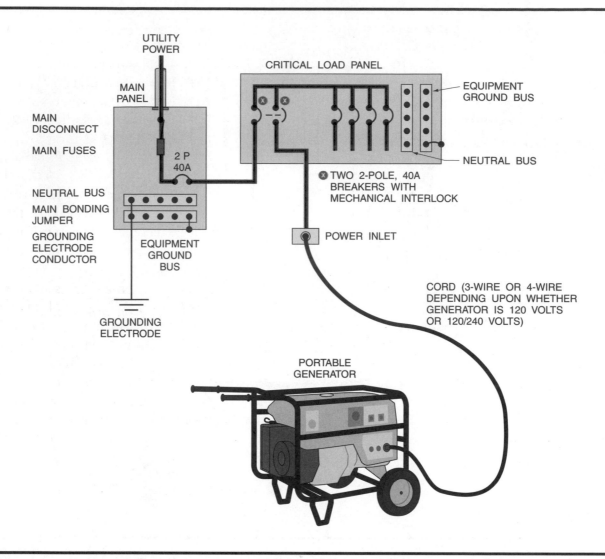

UTILITY
POWER

MAIN
PANEL

CRITICAL LOAD PANEL

EQUIPMENT
GROUND BUS

MAIN
DISCONNECT

MAIN FUSES

2 P
40A

NEUTRAL BUS

NEUTRAL BUS
MAIN BONDING
JUMPER

GROUNDING
ELECTRODE
CONDUCTOR

EQUIPMENT
GROUND
BUS

TWO 2-POLE, 40A
BREAKERS WITH
MECHANICAL INTERLOCK

POWER INLET

CORD (3-WIRE OR 4-WIRE
DEPENDING UPON WHETHER
GENERATOR IS 120 VOLTS
OR 120/240 VOLTS)

GROUNDING
ELECTRODE

PORTABLE
GENERATOR

Figure 32-3 A critical load panel served by normal power from the main service panel. A portable generator is shown plugged into the power inlet. To switch to standby power, the circuit breakers in the critical load panel must be manually turned to the standby position. These breakers are equipped with a mechanical interlock so there will be no feedback of electricity between the generator power and the utility normal power.

receptacle mounted as part of and just below the generator panel. In this case, the flexible cord is run through an open window or door to this receptacle from the generator located outside of the house.

During normal operation, this generator panel is fed from a two-pole circuit breaker in the main panel. This two-pole breaker might be rated 30, 40, 50, or 60 amperes, depending upon the number of critical branch-circuits to be supplied. When utility power is lost, this generator panel is fed from the generator. When the electric utility loses power, you must *manually* turn the "transfer switch" from its normal power position to its temporary power posi-

tion. One manufacturer of electrical equipment furnishes a panel that contains two 30-ampere, two-pole circuit breakers that are mechanically interlocked. These serve as the transfer switching means. Only one breaker can be in the "ON" position at the same time. One of these breakers brings the normal power to the generator panel. The second breaker brings the power from the generator to the generator panel. When one breaker is turned "OFF," the other is turned "ON." This panel will have four or more branch-circuit breakers.

With this system, the procedure generally is to first plug the four-wire cord into the polarized twistlock receptacle on the generator, plug the other end into the power inlet receptacle, and then start the generator, which is always located outdoors. The transfer switch in the generator panel is then switched to the "GEN" position.

When normal power is restored, turn the transfer switch in the generator panel back to its normal position, turn off the generator according to the manufacturer's instructions, unplug the power cord from the generator, and finally, unplug the other end of the cord from the power inlet receptacle.

Important: If you intend to have a GFCI breaker-protected branch-circuit transferred to the generator panel, be sure to connect this circuit to a GFCI breaker in the generator panel.

Because GFCIs and AFCIs require a neutral conductor, make sure the neutral conductor is properly connected between the main panel neutral bus and the neutral bus in the critical panel.

The neutral bus in the generator panel is insulated from the generator panel enclosure.

The ground bus in the generator panel is bonded to the generator panel enclosure.

Standby generators of this type can have run times as long as 16 hours. Again, it depends upon the loading. The manufacturer's installation and operating instructions will provide this information.

If you are not home when the power outage occurs, you are out of luck!

The Top of The Line

Standby power served by a permanently installed generator is truly standby power. These types of generator systems are covered under UL Standard 2200, *Stationary Engine Generator Assemblies*.

Standby power is a system that allows you to safely provide electrical power to your home in the event of a power outage. Instead of running extension cords from a portable generator located outside of the house to whatever critical loads (lights, sump pumps, refrigerators, etc.), you set up the critical branch-circuit wiring in a generator panel in such a manner that all that is necessary to do in the event of a power outage is to transfer the critical loads from normal utility power to generator power. The diagrams in this unit show this.

The ultimate and most expensive in standby power systems is one where the homeowner has

nothing to do in the event of a power outage. Standby power comes on when the normal utility power goes off. There is complete peace of mind! The generator for this type of system is permanently installed on a concrete pad at a suitable location outside of the home, convenient to making up all of the electrical connections to serve the selected critical loads. These systems generally run on natural gas (Figure 32-4), or propane (liquid petroleum gas—LPG) and in rare cases, on diesel fuel. All the conductors between the generator, the transfer switch, and the generator panel installed for the selected critical loads are permanently wired. The automatic transfer switch monitors the incoming electric utility's voltage. Should a power outage occur, the standby generator automatically starts, the transfer switch "transfers," and the critical loads continue to function. When normal power is restored, the transfer switch automatically "transfers" back to the normal power source, and the generator shuts down.

Since the fuel supply is a permanently installed natural gas line, these systems can run indefinitely in conformance to the manufacturer's specifications.

To be totally automatic, a solid-state controller that is part of the transfer switch immediately senses the loss of normal power and "tells" the standby generator to start, and to "tell" the transfer switch to transfer from normal power to standby power. This controller senses when normal power is restored, allowing the transfer switch to return to normal power and to shut off the standby generator. All of the circuitry is reset, and it is now ready for the next power interruption.

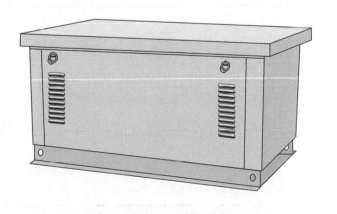

Figure 32-4 A natural gas-driven generator.

WIRING DIAGRAMS FOR A TYPICAL STANDBY GENERATOR

For simplicity, the diagrams in Figure 32-5 and Figure 32-6 are one-line diagrams. Actual wiring consists of two ungrounded conductors and one grounded "neutral" conductor. Equipment grounding of all of the components is accomplished through the metal raceways that interconnect the components. See *250.118.*

The neutral bus in a panel that serves critical loads must *not* be connected to the metal panel enclosure. Connecting the neutral, the grounding electrode conductor, and the equipment grounding conductors together is permitted only in the main service panel. If the neutral of the generator were to be connected to the metal frame of the generator, there would be a parallel path for the neutral return current through the neutral conductor, and through the metal raceway. Not a good situation.

A sign must be placed at the service-entrance main panel indicating where the standby generator is located, and what type of standby power it is. See *702.8(A).*

The total time for complete transfer to standby power is approximately 45–60 seconds.

To further give the homeowner assurance that the standby generator will operate when called upon, some systems provide automatic "exercising" of the system periodically, such as once every 7 or 14 days for a run time of 7 to 15 minutes.

Warning: When an automatic type of standby power system is in place and set in the automatic mode, the engine may crank and start at any time without warning. This would occur when the utility

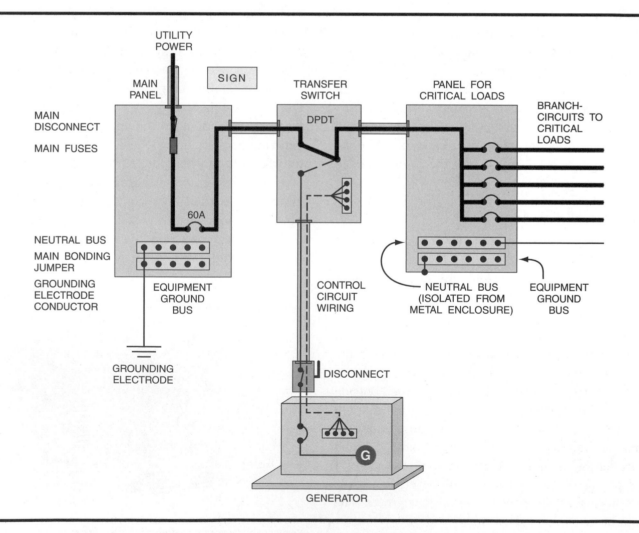

Figure 32-5   A natural gas-driven generator connected through a transfer switch. The transfer switch is in the normal power position. Power to the critical load panel is through the main panel, then through the transfer switch.

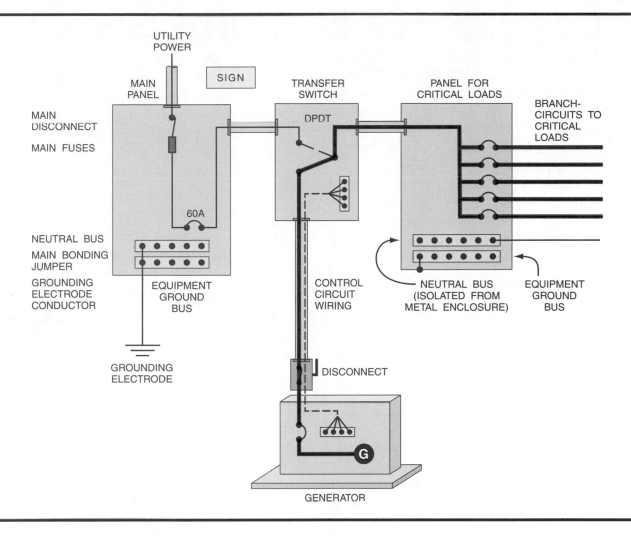

Figure 32-6 A natural gas-driven generator connected through a transfer switch. The transfer switch is in the standby position, feeding power to the critical load panel.

power supply is lost. To prevent possible injury, be alert, aware, and very careful when working on the standby generator equipment or on the transfer switch. Always turn the generator disconnect to the "OFF" position, then lock out and tag out the switch, warning others not to turn the switch back "ON." In the main panel, turn it "OFF," then lock out and tag out the circuit breaker feeding the transfer switch.

TRANSFER SWITCHES

A transfer switch shifts electrical power from the normal utility power source to the standby power source. A transfer switch isolates the utility source from the standby source in such a way that there is no feedback from the generator to the utility's system, or vice versa. When normal power is restored,

the transfer switch resets itself, and is ready for the next power outage.

Transfer switches for typical residential applications might have ratings of 40 to 200 amperes. Some have wattmeters for balancing generator loads.

For home installations, transfer switches are double-pole, double-throw (DPDT). This type of transfer switch does not break the "neutral" conductor. Technically speaking, the *NEC®* refers to this type of system as a *nonseparately derived system*. A nonseparately derived system is properly grounded through the grounding electrode system of the normal premise wiring. A system where the neutral is disconnected by the transfer switch is referred to as a *separately derived system*, and must be regrounded according to *250.30*. Separately derived systems are common in commercial and industrial installations.

If a transfer switch is connected on the line side of the main service disconnect, it must be listed as being suitable for use as service equipment. In Figure 32-5 and Figure 32-6, the transfer switch is not on the line side of the main service disconnect, and therefore does not have to be listed for use as service equipment.

A transfer switch must "break-before-make." Without the "break-before-make" feature, a dangerous situation is present and could lead to destruction of the generator, personal injury, or death. If the transfer switch does not separate the utility line from the standby power line, utility workers working on the line could be seriously injured. For low-cost, non-automatic transfer systems, the transferring means might consist of two two-pole circuit breakers that are mechanically interlocked so both cannot be "ON" at the same time.

The rating of the transfer switch must be capable of safely handling the load to be served, otherwise the transfer switch could get so hot as to cause a fire. This is nothing new; it is the same hazard as overloading a conductor. The transfer switch must also be capable of safely interrupting the available fault-current that the generator is capable of delivering.

A typical transfer switch might take 10–15 seconds to start the generator. This eliminates nuisance start ups during momentary utility power outages. After start up, another 10–15 second delay is provided to allow the voltage to stabilize. After start-up and warm-up, full transfer takes place. Similar time delay features are used for the return to normal power. Some systems can do the full transfer in a few seconds.

Only UL-listed transfer switches are permitted to be used. The UL Product Standard is 1008, *Transfer Switch Equipment*.

DISCONNECTING MEANS

A disconnecting means is required for generators, *NEC® 445.18*. This section states that "Generators shall be equipped with a disconnect by means of which the generator and all protective devices and control apparatus are able to be disconnected entirely from the circuits supplied by the generator," except where

1. The driving means for the generator can be readily shut down; and

2. The generator is not arranged to operate in parallel with another generator or other source of voltage.

   ▶*NEC® 225.31, Exception 5* states that *"For outdoor emergency legally required standby or optional standby generator sets, the disconnecting means, when listed as suitable for use as service equipment, shall be permitted to be located at the generator set."*◀

NEC® 225.34 requires that all disconnecting means shall be grouped.

NEC® 230.72(A) also requires that all disconnecting means shall be grouped. If a disconnect is located outside, it must be on the building or within sight of the building.

Where the normal power main service disconnecting means is located outside of the home and the standby generator is located nearby, the electrical inspector may permit the disconnect at the generator to be in conformance to *NEC® 225.31, Exception 5*.

Where the normal power main service disconnecting means is located inside of the home, some electrical inspectors will require that a disconnect for the standby power from the generator be installed near the main service equipment so as to conform to *225.34* and *230.72(A)*.

Having one disconnecting means inside the home and another outside of the home presents quite a challenge for firemen or others wanting to totally shut off the power to the home.

CONDUCTOR SIZE FROM STANDBY GENERATOR

The conductors that run from the standby generator to the transfer switch and to the generator panel that contains the branch-circuit overcurrent devices shall be sized not less than 115 percent of the generator's nameplate current rating. See *NEC® 445.13*.

EXAMPLE: a 7500-watt, single-phase, 120/240-volt generator.

Step 1. Calculate the generator's current rating.

$$\text{Amperes} = \frac{\text{Watts}}{\text{Volts}} = \frac{7500}{240} = 31.25 \text{ Amperes}$$

Step 2. $31.25 \times 1.15 = 35.94$ amperes.

Step 3. Refer to *NEC® Table 310.16* to determine the minimum size conductors.

In the 140°F (60°C) column of *Table 310.16*, we find that an 8 AWG copper conductor has an allowable ampacity of 40 amperes. The discussion on when to use the 140°F (60°C) and when to use the 167°F (75°C) columns of *NEC® Table 310.16* is covered in Unit 4.

The neutral conductor is permitted to be sized according to *220.22*, which under certain conditions allows the neutral conductor to be smaller than the phase conductors. This information is found in *NEC® 445.13*.

GENERATOR SIZING RECOMMENDATIONS

No matter what type or brand of generator you purchase, you will need to size it properly. There is no better way to do this than to follow the manufacturer's recommendations. Generators for home use are generally rated in watts. The manufacturer of the generator has taken into consideration that watts = volts × amperes. Some manufacturers suggest that after adding up all of the loads to be picked up by the generator, add another 20 percent capacity for future loads.

Wattage ratings of appliances vary greatly, as shown in Table 32-1. Resistive loads such as toasters, heaters, and lighting do not have an initial high inrush surge of current. Motors, on the other hand, do have a high inrush surge of starting current, which lasts for only a few seconds until the motor gets up to speed. This inrush must be taken into consideration when selecting a generator. Here are some typical *approximate* values for household loads. For heating-type appliances, the typical operating wattage values are used. For motor-operated appliances, the starting wattage values should be used. Verify actual wattage ratings by checking the nameplate on the appliance. Verify that the connected loads do not exceed the generator's marked capacity.

Most generators are capable of handling a momentary "inrush" of an extra 50 percent of their rating. Here again, check the manufacturer's specifications.

THE *NATIONAL ELECTRICAL CODE®* REQUIREMENTS

The *NEC®* addresses optional standby systems in *Article 702*. This includes those systems that are "permanently installed in their entirety," and those systems that are intended for "connection to a premises wiring system from a portable alternate power supply."

- *NEC® 702.4*: All equipment shall be approved for the intended use.

- *NEC® 702.5*: All of the equipment must have the capacity and rating for the loads that will be supplied by the equipment. The equipment shall be suitable for the maximum available fault-current at its terminals.

- *NEC® 702.6*: Transfer equipment shall be suitable for the intended use and designed to prevent the interconnection of the generator and the normal utility supply.

- *NEC® 702.8(A)*: A sign is required at the service-entrance equipment to indicate the type and location of the standby generator.

- *NEC® 702.7*: Some systems, where practical, require audible and visual signal devices to indicate when the normal power system has been transferred to the standby system. The manufacturer of the equipment provides these features.

- *NEC® 702.8(B)*: ▶Where the grounded circuit conductor connected to the optional standby generator is connected to a grounding electrode conductor at a location remote from the standby source, there shall be a sign at the grounding location that shall identify all standby and normal sources connected at that location.◀ For most residential standby systems, this connection is made on the neutral bus in the generator panel. This neutral bus in turn has a conductor run to the neutral bus in the main panel, where the neutral bus and equipment grounding bus are connected together, and a grounding electrode conductor is connected and run to the grounding electrode of the system. The neutral bus in the generator panel is isolated from the panel enclosure. The equipment grounding bus in the generator panel is bonded to the panel enclosure.

- *NEC® 702.9:* It is permissible to run standby power conductors and normal power conductors in the same raceway or enclosure.

	TYPICAL OPERATING WATTAGE REQUIREMENTS	STARTING WATTAGE REQUIREMENTS
Blender	600	700
Broiler	1600	1600
Central Air Conditioner		
10,000 BTU	1500	4500
20,000 BTU	2500	7500
24,000 BTU	3800	11,000
32,000 BTU	5000	15,000
40,000 BTU	6000	18,000
Coffee Maker	900–1750	900–1750
CD Player	50–100	50–100
Clothes Dryer		
Electric	5000 @ 240 volts	6500
Gas	700	2200
Computer	300–800	300–800
Curlers	50	50
Dehumidifier	250	350
Dishwasher	1500	2500
Electric		
Drill	250–750	300–900
Fry Pan	1300	1300
Blanket	50–200	50–200
Space Heater	1650	1650
Water Heater	5000–8000 @ 240 volts	5000–8000 @ 240 volts
Electric Range		
6-in. surface unit	1500	1500
8-in. surface unit	2100	2100
Oven	4500	4500
Lights	Add wattage of bulbs	Add wattage of bulbs
Furnace Fan		
1/8 Horsepower Motor	300	900
1/6 Horsepower Motor	500	1500
1/4 Horsepower Motor	600	1800
1/3 Horsepower Motor	700	2100
1/2 Horsepower Motor	875	2650
Garage Door Opener		
1/3 Horsepower	725	1400
1/4 Horsepower	550	1100
Hair Dryers	300–1650	300–1650
Iron	1200	1200
Microwave Oven	700–1500	1000–2300
Radio	15 to 500 (with components)	15 to 500
Refrigerator	700–1000	2200
Security, Home	25–100	25–100
Sump Pump		
1/3 Horsepower	800	2400
1/2 Horsepower	1050	3150
Television	300	300
Toaster		
4-Slice	1500	1500
2-Slice	950	950

Table 32-1   Wattage ratings of appliances (*continues*)

	TYPICAL OPERATING WATTAGE REQUIREMENTS	STARTING WATTAGE REQUIREMENTS
Automatic Washer	1150	3400
Well Pump Motor ⅓ Horsepower ½ Horsepower 1 Horsepower	800 1050 1920 @ 240 volts	2400 3150 5500
Vacuum Cleaner	1100	1600
VCR	150–250	150–250
Window Fan	200	300

Table 32-1 *(continued)*

- *NEC® 702.10(B):* Where a portable optional standby source is used as a nonseparately derived system, the equipment grounding conductor shall be bonded to the system grounding electrode. The equipment grounding conductor (green or bare) from the standby generator to the generator panel is connected to the equipment grounding bus in the generator panel. The equipment grounding bus in the generator panel is bonded to the generator panel enclosure. In turn, the equipment grounding bus in the generator panel is bonded back to the main panel either by a metallic wiring method, or by a separate equipment grounding conductor. At the main panel, the grounding electrode conductor is connected to both the equipment ground bus and the neutral bus.

Permits

Permanently installed generators require a considerable amount of electrical work and gas line piping. The installation will probably require applying for a permit so that proper inspection by the authorities can be made. Check with your local building official.

Sound Level

Because generators produce a certain level of "noise" (decibels) when running, you will want to check with your building authorities to make sure the generator you choose is in compliance with local Codes relative to any sound ordinance that might be applicable. Sound level information is provided by the manufacturer in their descriptive literature.

REVIEW

Note: Refer to the Code or the plans where necessary.

1. The basic safety rule when working with electricity is to _____

   _____ .

2. Where would the logical location be for running a portable generator?

3. The best advice to follow is to always use _____ equipment.
   Insert the appropriate word in the blank space: listed, cheapest, smallest.

4. How would you define the term "standby power?" _____

   _____

5. Describe in simple terms the three types of standby power systems.

_____

_____

_____

_____

_____

6. Is it permitted to ground the neutral conductor of the standby generator to the metal enclosure of the generator, transfer switch, or critical (generator) panel? Explain.

_____

_____

_____

7. Briefly explain the function of a transfer switch.

_____

_____

_____

8. When a transfer switch transfers to standby power, the electrical connection inside the switch (a) maintains connection to the normal power supply as it makes connection to the standby power supply, or (b) breaks the connection to the normal power supply before it makes connection to the standby power supply. Circle (a) or (b) as the correct answer.

9. A typical transfer switch for residential application is (a) a double-pole, double-throw switch (DPDT) (b) a single-pole, double-throw switch (SPDT) or (c) a double-pole, single-throw switch (DPST). Circle the correct answer.

10. The technical *NEC*® definition of a system in which the neutral conductor is not switched is referred to as (a) a separately derived system (b) a nonseparately derived system. Circle the letter indicating the correct definition.

11. The *NEC*® requires that a _____ means be provided for standby power systems. In the case of portable gen sets, this might be as simple as pulling out the plug on the extension cord that is plugged into the receptacle on the gen set. For permanently installed standby power generators, this might be on the gen set and/or separately provided inside or outside of the home.

12. The conductors running from a permanently installed standby power generator to the transfer switch shall not be less than (100%) (115%) (125%) of the generator's output rating.

Specifications for Electrical Work—Single-Family Dwelling

1. **GENERAL:**   The "General Clause and Conditions" shall be and are hereby made a part of this division.

2. **SCOPE:**   The electrical contractor shall furnish and install a complete electrical system as shown on the drawings and/or in the specifications. Where there is no mention of the responsible party to furnish, install, or wire for a specific item on the electrical drawings, the electrical contractor will be responsible completely for all purchases and labor for a complete operating system for this item.

3. **WORKMANSHIP:**   All work shall be executed in a neat and workmanlike manner. All exposed conduits shall be routed parallel or perpendicular to walls and structural members. Junction boxes shall be securely fastened, set true and plumb, and flush with finished surface when wiring method is concealed.

4. **LOCATION OF OUTLETS:**   The electrical contractor shall verify location, heights, outlet and switch arrangements, and equipment prior to rough-in. No additions to the contract sum will be permitted for outlets in wrong locations, in conflict with other work, and so on. The owner reserves the right to relocate any device up to 10 ft (3.0 m) prior to rough-in, without any charge by the electrical contractor.

5. **CODES:**   The electrical installation is to be in accordance with the latest edition of the *National Electrical Code,*® all local electrical codes, and the utility company's requirements.

6. **MATERIALS:**   All materials shall be new and shall be listed and bear the appropriate label of Underwriters Laboratories, Inc., or another nationally recognized testing laboratory for the specific purpose. The material shall be of the size and type specified on the drawings and/or in the specifications.

7. **WIRING METHOD:**   Wiring, unless otherwise specified, shall be nonmetallic-sheathed cable, armored cable, or electrical metallic tubing (EMT), adequately sized and installed according to the latest edition of the *National Electrical Code*® and local ordinances.

8. **PERMITS AND INSPECTION FEES:**   The electrical contractor shall pay for all permit fees, plan review fees, license fees, inspection fees, and taxes applicable to the electrical installation and shall be included in the base bid as part of this contract.

9. **TEMPORARY WIRING:**   The electrical contractor shall furnish and install all temporary wiring for handheld tools and construction lighting per latest OSHA standards and *Article 527, NEC,*® and include all costs in base bid.

10. **WORKSHOP:**   Workshop wiring is to be installed in EMT using steel compression gland fittings.

11. **NUMBER OF OUTLETS PER CIRCUIT:**   In general, not more than ten (10) lighting and/or receptacle outlets shall be connected to any one lighting branch-circuit. Exceptions may be made in the case of low-current-consuming outlets.

12. **CONDUCTOR SIZE:**   General lighting branch-circuits shall be 14 AWG copper protected by 15-ampere overcurrent devices. Small appliance circuits shall be 12 AWG copper protected by 20-ampere overcurrent devices.

   All other circuits: conductors and overcurrent devices as required by the Code.

13. **LOAD BALANCING:**   The electrical contractor shall connect all loads, branch-circuits, and feeders per the Panel Schedule, but shall verify and modify these connections as required to balance connected and computed loads to within 10 percent variation.

14. **SPARE CONDUITS:**   Furnish and install two empty trade size 1 EMT conduits between workshop and attic for future use.

15. **GUARANTEE OF INSTALLATION:**   The electrical contractor shall guarantee all work and materials for a period of one full year after final acceptance by the architect/engineer, electrical inspector, and owner.

16. **APPLIANCE CONNECTIONS:**   The electrical contractor shall furnish all wiring materials and make all final electrical connections for all permanently installed appliances such as, but not limited to, the furnace, water heater, water pump, built-in ovens and ranges, food waste disposer, dishwasher, and clothes dryer.

   These appliances are to be furnished by the owner.

17. **CHIMES:**   Furnish and install two (2) two-tone door chimes where indicated on the plans, complete with two (2) push buttons and a suitable chime transformer. Allow $150.00 for above items. Chimes and buttons to be selected by owner.

18. **DIMMERS:**   Furnish and install dimmer switches where indicated.

19. **EXHAUST FANS:**   Furnish, install, and provide connections for all exhaust fans indicated on the plans, including, but not limited to, ducts, louvers, trims, speed controls, and lamps. Included are the recreation room, laundry, rear entry powder room, range hood, and bedroom hall ceiling fans. Allow a sum of $700.00 in the base bid for this. This allowance does not include the two bathroom heat/vent/light units.

20. **LUMINAIRES (FIXTURES):**   A luminaire (fixture) allowance of $1,500.00 shall be included in the electrical contractor's bid for all surface-mounted luminaires (fixtures) and post lanterns, not including the ceiling paddle fans. Luminaires (fixtures) to be selected by owner. This allowance includes the three (3) medicine cabinets.

   The ceiling paddle fans, speed controls, and lamps for same shall be furnished by the owner and installed by the contractor.

   The electrical contractor shall:

   a. furnish and install all incandescent and fluorescent recessed luminaires (fixtures).

   b. install all surface, recessed, track, strip, pendant, and hanging luminaires (fixtures), and post lanterns.

   c. furnish and install two 2-lamp fluorescent luminaires (fixtures) above the workbench.

   d. furnish and install all lamps except as noted. Fluorescent lamps shall be Watt-Mizer F40 SP30/RS/WM or equivalent. Incandescent lamps shall be energy-efficient Watt-Mizer lamps or equivalent. Lamps for ceiling paddle fans to be furnished by owner.

   e. furnish and install all porcelain pull-chain and keyless lampholders.

   This luminaire (fixture) allowance does not include the two bathroom ceiling heat/vent/light units. See Clause 21.

21. **HEAT/VENT/LIGHT CEILING UNITS:** Furnish and install two heat/vent/light units where indicated on the plans, complete with switch assembly, ducts, and louvers required to perform the heating, venting, and lighting operations as recommended by the manufacturer.

22. **PLUG-IN STRIP:** Where noted in the workshop, furnish and install a multioutlet assembly with six outlets.

23. **SWITCHES, RECEPTACLES, AND FACEPLATES:** All flush switches shall be of the quiet ac-rated toggle type. They shall be mounted 46 in. (1.15 m) to center above the finished floor unless otherwise noted.

    Receptacle outlets shall be mounted 12 in. (300 mm) to center above the finished floor unless otherwise noted. All convenience receptacles shall be of the grounding type. Furnish and install, where indicated, ground-fault circuit interrupter (GFCI) receptacles to provide ground-fault circuit protection as required by the *National Electrical Code.* All wiring devices are to be provided with ivory handles or faces and shall be trimmed with ivory faceplates except in the kitchen, where chrome-plated steel faceplates shall be used.

    Receptacle outlets, where indicated, shall be of the split-circuit design.

24. **TELEVISION OUTLETS:** Furnish and install 4-in. square, 1½-in.-deep outlet boxes with single-gang raised plaster covers at each television outlet where noted on the plans. Mount at the same height as receptacle outlets. Furnish and install 75-ohm coaxial cable to each television outlet from a point in the workshop near the main service-entrance switch. Allow 6 ft (1.8 m) of cable in workshop. Furnish and install television plug-in jacks at each location. Faceplates are to match other faceplates in home. All remaining work done by others.

25. **TELEPHONES:** Furnish and install a 3-in.-deep device box or 4-in. square box, 1½ in. deep, with a suitable single-gang raised plaster cover at each telephone location, as indicated on the plans.

    Furnish and install four-conductor 18 AWG copper telephone cables from the telephone company's point of demarcation near the service-entrance panel to each designated telephone location. Terminate in a proper modular jack, complete with faceplates that match the electrical device faceplates. Allow 6 ft (1.8 m) of cable to hang below ceiling joists. Telephone company to furnish, install, and connect any and all equipment (including grounding connection) up to and including their Standard Network Interface (SNI)device.

    Installation shall be in accordance with any and all applicable *National Electrical Code®* and local code regulations.

26. **MAIN SERVICE PANEL:** Furnish and install in the Workshop where indicated on the plans one (1) 200-ampere, 120/240-volt, single-phase, three-wire load center with a 200-ampere main circuit breaker. Load center shall be series-rated for 22,000 amperes available fault current.

    Furnish and install all active and spare breakers as indicated on the Panel Schedule.

27. **SERVICE-ENTRANCE UNDERGROUND LATERAL CONDUCTORS:** To be furnished and installed by the utility. Meter equipment (pedestal type) to be furnished by the utility and installed by the electrical contractor where indicated on plans. Electrical contractor to furnish and install all panels, conduits, fittings, conductors, and other materials required to complete the service-entrance installation from the demarcation point of the utility's equipment to and including the Main Panel.

28. **SERVICE-ENTRANCE CONDUCTORS:** Service-entrance conductors supplied by the electrical contractor shall be two 2/0 AWG THW, THHW, THWN, or XHHW phase conductors and one 1 AWG bare neutral conductor. Install trade size 1½ EMT from Main Panel A to the meter pedestal.

29. **BONDING AND GROUNDING:**   Bond and ground service-entrance equipment in accordance with the latest edition of the *National Electrical Code,*®local, and utility code requirements. Install one 4 AWG copper grounding electrode conductor.

30. **SUBPANEL:**   Furnish and install in the Recreation Room where indicated on the plans one (1) 20-circuit, 125-ampere, 120/240-volt, single-phase, three-wire MLO load center. Load center shall be series-rated for 22,000 amperes available fault-current.

    Furnish and install all active and spare breakers as indicated on the Panel Schedule.

    Feed load center with three 3 AWG THHN or THWN conductors from a 100-ampere, two-pole breaker in Main Panel A. Install conductors in trade size 1 EMT.

31. **CIRCUIT IDENTIFICATION:**   All panelboards shall be furnished with typed-card directories with proper designation of the branch-circuit loads, feeder loads, and equipment served. The directories shall be located in the panel in a holder for clear viewing.

32. **SEALING PENETRATIONS:**   The electrical contractor shall seal and weatherproof all penetrations through foundations, exterior walls, and roofs.

33. **FINAL SITE CLEAN-UP:**   Upon completion of the installation, the electrical contractor shall review and check the entire installation, clean equipment and devices, and remove surplus materials and rubbish from the owner's property, leaving the work in neat and clean order and in complete working condition. The electrical contractor shall be responsible for the removal of any cartons, debris, and rubbish for equipment installed by the electrical contractor, including equipment furnished by the owner or others and removed from the carton by the electrical contractor.

34. **SMOKE DETECTORS:**   Furnish and install all smoke detectors and associated wiring per manufacturer's instructions and all codes. See Clause 5. Detectors to be of the ac/dc type. Interconnect to provide simultaneous signaling.

35. **ALTERNATIVE BID   Low-Voltage, Remote-Control System:**   The electrical contractor shall submit an alternative bid on the following:

    Furnish and install a complete low-voltage remote-control system to accomplish the same results as would be obtained with the conventional switching arrangement as indicated on the electrical plans.

    In addition, furnish and install two eight-position master selector switches, one in the master bedroom and one in the front hall or as directed by the architect or owner. Outlets to be controlled by this switch to be selected by the owner.

    All low-voltage wiring to conform to the *National Electrical Code.*®

36. **SPECIAL-PURPOSE OUTLETS:**   Install, provide, and connect all wiring for all special-purpose outlets. Upon completion of the job, all luminaires (fixtures) and appliances shall be operating properly. See plans and other sections of the specifications for information as to who is to furnish the luminaires (fixtures) and appliances.

37. ▶Furnish and install ground-fault circuit interrupter (GFCI) and arc-fault circuit interrupter (AFCI) protection as required by the latest edition of the *National Electrical Code.*®◄

SCHEDULE OF SPECIAL-PURPOSE OUTLETS

SYMBOL	DESCRIPTION	VOLTS	HORSE-POWER	APPLIANCE AMPERE RATING	TOTAL APPLIANCE WATTAGE RATING (OR VA)	CIRCUIT AMPERE RATING	POLES	WIRE SIZE THHN	CIRCUIT NUMBER	COMMENTS
▲A	Hydromassage tub, master bedroom	120	½	10	1200	15	1	14	A9	Connect to separate circuit. Class "A" GFCI circuit breaker.
▲B	Water pump	240	1	8	1920	20	2	12	A(5–7)	Run circuit to disconnect switch on wall adjacent to pump. Install 10-ampere dual-element fuses as "back-up" protection. Pump controller has integral overload protection.
▲C	Water heater: top element 2000 W. Bottom element 3000 W.	240	–0–	12.5	3000	20	2	12	A(6–8)	Connected for limited demand. Both elements cannot operate at the same time.
▲D	Dryer	120/240	120 V ⅙ Motor Only	23.75	5700 Total	30	2	10	B(1–3)	Provide flush-mounted 30-A dryer receptacle.
▲E	Overhead garage door opener	120	¼	5.8	696	15	1	14	B14	Unit comes with 3-W cord. Provide box-cover unit (fuse/switch). Install Fustat Type S fuse, 8 amperes. Unit has integral protection. Fustat fuses are additional "back-up" protection. Connect to garage lighting circuit.
▲F	Wall-mounted oven	120/240	–0–	27.5	6600	30	2	10	B(6–8)	
▲G	Countertop range	120/240	–0–	31	7450	40	2	8	B(2–4)	
▲H	Food waste disposer	120	⅓	7.2	864	20	1	12	B19	Controlled by S.P. switch on wall to the right of sink.
▲I	Dishwasher	120	⅓ Motor Only	Motor 7.20 Htr. 8.33 Total 15.53	864 1000 W 1864	20	1	12	B5	
▲J	Heat/vent/light master B.R. bath	120	–0–	12.5	1500	20	1	12	A12	
▲K	Heat/vent/light front B.R. bath	120	–0–	12.5	1500	20	1	12	A11	
▲L	Attic exhaust fan	120	¼	5.8	696	15	1	14	A10	Run circuit to 4-in. square box. Locate near fan in attic. Provide box-cover unit. Install Fustat Type S fuse, 8 amperes. Unit has integral protection. Fustat fuses provide additional back-up protection.
▲M	Electric furnace	240	⅓ Motor Only	Motor 3.5 Htr. 50.7 Total 54.2	13000	70	2	4	A(1–3)	The overcurrent device and branch-circuit conductors shall not be less than 125% of the total load of the heaters and motor. 54.2 × 1.25 = 67.75 (424.3(B)).
▲N	Air conditioner	240	–0–	26.4	6336	45	2	10	A(2–4)	
▲O	Freezer	120	¼	5.8	696	15	1	14	A13	Install single receptacle outlet. Do not provide GFCI protection.

USEFUL FORMULAS

TO FIND	SINGLE PHASE	THREE PHASE	DIRECT CURRENT
AMPERES when kVA is known	$\dfrac{kVA \times 1000}{E}$	$\dfrac{kVA \times 1000}{E \times 1.73}$	not applicable
AMPERES when horsepower is known	$\dfrac{HP \times 746}{E \times \% \text{ eff.} \times pf}$	$\dfrac{HP \times 746}{E \times 1.73 \times \% \text{ eff.} \times pf}$	$\dfrac{HP \times 746}{E \times \% \text{ eff.}}$
AMPERES when kilowatts are known	$\dfrac{kW \times 1000}{E \times pf}$	$\dfrac{kW \times 1000}{E \times 1.73 \times pf}$	$\dfrac{kW \times 1000}{E}$
HORSEPOWER	$\dfrac{I \times E \times \% \text{ eff.} \times pf}{746}$	$\dfrac{I \times E\ 1.73 \times \% \text{ eff.} \times pf}{746}$	$\dfrac{I \times E \times \% \text{ eff.}}{746}$
KILOVOLT AMPERES	$\dfrac{I \times E}{1000}$	$\dfrac{I \times E \times 1.73}{1000}$	not applicable
KILOWATTS	$\dfrac{I \times E \times pf}{1000}$	$\dfrac{I \times E \times 1.73 \times pf}{1000}$	$\dfrac{I \times E}{1000}$
WATTS	$E \times I \times pf$	$E \times I \times 1.73 \times pf$	$E \times I$

$$\text{ENERGY EFFICIENCY} = \frac{\text{Load Horsepower} \times 746}{\text{Load Input kVA} \times 1000}$$

$$\text{POWER FACTOR (pf)} = \frac{\text{Power Consumed}}{\text{Apparent Power}} = \frac{W}{VA} = \frac{kW}{kVA} = \cos\varnothing$$

I = Amperes	E = Volts	kW = Kilowatts	kVA = Kilovolt-amperes
HP = Horsepower	% eff. = Percent Efficiency		pf = Power Factor
	e.g., 90% eff. is 0.90		e.g., 95% pf is 0.95

EQUATIONS BASED ON OHM'S LAW:

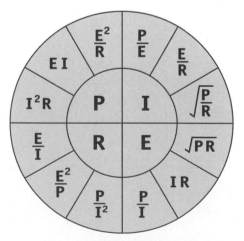

P = POWER, IN WATTS
I = CURRENT, IN AMPERES
R = RESISTANCE, IN OHMS
E = ELECTROMOTIVE FORCE, IN VOLTS

METRIC SYSTEM OF MEASUREMENT

The *National Electrical Code®* contains both English (inch-pound) and metric measurements. Whenever possible, use the measurements found in the *NEC®* instead of making your own conversions.

The following table provides useful conversions of English Customary terms to SI terms, and SI terms to English Customary terms. The metric system is known as the International System of Units (SI), taken from the French *Le Système International d'Unites*. Whenever the slant line (/) is found, say it as "per." Be practical when using the values in the following table. Present-day calculators and computers provide many "places" beyond the decimal point. You must decide how accurate your results must be when performing a calculation. When converting centimeters to feet, for example, one could be reasonably accurate by rounding the value of 0.03281 to 0.033. When a manufacturer describes a product's measurement using metrics, but does not physically change the product, it is called a *soft conversion* (mathematically identical). When a manufacturer actually changes the physical measurements of the product to standard metric size or a rational whole number of metric units, it is called a *hard conversion* (rounded). When rounding off numbers, be sure to round off in such a manner that the final result does not violate a "maximum" or "minimum" value, such as might be required by the *National Electrical Code.®* The rule for discarding digits is that if the digits to be discarded begin with a 5 or more, increase the last digit retained by one. For example: If rounded to three digits, 6.3745 would become 6.37. If rounded to four digits, it would become 6.375. The following information has been developed from the latest U.S. Department of Commerce, National Institute of Standards and Technology publications covering the metric system.

Multiply This Unit(s)	By This Factor	To Obtain This Unit(s)
acre	4 046.9	square meters (m²)
acre	43 560	square feet (ft²)
ampere hour	3 600	coulombs (C)
angstrom	0.1	nanometer (nm)
atmosphere	101.325	kilopascals (kPa)
atmosphere	33.9	feet of water (at 4°C)
atmosphere	29.92	inches of mercury (at 0°C)
atmosphere	0.76	meter of mercury (at 0°C)
atmosphere	0.007 348	ton per square inch
atmosphere	1.058	tons per square foot
atmosphere	1.0333	kilograms per square centimeter
atmosphere	10 333	kilograms per square meter
atmosphere	14.7	pounds per square inch
bar	100	kilopascals (kPa)
barrel (oil, 42 U.S. gallons)	0.158 987 3	cubic meter (m³)
barrel (oil, 42 U.S. gallons)	158.987 3	liters (L)
board foot	0.002 359 737	cubic meter (m³)
bushel	0.035 239 07	cubic meter (m³)
Btu	778.16	foot-pounds
Btu	252	gram-calories
Btu	0.000 393 1	horsepower-hour
Btu	1 054.8	joules (J)
Btu	1.055 056	kilojoules (kJ)
Btu	0.000 293 1	kilowatt-hour (kWh)
Btu per hour	0.000 393 1	horsepower (hp)
Btu per hour	0.293 071 1	watt (W)
Btu per degree Fahrenheit	1.899 108	Kilojoules per kelvin (kJ/K)
Btu per pound	2.326	Kilojoules per kilogram (kJ/kg)
Btu per second	1.055 056	kilowatts (kW)
calorie	4.184	joules (J)
calorie, gram	0.003 968 3	Btu
candela per feet squared (cd/ft²)	10.763 9	candelas per meter squared (cd/m²)

Note: The former term *candlepower* has been replaced with the term *candela*.

candela per meter squared (cd/m²)	0.092 903	candela per feet squared (cd/ft²)
candela per meter squared (cd/m²)	0.291 864	footlambert*
candela per square inch (cd/in²)	1 550.003	candelas per meter squared (cd/m²)

Celsius = (Fahrenheit − 32) × 5/9
Celsius = (Fahrenheit − 32) × 0.555555

Multiply This Unit(s)	By This Factor	To Obtain This Unit(s)
Celsius = (0.556 × Fahrenheit) − 17.8		

Note: The term *centigrade* was officially discontinued in 1948, and was replaced by the term *Celsius*. The term *centigrade* may still be found in some publications.

Multiply This Unit(s)	By This Factor	To Obtain This Unit(s)
centimeter (cm)	0.032 81	foot (ft)
centimeter (cm)	0.393 7	inch (in.)
centimeter (cm)	0.01	meter (m)
centimeter (cm)	10	millimeters (mm)
centimeter (cm)	393.7	mils
centimeter (cm)	0.010 94	yard
circular mil	0.000 005 067	square centimeter (cm²)
circular mil	0.785 4	square mil
circular mil	0.000 000 785 4	square inch (in²)
cubic centimeter (cm³)	0.061 02	cubic inch (in³)
cubic foot per second	0.028 316 85	cubic meter per second (m³/s)
cubic foot per minute	0.000 471 947	cubic meter per second (m³/s)
cubic foot per minute	0.471 947	liter per second (L/s)
cubic inch (in³)	16.39	cubic centimeters (cm³)
cubic inch (in³)	0.000 578 7	cubic foot (ft³)
cubic meter (m³)	35.31	cubic feet (ft³)
cubic yard per minute	12.742 58	liters per second (L/s)
cup (c)	0.236 56	liter (L)
decimeter	0.1	meter (m)
dekameter	10	meters (m)
Fahrenheit = (9/5 Celsius) + 32		
Fahrenheit = (Celsius × 1.8) + 32		
fathom	1.828 804	meters (m)
fathom	6.0	feet
foot	30.48	centimeters (cm)
foot	12	inches
foot	0.000 304 8	kilometer (km)
foot	0.304 8	meter (m)
foot	0.000 189 4	mile (statute)
foot	304.8	millimeters (mm)
foot	12 000	mils
foot	0.333 33	yard
foot, cubic (ft³)	0.028 316 85	cubic meter (m³)
foot, cubic (ft³)	28.316 85	liters (L)
foot, board	0.002 359 737	cubic meter (m³)
cubic feet per second (ft³/s)	0.028 316 85	cubic meter per second (m³/s)
cubic feet per minute (ft³/min)	0.000 471 947	cubic meter per second (m³/s)
cubic feet per minute (ft³/min)	0.471 947	liter per second (L/s)
foot, square (ft²)	0.092 903	square meter (m²)
footcandle	10.763 91	lux (lx)
footlambert*	3.426 259	candelas per square meter (cd/m²)
foot of water	2.988 98	kilopascals (kPa)
foot pound	0.001 286	Btu
foot pound-force	1 055.06	joules (J)
foot pound-force per second	1.355 818	joules (J)
foot pound-force per second	1.355 818	watts (W)
foot per second	0.304 8	meter per second (m/s)
foot per second squared	0.304 8	meter per second squared (m/s²)
gallon (U.S. liquid)	3.785 412	liters (L)
gallons per day	3.785 412	liters per day (L/d)
gallons per hour	1.051 50	milliliters per second (mL/s)
gallons per minute	0.063 090 2	liter per second (L/s)
gauss	6.452	lines per square inch
gauss	0.1	millitesla (mT)
gauss	0.000 000 064 52	weber per square inch
grain	64.798 91	milligrams (mg)
gram (g) (a little more than the weight of a paper clip)	0.035 274	ounce (avoirdupois)
gram (g)	0.002 204 6	pound (avoirdupois)
gram per meter (g/m)	3.547 99	pounds per mile (lb/mile)
grams per square meter (g/m²)	0.003 277 06	ounces per square foot (oz/ft²)
grams per square meter (g/m²)	0.029 494	ounces per square yard (oz/yd²)
gravity (standard acceleration)	9.806 65	meters per second squared (m/s²)
quart (U.S. liquid)	0.946 352 9	liter (L)
horsepower (550 ft•lbf/s)	0.745 7	kilowatt (kW)

Multiply This Unit(s)	By This Factor	To Obtain This Unit(s)
horsepower	745.7	watts (W)
horsepower hours	2.684 520	megajoules (MJ)
inch per second squared (in./s²)	0.025 4	meter per second squared (m/s²)
inch	2.54	centimeters (cm)
inch	0.254	decimeter (dm)
inch	0.083 3	feet
inch	0.025 4	meter (m)
inch	25.4	millimeters (mm)
inch	1 000	mils
inch	0.027 78	yard
inch, cubic (in³)	16 387.1	cubic millimeters (mm³)
inch, cubic (in³)	16.387 06	cubic centimeters (cm³)
inch, cubic (in³)	645.16	square millimeters (mm²)
inches of mercury	3.386 38	kilopascals (kPa)
inches of mercury	0.033 42	atmosphere
inches of mercury	1.133	feet of water
inches of water	0.248 84	kilopascal (kPa)
inches of water	0.073 55	inch of mercury
joule (J)	0.737 562	foot pound-force (ft•lbf)
kilocandela per meter squared (kcd/m²)	0.314 159	lambert*
kilogram (kg)	2.204 62	pounds (avoirdupois)
kilogram (kg)	35.274	ounces (avoirdupois)
kilogram per meter (kg/m)	0.671 969	pound per foot (lb/ft)
kilogram per square meter (kg/m²)	0.204 816	pound per square foot (lb/ft²)
kilogram meter squared (kg•m²)	23.730 4	pounds foot squared (lb•ft²)
kilogram meter squared (kg•m²)	3 417.17	pounds inch squared (lb•in²)
kilogram per cubic meter (kg/m³)	0.062 428	pound per cubic foot (lb/ft³)
kilogram per cubic meter (kg/m³)	1.685 56	pound per cubic yard (lb/yd³)
kilogram per second (kg/s)	2.204 62	pounds per second (lb/s)
kilojoule (kJ)	0.947 817	Btu
kilometer (km)	1 000	meters (m)
kilometer (km)	0.621 371	mile (statute)
kilometer (km)	1 000 000	millimeters (mm)
kilometer (km)	1 093.6	yards
kilometer per hour (km/h)	0.621 371	mile per hour (mph)
kilometer squared (km²)	0.386 101	square mile (mile²)
kilopound-force per square inch	6.894 757	megapascals (MPa)
kilowatts (kW)	56.921	Btus per minute
kilowatts (kW)	1.341 02	horsepower (hp)
kilowatts (kW)	1 000	watts (W)
kilowatt-hour (kWh)	3 413	Btus
kilowatt-hour (kWh)	3.6	megajoules (MJ)
knots	1.852	kilometers per hour (km/h)
lamberts*	3 183.099	candelas per square meter (cd/m²)
lamberts*	3.183 01	kilocandelas per square meter (kcd/m²)
liter (L)	0.035 314 7	cubic foot (ft³)
liter (L)	0.264 172	gallon (U.S. liquid)
liter (L)	2.113	pints (U.S. liquid)
liter (L)	1.056 69	quarts (U.S. liquid)
liter per second (L/s)	2.118 88	cubic feet per minute (ft³/min)
liter per second (L/s)	15.850 3	gallons per minute (gal/min)
liter per second (L/s)	951.022	gallons per hour (gal/hr)
lumen per square foot (lm/ft²)	10.763 9	lux (lx)
lumen per square foot (lm/ft²)	1.0	footcandles
lumen per square foot (lm/ft²)	10.763	lumens per square meter (lm/m²)
lumen per square meter (lm/m²)	1.0	lux (lx)
lux (lx)	0.092 903	lumen per square foot (foot-candle)
maxwell	10	nanowebers (nWb)
megajoule (MJ)	0.277 778	kilowatt hours (kWh)
meter (m)	100	centimeters (cm)
meter (m)	0.546 81	fathom
meter (m)	3.2809	feet
meter (m)	39.37	inches
meter (m)	0.001	kilometer (km)
meter (m)	0.000 621 4	mile (statute)
meter (m)	1 000	millimeters (mm)

Multiply This Unit(s)	By This Factor	To Obtain This Unit(s)
meter (m)	1.093 61	yards
meter, cubic (m³)	1.307 95	cubic yards (yd³)
meter, cubic (m³)	35.314 7	cubic feet (ft³)
meter, cubic (m³)	423.776	board feet
meter per second (m/s)	3.280 84	feet per second (ft/s)
meter per second (m/s)	2.236 94	miles per hour (mph)
meter, squared (m²)	1.195 99	square yards (yd²)
meter, squared (m²)	10.763 9	square feet (ft²)
mho per centimeter (mho/cm)	100	siemens per meter (S/m)

Note: The older term *mho* has been replaced with *siemens*. The term *mho* may still be found in some publications.

micro inch	0.025 4	micrometer (μm)
mil	0.002 54	centimeters
mil	0.000 083 33	feet
mil	0.001	inches
mil	0.000 000 025 40	kilometers
mil	0.000 027 78	yards
mil	25.4	micrometers (μm)
mil	0.025 4	millimeter (mm)
miles per hour	1.609 344	kilometers per hour (km/h)
miles per hour	0.447 04	meter per second (m/s)
miles per gallon	0.425 143 7	kilometer per liter (km/L)
miles	1.609 344	kilometers (km)
miles	5 280	feet
miles	1 609	meters (m)
miles	1 760	yards
miles (nautical)	1.852	kilometers (km)
miles squared	2.590 000	kilometers squared (km²)
millibar	0.1	kilopascal (kPa)
milliliter (mL)	0.061 023 7	cubic inch (in³)
milliliter (mL)	0.033 814	fluid ounce (U.S.)
millimeter (about the thickness of a dime)	0.1	centimeter (cm)
millimeter (mm)	0.003 280 8	foot
millimeter (mm)	0.039 370 1	inch
millimeter (mm)	0.001	meter (m)
millimeter (mm)	39.37	mils
millimeter (mm)	0.001 094	yard
millimeter squared (mm²)	0.001 550	square inch (in²)
millimeter cubed (mm³)	0.000 061 023 7	cubic inches (in³)
millimeter of mercury	0.133 322 4	kilopascal (kPa)
ohm	0.000 001	megohm
ohm	1 000 000	micro ohms
ohm circular mil per foot	1.662 426	nano ohms meter (nΩ•m)
oersted	79.577 47	amperes per meter (A/m)
ounce (avoirdupois)	28.349 52	grams (g)
ounce (avoirdupois)	0.062 5	pound (avoirdupois)
ounce, fluid	29.573 53	milliliters (mL)
ounce (troy)	31.103 48	grams (g)
ounce per foot squared (oz/ft²)	305.152	grams per meter squared (g/m²)
ounces per gallon (U.S. liquid)	7.489 152	grams per liter (g/L)
ounce per yard squared (oz/yd²)	33.905 7	grams per meter squared (g/m²)
pica	4.217 5	millimeters (mm)
pint (U.S. liquid)	0.473 176 5	liter (L)
pint (U.S. liquid)	473.177	milliliters (mL)
pound (avoirdupois)	453.592	grams (g)
pound (avoirdupois)	0.453 592	kilograms (kg)
pound (avoirdupois)	16	ounces (avoirdupois)
poundal	0.138 255	newton (N)
pound foot (lb•ft)	0.138 255	kilogram meter (kg•m)
pound foot per second	0.138 255	kilogram meter per second (kg•m/s)
pound foot squared (lb•ft²)	0.042 140 1	kilogram meter squared (kg•m²)
pound-force	4.448 222	newtons (N)
pound-force foot	1.355 818	newton meters (N•m)
pound-force inch	0.112 984 8	newton meter (N•m)
pound-force per square inch	6.894 757	kilopascals (kPA)
pound-force per square foot	0.047 880 26	kilopascal (kPa)

Multiply This Unit(s)	By This Factor	To Obtain This Unit(s)
pound per cubic foot (lb/ft³)	16.018 46	kilograms per cubic meter (kg/m³)
pound per foot (lb/ft)	1.488 16	kilograms per meter (kg/m)
pound per foot squared (lb/ft²)	4.882 43	kilograms per meter squared (kg/m²)
pound per foot cubed (lb/ft³)	16.018 5	kilograms per meter cubed (kg/m³)
pound per gallon (U.S. liquid)	119.826 4	grams per liter (g/L)
pound per second (lb/s)	0.453 592	kilogram per second (kg/s)
pound inch squared (lb•in²)	292.640	kilograms millimeter squared (kg•mm²)
pound per mile	0.281 849	gram per meter (g/m)
pound square foot (lb•ft²)	0.042 140 11	kilogram square meter (kg/m²)
pound per cubic yard (lb/yd³)	0.593 276	kilogram per cubic meter (kg/m³)
quart (U.S. liquid)	946.353	milliliters (mL)
square centimeter (cm²)	197 300	circular mils
square centimeter (cm²)	0.001 076	square foot (ft²)
square centimeter (cm²)	0.155	square inch (in²)
square centimeter (cm²)	0.000 1	square meter (m²)
square centimeter (cm²)	0.000 119 6	square yard (yd²)
square foot (ft²)	144	square inches (in²)
square foot (ft²)	0.092 903 04	square meter (m²)
square foot (ft²)	0.111 1	square yard (yd²)
square inch (in²)	1 273 000	circular mils
square inch (in²)	6.4516	square centimeters (cm²)
square inch (in²)	0.006 944	square foot (ft²)
square inch (in²)	645.16	square millimeters (mm²)
square inch (in²)	1 000 000	square mils (m²)
square meter (m²)	10.764	square feet (ft²)
square meter (m²)	1 550	square inches (in²)
square meter (m²)	0.000 000 386 1	square mile
square meter (m²)	1.196	square yards (yd²)
square mil	1.273	circular mils
square mil	0.000 001	square inch (in²)
square mile	2.589 988	square kilometers (km²)
square millimeter (mm²)	1 973	circular mils
square yard (yd²)	0.836 127 4	square meter (m²)
tablespoon (tbsp)	14.786 75	milliliters (mL)
teaspoon (tsp)	4.928 916 7	milliliters (mL)
therm	105.480 4	megajoules (MJ)
ton (long) (2 240 lb)	1 016.047	kilograms (kg)
ton (long) (2 240 lb)	1.016 047	metric tons (t)
ton, metric	2 204.62	pounds (avoirdupois)
ton, metric	1.102 31	tons, short (2 000 lb)
ton, refrigeration	12 000	Btus per hour
ton, refrigeration	4.716 095 9	horsepower-hours
ton, refrigeration	3.516 85	kilowatts (kW)
ton (short) (2 000 lb)	907.185	kilograms (kg)
ton (short) (2 000 lb)	0.907 185	metric ton (t)
ton per cubic meter (t/m³)	0.842 778	ton per cubic yard (ton/yd³)
ton per cubic yard (ton/yd³)	1.186 55	tons per cubic meter (ton/m³)
torr	133.322 4	pascals (Pa)
watt (W)	3.412 14	Btus per hour (Btu/hr)
watt (W)	0.001 341	horsepower
watt (W)	0.001	kilowatt (kW)
watt-hour (Wh)	3.413	Btus
watt-hour (Wh)	0.001 341	horsepower-hour
watt-hour (Wh)	0.001	kilowatt-hour (kW/h)
yard	91.44	centimeters (cm)
yard	3	feet
yard	36	inches
yard	0.000 914 4	kilometer (km)
yard	0.914 4	meter (m)
yard	914.4	millimeters
yard, cubic (yd³)	0.764 55	cubic meter (m³)
yard, squared (yd²)	0.836 127	meter squared (m²)

* These terms are no longer used, but may still be found in some publications.

SINGLE-FAMILY DWELLING SERVICE-ENTRANCE CALCULATIONS

1. **General Lighting Load (*220.3(A)*).**

   _____ ft² @ 3 VA per ft²   = _____ VA

2. **Minimum Number of Lighting Branch-Circuits.**

   $\dfrac{\text{Line 1}}{120} = \dfrac{}{120}$ = _____ amperes

   then, $\dfrac{\text{amperes}}{15} = \dfrac{}{15}$ = _____ 15-ampere branch-circuits

3. **Small Appliance Load [*210.11(C)(1)* and *220.16(A)*].**
   (Minimum of two 20-ampere branch-circuits)

   _____ branch-circuits @ 1500 VA each   = _____ VA

4. **Laundry Branch-Circuit [*210.11(C)(2)* and *220.16(B)*].**
   (Minimum of one 20-ampere branch-circuit)

   _____ branch-circuit(s) @ 1500 VA each   = _____ VA

5. **Total General Lighting, Small Appliance, and Laundry Load.**
   Lines 1 + 3 + 4   = _____ VA

6. **Net Computed General Lighting, Small Appliance, and Laundry Loads (less ranges, ovens, and "fastened-in-place" appliances). Apply demand factors from *Table 220.11*.**

   a. First 3000 VA @ 100%   = ____3000____ VA

   b. Line 5 _____ − 3000 = _____ @ 35% = _____ VA

   Total a + b = _____ VA

7. **Electric Range, Wall-Mounted Ovens, Counter-Mounted Cooking Units (*Table 220.19*).**   = _____ VA

8. **Electric Clothes Dryer (*Table 220.18*).**   = _____ VA

9. **Electric Furnace (*220.15*).**
   Air Conditioner, Heat Pump (*Article 440*).
   (Enter largest value, *220.21*)   = _____ VA

10. **Net Computed General Lighting, Small Appliance, Laundry, Ranges, Ovens, Cooktop Units, HVAC.**

    Lines 6 + 7 + 8 + 9   = _____ VA

11. List "Fastened-in-Place" Appliances *in addition* to Electric Ranges, Electric Clothes Dryers, Electric Space Heating, and Air-Conditioning Equipment.

Appliance		VA Load
Water heater:	= _____ VA	
Dishwasher:	= _____ VA	

    Motor: _____ VA

    Heater: _____ VA (watts)

    Enter largest value, *220.21.*

Garage door opener: = _____ VA

Food waste disposer: = _____ VA

Water pump: = _____ VA

Gas-fired furnace: = _____ VA

Sump pump: = _____ VA

Other: _____   = _____ VA

_____   = _____ VA

_____   = _____ VA

_____   = _____ VA

                     Total   = _____ VA

12. **Apply 75% Demand Factor (*220.17*) if Four or More "Fastened-in-Place" Appliances. If Less Than Four, Figure @ 100%.** Do not include electric ranges, electric clothes dryers, electric space heating, or air-conditioning equipment.

    Line 11 Total: _____ × 0.75   = _____ VA

13. **Total Computed Load (Lighting, Small Appliance, Ranges, Dryer, HVAC, "Fastened-in-Place" Appliances).**

    Line 10 _____ + Line 12 _____   = _____ VA

14. **Add 25% of Largest Motor (*220.14*).**

    _____ × 0.25   = _____ VA

15. **Grand Total Line 13 + Line 14.**   = _____ VA

16. **Minimum Ampacity for Ungrounded Service-Entrance Conductors.**

$$\text{Amperes} = \frac{\text{Line 15}}{240} = \frac{\underline{\hspace{2cm}}}{240} \qquad = \underline{\hspace{2cm}} \text{ amperes}$$

17. Minimum Ampacity for Neutral Service-Entrance Conductor, *220.22* and *310.15(B)(6)*. Do Not Include Straight 240-Volt Loads.

    a. Line 6:   = _____ VA

    b. Line 7: _____ @ 0.70   = _____ VA

    c. Line 8: _____ @ 0.70   = _____ VA

    d. Line 11: (Include only 120-volt loads.)

        _____   _____ VA

        _____   _____ VA

        _____   _____ VA

        _____   _____ VA

        _____   _____ VA

                 Total   _____ VA

    e. Line d total @ 75% demand factor if four or more, otherwise use 100%.

        _____ × 0.75   = _____ VA

    f. Add 25% of largest 120-volt motor.

        _____ × 0.25   = _____ VA

                      Total   = _____ VA

    g. Total a + b + c + e + f.   = _____ VA

$$\text{Amperes} = \frac{\text{Line g}}{240} = \frac{}{240} \qquad = \underline{} \text{ amperes}$$

18. Ungrounded Conductor Size (copper).   _____ AWG

    Note: *Table 310.15(B)(6)* may be used only for 120/240-volt, three-wire, residential single-phase service-entrance conductors, service lateral conductors, and feeder conductors that serve as the main power feeder to a dwelling unit.

19. Neutral Conductor Size (copper) (*220.22*).   _____ AWG

    Note: *NEC® 310.15(B)(6)* permits the neutral conductor to be smaller than the ungrounded "hot" conductors if the requirement of *215.2, 220.22,* and *230.42* are met. When bare conductors are used with insulated conductors, the conductors' ampacity is based on the ampacity of the other insulated conductors in the raceway, *310.15(B)(3)*. The neutral conductor must not be smaller than the grounding electrode conductor, *250.24(B)(1)*.

20. Grounding Electrode Conductor Size (copper) (*Table 250.66*).   _____ AWG

21. Raceway Size.   _____ Trade Size

    Obtain dimensional data from *Table 1, Table 4, Table 5,* and *Table 8, Chapter 9, NEC.®*

MEMBERSHIP APPLICATION

for
Inspector Member
Associate Member
Student Member

Annual Membership Dues $60
Students $50

International Association of Electrical Inspectors' Major Objectives

- Formulation of standards for safe installation and use of electrical materials, devices, and appliances.
- Promotion of uniform understanding and application of the National Electrical Code®, other codes, and any adopted electrical codes in other countries.
- Promotion of uniform administrative ordinances and inspection methods.
- Collection and dissemination of information relative to the safe use of electricity.
- Representation of electrical inspectors in the electrical industry, nationally, and internationally.
- Cooperation with national and international organizations in the further development of the electrical industry.
- Promotion of cooperation among inspectors, inspection departments of city, county and state at national and international levels, the electrical industry, and the public.

Services from the IAEI for you and your organization include all this and more:

- Subscription to *IAEI News* - Bimonthly magazine with latest news on *Code* issues
- Local Chapter or Division meetings - New products, programs, national experts answer questions on local problems
- Section meetings to promote education and cooperation
- Promote advancement of electrical industry
- Save TIME and MONEY by having the latest information

Contact the IAEI Customer Service Department for information on our special membership categories - Section, National, and International Member; Sustaining Member (Bronze, Silver, Gold, or Platinum); and Inspection Agency Member.

PLEASE PRINT

Name - Last   First   M.I.

Chapter, where you live or work, if known

Title

(Division, where appropriate)

Employer

For Office Use Only   Section No.   Chapter No.   Division No.

Address of Applicant

If previous member, give last membership number

City   State or Province   Zip or Postal *Code*

and last year of membership

19____

Student applicants give school attending **

Graduation Date   /   /

Endorsed by

R A Y   C   M U L L I N

Endorser's Membership Number

5 3 3 8 0 0

Applicant's Signature

(Area *Code*) Telephone Number   -   -

☐ Check ☐ Money Order ☐ Diners Club   Name on Card   Charge Card Number   Exp. Date
☐ MasterCard ☐ Visa ☐ Amex ☐ Discover

☐ Inspector $60   ☐ Chief Electrical Inspector * $60   ☐ Associate $60   ☐ Student ** $50

Inspector Members and Chief Electrical Inspector Members MUST sign below:

☐ I, _____, meet the qualifications for inspector member type as described below.

Inspector members must regularly make electrical inspections for preventing injury to persons or damage to property on behalf of a governmental agency, insurance agency, rating bureau, recognized testing laboratory or electric light and power company.

*Chief electrical inspector members must regularly provide technical supervision of electrical inspectors.

**Student member must be currently enrolled in an approved college, university, vocational technical school or trade school specializing in electrical training or approved electrical apprenticeship school.

Mail to: IAEI, P. O. Box 830848 Richardson, TX 75083-0848   For information call: (972) 235-1455 (8-5 CT)

The following is a list of terms used extensively in the electrical trades. These terms are unique to the electrical industry. Words marked with an * are defined in the *National Electrical Code.*®

Accent Lighting: Directional lighting to emphasize a particular object or to draw attention to a part of the field in view.

Accessible (equipment)*: Admitting close approach: not guarded by locked doors, elevation, or other effective means.

Accessible (wiring methods)*: Capable of being removed or exposed without damaging the building structure or finish, or not permanently closed in by the structure or finish of the building.

Accessible, Readily*: Capable of being reached quickly for operation, renewal, or inspections, without requiring those to whom ready access is requisite to climb over or remove obstacles or to resort to portable ladders, chairs, and so forth.

Addendum: Modification (change) made to the construction documents (plans and specifications) during the bidding period.

Adjustment Factor: A multiplier (penalty) that is applied to conductors when there are more than three current-carrying conductors in a raceway or cable. These multipliers are found in *Table 310.15(B)(2)*. An *adjustment factor* is also referred to as a *derating factor*.

AL/CU: Terminal marking on switches and receptacles rated 30 amperes and greater suitable for use with aluminum, copper, and copper-clad aluminum conductors. If not marked, suitable for copper conductors only.

Ambient Lighting: Lighting throughout an area that produces general illumination.

Ambient Temperature: The environmental temperature surrounding the object under consideration.

American National Standards Institute (ANSI): An organization that identifies industrial and public requirements for national consensus standards and coordinates and manages their development, resolves national standards problems, and ensures effective participation in international standardization. ANSI does not itself develop standards. Rather, it facilitates development by establishing consensus among qualified groups. ANSI ensures that the guiding principles—consensus, due process, and openness—are followed.

Ampacity*: The current, in amperes, that a conductor can carry continuously under the conditions of use without exceeding its temperature rating.

Ampere: The measurement of intensity of rate of flow of electrons in an electric circuit. An ampere is the amount of current that will flow through a resistance of one ohm under a pressure of one volt.

Appliance*: Utilization equipment, generally other than industrial, normally built in standardized sizes or types, that is installed or connected as a unit to perform one or more functions such as clothes washing, air conditioning, food mixing, deep frying, and so forth.

Approved*: Acceptable to the authority having jurisdiction (AHJ).

Arc-Fault Circuit Interrupter (AFCI)*: A device intended to provide protection from the effects of arc faults by recognizing characteristics unique to arcing and by functioning to de-energize a circuit when an arcing fault is detected.

Architectural Drawing: A line drawing showing plan and/or elevation views of the proposed building for the purpose of showing the overall appearance of the building.

As-Built Plans: When required, a modified set of working drawings that are prepared for a construction project that includes all variances from the original working drawings that occurred during the project construction.

►**Authority Having Jurisdiction (AHJ):** The organization, office, or individual responsible for approving equipment, materials, an installation, or a procedure.◄ An AHJ is usually a governmental body that has legal jurisdiction over electrical installations. The AHJ has the responsibility for making interpretations of the rules, for deciding upon the approval of equipment and materials, and for granting special permission if required. Where a specific interpretation or deviation from the *NEC®* is given, get it in writing. See *90.4* of the *NEC®.*

Average Rated Life (lamp): How long it takes to burn out the lamp, for example, a 60-watt lamp is rated 1000 hours. The 1000-hour rating is based on the point in time when 50 percent of a test batch of lamps burn out and 50 percent are still burning.

AWG: The American Wire Gauge. Previously known as the Brown & Sharpe Gauge. The smaller the AWG number, the larger the conductor. There are 40 electrical conductor sizes from 36 AWG through 0000 AWG. Starting from the smallest size, 36 AWG, each successive size is approximately 1.26 times larger than the previous AWG size. *Table 8* of *Chapter 9* in the *NEC®* lists conductor sizes commonly found in the electrical industry. Conductor sizes through 0000 AWG are shown using the AWG designation. Conductors larger than 0000 AWG are shown in circular mil area. The 1.26 relationship between sizes is easily confirmed in *Table 8* by checking conductor sizes 4 AWG through 0000 AWG.

Backlight: Illumination from behind a subject directed substantially parallel to a vertical plane through the optical axis of the camera.

Ballast: Used for energizing fluorescent lamps. It is constructed of a laminated core and coil windings or solid-state electronic components.

Ballast Factor: The percentage of rated light output that can be achieved with a specific ballast.

Bare Lamp: A light source with no shielding.

Black Line Print: Another term for a photocopy or computer-aided drawing (CAD).

Blueprints: A term used to refer to a plan or plans. A photographic print in white on a bright blue ground or blue on a white ground used especially for copying maps, mechanical drawings, and architects' plans. See text for further explanation.

Bonding*: The permanent joining of metallic parts to form an electrically conductive path that will ensure electrical continuity and the capacity to conduct safely any current likely to be imposed.

Bonding Jumper*: A reliable conductor to ensure the required electrical conductivity between metal parts required to be electrically connected.

Bonding Jumper, Circuit*: The connection between portions of a conductor in a circuit to maintain the required ampacity of the circuit.

Bonding Jumper, Equipment*: The connection between two or more portions of the equipment grounding conductor.

Bonding Jumper, Main*: The connection between the grounded circuit conductor and the equipment grounding conductor at the service.

Branch-Circuit*: The circuit conductors between the final overcurrent device protecting the circuit and the outlet(s).

Branch-Circuit, Appliance*: A branch-circuit supplying energy to one or more outlets to which appliances are to be connected; such circuits are to have no permanently connected luminaires (lighting fixtures) not a part of an appliance.

Branch-Circuit, General Purpose*: A branch-circuit that supplies a number of outlets for lighting and appliances.

Branch-Circuit, Individual*: A branch-circuit that supplies only one utilization equipment.

▶**Branch-Circuit—Multiwire*:** *"A branch-circuit consisting of two or more ungrounded conductors having a voltage between them, and a grounded conductor having equal voltage between it and each ungrounded conductor of the circuit and that is connected to the neutral or grounded conductor of the system."* ◀

Branch-Circuit Selection Current*: Branch-circuit selection current is the value in amperes to be used instead of the rated-load current in determining the ratings of motor branch-circuit conductors, disconnecting means, controllers, and branch-circuit short-circuit and ground-fault protective devices wherever the running overload protective device permits a sustained current greater than the specified percentage of the rated-load current. The value of branch-circuit selection current will always be equal to or greater than the marked rated-load current.

Brightness: In common usage, the term "brightness" usually refers to the strength of sensation that results from viewing surfaces or spaces from which light comes to the eye.

Candela: The international (SI) unit of luminous intensity. Formerly referred to as a "candle," as in "candlepower."

Candlepower: A measure of intensity mathematically related to lumens.

Carrier: A wave having at least one characteristic that may be varied from a known reference value by modulation.

Carrier Current: The current associated with a carrier wave.

Circuit Breaker*: A device designed to open and close a circuit by nonautomatic means and to open the circuit automatically on a predetermined overcurrent without damage to itself when properly applied within its rating.

CC: A marking on wire connectors and soldering lugs indicating they are suitable for use with copper-clad aluminum conductors only.

CC/CU: A marking on wire connectors and soldering lugs indicating they are suitable for use with copper or copper-clad aluminum conductors only.

CU: A marking on wire connectors and soldering lugs indicating they are suitable for use with copper conductors only.

CO/ALR: Terminal marking on switches and receptacles rated 15 and 20 amperes that are suitable for use with aluminum, copper, and copper-clad aluminum conductors. If not marked, they are suitable for copper or copper-clad conductors only.

Conductor*:

> **Bare:** A conductor having no covering or electrical insulation whatsoever. (See "Conductor, Covered.")
>
> **Covered:** A conductor encased within material of a composition or thickness that is not recognized by this Code as electrical insulation. (See "Conductor, Bare.")
>
> **Insulated:** A conductor encased within material of a composition and thickness that is recognized by this Code as electrical insulation.

Construction Documents: A term used to represent all drawings, specifications, addenda, and other pertinent construction information associated with the construction of a specific project.

Construction Drawing: See Drawings.

Construction Types:

> **Type I:** Fire resistive construction. A building constructed of noncombustible materials such as reinforced concrete, brick, stone, etc., and having any metal members properly "fireproofed" with major structural members designed to withstand collapse and to prevent the spread of fire. Type I construction is subdivided into Types IA and IB.

Type II: Noncombustible construction. A building having all structural members, including walls, floors, and roofs, of noncumbustible materials such as reinforced concrete, brick, stone, etc., and not qualifying as fire-resistive construction. Type II construction is divided into Types IIA, IIB, and IIC.

Type III: Construction where the exterior walls are of concrete, masonry or other noncombustible material. The interior structural members are constructed of any approved materials such as wood or other combustible materials. Type III construction is subdivided into Types IIIA and IIIB.

Type IV: Construction where the exterior walls are constructed of approved noncombustible materials. The interior structural members are of solid or laminated wood. Type V construction is divided into Types VA and VB.

Type V: Construction where the exterior walls, loadbearing walls, partitions, floors, and roofs are constructed of any approved material. This is wood construction as found in typical residential construction.

Continuous Load*:   A load where the maximum current is expected to continue for three hours or more.

Contract Modifications: After the agreement has been signed, any additions, deletions, or modifications of the work to be done are accomplished by change order, supplemental instruction, and field order. They can be issued at any time during the contract period.

Correction Factor: A multiplier (penalty) that is applied to conductors when the ambient temperature is greater than 86°F (30°C) . These multipliers are found at the bottom of *Table 310.16* of the *NEC.*®

CSA: This is the Canadian Standards Association that develops safety and performance standards in Canada for electrical products, similar to but not always identical to those of (Underwriters Laboratories) UL in the United States.

Current: The flow of electrons through an electrical circuit, measured in amperes.

Derating Factor: A multiplier (penalty) that is applied to conductors when there are more than three current-carrying conductors in a raceway or cable. These multipliers are found in *Table 310.15(B)(2)*. A *derating factor* is also referred to as an *adjustment factor*.

Details: Plans, elevations, or sections that provide more specific information about a portion of a project component or element than smaller scale drawings.

Diagrams: Nonscaled views showing arrangements of special system components and connections not possible to clearly show in scaled views. A *schematic diagram* shows circuit components and their electrical connections without regard to actual physical locations. A *wiring diagram* shows circuit components and the actual electrical connections.

Dimmer: A switch with components that permits variable control of lighting intensity. Some dimmers have electronic components; others have core and coil (transformer) components.

Direct Glare: Glare resulting from high luminances or insufficiently shielded light sources in the field of view. Usually associated with bright areas, such as luminaires (fixtures), ceilings, and windows that are outside the visual task of the region being viewed.

Disconnecting Means*:   A device, or group of devices, or other means by which the conductors of a circuit can be disconnected from their source of supply.

Downlight: A small, direct lighting unit that guides the light downward. It can be recessed, surface-mounted, or suspended.

Drawings: (1) A term used to represent that portion of the contract documents that graphically illustrates the design, location, and dimensions of the components and elements contained in a specific project. (2) A line drawing.

Dry Niche Luminaire (lighting fixture)*:   A luminaire (lighting fixture) intended for installation in the wall of a pool or fountain in a niche that is sealed against the entry of pool water.

Dwelling Unit*:   One or more rooms for the use of one or more persons as a housekeeping unit with space for eating, living, and sleeping, and permanent provisions for cooking and sanitation.

Efficacy: The total amount of light energy (lumens) emitted from a light source divided by the total lamp and ballast power (watts) input, expressed in lumens per watt.

Elevations: Views of vertical planes, showing components in their vertical relationship, viewed perpendicularly from a selected vertical plane.

Feeder*:   All circuit conductors between the service equipment, the source of a separately derived system, or other power-supply source and the final branch-circuit overcurrent device.

Fill Light: Supplementary illumination to reduce shadow or contract range.

Fine Print Note (FPN)*:   Explanatory material is in the form of Fine Print Notes (FPN). Fine Print Notes are informational only, and are not enforceable as a requirement by the Code.

Fire Rating: The classification indicating in time (hours) the ability of a structure or component to withstand fire conditions.

Flame Detector: A radiant energy-sensing fire detector that detects the radiant energy emitted by a flame.

Fluorescent Lamp: A lamp in which electric discharge of ultraviolet energy excites a fluorescing coating (phosphor) and transforms some of that energy to visible light.

Footcandle: The unit used to measure how much total light is reaching a surface. One lumen falling on one ft^2 of surface produces an illumination of one footcandle.

Fully Rated System: Panelboards are marked with their short-circuit rating in RMS symmetrical amperes. In a fully-rated system, the panelboard short-circuit current rating will be equal to the lowest interrupting rating of any branch-circuit breaker or fuse installed. All devices installed shall have an interrupting rating greater than or equal to the specified available fault-current. (See "Series-Rated System").

Fuse: An overcurrent protective device with a fusible link that operates to open the circuit on an overcurrent condition.

General Lighting: Lighting designed to provide a substantially uniform level of illumination throughout an area, exclusive of any provision for special lighting.

Glare: The sensation produced by luminance within the visual field that is sufficiently greater than the luminance to which the eyes are adapted to cause annoyance, discomfort, or loss of visual performance and visibility.

Ground*:   A conducting connection, whether intentional or accidental, between an electrical circuit or equipment and the earth, or to some conducting body that serves in place of the earth.

Grounded*:   Connected to earth or to some conducting body that serves in place of the earth.

Grounded Conductor*:   A system or circuit conductor that is intentionally grounded.

Grounded, Effectively*:   Intentionally connected to earth through a ground connection or connections of sufficiently low impedance and having sufficient current-carrying capacity to prevent the buildup of voltages that may result in undue hazards to connected equipment or to persons.

▶**Ground-Fault*:**   An unintentional, electrically conducting connection between an ungrounded conductor of an electrical circuit and the normally noncurrent-carrying conductors, metallic enclosures, metallic raceways, metallic equipment, or earth.◀

▶**Ground-Fault Circuit-Interrupter (GFCI)*:**   A device intended for the protection of personnel that functions to de-energize a circuit or portion thereof within an established period of time when a current to ground exceeds the values established for a Class A device.◀

►**Ground-Fault Current Path*:**   An electrically conductive path from the point of a ground-fault on a wiring system through normally noncurrent-carrying conductors, equipment, or the earth to the electrical supply source.◄

►**Ground-Fault Current Path, Effective*:**   An intentionally constructed, permanent, low-impedance electrically conductive path designed and intended to carry current under ground-fault conditions from the point of a ground-fault on a wiring system to the electrical supply source.◄

Ground-Fault Protection of Equipment (GFPE)*:   A system intended to provide protection of equipment from damaging line-to-ground fault currents by operating to cause a disconnecting means to open all ungrounded conductors of the faulted circuit. This protection is provided at current levels less than those required to protect conductors from damage through the operation of a supply circuit overcurrent device.

Grounding Conductor*:   A conductor used to connect equipment or the grounded circuit of a wiring system to a grounding electrode or electrodes.

Grounding Conductor, Equipment*:   The conductor used to connect the noncurrent-carrying metal parts of equipment, raceways, and other enclosures to the system grounded conductor, the grounding electrode conductor, or both, at the service equipment or at the source of a separately derived system.

►**Grounding Electrode Conductor*:**   The conductor used to connect the grounding electrode(s) to the equipment grounding conductor, to the grounded conductor, or to both, at the service, at each building or structure where supplied from a common service, or at the source of a separately derived system.◄

HACR: A circuit breaker that has been tested and found suitable for use on heating, air-conditioning, and refrigeration equipment. The circuit breaker is marked with the letters HACR. Equipment that is suitable for use with HACR-type circuit breakers will be marked for use with HACR-type circuit breakers.

Halogen Lamp: An incandescent lamp containing a halogen gas that recycles tungsten back onto the tungsten filament surface. Without the halogen gas, the tungsten would normally be deposited onto the bulb wall.

Heat Alarm: A single or multiple station alarm responsive to heat.

Hermetic Refrigerant Motor-Compressor: A combination consisting of a compressor and motor, both of which are enclosed in the same housing, with no external shaft or shaft seals; the motor operating in the refrigerant.

High Intensity Discharge Lamp (HID):   A general term for a mercury, metal-halide, or high-pressure sodium lamp.

High-Pressure Sodium Lamp: A high-intensity discharge light source in which the light is primarily produced by the radiation from sodium vapor.

Household Fire Alarm System: A system of devices that produces an alarm signal in the household for the purpose of notifying the occupants of the presence of a fire so that they will evacuate the premises.

Identified* (conductor):   The identified conductor is the insulated grounded conductor.
- For sizes 6 AWG or smaller, the insulated grounded conductor shall be identified by a continuous white or gray outer finish or by three continuous white stripes on other than green insulation along its entire length.
- For sizes larger than 6 AWG, the insulated grounded conductor shall be identified either by a continuous white or gray outer finish or by three continuous white stripes on other than green insulation along its entire length, or at the time of installation by a distinctive white marking at its terminations.

Identified* (equipment):   Recognizable as suitable for the specific purpose, function, use, environment, application, and so on, where described in a particular Code requirement.

Identified* (terminal):   The identification of terminals to which a grounded conductor is to be connected shall be substantially white in color. The identification of other terminals shall be of a readily distinguishable different color.

IEC: The International Electrotechnical Commission is a worldwide standards organization. These standards differ from those of Underwriters Laboratories. Some electrical equipment might conform to a specific IEC Standard, but may not conform to the UL Standard for the same item.

Illuminance: The amount of light energy (lumens) distributed over a specific area expressed as footcandles (lumens/ft^2) or lux (lumens/m^2).

Illumination: The act of illuminating or the state of being illuminated.

Immersion Detection Circuit Interrupter (IDCI):   A device integral with grooming appliances that will shut off the appliance when the appliance is dropped in water.

Incandescent Filament Lamp: A lamp that provides light when a filament is heated to incandescence by an electric current. Incandescent lamps are the oldest form of electric lighting technology.

Indirect Lighting: Lighting by luminaires (fixtures) that distribute 90 to 100 percent of the emitted light upward.

Inductive Load: A load that is made up of coiled or wound wire that creates a magnetic field when energized. Transformers, core and coil ballasts, motors, and solenoids are examples of inductive loads.

In Sight: Where this Code specifies that one equipment shall be "in sight from," "within sight from," or "within sight," and so on, of another equipment, the specified equipment is to be visible and not more than 50 ft (15 m) from the other.

Instant Start: A circuit used to start specially designed fluorescent lamps without the aid of a starter. To strike the arc instantly, the circuit utilizes higher open-circuit voltage than is required for the same length preheat lamp.

International Association of Electrical Inspectors (IAEI):   A not-for-profit and educational organization cooperating in the formulation and uniform application of standards for the safe installation and use of electricity, and collecting and disseminating information relative thereto. The IAEI is made up of electrical inspectors, electrical contractors, electrical apprentices, manufacturers, and governmental agencies.

Interrupting Rating*:   The highest current at rated voltage that a device is intended to interrupt under standard test conditions.

Isolated Ground Receptacle: A grounding type device in which the equipment ground contact and terminal are electrically isolated from the receptacle mounting means.

Kilowatt: One thousand watts equals one kilowatt.

Kilowatt Hour: One thousand watts of power consumed in one hour. Ten 100-watt lamps burning for ten hours is one kilowatt hour. Two 500-watt electric heaters operated for two hours is one kilowatt hour.

Labeled*:   Equipment or materials to which has been attached a label, symbol, or other identifying mark of an organization that is acceptable to the authority having jurisdiction (AHJ) and concerned with product evaluation that maintains periodic inspection of production of labeled equipment or materials and by whose labeling the manufacturer indicates compliance with appropriate standards or performance in a specified manner.

Lamp: A generic term for a man-made source of light.

Light: The term generally applied to the visible energy from a source. Light is usually measured in lumens or candlepower. When light strikes a surface, it is either absorbed, reflected, or transmitted.

Lighting Outlet*:   An outlet intended for the direct connection of a lampholder, a luminaire (fixture), or a pendant cord terminating in a lampholder.

▶**Listed*:** Equipment, materials, or services included in a list published by an organization that is acceptable to the authority having jurisdiction (AHJ) and concerned with evaluation of products or services, that maintains periodic inspection of production of listed equipment or materials or periodic evaluation of services, and whose listing states that either the equipment, material, or services meets appropriate designated standards or has been tested and found suitable for a specified purpose. ◀

Load: The electric power used by devices connected to an electrical system. Loads can be figured in amperes, volt-amperes, kilovolt-amperes, or kilowatts. Loads can be intermittent, continuous intermittent, periodic, short-time, or varying. See the definition of "Duty" in the *NEC*.

Load Center: A common name for residential panelboards. A load center may not be as deep as a panelboard, and generally does not contain relays or other accessories as are available for panelboards. Circuit breakers "plug in" as opposed to the "bolt-in" types used in panelboards. Manufacturers' catalogs will show both load centers and panelboards. The UL standards do not differentiate.

Locked Rotor Current (LRC): The steady-state current taken from the line with the rotor locked and with rated voltage and frequency applied to the motor.

Low-Pressure Sodium Lamp: A discharge lamp in which light is produced from sodium gas operating at a partial pressure.

Lumens: The SI unit of luminous flux. The units of light energy emitted from the light source.

▶**Luminaire (fixture)*:** A complete lighting unit consisting of a lamp or lamps together with the parts designed to distribute the light, to position and protect the lamps and ballast (where applicable), and to connect the lamps to the power supply. ◀ Prior to the *National Electrical Code®* adopting the International System (SI) definition of *luminaire*, the commonly used term in the United States was and in most instances still is "lighting fixture." It will take years for the electrical industry to totally change and feel comfortable with the term *luminaire*.

Luminaire (fixture) Efficiency: The total lumen output of the luminaire (fixture) divided by the rated lumens of the lamps inside the luminaire (fixture).

Lux: The SI (International System) unit of illumination. One lumen uniformly distributed over an area one square meter in size.

Mandatory Rules: Required. The terms *shall* or *shall not* are used when a Code statement is mandatory.

Maximum Continuous Current (MCC): A value that is determined by the manufacturer of hermetic refrigerant motor compressors under high load (high refrigerant pressure) conditions. This value is established in the *NEC®* as being no greater than 156 percent of the marked rated amperes or the branch-circuit selection current.

Maximum Overcurrent Protection (MOP): A term used with equipment that has a hermetic motor compressor(s). MOP is the maximum ampere rating for the equipment's branch-circuit overcurrent protective device. The ampere rating is determined by the manufacturer, and is marked on the nameplate of the equipment. The nameplate will also indicate if the overcurrent device can be an HACR-type circuit breaker, a fuse, or either.

Mercury Lamp: A high-intensity discharge light source in which radiation from the mercury vapor produces visible light.

Metal-Halide Lamp: A high-intensity discharge light source in which the light is produced by the radiation from mercury, together with the halides of metals such as sodium and candium.

Minimum Circuit Ampacity (MCA): A term used with equipment that has a hermetic motor compressor(s). MCA is the minimum ampere rating value used to determine the proper size conductors, disconnecting means, and controllers. MCA is determined by the manufacturer, and is marked on the nameplate of the equipment.

National Electrical Code® (NEC®): The electrical code published by the National Fire Protection Association. This Code provides for practical safeguarding of persons and property from hazards arising from the use of electricity. It does not become law until adopted by federal, state, and local laws and regulations.

National Electrical Manufacturers Association (NEMA): NEMA is a trade organization made up of many manufacturers of electrical equipment. They develop and promote standards for electrical equipment.

National Fire Protection Association (NFPA): Located in Quincy, MA. The NFPA is an international Standards Making Organization dedicated to the protection of people from the ravages of fire and electric shock. The NFPA is responsible for developing and writing the *National Electrical Code*, *The Sprinkler Code*, *The Life Safety Code*, *The National Fire Alarm Code*, and over 295 other codes, standards, and recommended practices. The NFPA may be contacted at (800) 344-3555 or on their Web site at www.nfpa.org.

Nationally Recognized Testing Laboratory (NRTL): The term used to define a testing laboratory that has been recognized by OSHA. For example, Underwriters Laboratories (UL).

Neutral: In residential wiring, the neutral conductor is the grounded conductor in a circuit consisting of three or more conductors. There is no neutral conductor in a two-wire circuit, although electricians many times refer to the white grounded conductor as the neutral.

The voltages from the ungrounded conductors to the grounded conductor are of equal magnitude.

See the definition of "Multiwire Branch-Circuit" in the *NEC*.

No Niche Luminaire (lighting fixture)*: A luminaire (lighting fixture) intended for installation above or below the water without a niche.

Noncoincidental Loads: Loads that are not likely to be on at the same time. Heating and cooling loads would not operate at the same time. See *220.21* of the *NEC*.

Notations: Words found on plans to describe something.

Occupational Safety and Health Act (OSHA): This is the code of federal regulations developed by the Occupational Safety and Health Administration, U.S. Department of Labor. The electrical regulations are covered in *Part 1910, Subpart S*. In order to shorten and simplify the regulations, *Part 1920, Subpart S* contains only the most common performance requirements of the *NEC*. The *NEC*® must still be referred to in conjunction with OSHA regulations.

Ohm: A unit of measure for electric resistance. An ohm is the amount of resistance that will allow one ampere to flow under a pressure of one volt.

Outlet*: A point on the wiring system at which current is taken to supply utilization equipment.

Overcurrent*: Any current in excess of the rated current of equipment or the ampacity of a conductor. It may result from overload, short circuit, or ground-fault.

Overcurrent Device: Also referred to as an overcurrent protection device. A form of protection that operates when current exceeds a predetermined value. Common forms of overcurrent devices are circuit breakers, fuses, and thermal overload elements found in motor controllers.

Overload*: Operation of equipment in excess of the normal, full-load rating, or of a conductor in excess of rated ampacity that, when it persists for a sufficient length of time, would cause damage or dangerous overheating. The current flow is contained in its normal intended path. A fault, such as a short circuit or a ground-fault, is not an overload.

Panelboard*: A single panel or group of panel units designed for assembly in the form of a single panel, including buses and automatic overcurrent devices; equipped with or without switches for the control of light, heat, or power circuits; designed to be placed in a cabinet or cutout box placed in or against a wall or partition and accessible only from the front.

Permissive Rules: Allowed but not required. Terms such as *shall be permitted* or *shall not be required* are used when a Code statement is permissive.

Photometry: The pattern and amount of light that is emitted from a luminaire (fixture), normally represented as a cross-section through the luminaire (fixture) distribution pattern.

Plan: (1) A line drawing (by floor) representing the horizontal geometrical section of the walls of a building. The section (a horizontal plane) is taken at an elevation to include the relative positions of the walls, partitions, windows, doors, chimneys, columns, pilasters, etc. (2) A plan can be thought of as cutting a horizontal section through a building at an eye level elevation.

Plans: A term used to represent all drawings including sections and details, and any supplemental drawings for complete execution of a specific project.

Power Supply: A source of electrical operating power including the circuits and terminations connecting it to the dependent system components.

Preheat Fluorescent Lamp Circuit: A circuit used on fluorescent lamps wherein the electrodes are heated or warmed to a glow stage by an auxiliary switch or starter before the lamps are lighted.

Prints: A term used to refer to a plan or plans.

Qualified Person*: One who has skills and knowledge related to the construction and operation of the equipment and has received safety training on the hazards involved.

Raceway*: An enclosed channel of metallic or nonmetallic materials designed expressly for holding wires, cables, or busbars, with additional functions as permitted in this Code. Raceways include, but are not limited to, rigid metal conduit, rigid nonmetallic conduit, intermediate metal conduit, liquidtight flexible conduit, flexible metallic tubing, flexible metal conduit, electrical nonmetallic tubing, electrical metallic tubing, underfloor raceways, cellular concrete floor raceways, cellular metal floor raceways, surface raceways, wireways, and busways.

Rapid Start Fluorescent Lamp Circuit: A circuit designed to start lamps by continuously heating or preheating the electrodes. Lamps must be designed for this type of circuit. This is the modern version of the "trigger start" system. In a rapid-start two-lamp circuit, one end of each lamp is connected to a separate starting winding.

Rate of Rise Detector: A device that responds when the temperature rises at a rate exceeding a predetermined value.

Rated-Load Current: The rated-load current for a hermetic refrigerant motor-compressor is the current resulting when the motor-compressor is operated at the rated load, rated voltage, and rated frequency of the equipment it serves.

Receptacle*: A receptacle is a contact device installed at the outlet for the connection of an attachment plug. A single receptacle is a single contact device with no other contact device on the same yoke. A multiple receptacle is two or more contact devices on the same yoke.

Receptacle Outlet*: An outlet where one or more receptacles are installed.

Resistive Load: An electric load that opposes the flow of current. A resistive load does not contain cores or coils of wire. Some examples of resistive loads are the electric heating elements in an electric range, ceiling heat cables, and electric baseboard heaters.

Root-Mean-Square (RMS): The square root of the average of the square of the instantaneous values of current or voltage. For example, the RMS value of voltage, line to neutral, in a home is 120 volts. During each electrical cycle, the voltage rises from zero to a peak value ($120 \times 1.4142 = 169.7$ volts), back through zero to a negative peak value, then back to zero. The RMS value is 0.707 of the peak value ($169.7 \times 0.707 = 120$ volts). The root-mean-square value is what an electrician reads on an ammeter or voltmeter.

Schedules: Tables or charts that include data about materials, products, and equipment.

Sections: Views of vertical cuts through and perpendicular to components, showing their detailed arrangement.

Series-Rated System: Panelboards are marked with their short-circuit rating in RMS symmetrical amperes. A series-rated panelboard will be determined by the main circuit breaker or fuse, and branch-circuit breaker combination tested in accordance to UL *Standard 489*. The series-rating will be less than or equal to the interrupting rating of the main overcurrent device, and greater than the interrupting rating of the branch-circuit overcurrent devices. (See "Fully-Rated System.")

Service*: The conductors and equipment for delivering energy from the serving utility to the wiring system of the premises served.

Service Conductors*: The conductors from the service point to the service disconnecting means.

Service Drop*: The overhead service conductors from the last pole or other aerial support to and including the splices, if any, connecting to the service-entrance conductors at the building or other structure.

Service Equipment*: The necessary equipment, usually consisting of a circuit breaker or switch and fuses, and their accessories, located near the point of entrance of supply conductors to a building or other structure, or an otherwise defined area, and intended to constitute the main control and means of cutoff of the supply.

Service Lateral*: The underground service conductors between the street main, including any risers at a pole or other structure or from transformers, and the first point of connection to the service-entrance conductors in a terminal box or meter or other enclosure with adequate space, inside or outside the building wall. Where there is no terminal box, meter, or other enclosure with adequate space, the point of connection shall be considered to be the point of entrance of the service conductors into the building.

Service Point*: Service point is the point of connection between the facilities of the serving utility and the premises wiring.

Shall: Indicates a mandatory requirement.

Short Circuit: A connection between any two or more conductors of an electrical system in such a way as to significantly reduce the impedance of the circuit. The current flow is outside of its intended path, thus the term "short circuit." A short circuit is also referred to as a fault.

Should: Indicates a recommendation or that which is advised but not required.

Smoke Alarm: A single or multiple station alarm responsive to smoke.

Smoke Detector: A device that detects visible or invisible particles of combustion.

Specifications: Text setting forth details such as description, size, quality, performance, workmanship, and so forth. Specifications that pertain to all of the construction trades involved might be subdivided into "General Conditions" and "Supplemental General Conditions." Further subdividing the specifications might be specific requirements for the various contractors such as electrical, plumbing, heating, masonry, and so forth. Typically, the electrical specifications are found in Division 16.

Split-Circuit Receptacle: A receptacle that can be connected to two branch-circuits. These receptacles may also be used so that one receptacle is live at all times, and the other receptacle is controlled by a switch. The terminals on these receptacles usually have breakaway tabs so the receptacle can be used either as a split-circuit receptacle or as a standard receptacle.

Standard Network Interface (SNI): A device usually installed by the telephone company at the demarcation point where their service leaves off and the customer's service takes over. This is similar to the service point for electrical systems.

Starter: A device used in conjunction with a ballast for the purpose of starting an electric discharge lamp.

Surface-Mounted Luminaire (fixture):   A luminaire (fixture) mounted directly on the ceiling.

Surge Suppressor: A device that limits peak voltage to a predetermined value when voltage spikes or surges appear on the connected line.

Suspended (pendant) Luminaire (fixture):   A luminaire (fixture) hung from a ceiling by supports.

Switches*:

General-Use Snap Switch: A form of general-use switch constructed so that it can be installed in device boxes or on box covers, or otherwise used in conjunction with wiring systems recognized by this Code.

General-Use Switch: A switch intended for use in general distribution and branch-circuits. It is rated in amperes, and it is capable of interrupting its rated current at its rated voltage.

Motor-Circuit Switch: A switch, rated in horsepower, capable of interrupting the maximum operating overload current of a motor of the same horsepower rating as the switch at the rated voltage.

Symbols: Graphic representations that stand for or represent other things. A symbol is a simple way to show such things as lighting outlets, switches, and receptacles on an electrical plan. The American Institute of Architects has developed a very comprehensive set of symbols that represent just about everything used by all building trades. When an item cannot be shown using a symbol, then a more detailed explanation using a notation or inclusion in the specifications is necessary.

Task Lighting: Lighting directed to a specific surface or area that provides illumination for visual tasks.

Terminal: A screw or a quick-connect device where a conductor(s) is intended to be connected.

Thermocouple: A pair of dissimilar conductors so joined at two points that an electromotive force is developed by the thermoelectric effects when the junctions are at different temperatures.

Thermopile: More than one thermocouple connected together. The connections may be series, parallel, or both.

Transient Voltage Surge Suppressor (TVSS):   A device that clamps transient voltages by absorbing the major portion of the energy (joules) created by the surge, allowing only a small, safe amount of energy to enter the actual connected load.

Troffer: A recessed lighting unit, usually long and installed with the opening flush with the ceiling. The term is derived from "trough" and "coffer."

UBC (Uniform Building Code):   The Uniform Building Code is one of the family of codes and related publications published by the International Conference of Building Officials (ICBO) and other organizations, such as the International Association of Plumbing and Mechanical Officials (IAPMO) and the National Fire Protection Association (NFPA), which have similar goals as far as code publications are concerned. The Uniform Building Code is designed to be compatible with these other codes, as together they make up the enforcement tools of a jurisdiction.

UL: Underwriters Laboratories (UL) is an independent not-for-profit organization that develops standards and tests electrical equipment to these standards.

UL-Listed: Indicates that an item has been tested and approved to the standards established by UL for that particular item. The UL Listing Mark may appear in various forms, such as the letters UL in a circle. If the product is too small for the marking to be applied to the product, the marking must appear on the smallest unit container in which the product is packaged.

UL-Recognized: Refers to a product that is incomplete in construction features or limited in performance capabilities. A "Recognized" product is intended to be used as a component part of equipment that has been "listed." A "Recognized" product must not be used by itself. A UL product may contain a number of components that have been "Recognized."

Ungrounded Conductor: The conductor of an electrical system that is not intentionally connected to ground. This conductor is referred to as the "hot" or "live" conductor.

Volt: The difference of electric potential between two points of a conductor carrying a constant current of one ampere, when the power dissipated between these points is equal to one watt. A voltage of one volt can push one ampere through a resistance of one ohm.

Voltage (of a circuit)*: The greatest root-mean-square (effective) difference of potential between any two conductors of the circuit concerned.

Voltage (nominal)*: A nominal value assigned to a circuit or system for the purpose of conveniently designating its voltage class (e.g., 120/240 volts, 480Y/277 volts, 600 volts).

   The actual voltage at which a circuit operates can vary from the nominal within a range that permits satisfactory operation of equipment.

Voltage to Ground*: For grounded circuits, the voltage between the given conductor and that point or conductor of the circuit that is grounded; for ungrounded circuits, the greatest voltage between the given conductor and any other conductor of the circuit.

Voltage Drop: Also referred to as IR drop. Voltage drop is most commonly associated with conductors. A conductor has resistance; when current is flowing through the conductor, a voltage drop will be experienced across the conductor. Voltage drop across a conductor can be calculated using Ohm's law: $E = IR$.

Volt-ampere: A unit of power determined by multiplying the voltage and current in a circuit. A 120-volt circuit carrying 1 ampere is 120 volt-amperes.

Watt: A measure of true power. A watt is the power required to do work at the rate of 1 joule per second. Wattage is determined by multiplying voltage times amperes times the power factor of the circuit: $W = E \times I \times PF$.

Watertight: Constructed so that moisture will not enter the enclosure under specified test conditions.

Weatherproof*: Constructed or protected so that exposure to the weather will not interfere with successful operation.

   (FPN): Rainproof, raintight, or watertight equipment can fulfill the requirements for weatherproof where varying weather conditions other than wetness, such as snow, ice, dust, or temperature extremes, are not a factor.

Wet Niche Luminaire (lighting fixture)*: A luminaire (lighting fixture) intended for installation in a forming shell mounted in a pool.

Working Drawing(s): A drawing sufficiently complete with plan and section views, dimensions, details, and notes, so that whatever is shown can be constructed and/or replicated without instructions but subject to clarifications, see drawings.

The following list of Internet World Wide Web sites has been included for your convenience. These are current as of time of printing. Web sites are a "moving target" with continual changes. We do our utmost to keep these web sites current. If you are aware of web sites that should be added, deleted, or that are different than shown, please let us know. Most manufacturers will provide catalogs and technical literature upon request either electronically or by snail-mail. You will be amazed at the amount of electrical and electronic equipment information that is available on these web sites.

Web sites marked with an asterisk (*) are closely related to the "home automation" market.

Electrical Industry (This web site links to over 800 sites related to the electrical industry.)	www.electricalsearch.com
Electrical Industry (This web site provides current electrical industry news, as well as links to manufacturers, Code tips, and products.)	www.electricalzone.com
Electrical Industry (This web site offers over 6200 links to the entire electrical industry.)	www.electric-find.com
Electrical Industry (A great web site for electrical industry professionals. Excellent articles on the Code.)	www. econline.com
National Fire Protection Association (A web site of Code questions, newsletters, and articles.)	www.necdirect.org

ACCUBID	www.accubid.com
Acme Electric Corporation	www.acmepowerdist.com
Acronym/Abbreviation finder	www.acronym.com
ActiveHome	www.X10-beta.com/ActiveHome/
Adalet	www.adalet.com
Ademco*	www.ademco.com
Advance Transformer Co.	www.advancetransformer.com
Advanced Cable Ties	www.advancedcableties.com
AEMC Instruments	www.aemc.com
AFC Cable Systems	www.afcweb.com
Alpha Wire Co.	www.alphawire.com
Alflex	www.alflex.com
Allen-Bradley	www.automation.rockwell.com
or	www.ab.com
Allied Moulded Products	www.alliedmoulded.com
Allied Support Products	www.alliedsupport.com
Allied Tube & Conduit	www.alliedtube.com
Alpha Wire	www.alphawire.com

ALTO Lamps	www.altolamp.com
The Aluminum Association, Inc.	www.aluminum.org
American Council for an Energy-Efficient Economy	www.aceee.org
American Gas Association	www.aga.org
American Institute of Architects	www.aiaonline.com
American Insulated Wire Corp.	www.aiwc.com
American Lighting Association	www.americanlightingassoc.com
American National Standards Institute	www.ansi.org
American Pipe & Plastics	www.tier.net/ampipe.com
American Power Conversion	www.apcc.com
American Public Power Assoc.	www.Appanet.org
American Society of Heating, Refrigerating, and Air-Conditioning Engineers	www.ashrae.org
American Society for Testing and Materials	www.astm.org
American Society of Mechanical Engineers	www.ASME.org
American Society of Safety Engrs.	www.asse.org
AMP*	www.amp.com
Amprobe	www.amprobe.com
Angelo	www.angelobrothers.com
Anixter (data & telecommunications cabling)*	www.anixter.com
Appleton Electric Co.	www.appleton.com
Arlington Industries	www.aifittings.com
Arrow Fasteners	www.arrowfastener.com
Arrow Hart Wiring Devices	www.arrowhart.com
Assisted Living: The World of Assisted Technology	www.abledata.com
Associated Builders & Contractors	www.abc.org
Association of Cabling Professions	www.wireville.com
Association of Home Appliance Manufacturers	www.aham.org
Automated Home Technologies*	www.autohometech.com
Automatic Switch Co. (ASCO)	www.asco.com
Avaya HomeStar*	http://connectivity1.avaya.com/homestar/
AVO International	www.avointl.com
B-Line Systems, Inc.	www.B-Line.com
Baldor Motors & Drives	www.baldor.com
R. W. Beckett Corporation (oil burner technical info.)	www.beckettcorp.com
Belden Wire & Cable Co.	www.belden.com
Bell	www.hubbell-bell.com
Bender	www.bender.org
Benfield International Corp.	www.benfieldintl.com
Best Power	www.bestpower.com
BICC General	www.generalcable.com
BICSI: (formerly Building Industry Consulting Services, International)*	www.bicsi.org
BidStreet	www.BidStreetUSA.com
The Blue Book	www.thebluebook.com
Bodine Electric Company	www.bodine-electric.com
BOCA	www.bocai.org

boltswitch, Inc.	www.boltswitch.com
Brady Worldwide, Inc.	www.whbrady.com
Brand-Rex	www.biccgeneral.com
Bridgeport Fittings, Inc.	www.bptfittings.com
Broan-NuTone	www.broan.com
Bryant Electric	www.hubbell-bryant.com
BuildNet	www.buildnet.com
BuildPoint	www.buildpoint.com
Burndy Electrical Products	www.burndy.com
or	www.fciconnect.com
Bussmann, Cooper Industries	www.bussmann.com
Cable Systems International	www.csi-cables.com
CABLOFIL, Inc.	www.cablofil.com
Caddy	www.erico.com
Cadweld	www.erico.com
Canadian Standards Association	www.csa-international.org
Cantex	www.cantexinc.com
Carlon, A Lamson & Sessions Co.	www.carlon.com
Carol Cable	www.biccgeneral.com
Casablanca Fan Company	www.casablancafanco.com
Caterpillar Engine Division	www.cat-engines.com
Caterpillar Generator Sets	www.cat-electricpower.com
CEBus Industry Council*	www.cebus.org
CEE News magazine	www.ceenews.com/
CEPro*	www.ce-pro.com
Certified Ballast Manufacturers	www.certbal.org
Channellock, Inc.	www.channellock.com
ChannelPlus*	www.channelplus.com
Chromalox	www.chromalox.com
Clifford of Vermont*	www.cliffordvt.com
Coleman Cable	www.colemancable.com
Columbia Lighting	www.columbia-ltg.com
CommScope* (coax cable mfg.)	www.commscope.com
Computer and Telecommunications Page*	www.cmpcmm.com/cc/
Construction Specifications Institute	www.csinet.org
Construction Terms/Glossary	www.constructionplace.com/glossary.html
Consumer Electronic Manufacturers	
Association (CEMA)*	www.techhome.org
Consumer Product Safety Commission (CPSC)	www.cpsc.gov
Continental Automated Building	
Association (CABA)*	www.caba.org
Cooper Hand Tools	www.coopertools.com
Cooper Lighting	www.cooperlighting.com
Cooper Power Systems	www.cooperpower.com
Copper Development Assoc., Inc.	www.copper.org
Crouse-Hinds	www.crouse-hinds.com
Crouse-Hinds (hazardous locations)	www.zonequest.com
Custom Electronic Design &	
Installation Association*	www.CEDIA.org
Custom Electronic Professional*	www.ce-pro.com

Cutler-Hammer	www.cutler-hammer.com
Cutler-Hammer safety web site	www.homesafetycentral.com
Daniel Woodhead Company	www.danielwoodhead.com
Delmar, a division of Thomson Learning	www.delmar.com
F. W. Dodge	www.fwdodge.com
DOMOSYS*	www.domosys.com
Dremel	www.dremel.com
Dual-Lite	www.dual-lite.com
Dynacom Corporation	www.dynacomcorp.com
Eagle Electric	www.eagle-electric.com
or	www.cooperindustries.com
Eaton/Cutler-Hammer	www.cutler-hammer.com
EC & M magazine	www.ecmweb.com
Eclipse	www.eclipseinc.com
Edison Electric Institute	www.eei.org
Edwards Signaling Products	www.edwards-signals.com
ELAN Home Systems*	www.elanhomesystems.com
Electric Web	www.electricwebsite.net
Electri-Flex	www.electriflex.com
Electrical Apparatus Service Assoc.	www.easa.com
Electrical Construction & Maintenance (Has a list of over 2000 manufacturers.)	www.ecmweb.com
Electrical Contractor magazine	www.ecmag.com
Electrical Contracting & Engineering News	www.ecenmag.com
The Electrical Contractor Network	www.electrical-contractor.net/
Electrical Designers Reference	www.edreference.com
The Electrical Distributor magazine	www.tedmag.com
Electrical Zone	www.electricalzone.com
Electrician.com (formerly known as Newton's International Electric Journal)	www.electrician.com
Electrician's Web	www.electriciansweb.com
Electric Smarts	www.Electricsmarts.com
Electro-Test, Inc.	www.electro-test.com
Electronic House magazine*	www.electronichouse.com
Electronic Industry Alliance*	www.eia.org
Electronic Wire & Cable	www.e-wire.cable.com
Emerson	www.emersonfans.com
Encore Wire Corp.	www.encorewire.com
Energy Efficiency & Renewable Energy Network (U.S. Dept. of Energy)	www.eren.doe.gov
Energy Star	www.energystar.gov
Environmental Protection Agency	www.epa.gov
Erico, Inc.	www.erico.com
Essex Group*	www.essexgroup.com
Estimation, Inc.	www.estimation.com
ETL Semko	www.etlsemko.com
Exide Electronics	www.exide.com
Factory Mutual Insurance Co.	www.factorymutual.com
Faraday	www.faradayllc.com
FCI/Burndy (telecom connectors)	www.fciconnect.com

Federal Communications Commission*	www.fcc.gov
Federal Pacific Co.	www.electro-mechanical.com/fpc.html
Federal Signal Corp.	www.federalsignal.com
Ferraz	www.ferraz.com
Ferraz/Shawmut	www.ferrazshawmut.com
Fiber Optic Association, Inc.*	www.world.std.com/~foa/
Fibertray	www.ditel.net
First Alert	www.FirstAlert.com
Flex-Core	www.flex-core.com
Fluke Corporation	www.fluke.com
Future Smart Networks*	www.futuresmart.com
Fyrnetics	www.fyrnetics.com
Gardner Bender	www.gardnerbender.com
GE Electrical Distribution & Control	www.ge.com/edc
GE Industrial Systems	www.geindustrial.com
GE Lighting	www.gelighting.com
GE Total Lighting Control	See Horton Control
General Cable	www.generalcable.com
Generators:	
Coleman Powermate	www.colemanpowermate.com
Cummins Power Generation	www.onan.com
Cummins Power Generation	www.cumminspowergeneration.com
GE	www.ge.com
Generac Power Systems	www.generac.com
Gen-X	www.gcnxnow.com
Kohler Generators	www.kohlergenerators.com
Onan Corporation	www.onan.com
Reliance Electric	www.rcliancc.com
Genesis Cable Systems*	www.genesiscable.com
GESmart	www.GESmart.com
Global Engineering Documents	www.global.ihs.com
Gould Shawmut	www.gouldshawmut.com
Grainger	www.grainger.com
Grayline, Inc.	www.graylineinc.com
Greenlee Textron, Inc.	www.greenlee.textron.com
Greyfox Systems*	www.greyfox.com
Grinnell Fire Protection Systems	www.grinnellfire.com
Guth Lighting	www.guth.com
Halo (Cooper Lighting)	www.cooperlighting.com
Halex	www.halexco.com
Handicapped Suggestions	
The World of Assisted Technology	www.abledata.com
Hatch Transformers, Inc	www.hatchtransformers.com
Heavy Duty	www.heavydutytool.com
Helix/HiTemp Cables, Inc.*	www.drakeusa.com
Hewlett Packard	www.wirescope.com/entry
Heyco Molded Products, Inc.	www.heyco.com
Hilti	www.hilti.com
HIOKI	www.hioki.co.jp/
Hitachi Cable Manchester, Inc.*	www.hcm.hitachi.com

Hoffman Enclosures, Inc.	www.hoffmanonline.com
Holophane Corporation	www.holophane.com
Holt, Mike Enterprises	www.mikeholt.com
Home Automation and Networking Association*	www.homeautomation.org
Home Automation Links*	www.handilink.com
Home Automation Links*	www.asihome.com
Home Automation Systems*	www.smarthome.com
Home Controls, Inc.*	www.homecontrols.com
Home Director, Inc.*	www.homedirector.net
HomeStar Wiring Systems*	http://connectivity1.avaya.com/homestar/
Homestore*	www.homestore.com
Home Team Network*	www.hometeam.com/
Home Toys*	www.hometoys.com
Honeywell	www.Honeywell.com
Horton Controls (was GE Total Lighting Control)	www.hortoncontrols.com
Housing Urban Development	www.HUD.gov.
Houston Wire & Cable Company	www.houwire.com
Hubbell Electrical Products	www.hubbell-bell.com
	www.hubbell-bryant.com
	www.hubbell-killark.com
	www.hubbell-raco.com
	www.hubbell-wiegmann.com
Hubbell Incorporated	www.hubbell.com
Hubbell Lighting Incorporated	www.hubbell-ltg.com
Hubbell Premise Wiring	www.hubbell-premise.com
Hubbell Wiring Devices	www.hubbell-wiring.com
Hunt Dimmers	www.huntdimming.com
Hunter Fans	www.hunterfan.com
Husky	www.mphusky.com
ICBO	www.icbo.org
Ideal Industries, Inc.	www.idealindustries.com
Illuminating Engineering Society of North America	www.iesna.org
ILSCO	www.ilsco.com
Independent Electrical Contractors	www.ieci.org
Intel's Home Network	www.intel.com/anypoint
Intermatic	www.intermatic.com
International Association of Electrical Inspectors (IAEI)	www.iaei.com
International Brotherhood of Electrical Workers	www.ibew.org
International Code Council (ICC)	www.intlcode.org
International Dark-Sky Association	www.darksky.org
International Electrotechnical Commission (IEC)	www.iec.ch/
International Organization for Standardization (ISO)	www.iso.ch
Intertec Testing Services (ETL)	www.etlsemko.com
Intertec Testing Services (ITS)	www.itsqs.com

inter.Light, Inc.	www.lightsearch.com
Institute of Electrical and Electronic Engineers, Inc.	www.ieee.org
Instrument Society of America	www.isa.org
Insulated Tools (Certified Insulated Products)	www.insulatedtools.com
Jacuzzi, Inc.	www.jacuzzi.com
Jefferson Electric	www.jeffersonelectric.com
Johnson Controls	www.johnsoncontrols.com
Joslyn	www.jesc.com
Juno Lighting, Inc.	www.junolighting.com
KatoLight	www.katolight.com
Kenall Lighting	www.kenall.com
King Safety Products	www.kingsafety.com
King Wire, Inc.	www.kingwire.com
Klein Tools	www.klein-tools.com
Leviton Mfg. Co., Inc.	www.leviton.com
Leviton Telcom*	www.levitontelcom.com
Lew Electric	www.lewelectric.com
Liebert	www.liebert.com
Lighting Design Forum	www.qualitylight.com
Lighting Research Center	www.lrc.rpi.edu
Lightolier	www.lightolier.com
Litetouch*	www.litetouch.com
Lithonia Lighting	www.lithonia.com
Littelfuse, Inc.	www.littelfuse.com
LonWorks*	www.lonworks.echelon.com
Lucent Technologies*	www.lucent.com
Lumisistemas, S.A. de C.V	www.lumisistemas.axasa.com
Lutron Electronics Co., Inc.	www.lutron.com
Lutron Home Theater*	www.ultimatehometheater.com
MagneTek Lighting	www.distributors.magnetek.com
MagneTek Lighting Products	www.magnetek.com
MagneTek (Motors)	www.magnetek.com
Manhattan/CDT	www.manhattancdt.com
MapleChase (Paragon timers)	www.maplechase.com
McGill	www.mcgillelectrical.com
Meggers and Other Test Meters	www.avointl.com
MET Laboratories, Inc.	www.metlabs.com
MI Cable Company	www.micable.com
Midwest Electric Products, Inc.	www.midwestelectric.com
Milbank Manufacturing Co.	www.milbankmfg.com
Milwaukee Electric Tool	www.mil-electric-tool.com
Minerallac	www.minerallac.com
Molex, Inc.	www.molex.com
Motor Control	www.motorcontrol.com
Motorola Lighting, Inc.	www.motorola.com
MYTECH	www.lightswitch.com
National Association of Electrical Distributors	www.naed.org
National Association of Home Builders	www.nahb.org

National Association of Radio and Telecommunications Engineers, Inc.	www.narte.com
National Electrical Contractors Association (NECA)	www.necanet.org
National Electrical Manufacturers Association (NEMA)	www.nema.org
National Electrical Safety Foundation	www.nesf.org
National Fire Protection Association	www.nfpa.org
National Fire Protection Association *National Electrical Code®* Information	www.necdirect.org
National Institute of Standards and Technology (NIST)	www.nist.gov
National Joint Apprenticeship and Training Committee (NJATC)	www.njatc.org
National Lighting Bureau	www.nlb.org
National Resource for Global Standards	www.nssn.org
National Safety Council	www.nsc.org
National Spa and Pool Institute	www.nspi.org
National Systems Contractors Association*	www.nsca.org
NCI Products, Inc.	www.nciproducts.com
Nora Lighting	www.noralighting.com
NuTone	www.nutone.com
Occupational Safety & Health Administration	www.osha.gov
Occupation Health & Safety	www.ohsonline.com
Okonite Co	www.okonite.com
Olflex Cable	www.olflex.com
OnQ*	www.onqtech.com/
Optical Cable Corporation*	www.occfiber.com
Ortronics, Inc.	www.ortronics.com
Osram Sylvania Products, Inc.	www.sylvania.com
O-Z /Gedney	www.o-zgedney.com
Panasonic Lighting	www.panasonic.com/lighting
Panasonic Technologies	www.panasonic.com
Panduit Corporation	www.panduit.com
Paragon Electric Company	See MapleChase
Pass and Seymour, Legrand	www.passandseymour.com
Pelican Rope Works	www.pelicanrope.com
Penn-Union	www.penn-union.com
Phase-A-Matic, Inc.	www.phase-a-matic.com
Philtek Power Corporation	www.philtek.com
Philips	www.philips.com
Philips Lamps	www.ALTOlamp.com
Philips Lighting Company	www.lighting.philips.com
Phoenix Contact	www.phoenixcon.com
Pigeon Hole, The	www.thepigeonhole.com
Plant Engineering On-Line	www.plantengineering.com
Popular Home Automation*	www.pophome.com
Porter Cable	www.porter-cable.com
Power Conversion Systems	www.ups-power.com
Power Maintenance International	www.pmiservice.com

Power-Sonic Corporation	www.power-sonic.com
Power Quality Monitoring	www.pqmonitoring.com
Premise Wiring Products	www.unicomlink.com
Prescolite	www.prescolite.com
Pringle Electrical Mfg. Co.	www.pringle-elec.com
Progress Lighting	www.progresslighting.com
Radix Wire Company	www.radix-wire.com
RACO	www.racoinc.com
Rayovac Corporation	www.rayovac.com
Regal Manufacturing Co.	www.regalfittings.com
Refrigeration Service Engineering Society	www.rses.org
Reliable Power Meters	www.reliablemeters.com
Remke Industries, Inc.	www.remke.com
Robertson Transformer Co.	www.robertsontransformer.com
Robicon	www.robicon.com
Robroy Industries – Conduit Div.	www.robroy.com
Rome Cable	www.romecable.com
Romex Cable	www.biccgeneral.com
RotoZip Tool Corp.	www.rotozip.com
Russelectric	www.russelectric.com
Safety Technology International, Inc.	www.sti-usa.com
S&C Electric Company	www.sandc.com
Sea Gull Lighting	www.seagulllighting.com
Seatek, Inc.	www.SeatekCo.com
The Siemon Company*	www.siemon.com
Sensor Switch, Inc.	www.sensorswitchinc.com
Shat-R-Shield	www.shat-r-shield.com
Siemens Energy & Automation	www.aut.sea.siemens.com
Siemens-Furnas Controls	www.furnas.com
Siemon*	www.siemon.com
Silent Knight	www.silentknight.com
Simplex	www.simplexnet.com
Simpson Electric Co	www.simpsonelectric.com
A.W. Sperry Instruments, Inc.	www.awsperry.com
Sli Lighting	www.powerlighting.com
Smart Corporation*	www.smartcorporation.com
Smart Electrical	www.SmartElectrical.com
Smart Home*	www.smarthome.com
Smart Home Pro*	www.smarthomepro.com
Smart House, Inc.*	www.smart-house.com
Society of Cable Telecommunications Engineers*	www.scte.org
Sola/Hevi Duty	www.sola-hevi-duty.com
Southwire Co.	www.southwire.com
SBCCI	www.sbcci.org
SP Products, Inc.	www.spproducts.com
Square D, Groupe Schneider	www.squared.com
Square D (Homeline)	www.squared.com/retail
Square D, Power Logic	www.powerlogic.com
Steel Tube Institute	www.steeltubeinstitute.org
Sunbelt Transformers	www.sunbeltusa.com

Superior Electric	www.superiorelectric.com
Superior Essex Group	www.superioressex.com
Suttle Telephone Apparatus	www.suttleapp.com
Sylvania	www.sylvania.com
Technical Consumer Products, Inc.	www.springlamp.com
Technology Information Center*	www.tickit.com
Telecommunications Industry Association (TIA)*	www.tia.org
Terms/Construction/Glossary	www.constructionplace.com/glossary.html
This Old House	www.thisoldhouse.org
Thomas & Betts Corporation	www.tnb.com
Thomas Lighting	www.thomaslighting.com
Thomas Register	www.thomasregister.com
Thompson Consumer Electronics*	www.rca-electronics.com
3M Electrical Products	www.3M.com/elpd
Tork	www.tork.com
Trade Service Systems	www.tradeservice.com
Triplett Corporation	www.triplett.com
Tripp Lite	www.tripplite.com
Trade Service	www.tra-ser.com
Tyton Hellermann	www.tytonhellermann.com
UEI Test Equipment	www.ueitest.com
UL Standards Dept.	http://ulstandardsinfonet.ul.com
Underwriters Laboratories Hazardous Locations	www.ul.com/hazloc/
Underwriters Laboratories (Code Authority)	www.ul.com/auth/index.html
Underwriters Laboratories	www.ul.com
Unistrut Corp.	www.unistrut.com
Unity Manufacturing	www.unitymfg.com
USA Wire and Cable, Inc.	www.usawire-cable.com
US Department of Energy (DOE)	www.doe.gov
US Environmental Protection Agency (EPA)	www.epa.gov
Ustec*	www.ustecnet.com
Versa-Duct	www.versaduct.com
Watt Stopper	www.wattstopper.com
Waukesha Electric Systems	www.waukeshaelectric.com
Wheatland Tube Company	www.wheatland.com
Wireless Industry*	www.wireless.about.com
Wireless Thermostats	www.wirelessthermostats.com
Wiremold	www.wiremold.com
Wiring America's Homes	www.connectedhome.org
Woodhead Industries	www.danielwoodhead.com
X-10 Pro Automation Products*	www.x10pro.com
Zenith Controls, Inc.	www.zenithcontrols.com
Zircon	www.zircon.com

CODE INDEX

Note: Page numbers in **bold** reference Figures

UL STANDARDS

INDEX

Note: Page numbers in **bold type** reference non-text material, such as tables and figures.